THE IMPACT OF VLBI ON ASTROPHYSICS AND GEOPHYSICS

Sponsored by
International Astronomical Union
International Union of Radio Sciences
American Geophysical Union

with support from the
National Aeronautics and Space Administration
National Science Foundation
Smithsonian Institution
Interferometrics, Inc.
Signatron, Inc.

hosted by
Harvard–Smithsonian Center for Astrophysics

INTERNATIONAL ASTRONOMICAL UNION

UNION ASTRONOMIQUE INTERNATIONALE

THE IMPACT OF VLBI ON ASTROPHYSICS AND GEOPHYSICS

PROCEEDINGS OF THE 129TH SYMPOSIUM OF THE
INTERNATIONAL ASTRONOMICAL UNION
HELD IN CAMBRIDGE, MASSACHUSETTS, U.S.A.,
MAY 10-15, 1987

EDITED BY

M. J. REID

and

J. M. MORAN

Harvard-Smithsonian Center for Astrophysics,
Cambridge, Massachusetts, U.S.A.

KLUWER ACADEMIC PUBLISHERS

DORDRECHT / BOSTON / LONDON

Library of Congress Cataloging in Publication Data CIP

International Astronomical Union. Symposium (129th : 1987 : Cambridge, Mass.)
The impact of VLBI on astrophysics and geophysics.

Includes bibliographies and indexes.
1. Radio astrophysics--Congresses. 2. Geophysics--Congresses. 3. Astrometry--Congresses. 4. Very long baseline interferometry--Congresses. I. Reid, Mark Jonathan, 1948- . II. Moran, James M. III. Title.
QB462.5.I57 1987 522'.682 88-3175

ISBN 90-277-2704-X (HB)
ISBN 90-277-2705-8 (PB)

Published on behalf of
the International Astronomical Union
by
Kluwer Academic Publishers, P.O. Box 17, 3300 AA Dordrecht, The Netherlands.

Kluwer Academic Publishers incorporates
the publishing programmes of
D. Reidel, Martinus Nijhoff, Dr W. Junk and MTP Press.

Sold and distributed in the U.S.A. and Canada
by Kluwer Academic Publishers,
101 Philip Drive, Norwell, MA 02061, U.S.A.

In all other countries, sold and distributed
by Kluwer Academic Publishers Group,
P.O. Box 322, 3300 AH Dordrecht, The Netherlands.

Printed in The Netherlands

Dedication

The radio astronomy community was saddened by the untimely death on August 20, 1987 of Alan Moffet, Professor of Radio Astronomy at the California Institute of Technology. Alan was actively involved with radio interferometry from its early days and he participated in the development of VLBI techniques for nearly two decades. To all who knew him his quick mind and technical expertise will be missed. We dedicate these proceedings to his memory.

Table of Contents

Extragalactic

Galactic

Astrometry

Geophysics

Instrumentation and Analysis

Preface

The symposium "The Impact of VLBI on Astrophysics and Geophysics" was held in Cambridge, Massachusetts, USA, from May 10 to May 15, 1987. 268 participants from 23 countries attended the meeting, and 182 papers were presented, including 23 invited review papers, 45 other oral contributions, and 114 poster papers.

Conferences on scientific results of VLBI have been held at approximately four-year intervals since 1970. Previous meetings were: Charlottesville, VA 1970, Pasadena, CA 1974, Heidelberg 1978, Bologna 1983. The present meeting was held on the twentieth anniversary of the first coherent VLBI experiments conducted in 1967 by groups in Canada and the U.S. and also on the 100th anniversary of the discovery of radio waves by Hertz and of the Michelson-Morley experiment.

There were 18 scientific sessions: 6 devoted to extragalactic astronomy; 3 to galactic astronomy; 1 to astrometry; 4 to geophysics; and 3 to instrumentation and analysis. Major topics of discussion included: superluminal radio sources; gravitational lenses; steep spectrum sources; low frequency variables, polarization; distance measurements; cosmic masers; radio stars; interstellar scattering; reference frames; plate tectonics; earth orientation; propagation effects; orbiting VLBI; new telescopes and arrays; and image formation and processing techniques. At the end of most sessions, time was set aside for an oral review of a group of poster papers by a knowledgeable reviewer.

The symposium was sponsored by the International Astronomical Union (Symposium 129), International Union of Radio Sciences, and American Geophysical Union, and additional support was received from the National Aeronautics and Space Administration, National Science Foundation, Smithsonian Institution, Interferometrics Corporation, and Signatron Corporation. With this support, travel grants were provided to 45 participants.

Excursions were made to a Red Sox baseball game; a performance by the Boston Pops orchestra; Plimouth Plantation; Haystack Observatory; and the Museum of Fine Arts. A banquet was held at the Museum of Science in Boston, in the midst of a special exhibit on Science in India that included a model of the Ooty array.

We thank Carolann Barrett, Marilyn Bibeau, and Cristina Brigham, whose diligent efforts made the meeting and these proceedings possible.

Mark Reid
James Moran

1. J. Ulvestad 2. H. Aller 3. R. Preston 4. R. Gaume 5. M. Aller 6. T. Menon 7. T. Legg 8. G. Langston 9. M. Salvati 10. D. Hough 11. J. Galt 12. E. Fomalont 13. K. Weiler 14. C. Schalinski 15. R. Dewey 16. J. Hewitt 17. A. Rogers 18. D. Jones 19. R. Coates 20. R. Predmore 21. M. Morimoto 22. G. White 23. G. Levy 24. P. Biermann 25. L. Molnar 26. J. Cordes 27. D. Jauncey 28. J. Romney 29. A. Marscher 30. P. Hughes 31. J. Broderick 32. K. Kellermann 33. S. Broderick 34. D. Gordon 35. J. Schmelz 36. M. Inoue 37. G. Pooley 38. R. Wielebinski 39. R. Potash 40. F. Schwab 41. A. Wolszczan 42. J. Benson 43. A. Witzel

1. P. Kaufmann 2. C. Boucher 3. D. Fort 4. G. Nicolson 5. N. Broten 6. J. Dreher 7. J. Johansson 8. I. Shapiro 9. J. Marcaide 10. A. Baudry 11. G. Pilbratt 12. T. Clark 13. W. vanBreugel 14. G. Lundqvist 15. P. Crane 16. G. Elgered 17. S. Neff 18. J. Hooimeyer 19. M. Eubanks 20. N. Renzetti 21. M. Ratner 22. G. deBruyn 23. R. Gordon 24. C. Yoder 25. S. Gorgolewski 26. O. Sovers 27. B. Buerki 28. F. Pacini 29. P. Barthel 30. T. Herring 31. P. Shastri 32. G. Comoretto 33. F. Mantovani 34. D. Robertson

1. C. Walker 2. Y. Xia 3. T. Wan 5. D. Matsakis 6. N. Kylafis 7. M. Pearlman 8. Z. Qian 9. D. Meier 10. A. Kus 11. T. Maccacaro 12. R. Fanti 13. R. Ekers 14. G. Giovanni 15. G. Setti 16. V. Kulkarni 17. J. Wright 18. R. Porcas 19. A. Bombanati 20. C. Fanti 21. R. Perley 22. S. Wu 23. A. Whitney 24. P. Tomasi 25. T. Venturi 26. P. Charlot 27. L. Gregorini 29. J. Spencer 31. W. Carter 32. L. Padrielli 33. G. Petit 34. R. Treuhaft 35. T. Fischetti 36. W. Alef 37. A. Caporali 38. T. Pearson 39. R. Davis 40. M. Birkinshaw 41. V. Slysh 42. J. Zensus 43. G. Perola 44. S. Unwin 46. K. Kingham 47. J. Popelar 48. P. Parma 49. H. Chu 50. G. Tang 51. D. Shaffer 52. N. Kardashev 53. J. Moran 54. J. Wahr 55. A. Babcock 56. A. Tlamicha 57. M. Fujishita 58. X. Zheng 59. R. Rusk

1. P. Elosegui 2. S. Qian 3. A. Kemball 4. D. Dong 5. M. Reid 6. X. Tan
7. J. Campbell 8. A. Alberdi 9. M. Sarasso 10. B. Pernice 11. B. Corey
12. G. Verrone 13. R. Kollgaard 14. D. Salstein 16. R. Langley 17. J. Wardle
18. W. Cannon 19. D. Roberts 20. W. Cotton 21. J. Gomez-Gonzalez
22. A. Bogolomov 24. L. Litvinenko 26. B. Dennison 27. S. Spangler
28. P. Diamond 29. F. Briggs 31. C. Lonsdale 32. W. Watson 33. T. Muxlow

1. P. Thaddeus 2. A. Readhead 3. M. Wright 4. R. Vermeulen 5. M. Standish 6. R. Schilizzi 7. R. Nan 8. B. Doubinsky 9. D. Gabuzda 10. C. Gwinn 11. C. Flatters 12. P. Pismas 13. J. Ray 14. K. Johnston 15. T. Cawthorne 16. S. Ananthrakrishnan 17. L. Brown 18. W. Cobb 19. J. Lestrade 20. N. Vandenberg 21. R. Simon 22. J. Garcia-Barreto 23. R. Spencer 24. M. Cohen 25. D. Florkowski 26. E. Himwich 27. B. Burke 28. C. Ma 29. C. O'Dea 30. L. Bååth 31. D. Backer 32. S. Zisk 33. J. Marr 34. A. Mallama 35. R. Linfield 36. E. Preuss 37. A. Anderson 38. J. Biretta 39. P. Wilkinson 40. J. Ball 41. B. Rönnäng 42. R. Thompson 43. J. Wrobel 45. G. Tofani 46. K. Lind 47. M. Felli 48. R. Phillips 49. R. Norris 50. J. Waak 51. L. Oster 52. A. Niell 53. R. Cappalo 54. J. Yen

List of Participants

Antonio Alberdi, Inst. de Astrof. de Andalucia, Granada, Spain
Walter Alef, MPIFR, Bonn, Federal Republic of Germany
Hugh D. Aller, University of Michigan, Ann Arbor, MI USA
Margo F. Aller, University of Michigan, Ann Arbor, MI USA
S. Ananthakrishnan, Tata Inst. of Fund. Res., Ootacamund, India
Allen Joel Anderson, University of Uppsala, Uppsala, Sweden
Lars B. Bååth, Onsala Space Observatory, Onsala, Sweden
Alice K. Babcock, U.S. Naval Observatory, Washington, DC USA
Donald C. Backer, University of California, Berkeley, CA USA
John A. Ball, Haystack Observatory, Westford, MA USA
Norbert H. Bartel, Center for Astrophysics, Cambridge, MA USA
Peter D. Barthel, California Institute of Technology, Pasadena, CA USA
Alain Baudry, Observatoire de Bordeaux, Floriac, France
John M. Benson, NRAO, Charlottesville, VA USA
Peter Biermann, MPIFR, Bonn, Federal Republic of Germany
John A. Biretta, Center for Astrophysics, Cambridge, MA USA
Mark Birkinshaw, Harvard University, Cambridge, MA USA
Eric E. Bloemhof, Center for Astrophysics, Cambridge, MA USA
A.F. Bogomolov, Power Engineering Institute, Moscow, USSR
Alberto Bombonati, Instituto di Radioastronomia, Bologna, Italy
Claude Boucher, Institut Geographique National, Saint Mande, France
Franklin H. Briggs, University of Pittsburgh, Pittsburgh, PA USA
John J. Broderick, Virginia Polytechnic Inst., Blacksburg, VA USA
Norman W. Broten, National Research Council, Ottawa, Ontario, Canada
Leslie F. Brown, Brandeis University, Waltham, MA USA
Beat Buerki, Swiss Federal Institute of Technology, Zurich, Switzerland
Bernard F. Burke, MIT, Cambridge, MA USA
James Campbell, Geodetic Institute, University of Bonn, Bonn, FRG
Wayne H. Cannon, York University, Toronto, Ontario, Canada
Alessandro Caporali, University of Padova, Padova, Italy
Roger J. Cappallo, Haystack Observatory, Westford, MA USA
William E. Carter, National Geodetic Survey, Rockville, MD USA
Marco Catarzi, Osservatorio Astrofisico di Arcetri, Firenze, Italy
Timothy V. Cawthorne, Brandeis University, Waltham, MA USA
Mark J. Chandler, Allied Bendix Aerospace, Columbia, MD USA
Patrick Charlot, Institut Geographique National, Saint Mande, France
Han Shu Chu, Purple Mountain Obs., Nanjing, People's Republic of China
Thomas A. Clark, NASA/GSFC, Greenbelt, MD USA
Mark J. Claussen, FCRAO, Amherst, MA USA
Robert J. Coates, NASA/GSFC, Greenbelt, MD USA
Wesley K. Cobb, Brandeis University, Waltham, MA USA
Marshall H. Cohen, California Institute of Technology, Pasadena, CA USA
Giovanni Comoretto, Osservatorio Astrofisico di Arcetri, Firenze, Italy
James M. Cordes, Cornell University, Ithaca, NY USA
Brian E. Corey, Haystack Observatory, Westford, MA USA
James C. Cornell, Center for Astrophysics, Cambridge, MA USA
William D. Cotton, NRAO, Charlottesville, VA USA
Charles C. Counselman III, MIT, Cambridge, MA USA

Patrick C. Crane, NRAO, Socorro, NM USA
James L. Davis, Center for Astrophysics, Cambridge, MA USA
Richard J. Davis, University of Manchester, Cheshire, United Kingdom
A.G. deBruyn, Radio Sterrewacht Dwingeloo, Dwingeloo, Netherlands
Brian K. Dennison, Virginia Tech, Blacksburg, VA USA
Hans DeRuiter, Instituto di Radioastronomia, Bologna, Italy
Rachel J. Dewey, Cornell University, Ithaca, NY USA
Philip J. Diamond, NRAO, Charlottesville, VA USA
Dahan Dong, 166 Brookline Street, Cambridge, MA USA
Boris A. Doubinsky, I.R.E., Moscow, USSR
John W. Dreher, MIT, Cambridge, MA USA
Peter J. Duffett-Smith, MRAO, Cambridge, United Kingdom
Ronald D. Ekers, NRAO, Socorro, NM USA
Gunnar K. Elgered, Center for Astrophysics, Cambridge, MA USA
Pedro Elosegui, Inst. de Astrof. de Andalucia, Granada, Spain
Thomas Marshall Eubanks, Jet Propulsion Laboratory, Pasadena, CA USA
Emilio E. Falco, Center for Astrophysics, Cambridge, MA USA
Carla Fanti, Istituto di Radioastronomia, Bologna, Italy
Roberto Fanti, Istituto di Radioastronomia, Bologna, Italy
Marcello Felli, Osservatorio Astrofisico di Arcetri, Firenze, Italy
Attilio Ferrari, Universita di Torino, Torino, Italy
Ralph L. Fiedler, Naval Research Laboratory, Washington, DC USA
Thomas L. Fischetti, Technology Mgmt. Consult., Silver Spring, MD USA
Christopher Flatters, MPIFR, Bonn, Federal Republic of Germany
David R. Florkowski, US Naval Observatory, Washington, DC USA
Edward B. Fomalont, NRAO, Charlottesville, VA USA
David N. Fort, National Research Council of Canada, Ottawa, Ontario, Canada
Mitsumi Fujishita, Intl. Latitude Obs. of Mizusawa, Mizusawa, Japan
Denise C. Gabuzda, Brandeis University, Waltham, MA USA
John Galt, Dominion Radio Astrophysical Observatory, Pendicton, BC, Canada
Guido Garay, Universidad de Chile, Santiago, Chile
J. A. Garcia-Barreto, Instituto de Astronomia, UNAM, Ensenada, BC, Mexico
Ralph A. Gaume, University of Michigan, Ann Arbor, MI USA
Gabriele Giovannini, Istituto di Radioastronomia, Bologna, Italy
Jesus Gomez-Gonzalez, Centro Astronomico de Yebes, Guadalajara, Spain
David Gordon, NASA/GSFC, Greenbelt, MD USA
Richard Gordon, Northwestern University, Evanston, IL USA
Marc V. Gorenstein, Center for Astrophysics, Cambridge, MA USA
Stanislaw Gorgolewski, Torun Radio Astronomy Observatory, Torun, Poland
Lincoln J. Greenhill, Center for Astrophysics, Cambridge, MA USA
Loretta Gregorini, Istituto di Radioastronomia, Bologna, Italy
Carl R. Gwinn, Center for Astrophysics, Cambridge, MA USA
Bradford H. Hager, California Institute of Technology, Pasadena, CA USA
Philip E. Hardee, University of Alabama, Tuscaloosa, AL USA
Aubrey D. Haschick, Haystack Observatory, Westford, MA USA
Michael B. Heflin, MIT, Cambridge, MA USA
Paul D. Hemenway, University of Texas, Austin, TX USA
Thomas A. Herring, Center for Astrophysics, Cambridge, MA USA
Jacqueline N. Hewitt, Haystack Observatory, Westford, MA USA
W. Edwin Himwich, Interferometrics, Inc., Vienna, VA USA

Hans F. Hinteregger, Haystack Observatory, Westford, MA USA
Hisashi Hirabayashi, Nobeyama Radio Observatory, Nagano, Japan
Jeannette R. Hooimeyer, Sterrewacht Leiden, Leiden, Netherlands
David H. Hough, Jet Propulsion Laboratory, Pasadena, CA USA
Philip A. Hughes, University of Michigan, Ann Arbor, MI USA
Makoto Inoue, Nobeyama Radio Observatory, Nagano, Japan
David L. Jauncey, CSIRO, Epping, NSW, Australia
Jan M. Johansson, Onsala Space Observatory, Onsala, Sweden
Kenneth J. Johnston, Naval Research Laboratory, Washington, DC USA
Dayton L. Jones, Jet Propulsion Laboratory, Pasadena, CA USA
Thomas H. Jordan, MIT, Cambridge, MA USA
Nicholai S. Kardashev, Space Research Institute, Moscow, USSR
Pierre Kaufmann, INPE, Sao Paulo, Brazil
Kenneth I. Kellermann, NRAO, Charlottesville, VA USA
Athol J. Kemball, Hartbeesthoek Radio Ast. Obs., Johannesburg, South Africa
Robert W. King, MIT, Cambridge, MA USA
Kerry Kingham, US Naval Observatory, Washington, DC USA
Ronald Kollgaard, Brandeis University, Waltham, MA USA
Peter M. Kroger, Jet Propulsion Laboratory, Pasadena, CA USA
Vasant K. Kulkarni, Tata Institute of Fund. Res., Ootacamund, India
A. J. Kus, Torun Radio Astronomy Observatory, Torun, Poland
Nikolaos Kylafis, University of Crete, Crete, Greece
Richard B. Langley, University of New Brunswick, New Brunswick, Canada
Ann C. Lazzarin, Brandeis University, Waltham, MA USA
Thomas H. Legg, Natl. Research Council of Canada, Ottawa, Ontario, Canada
Jean-Francois Lestrade, Jet Propulsion Laboratory, Pasadena, CA USA
Gerald S. Levy, Jet Propulsion Laboratory, Pasadena, CA USA
A. Edward Lilley, Center for Astrophysics, Cambridge, MA USA
Kevin R. Lind, NRAO, Charlottesville, VA USA
Roger Linfield, Jet Propulsion Laboratory, Pasadena, CA USA
L.N. Litvinenko, Radioastronomy Institute, Latvia, USSR
Colin J. Lonsdale, Haystack Observatory, Westford, MA USA
Goran Lundqvist, Interferometrics, Inc., Vienna, VA USA
Chopo Ma, NASA/GSFC, Greenbelt, MD USA
Anthony Mallama, SAR/GSFC, Greenbelt, MD USA
Franco Mantovani, Istituto di Radioastronomia, Bologna, Italy
Juan Maria Marcaide, Inst. de Astrof. de Andalucia, Granada, Spain
Jonathan Marr, University of California, Berkeley, CA USA
Alan P. Marscher, Boston University, Boston, MA USA
Demetrios N. Matsakis, U.S. Naval Observatory, Washington, DC USA
Leonid I. Matveyenko, Space Research Institute, Moscow, USSR
Peter R. McCullough, University of California, Berkeley, CA USA
Gordon C. McIntosh, University of Massachusetts, Amherst, MA USA
David L. Meier, Jet Propulsion Laboratory, Pasadena, CA USA
T.K. Menon, University of British Columbia, Vancouver, BC, Canada
Lawrence A. Molnar, Center for Astrophysics, Cambridge, MA USA
James M. Moran, Center for Astrophysics, Cambridge, MA USA
Masaki Morimoto, Nobeyama Radio Observatory, Nagano, Japan
Mark Murray, MIT, Cambridge, MA USA
Robert L. Mutel, University of Iowa, Iowa City, IA USA

T.W.B. Muxlow, University of Manchester, NRAL, Cheshire, United Kingdom
Rendong Nan, Radio Sterrewacht Dwingeloo, Dwingeloo, Netherlands
Susan G. Neff, NASA/GSFC, Greenbelt, MD USA
G.D. Nicolson, Hartebeesthoek Radio Ast. Obs., Johannesburg, South Africa
Arthur E. Niell, Haystack Observatory, Westford, MA USA
Ray P. Norris, CSIRO, Epping, NSW, Australia
Christopher P. O'Dea, NRAO, Charlottesville, VA USA
Ludwig F. Oster, National Science Foundation, Washington, DC USA
Franco Pacini, Osservatorio Astrofisico di Arcetri, Firenze, Italy
Lucia Padrielli, Istituto di Radioastronomia, Bologna, Italy
Paola Parma, Istituto di Radioastronomia, Bologna, Italy
Michael Pearlman, Center for Astrophysics, Cambridge, MA USA
Timothy J. Pearson, California Institute of Technology, Pasadena, CA USA
Richard A. Perley, NRAO, Socorro, NM USA
Bartolomeo Pernice, Universita di Bari, Bari, Italy
Giuseppe Cesare Perola, Istituto Astronomico, Rome, Italy
Gerard Petit, Institut Geographique National, Saint Mande, France
William Petrachenko, Haystack Observatory, Westford, MA USA
Robert B. Phillips, Haystack Observatory, Westford, MA USA
Goran Pilbratt, European Space Agency, Noordwijk, Netherlands
Paris Pismis, Instituto de Astronomia, UNAM, Mexico City, Mexico
Guy G. Pooley, MRAO, Cambridge, United Kingdom
J. Popelar, Earth Physics Branch, E.M.R., Ottawa, Ontario, Canada
Richard W. Porcas, MPIFR, Bonn, Federal Republic of Germany
Robert Potash, Interferometrics, Inc., Vienna, VA USA
C. Read Predmore, FCRAO, University of Massachusetts, Amherst, MA USA
Robert A. Preston, Jet Propulsion Laboratory, Pasadena, CA USA
Eugen Preuss, MPIFR, Bonn, Federal Republic of Germany
Shanjie Qian, Beijing Astronom. Obs., Beijing, People's Republic of China
Zhihan Qian, Shanghai Observatory, Shanghai, People's Republic of China
Michael Ratner, Center for Astrophysics, Cambridge, MA USA
James Ray, Interferometrics, Inc., NASA/GSFC, Greenbelt, MD USA
Anthony C.S. Readhead, California Institute of Technology, Pasadena, CA USA
Mark J. Reid, Center for Astrophysics, Cambridge, MA USA
Nicholas A. Renzetti, Jet Propulsion Laboratory, Pasadena, CA USA
Barney Rickett, University of California, La Jolla, CA USA
David H. Roberts, Brandeis University, Waltham, MA USA
Douglas S. Robertson, NGS/NOAA, Rockville, MD USA
Alan E. E. Rogers, Haystack Observatory, Westford, MA USA
Jonathan D. Romney, NRAO, Charlottesville, VA USA
Bernt O. Rönnäng, Onsala Space Observatory, Onsala, Sweden
Raymond E. Rusk, University of Toronto, Toronto, Ontario, Canada
James W. Ryan, NASA/GSFC, Greenbelt, MD USA
Joseph E. Salah, Haystack Observatory, Westford, MA USA
Jarir Saleh, Israel Institute of Technology, Haifa, Israel
David A. Salstein, Atmos. & Environmental Res. Inc., Cambridge, MA USA
Marco Salvati, Osservatorio Astrofisico di Arcetri, Firenze, Italy
Maria Sarasso, Osservatorio Astron. di Torino, Pino Torinese, Italy
Jeanne M. Sauber, MIT, Cambridge, MA USA
Cornelius J. Schalinski, MPIFR, Bonn, Federal Republic of Germany

Richard T. Schilizzi, Neth. Fndn. for Radio Astronomy, Dwingloo, Netherlands
Joan Schmelz, Applied Research Corporation, Landover, MD USA
Matthew H. Schneps, Center for Astrophysics, Cambridge, MA USA
Frederic R. Schwab, NRAO, Charlottesville, VA USA
George A. Seielstad, NRAO, Green Bank, WV USA
Naosuke Sekiguchi, Tokyo Astronomical Observatory, Tokyo, Japan
Giancarlo Setti, European Southern Observatory, Munich, FRG
David B. Shaffer, Interferometrics, Inc., NASA/GSFC, Greenbelt, MD USA
Michael Shao, Center for Astrophysics, Cambridge, MA USA
Irwin I. Shapiro, Center for Astrophysics, Cambridge, MA USA
Prajval Shastri, Tata Institute of Fund. Res., Bangalore, India
Richard S. Simon, Naval Research Laboratory, Washington, DC USA
Vjacheslav I. Slysh, Space Research Institute, Moscow, USSR
Ojars J. Sovers, Jet Propulsion Laboratory, Pasadena, CA USA
Steven R. Spangler, University of Iowa, Iowa City, IA USA
John H. Spencer, Naval Research Laboratory, Washington, DC USA
Ralph E. Spencer, University of Manchester, NRAL, Cheshire, United Kingdom
E. Myles Standish, Jet Propulsion Laboratory, Pasadena, CA USA
William E. Strange, National Geodetic Survey, Rockville, MD USA
Alan J. Stratton, Signatron, Inc., Lexington, MA USA
Laurence G. Taff, Lincoln Laboratory, Lexington, MA USA
Xin Tan, University of New Brunswick, Fredericton, New Brunswick, Canada
Guoqiang Tang, Onsala Space Observatory, Onsala, Sweden
Guy Tarnstrom, MIT, Cambridge, MA USA
Patrick Thaddeus, Center for Astrophysics, Cambridge, MA USA
A. Richard Thompson, NRAO, Charlottesville, VA USA
Antonin Tlamicha, Astronom. Inst., Acad. of Sci., Ondrejov, Czechoslovakia
Gianni Tofani, Osservatorio Astrofisico di Arcetri, Firenze, Italy
Paolo Tomasi, Istituto di Radioastronomia, Bologna, Italy
Robert N. Treuhaft, Jet Propulsion Laboratory, Pasadena, CA USA
Edwin L. Turner, Princeton University, Princeton, NJ USA
James S. Ulvestad, Jet Propulsion Laboratory, Pasadena, CA USA
Stephen C. Unwin, California Institute of Technology, Pasadena, CA USA
Wil J.M. van Breugel, University of California, Berkeley, CA USA
Nancy R. Vandenberg, Interferometrics, Inc., Vienna, VA USA
Paul A. Vanden Bout, NRAO, Charlottesville, VA USA
Tiziana Venturi, Istituto di Radioastronomia, Bologna, Italy
Rene C. Vermeulen, Leiden Observatory, Leiden, Netherlands
Grazia Verrone, Universita di Bari, Bari, Italy
Robert F. C. Vessot, Center for Astrophysics, Cambridge, MA USA
Serge Volonte, European Space Agency, Paris, France
John A. Waak, Naval Research Laboratory, Washington, DC USA
John M. Wahr, University of Colorado, Boulder, CO USA
R. Craig Walker, NRAO, Socorro, NM USA
Tongshan Wan, Shanghai Observatory, Shanghai, People's Republic of China
John F.C. Wardle, Brandeis University, Waltham, MA USA
William D. Watson, University of Illinois, Urbana, IL USA
John C. Webber, Haystack Observatory, Westford, MA USA
Kurt W. Weiler, Naval Research Laboratory, Washington, DC USA
Sander Weinreb, NRAO, Green Bank, WV USA

Graeme L. White, CSIRO, Epping, NSW, Australia
Alan R. Whitney, Haystack Observatory, Westford, MA USA
Richard Wielebinski, MPIFR, Bonn, Federal Republic of Germany
Rick Wietfeldt, York University, Toronto, Ontario, Canada
Peter N. Wilkinson, University of Manchester, Cheshire, United Kingdom
Arno Witzel, MPIFR, Bonn, Federal Republic of Germany
Aleksander Wolszczan, NAIC, Arecibo Observatory, Arecibo, PR USA
Diana M. Worrall, Center for Astrophysics, Cambridge, MA USA
James P. Wright, National Science Foundation, Washington, DC USA
Melvyn C.H. Wright, University of California, Berkeley, CA USA
Joan M. Wrobel, New Mexico Tech., Socorro, NM USA
Shengyin Wu, Beiging Astron. Obs., Beijing, People's Republic of China
Yifie Xia, Nanjing University, Nanjing, People's Republic of China
Jui L. Yen, University of Toronto, Toronto, Ontario, Canada
Charles F. Yoder, Jet Propulsion Laboratory, Pasadena, CA USA
John Kar-Kin Zao, University of Toronto, Toronto, Ontario, Canada
Marguerite Zarrillo, Brandeis University, Waltham, MA USA
J. Anton Zensus, California Institute of Technology, Pasadena, CA USA
Xingwu Zheng, Nanjing University, Nanjing, People's Republic of China
Stanley H. Zisk, Haystack Observatory, Westford, MA USA

HERTZ AND MICHELSON

Irwin I. Shapiro
Harvard-Smithsonian Center for Astrophysics
Cambridge, MA, USA

1. INTRODUCTION

I have mixed feelings about giving the opening talk at this symposium. On the one hand I am delighted to welcome you to this conference on the impact of VLBI on astrophysics and geophysics. On the other hand I am very sad that Ed Purcell is bothered by a bad back and cannot give the introduction as originally scheduled. Ed is one of the truly great of the world's physicists and one of the nicest persons ever born. My substituting for Ed is somewhat like Bill Buckner pinch hitting for Lou Gehrig. (Those who do not understand this reference to American baseball might have a better appreciation after viewing tonight's Red Sox game.)

We originally intended this symposium to be a celebration of the 20th anniversary of the first successful VLBI experiments. However, while beginning to plan for this event, I realized that 1987 would be the 100th anniversary of the discovery of radio waves by Heinrich Hertz. A few days ago it occurred to me that 1987 was also the 100th anniversary of the Michelson-Morley experiment. These developments together, of course, supply the two ingredients of VLBI -- radio waves and interferometry. Hence, it seemed appropriate to dedicate the conference to both Heinrich Hertz and Albert Michelson. I wanted to invite both, but I was a little late getting out the invitations.

To honor these two gentlemen in absentia I will go over a few -- very few -- of the highlights of their discoveries and of their lives.

2. HERTZ

Like our George Washington, Hertz (Fig. 1) was born on 22 February, a child prodigy who changed from a career in engineering to one in physics, just as did Purcell. Like Mozart, but unlike Purcell, Hertz died very young, of a bone malignancy in his head -- a terrible tragedy for his family and for science.

In the 1870s and 1880s a major question in physics concerned electrodynamics. Maxwell's theory was not yet well accepted. In England, yes; on the continent, hardly at all. Competing on the continent were the so-called action-at-a-distance theories of Newmann and Weber. Experimental distinctions were not easy to come by.

M. J. Reid and J. M. Moran (eds.), The Impact of VLBI on Astrophysics and Geophysics, 1–6.

The conflict concerned primarily the consequences of open circuits which Maxwell resolved with his ingenious fourth equation that modified Ampere's law by introducing the displacement current and thereby insuring conservation of charge. One question concerned the possible generation in space of electric waves, predicted by Maxwell's theory, but not by the others. While at the Technische Hochschule in Karlsruhe, Hertz devised a method[1] to attempt to detect such electric waves ("radio waves" to us). He used for his primary (transmitter) circuit an induction-coil, spark-gap dipole arrangement which, as Purcell deduced, radiated ~10 watts at ~60 MHz. The oscillations were set up in this spark gap by means of "the most powerful discharges that could be obtained from a large induction coil." His secondary (receiver) circuit was a circle of wire 35 cm in radius with a dielectric "handle" and a short spark gap, adjustable with a micrometer screw for ease in rotating it and in placing it at any desired position. For some experiments, Hertz modified the secondary to be very nearly in resonance with the primary through soldering small pieces of sheet metal to the poles to change the capacitance so as to maximize the spark length in a given position and orientation. By inducing repeated discharges and measuring the effects at many different places around the room with his receiver, Hertz was able to establish the presence of standing waves and to measure their length. Calculating frequency (see above) from the inductance and capacitance of the relevant components in the transmitter, much as physics undergraduate students would now do, Hertz then deduced the velocity of the radiation, a quite respectable 320,000 (+ 50% - 30%) km s^{-1}. At the end, Hertz noted how it was "certainly remarkable that proof of the finite rate of propagation should have been first brought forward in the case of a force which diminishes in inverse proportion to the distance, and not the square of the distance."

Fig. 1 Heinrich Hertz (1857-1894), in his prime, circa 1887.

To investigate these waves further, Hertz needed higher frequency radiation to make use of concave parabolic mirrors, while avoiding having his experiments distorted by the "disproportion between the length (~5 m) of the waves (previously) used and the dimensions which (he) was able .... to give to the mirror." By pushing the then state of the art to its limit, Hertz decreased the wavelength of the radiation to ~30 cm, a value, in fact, not too far from that used by the Canadian group in their first VLBI experiment in 1967. Hertz then constructed two identical zinc, concave, cylindrical, 12.5-cm focal length, paraboloids, forming them on a wooden template. The spark gap from the primary was placed on the focal line of one of the mirrors and the secondary gap was placed, in effect, on the focal line of the other, although the wires were arranged so that the spark gap was behind the mirror and could be adjusted and examined "without obstructing the course of the waves." In some cases, Hertz dealt with minute sparks -- only a few hundredths of a millimeter in length.

By placing "blocking" conductors at various places, Hertz established beyond a shadow of a doubt that the propagation was rectilinear; he also noted the lack of a sharp geometric shadow to the rays and the correspondence to diffraction, although he could not discern maxima and minima at the edges of the shadows. To demonstrate reflection, Hertz used 2m x 2m plain vertical "walls" of zinc, positioned in various ways with respect to the two mirrors. He determined the position of the wavefront with respect to the rays before and after reflection, in addition to the plane of oscillation before and after, all as a function of the angle of incidence. He, of course, established the equality of the angle of incidence and reflection, and noted that the reflection was more nearly "regular" than diffuse. He even placed his transmitter and receiver in adjoining rooms, with the reflector situated in the doorway between, to study large angles of incidence.

To investigate refraction Hertz made a large prism in the shape of an isosceles triangle, 1.2 m on a side, with about a 30° refracting angle. He made this prism from hard pitch, cast in wooden boxes, and blocked passage of any rays not going through the prism with conducting screens. For the refractive index, he obtained 1.7; for light, it was known to be between 1.5 and 1.6; the difference was within Hertz's (uncertain) estimate of the uncertainty.

Hertz also demonstrated that the electric waves were transverse and plane polarized. In particular, he rotated the "receiving" mirror until its focal line was perpendicular to that of the transmitter, and watched the sparks in the receiver gradually become more feeble and then disappear for the right-angle configuration. He further demonstrated polarization properties by interposing a screen of parallel copper wires; for example, with the wires making a 45° angle with the mutually perpendicular focal lines of the two mirrors, he was again able to observe sparks in the secondary and noted the analogy to "the brightening of the dark field of two crossed Nicols by the interposition of a crystalline plate in a suitable position." Hertz also established to his satisfaction that electric and magnetic oscillations in the wavefront were perpendicular to each other.

For Hertz, these experiments, in toto, removed any doubt about electric waves being the same as light waves, save for having far longer wavelengths. This grand synthesis of electromagnetic and light phenomena was of profound importance for physics.

Hertz's extraordinary thoroughness was perhaps nowhere better illustrated than in his discovery of the photoelectric effect. In his experiments on electric waves, he at times enclosed the (secondary) spark coil in a dark case to observe more easily the induced spark. The dielectric case he chose was, of course, transparent to electric waves. To Hertz's surprise, "the maximum spark-length became decidedly smaller inside the case" than when the coil was exposed directly to the primary. Removing the case in parts, he deduced that only the portion interposed between the secondary and the primary produced the effect. He continued with a painstakingingly thorough, exhaustive investigation of all conceivably relevant variables that might influence the effect; he deduced, for example, that its cause must travel in straight lines. By studying various interposed materials -- solid, liquid, and gaseous -- he found which ones did, and which did not, hinder this travel, and how the effects depended on the thickness of the interposed material. He even demonstrated that the effect could be produced with reflectors and refractors appropriately placed between the primary and the appropriately shielded secondary. After nearly 20 different sets of experiments, he concluded that ultraviolet light was the cause of this increase in the spark length.

Hertz was also no slouch as a theorist and, near the end of his life, while ill, concentrated on theory. He developed the modern form of Maxwell's equations, if not the vector and tensor notations. Maxwell's original formulation was rather abstruse from our point of view, being entangled with the then pervasive mechanical model of the world. Hertz stripped away the irrelevancies, including the attempted rationalizations in terms of the mechanical view, and gave those equations nearly the form we see on "T" shirts. His philosophy was also very modern; to the question "what is Maxwell's theory?," he responded that he knew no shorter or more definitive answer than "Maxwell's theory is Maxwell's set of equations."

Purcell, as a young graduate student, had gone to the Technische Hochschule in Karlsruhe and was in the very room in which Hertz's first experiments on radio waves were performed; he recalls clearly the inscription on the plaque there commemorating Hertz's achievements. Curiously, Purcell, who was trained in electrical engineering as an undergraduate, had not even seen Maxwell's equations, *per se*, when he went to Karlsruhe, and the "facts" of the fourth equation were then unknown to him. Needless to say, that condition neither lasted much longer nor hindered his later enormous contributions to physics.

3. MICHELSON

Now let me turn to Albert Michelson (Fig. 2). The Polish-born son of immigrants to the US[2], Michelson started his career at the Naval Academy in Annapolis, but was soon drawn to physics and, especially, optics. He was asked, at 25, to begin teaching an advanced physics course with a demonstration of the measurement of the speed of light by Foucault's rotating mirror method. Within a short time he saw how to improve this method significantly and, within about a year, had built the needed equipment and obtained a result: 299,940 km s^{-1} -- the first of his many improvements on the measurement of that fundamental quantity. His work brought him to the attention of the already renowned Simon Newcomb, and Michelson's scientific career was set by a transfer to the Nautical Almanac Office from the Naval Academy.

Maxwell, shortly before his death in 1879, wrote to David Todd, the then director of the Nautical Almanac Office, asking about the possibility of a celestial measurement of the aether drift. Maxwell was not sanguine, however, about the prospects for measuring any quantity proportional to "the square of the ratio of the earth's velocity to that of light." Michelson, intrigued, took up the challenge; first he felt he needed advanced training in physics and so went to the University of Berlin. He developed his first interferometer in Europe, inspired by Fizeau's (1859) experiments, and performed his first aether drift experiment in 1881 in Germany. He was reportedly disappointed at the null result. In his report on this experiment, however, he had made a small, but embarrassing, mistake in calculation, immediately recognized and corrected independently by André Potier and by Hendrik Lorentz. This error -- the neglect of the effect of the earth's motion on the interferometer arm perpendicular to that of the motion -- did not affect, qualitatively, the null result, but did cast a bit of a cloud over its "credibility." Michelson's null result was apparently not taken very seriously in the physics community. Nonetheless, he wished to pursue and to improve his experiment. Even then, such undertakings required raising funds. Luckily, Michelson was quite adept at such endeavors. He had secured funds for his first experiment, with Newcomb's help, from Alexander Graham Bell's Volta Foundation; for his later experiment, he received an endorsement from Lord Rayleigh for a repetition of the experiment and recognized immediately its value, and used it, in raising the needed funds. The experiment, undertaken with Edward Morley, was carried out while Michelson was at the Case School of Applied Science (now part of Case-Western Reserve). The result, announced in 1887, almost exactly 100 years ago today, was now taken quite seriously by the world's scientists; it showed that any drift of the earth through the aether, if it existed, was less than 1/6 the earth's orbital velocity. Thus did interferometry first make its lasting mark.

Fig. 2 Albert A. Michelson (1852-1931), in 1887, at the time of THE experiment.

Not long thereafter, c. 1890, Michelson tried to apply his technique to observations of the Galilean satellites of Jupiter. At that time there was also a mini-controversy raging over inconsistencies between different determinations of the sizes of these satellites. Michelson wrote a low-key letter to one of my professional ancestors here at the Harvard College Observatory, requesting permission to use the Great Refractor (15" diameter) to try out these interferometric measurements. After much procrastination for reasons lost in the mist of history, E. C. Pickering gave Michelson his chance. However, apparently a combination of delays, instrument problems, and bad seeing caused the endeavor to fail. It is not clear which was the most important cause. Michelson later succeeded in these measurements, in better seeing, at the Lick Observatory in California. As we all know, he went on early in this century to make important interferometric determinations, with Francis

Pease, of stellar diameters, using primarily the 100" diameter reflecting telescope on Mt. Wilson, the site of current advances in the application of optical interferometry to astrometry.

Michelson was respected throughout the physics and astronomy communities as a superb experimentalist. It was widely felt that only he could make an astronomical interferometer work and, perhaps because of that mystique, such interferometers were neither developed nor used by any other astronomers for many years.

4. HISTORICAL CONNECTIONS BETWEEN HERTZ AND MICHELSON

Hertz and Michelson both worked under Herman von Helmholtz in Berlin in the early 1880s. Although I found no definitive reference to their knowing one another, it could hardly have been otherwise: Von Helmholtz's group was rather small.

When Clark University in Worcester, MA was founded with the intent to have the finest obtainable faculty in science, both Hertz and Michelson were invited to join. Michelson went; Hertz did not. The reason proffered for Hertz's declination of the invitation, it is said, was his then recent marriage.

Both Hertz and Michelson measured the speed of light, although at vastly different wavelengths and with vastly different accuracy. Both were also connected with Albert Einstein's "wunder yahr" of 1905: Hertz through his discovery of the photoelectric effect and the establishment of the equivalence between radio and light waves, and Michelson (more tenuously) through the tight experimental limit he placed on any motion of the earth through the aether.

Hertz and Michelson both won the Rumford Prize of the Royal Society, albeit not in the same year. Michelson went on to win the Nobel Prize; Hertz doubtless would have won this prize, too, had he not died before its establishment. Michelson's Nobel citation did not even mention the aether-drift experiments; it emphasized his interferometer-based experiments directed toward the use of a wavelength of light as a universal standard of length. Hertz would most likely have been cited for his experiments on electric waves.

The first successful VLBI observations were also recognized by the Rumford Award of the American Academy of Arts and Sciences in 1971. The prize was split 21 ways, and many of the recipients are here today to help honor the 20th anniversary of this achievement.

The VLBI community has now proliferated from the original 21 to casts of hundreds, if not thousands. With all of you, I look forward to hearing about the fascinating results recently obtained through the use of VLBI, in astrophysics and geophysics. This conference is probably unique in bringing together experts in fields as diverse as earthquake hazard reduction and quasar kinematics. VLBI provides the glue that should keep these various fields together for their mutual benefit.

5. REFERENCES

1. Hertz, H. Electric Waves (Dover, New York, 1962).
2. Livingston, D.M. The Master of Light (Scribner's, New York, 1973).

SUPERLUMINAL RADIO SOURCES

J. A. Zensus and T. J. Pearson
Owens Valley Radio Observatory
California Institute of Technology
Pasadena, California 91125
USA

ABSTRACT There are now at least 23 radio sources in which apparent superluminal motion has been found. The effect is not limited to the "classic" examples like 3C 345, but is found in objects with a wide range of characteristic properties. We review the observations of the phenomenon and present an up-to-date list of the known superluminal sources.

1. INTRODUCTION

The discovery of superluminal sources in 1970–71 is one of the most exciting results obtained in the twenty-year history of VLBI. It provided a remarkable confirmation of Martin Rees's prediction that "an object moving relativistically in suitable directions may appear to a distant observer to have a transverse velocity much greater than c" (Rees 1966).

The development of the study of superluminal sources closely parallels that of VLBI. Rees's prediction of superluminal motion was made in response to the discovery of rapid variations in the radio flux of quasars (e.g., Dent 1965). Soon after the introduction of VLBI (Broten *et al.* 1967; Bare *et al.* 1967), variations in flux density were found to be accompanied by variations in source structure, which in some cases appeared to require faster-than-light expansion (e.g., Moffet *et al.* 1972). The first clear indication of superluminal expansion came when long-track observations of the quasars 3C 273 and 3C 279 revealed sharp visibility minima which moved in hour-angle between observing sessions (Knight *et al.* 1971; Whitney *et al.* 1971; Cohen *et al.* 1971). In the case of 3C 279, the changes were most simply interpreted as due to an increase in the separation of a symmetric double source with a speed of about $10c$.

Four such superluminal sources (3C 120, 3C 273, 3C 279, and 3C 345) were monitored in the period 1971–1977 (e.g., Cohen *et al.* 1977; Cotton *et al.* 1979); but some skepticism about the phenomenon persisted until real images of the sources were made using the structure information contained in the closure phase (Rogers *et al.* 1974; Readhead and Wilkinson 1978). For example, the sequence of images of 3C 273 by Pearson *et al.* (1981) won a number of converts. Now, the basic phenomenon is no longer in doubt, and observers concentrate on unravelling the details of the trajectories and evolution of the components and on studying the statistics of superluminal sources, in the hope of determining the physical processes at work in the sources and of testing the predictions of models of extragalactic radio sources.

The Mark-III VLBI system has produced further discoveries. Its increased sen-

M. J. Reid and J. M. Moran (eds.), The Impact of VLBI on Astrophysics and Geophysics, 7–16.

sitivity has allowed the detection of sources as weak as 50 mJy, and is essential for the study of superluminal motion in the weak cores of extended, lobe-dominated quasars. It has also permitted phase-referencing which can provide an absolute frame of reference for superluminal motion. VLBI imaging of polarization promises to be a very productive new technique. We confidently expect that the next technical steps forward, the Very Long Baseline Array and mm and space VLBI, will also provide new surprises and insights on the superluminal phenomenon.

Earlier reviews of observational results on superluminal sources include the articles by Cohen *et al.* (1977), Cohen and Unwin (1982, 1984), Kellermann and Pauliny-Toth (1981), Cohen (1986), and Porcas (1984, 1987). For a theoretical perspective we refer to, e.g., Scheuer (1984), Rees (1984), Phinney (1985), and Blandford (1984, 1987). Here we summarize recent observations and present a compilation of all known superluminal sources, without preempting the multitude of detailed results discussed at this Symposium. A more detailed account of the field can be found in the proceedings of the Workshop on Superluminal Radio Sources, held at Big Bear Solar Observatory in October 1986 (Zensus and Pearson 1987).

2. CURRENT OBSERVATIONAL CAPABILITIES

In spite of the great progress made in VLBI over the last 20 years, our understanding of superluminal sources is still limited by the instruments with which we study them.

VLBI images can now be made routinely at wavelengths $\lambda = 50$, 18, 6, 2.8, and 1.3 cm, using as many as 18 antennas. The maximum resolution (FWHM) is $\sim 0.5\lambda/D$; for baseline length $D = 8200$ km (Europe–California) and $\lambda = 1.3$ cm this is 0.2 milliarc-seconds (mas). Most of the available images have been made at 6 cm and have resolution ~ 1 mas. Shorter wavelengths (7 mm, 3 mm) have been used successfully but are not yet routine, and image-forming capability is crude.

In terms of linear resolution, 1 mas ranges from about 0.25 pc in nearby active galactic nuclei to about 4 pc in most quasars. For comparison, the Schwarzschild radius of a $10^9 M_\odot$ black hole is $\sim 10^{-4}$ pc, the size of the "broad emission line region" in active nuclei is typically ~ 1 pc, and the size of the "narrow emission line region" is typically ~ 100 pc. Thus VLBI can resolve the broad-line region, but it is still some way from resolving the "central engine".

When making VLBI images, self-calibration techniques are used to circumvent atmospheric and oscillator phase and amplitude instability (e.g., Pearson and Readhead 1984*a*). These techniques require a source to be detectable in the coherence time (in the range 30 s to 10 min, depending on frequency and weather conditions), so that observations are restricted to strong sources, brighter than ~ 0.5 Jy for the Mark II recording system (bandwidth 2 MHz), or ~ 0.1 Jy for the Mark III system (bandwidth up to 112 MHz). When self-calibration is used, absolute position information is lost, which makes it difficult to register images made at different wavelengths or epochs.

The dynamic range achieved in VLBI images (the ratio of the strongest feature in the image to the noise level) ranges from about 20 : 1 to 1000 : 1, and is limited by (*a*) sensitivity (a few millijanskys per beam area), (*b*) "non-closing" calibration errors which cannot be removed by self-calibration, and (*c*) inadequate coverage of the (u, v) plane (i.e., missing Fourier components of the image). All of these problems should be alleviated when the Very Long Baseline Array and other new arrays are completed in the 1990s.

3. APPARENT VELOCITY

What is measured in superluminal sources is the rate of change of the apparent separation of two features in the brightness distribution. To convert this angular velocity μ into an apparent linear velocity $v_{\rm app} = c\beta_{\rm app}$ (projected on the sky), the distance to the source is needed, and for almost all the known superluminal sources this must be inferred from the redshift, z. In the standard Friedmann–Robertson–Walker model, parameterized by the Hubble constant H_0 and the deceleration parameter q_0, the apparent transverse velocity $\beta_{\rm app}$ is related to the observed angular velocity μ by

$$\beta_{\rm app} = \frac{\mu z}{H_0(1+z)} \left(\frac{1+\sqrt{1+2q_0 z}+z}{1+\sqrt{1+2q_0 z}+q_0 z} \right). \qquad (1)$$

The values of H_0 and q_0 are unfortunately still subject to debate. For consistency, we shall use $H_0 = 100h$ km s^{-1} Mpc^{-1}and $q_0 = 0.5$ in this review. If μ is measured in milliarcseconds per year, the appropriate numerical factor in equation (1) is $H_0 = 2.11 \times 10^{-2} h$ mas yr^{-1}.

4. 3C 345: A SUPERLUMINAL ARCHETYPE

The quasar 3C 345 is in many respects the best-studied superluminal source owing to its high radio flux density and its high declination (39°) which allow routine Mark-II VLBI monitoring with many stations and good (u,v)-coverage. Figure 1 shows a sequence of hybrid maps of this source made at 10.7 GHz (Biretta, Moore, and Cohen 1986).

The compact structure of 3C 345 has the typical asymmetric "core-jet" character, with several "jet" components separating superluminally from the "core" D. The identification of core and jet components in superluminal sources is usually based on the finding that one component, the core, is compact, located at one end of the structure, and has a flat radio spectrum, whereas the jet components tend to be resolved and have steeper spectra. All too often, however, such a distinction is based on an educated guess rather than firm observational evidence. A related problem is the difficulty of uniquely identifying components from one epoch to another (e.g., Blandford 1987). In 3C 345, the phase-referencing observations of Bartel *et al.* (1986) have shown that the core indeed appears stationary on the sky and that any motion with respect to the nearby quasar NRAO 512 is smaller than 0.2 mas per year. At least three superluminal components (C2, C3, C4) in 3C 345 have now been monitored for several years. Their properties are compared with the predictions of a synchrotron self-Compton model by Biretta and Cohen at this Symposium. Here we list some of the observational results.

1. Component sizes and separation from the core can be frequency-dependent, as seen for C3 in 3C 345.
2. Superluminal components in a given source do not always all move with the same velocity. This is seen not only in 3C345, but also in 3C 120 and 3C 279, and several other sources.
3. Acceleration is seen for component C4 in 3C 345 with changes in both magnitude and direction of the superluminal velocity.
4. The size of the superluminal components in 3C 345 increases roughly linearly with separation from the core, corresponding to a projected jet opening angle of 26°.
5. An important question is whether individual components follow the same path or are moving in different directions. There is some evidence that components in 3C 345 (and also in 3C 273) follow a common curved path, but this is still tentative.

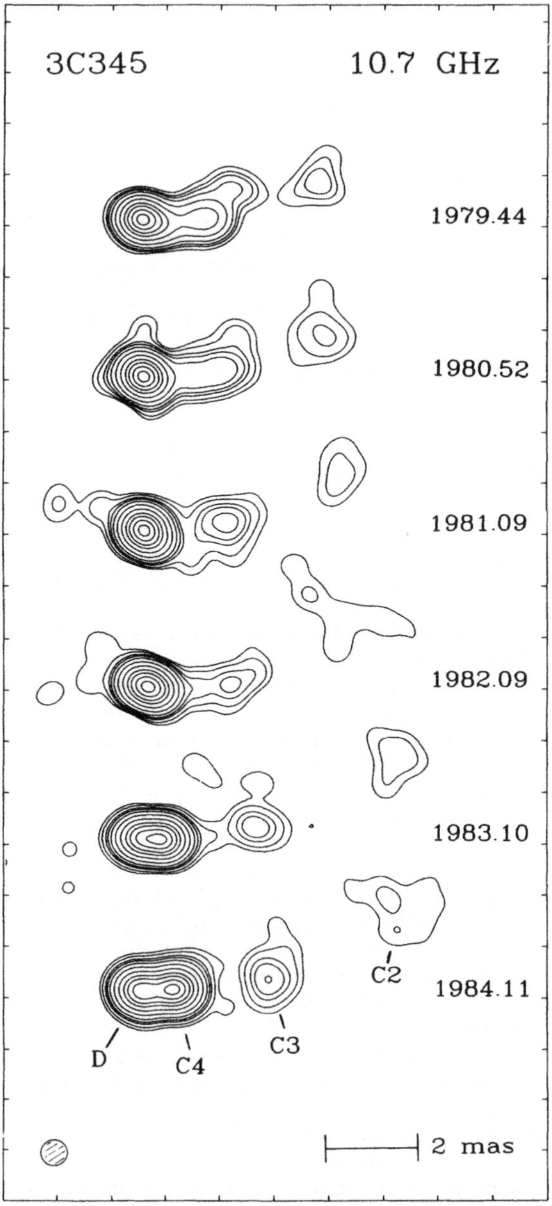

Figure 1 A sequence of hybrid maps of 3C 345 at 10.7 GHz (from Biretta, Moore, and Cohen 1986). North is at the top; tick marks are 1.2 mas apart. Contour levels are 0.5, 1, 2, 3, 5, 10, 20, 35, 50, 70, and 90% of the peak intensity; lower contours have been omitted in some of the maps. The FWHM of the Gaussian restoring beam is 0.6 mas (*hatched circle*).

6. The structure in 3C 345 is curved northward in the direction of the extended emission regions located 1–3 arcsec from the core. Such curvature is quite common. The early finding that curvature and also misalignments with the large-scale symmetry axis are more pronounced in core-dominated sources has been confirmed.

Although some of the above properties have been seen only in 3C 345 it is likely that this object is typical of the class of superluminal sources that contain strong, core-dominated sources as found in radio samples selected at high frequencies.

5. SEARCHING FOR SUPERLUMINAL SOURCES

The first few superluminal sources were discovered accidentally in observations of the brightest compact sources in the sky. These include 3C 345, 3C 273, 3C 279, and 3C 120. In one case, NRAO 140, superluminal motion was successfully predicted based on its low X-ray intensity. Marscher (1987) has recently cautioned that such predictions may be much weaker than previously thought, owing to ill-defined parameters in the assumed synchrotron self-Compton models.

Recently, a number of systematic searches for superluminal sources have been undertaken. Such programs are essential for determining the frequency of occurrence of superluminal motion and the distribution of apparent velocities. Witzel and his colleagues (Witzel 1987; Eckart *et al.* 1986, 1987) have selected a sample of 13 sources from the NRAO-MPIfR 5-GHz S5 survey, with flux densities $\geq$ 1 Jy and flat spectra. They have studied these sources intensively in the X-ray, optical, and radio wavebands, including multi-epoch VLBI at several frequencies. Three of the sources with known redshift show superluminal motion (0212+735, 1150+812, 1928+738) and two more are superluminal candidates. Four sources without redshifts show apparent expansion which would be superluminal if, as seems probable, the redshifts exceed 0.3. There is also evidence from a comparison of measured and predicted inverse-Compton X-ray fluxes for bulk relativistic motion in 12 of the 13 sources. They conclude that bulk relativistic motion is a common phenomenon in core-dominated sources.

Pearson, Readhead, and Barthel (1987; see also Pearson and Readhead 1984*b*) are studying a larger sample of 65 sources, also selected at 5 GHz; the selection criterion was flux density $\geq$ 1.3 Jy, with no spectral selection. Forty-five of these sources are strong enough to map with Mark-II VLBI. Although the observations are not yet complete, nine sources in this sample are already known to be superluminal; four of these are new (0850+581, 0906+430, 1642+690, and 1928+738). Superluminal sources are found in all types of source except the "compact double" sources (Phillips and Mutel 1982; Hodges and Mutel 1987).

Both of the above programs have selected strong sources. If the beaming theories are correct, sources in such flux-density limited samples are not randomly oriented but are preferentially beamed towards the observer. To avoid this bias, several groups have attempted to define orientation-unbiased samples for study, selecting sources by the flux density of the extended lobes, which are assumed to be unbeamed, rather than the compact core. Hough and Readhead (1987) are studying an unbiased sample of 26 double-lobed quasars using the Mark-III system at 10.7 GHz. So far they have studied six sources and found two of them to be superluminal (3C 245 and 3C 263). Zensus and Porcas (1987) have selected a similar sample of 30 quasars, based on the flux density at 966 MHz where the core is not a major contributor. Of the five sources they have studied closely, three are superluminal (3C 179, 3C 263, and 1951+498). The results from both studies indicate a trend that velocities in lobe-dominated superluminals are smaller than in core-dominated objects, in agreement with the expectations of beaming models.

Table 1 Superluminal Radio Sources

Source		z	Component	μ [mas yr^{-1}]	$\beta_{\rm app}h$	Ref.
0212+735		2.367	–	0.09[a]	3.9	1
0333+321	NRAO 140	1.258	B	0.15	4.8	2
0430+052	3C 120	0.033	A	1.35	2.1	3
			B	2.53	3.9	3
			C	2.47	3.8	3
			D	2.66	4.1	3
			E	2.54	3.9	3
0723+679	3C 179	0.846	–	0.19	4.8	4
0735+178		0.424	NE	0.18	2.8	5
0850+581		1.322	–	0.12	3.9	6
0851+202	OJ 287	0.306	SW1–2	0.28[b]	3.3	7
0906+430	3C 216	0.669	–	0.11	2.4	8
0923+392	4C 39.25	0.699	a–c	<0.006[c]	<0.1	9
			b	0.16	3.5	9
1040+123	3C 245	1.029	–	0.11[d]	3.1	10
1137+660	3C 263	0.652	–	0.06	1.3	11
1150+812		1.25	–	0.13	4.1	1
1226+023	3C 273	0.158	C3	0.79	5.3	12
			C4	0.99	6.6	12
			C5	1.20	8.1	13
			C7a	0.76	5.1	13
1253–055	3C 279	0.538	?	0.5[e]	9.2	14
			B2	0.11	2.0	15
1641+399	3C 345	0.595	C2	0.48	9.5	16
			C3	0.30	5.9	16
			C4	0.07,0.3	1.4,5.9	16
1642+690		0.751	–	0.34	7.9	17
1721+343	4C 34.47	0.206	–	0.36	3.1	18
1901+319	3C 395	0.635	1–2	<0.06[c]	<1.2	19
			3	0.64[d]	13.2	19
1928+738		0.302	A1–4	0.6	7.0	20
1951+498		0.466	–	~0.07[d]	~1.2	21
2200+420	BL Lac	0.0695	1–4	~0.76[f]	~2.4	22
2230+114	CTA 102	1.037	–	~0.65[g]	~18.5	23
2251+158	3C 454.3	0.859	2	<0.05	<1.3	24
			4	0.35[h]	8.9	24

Among other programs to search for superluminal sources, Bååth (1987) has studied several low-frequency variable sources, and found one superluminal (CTA 102 = 2230+114). This source has an unusually high expansion speed, $(18 \pm 4)h^{-1}c$, which may be related to the high Lorentz factors required to explain low-frequency variability as an intrinsic effect. The discovery of superluminal motion lends some support to the idea that the variability is indeed intrinsic rather than due to refractive scintillation in the interstellar medium. Bååth (1984) has also studied a number of BL-Lac objects and has found one of these to be superluminal (0735+178).

6. INVENTORY

At the last IAU meeting on VLBI (IAU Symposium No. 110 in Bologna), eight sources were known to show apparent superluminal motion (Cohen and Unwin 1984), but since then the number has increased rapidly and is now at least 23. In Table 1 we have assembled a list of all superluminal sources known at the time of writing (May 1987), including redshift, apparent angular velocity, and inferred apparent velocity. The list includes the "classic" sources, the sources discovered in the surveys described above, and a few more individual

Table 2 Subluminal and Stationary Sources

Source		z	Component	μ [mas yr^{-1}]	$\beta_{app}h$	Ref.
0153+744		2.33	–	<0.03[i]	<1.3	1
0316+413	3C 84	0.0172	–	0.24	0.2	25
0710+439		0.517	B	<0.04[j]	<0.7	8
			C	<0.07[j]	<1.3	8
0711+356		1.62	–	−0.05	−1.8	26
1228+127	M 87	0.004	–	<0.3	<0.1	27
1637+826	NGC 6251	0.023	–	<0.3	<0.3	28
1934−638		0.183	–	<0.4[k]	<3.1	29
2021+614		0.2266	–	<0.04	<0.4	26
2134+004		1.936	–	<0.01[h]	<0.4	30

Notes to Tables 1 and 2: [a] no published sequence of maps available yet. [b] polarization data. [c] limit on velocity between stationary components. [d] well-separated double structure only at 2 epochs. [e] one-baseline results. [f] average of four events; possibly deceleration. [g] maps made at 932 MHz; well-separated components only at 2 epochs. [h] complex structure. [i] upper limit on separation of 12-mas double structure. [j] 1-σ limit; motion relative to component A. [k] 12-yr limit.

References in Tables 1 and 2: 1. Witzel 1987. 2. Marscher and Broderick 1985. 3. Walker, Benson, and Unwin 1987. 4. Porcas 1987. 5. Bååth 1984. 6. Barthel *et al.* 1986. 7. Roberts and Wardle 1987. 8. Pearson, Readhead, and Barthel 1987. 9. Shaffer *et al.* 1987. 10. Hough and Readhead 1987. 11. Zensus, Hough, and Porcas 1987. 12. Unwin *et al.* 1985. 13. Cohen *et al.* 1987. 14. Cotton *et al.* 1979. 15. Unwin 1987. 16. Biretta, Moore, and Cohen 1986. 17. Pearson *et al.* 1986. 18. Barthel 1987. 19. Simon *et al.* 1987. 20. Eckart *et al.* 1985. 21. Zensus and Porcas 1987. 22. Mutel and Phillips 1987. 23. Bååth 1987. 24. Pauliny-Toth 1987. 25. Romney *et al.* 1984. 26. Readhead, Pearson, and Unwin 1984. 27. Schmitt and Reid 1985. 28. Jones 1986. 29. A. K. Tzioumis, private communication. 30. Pauliny-Toth *et al.* 1984.

objects. We have tried to restrict the list to sources mapped at several epochs, where component motion can be clearly followed from one epoch to the next, and an accurate proper motion can be measured. Few authors are brave enough to attach error estimates to their measurements, so we have not included any in the table.

In Table 2, we have listed a miscellaneous collection of additional sources. These include a few which have been studied well enough to establish definite upper limits on any motion. Some of the upper limits are subluminal; but only two sources, NGC 1275 and M 87, have measured non-zero subluminal velocities. Comments on these sources are included in the notes below the table. There are also several sources which have shown "superluminal behavior" but for which we think it is premature to estimate a velocity (3C 446, 0235+164, 3C 147, 3C 371).

7. SOME INTERESTING SOURCES

The galaxy 3C 120 is the closest superluminal source and the detailed observations by Walker, Benson, and Unwin (1987) have revealed a jet with continuous properties on scales from $0.5h^{-1}$ pc to more than $100h^{-1}$ kpc, indicating that the jet characteristics are defined in the regions of the superluminal components.

The quasar 3C 179 stands out from the small group of established superluminal sources. This source is the first classical double radio source with a relatively weak core in which superluminal motion was found. Although much weaker, the components in this source exhibit properties similar to those in 3C 345 (Porcas 1987). It seems likely that 3C 179 is typical of the weak cores that are now being studied in well-defined samples.

The quasar 4C 39.25 was earlier considered the standard example of a source that does not show superluminal components, but rather has a stationary double structure. Subsequent observations indicated an apparent superluminal contraction, but it has now become clear that we see a superluminal component moving between two stationary components (Shaffer *et al.* 1987). It is interesting that none of these components appears to have the typical properties of a core. A similar situation is present in the quasar 3C 395, but new results shown at this Symposium by Simon *et al.* indicate that the superluminal motion has stopped. Deceleration of components to rest is also seen in BL Lac (Mutel and Phillips 1987). There is still no clear evidence in any source for contractions, although the identification of changes as expansion is not in all cases unique.

The quasar 3C 454.3 appears to be a good example of an object that exhibits rapid brightness changes in a stationary structure, in this case together with superluminally moving components (Pauliny-Toth 1987).

8. CONCLUSIONS

Superluminal motion is now established as a common phenomenon in active galactic nuclei. The available observational results are beginning to constrain the parameters of physical models for individual sources, and to test more general source models (e.g., Cohen *et al.*, this Symposium). While relativistic beaming clearly does play a role in compact radio sources (e.g., Blandford 1987), difficulties such as as the problem of exceedingly large linear sizes of superluminal sources (e.g., Schilizzi and de Bruyn 1983; Browne 1987; Barthel *et al.*, this Symposium) seem to require modifications of the most naïve beaming theories. Future work will increasingly focus on the comparison of superluminal sources with other classes of object in the quest for unified models of extragalactic sources. For example, a connection between superluminal sources and blazars is being discussed (e.g., Impey 1987; Rudnick 1987).

REFERENCES

Bååth, L. B. 1984, in *IAU Symposium 110, VLBI and Compact Radio Sources*, ed. R. Fanti, K. Kellermann, and G. Setti (Dordrecht: Reidel), p. 127.

Bååth, L. B. 1987, in *Superluminal Radio Sources*, ed. J. A. Zensus and T. J. Pearson (Cambridge: Cambridge University Press), p. 206.

Bare, C., Clark, B. G., Kellermann, K. I., Cohen, M. H., and Jauncey, D. L. 1967, *Science*, **157**, 189.

Bartel, N., Herring, T. A., Ratner, M. I., Shapiro, I. I., and Corey, B. E. 1986, *Nature*, **319**, 733.

Barthel, P. D. 1987, in *Superluminal Radio Sources*, ed. J. A. Zensus and T. J. Pearson (Cambridge: Cambridge University Press), p. 148.

Barthel, P. D., Pearson, T. J., Readhead, A. C. S., and Canzian, B. J. 1986, *Astrophys. J. (Letters)*, **310**, L7.

Biretta, J. A., Moore, R. L., and Cohen, M. H. 1986, *Astrophys. J.*, **308**, 93.

Blandford, R. D. 1984, in *IAU Symposium 110, VLBI and Compact Radio Sources*, ed. R. Fanti, K. Kellermann, and G. Setti (Dordrecht: Reidel), p. 215.

Blandford, R. D. 1987, in *Superluminal Radio Sources*, ed. J. A. Zensus and T. J. Pearson (Cambridge: Cambridge University Press), p. 310.

Broten, N. W., Legg, T. H., Locke, J. L., McLeish, C. W., Richards, R. S., Chisholm, R. M., Gush, H. P., Yen, J. L., and Galt, J. A. 1967, *Nature*, **215**, 38.

Browne, I. W. A. 1987, in *Superluminal Radio Sources*, ed. J. A. Zensus and T. J. Pearson (Cambridge: Cambridge University Press), p. 129.

Cohen, M. H. 1986, in *Highlights of Modern Astrophysics*, ed. S. L. Shapiro and S. A. Teukolsky (New York: Wiley), p. 299.

Cohen, M. H., Cannon, W., Purcell, G. H., Shaffer, D. B., Broderick, J. J., Kellermann, K. I., and Jauncey, D. L. 1971, *Astrophys. J.*, **170**, 207.

Cohen, M. H., Kellermann, K. I., Shaffer, D. B., Linfield, R. P., Moffet, A. T., Romney, J. D., Seielstad, G. A., Pauliny-Toth, I. I. K., Preuss, E., Witzel, A., Schilizzi, R. T., and Geldzahler, B. J. 1977, *Nature*, **268**, 405.

Cohen, M. H., and Unwin, S. C. 1982, in *IAU Symposium 110, VLBI and Compact Radio Sources*, ed. R. Fanti, K. Kellermann, and G. Setti (Dordrecht: Reidel), p. 95.

Cohen, M. H., and Unwin, S. C. 1984, in *IAU Symposium 97, Extragalactic Radio Sources*, ed. D. S. Heeschen and C. M. Wade (Dordrecht: Reidel), p. 345.

Cohen, M. H., Zensus, J. A., Biretta, J. A., Comoretto, G., Kaufmann, P., and Abraham, Z. 1987, *Astrophys. J. (Letters)*, **315**, L89.

Cotton, W. D., Counselman III, C. C., Geller, R. B., Shapiro, I. I., Wittels, J. J., Hinteregger, H. F., Knight, C. A., Rogers, A. E. E., Whitney, A. R., and Clark, T. A. 1979, *Astrophys. J. (Letters)*, **229**, L115.

Dent, W. A. 1965, *Science*, **148**, 1458.

Eckart, A., Witzel, A., Biermann, P., Johnston, K. J., Simon, R., Schalinski, C., and Kühr, H. 1986, *Astron. Astrophys.*, **168**, 17.

Eckart, A., Witzel, A., Biermann, P., Johnston, K. J., Simon, R., Schalinski, C., and Kühr, H. 1987, *Astron. Astrophys. Suppl.*, **67**, 121.

Eckart, A., Witzel, A., Biermann, P., Pearson, T. J., Readhead, A. C. S., and Johnston, K. J. 1985, *Astrophys. J. (Letters)*, **296**, L23.

Hodges, M. W., and Mutel, R. L. 1987, in *Superluminal Radio Sources*, ed. J. A. Zensus and T. J. Pearson (Cambridge: Cambridge University Press), p. 168.

Hough, D. H., and Readhead, A. C. S. 1987, in *Superluminal Radio Sources*, ed. J. A. Zensus and T. J. Pearson (Cambridge: Cambridge University Press), p. 114.

Impey, C. 1987, in *Superluminal Radio Sources*, ed. J. A. Zensus and T. J. Pearson (Cambridge: Cambridge University Press), p. 233.

Jones, D. L. 1986, *Astrophys. J. (Letters)*, **309**, L5.

Kellermann, K. I., and Pauliny-Toth, I. I. K. 1981, *Ann. Rev. Astron. Astrophys.*, **19**, 373.

Knight, C. A., Robertson, D. S., Rogers, A. E. E., Shapiro, I. I., Whitney, A. R., Clark, T. A., Goldstein, R. M., Marandino, G. E., and Vandenberg, N. R. 1971, *Science*, **172**, 52.

Marscher, A. P. 1987, in *Superluminal Radio Sources*, ed. J. A. Zensus and T. J. Pearson (Cambridge: Cambridge University Press), p. 280.

Marscher, A. P., and Broderick, J. J. 1985, *Astrophys. J.*, **290**, 735.

Moffet, A. T., Gubbay, J., Robertson, D. S., and Legg, A. J. 1972, in *IAU Symposium 44, External Galaxies and Quasi-Stellar Objects*, ed. D. S. Evans (Dordrecht: Reidel), p. 228.

Mutel, R. L., and Phillips, R. B. 1987, in *Superluminal Radio Sources*, ed. J. A. Zensus and T. J. Pearson (Cambridge: Cambridge University Press), p. 60.

Pauliny-Toth, I. I. K. 1987, in *Superluminal Radio Sources*, ed. J. A. Zensus and T. J. Pearson (Cambridge: Cambridge University Press), p. 55.

Pauliny-Toth, I. I. K., Porcas, R. W., Zensus, A., and Kellermann, K. I. 1984, in *IAU Symposium 110, VLBI and Compact Radio Sources*, ed. R. Fanti, K. Kellermann, and G. Setti (Dordrecht: Reidel), p. 149.

Pearson, T. J., Barthel, P. D., Lawrence, C. R., and Readhead, A. C. S. 1986, *Astrophys. J. (Letters)*, **300**, L25.

Pearson, T. J., and Readhead, A. C. S. 1984*a*, *Ann. Rev. Astron. Astrophys.*, **22**, 97.

Pearson, T. J., and Readhead, A. C. S. 1984*b*, in *IAU Symposium 110, VLBI and Compact Radio Sources*, ed. R. Fanti, K. Kellermann, and G. Setti (Dordrecht: Reidel), p. 15.

Pearson, T. J., Readhead, A. C. S., and Barthel, P. D. 1987, in *Superluminal Radio Sources*, ed. J. A. Zensus and T. J. Pearson (Cambridge: Cambridge University Press), p. 94.

Pearson, T. J., Unwin, S. C., Cohen, M. H., Linfield, R. P., Readhead, A. C. S., Seielstad, G. A., Simon, R. S., and Walker, R. C. 1981, *Nature*, **290**, 365.

Phillips, R. B., and Mutel, R. L. 1982, *Astron. Astrophys.*, **106**, 21.

Phinney, E. S. 1985, in *Astrophysics of Active Galaxies and Quasi-Stellar Objects*, ed. J. S. Miller (Mill Valley: University Science Books), p. 453.

Porcas, R. W. 1984, in *Active Galactic Nuclei*, ed. J. E. Dyson (Manchester: Manchester University Press), p. 20.

Porcas, R. W. 1987, in *Superluminal Radio Sources*, ed. J. A. Zensus and T. J. Pearson (Cambridge: Cambridge University Press), p. 12.

Readhead, A. C. S., Pearson, T. J., and Unwin, S. C. 1984, in *IAU Symposium 110, VLBI and Compact Radio Sources*, ed. R. Fanti, K. Kellermann, and G. Setti (Dordrecht: Reidel), p. 131.

Readhead, A. C. S., and Wilkinson, P. N. 1978, *Astrophys. J.*, **223**, 25.

Rees, M. J. 1966, *Nature*, **211**, 468.

Rees, M. J. 1984, in *IAU Symposium 110, VLBI and Compact Radio Sources*, ed. R. Fanti, K. Kellermann, and G. Setti (Dordrecht: Reidel), p. 207.

Roberts, D. H., and Wardle, J. F. C. 1987, in *Superluminal Radio Sources*, ed. J. A. Zensus and T. J. Pearson (Cambridge: Cambridge University Press), p. 193.

Rogers, A. E. E., Hinteregger, H. F., Whitney, A. R., Counselman, C. C., Shapiro, I. I., Wittels, J. J., Klemperer, W. K., Warnock, W. W., Clark, T. A., Hutton, L. K., Marandino, G. E., Rönnäng, B. O., and Rydbeck, O. E. H. 1974, *Astrophys. J.*, **193**, 293.

Romney, J. D., Alef, W., Pauliny-Toth, I. I. K., Preuss, E., and Kellermann, K. I. 1984, in *IAU Symposium 110, VLBI and Compact Radio Sources*, ed. R. Fanti, K. Kellermann, and G. Setti (Dordrecht: Reidel), p. 137.

Rudnick, L. 1987, in *Superluminal Radio Sources*, ed. J. A. Zensus and T. J. Pearson (Cambridge: Cambridge University Press), p. 217.

Scheuer, P. A. G. 1984, in *IAU Symposium 110, VLBI and Compact Radio Sources*, ed. R. Fanti, K. Kellermann, and G. Setti (Dordrecht: Reidel), p. 197.

Schilizzi, R. T., and de Bruyn, A. G. 1983, *Nature*, **303**, 26.

Schmitt, J. H. M. M., and Reid, J. M. 1985, *Astrophys. J.*, **289**, 120.

Shaffer, D. B., Marscher, A. P., Marcaide, J., and Romney, J. D. 1987, *Astrophys. J. (Letters)*, **314**, L1.

Simon, R. S., Johnston, K. J., Hall, J., Spencer, J. H., and Waak, J. A. 1987, in *Superluminal Radio Sources*, ed. J. A. Zensus and T. J. Pearson (Cambridge: Cambridge University Press), p. 72.

Unwin, S. C. 1987, in *Superluminal Radio Sources*, ed. J. A. Zensus and T. J. Pearson (Cambridge: Cambridge University Press), p. 34.

Unwin, S. C., Cohen, M. H., Biretta, J. A., Pearson, T. J., Seielstad, G. A., Walker, R. C., Simon, R. S., and Linfield, R. P. 1985, *Astrophys. J.*, **289**, 109.

Walker, R. C., Benson, J. M., and Unwin, S. C. 1987, in *Superluminal Radio Sources*, ed. J. A. Zensus and T. J. Pearson (Cambridge: Cambridge University Press), p. 48.

Whitney, A. R., Shapiro, I. I., Rogers, A. E. E., Robertson, D. S., Knight, C. A., Clark, T. A., Goldstein, R. M., Marandino, G. E., and Vandenberg, N. R. 1971, *Science*, **173**, 225.

Witzel, A. 1987, in *Superluminal Radio Sources*, ed. J. A. Zensus and T. J. Pearson (Cambridge: Cambridge University Press), p. 83.

Zensus, J. A., Hough, D. H., and Porcas, R. W. 1987, *Nature*, **325**, 36.

Zensus, J. A., and Pearson, T. J. (ed.) 1987, *Superluminal Radio Sources* (Cambridge: Cambridge University Press).

Zensus, J. A., and Porcas, R. W. 1987, in *Superluminal Radio Sources*, ed. J. A. Zensus and T. J. Pearson (Cambridge: Cambridge University Press), p. 126.

SUPERLUMINAL MOTION IN THE LARGEST KNOWN QUASAR

P. D. Barthel[1,2], R. T. Schilizzi[3], G. K. Miley[2,4], E.Preuss[5], J. R. Hooimeyer[2]

1. California Institute of Technology; 2. Leiden Observatory; 3. Netherlands Foundation for Radio Astronomy; 4. Space Telescope Science Institute; 5. Max-Planck-Institut für Radioastronomie

ABSTRACT. We discuss VLBI observations of the bright core in the large, triple radio source 4C 34.37, associated with a quasar at $z = 0.206$. We have discovered superluminal motion in this quasar, which is the largest quasar known. We comment on the implications of these findings, and subsequently examine the projected size distributions of lobe and core dominated quasars in general. The results of these investigations lead us to conclude that neither superluminal motion, nor core prominence can simply be translated into an orientation of the overall radio source.

Superluminal motion, (curved) one-sided parsec scale jets, and rapid variability in compact radio sources have found a more or less satisfactory explanation in relativistic effects, resulting from a relativistic beam observed from a suitable direction (e.g., Zensus & Pearson, these Proceedings). Along these lines of thought, attempts have also been made to unify the properties of extended, lobe dominated, and compact, core dominated radio sources (e.g., Orr & Browne 1982), in which the morphologies and relative occurrences of these sources are explained as purely an orientation effect. Defining the simple beaming model as having unidirectional, uniform velocity, bulk relativistic motion in a jet along the overall source axis, we will in the following comment on its applicability, to explain superluminal motion and as the basis for unifying schemes.

To test the model, the occurrence of superluminal motion in the cores of large extended triple quasars, supposedly oriented near the sky plane, is of prime importance. For example, in a randomly oriented sample, one out of four extended quasars should be within 15° of the sky plane, whereas according to the simple beaming model an apparent velocity of $2c$ is impossible in this source orientation. Motivated by an increasing number of superluminal quasars with extended morphologies Barthel *et al.* (1986) pointed out that the simple beaming hypothesis might be inadequate to explain the observations. In the meantime, a project to study the structure and evolution of the core morphologies in very large triple quasars (Barthel *et al.* 1984) has resulted in another superluminal quasar, which happens to be the largest quasar known ($560/h$ kpc, $H_0 = 100h$ km s^{-1} Mpc^{-1} and $q_0 = 0.5$). Fig. 1 shows a montage of the large and small scale morphology of 4C 34.47, associated with the quasar 1721+343 at redshift $z = 0.206$. More maps can be found in Barthel (1987). Combining 5 and 10.7 GHz VLBI data we find that the northernmost core component has an inverted radio spectrum, whereas the other components have steep spectral indices. We therefore identify this northernmost component with the actual radio core, and measure $\beta_{app} = 3.1/h$ for the two components south of the core: 4C 34.47 is a superluminal quasar. With $H_0 = 100$ this observation implies that the quasar lies within a 36° cone with the line of sight. The chance for a randomly oriented quasar to be within this cone is 19%. Unless 4C 34.47 is a freak, we would therefore expect at least five larger quasars. We

M. J. Reid and J. M. Moran (eds.), The Impact of VLBI on Astrophysics and Geophysics, 17–18.

conclude that superluminal motion cannot simply be translated into an orientation of the overall radio source.

This 'uncomfortable' (Barthel 1987) situation led us to investigate the extent of another prediction of the simple beaming model: using similar projection arguments as above, core dominated quasars should be smaller in projected linear size than lobe dominated quasars. To this end we extracted quasars with known flux densities for their core and extended emission, and available radio maps, from the compilation of Wills & Browne (1986). Based on their core prominence, we divided these sources in two samples: a core dominated sample having $0 \leq \log \mathrm{R} < 1$, and a lobe dominated sample having $-1 \leq \log \mathrm{R} < 0$. Median values for the luminosities of the unbeamed (lobe) emission were the same for both samples (important criterion – we don't want to compare apples and pears!). It is easy to show that the simple beaming model ($\mathrm{R}_T = 0.024$ and $\gamma = 5$) predicts that our core dominated sample should have half the projected linear extent of the lobe dominated sample. The actual size distributions, shown in Fig. 2, indicate that this is definitely not the case: the two distributions are statistically undistinguishable. Apparently both lobe and core dominated quasars can display compact and extended morphologies. Similar results have been obtained by Murphy (Ph. D. thesis, as reported in Browne, 1987). We conclude that core prominence cannot simply be translated into an orientation of the overall source.

In summary, it seems hard to reconcile both superluminal motion and core prominence with a preferred orientation of the overall radio source axis. Maintaining the beaming hypothesis, we have to conclude that superluminal motion and flux boosting can be observed from a fairly wide range of directions, and a wide jet cone angle or oblique shocks may be relevant in this respect. The question as to the existence of ***unbeamed*** quasars was addressed at length in Barthel (1987). On the other hand, orientation independent mechanisms such as screens or intersecting shocks may apply, and core prominence may have an intrinsic, physical origin. Conclusive statements will have to be based on statistical studies of large, carefully matched samples of core and lobe dominated quasars.

References

Barthel, P.D., Miley, G.K., Schilizzi, R.T., Preuss, E. 1984: Astron. Ap. **148**, 243.

Barthel, P.D., Pearson, T.J., Readhead, A.C.S., Canzian, B.J. 1986: Ap. J. (Letters) **310**, L7.

Barthel, P.D. 1987: in *Superluminal Radio Sources*, ed. Zensus & Pearson (Cambridge University Press), p. 148.

Browne, I.W.A. 1987: in *Superluminal Radio Sources*, ed. Zensus & Pearson (Cambridge University Press), p. 129.

Orr, M.J.L., Browne, I.W.A. 1982: M. N. R. A. S. **200**, 1067.

Wills, B.J., Browne, I.W.A. 1986: Ap. J. **302**, 56.

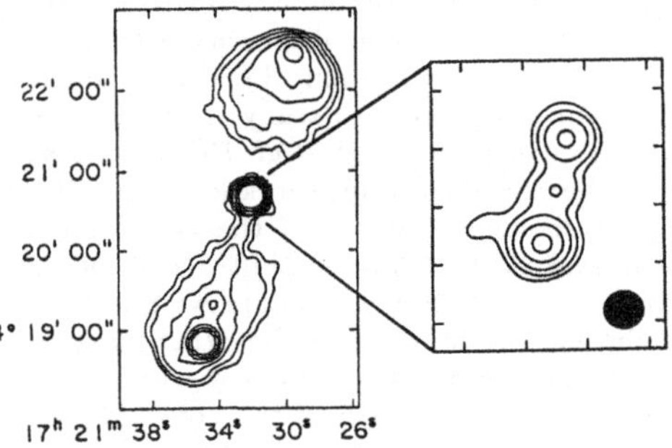

Fig. 1. Large and small scale morphology of 4C 34.47

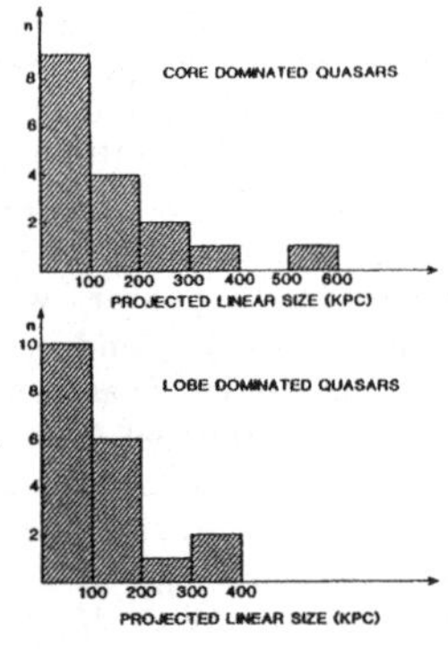

Fig. 2. Projected linear size distributions

INVESTIGATIONS OF 3C345

J. A. Biretta
Harvard - Smithsonian Center for Astrophysics

M. H. Cohen
California Institute of Technology

The well known superluminal source 3C345 is associated with a 16th magnitude quasar at z=0.595. The optical continuum is highly polarized and shows violent variations. At x-ray frequencies it is a weak source; while at radio frequencies it is bright and variable. The radio structure consists of a 200 kpc halo, a 20 kpc jet, and a very bright pc scale region which shows superluminal motion. Recent MkII VLBI monitoring (Biretta, Moore, Cohen 1986) at 2, 5, 11, and 22 GHz has provided accurate spectra, sizes, expansion rates, velocities, and trajectories for the superluminal knots and the "core." Here we consider physical conditions in the emission regions and models for the kinematics.

To derive physical conditions in the emission regions, and properties of the jet, we apply a synchrotron self-Compton model to observations. We assume the radio emission is incoherent synchrotron emission, and that inverse Compton scattering will produce x-rays. The "core" component is modeled as an inhomogeneous conical jet (Königl 1981) with magnetic fields and relativistic electron densities varying smoothly as $B(r) = B(r_o)\, r^{-m}$ and $N(E,r) = KE^{-p} r^{-n}$, where r is the distance along the jet. The superluminal knots are modeled as homogeneous spheres with a uniform electron density and (well tangled) magnetic field. For the "core" we assume the conical pattern is fixed relative to the "central engine" and the fluid moves with a Lorentz factor γ; for the knots we assume both the spherical pattern and fluid move with γ (Lind and Blandford 1985). The model is constrained by the radio spectra and component sizes, and by the total x-ray flux.

Results of the model can be summarized as follows. Each of the emission regions must have a relativistic Doppler shift factor δ significantly greater than unity, or the model x-ray flux will exceed the observed x-ray flux. The "core" has the largest limit $\delta \gtrsim 10$ and knot C3 has the smallest $\delta \gtrsim 3$. This is evidence for relativistic motion in the source, and is independent of the observed superluminal motions. The magnetic fields can be estimated from the turnover in the synchrotron spectrum for knots C4, C3, and C2, and we find that the magnetic field falls off as $B \propto r^{-1.0\pm0.5}$ along the jet. For the "core" component we cannot derive the magnetic field and particle density functions independently. However, if

M. J. Reid and J. M. Moran (eds.), The Impact of VLBI on Astrophysics and Geophysics, 19–20.

we assume the number of relativistic particles is conserved along the jet ($n = 2$) we then find $m = 1$ ($B \propto r^{-1}$) in the "core" as well. The pressures in the knots appear to exceed the external (narrow line region) pressures, unless the Doppler shift factor $\delta \sim 15$, which is much larger than the minimum values. Also, the particle energy densities exceed those of the magnetic field, unless $\delta \gtrsim 15$. This model predicts that the position of the "core" will shift with frequency. At any given frequency the "core" will be brightest at the optical depth $\tau \sim 1$ point; at high frequencies this point occurs near the apex of the jet (near the central engine), whereas at low frequencies it occurs farther from the apex. While we do not have absolute positions for the "core," we can measure its position relative to the knots. For example, the model predicts that "core" will be 0.27mas farther from the knots at 5 GHz than at 11 GHz, and in fact we observe that knot C3 is 0.30 ± 0.04mas farther from the "core" at 5 GHz than at 11 GHz. This and a similar agreement found between 22 and 11 GHz lends support to this model.

The superluminal knots in 3C345 display interesting kinematics which have not yet been seen in other sources. Knot C4 was observed to change position angle from $-135°$ to $-87°$ relative to the "core" as it moved 0.3 to 0.8 mas away from the "core." Components C3, C2, C1, and the 3" jet continue the northward curvature, and are at position angle $-86°$, $-74°$, $-64°$, and $-31°$, respectively. Component C4 accelerates from v/c=1.3 to 6.5 as it moves away from the "core," while knots C3 and C2 have v/c = 6.0 and 9.5, respectively.

Precession of the central engine (Begelman, Blandford, and Rees 1980) has been proposed to explain the curvature seen in mas scale jets, but this mechanism fails in 3C345 since it cannot explain the changing position angle and acceleration of C4. Furthermore, the required rapid precession would be difficult to obtain (Biretta, *et al.* 1985). The kinematics can instead be explained as a bending or twisting of the jet by an external medium. If the jet has a non-relativistic equation of state, its bulk velocity will remain approximately constant as it move through the external medium. The observed kinematics could then be explained by a $\gamma \sim 10$ jet which is initially aligned at an angle $\theta \lesssim 1/2°$ to the line of sight, and then bends away from the line of sight to $\theta \sim 4°$. The bending might result from a large pressure gradient in the external medium, or from a Kelvin-Helmholtz instability (Hardee, 1987). However, such a trajectory seems improbable since the initial θ is so small. A larger initial θ is possible if the jet has a relativistic equation of state. Such a jet would probably consist of electrons and positrons, and would accelerate as the external pressure decreased (Smith and Norman 1981). A possible trajectory might have $\gamma \sim 5$ and $\theta \sim 2°$ near the "core" and then accelerate to $\gamma \sim 10$ with $\theta \sim 4°$. This has the advantage that the initial θ need not be very small, and the jet can be bent by smaller pressure gradients. Further study of hydrodynamic acceleration in relativistic jets is needed.

References

Begelman, M.C., Blandford, R.D., and Rees, M.J.: *Nature* **287**, 307 (1980).
Biretta, J.A., *et al.* : *Ap.J.* **292**, L5 (1985).
Biretta, J.A., Moore, R.L., and Cohen, M.H.: *Ap.J.* **308**, 93 (1986).
Hardee, P.E.: this conference (1987).
Königl, A.: *Ap.J.* **243**, 700 (1981).
Lind, K.R. and Blandford, R.D.: *Ap.J.* **295**, 358 (1985).
Smith, M.D. and Norman, C.A.: *M.N.R.A.S.* **194**, 771 (1981).

RECENT OBSERVATIONS AND SUPERLUMINAL MONITORING OF THE QUASAR 3C 395

R. S. Simon, K. J. Johnston, and J. H. Spencer
E. O. Hulburt Center for Space Research
Naval Research Laboratory, Washington, D.C. 20375-5000

Until recently, the compact structure in 3C 395 could be described as follows (Simon et al. 1987):

(1) A bright, unresolved, flat spectrum core to the northwest. The core brightened by about 0.4 Jy between 1983.3 and 1985.4, from 0.6 Jy to 1.0 Jy.

(2) A moderately resolved component 15.8+/-0.2 mas to the southwest along P.A.118°. The position of this component relative to the core has changed by less than 0.2 mas in 6 years and the flux density of this component has remained roughly constant.

(3) A component which was moving rapidly away from the core towards component (2) with a proper motion of 0.64+/-0.1 mas yr^{-1}. For 3C 395 (z=0.635), this implies (v/c)~20c for H_o=100 km/s/Mpc and q_o=0.5.

Thus 3C 395 appeared to show typical superluminal motion near its core, yet have an apparently stationary component at a projected distance of ~60pc.

In an 11-station experiment in November 1986 we re-observed 3C 395 and found a surprising result: the relative motion between components (1) and (3) in 3C 395 had apparently ceased between 1985.4 and 1986.9. In Fig. 1 we show the 4 maps at 6 cm. In Fig. 2 we plot the relative separation between the core and the moving component. The dashed line in Fig. 2 is a linear fit to the measured separation in the first three maps at 6cm.

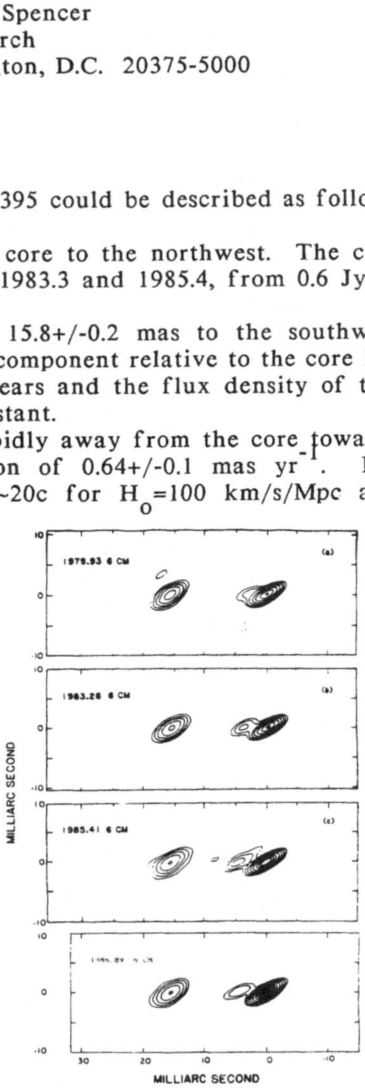

FIGURE 1: 6 cm images of 3C 395 for 4 epochs: 1979.9, 1983.3, and 1985.4 and 1986.9. All four maps have been rotated clockwise by 28 degrees and have been convolved with a beam 3x1 mas at -30°.

M. J. Reid and J. M. Moran (eds.), The Impact of VLBI on Astrophysics and Geophysics, 21–22.

The core of 3C 395 has also brightened further from 1985.4 to 1986.9 by 0.2 Jy to 1.2 Jy. The core has roughly tripled in brightness since 1979.9.

There seem to be two possible explanations for the behavior we see in 3C 395. First, the moving component in 3C 395 may have actually slowed relative to the core. If so, this would be the first time a clearly defined component in a superluminal source has been seen to slow down. This would strongly suggest that the ambient conditions in 3C 395 change at this distance from the core and that the conditions which have created the stationary component at ~60pc may extend in as close as 15pc to the core (in projected distance).

A second possibility is that a new superluminal component has been ejected from the core of 3C 395 and that the separation measured is to its position and not to the nominal core position. This possibility would give a natural explanation for the observed flux density increase in the core and the apparent halt in superluminal motion in 3C 395. The discrepancy between the observed and linearly extrapolated positions of the moving component is ~1 milliarcsec, or more than a beam diameter, so this explanation faces some difficulties. However, in the most recent image from November 1986, the core of 3C 395 is slightly resolved on the SE side, implying that there is indeed a new component emerging from the core. If the separation between the true core and the stationary component is assumed constant, the apparent core position observed in November 1986 is less than ~0.3 mas from its position in earlier images.

There is limited evidence that the separations between the core and the other components in 3C 395 may be systematically lower at 18cm than they are at 6cm (Simon et al. 1987). If this turns out to be the case, it would suggest that the observations of Biretta et al. (1987) of similar behavior in 3C 345 may be a common property of superluminal sources.

It has been suggested (Biermann et al. 1987, in preparation) that stationary components in superluminal sources may be regions where the pressure in an expanding jet falls substantially below the external pressure so that a stationary, reconfinement shock forms. It should be possible to test this hypothesis in the case of 3C 395 by determining precisely the way the milliarcsec structure in 3C 395 is connected to the larger scale emission. At present, the most recent (Nov. 1986) observations of 3C 395 show that the edge of the stationary component closest to the core is ~unresolved whereas in the opposite direction the emission fades more gradually.

REFERENCES

Biermann, P., et al. 1987. In preparation.

Biretta, J. A.,, Moore, R. L., and Cohen, M. H. 1986. Ap. J. 308, 93.

Simon R. S., Hall, J., Johnston, K. J., Spencer, J. H., Waak, J., and Mutel, R. L. 1987. Ap. J. Lett. (submitted).

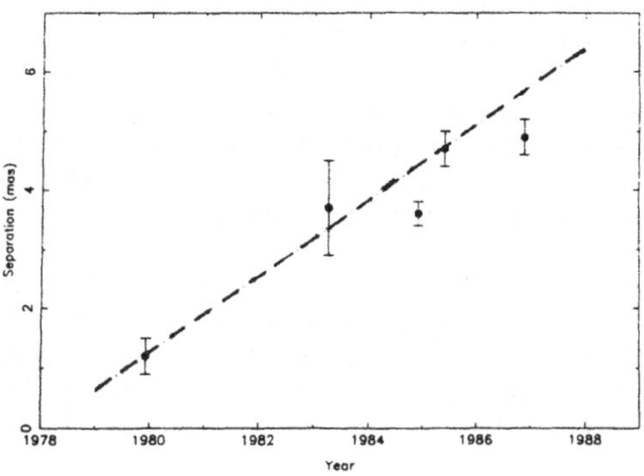

FIGURE 2: Separation as a function of time between the superluminal component and the core. The dashed line is the best linear fit to the first 3 6cm epochs. The point at 1984.9 was determined from model fitting at 18cm.

EXPANDING QUASARS AND THE EXPANSION OF THE UNIVERSE

M. H. Cohen, P. D. Barthel, T. J. Pearson, and J. A. Zensus
Owens Valley Radio Observatory
California Institute of Technology
Pasadena, California 91125
USA

The μ–z diagram (Figure 1) plots the observed internal proper motion μ versus redshift z for 32 extragalactic radio sources associated with active galactic nuclei. The observed points fall below an upper bound which decreases with redshift; there is a statistically significant anticorrelation between redshift and internal proper motion.

We only include sources for which measurements of internal motion are published (see the table of Zensus and Pearson in this volume). The selection is biased towards sources with large proper motions, which tend to be monitored intensively, but we are not aware of any selection biases which would introduce the observed correlation. The filled circles represent sources with well-established measurements, i.e., based on maps at three or more epochs where the components are well identified. The crosses represent sources with only limited observations or ill-defined proper motions.

The correlation between redshift and proper motion provides a powerful test of models of expanding sources and of theories of the origin of the redshift. We discuss here only the currently popular relativistic beam model, with standard Friedmann cosmology. This gives a reasonable fit to the data. Other models are discussed by Cohen *et al.* (in preparation).

Assume that z measures distance in a Friedmann universe, and that we are dealing with thin randomly-oriented jets containing luminous blobs all moving with the same γ. Then the observed proper motions will have an upper envelope given by

$$\mu \approx 2.11 \times 10^{-4} \gamma H_0 g(q_0, z) \text{ mas yr}^{-1}, \tag{1}$$

for $\gamma^2 \gg 1$, where $g(q_0, z)$ is given by

$$g(q_0, z) = q_0^2 (1+z) \left(q_0 z + (q_0 - 1)(\sqrt{1 + 2q_0 z} - 1) \right)^{-1} . \tag{2}$$

The solid line shows this envelope for $\gamma H_0 = 900$ km s^{-1} Mpc^{-1}, and $q_0 = 0.5$. This curve could reasonably well form an upper limit to the data. The two high points (CTA 102 and 3C 395) need confirmation, but if taken at face value, they would require $\gamma H_0 \sim 1800$.

In a randomly oriented sample of objects, the distribution of points below the upper envelope is predicted by the model; e.g., half the sources lie closer to the line of sight than $\theta = 60°$. The dashed line shows this median $\mu(z)$ for the particular case $\gamma = 4$ and $H_0 = 100$ km s^{-1} Mpc^{-1}, $q_0 = 0.5$. The position of the median line does not change much if $\gamma > 4$, while for a smaller Hubble constant, it moves down. Thus we would expect

M. J. Reid and J. M. Moran (eds.), The Impact of VLBI on Astrophysics and Geophysics, 23–24.

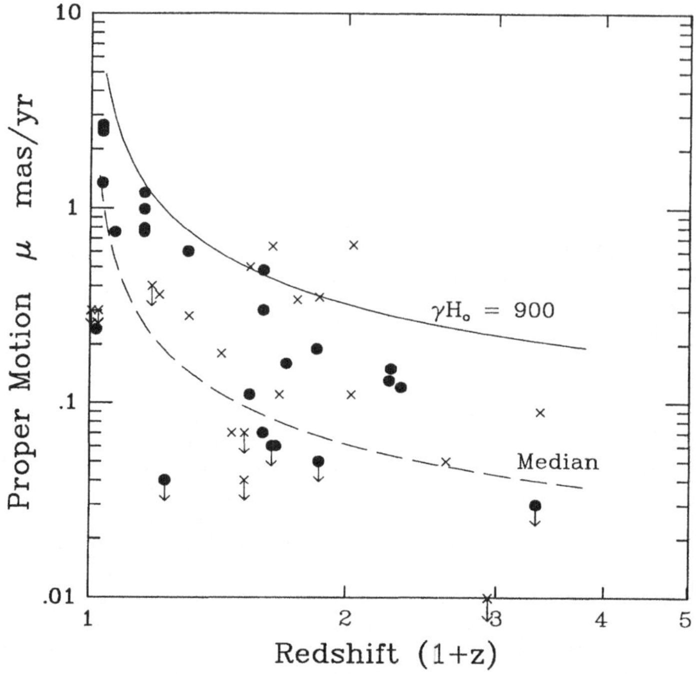

Figure 1 The μ–z diagram. The filled circles and crosses show observed sources; the solid line is an upper bound in Friedmann cosmology for a particular value of γ and H_0; the dashed line shows the median proper motion in an orientation-unbiased sample.

less than half of the points to lie above the dashed line. This is obviously not the case, and we conclude that the sources in the Figure must be predominantly aligned in our direction.

An earlier version of this work was published by Cohen 1987, in *Superluminal Radio Sources*, ed. Zensus and Pearson (Cambridge: Cambridge University Press).

THE LARGE SCALE STRUCTURE OF SUPERLUMINAL RADIO SOURCES

R. S. Simon and K. J. Johnston
E. O. Hulburt Center for Space Research
Naval Research Laboratory, Washington, D.C. 20375-5000

The standard model for apparent superluminal motion is that of relativistic motion in a jet pointed nearly along the line of sight. Variants of this model give a natural explanation for the extremely asymmetric morphology observed in the radio cores of these sources. Despite this, examination of the large scale structure in superluminal radio sources strongly suggests that large scale symmetrical structure with two large radio lobes placed on either side of the compact emission is a common phenomenon among these objects. When a comparison is done between the large scale structure typically seen in double quasars and that seen in superluminal sources, it is found that the de-projected physical sizes estimated for superluminal quasars are similar to the size scales observed in the double quasars. Large scale radio structure is probably as common among apparent superluminal sources as it is in radio quasars.

In Table 1 we present a summary of superluminal sources and their observed angular sizes, along with apparent transverse velocities and de-projected sizes derived from their measured proper motions. For a few of the sources in Table 1, the de-projected sizes are significantly larger than the sizes observed in double quasars. This implies that simple de-projection over-estimates the physical sizes of some superluminal quasars, and that for those cases it appears that the angle to the line of sight of the superluminal motion is much less than the angle to the line of sight of the axis defined by the large-scale radio lobes.

For example, in the radio quasar 1928+738 the superluminal motion in the core implies that the core is oriented less than 15° to line of sight. However, the de-projected angular size would then be about 850 kpc for that projection angle. Unless 1928+738 is substantially larger than other double radio quasars, the extremely large angular size of 1928+738 suggests that the large scale structure is significantly misaligned to the line of sight at an angle much greater than 15°.

REFERENCES

(1) Walker, R. C., Benson, J. M., and Unwin, S. C. 1987, Ap. J. 316, 546.
(2) Schilizzi, R. T. and de Bruyn, A. G. 1983, Nature, 303, 26.
(3) Porcas, R. W. 1984, in IAU Symposium #110, VLBI and Compact Radio Sources, ed. Fanti, Kellermann, and Setti (Dordrecht: Reidel), 157.
(4) Pearson, T. J., Perley, R. A., and Readhead, A. C. S. 1985, Astron. J., 90, 738.

M. J. Reid and J. M. Moran (eds.), The Impact of VLBI on Astrophysics and Geophysics, 25–26.

(5) Pearson, T. J., Readhead, A. C. S., and Barthel, P. 1987, in "Superluminal Radio Sources", ed. Zensus, J. A., and Pearson, T. J. (Cambridge: Cambridge Univ. Press), 94.

(6) Hough D. H., and Readhead, A. C. S. 1987, in "Superluminal Radio Sources", ed. Zensus, J. A., and Pearson, T. J. (Cambridge: Cambridge Univ. Press), 114.

(7) Cohen, M. H. and Unwin, S. C. 1984, in IAU Symposium 110, "VLBI and Compact Radio Sources" ed R. Fanti, K. Kellermann, and G. Setti (Dordrecht:Reidel), 95.

(8) Biretta, J. A., Moore, R. L., and Cohen, M. H. (1986). Ap. J. 308, 93.

(9) Pearson, T. J., Barthel, P. D., Lawrence, C. R., and Readhead, A. C. R. S., 1986, Ap. J. (Lett.), 300, L25.

(10) Browne, I. W. A. and Orr, M. J. L. 1981, in "Optical Jets in Galaxies", Proc. of ESO/ESA workshop ESA SP-162 (Noordwijk: European Space Agency), 87.

(11) Barthel, P. D. 1987, in "Superluminal Radio Sources", ed. Zensus, J. A., and Pearson, T. J. (Cambridge: Cambridge Univ. Press), 148.

(12) Eckart, A., Witzel, A., Biermann, P., Pearson, T. J., Readhead, A. C. S., and Johnston, K. J. 1985, Ap. J. (Lett.), 296, L23.

(13) Johnston, K. J., Simon, R. S., Eckart, A., Biermann, P., Schalinski, C., Witzel, A., and Strom, R. G. 1987, Ap. J. (Lett.), 313, L85.

(14) Waak, J. A., Spencer, J. H., Johnston, K. J., and Simon, R. S. 1985, A.J., 90, 1989.

(15) Perley, R., Fomalont, E., and Johnston, K. J. 1982, Ap. J. (Lett.), 255, L93.

(16) Simon, R. S., Hall, J., Johnston, K. J., Spencer, J. H., Waak, J. A., and Mutel, R. L. 1987, Ap. J. (Lett.), submitted.

TABLE 1: Angular Sizes of Superluminal Sources

Source	Morphology[a]	z	Size arc sec	v/c[b]	$Theta_{max}$[c] degrees	Minimum Apparent size h^{-1} kpc	Deprojected size[d] h^{-1} kpc	Ref
3120	C	0.033	720	2.1	51	370	424	(1)
NRA0140	D	1.258	11	5.4	21	61	170	(2)
3C179	C/D	0.846	16	4.2	27	80	176	(3)
3C216	C	0.668	2	2.5	44	9	13	(4,5)
3C245	D	1.029	9	3.1	36	48	82	(6)
3C263	D	0.652	44	1.3	75	199	206	(6)
3C273	A	0.158	20	5.3	21	36	100	(1)
3C279	C/D	0.538	18	3.5	32	74	140	(1)
3C345	D	0.595	20	8.2	3	85	1620	(2,7,8)
1642+690	D	0.75	11	9.3	12.3	53	249	(9,10)
4C34.47	D	0.206	180	3.2	35	406	713	(11)
1928+738	D	0.302	80	7.5	15.1	235	848	(12,13)
3C395	D	0.635	1	10.5	10.9	4	21	(14,15,16)
BL Lac	C	0.07	25	2.0	53	22	28	(2,6)

NOTES TO TABLE 1

[a] A = Asymmetric, C = Complex, D = Double

[b] We assume $H_o = 100\ h^{-1}$ km s^{-1} Mpc^{-1} and $q_o = 0.5$.

[c] Assumes that large scale emission is misaligned to the line of sight by the angle predicted by the apparent superluminal motion. No correction is made for the presence of radio lobes, so this is the maximum size for the source.

[d] $Theta_{max} = 2 \cot^{-1} (v/c)$ for all sources except 3C345 for which additional constraints are available (Cohen and Unwin 1984). If v/c were the only constraint, $Theta_{max} = 14$ for 3C345 and the deprojected size is > 351 h^{-1} kpc.

VLBI OBSERVATIONS OF THE RADIO JET IN 3C 273

S. C. Unwin
California Institute of Technology, Pasadena, U.S.A.

R. J. Davis
Nuffield Radio Astronomy Laboratories, Jodrell Bank, U.K.

ABSTRACT We present a new high dynamic range map of the quasar 3C 273, made from observations with a VLBI network of 12 telescopes. This new map at 18 cm wavelength has one of the highest dynamic ranges yet achieved with VLBI, and it shows the 'jet' extending to at least 180 milliarcsec, or 330 pc from the nucleus of the quasar. Strong limits can be placed on the brightness of any 'counter-jet' on kiloparsec-scales, as no emission is visible on the opposite side of the 'core'. Combining with other VLBI, VLA and MERLIN maps shows that the jet is visible and continuous over a very large range of scales, from 1 pc to 40 kpc.

1. VLBI MAP

The new map, shown in Figure 1, was made using the Jodrell Bank **OLAF** mapping software (R.G. Noble, A. Bailey and T.W.B. Muxlow, unpublished), which corrects for errors in the data due to (*a*) station-dependent amplitudes and phases, and (*b*) baseline-dependent terms. Station corrections are part of normal 'hybrid' mapping, but for high SNR data, as in this case, the map quality is limited by baseline errors, and correcting them makes a dramatic improvement.

Using this data we have achieved one of the highest dynamic ranges of any VLBI map made to date, over 7000:1 (defined as peak brightness to rms noise level). Away from the map center, the noise level is 1.2 mJy/beam, rising to 10 mJy/beam at the largest feature believed to be an artifact, just east of the core. Even this feature is only 0.1 % of the peak brightness. Although the resolution is poorer in the N–S direction, the jet is unresolved on our map at all points along its length.

'Superluminal' components seen on VLBI maps at higher frequencies and higher resolution (*e.g.,* Unwin *et al.* 1985) are identified as 'knots' on our map out to about 20 milliarcsec from the core, and the slight jet curvature is consistent with the positions of those features. There are indications the jet extends beyond the area plotted, but shorter baselines and better self-calibration are needed to map this weak emission reliably.

Some components further down the jet are sufficiently well-defined that proper motions could be detected between them in a few years. We therefore plan to repeat the observation within the next year or so, to search for differences between the two maps. With a map of similar quality, detection of proper motions similar to those seen in the close-in components will be possible. A direct measurement at larger radii would be a critical test of the Davis, Muxlow, and Conway (1986) model of the jet morphology, which seeks to explain the wiggles seen in the ridge-line of the jet in terms of precession of the jet axis.

M. J. Reid and J. M. Moran (eds.), The Impact of VLBI on Astrophysics and Geophysics, 27–28.

2. DISCUSSION

The brightness of the jet along the ridge line decays rapidly with distance from the nucleus, in a power-law fashion, out to the first large-scale knot at 7 arcsec. A smooth relation over three orders of magnitude in radius suggests that the physical parameters of the jet also follow simple power laws, as found for 3C 120 by Walker, Benson, and Unwin (1987).

One possible model which reproduces the main features of the source over a wide range of scales is one in which the radio emission close to the core is caused by internal turbulence, or shocks, in a flow which moves out rapidly. The decay in jet brightness is caused by loss of energy by radiation, leaving power only in the bulk flow. Further out, the jet gradually slows down, and becomes more visible again, by interaction with the confining external medium. This model contrasts somewhat with the 'standard' model, in that the jet is moving slowly by the time it reaches the hotspot at the end of the head.

No 'counter-jet' is seen on the opposite side of the core at any resolution in our new data, or indeed, of any other observation of this object in any waveband. Our map yields a limit of several 100:1 on the jet:counter-jet ratio on scales up to about 0.1 arcsec. Since the outer 'head' appears to be genuinely one-sided, and is fed with energetic particles by the jet, it seems that 3C 273 represents the best case yet for an intrinsically one-sided jet.

These observations were conducted by the US VLBI Network Users Group, and we thank all those who help make global VLBI possible. This research was supported by the NSF, via grant 85-09822 to the Owens Valley Radio Observatory.

3. REFERENCES

Unwin, S.C., Cohen, M.H., Biretta, J.A., Pearson, T.J., Seielstad, G.A., Walker, R.C., Simon, R.S., and Linfield, R.P.: 1985, *Astrophys. J.* **289**, 109.
Davis, R.J., Muxlow, T.W.B., and Conway, R.G.: 1985, *Nature* **318**, 343.
Walker, R.C., Benson, J.M., and Unwin, S.C.: 1987, *Astrophys. J.* **316**, 546.

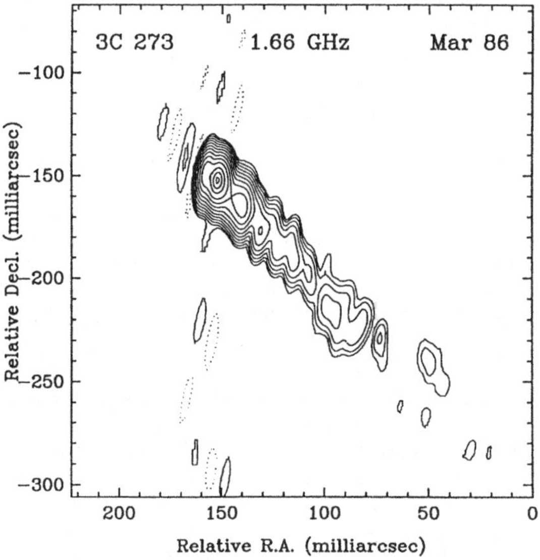

FIGURE 1. VLBI map of 3C 273 at 18 cm wavelength. The restoring beam is 12 × 4 milliarcsec, and contours are drawn at –0.05, 0.05, 0.1, 0.2, 0.4, 0.8, 1.6, 3.2, 6.4, 13, 25, 50, 75, and 90 % of the peak (negative contours dashed).

SUPERLUMINAL MOTIONS AND LARGE SCALE STRUCTURE IN 3C 120

R. C. Walker[1], J. M. Benson[2], and S. C. Unwin[3]

[1] *National Radio Astronomy Observatory*, Socorro, N.M., U.S.A.*
[2] *National Radio Astronomy Observatory*, Charlottesville, VA, U.S.A.*
[3] *California Institute of Technology, Pasadena, CA, U.S.A.*

ABSTRACT. The jet in the superluminal source 3C 120 has been observed on all scales from under a parsec to about 100 kpc. The jet brightness decays rather smoothly with core distance, indicating that the basic characteristics of the jet are established on parsec or smaller scales and evolve simply with distance.

3C 120 is the closest of the superluminal sources and has the highest angular expansion rates. It is associated with a galaxy that is usually classified as a Seyfert, although it is highly disturbed by the nuclear source and its morphological type is not clear. (cf. Baldwin *et al.* 1980). The parsec scale structure of 3C 120 has been monitored extensively since the early 1970's (Walker *et al.* 1982 and references therein). Clear epochs of expansion were seen in 1972–1974 and in 1979 with angular rates of about 1.5 mas per year. Between 1980 and 1983, four distinct knots were seen to emerge from the core and move away at a rate of about 2.5 mas yr^{-1} (about 4 c for $H_o = 100$ km $s^{-1}Mpc^{-1}$ and a redshift of 0.033). The knots were resolved from the core at about 1 mas and decayed as they moved out until, at roughly 6 mas, limited dynamic range made them hard to follow. Some details of these motions can be found in Walker, Benson, and Unwin (1987).

Extensive VLBI and VLA observations of the radio structure of 3C 120 on scales larger than that observed in the VLBI monitoring runs have been made (Walker, Benson, and Unwin 1987). They show that there is a continuous jet structure from the parsec scales, where the superluminal motions are seen, to over 100 kpc, well outside of the optical galaxy. In addition, diffuse structures are seen with an overall extent of at least 400 kpc in projection. The jet is one sided and has a longitudinal magnetic field, characteristics typical of the more powerful radio sources (Bridle and Perley 1984). On the side opposite the jet, at about 60 kpc, there is an edge-brightened lobe. Any counterjet that feeds this lobe is at least 100 times weaker in the radio than the observed jet. The larger scale structures are very confused. This may be the result of the projection effects that are expected in the line-of-sight models for superluminal motions.

Perhaps the most significant result of the observations of the structure of 3C 120 is the simple evolution of brightness that was found. Figure 1 shows the brightness as a function of jet width. It covers about 3.5 orders of magnitude in core distance and 8 orders of magnitude in brightness. The data are fit well by a power law. The simple evolution of the jet suggests that its basic character is established on parsec scales and it evolves simply, perhaps in response to a gradient of external pressure, as it travels out. The source was actually observed over 5 orders of magnitude in core distance and the simple evolution seems to hold over the full range. However the jet width on small

* *The National Radio Astronomy Observatory is operated by Associated Universities Incorporated under contract to the National Science Foundation.*

M. J. Reid and J. M. Moran (eds.), The Impact of VLBI on Astrophysics and Geophysics, 29–30.

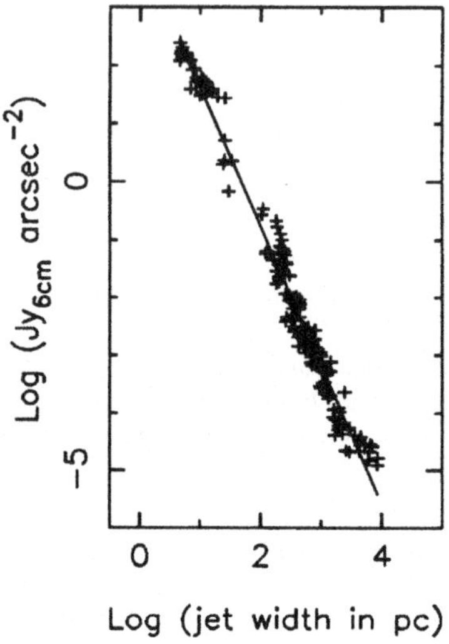

Figure 1: The brightness (Jy per square arcsecond) of the 3C 120 jet vs. the width of the jet in parsecs. The solid line represents a power law with an index of -2.40 ± 0.08 that is the result of a least squares fit to the data.

scales and the brightness on large scales could not be measured reliably so the full range is not used in Figure 1.

With the brightness information shown in Figure 1, the physical parameters of the jet can be calculated under the assumption that there is equipartition between the energy in relativistic particles and magnetic field. Some results are:

Minimum energy density: $U_{min} = 1.5 \times 10^{-10}\ r^{-1.95}$erg cm^{-3}

Magnetic field: $B = 4.1 \times 10^{-5} r^{-0.97}$ gauss

Pressure: $p/k = 3.7 \times 10^{5} r^{-1.95}$ cm^{-3}deg,

where the pressure is given in units of density times temperature, and r is the FWHM of the jet in arc seconds.

The pressure is about what is expected for the external medium on kpc and larger scales but is somewhat lower that what is expected in the broad and narrow line regions on small scales. The evolution of brightness is not consistent with a flux conserving, adiabatic expansion of a jet with a longitudinal field even with slowing of the bulk velocity (Bridle and Perley 1984) — the velocity would become too low to reach the outer structures. The data appear to be most consistent with a light jet, containing little or no thermal material, that is relativistic at least on small scales and perhaps on all scales. For further details, see Walker, Benson, and Unwin (1987).

REFERENCES

Baldwin, J. A., Carswell, R. F., Wampler, E. J., Smith, H. E., Burbidge, E. M., and Boksenberg, A. 1980, *Ap. J.*, **236**, 388.

Bridle, A. H., and Perley, R. A. 1984, *Ann. Rev. Astron. Ap.*, **22**, 319.

Walker, R. C., Benson, J. M., and Unwin, S. C. 1987, *Ap. J.*, **316**, 546.

Walker, R. C., Benson, J. M., and Unwin, S. C. 1987, in *Superluminal Radio Sources*, ed. J. A. Zensus and T. J. Pearson (Cambridge: Cambridge University Press), p. 48.

HIGH DYNAMIC RANGE VLBI OBSERVATIONS OF THE 3C120 RADIO JET

J.M. Benson[1], R.C. Walker[1], T.W.B. Muxlow[2]
P.N. Wilkinson[2], R. Booth[3], G. Pilbratt[3], S.C. Unwin[4]
[1]NRAO, Charlottesville, VA
[2]NRAL, Jodrell Bank, Macclesfield, UK
[3]Onsala Space Obs., Onsala, Sweden
[4]OVRO, Caltech, Pasadena, CA

3C 120 is a nearby radio galaxy (z = 0.033) that shows superluminal motions in its radio jet within one parsec of the nucleus. The apparent velocities of the moving features are approximately 1.5 mas/yr. This object is of particular interest because it is clearly identified with an N-type galaxy, unlike all other superluminal radio sources which are only seen in quasars and BL Lac objects. Because of the close proximity of 3C 120, we have been able to make a series of VLBI and VLA images at stepped resolutions, and thereby map the radio jet continuously from the radio nucleus (0.5 h^{-1} parsec) into the diffuse 400 h^{-1} kpc structure.

We observed 3C 120 at two epochs with VLBI arrays consisting of 14 and 18 stations. The latter was the so-called 'World Radio Array' experiment. The observations were conducted at 18 cm. At the time of the second VLBI run, 18 cm observations were also obtained with the MERLIN array. The large number of stations and carefully organized calibration observations enabled us to make radio images having dynamic ranges of nearly 1000:1 (peak flux to off-source rms). The off-source noise level in the Apr 1984 map is 0.5 mJy/beam, roughly twice the theoretically derived noise level.

There are four basic results from the maps shown in the figure. 1). We see, in both VLBI maps, a well collimated jet running eastward from the radio core for about 75 h^{-1} pc. Beyond that distance, the jet suddenly expands into an opening angle of 30 degrees. The superluminal components, whose motions we monitor at 6 cm, are blended together in the region within 20 mas of the core. 2). The 1982 and 1984 images both distinctly show emission knots and twists in the jet before it begins to expand. The individual features are clearly seen in both epochs and are essentially spatially coincident. 3). The emission knot lying between 40 and 60 mas from the core may have moved outward at approximately 1.2 mas/yr, nearly the same velocity as the superluminal features near the radio core. The jet may therefore exhibit superluminal velocities up to the point where it begins to expand. 4). The VLBI and MERLIN images taken together provide a link

M. J. Reid and J. M. Moran (eds.), The Impact of VLBI on Astrophysics and Geophysics, 31–32.

in spatial resolution between the superluminal region (10 h^{-1} pc) and the kiloparsec features which have been completely imaged by the VLA. The radio jet is continuously connected from the radio core to the most distant 400 h^{-1} kpc emission region.

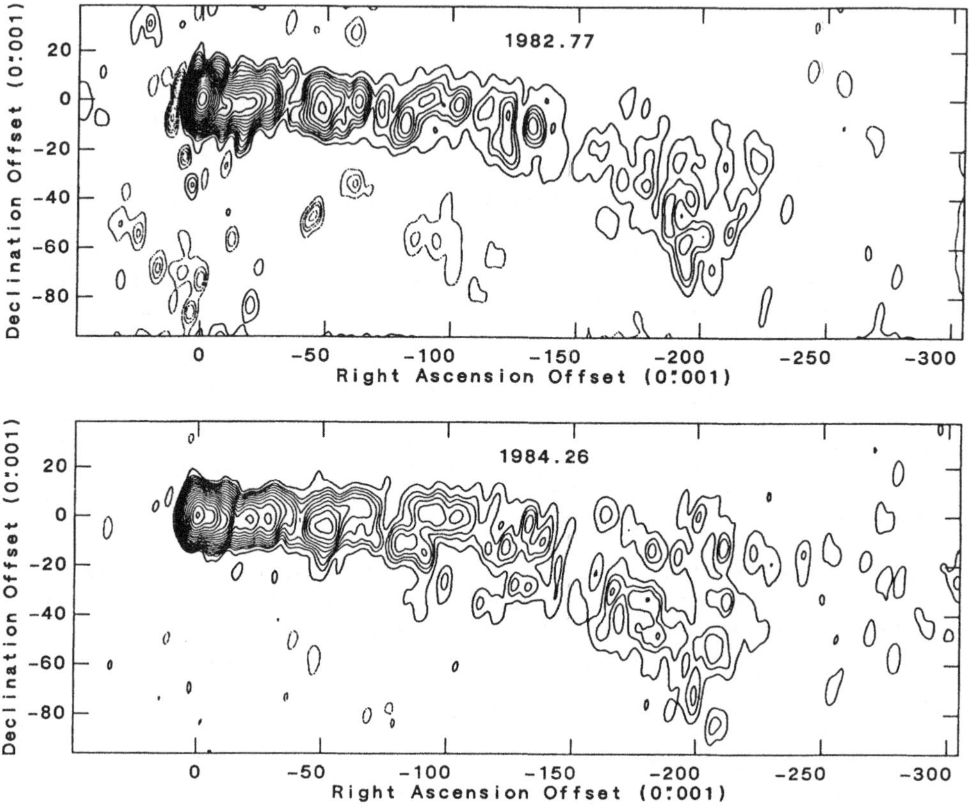

Fig. 1. Full resolution maps of 3C 120 at 18 cm in October 1982 and April 1984. The convolving beam is 3 by 6 milliarcsec in position angle -10 deg. The contour levels are logarithmic : -6.2, -4.5, -3.2, -2.3, -1.2, 1.2, 2.3, 3.2, 4.5, 6.2, 8.6, 12.0, 16.7, 23.2, 32.2, 44.8, 62.2, 86.4, 120.0, 166.9, 231.6, 321.6, 447.6, 621.6, 864.0, 1666.8 milliJanksies.

3C 273 AND DA 193 MAPPED WITH CRUSTAL DYNAMICS VLBI DATA

Patrick Charlot[(1)], Jean-François Lestrade [(2)], Claude Boucher[(1)]

[(1)]Institut Géographique National
2 Avenue Pasteur, F-94160, Saint-Mandé, France.
[(2)]Bureau des Longitudes
77 Avenue Denfert-Rochereau, F-75014, Paris, France.

ABSTRACT. Hybrid maps of 3C 273 and DA 193 at 2.3 and 8.4 GHz have been produced with VLBI data acquired during a Crustal Dynamics campaign. By comparing to maps at previous epochs, superluminal components of 3C 273 are clearly detected. DA 193 is relatively compact. Structure corrections for the VLBI delay used in astrometry have been estimated from these maps. Some corrections are significant when compared with the precision of models used in astrometry and geodesy.

The Crustal Dynamics Program (CDP) kindly provided us with the VLBI data acquired during the experiment conducted on North Pacific baselines on 1985 May 15 with a 6-station VLBI array (Mojave, Vandenberg, Hatcreek, Gilmore Creek, Kauai, Kashima). Figure 1 shows the u-v plane for 3C 273 at 8.4 GHz. It is interesting that such coverage is more uniform than those provided by the other networks. Figure 2 shows the hybrid maps of DA 193 and 3C 273 made with these data.

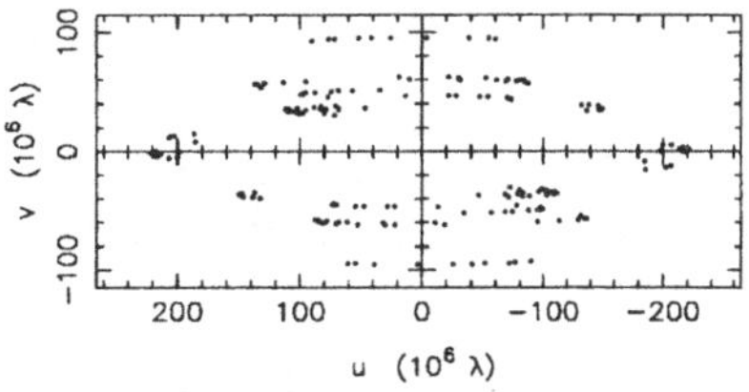

Figure 1. The u-v plane for 3C 273 at 8.4 GHz.

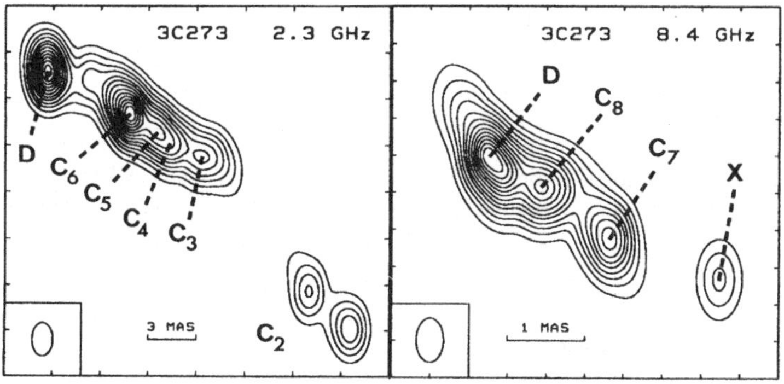

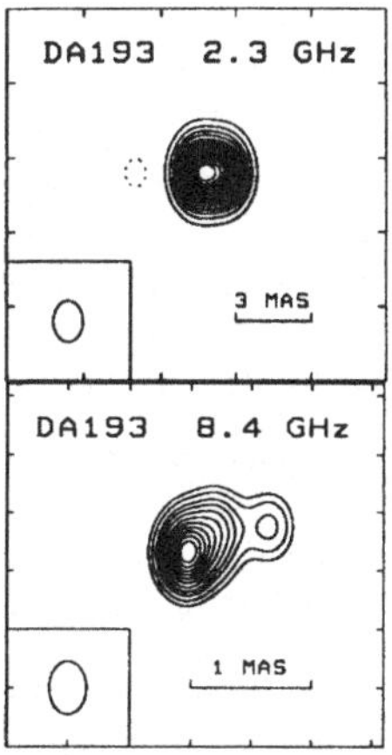

Figure 2. Hybrid maps of 3C 273 and DA 193 at 2.3 and 8.4 GHz. Contour levels: 1, 3, 6, 10, 16, 22, 30, 40, 50, 60, 70, 80, 90 % of the peak brightness (1 and 3 % not drawn for DA 193 at 8.4 GHz).

M. J. Reid and J. M. Moran (eds.), The Impact of VLBI on Astrophysics and Geophysics, 33–34.

Our maps of DA 193 are similar to the one found by Spangler *et al* (1983) at 5 GHz at epoch 1981.7. However we find a slight extension in the N-W direction that is not seen in their map. Our maps of 3C 273 are consistent with previous maps published by Unwin *et al* (1985), hereafter U85, Biretta *et al* (1985), and Cohen *et al* (1987), hereafter C87. The separations and proper motions of the components of 3C 273 are given by U85 and C87. We have used these values to predict the separations of the components at our epoch of observation (1985.37) in order to identify them. This comparison is summarized in Table I.

	Map at 2.3 GHz					Map at 8.4 GHz		
	C_2	C_3	C_4	C_5	C_6	X	C_7	C_8
U85 or C87 at 1985.37	20-25	11.9 ±0.5	9.3 ±2.2	7.8 ±0.2		3.6 ±0.2	2.1 ±0.2	0.7 ±0.2
Our maps	21-26	11.1	9.1	7.8	5.9	3.4	1.9	0.7

Table I. Separations between the core D and the components of 3C 273 (milliarcseconds).

In our 2.3 GHz map, the strongest component is assumed to be C_6. This component was first seen in 1981.10 at 10.7 GHz (U85) when its separation from D was 1.1 milliarcsecond. It is not seen in later maps at 10.7 GHz (C87), but it is visible in our 2.3 GHz map with a separation of 5.9 milliarcseconds from D. This implies a proper motion of 1.1 milliarcsecond per year. This proper motion is consistent with the measured proper motions of the other components. The weak emission region located between 21 and 26 milliarcseconds from the core may correspond to the fading component C_2.

The maps above can be used to derive astrometric corrections for the VLBI delays in order to refer the position of each source to a specific feature of its morphology (Thomas, 1980). Table II gives the magnitudes of these corrections at 2.3 and 8.4 GHz in the cases of the complex source 3C 273 and of the simple source DA 193. These corrections must be included for modelling the VLBI delay at the 0.1 nanosecond level.

	3C 273		DA 193	
	2.3 GHz	8.4 GHz	2.3 GHz	8.4 GHz
Maxi	1.44	0.44	< 0.01	0.02
Mean	0.38	0.08	< 0.01	0.01
RMS	0.57	0.15	< 0.01	0.01

Table II. Structure corrections in the delays (nanoseconds) for 3C 273 and DA 193 during the North Pacific CDP campaign on 1985 May 15.

Acknowledgements:

We are grateful to the Crustal Dynamics Program for providing us with the VLBI data analysed in this report. The Caltech VLBI package has been used to map the structures of the sources presented in this report.

REFERENCES

Biretta *et al*, 1985, *Astrophys. J.*, **292**: L5-L8.
Cohen *et al*, 1987, *Astrophys. J.*, **315**: L89-L92 (C87).
Spangler *et al*, 1983, *Astrophys. J.*, **271**: 44-50.
Thomas, J. B., 1980, *NASA-JPL Publication* **80-84**.
Unwin *et al*, 1985, *Astrophys. J.*, **289**: 109-119 (U85).

5 YEARS OF VLBI AND X-RAY OBSERVATIONS OF NRAO 140

Alan P. Marscher
Department of Astronomy, Boston University
725 Commonwealth Ave.
Boston, MA 02215
U.S.A.

ABSTRACT. The author and his collaborators have observed NRAO 140 twice at X-ray energies and numerous times with multifrequency VLBI. The VLBI observations reveal a knotty jet structure with superluminal motion of the innermost two knots relative to the core. The VLBI core decreased by about a factor of 2 in flux density between 1980 and late 1984. The X-ray flux also declined by about the same factor during this period. Monitoring at 18 cm during periods of low-frequency variability has revealed pronounced changes in the relative brightnesses of the components of the source while the total flux density has varied by $\lesssim$ 10%.

1. INTRODUCTION.

Advances in our understanding of the quasar phenomenon have been accomplished through both statistical studies of large samples of objects and detailed analyses of the emission from single objects. The quasar NRAO 140 (z=1.26) is a particularly good choice for the latter type of investigation, since it is among the strongest X-ray quasars and exhibits many interesting of radio properties apparent superluminal motion, core-jet morphology, and variability at both high and low frequencies. The author and his collaborators have been observing NRAO 140 with X-ray satellites and VLBI arrays since 1980. The results impact our current thoughts on how to explain the X-ray and radio emission from quasars.

2. X-RAYS.

The author has observed NRAO 140 in 1980 (with *HEAO 2*) and 1985 (with *EXOSAT*). Upon combining these results with the X-ray measurements by other authors, it is evident that the X-ray flux density decreased by a factor of 2 between 1978 and 1985. This is coincident with a similar decline in the intensity of the radio core flux density at 2.8 cm (see below). This behavior is consistent with inferences drawn from the previously reported strong correlation between high-frequency radio flux density and X-ray flux in compact radio sources (*e.g.*, Owen, Helfand, and Spangler 1981). One would not, however, expect a one-to-one correspondence between X-ray and radio core flux density if the X-rays originate through inverse Compton scattering in the core except under special circumstances (Marscher 1987a).

The absorption column density (expressed in terms of the amount of cold

M. J. Reid and J. M. Moran (eds.), The Impact of VLBI on Astrophysics and Geophysics, 35–36.

hydrogen gas along the line-of-sight), as determined by the low-energy cutoff in the X-ray spectrum, is consistent with the galactic value of $2 \times 10^{21}\text{cm}^{-2}$ in 1979 and 1985. In 1980, however, it exceeded $6 \times 10^{21}\text{cm}^{-2}$ (90% confidence limit) (Marscher 1987b). The excess absorption could have been produced either by a solar-system sized globule of density $\sim 10^{9}\text{cm}^{-3}$ or by a cloud in the quasar, with the latter case requiring a column density 9 times greater owing to the strong energy dependence of the absorption cross-section and the redshift of the quasar. The former interpretation is made more plausible by the fact that NRAO 140 lies behind the outer edge of a giant molecular cloud in Perseus.

3. VLBI.

The 6 cm VLBI observations reveal that the quasar has a core-jet structure, with 6 components evident on the map. The opening half-angle of the jet is $\sim 3°$, which implies a true opening half-angle less than half a degree if the jet is pointing nearly along the line-of-sight. Superluminal motion of the innermost knots is revealed at 2.8 cm. The core at 2.8 cm decreased in flux density by about a factor of 2 between 1980 and late 1984. A spectral dissection using multifrequency VLBI performed in mid-1981 and 1984 illustrates how the various self-absorbed components contribute to an overall flat spectrum. Analysis of two of the components reveals a "Compton problem," which refers to an excess of self-Compton X-ray flux predicted over that observed (Marscher and Broderick 1981; Marscher *et al.* 1987a).

NRAO 140 is a well known low-frequency variable, with an outburst which peaked in mid-1981. The author and his collaborators have observed NRAO 140 using VLBI at 18 cm in 1981, 1984, and 1986. Changes in the visibilities indicate that a particular component decreased significantly in brightness at 18 cm between the first two epochs, as the low-frequency outburst faded (Marscher *et al.* 1987a). The 1986 map (Marscher *et al.* 1987b) shows that the southeastern half of the source declined further in brightness while the northwestern half increased in flux density. The total flux density at 18 cm fluctuated only slightly (by ± 0.15 Jy) during this period. The "mid-frequency gap" in variability seems therefore to be another conspiracy, in this case resulting in nearly constant total flux density at 18 cm while the flux densities of the individual components fluctuate considerably. It is not clear whether the refractive scintillation model can explain the near constancy of the total flux density at 18 cm. Also, the size of the variable component measured at 6 cm indicates that it has a "Compton problem" which implies the existence of relativistic motion, which by itself can also explain the low-frequency variability.

The X-ray observations of NRAO 140 were part of the *HEAO-2* and *EXOSAT* Guest Investigator Programs, with financial support provided in part by NASA. The VLBI observations were supported in part by the National Science Foundation.

REFERENCES.

Marscher, A.P. 1987a, *Ap. J.*, submitted.
Marscher, A.P. 1987b, *Ap. J. (Letters)*, submitted.
Marscher, A.P., and Broderick, J.J. 1981, *Ap. J. (Letters)*, **247**, L49.
Marscher, A.P., Broderick, J.J., Padrielli, L., Bartel, N., and Romney, J.D. 1987a, *Ap. J.*, **319**, in press.
Marscher, A.P., Rickett, B.J., Padrielli, L., Bartel, N., and Romney, J.D. 1987b, in preparation.
Owen, F.N., Helfand, D.J., and Spangler, S.R. 1981, *Ap. J. (Letters)*, **250**, L55.

SUB-LUMINAL MOTION IN M87

J. A. Biretta, M. J. Reid
Harvard - Smithsonian Center for Astrophysics

W. Junor, R. Spencer, T. Muxlow
Nuffield Radio Astronomy Laboratories

The galaxy M87 (3C274, NGC4486) contains one of the nearest jet sources in the northern hemisphere, and one of our best opportunities to study a jet in detail. We have made MkII VLBI observations of the nuclear jet at 1.66 GHz using 18 stations of the "World Array." The array included Arecibo which greatly improved the north-south resolution. The data were global fringe fitted, and self-calibrated to remove both antenna and baseline amplitude and phase errors. The resulting map (Fig. 1) has a dynamic range of 2300:1 and reveals a very complex structure. At distances $\geq$50 mas from the core, the jet is resolved across its width and appears limb-brightened. The brightness centroid oscillates from side to side by an amount roughly equal to the jet's width; this may be caused by brightness variations at the edges of a limb-brightened jet. Within 50 mas of the core the opening angle (at the FWHM points) is $6.5° \pm 0.3°$; this is larger than that of the 20" jet, and suggests there is recollimation between the pc and kpc scales. The jet is aligned within 1° of the 20" jet. There is no evidence for a counter-jet; the jet/counter-jet brightness ratio is $R \geq 200$.

Comparison of this map with reanalyzed data from 1980 and 1982 (Schmitt and Reid 1985) shows a prominent feature (labeled N2) moving outward at 1.1 ± 0.3 mas/year, or $0.34 \pm 0.09c$ for a distance of 20 Mpc. A weaker feature, N1, also shows evidence for motion. Future observations will give more accurate velocities and show whether the side to side oscillation pattern also moves outward. Sub-luminal motion has been detected in only one other active galactic nuclei, 3C84, which has similar speeds (Marr, et al. 1987). Both sources have 18cm total powers which are much lower than those of the super-luminal quasars.

The observed proper motion β_{obs} and counter-jet ratio R can be used to constrain the geometry and intrinsic velocity of the jet. Assuming (1) the observed feature and jet fluid move with the same speed, and (2) that there is a counter jet with an intrinsic brightness equal to that of the observed jet, we obtain limits of $\theta \leq 5.2°$ for the angle to the line of sight, and $\beta \geq 0.8$ for the fluid speed. Such a small angle to the line of sight seems very unlikely for several reasons. (a) The enclosed solid angle is very small. (b) The deprojected opening angle for the 20" jet would be $\leq 0.3°$, which is much smaller any other radio source of similar luminosity (Bridle, 1984). And (c) the sharp feature seen in Knot A (Biretta, Owen, Hardee 1983) is difficult to obtain for small θ. Together this suggests the counter-jet is either intrinsically weak or non-existent, or that the jet fluid has a much higher velocity than the observed feature (Lind and Blandford 1985).

M. J. Reid and J. M. Moran (eds.), The Impact of VLBI on Astrophysics and Geophysics, 37–38.

References

Bridle, A.H.: *A.J.* **89**, 979 (1984).
Biretta, J.A., Owen, F.N., Hardee, P.E.: *Ap.J. Lett.* **274**, L27 (1983).
Lind, K.R. and Blandford, R.D.: *Ap.J.* **295**, 358 (1985).
Marr, J.M., *et al.*: this conference (1987).
Schmitt, J.M. and Reid, M.J.: *Ap.J.* **289**, 120 (1985).

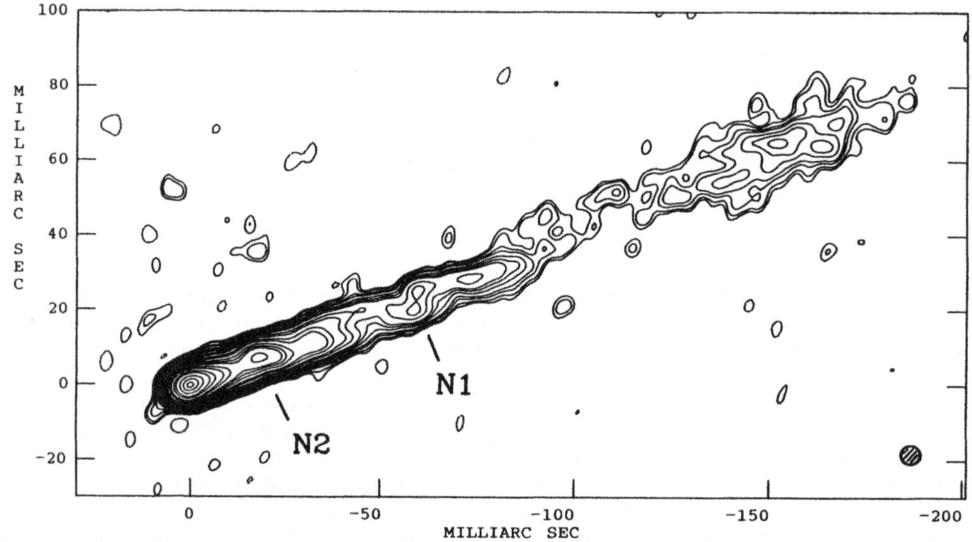

Fig. 1. Map of M87 nucleus for 1984.26. Contours are ± .15, .2, .3, .4, .6, .8, 1, 1.5, 2, 3, 4, 6, 8, 10, 15, 20, 30, 40, 60, 80, and 95 percent of the peak, and the beam has a 5 mas FWHM.

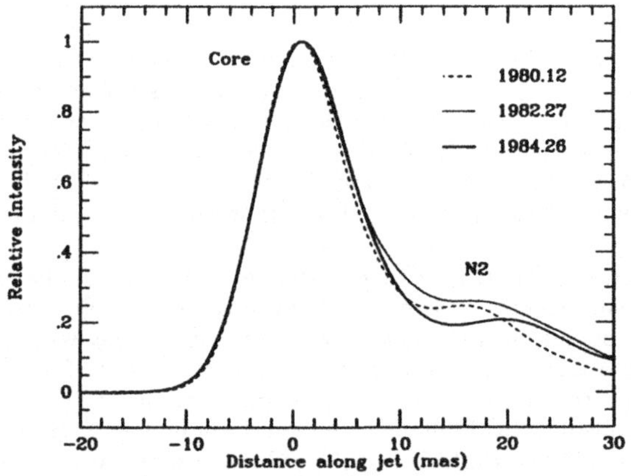

Fig. 2. Intensity profiles taken along P.A. -70 degrees. Peak intensities are normalized. All epochs have been convolved to the same 8 mas FWHM beam.

UNIFORM APPARENT SUPERLUMINAL MOTION IN 4C39.25

C.J.Schalinski
Max Planck Institut für Radioastronomie, Bonn, RFG

A.Alberdi, P.Elósegui, and J.M.Marcaide
Instituto de Astrofísica de Andalucía, Granada, Spain

The radiosource 4C39.25 (0923+392) is an 18 mag quasar with a redshift of 0.699. Throughout the seventies its compact structure at centimetric wavelengths consisted of 2 components separated by an angle of about 2 milliarcseconds (mas). In the eighties the structure has been found to consist of 3 components (Marcaide et al.,1985; Shaffer et al.,1987). The angular separation between the "a" and "c" components (see Fig. 1, which shows our most recent map of this source made at a wavelength of 2.8 cm) has remained constant at about 2 mas, while component "b" is separating from component "c" at a rate which we estimate from 2.8 cm observations as 0.18 ± 0.01 mas/yr, and which translates into an apparent linear separation velocity of 4.0 ± 0.2 c (H_0=100 km/s/Mpc, q_0=0.5, c is the speed of light).

In Figure 2 we present maps obtained at 3.6 cm using data from geodetic IRIS-experiments (see Schalinski et al., this volume). The proper motion of component "b" relative to component "c" which we estimate at this wavelength is compatible with that estimated from 2.8 cm observations. It should be noted that this relative proper motion is very uniform. Component "b" is rapidly brightening, component "c" is progressively dimming, whereas the flux density of component "a" appears to remain rather constant in time. Detailed considerations about the evolution of each one of the three components will be presented elsewhere.

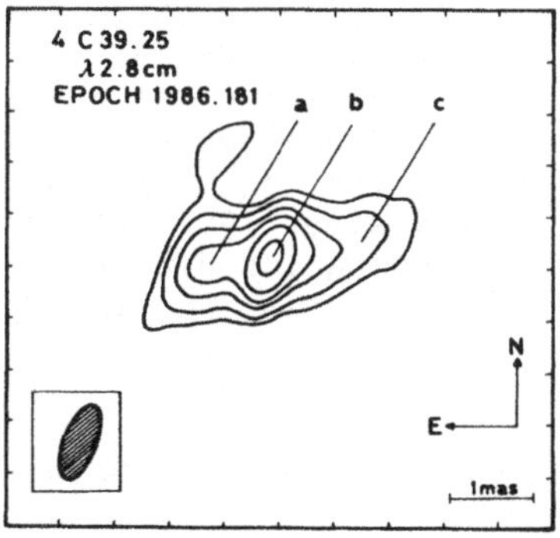

Figure 1. Hybrid map of quasar 4C39.25 made at 2.8 cm. Component "b" is moving superluminally to the east with respect to the other components. Contours:5,10,20, 30,50,80% of peak.

M. J. Reid and J. M. Moran (eds.), The Impact of VLBI on Astrophysics and Geophysics, 39–40.

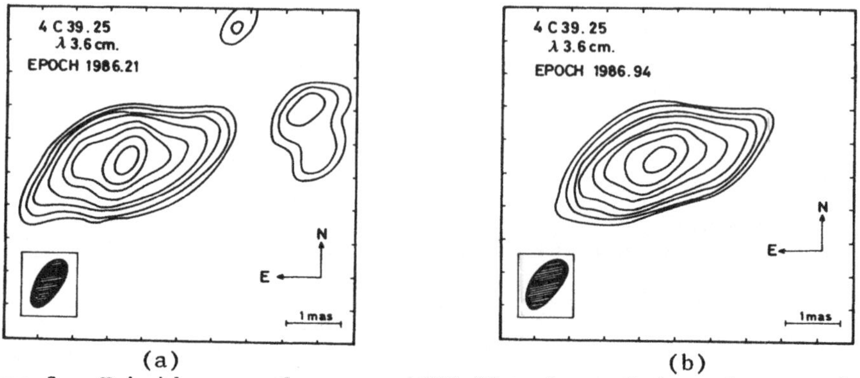

Figure 2. Hybrid maps of quasar 4C39.25 made at 3.6 cm from geodetic IRIS experiments. Contours: 2,3,5,10,20,30,50,80% of peak.

Independent evidence for bulk relativistic motion in this source appears to come from the excess of calculated inverse-Compton flux density over the observed X-ray flux density, indicating a minimum Doppler-factor of 4.2 (Biermann et al.,1987). These authors find 4C39.25 to be the source with the strongest discrepancy between expected and observed X-ray flux density among those in a complete sample of 56 flat spectrum radio sources (Zensus et al.,1984).

Like the source 3C395 (Waak et al.,1985), 4C39.25 displays apparent superluminal motion relative to other parts of the compact structure which appear stationary with respect to each other. The evolution of 4C39.25 is such that unless something new, and unusual, again happens the source may be around 1991 -when we expect the apparent collision of component "b" with component "a"- an excellent geodetic and astrometric calibrator, since by that time the component "c" may be very dimm. A careful monitoring of the "collision" process before and after 1991 should provide new clues about the physics of the very compact radio sources.

REFERENCES

Biermann,P.L.,Kuehr,H.,Snyder,W.A.,Zensus,J.A.,1987:Astron. Astrophys., in press

Marcaide,J.M.,Bartel,N.,Gorenstein,M.V.,Shapiro,I.I.,Corey,B.E., Rogers,A.E.E.,Webber,J.C.,Clark,T.A.,Romney,J.D.,Preston,R.A., 1987:Nature,314,424

Shaffer,D.B.,Marscher,A.P.,Marcaide,J.M.,Romney,J.D.,1987:Ap.J.Lett., 314,L1

Waak,J.A.,Spencer,J.H.,Johnston,K.J.,Simon,R.S.,1985:A.J.,90,1989

Zensus,J.A.,Porcas,R.W.,Pauliny-Toth,I.I.K.,1984:Astron.Astrophys., 133, 27

THE STRUCTURE OF 4C39.25 ON INTERMEDIATE SCALES

S. Y. Wu (*)
I. I. K. Pauliny-Toth and R. W. Porcas (**)
(*) Beijing Astronomical Observatory
Academia Sinica
Beijing 100080, P. R. China
(**) Max-Planck-Institute for Radio Astronomy
Auf dem Huegel 69
D-5300 Bonn 1, F. R. Germany

ABSTRACT. VLBI observations at 18 and 6 cm with European baselines of lengths up to 15 and 23 million wavelengths respectively have been used to search for structure in the quasar 4C39.25 on scales of 10 to 100 milliarcsec. No significant structure on these scales brighter than 5 percent of the peak brightness has been detected. The core itself, however appears to be extended on a scale of 2 to 4 milliarcsec and variable in flux density.

1. INTRODUCTION AND OBSERVATIONS

4C39.25, a quasar with redshift 0.698, is one of the ten strongest compact sources at centimetre wavelengths. Recent observations (Marcaide et al. 1985, Shaffer and Marcher 1985) have revealed a weak, probably superluminal feature E between 2 stationary components with flux density increasing year by year. A larger component B with a size of about 3 mas beside the strong double was suggested by Baath et al. (1980).

VLBI observations at 18 and 6 cm with European baselines have been made to search for structure in the quasar on scales of 10 to 100 mas between March 30 and April 8, 1984. The calibration has been checked by using observations of the unresolved sources OQ208 (6 cm) and DA193 (18 cm). The calibration was less accurate than expected, owing to the poor weather conditions during the observations, and the error in fringe amplitude is estimated to be less than 10 percent.

2. RESULTS AND DISCUSSION

The hybrid maps at two frequencies show that the emission originates mostly in a bright, nearly unresolved core. The fringe amplitudes show peak-to-peak variations of about 20 percent on the longest baseline, indicating that the core is slightly resolved, and is more compact at

M. J. Reid and J. M. Moran (eds.), The Impact of VLBI on Astrophysics and Geophysics, 41–42.

6 cm than at 18 cm. There is evidence for a weak extension to the northwest, although this is at a level of only 5 percent of the peak brightness. It is also present (at level of 1%) at 5 GHz and is thus probably real. There are no other extended features stronger than 2 percent of the maximum in the maps.

The parameters of the features obtained from model fitting are shown in Table I. The core appears to be elongated in directions consistent with that of the compact double (-81 deg.) and of the arcsec structure (-100 deg.; Browne et al. 1982) respectively. The results imply a steep-spectrum feature with a size of 3.6 mas at 18 cm, elongated in P.A.=-105 deg. surrounding the core. It seems to be component B suggested by Baath et al..

The weak extended feature with 16 and 18 mas separation to the core and P.A.=-26 deg. has a spectral index of about -0.63, and a size of about 10 mas. It perhaps is only structure on the intermediate scales revealed in these observations.

Table I. Structural parameters deduced from the model fitting

Frequency	Total flux density	Parameters of the features (1984.25)					
		Flux density	Separ. from core	P.A. of separ.	Major axis	Axial ratio	P.A. of major axis
(GHz)	(Jy)	(Jy)	(mas)	(deg.)	(mas)		(deg.)
5	5.9	5.1	0	0	2.0	0.2	-81
		0.1	16	-26	7	0.4	-33
1.66	3.3	2.7	0	0	3.6	0.5	-105
		0.2	18	-26	11	0.5	-95

4C39.25 shows slow variations in its flux density at centimetre wavelengths, on a time-scale of years. At 5 GHz, the flux density has decreased from 9.0 Jy at epoch 1974.1 (Pauliny-Toth et al. 1981) to 5.9 Jy at 1984.25. At 1.66 GHz, however, the flux density has remained constant at about 3.3 Jy over the same interval (Baath et al. and our data). The fact that there has been no significant change at 1.66 GHz in spite of the considerable decrease at 5 GHz also supports the existence of feature B. As Baath et al. indicated that about 80 percent of the emission at 1.66 GHz should arise from component B and the arcsec structure. The observed variation of flux density arises mostly from the compact triplex.

REFERENCES

Baath, L. B., et al.: 1980, Astron. astrophys. 86, 364
Browne, I. W. A., et al.: 1982, Nature 299, 788
Marcaide, J. M., et al.: 1985, Nature 314, 424
Pauliny-Toth, I. I. K., et al.: 1981, Astron. J. 86, 371
Shaffer, D. B., et al.: 1985, Bull. Am. Astron. Soc. 17, 608

4C39.25 – SUPERLUMINAL MOTION BETWEEN STATIONARY COMPONENTS

D. B. Shaffer
Interferometrics Inc.
8150 Leesburg Pike
Vienna, Virginia 22180 USA
and
A. P. Marscher
Department of Astronomy
Boston University
725 Commonwealth Avenue
Boston, Massachusetts 02215 USA

ABSTRACT. During the 1970's, the compact radio structure of 4C39.25 consisted of a simple double source, whose separation did not change. A new component, which probably originated in mid-1979, is now seen to be moving superluminally between the original, still stationary, components. It is not clear which component of the source is the core, nor how the moving component will interact with the stationary component that it is now approaching on a collision course.

From 1972 until 1979, the compact radio structure of the quasar 4C39.25 (z=0.699) consisted of an almost east/west (position angle 98°) double source, whose separation was about 2.0 mas at frequencies from 5 to 15 GHz (Shaffer *et al.* 1977 and 1987). In late 1979, the structure of the westernmost component of the source began to change, and by 1982 a new component was clearly evident between the two original components. Since then, the source structure has consisted of three distinct components. The central component is moving eastward with an angular velocity of 0.16 ± 0.02 mas year^{-1} with respect to the two outer components, which have the original 2.0 mas separation. The corresponding velocity is $3.5h^{-1}c$ for $q_0 = 0.5$ and $H_0 = 100h$ km s^{-1} Mpc^{-1}. The epoch of zero separation of the central component from the westernmost component of the original double is about 1979.5. This time coincides with a plateau in the long-term decline of the flux density of the source (Aller *et al.* 1985), perhaps indicative of a weak outburst.

Shaffer *et al.* (1987) show maps of the structure of 4C39.25 at 8.4 and 10.7 GHz, from 1979 through 1985. Our 1986 maps at 8.4 GHz show that the motion of the central component has continued at the same rate. Observations at 22 GHz (Marscher *et al.* 1987) also show a triple source, whose central component is the brightest and most compact.

The structure and its variations in 4C39.25 are quite unlike those seen in any other source, and it is not clear how to interpret these observations. For instance, we are not certain which component is the nucleus of the source. With no other information, one would be inclined to call the central component the core because it is so strong and compact at 1.3 cm. Such characteristics usually identify the nuclear component. Alternatively, since the central component is seen to be moving to the east, away from the western component, one would argue that the western component is the core, by analogy to all other superluminal sources, since that is where the moving component seems to have originated.

M. J. Reid and J. M. Moran (eds.), The Impact of VLBI on Astrophysics and Geophysics, 43–44.

On the arcsecond scale, 4C39.25 shows a two-sided jet, with the brightest component on the east (Browne *et al.* 1982). In almost all other sources, the superluminal components move out towards the brightest outer features (3C120 or 3C273, for example). If 4C39.25 is similar, its extended structure is consistent with the western component being the core. The source clearly has had two-sided ejection in the past, but we can not tell if we are seeing only one-sided activity now or just the beamed side of a two-sided jet.

We can not even tell for sure which component(s) is (are) really moving. We do not have any strong limits on the overall motion of the source, except that if its position had changed by more than about one or two milliarcseconds over the last five years, we would have detected the motion with NASA geodesy observations and analysis. It is possible that the central component is stationary, and that the two outer components are moving with the same velocity, or that all three components are moving but with different velocities.

Both the outer components could be stationary features in a relativistic jet. Depending on its environment, such a jet can alternately expand and contract, showing compact hot spots where the contraction reaches a local maximum. In this case, the central component would then be a shock or instability caught up in the flow.

It is possible, although we feel it is unlikely, that the outer components are gravitational lens images of the same compact core. In this case, both "cores" should be seen to eject the same component, although not necessarily at the same time because of path length differences for the two images. There is no evidence for lensing in the arcsecond structure of the source (Browne *et al.*).

Which, if any, of these models is correct may be revealed as we follow the evolution of 4C39.25. The central component is on a collision course with the eastern component. What happens when (or if) a collision occurs should provide considerable insight into the nature of radio jets and the composition of superluminal material. The actual encounter, due in 1991 at the present pace, may show one of several possible results: If the eastern component is a stationary, compact lobe being fed by the jet or caused by an obstruction in the jet flow, the moving component should stop when it gets there, and the eastern end of the source will probably brighten. If the eastern component is a contraction in an undulating jet, the moving component should pass through it at constant velocity, although there may be a brightening as the moving material is compressed. If the components are just images, they should show no interaction at all. We are continuing our observations of this fascinating object, waiting to see what happens.

REFERENCES

Aller, H. D., Aller, M. F., Latimer, G. E., and Hodge, P. E. 1985, *Ap. J. Suppl.*, **59**, 513.

Browne, I. W. A., Clark, R. R., Moore, P. K., Muxlow, T. W. B., Wilkinson, P. N., Cohen, M. H., and Porcas, R. W. 1982, *Nature*, **299**, 788.

Marscher, A. P., Shaffer, D. B., Booth, R. S., and Geldzahler, B. J. 1987, *Ap. J. Letters*, submitted.

Shaffer, D. B. *et al.*, 1977, *Ap. J.*, **218**, 353.

Shaffer, D. B., Marscher, A. P., Marcaide, J., and Romney, J. D. 1987, *Ap. J. Letters*, **314**, L1.

SUPERLUMINAL MOTION OF 3C273B AND THE APPARENT RIGHT ASCENSION SHIFT

N.Kawaguchi and Y.Takahashi
Kashima Space Research Center,
Radio Research Laboratory
893-1, Hirai, Kashima, Ibaraki 314
Japan

ABSTRACT. The effect of the superluminal motion of 3C273b was observed on the fringe amplitudes obtained in VLBI experiments which have been conducted on the baselines between Japan and United States since 1984 under the Crustal Dynamics Project. The speed was 0.7 mas/year and the direction was -83 degrees in the position angle. It was also confirmed as the apparent right ascention shift of the source.

1. Observations

The VLBI observations under the Crustal Dynamics Project have been made since 1984 with stations in Japan and United States, Kashima station in Japan, Mojave station in California, Gilcreek station in Alaska, and Kauai and Kwajalein stations on the Pacific plate, and many good fringes were obtained at 2 GHz and 8 GHz.

2. Fringe Amplitudes

We made analysis on the fringe amplitudes of 3C273b obtained on the Kashima-Mojave, Kashima-Kauai and Kauai-Mojave baselines at 8 GHz and found large variation of the amplitude which seems like an interferometer responce to the superluminal motion of 3C273b. From the best fitted theoretical curve to the observations, we derived the speed of the superluminal motion and the direction as shown in Figure 1. We also got the starting epoch of the evolution of 1979.8 which suggests the Jet knot we detected was C6 named by Unwin et al. [1].

3. Right Asension Shift

The superluminal motion of the Jet knot towards the West as was observed in the fringe amplitudes may cause the apparent shift of the source position. We derived the position of 3C273b from precise delay observations by making 0552+398 a fixed reference and found the decrease of the right ascension as shown in Figure 2. No declination shift was observed. This supports the results of the amplitude analysis, but we need much more data to derive a firm conclusion.

M. J. Reid and J. M. Moran (eds.), The Impact of VLBI on Astrophysics and Geophysics, 45–46.

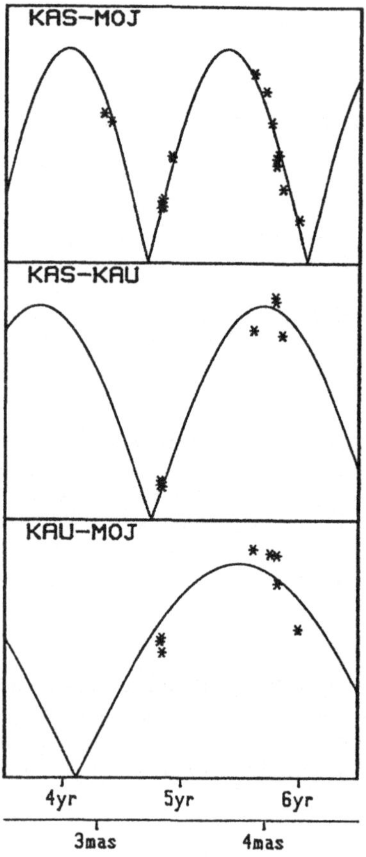

REFERENCE

[1] Unwin, S.C., et al. 1985, Ap. J.,**289**, 109

Figure 1. Observed variation of fringe amplitudes of 3C273b at 8 GHz and the best fitted theoretical curve which assumes that the evolution started in 1979.8 at the speed of 0.70 mas/year to the position angle of -83 degree.

KAS : Kashima in Japan
MOJ : Mojave in California
KAU : Kauai in Hawaii Islands

Figure 2. Observed right ascension shift. The decrease in the right ascension and no change in the declination means that the apparent position of 3C273b shifted towards the west. The slope was -0.83 ± 0.57 mas/year.

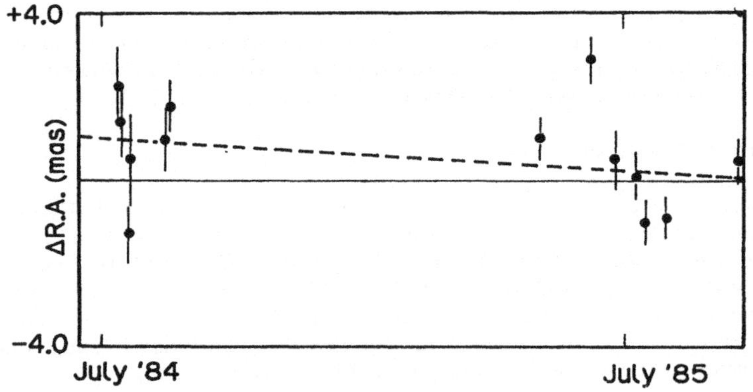

ELECTRODYNAMICS OF THE CENTRAL REGIONS OF ACTIVE GALACTIC NUCLEI

M. Salvati
Osservatorio Astrofisico di Arcetri
L. E. Fermi 5
I-50125 Firenze
Italy

ABSTRACT. We review the information available on some bulk properties of the non-thermal emission regions in Active Galactic Nuclei, such as streaming motions, nature of composing particles, brightness temperature in the proper frame. Most of the information concerns the compact radio emission at relatively large distances from the center, whereas most of the action is thought to occur in the innermost regions which are observed at higher frequencies. We investigate whether these bulk properties can be understood within a general scheme for energy transport and particle acceleration around a magnetized rotating body.

1. INTRODUCTION

Investigating the electrodynamics of Active Galactic Nuclei (AGN's) by means of radio observations has an inherent difficulty. This is illustrated in Fig. 1, which shows a composite spectrum of the QSO 3C273 in the form of a plot of νF_ν - i.e., flux per logarithmic frequency interval - versus ν; plots of this kind give an indication on how the total power is distributed among the various spectral domains. The distribution is remarkably flat from the infrared up to the gamma rays; the exception is represented by the radio frequencies, whose contribution to the total luminosity is only a minor one.

The situation has been summarized by saying that, as far as the overall AGN emission is concerned, the radio emission is just "smoke", and the "fire" has to be looked for at higher frequencies. In the present paper we address precisely the problem of learning about the "fire" by studying the "smoke".

2. RELATIVISTIC MOTIONS AND COMPACTNESS

For our purposes, the two most important pieces of information derived from radio studies of AGN's are superluminal motions and rapid flux variations. By

M. J. Reid and J. M. Moran (eds.), The Impact of VLBI on Astrophysics and Geophysics, 47–54.

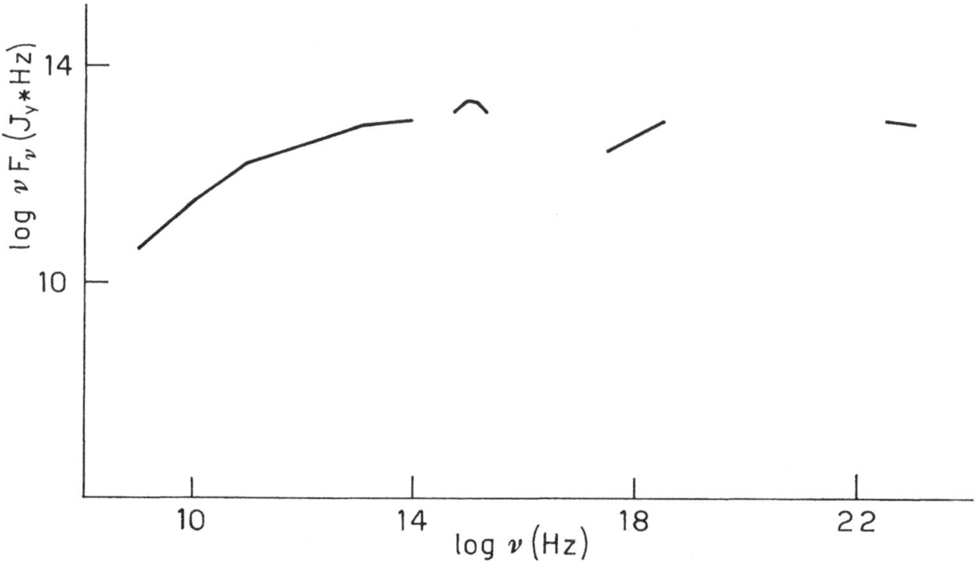

Figure 1. Composite spectrum of 3C273, showing the relative contribution of the various spectral regions to the total luminosity.

superluminal motions one means angular displacements of milliarcsec radio features, which imply faster-than-light transverse velocities at the distances indicated by the optical redshifts; these displacements have been firmly established by modern VLBI techniques, and cannot be explained away as due to insufficient sampling of the Fourier plane (Cohen and Unwin 1984).

Among the numerous interpretations which have been put forward, the one which encounters the largest degree of consensus involves the bulk motion of the emitting plasma at a speed very close to the speed of light along a trajectory very close to the line of sight (Blandford *et al.* 1977). This approach has some difficulties of its own, and its decisive advantage over the competing ones is that it provides a powerful unifying scheme for several other phenomena (Begelman *et al.* 1984).

Additional evidence for the occurence of relativistic bulk motions in AGN's derives from rapid flux variability. The observational situation is rather complicated, especially at low radio frequencies where the importance of propagation effects in the interstellar medium has been recognized only recently (Padrielli 1984; Rickett 1986). At any rate, in at least some instances the observed variations appear to be intrinsic to the source; then, using the well-known causality argument to infer the source size r from the variability time scale Δt, $r < c\ \Delta t$, one runs into serious difficulties in building up a consistent synchrotron model.

In their simplest form, models of this kind require only a few ingredients (see, e.g., Rybicki and Lightman 1979). The size of the emitting region, r, is either measured directly by means of inteferometric observations, or is deduced from the time scale Δt, as indicated above. The magnetic field, B, is measured or bounded from above according to the presence or absence of a synchrotron self-absorption turn-over at low frequencies. The energy distribution of the relativistic electrons and their total number, N, follow from the spectral distribution and total luminosity of the radio emission, respectively. The self-consistent solution for the overall spectrum is a pure synchrotron spectrum only in the limit of a very dilute source. In the opposite limit of a very compact source one has a synchrotron-self-Compton spectrum (SSC): here a cascade of Compton components is included, produced in succession by the same relativistic electrons which emit the "seed" synchrotron photons. The parameter governing the relative importance of any two successive orders is

$$A = \sigma_T N \gamma^2/r^2 \qquad (1)$$

where γ is a suitable average of the particle Lorentz factor. The luminosity contributed by the i-th element in the cascade is

$$L_i \approx c\ B^2\ r^2\ A^{i+1} \qquad (2)$$

which includes the synchrotron luminosity, L_s, for i = 0. When A is larger or smaller than unity, the overall luminosity is dominated by the high or low frequency emission, respectively. In radio astronomical terms the same notion is expressed by means of the synchrotron brightness temperature, T_b: A being larger or smaller than unity is equivalent to T_b being larger or smaller than a critical value $T_c \approx 10^{12}$ °K (Kellermann and Pauliny-Toth 1969). If SSC is relevant to the continuum of AGN's, then the flatness illustrated in Fig. 1 indicates an effective brightness temperature close to the critical value.

Fast variations around a given radio luminosity imply small r's and large apparent T_b's; these in turn imply large luminosities at higher frequencies - typically in the X-ray range -, in contradiction with the luminosities which are actually observed. Bulk motions at relativistic speeds (Rees 1967) allow large r's for the same Δt's, and entail an effective T_b arbitrarily lower than the apparent one.

We are thus led to the conclusion that relativistic flows are an essential ingredient of AGN electrodynamics, and that the conditions are suitable for the SSC process to play an important role.

From a theoretical point of view SSC is appealing since it naturally produces a wide-band spectrum, without postulating a wide-band distribution of radiators. Also, it always comes with compact synchrotron sources, and avoids the introduction of further ad hoc hypotheses.

3. POWER SUPPLY VIA POYNTING FLUX

In the preceding arguments the magnitude of the relativistic effects depends on the Doppler factor, $\mathcal{D}$, of the bulk motion relative to the observer; even if a universal value of the Lorentz factor, Γ, were enforced by some acceleration mechanism, a variety of line-of-sight angles would still result in a variety of $\mathcal{D}$'s ; there is no reason why several apparent brightness temperatures and several Doppler factors should conspire so as to produce a rather narrow distribution of effective brightness temperatures, clustered around

T_c. One is led to the conclusion that some kind of universal radiation process is at work in the proper frame of the emitting plasma, endowed with a self-regulating mechanism which keeps T_b at around T_c. Once the luminosity ratios among the various spectral domains have been set at around unity in the proper frame, an arbitray Lorentz transformation to the observer frame leaves them unchanged, and only affects the variability time scales.

In our formalism the relative weight of the spectral orders is given by A, defined in Eq. 1. When A = 1, Eq. 2 shows that the value of the luminosity common to all orders is $L_B = c\ B^2 r^2$, that is, the energy density in magnetic field flowing across the source boundaries at speed c. The primary power supply appears to be in the form of electromagnetic fields, which are converted with high efficiency into observable radiation.

A plausible scenario where these conditions could be met has been suggested by Pacini and Salvati (1978): they assumed that the active region is energized by large scale electromagnetic fields with $E \approx B$, $\mathbf{E} \perp \mathbf{B}$, so that the Poynting vector flux $\approx c\,B^2 r^2$. The electromagnetic power is spent in maintaining a suitable number of electrons at relativistic energies, at balance with their radiative losses; the gains-vs-losses balance guarantees a high radiative efficiency, and the overall luminosity has to be equal to L_B.

Furthermore, when the process is roughly half-way, the relativistic electrons see a comparable amount of energy density in magnetic field and target synchrotron photons. Hence we have

$$L_S/(c\,r^2) \approx B^2, \quad L_S/(c\,r^2) \approx N\ c\ \sigma_T\ \gamma^2\ B^2/(c\,r^2) = A\ B^2$$
$$A \approx 1 \qquad (3)$$

and this is the condition required for the individual orders to be comparable to one another.

In the following we will show how the proposed scheme is consistent with global pictures currently proposed for the central regions of Active Galactic Nuclei; it is also amenable to a quantitative description, provided that a number of approximations are accepted; and, finally, it can be proven very easily to be Lorentz-invariant.

4. A MODEL

According to recent studies of AGN continua, several different processes are at work simultaneously (Malkan and Sargent 1982; Netzer 1985). Very crudely, one can distinguish a thermal bump, which is responsible for most of the ionizing radiation, and is related to at least part of the X-rays via non-relativistic comptonization; this is superimposed on a power law, presumably of a non-thermal origin, which runs from the submillimeter region to the hard X-rays; the radio emission in radio loud objects is an extension of this power law, coming from larger and less compact regions, and part of the infrared might be due to reprocessing by dust.

The physical picture is similarly complicated (e.g., Rees 1984). While the prime mover is generally identified with a mass-accreting potential well, different components are thought to be relevant to different parts of the spectrum. The UV bump and the thermal X-rays are probably due to an accretion disk and its corona. If a magnetic field threads the disk, or the disk and the black hole horizon (Blandford 1976; Blanford and Znajek 1977), part of the power could be extracted by electromagnetic stresses, rather than viscous stresses; in this case the vicinity of the hole will be pervaded by large scale electromagnetic fields organized in a wind-like geometry, suitable - in our scenario - for the production of the non-thermal power-law. The wind would be eventually collimated, perhaps by magnetic pinch, and at the appropriate distance it would become observable in the radio domain.

Close to the black hole, at distances sufficiently smaller than the scale of the collimating processes, the geometry is expected to resemble the far zone of an aligned "pulsar" (Goldreich and Julian 1969). An artist's view of the proposed arrangement is given in Fig. 2. A dipolar poloidal magnetic field induces a quadrupolar charge distribution, which sustains a quadrupolar poloidal electric field. The charges move radially outward at practically the speed of light, and generate currents which sustain a toroidal magnetic field having opposite signs in the two hemispheres. Since the charges are moving

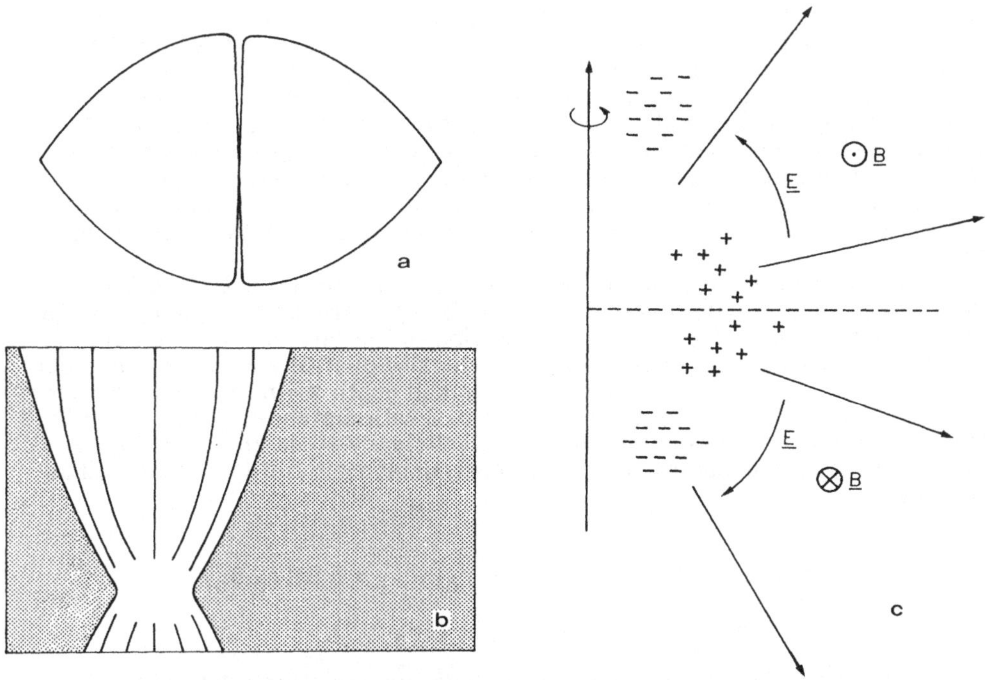

Figure 2. Schematic diagram indicating location and geometry of the proposed non-thermal source. Panels a, b, and c are a zooming sequence towards the innermost AGN regions. They show respectively: a thick accretion disk; the region of funnel formation; the field and charge arrangement in a relativistic wind. From Calvani and Nobili (1983), Phinney (1983), and Goldreich and Julian (1969), with modifications.

with speed $\approx$ c , the E and B fields are very nearly equal

$$\rho_e \approx E/r, \quad j \approx c\,\rho_e \approx c\,E/r, \quad B \approx r\,j/c \approx E \qquad (4)$$

Then the drift velocity is close to c and is directed radially outward, and an approximate dynamical consistency is ensured. The large scale electromagnetic fields make the dominant contribution to the energy density and the energy flux. The carriers are irrelevant as far as the global energetics are concerned, and their nature and total number cannot be determined without including the boundary conditions at injection and solving the acceleration problem. For the sake of simplicity, and in agreement with previous suggestions, we consider a plasma composed solely of electrons and positrons.

Our basic assumption is that the radial outflow of the particles is somehow perturbed. One can imagine "collisions" with heavy "scattering centers", or with the waves of a turbulent substratum. We postulate that the effect of such collisions is to remove momentum without changing the energy; the momentum removal is described by means of a scale length, S, which enters the momentum equation via a term $m\,c^3/S$. The collisions decrease the bulk velocity of the outflow, and the surplus energy goes into particle gyrations; these can be relativistic, with Lorentz factor γ. Since the $\mathbf{v} \times \mathbf{B}$ force is decreased, the **E** force brings opposite charges together in the meridian plane; this means closing some current loops, and E and B diminish together; the bulk motion is accelerated anew because of the drift along **E** and equilibrium is reestablished at a lower value of the Poynting vector flux. Of course the process goes on continuously, and the energy which is being fed into disordered particle motions is available for radiation. The details of the formalism will be given elsewhere; here we list the free parameters of the model, and indicate the – potentially – observable quantities which depend on them.

At a fiducial radius r_0 one has to assign: i) the fields E_0 and B_0: these determine the available power supply $c\,E_0 B_0\,r_0^2$, the drift velocity $c\;E_0/B_0$, and the bulk Lorentz factor Γ corresponding to the latter; ii) the particle number density n_0: this determines the density profile via particle number conservation, $n = n_0(r_0/r)^2$, and the Thomson depth $\tau = n_0\,r_0\,\sigma_T$; the quantities S_0 and k, describing the behaviour of S with distance, $S = S_0\,(r_0/r)^{k+1}$: these concur in the determination of the efficiency ε, as shown in Eq. 6 below. Then, the consumption of the electromagnetic field is described as follows

$$\begin{aligned} &d\,y^2/\,d\,x = -\,\sigma\,x^{-k-1}, \qquad x = r/r_0, \qquad y = r\,B/(r_0\,B_0) \\ &\sigma = 4\,\pi\,m\,c^2\;n_0\,r_0\,\Gamma^2\,(1-1/\Gamma^2)^{\frac{1}{2}}/(S_0\,B_0^{\,2}) \end{aligned} \qquad (5)$$

By integrating up to infinity, one easily finds the radiative efficiency

$$\varepsilon = \sigma/k \qquad (6)$$

The comoving Lorentz factor γ is a function of position, and is determined by the requirement that the energy input – depending on S – be equal to the radiative losses. An exact solution involves a careful treatment of the transfer problem; order-of-magnitude wise, however, one finds

$$\gamma \approx \Gamma \ (\varepsilon/\tau)^{\frac{1}{2}} \tag{7}$$

When ε approaches unity, νF_ν becomes flat, as expected. Simple arguments can be given showing that this result is indeed independent of Γ: the inverse Compton emission is proportional to Γ^{-4} because of the angular factor introduced by the relativistic expansion, but the synchrotron emission L_S has the same dependence. Very high Γ's require E and B fields very nearly equal, and the residual B field in the comoving frame is very small; then, taking advantage of the Lorentz invariance of the radiated power, one can write $L_S \propto (\gamma B/\Gamma)^2 = (\gamma\Gamma)^2 \ B^2 \Gamma^{-4}$, which proves our assertion, since $\gamma\Gamma$ is the particle Lorentz factor in the observer frame.

5. CONCLUSIONS

Besides the general notions of relativistic velocities and SSC, the comparison between our model and radio observations of AGN's is rather difficult. It is plausible that the radio emitting plasma is the same plasma of the wind which we have been discussing, but the radio emission takes place at such a large distance that a good deal of extrapolation is needed to link to the latter the properties measured in the former.

However, as the observing frequency increases and approaches the submillimeter range, the emitting region comes closer and closer to the center. In the pictorial language of the Introduction, increasing the observing frequency is equivalent to going down the "chimney" toward the real "fire". There is evidence, not only from preliminary synthetic spectra, but also from direct observations (Owen *et al.* 1981), that at the highest radio frequencies one already touches the tip of the core power law. It is obvious that systematic studies of this spectral region, including perhaps VLBI mapping, would directly affect our knowledge of the "fire", and would put on a firmer footing theories as the one presented here.

REFERENCES

Begelman, M.C., Blandford, R.D., and Rees, M.J. 1984, *Rev. Mod. Phys.*, **56**, 255
Blandford, R.D. 1976, *M.N.R.A.S.*, **176**, 465
Blandford, R.D., McKee, C.E., and Rees, M.J. 1977, *Nature*, **267**, 211
Blandford, R.D., and Znajek, R.L. 1977, *M.N.R.A.S.*, **179**, 433
Calvani, M., and Nobili, L. 1983, in *Astrophysical Jets*, eds. A. Ferrari and A.G. Pacholczyk (Dordrecht: Reidel), p. 189
Cohen, M.H., and Unwin, S.C. 1984, in *VLBI and Compact Radio Sources*, IAU Symp. no. 110, eds. R. Fanti, K. Kellermann, and G. Setti (Dordrecht: Reidel), p. 95
Goldreich, P., and Julian, W.H. 1969, *Ap. J.*, **157**, 869
Kellermann, K.I., and Pauliny-Toth, I.I.K. 1969, *Ap. J. (Letters)*, **155**, L71

Malkan, M.A., and Sargent, W.L.W. 1982, *Ap. J.*, **254**, 22
Netzer, H. 1985, *Ap. J.*, **289**, 451
Owen, F.N., Helfand, D.J., and Spangler, R.S. 1981, *Ap. J. (Letters)*, **250**, L55
Pacini, F., and Salvati, M. 1978, *Ap. J. (Letters)*, **225**, L99
Padrielli, L. 1984, in *VLBI and Compact Radio Sources*, IAU Symp. no. 110, eds. R. Fanti, K. Kellermann, and G. Setti (Dordrecht: Reidel), p. 169
Phinney, E.S. 1983, Ph.D. Thesis, University of Cambridge.
Rickett, B.J. 1986, *Ap. J.*, **307**, 56
Rees, M.J. 1967, *M.N.R.A.S.*, **135**, 34
Rees, M.J. 1984, *Ann. Rev. Astr. Ap.*, **22**, 472
Rybicki, G.B., and Lightman, A.P. 1979, *Radiative Processes in Astrophysics* (New York: Wiley)

KINEMATIC AND DYNAMICAL MODELS OF SMALL STRUCTURES IN RADIO SOURCES

Kevin R. Lind
National Radio Astronomy Observatory
Edgemont Road
Charlottesville VA 22903-2475

ABSTRACT.

Models of the small-scale structures in radio sources are limited by the resolution, dynamic range, and sensitivity of the observations upon which they are based. However, evidence from variability, brightness, and proper motion observations strongly support the contention that these objects involve directed outflow or ejection of material at relativistic speeds. The evidence for relativistic motion will be discussed, and a variety of models involving both bulk and pattern motion will be analyzed in terms of spatial and temporal characteristics which may distinguish between models. The possibilities and hazards of generalizing nonrelativistic jet models to the relativistic case will also be briefly discussed, especially regarding possible observational signatures of magnetized versus unmagnetized jets.

INTRODUCTION

Theoretical modeling of small-scale structure in radio sources is made difficult by the limited number of constraints obtainable from the available data. The only possible imaging technique at present is VLBI, which is plagued by sensitivity and calibration limitations, and only provides a minimal level of resolution for serious theoretical work. Likewise, studies of short-period variability over all frequency regimes give limited information on the size and velocities of the emitting regions, but do not really place them accurately within the structure of the compact source. Confirmation of theory, apart from the generation of theories, requires considerable cleverness and caution. It is my intent to discuss the present observational and theoretical climate for modeling of small-scale structure, given the assumption that there are indeed relativistic velocities involved. Evidence from VLBI observations will be stressed, and shall be evaluated and discussed in terms of both theoretical models and observational format from which these results may best be derived.

M. J. Reid and J. M. Moran (eds.), The Impact of VLBI on Astrophysics and Geophysics, 55–64.

Kinematic models will be used to demonstrate the variety of possible results given the basic theory, and some discussion of dynamics shall be made in terms of possible qualitative results obtainable from present observations.

OBSERVATIONS

The structures of radio sources observed using VLBI typically fall into four classes: very compact, core-jet, compact double, and steep spectrum. The very compact sources are almost point sources, and hence provide little data apart from size. Unfortunately, this resolution is insufficient to determine whether or not the source exceeds the 10^{12} K brightness limit for synchrotron emission, although interstellar scintillation at radio wavelengths and X-ray variability suggest that this limit is, indeed, exceeded. Core-jet sources, on the other hand, have a definite core with some type of structure extending along some more-or-less well-defined axis. The compact doubles exhibit two more-or-less equal components with little or no change in their separation. The steep-spectrum objects are often rather complex, not necessarily possessing a strong axis of symmetry. These cores also vary in strength, with strong cores typically associated with core-jet sources at the centers of quasars, and the weaker cores associated with a variety of types associated with radio galaxies; even powerful radio galaxies such as Cygnus A tend to have weak compact cores.

The physical resolution scales of these observations vary with frequency and distance, from roughly 0.01 pc for the nearest Seyfert detected using VLBI to about 10 pc for the farthest quasar. Since the predicted size scale for the central engines of radio jets is of the order of a few A.U., this does not even approach the minimum characteristic length scale associated with models of these regions. Thus, the observed structures should be characteristic of the intrinsic power and composition of the jets, as well as the surrounding medium, and depend on the initial conditions at the central engine only insofar as they influence the gross properties of the jet. In other words, the properties of the radio images and the properties of the objects which generate them can only be linked indirectly, by theoretical modeling combined with short timescale variability studies. Conversely, the jets can be studied as if injected by a vaguely-specified ''nozzle'', as objects apart from the region in which they were generated.

A further observational constraint is the limited number of pixels per structure in VLBI images. In particular, very little internal detail of the individual components is possible, and, in the case of jetlike structure, no structural detail transverse to the jet can be resolved. Therefore, there is no reliable information as to the shapes of the structures composing the VLBI images, and hence it is difficult to distinguish between jets, plasmoids, or other types of structures. In fact, the only truly reliable parameters in VLBI maps are the relative flux distribution, and in particular the positions and relative brightnesses of isolated conponents, and the overall structure of continuous emission regions. The relative flux distribution can be tied to some absolute scale by careful calibration, but this is difficult and is generally unreliable by a few percent even if extreme care is taken.

The class of objects which I shall discuss today shall be the core-jet sources. Within the limits discussed above, we can measure three properties to reasonable precision: absolute flux (i.e., variability) at a variety of wavelengths but limited resolution; overall symmetry; and positions of relatively strong isolated components.

Variability fractions and time scales vary considerably. Some sources have X-ray variability time scales on the order of hours, and vary by a large fraction of their average flux, whereas some vary slowly if at all. This variety can be contrasted to the overall structure, which tends to be one-sided and, in the cases where large-scale structure is also present, appears to correlate with the sidedness of the large-scale structure. The positions of isolated structures in many sources are observed to vary so as to give the appearance of faster-than-light, or superluminal, motion. The short variability time scales, one-sidedness, and superluminal motion can be explained by relativistic motion.

BASIC THEORY

Consider a fixed point in space and an emitter moving relative to that emitter at some speed and viewing angle to θ, and define $\beta = v/c$, where c is the speed of light. Let the separation between them at time $t = 0$, r_0, be such that $r_0/R << 1$, and the final time of observation t_f such that $ct_f << R$. The light emitted from the initial and final positions of the emitter travels along parallel lines to the observer. The light emitted at time $t = 0$ arrives at the observer's position at time $Rcos\theta/c$; the light emitted at time t_f arrives at the observer's position at time $t_f + (R/c - \beta t_f)cos\theta$. Thus, the difference in arrival time, $\Delta t_{obs} = (1 - \beta\cos\theta)t_f$, can be considerably less than the physical time interval t_f At the same time, the projected physical separation between the two positions is simply $\beta sin\theta ct_f$. The apparent velocity of separation, $v_{app} = \beta csin\theta/(1 - \beta\cos\theta)$, may exceed the speed of light c for $\beta > 1/\sqrt{2}$. The maximum separation velocity $\gamma\beta c$, where $\gamma = (1 - \beta^2)^{-1/2}$, occurs at $\cos\theta = v/c$; thus, almost all sources which exhibit superluminal motion should have their jets aligned close to the line of sight. Given the high percentage of compact core-jet sources which now appear to exhibit superluminal motion, it appears that either superluminal motion is not due to relativistic motion along a well-defined direction, or there is a selection effect which enhances the number of observable sources which have jets which are nearly aligned with the observer.

This enhancement is due to the beaming of the flux from an emitter moving at relativistic velocities. It can be shown that, for an optically thin isolated component, $S(\theta) = S_{90}(1 - \beta\cos\theta)^{-(3+\alpha)}$, where S_{90} is the flux observed along viewing angle $\theta = 90°$, and α is the spectral index (Scheuer and Readhead 1979). For an optically thin jet, or an infinitely long line of emitting isolated components, the exponent changes from $3 + \alpha$ to $2 + \alpha$ because, since the pattern over which we integrate the flux is essentially fixed relative to the observer, the emitting volume in the frame of the observer is stationary relative to the observer, whereas for a moving isolated component it is not. A further demonstration of this is that the emission from an excitation moving through stationary gas --- that is, some disturbance which illuminates gas which is stationary relative to the observer --- goes as $(1 - \beta\cos\theta)^{-1}$, because of the difference in travel time between light emitted from the front and back of the moving emission region. In any case, the flux from an optically thin emitter is strongly enhanced in the direction of relativistic motion, which explains the large percentage of superluminal sources and the one-sidedness of the source; the ''counterjet'' can be much fainter than the jet even if intrinsically as bright. Note also that the spectral index for an optically thick source is 2.5, and so the optically thick flux is much more weakly beamed.

The beaming pattern, written above as a function of viewing angle θ, is traditionally presented as a number-flux relation P(x), i.e. as the probability of observing a source with flux ratio $S/S_{90} > x$. This is derived by assuming that the source orientations are randomly distributed relative to the observer, so that $dP \propto \cos\theta d\theta$, and integrating over all values of θ where $S/S_{90} > x$. The results for the standard models are shown in figure 1. These probability distributions may be changed by including evolution effects (such as an emitter which fades with distance from the core); this will be discussed later.

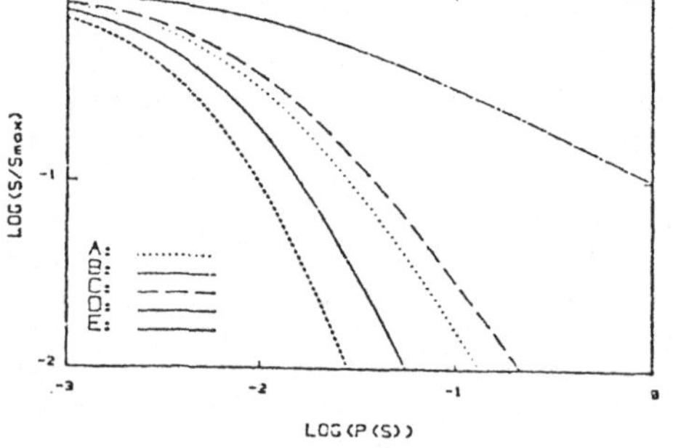

Figure 1

Another effect of relativistic motion can be seen by considering the arrival time of pulses emitted by a moving source at time intervals Δt_0 as measured by a comoving observer. The observed arrival times for the pulses are separated by a time interval $\Delta t_{obs} = \Delta t_0(1 - \beta \cos\theta)$. Thus, the time intervals are shorter for an emitter moving almost in the direction of the observer, and so the variability time scale is shorter. Also, the inferred size of the source, $c/\Delta t_{obs}$, is smaller than the size observed for a stationary source. Since the observed flux is larger than for a stationary source, and the inferred size is smaller than for a stationary unresolved source, then the brightness temperature inferred could be extremely large if it were assumed that the emitting region were stationary.

The point of this is, that the basic sense of the observations is consistent with the basic theory. However, the details are not necessarily consistent with the simple model presented here. The physical sources themselves are certainly much more complicated than this. One way to study the importance of these physical details is to select some simple models, such as relativistic shocks and plasmoids which fade with time, and examine the flux pattern in the frame of the observer for different intrinsic characteristics of the emitter, such as structure, shock strength, and opacity. This requires that the radiative transfer be carried out in full generality.

GENERAL THEORY

The general theory involving careful treatment of the radiative transfer is done by defining each of the relevant reference frames, and then transforming all quantities until everything is transformed into the reference frame of the observer (Lind and Blandford 1985). The three reference frames are (1) the comoving frame of the individual fluid elements, (2) the comoving frame of the emitting pattern, which is assumed to be stationary, and (3) the comoving frame of the observer. The relevant quantities are (1) emissivity and opacity, (2) shape and optical depth, and (3) flux. In detail, the emissivity and opacity are determined for each fluid element in its comoving frame, and then transformed to the comoving frame of the emitting pattern (e.g., the comoving frame of a relativistic shock which is itself moving relativistically relative to the observer. The radiative transfer equations are integrated along lines of sight in the pattern frame to obtain the optical depth and intensity. These are then transformed into the frame of the observer, and then the intensity in the observer's frame is integrated over to obtain the observed flux. All transformations are done assuming a power law for emissivity and opacity, and the flux calculated at a fixed frequency in the frame of the observer.

The results of these calculations for a planar relativistic shock deviate considerably from the case for a uniform optically thin sphere. Even the optically thin flux distribution deviates from that for an optically thin sphere, because of the internal relativistic velocities of the shock model. Furthermore, since optical depth depends strongly on orientation angle, $\tau = \tau_0(1-\beta\cos\theta)^{-(5/2+\alpha)}$, a high optical depth near the line of sight can suppress the flux toward the observer relative to the optically thin case, but may decrease so as to provide the strong optically thin beaming at larger angles. This reduces the observed effects of beaming. Similar results are derived for a variety of conical shocks, oriented both toward and away from the core. Even if we restrict ourselves to optically thin models, the distributions shown in figures 2 and 3, for pattern Lorentz factors of $\gamma = 7$ and $\gamma = 2$ respectively, demonstrate the large variety of possible flux distributions which may be associated with a given pattern Lorentz factor. Even given knowledge of the pattern Lorentz factor as measured by superluminal motion, some estimate of the orientation angle, and an accurate number-flux relation (equivalently, function of flux with angle), it is impossible to estimate the physical velocity of the jet without a detailed model of the jet structure.

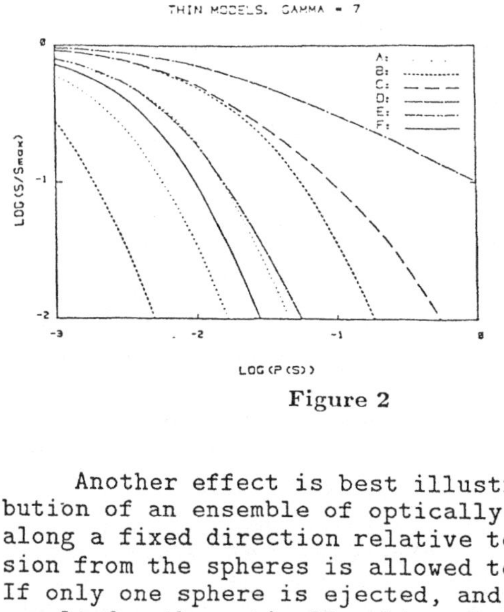

Figure 2

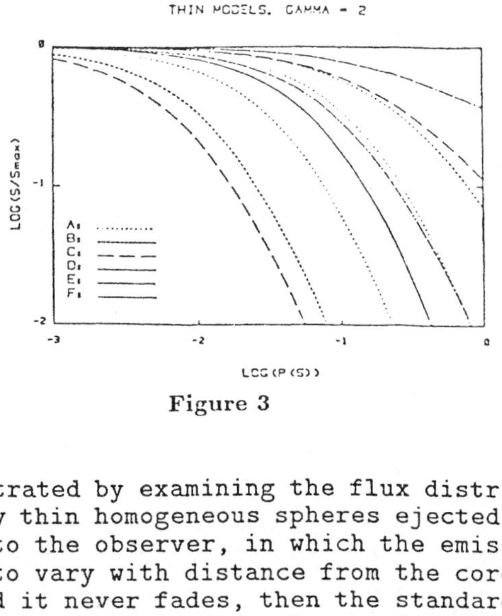

Figure 3

Another effect is best illustrated by examining the flux distribution of an ensemble of optically thin homogeneous spheres ejected along a fixed direction relative to the observer, in which the emission from the spheres is allowed to vary with distance from the core. If only one sphere is ejected, and it never fades, then the standard result for the optically thin sphere is obtained as discussed under the basic theory section. If a train of spheres have a fixed finite lifetime in the comoving frame of the observer Δt_l, and are ejected from the core at a fixed time interval of Δt_e, where $\Delta t_l >> \Delta t_e$, then the observed lifetime of a sphere is $\Delta t_{obs} = \Delta t_l(1-\beta\cos\theta)$, and at any given time the number of emitters observed is $\Delta t_{obs}/\Delta t_e \propto (1-\beta\cos\theta)$. The total flux is simply the number of emitters times their individual fluxes, and so the result for the continuous jet is then obtained.

Now, consider the case where $\Delta t_l \approx \Delta t_e$. There are two possibilities. If only one component is seen at a time, then the probability of seeing a source with a given flux S is then $dP \propto (1-\beta\cos\theta)\cos\theta d\theta$ rather than $dP \propto \cos\theta d\theta$ because of the shortened apparent lifetime of the emitting region. If a number of components are seen at one time, then the mean flux will be the same as that for the continuous jet, but the fractional variability $\Delta S/S = 1/N \propto (1-\beta\cos\theta)^{-1}$. Note that, the more nearly aligned the source axis and the observer, the larger the fraction by which the source varies. Recall also that, the more nearly aligned the source axis and the observer, the shorter the variability time scale. Thus, the most variable sources should also vary the fastest. This is consistent with observation. If the brightnesses of the spheres fall off as some function of distance form the core, then other flux distributions can be obtained. The variety of possible number-flux relations attainable with different evolution/ejection models is at least as great as the variety of number-flux relations attainable with different models for the emitting regions (as seen in figures 2 and 3), and again, seriously diminishes the predictive power of the number-flux relation in modeling the structures of the sources.

A final modification is to assume that the path followed by the emitting regions curves. For a single emitter moving out along this path, the effect is for the emitter to brighten and fade, depending on the angle relative to the observer, and for the observed proper motion to vary as well. For multiple emitters, the relative fluxes between them would vary as they moved along the curve. If all of the emitters were identical, the same spot would brighten and fade although the emitters were moving relativistically. In effect, the curve acts as a screen. A number of other screen models can be constructed, some of which would allow superluminal motion without beaming, some of which would allow beaming without superluminal motion, and some of which would allow both.

As a final refinement, allow a variety of emitters of varying structures to be ejected at varying intervals along slightly different angles or along curved paths. Because of the variations in intensity and optical depth with angle, a variety of different components may dominate the flux at any given viewing angle, and so, given the variety of functions associated with the ensemble of plausible models, the detailed number-flux relation --- or, the distribution of flux with angle --- cannot be determined unless the details of the source are accurately known. In short, the fitting of number-flux relations to theoretical curves in order to determine the velocity of the jet material is at best risky. Furthermore, the use of the proper motion of components to determine the jet velocity is also risky, since the pattern speed of a component need not be closely related to the velocity of the underlying jet. Given these conclusions, one may well wonder what can be done with the available data.

First, the existence of superluminal motion does indicate that there is relativistic motion of some sort going on, whether pattern motion or bulk motion. If there is pattern motion, it is either motion of some nonemitting jet or well-collimated beam of light against a screen, or there is some material present with a relativistic equation of state, through which the pattern may travel. Given material with a relativistic equation of state, it will eventually expand relativistically unless confined by the surrounding medium, or is already moving outward as a jet. Since almost all superluminal motion appears to involve motion away from the compact core, it appears that a relativistic jet is the best general underlying model.

Second, the classic superluminal sources eject new components at relatively long time intervals, and the components remain discrete and observable for long enough to be observed for several epochs, as has been done for a variety of superluminal sources. If this data is taken at several frequencies, and the data carefully calibrated, then the total flux and spectral index of a component can be monitored, and can be fit to theoretical models. This need not be done only for discrete components; since components appear in a more or less regular way following outbursts for a large number of these sources, and these outbursts can be well-separated, then the flux variations in the core can often be associated with the formation and propagation of one component, which may then be modeled. For example, Marscher and Gear (1985) constructed a very plausible model for a planar relativistic shock propagating in a relativistic jet of small opening angle, and proceeeded to fit it to an outburst in 3C273. Similarly, Hughes, Aller, and Aller (1985) have constructed models for the variations in flux and polarization caused by energy generation and subsequent cooling of material at a relativistic shock front and fit it to radio variability data for BL Lac, with considerable success. This confirms that at least some of the observed features in VLBI maps are relativistic shocks propagating through a faint underlying relativistic jet. This is further supported by high dynamic range imaging of a number of sources, such as 3C273 (Unwin and Davis 1987) and 3C371 (Lind 1987), which reveal a continuous underlying flux distribution with isolated components superimposed on top of it. The regularity in the falloff of the flux with radius for the above-mentioned map of 3C273, and for 3C120 (Walker, Benson, and Unwin 1987), suggest as well that either classic superluminals are intrinsically simple, or that beaming dominates such that one particular type of feature totally dominates the flux. In the first case, we can hope to model them; in the second, it is questionable.

Third, variability timescales, particularly in the X-ray, appear to indicate that much if not all of the flux from the compact core is indeed beamed. Given the large flux enhancement of an optically thin emitter moving at relativistic speeds over either a stationary or an optically thick emitter, it is not surprising that the compact core flux should come, not from the central core, but from some part of the outflow itself. In this regard, the inhomogeneous jet model of Blandford and Königl (1979) has proven to be in good agreement with the spectra and variability timescales of various compact cores, and suggests that the relativistic jet models are, in essence, correct.

RELATIVISTIC HYDRODYNAMICS

Numerical simulation of nonrelativistic jets has provided a number of interesting general results, but their usefulness for modeling of one-sided jets, both large and small-scale, is questionable because of the differences between relativistic and nonrelativistic fluid mechanics. For example, the factors of $\gamma = (1 - \beta^2)^{-1/2}$ in the momentum and energy equations provide for a large change in energy density for a small change in velocity, and can cause large pressure contrasts. On the other hand, there may well be some structural signatures which carry over from nonrelativistic to relativistic jets. An example of such a possible signature is the morphological difference between jets with strong and weak ordered toroidal fields. The weakly magnetized jet is simply terminated by a Mach disk structure, and the shocked jet material flows back into a surrounding cocoon (Lind 1987). The strongly magnetized jet recollimates into a broader, weaker jet which continues on (Clarke, Norman, and Burns 1987). These jets also demonstrate the complexity of ''real jets'', which may contain fixed and traveling planar and oblique shocks, as well as considerable vorticity and turbulence.

CONCLUSIONS

Inhomogeneous models of VLBI jets are at best difficult to construct because of the sparcity of the observations and the strong dependence of the observed flux distribution on model-dependent parameters. However, in many cases, which include the superluminal radio sources, the isolation of individual components allows the use of multifrequency imaging and flux monitoring to determine semi-quantitative models of individual outbursts or components. The precise modeling of compact radio sources requires high dynamic range imaging and frequent observation of variable sources --- attainable once the VLBA is built --- and theoretical modeling of relativistic flows. The latter is difficult numerically, but some qualitative results from present, nonrelativistic simulations may allow us to distinguish between general classes of theoretical models. Theoretical modeling of compact radio sources is indeed difficult and limited, but, with the use of all available data and cautious interpretation, progress can be made!

REFERENCES

Blandford, R. D. and Königl, A., 1979, Ap. J., **232**, 34.
Hughes, P.A., Aller, H.D., and Aller, M.F., 1985, Ap. J., **298**, 301.
Lind, K.R., 1985, thesis, California Institute of Technology.
Lind, K.R. and Blandford, R.D., 1985, Ap. J., **295**, 358.
Marscher, A.P. and Gear, W.K., 1985, Ap. J., **298**, 114.

Scheuer, P.A.G. and Readhead, A.C.S., 1979, Nature, **277**, 182.
Unwin, S.C. and Davis, R.J., 1987, these proceedings.
Walker, R.C., Benson, J.M., and Unwin, S.C., 1987, these proceedings.

RELATIVISTIC BEAMING MODELS AND VLBI OBSERVATIONS OF A COMPLETE SAMPLE OF RADIO SOURCES

A. C. S. Readhead, T. J. Pearson and P. D. Barthel
Owens Valley Radio Observatory
California Institute of Technology
Pasadena, California 91125
USA

1. INTRODUCTION

We have conducted a systematic study of the milliarcsecond structure of a complete, flux-density limited sample of strong radio sources selected at 5 GHz. We have made 5 GHz maps at two epochs of the 45 compact sources in the sample, and third-epoch observations are in progress. Our intention was to explore the full range of morphologies exhibited by compact radio sources, to search for new superluminal sources, and to determine how widespread such phenomena as parsec-scale jets, alignment of parsec-scale and kiloparsec-scale jets, and superluminal motion are. In addition, we hoped to use this well-defined sample for statistical tests of the beaming theories.

We have found that superluminal motion is common. Thus far it has been observed in all classes of object except the compact doubles, which comprise less than 10% of the objects in our complete sample (Pearson, Readhead and Barthel 1987). Examples of superluminal sources in our sample are the compact flat-spectrum objects 3C 345 and BL Lac, the compact steep-spectrum object 3C 216, and the extended triple sources 3C 179 and 1928+738. At least nine of the 65 sources in the complete sample are superluminal, and we can place subluminal limits on two of the compact double objects (0710+439 and 2021+614). In order to detect superluminal motion high quality observations at many epochs are needed. Based on the number of superluminal objects already detected in this survey, it seems likely that it will be only a matter of time and effort before superluminal motion is detected in 90% of the remaining sources.

The results of this survey will be discussed fully elsewhere. We concentrate here on some implications that these results have for relativistic beaming models.

2. THE SIMPLE BEAMING HYPOTHESIS

We make no attempt at a complete review of beaming theories here. There are a number of excellent recent discussions in the literature (Phinney 1985; Blandford 1987; Scheuer 1987) to which the reader is referred for the details of the arguments. However our observations of a complete sample do cast some interesting light on the subject, as discussed below.

M. J. Reid and J. M. Moran (eds.), The Impact of VLBI on Astrophysics and Geophysics, 65–70.

In its simplest form the beaming hypothesis makes the following assumptions:

i. Material is ejected from the nuclei of active galaxies with relativistic bulk velocity v, corresponding to Lorentz factor γ.
ii. This velocity is constant from the innermost region of the jet to the outer lobes.
iii. The velocity is the same for all active nuclei.
iv. The velocity is parallel to the axis of the jet, so that the emission cone has opening angle $\sim 1/\gamma$.
v. The jets are narrow (opening angle $< 1/\gamma$).
vi. The jets have very small intrinsic curvature over their entire length.
vii. The range of the ratio, R_i, of the intrinsic flux density (i.e., the flux density in the rest frame of the emission region) of the core to that of the outer lobes is much smaller than the variation of the observed ratio, R_o.
viii. The observed motions in the nuclear radio regions are causally connected, and are not simply pattern speeds.
ix. There is no orientation bias in the selection of optical quasars.

Of these assumptions (iii) is commonly dropped in beaming models and a range of velocities, which applies to all classes of active nuclei, is assumed.

There are five significant successes of the simple beaming models:

I. There is good evidence, based on three independent arguments, for relativistic motion in compact objects (superluminal motion, variability timescales and low X-ray luminosities of some compact objects; see e.g., Cohen and Unwin 1984).
II. Misalignments between the central components and outer components are greater in the core-dominated sources than in the lobe-dominated sources (Readhead *et al.* 1978).
III. The apparent velocities of extended double sources are lower than those of compact sources (Hough and Readhead 1987; Zensus and Porcas 1987).
IV. In powerful objects with both kpc-scale and pc-scale jets, the jets lie on the same side of the nucleus or they are connected (over 25 cases).
V. Classical triple objects with one-sided jets show less Faraday depolarization on the side coincident with the large-scale jet (about a dozen cases; Laing, private communication).

It has been clear from the start that this model is too simplistic (Scheuer and Readhead 1979; Blandford and Königl 1979), and the evidence against this simplest form of the model has increased steadily over the last decade:

A. The shape of the radio luminosity function is inconsistent with the predictions of simple beaming models, and there are too few radio quiet quasars or extended radio sources to account for the numbers of compact radio sources (Strittmatter *et al.* 1980; Phinney 1985; Kellermann *et al.* 1986).
B. The overall sizes of superluminal sources are too large if they are deprojected by the geometrical factor derived from the superluminal speeds (Schilizzi and de Bruyn 1983; Barthel *et al.* 1986; Browne 1987; Barthel, these proceedings).
C. Few, if any, two sided jets are observed in powerful radio sources.

3. IMPLICATIONS OF THE SURVEY FOR BEAMING THEORIES

It is clear that at least one of the assumptions—that the jets are narrow, with small emission cones, small intrinsic curvature over their entire length and that there is no orientation bias in the selection of optical objects—must be incorrect.

One of the most powerful arguments against the simplest model is that of the large deprojected sizes. However, it is difficult to discriminate between the alternative models on the basis of deprojected sizes alone. It has long been recognized that the observed differences in position angles provide a potentially powerful test. The first VLBI maps showed that there is a significant difference between the alignment of the inner and outer regions in core-dominated and lobe-dominated objects (Readhead *et al.* 1978). The direction of the superluminal motion is likely to be along the jets that we have already detected in many of these objects, and we can therefore use the present observations, together with VLA, MERLIN and WSRT observations of larger-scale structure, to measure variations in the apparent orientation of the jet axis.

We have looked at the change in position angle between the inner and outer components in the core-dominated objects in our sample, and the first significant result is that 1/3 of these objects, almost all of which have asymmetric milliarcsecond structure, have no detected large-scale structure at all. These have < 0.3 mJy in compact components within 6 arc seconds of the core (Perley 1982). Thus, although lower resolution VLA observations may reveal some weak extended features which were resolved out in Perley's observations, it appears that R_o exceeds 10^3 for these objects. Two of these objects (0212+735, 0454+844) have been the subject of very sensitive 20 cm VLA observations (Antonucci *et al.* 1986), and found to have $R_o > 2700$ and 1500, respectively. The lowest values of R_o are observed in double-lobed radio galaxies, where they can be as small as 10^{-3}. Thus we have to accommodate a total range in R_o of at least six orders of magnitude. One of the attractions of beaming models is that they can easily account for two to four orders of magnitude in R_o.

The second significant result is that we observe substantial misalignments in a number of objects (e.g., 3C 216, 0945+408, 1749+701, 1803+784, 1823+568).

Our results on position angle differences between small-scale and large-scale structure are shown in Figure 1 where we show the largest change in angle observed between the central component and features in the outer jet. We choose to plot the largest PA difference observed in each jet since it is the maximum curvature which should be used in any estimates of the deprojected length. Our distribution is broader than those obtained by Rusk and Rusk (1986) and Browne (1987), reflecting the fact that there is curvature along the large-scale jets.

Browne concluded, on the basis of the paucity of large misalignments in a sample of superluminal sources, that the intrinsic misalignment angles are rarely larger than the angle made by the core motion to the line of sight. His results showed that fewer than 20% of the sample had misalignments of greater than 90°. In Figure 1 41% of the objects have misalignments of greater than 90°. The results of Figure 1 can be tested against the simple beaming model (Moore *et al.* 1981; Readhead *et al.* 1983; Rusk and Rusk 1986). The chi-square test shows that simple beaming with an assumed intrinsic bend of 10° and $\gamma = 10$ cannot be ruled out, but that there is only a 13% probability of exceeding the value of chi square which obtains with an assumed bend of 11° and $\gamma = 5$. We wish to emphasize that for many of the objects included in Figure 1 only one or two comparatively

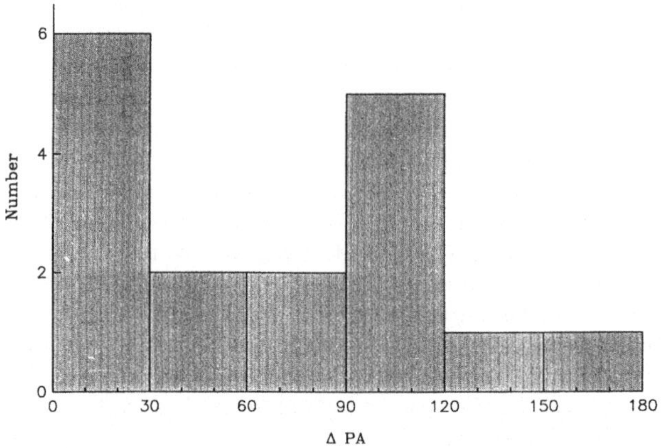

Figure 1 Histogram of the largest position angle differences ΔPA between the large-scale and small-scale features observed in the sources in the sample.

bright regions in the large-scale jet are visible. Higher dynamic range observations will likely reveal a larger fraction of the underlying jet, and some of these regions will show greater misalignments than we have observed thus far. Thus the misalignments plotted in Figure 1 are lower limits, and higher quality data are likely to be harder to reconcile with the simple beaming model. Furthermore, the tests assume that we have a complete sample, based on the core flux density, and as mentioned above, 1/3 of the objects are not included—these are the objects with high R_o which are expected to show the greatest curvature.

Since the observed misalignments are only marginally consistent with the simple model and we expect the evidence for misalignments to increase, we must consider alternatives to the simplest beaming models defined above.

If we admit intrinsic bends of order 30° in these objects is this alone sufficient or do we need, in addition, a broad ejection cone and/or a broad cone of emission at each position along the jet? Large intrinsic bends alone appear insufficient for the following reasons.

(a) A very large fraction of core-dominated objects are asymmetric core-jets. If intrinsic bends alone were responsible for increasing the overall emission cone angle in these objects we would expect to see a significant fraction of disembodied jets, i.e., jets with no flat-spectrum core at one end, and such objects are exceedingly rare. We would therefore be forced to postulate that the initial jet direction is closely aligned with the line of sight, and this leads to the problems with the luminosity distribution and relative numbers of faint objects mentioned above.

(b) Very large variations in the surface brightness of large-scale jets are not observed, although these would be expected on some models near bends in the jets. This expectation is relaxed if the emission cone is comparable in width to the intrinsic changes in jet direction.

We next consider whether broad ejection cones alone can explain the observations. Rees (1981) suggested a model with a broad ejection cone. In his model the "specific discharge" and/or the bulk velocity drops monotonically with angle from the axis of the jet. This ensures that the dominant superluminal effects are on one side of the core and appear in projection aligned with the approaching side. This model does not ensure that successive components follow the same path, and in addition, it is difficult to reconcile with the sharp decrease in surface brightness at the edges of the jets (see the map of M87 by Biretta *et al.* in these proceedings). Furthermore we would expect to see some components of brightness comparable to the main jet components moving at random angles relative to the jet axis. For all these reasons, a narrow ejection cone with a wide emission cone, caused by bulk motions at significant angles to the jet axis, seems to be favored by the observations.

It therefore appears that these objects must have both significant intrinsic curvature and broad emission cones. A promising approach is one in which shocks and instabilities in the jets give rise to a rather wide range of bulk speeds and directions of motion (Norman, Winkler and Smarr 1983; Blandford 1984; Lind and Blandford 1985; Phinney 1985), and therefore there is beamed emission over a much wider cone angle than the opening angle of the jet. On purely physical grounds this is an attractive possibility, and it is worth pursuing since, even in those objects for which the misalignment is large (e.g., 3C 216, 3C 309.1, 1823+568), the nuclear jets do join up with the larger-scale jets, often after bending through more than 90°. This fact, together with the fact that successive ejecta appear to follow the same path, suggests that the ejection cones in these objects are narrow ($\sim 1/\gamma$).

4. LARGE-SCALE COUNTERJETS

Until a few years ago no large-scale counterjets had been observed. However this is no longer the case. Simon *et al.* (1987) have published a map of 1928+738, a core-dominated double-lobed superluminal quasar which does have a counterjet, and in an ongoing program of mapping 3C quasars at the VLA Bridle and a large group of collaborators have recently found large-scale counterjets in 3C 9, 3C 68.1 and 3C 334 (private communication). Indeed, there are features in the Cygnus A map of Perley and Dreher (private communication) which look suspiciously like a counterjet.

The present incomplete statistics of large-scale counterjets argue against the beaming hypothesis (Wardle and Potash 1984; Barthel 1987 and these proceedings), and we may therefore be missing quasars with axes near the plane of the sky. This argument is strengthened if we admit large emission cones and intrinsic bends into the model since in some objects with symmetry axes near the plane of the sky there will be regions of the counterjet at angles significantly less than 90° to the line of sight. However the statistics are incomplete and we must await the results of VLA observations of complete samples.

5. CONCLUSIONS

We have summarized the arguments and presented some new evidence which suggest that significant modifications to the simplest beaming theories are required. In particular, it appears that all of the observations can be explained in terms of a model with a narrow ejection cone and large intrinsic curvature of the jet, in which some material moves at significant angles to the axis of the jet.

This work was supported in part by the National Science Foundation under grant AST 85-09822.

REFERENCES

Antonucci, R. R. J., Hickson, P., Olszewski, E. W., and Miller, J. S. 1986, *Astron. J.*, **92**, 1.

Barthel, P. D. 1987, in *Superluminal Radio Sources*, ed. J. A. Zensus and T. J. Pearson (Cambridge: Cambridge University Press), p. 148.

Barthel, P. D., Pearson, T. J., Readhead, A. C. S., and Canzian, B. J. 1986, *Astrophys. J. (Letters)*, **310**, L7.

Blandford, R. D. 1984, in *IAU Symposium 110, VLBI and Compact Radio Sources*, ed. R. Fanti, K. Kellermann, and G. Setti (Dordrecht: Reidel), p. 215.

Blandford, R. D. 1987, in *Superluminal Radio Sources*, ed. J. A. Zensus and T. J. Pearson (Cambridge: Cambridge University Press), p. 310.

Blandford, R. D., and Königl, A. 1979, *Astrophys. J.*, **232**, 34.

Browne, I. W. A. 1987, in *Superluminal Radio Sources*, ed. J. A. Zensus and T. J. Pearson (Cambridge: Cambridge University Press), p. 129.

Cohen, M. H., and Unwin, S. C. 1984, in *IAU Symposium 110, VLBI and Compact Radio Sources*, ed. R. Fanti, K. Kellermann, and G. Setti (Dordrecht: Reidel), p. 95.

Hough, D. H., and Readhead, A. C. S. 1987, in *Superluminal Radio Sources*, ed. J. A. Zensus and T. J. Pearson (Cambridge: Cambridge University Press), p. 114.

Kellermann, K., Sramek, R., Shaffer, D., Green, R., and Schmidt, M. 1986, in *IAU Symposium 119, Quasars*, ed. G. Swarup and V. K. Kapahi (Dordrecht: Reidel), p. 95.

Lind, K. R., and Blandford, R. D. 1985, *Astrophys. J.*, **295**, 358.

Moore, P. K., Browne, I. W. A., Daintree, E. J., Noble, R. G., and Walsh, D. 1981, *Monthly Notices Roy. Astron. Soc.*, **197**, 325.

Norman, M. L., Winkler, K.-H. A., and Smarr, L. 1983, in *Astrophysical Jets*, ed. A. Ferrari and A. G. Pacholczyk (Dordrecht: Reidel), p. 227.

Pearson, T. J., Readhead, A. C. S., and Barthel, P. D. 1987, in *Superluminal Radio Sources*, ed. J. A. Zensus and T. J. Pearson (Cambridge: Cambridge University Press), p. 94.

Perley, R. A. 1982, *Astron. J.*, **87**, 859.

Phinney, E. S. 1985, in *Astrophysics of Active Galaxies and Quasi-Stellar Objects*, ed. J. S. Miller (Mill Valley: University Science Books), p. 453.

Readhead, A. C. S., Cohen, M. H., Pearson, T. J., and Wilkinson, P. N. 1978, *Nature*, **276**, 768.

Readhead, A. C. S., Hough, D. H., Ewing, M. S., Walker, R. C., and Romney, J. D. 1983, *Astrophys. J.*, **265**, 107.

Rees, M. J. 1981, in *IAU Symposium 94, Origin of Cosmic Rays*, ed. G. Setti, G. Spada, and A. W. Wolfendale (Dordrecht: Reidel), p. 139.

Rusk, R., and Rusk, A. C. M. 1986, *Can. J. Phys.*, **64**, 440.

Scheuer, P. A. G. 1987, in *Superluminal Radio Sources*, ed. J. A. Zensus and T. J. Pearson (Cambridge: Cambridge University Press), p. 104.

Scheuer, P. A. G., and Readhead, A. C. S. 1979, *Nature*, **277**, 182.

Schilizzi, R. T., and de Bruyn, A. G. 1983, *Nature*, **303**, 26.

Simon, R. S., Johnston, K. J., Eckart, A., Biermann, P., Schalinski, C., Witzel, A., and Strom, R. G. 1987, in *Superluminal Radio Sources*, ed. J. A. Zensus and T. J. Pearson (Cambridge: Cambridge University Press), p. 155.

Strittmatter, P. A., Hill, P., Pauliny-Toth, I. I. K., Steppe, H., and Witzel, A. 1980, *Astron. Astrophys.*, **88**, L12.

Wardle, J. F. C., and Potash, R. I. 1984, in *Physics of Energy Transport in Extragalactic Radio Sources*, NRAO Workshop No. 9, ed. A. H. Bridle and J. A. Eilek (Green Bank: National Radio Astronomy Observatory), p. 30.

Zensus, J. A., and Porcas, R. W. 1987, in *Superluminal Radio Sources*, ed. J. A. Zensus and T. J. Pearson (Cambridge: Cambridge University Press), p. 126.

BULK RELATIVISTIC MOTION IN A COMPLETE SAMPLE OF FLAT SPECTRUM RADIO SOURCES

C.J. Schalinski, A. Witzel, Th.P. Krichbaum, C.A. Hummel, P.L. Biermann
Max-Planck-Institut für Radioastronomie, Bonn, F.R.G.

K.J. Johnston, R.S. Simon
E.O. Hulburt Center for Space Research, N.R.L., Washington, D.C., U.S.A.

As part of a multi-epoch and multi-wavelength study of the physics of core dominated radio sources we have investigated the occurrence of apparent superluminal motion and other indications for bulk relativistic motion (b.r.m) in a statistically complete, flux density limited ($S_{5GHz} \geq 1Jy$) sample of 13 objects with flat radio spectra ($\alpha_{5GHz}^{2.7GHz} \geq -0.5$, $S_\nu \sim \nu^\alpha$), $\delta \geq 70°$ and $|b_{II}| \geq 10°$. These sources come from the S5-survey (Kühr et al., 1981) and are optically identified as 7 quasars and 6 BL-Lac objects. They have been observed over a wide range of frequencies, from radio through X-rays (s. Eckart et al.,1986 and references therein). Radiomaps have been obtained at frequencies from 327 MHz to 22.2 GHz with resolutions from arcseconds to 0.2 mas, using the *VLA, MERLIN* and telescopes of the US- and European-VLBI networks. A recent status report on the VLBI-observations is given by Witzel (1987). In this paper we summarize the results on the direct evidence for b.r.m. in this sample as derived from repeated VLBI-observations at 5 GHz, as well as supporting evidence from SSC-calculations and flux density variability of the 5GHz VLBI core components (*Table 1*).

For the sources with unknown redshift we list in *col.6* of *Table 1* the "critical" redshifts beyond which the measured separation rate translates into $v > c$ ($H_0 = 100 km/s/Mpc$, $q_0 = 0.5$). Deep optical imaging (Kühr, priv. com.) gives evidence for

Table 1 :

SOURCE	*ID*	*N(VLBI)* [5GHz]	$d\theta/dt$ [mas/yr]	z	$z_{crit.}$	v_{app}/c h^{-1}	D_{SSC}^{min}	D_{var}^{min} $h^{2/3}$
0016 + 73	*Q*	1		1.76			8.0	
0153 + 74	*Q*	3	< 0.03	2.34		< 1.3		
0212 + 73	*BL*	3	0.09	2.37		3.9	2.4	1.4
0615 + 82	*Q*	2	0.05	0.71		1.1	1.3	
0836 + 71	*Q*	3	0.15	2.16		6.2	4.0	1.9
1039 + 81	*Q*	2	0.08	1.26		2.5	2.1	1.2
1150 + 81	*Q*	3	0.13	1.25		4.1	2.2	1.6
1803 + 78	*BL*	3	< 0.03	0.68[♮]		< 0.6	4.4	
1928 + 73	*Q*	5	0.60	0.30		7.0	1.7	1.2
0454 + 84	*BL*	2	0.15		0.16		2.1	
0716 + 71	*BL*	2	0.09		0.28		1.4	
1749 + 70	*BL*	2	0.10	> 0.7[♭]	0.25	> 2.2	1.3	
2007 + 77	*BL*	4	0.30		0.07		2.8	

redshifts from (♮) C.Lawrence (priv.com.),(♭) Wrobel et al. (this vol.)

M. J. Reid and J. M. Moran (eds.), The Impact of VLBI on Astrophysics and Geophysics, 71–72.

$z > 0.3$ in these sources. The results of *Table 1* indicate: {1} *10* out of *12* sources measured at least twice with VLB-interferometry at 5 GHz are candidates for apparent superluminal motion - 4 out of *6* sources with known z measured at least 3 times with VLBI show apparent faster than light motion -. {2} *12* out of the *13* sources show evidence for b.r.m. on the basis of SSC- calculations (s. Doppler-factors D in *col.8* of *Table 1*). {3} For the sources with known z this is in good agreement with the Doppler-factors (*col.9*), derived from the flux density variability of the 5 GHz core components adopting light travel time arguments (obviously, insufficient sampling has to be taken into account). A detailed discussion will follow (Schalinski et al.,in prep.). Thus, we conclude that b.r.m., especially apparent superluminal motion, is a common phenomenon among compact flat spectrum radio sources.

For 2 sources, *0153+74* and *1803+78*, we derived upper limits on v/c. In the case of the quasar *0153+74* (z=2.34) we detected 2 compact components *A* and *B* (s.*Fig.1*) - *A* has an inverted, *B* a flat spectrum between 1.6 and 5 GHz - with a separation of about 10 mas, and a "bridge" of steep spectrum components with a bending of $\sim 180°$. As *A* and *B* remained constant within the errors during almost 7 years, and neither SSC-calculations nor variability show evidence for Doppler-boosting, this source holds promise to be subluminal.

1803+78 (s.*Fig.1*: Schalinski et al.,this vol.) appears to be subluminal on the basis of observations at frequencies lower than 5 GHz, thus showing a discrepancy between expected inverse Compton- and observed X-ray flux densities. However, since the source was detected on transatlantic baselines at 43 GHz and with *IRAS* (s.Witzel,1987 for refs.), and recent 22.2 GHz-VLBI observations indicate the presence of a new component close to the core, relativistic effects cannot be excluded and require further studies at higher frequencies.

REFERENCES

Eckart A.,Witzel A.,Biermann P.,Johnston K.J.,Simon R.,Schalinski C.J., Kühr H., 1986 : Astron. & Astrophys., 168, 17

Kühr H.,Pauliny-Toth I.I.K.,Witzel A.,Schmidt J., 1981: A.J., 86, 854

Witzel A., 1987: in *Superluminal Radio Sources*, ed. J.A. Zensus and T.J. Pearson (Cambridge: Cambridge University Press),p. 83

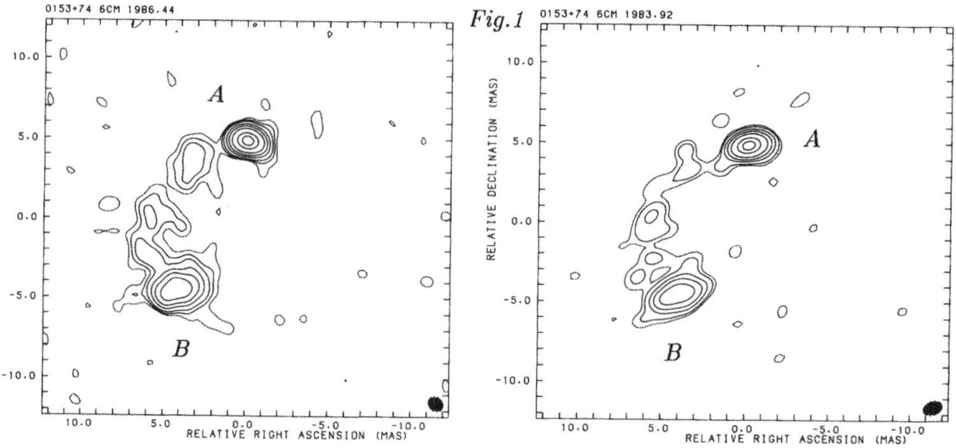

Fig.1

THE EVOLUTION OF COMPACT DOUBLE RADIO SOURCES

R.L. Mutel
Department of Physics and Astronomy
University of Iowa
Iowa City, IA 52242 USA

R.B. Phillips
Haystack Observatory
Westford, MA 01886 USA

ABSTRACT. The evolution of compact double radio sources is discussed using a model due to Carvahlo (1985). We suggest that for doubles associated with active *galaxies*, there is a plausible evolutionary sequence from compact doubles to steep-spectrum compact sources to extended (D1) doubles.

Most well-studied central radio sources are characterized by asymmetric 'core-jet' morphologies. The projected sizes are typically tens of parsecs, but there is almost always extended arcsecond structure along the jet axis extending well beyond the parent galaxy (e.g., Antonucci 1985). This implies a *minimum* age $t \sim 10^{6.5}$ y for a deprojected size $l \sim 1$ Mpc and relativistic bulk flow for the entire lifetime of the jet.

By contrast, the compact double (CD) sources (Phillips and Mutel 1982) have overall sizes $\lesssim 1$ kpc., implying a much younger evolutionary state. Their observed properties have been discussed elsewhere (Hodges and Mutel 1987), and will only be summarized here: well separated double morphology with no arcsecond structure, self-absorbed synchrotron spectra for each component, low polarization and variability, optically identified with faint galaxies and empty fields (*never quasars*). The CD sources bear a striking resemblance to extended double sources, particularly the luminous, edge-brightened Fanaroff-Riley Class II (FRII, also known as D1) objects such as 3C33 or Cygnus-A. They are also similar to many 'steep-spectrum compact' (SSC) sources whose dominant structure is often double, but with linear separations of 1 – 10 kpc. (Fanti, *et al.* 1985).

We have suggested that there is an evolutionary sequence connecting all three classes, *viz*, CD sources evolve into SSC sources, which eventually become extended doubles (Hodges and Mutel 1987). The lifetimes (and hence relative numbers) for each phase of evolution depend on the bulk expansion speed of the lobes. This is likely to be time variable, since the lobes pass through decreasing gas density as they move from near the galactic center to the intergalactic environment. If the lobes are pressure confined during the CD phase, the expansion speed is $v_{lobe} \sim 0.1c$ (Mutel, *et al.* 1985), implying a lifetime of $\tau \sim 10^4$ y.

Carvahlo (1985) has recently proposed a simple evolutionary model which is consistent with the hypothesis that CD sources are the progenitors of extended doubles. It is based on the model developed by Schueur (1974) in which the opposed radio lobes lose energy both by expansion and synchrotron radiation losses. The central engine supplies energy to the lobes at a constant rate until a characteristic time t_0, whence the lobe lu-

M. J. Reid and J. M. Moran (eds.), The Impact of VLBI on Astrophysics and Geophysics, 73–74.

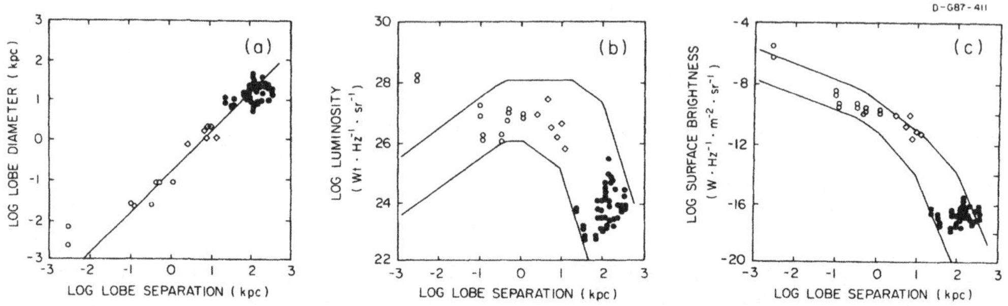

Figure 1 (*a*) Lobe diameter, (*b*) Radio luminosity, and (*c*) Surface brightness at 1.4 GHz of double radio sources as a function of lobe separation. The open circles (∘) refer to compact double (CD) sources, the diamonds (◇) refer to steep spectrum compact (SSC) sources, and the filled circles (•) refer to Fanaroff-Riley Class II (edge-brightened or D1) doubles. The lines are predicted evolutionary tracks using the model of Carvahlo (1985).

minosity and energy density begins to decrease as e^{-t/t_0}. He compares the expected size, luminosity, energy density, and surface brightness of the radio lobes as a function of linear separation for the model versus a sample of CD and FRII sources. Carvahlo finds that the observed properties of CD and FRII sources are consistent with a lobe velocity of $v \sim 0.2c$ and a turn-off time of $t_0 \sim 10^5$ y.

We have applied this model to include the SSC sources as an intermediate state. An example of the lobe size, radio luminosity, and surface brightness as a function of overall size is given in figure 1. The relatively good fit suggests that at least some of the SSC sources (those with dominant double structure) could be the evolutionary successors of CD sources and progenitors of extended doubles. Additional supportive evidence derives from the spectral turnover frequencies in the three groups — there is a systematic decrease in average peak frequency as would be expected for synchrotron sources suffering radiation and expansion losses ($\nu_{peak} \propto B^{-3}t^{-2}$). A systematic analysis of the evolution of double radio sources should include a detailed statistical analysis of the evolution of the lobe properties, relative number densities, and spectral properties. Unfortunately, there are as yet no systematic surveys of either CD or SSC sources. Nevertheless, the initial results using a relatively simple physical model are encouraging.

REFERENCES

Antonucci, R.R.J. 1985, in *Active Galactic Nuclei*, ed. J.E. Dyson (Manchester University Press:Manchester), p. 98

Carvahlo, J.C. 1985, *M. N. R. A. S.*, **215**, 463.

Fanti, C., Fanti, R., Parma, P., Schilizzi, R.T., and van Breugel, W. 1985, *Astr. Ap.*, **143**, 292.

Hodges, M.W. and Mutel, R.L. 1987,in *Superluminal Radio Sources*, ed. J.A. Zensus and T.J. Pearson (Cambridge University Press:Cambridge), p. 168

Mutel, R.L., Hodges, M.W., and Phillips, R.B. 1985, *Ap. J.*, **290**, 86.

Phillips, R.B. and Mutel, R.L. 1982, *Astr. Ap.*, **106**, 21.

Scheuer, P.A.G. 1974, *M. N. R. A. S.*, **166**, 513.

VLBI CORES IN A SAMPLE OF RADIO GALAXIES

G. Comoretto,[1] L. Feretti,[2,3] G. Giovannini [2,3]
1 - Osservatorio Astrofisico di Arcetri, L.go E. Fermi 5, Firenze, Italy
2 - Istituto di Radioastronomia, v. Irnerio 46, Bologna
3 - Dipartimento di Astronomia, v. Zamboni 33, Bologna, Italy

We present the first results of a statistical study of the milliarcsec structure in a complete sample of radio galaxies. We have selected from the B2 and 3CR samples of galaxies the sources which present, at the VLA or WSRT angular resolution, an unresolved core with a flux density at 5 GHz $S_c \geq 100$ mJy. The total sample consists of 30 radio galaxies, 17 from the B2 and 13 from the 3CR catalog. This complete sample covers a range of total radio power at 408 MHz log P = 23.5 - 26.5 W/Hz (low-intermediate luminosity). The radio structure of these sources on the arcsec-arcmin scale is well known, thanks to good dynamic range VLA and/or WSRT maps; a large variety of structures is present in the sample, from classical doubles to head-tail sources; flat, inverted and steep spectrum cores are also present.

The 12 radio galaxies with the strongest cores have been already observed by other authors with the VLBI, therefore we observed at 5 GHz the remaining 18 galaxies with the EVN array + Haystack antenna, using the Mark III recording system in mode B. As a first step, each source was observed with 1 to 8 scans of 13 minutes in order to have indications on the size and flux of the nuclear sources. Unambiguous mapping is not possible because of the limited u-v coverage, but crude indications about the core structure can be derived. Thanks to the transatlantic baselines, we were able to detect structure down to a resolution of 2 mas in E-W direction.

The results are summarized in table 1. We detected 15 out of the 18 sources. In 9 sources all the arcsec flux was detected on the VLBI scale, in 4 sources more than 70%; only in 2 sources less than 65% of the arcsec flux was detected. This result is different from that of Wehrle et al. (1984, Ap. J., **284**, 519), who observed on the baseline Goldstone-Madrid a complete sample of radio galaxies, selected according to the total flux at 408 MHz. They found that for most sources the flux in the VLBI component is only a fraction of the flux detected in the arcsec core. This implies that besides the compact feature some resolved structure, which can be missed in observations with only transatlantic baselines, is generally present in these sources.

M. J. Reid and J. M. Moran (eds.), The Impact of VLBI on Astrophysics and Geophysics, 75–76.

In our sample, of the 15 sources detected, 8 are unresolved and 7 show indication of an extended structure. The limited u-v coverage available allow us to derive only simple models. All sources show a structure of the type core-jet, with the parsec jet well aligned with the Kpc jet. In one source (0755+37) a jet-counterjet is strongly suggested both in the map and in the model.

The accurate mapping with better u-v coverage of all the detected sources is planned as a second step of this work.

TABLE 1

Summary of observations

Source name		Beam mas (deg)	Scans observed	Number of stations	Structure	S_{mas}/S_{as} %
0055+30	NGC315	6×2 (0)	6	6	one sided	100
0104+32	3C31	6×2 (0)	6	6	point	100
0220+43	3C66B	5×2 (60)	2	6	point	100
0222+36		6×2 (20)	2	6	extended?	65
0258+35	NGC1167	5×2 (25)	2	6	—	< 20
0331+39	4C39.12	6×2 (30)	1	6	point	85
0755+37	NGC2484	4×2 (0)	8	6	two sided	100
0836+29		4×2 (−15)	6	6	one sided	100
1037+30		4×2 (−15)	6	6	—	< 7
1142+20	3C264	8×2 (0)	4	6	one sided?	90
1144+35		6×2 (15)	2	6	one sided	variable
1322+36	NGC5141	4×2 (30)	2	6	point	50
1441+52	3C303	6×2 (−45)	3	6	point	100
1511+26	3C315	14×4 (30)	1	3	—	< 5
1626+39	3C338	7×2 (−15)	5	6	extended	100
1641+17	3C346	12×6 (−45)	2	5	point	70
1833+32	3C382	8×3 (0)	1	6	extended?	100
2249+39	3C452	3×2 (0)	3	6	one sided	100

COMPLETE SAMPLES OF FLAT SPECTRUM RADIO SOURCES FROM THE PARKES 2.7 GHZ SURVEY.

David L.Jauncey, G.L. White,B.R. Harvey, M.J. Batty and A.E. Wright,
Division of Radiophysics,CSIRO,Epping,NSW, Australia
Ann Savage,
UK Schmidt Telescope Unit, Coonabarabran,NSW
B.A.Peterson,W.L. Peters and J.E. Reynolds,
Mount Stromlo and Sidiing Spring Observatory, Canberra,ACT, Australia.
S.Gulkis,R.A. Preston and D.D. Morabito,
Jet Propulsion Laboratory,Pasadena,Ca,USA,
A.K.Tzioumis,
School of Physics, Sydney University,NSW.
J.J.Condon,
National Radio Astronomy Observatory, Charlottesville,Va,USA,
D.F. Malin,
Anglo-Australian Observatory,Epping,NSW.
G.D.Nicolson and A. Nothnagel,
Hartebeesthoek Radio Astronomy Observatory, South Africa.

ABSTRACT

We are investigating complete samples of southern hemisphere flat spectrum extra-galactic radio sources drawn from the Parkes 2.7 GHz Survey (see Bolton et al. 1979 and references therein). These samples are being used for a variety of investigations, including a determination of the space distribution and luminosity function of radio QSOs, their radio size distribution, as well as the structures of the individual sources. Accurate positions are being determined, as well, in order to establish an extra-galactic position reference frame in the southern hemisphere.

THE COMPLETE SAMPLES

The main component of this programme is a complete sample of some 400 sources satisfying the following criteria;

i) $S2.7 \geq 0.50$ Jy ii) Alpha $(2.7,5.0) \geq 0.5$,

iii) $10° \geq \text{dec} \geq -45°$, and iv) $|b| \geq 10°$

A second,strong source sample covers the southern sky outside of the Galactic Plane, and satisfies;

a) $S2.7 \geq 1.8$ Jy b) Alpha(2.7,5.0)or(.4,2.7)≥ 0.50,

c) Dec $\leq -20°$,and d) $|b| \geq 10°$.

Criterion c) was introduced to ensure that those compact peaked spectrum sources would also be included.

M. J. Reid and J. M. Moran (eds.), The Impact of VLBI on Astrophysics and Geophysics, 77–78.

HIGH RESOLUTION RADIO OBSERVATIONS

Single epoch VLB1 observations of more than half of the stronger sample have already been undertaken with the SHEVE array (see Preston et. al. this conference), and accurate, 0.1 arcsecond rms, radio positions measured for 10 of the stronger sources south of -30° that have bright optical counterparts. Comparison of the radio positions and optical positions measured with respect to reference stars from the Perth 70 Catalogue (Hog and van der Heide 1976) shows that at the level of 0.2 arcsec rms there appears to be no significant displacement between the radio and optical frames.

In addition,about half of the sources in the 0.5 Jy sample have been observed at 2.3 GHz with the Parkes-Tidbinbilla interferometer, the PTI, (see Norris, this conference, and Norris et al. 1985) in order to identify those showing structure on this 275 Km baseline, as well as a search for potential radio position calibrators.

OPTICAL IDENTIFICATIONS

The SERC J survey where available has been used for the identifications and the Palomar Sky Survey for the remainder. SERC J plates are being sought for all fields and AAT CCD photography is underway for the faint identifications. Accurate radio positions and maps have been obtained with the VLA at 6 cm wavelength for sources north of declination -45, in order to confirm identifications and to compare the optical and radio morphologies. Low resolution optical spectroscopy, mainly undertaken by us with the AAT, (see Jauncey et al. 1984, and references therein) has been completed for 70% of the identifications. The present redshift distribution shows a broad peak at about z=1.5, with a steady decrease in source numbers above this redshift. We see no evidence for a redshift cut-off above 3.0, and note that four sources have already been found in the 0.5 Jy sample with redshifts above 3.5.

SUMMARY

This is a major observational program parts of which have been underway for more than a decade. With limited VLB1 facilities available in the southern hemisphere, high resolution observations of the flat spectrum radio sources from the Parkes 2.7 GHz radio survey are proceeding slowly. In contrast, however, to similar programmes from the north, the present investigation has been able to compile extensive optical identifications, spectroscopy,deep photography and redshift determinations Observations presently in progress with the PTI are aimed at satistical angular size studies, while the development of the Long Baseline Array of the Australia Telescope in the coming years will eventually provide detailed milliarcsecond resolution maps.

REFERENCES

Bolton,J.G., Wright,A.E., & Savage,A (1979) Aust.J.Phys. Astrophys.Suppl., 46,1.

Hog,E.,& van der Heide,J.(1976) Abh Hamburger Sternw. Vol.IX.

Jauncey,D.L.,Batty ,M.J.,Wright,A.E.,Peterson,B.A. & Savage,A. (1984). Astrophys.J. 286,498.

Norris,R.P.,.Batty,M.J.,Kesterven,M.J.& Wellington,K.J (1985) Proc.Astr. Soc.Australia, 6, 137.

TESTS FOR A RELATIVISTIC BEAMING MODEL USING A VLBI SURVEY

Makoto Inoue
Nobeyama Radio Observatory
Nobeyama, Minamisaku, Nagano 384-13
Japan

ABSTRACT: We have examined a relativistic beaming model using a VLBI survey by Preston *et al.* (1985). Our statistical study of a ratio R of the flux density for the beamed compact core to that for unbeamed components of a source shows Lorentz factor γ to be $\cong$ 6 and R_T (R at transverse alignment) to be $\cong 10^{-2}$ for a sample of 222 QSO's. In addition, we find that a sample of 60 radio galaxies show the beaming effect with $\gamma \cong 4$. It should be emphasized here that the beaming effect strongly affects the source counts (Log N - Log S) especially at high frequencies.

Orr and Browne (1982) studied a statistical distribution of R and derived $\gamma \cong 5$ and $R_T = 0.024$. They showed that the distribution of R ranges $R_T \leqq R \leqq 2\gamma^4 R_T$. We can thus estimate R_T from the lower limit of R and γ from the range of R distribution.

In order to remove contribution of extended (unbeamed) component and to improve statistical accuracy, we used 222 QSO's (and probable QSO's) and 60 galaxies (and probable Galaxies) from a VLBI survey at 2.3 GHz (Preston *et al.* 1985), which covers entire sky except $|b| \leqq 10°$ with a completeness of the sample for $S \geqq 1.0$ Jy. Although its U-V coverage is limited, the baseline length of several tens of $10^6\lambda$ gives us the benefit of the study.

A distribution of R for QSO's (Figure 1 (a)) gives $\gamma \cong 6$ and $R_T \cong 10^{-2}$, which is consistent with the previous result by Orr and Browne (1982). The range of R for the steep spectrum ($\alpha < -0.5$; $S \propto \nu^{\alpha}$) QSO's (SSQ's) is one order of magnitude smaller than that of flat spectrum ($\alpha \geqq -0.5$) QSO's (FSQ's), while R_T seems to be the same with each other. This can be understood if SSQ's are a subset of QSO's which have smaller beaming effects with a common R_T for the other subset of FSQ's. Furthermore, Figure 1 (b) gives an evidence that galaxies also behave in a similar manner to QSO's; almost the same R_T as that of QSO's with $\gamma \cong 4$.

Figures 2 show that the source counts for total intensities (S_T) are flat for both QSO's and galaxies. On the other hand, they rapidly

M. J. Reid and J. M. Moran (eds.), The Impact of VLBI on Astrophysics and Geophysics, 79–80.

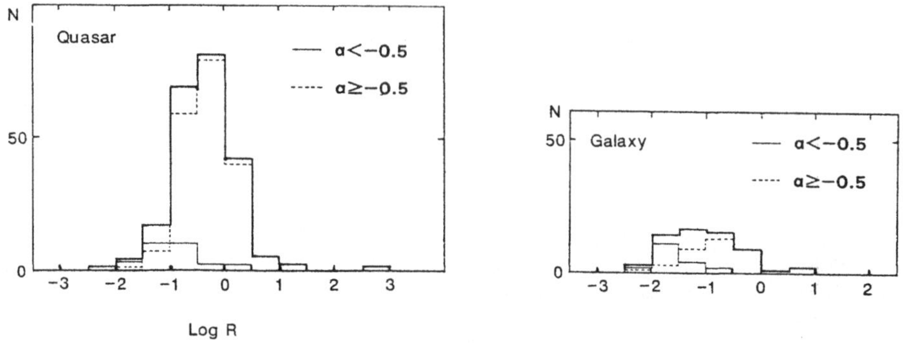

Figure 1. The R distribution for QSO's ((a); left) and for galaxies ((b); right).

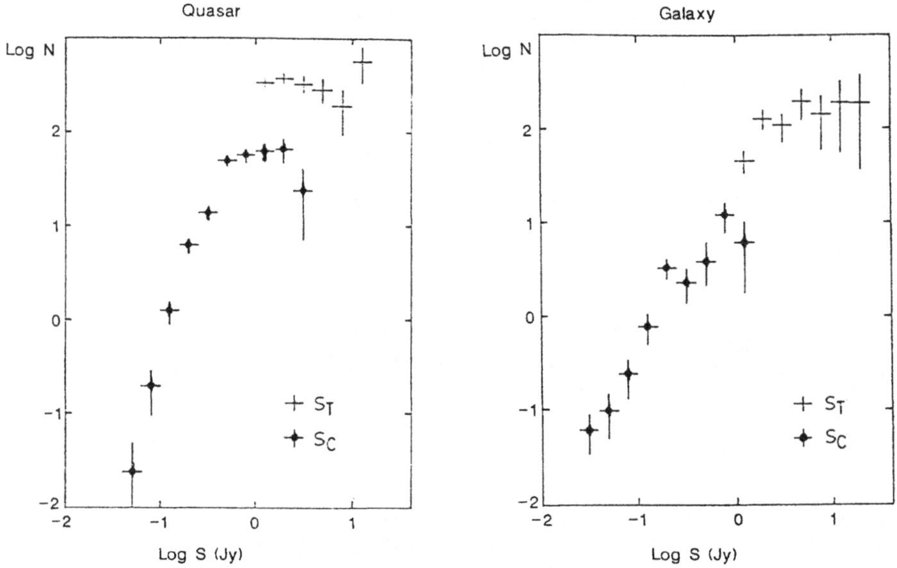

Figure 2. Same as Figure 1 but for the differential source counts.

decrease with decreasing correlated fluxes (S_C). Because of the beaming effect, the number of weaker S_C decreases and consequently that of stronger one increases. Thus Figures 2 (a) and (b) give another evidence for the beaming effect of both QSO's and galaxies. The difference of the inclination in Figures 2 again indicates the difference of γ. This effect will become dominant for high frequency (core-dominant) source counts.

REFERENCES

Orr, M.J.L. and Browne, I.W.A. 1986, *Mon. Not. R. astr. Soc.*, 200, 1067.
Preston, R.A. *et al.* 1985, *Astron. J.*, 90, 1599.

SHOCK MODELS OF TIME VARIABILITY AND SUPERLUMINAL MOTION IN COMPACT EXTRAGALACTIC RADIO SOURCES

Philip A. Hughes, Hugh D. Aller & Margo F. Aller
Radio Astronomy Observatory
Dennison Building
University of Michigan
Ann Arbor, Michigan 48109-1090

Following the success of a simple shock model for outbursts in BL Lacertae and 3C 279 (see Aller, Aller & Hughes, this meeting) we have constructed computer codes to study in detail the radiation from shocked, relativistic jets. These codes compute the transfer of synchrotron radiation, accounting for polarized emission and absorption, rotation, and mode conversion for a turbulent collimated flow with one or more shocks propagating parallel to the jet axis. We present results for a flow that evolves adiabatically, with the turbulence represented by a random component to the magnetic field within each computational cell, and with the shocks prescribed analytically following KönIgl (Phys. Fluids, 23, 1083, 1980). From the evolution of the total and polarized fluxes as a function of frequency, and from the corresponding projection of the source structure on the plane of the sky, we see that this type of model

a. is capable of explaining the variability of compact radio sources - see Aller, Aller & Hughes, this meeting,
b. highlights the care needed when interpreting VLBI maps, in that
   i) the component separations are frequency dependent (see Fig. 1)
   ii) the 'core' is not always the brightest component (see Fig. 1)
   iii) the Doppler boosting factor of the shocked flow is not directly related to the Lorentz factor derived from the apparent superluminal motion of a component
   iv) a multiplicity of components can give rise to both apparent contractions and accelerations,
c. clearly shows the link between time variability of compact sources and evolving VLBI structure, and suggests that both may be understood in terms of weak shocks that tap a small fraction of a jet's flow energy,
d. enables us to probe the physical conditions of the flow and the ambient material.

This work was supported in part by NSF Grant AST-8501093.

M. J. Reid and J. M. Moran (eds.), The Impact of VLBI on Astrophysics and Geophysics, 81–82.

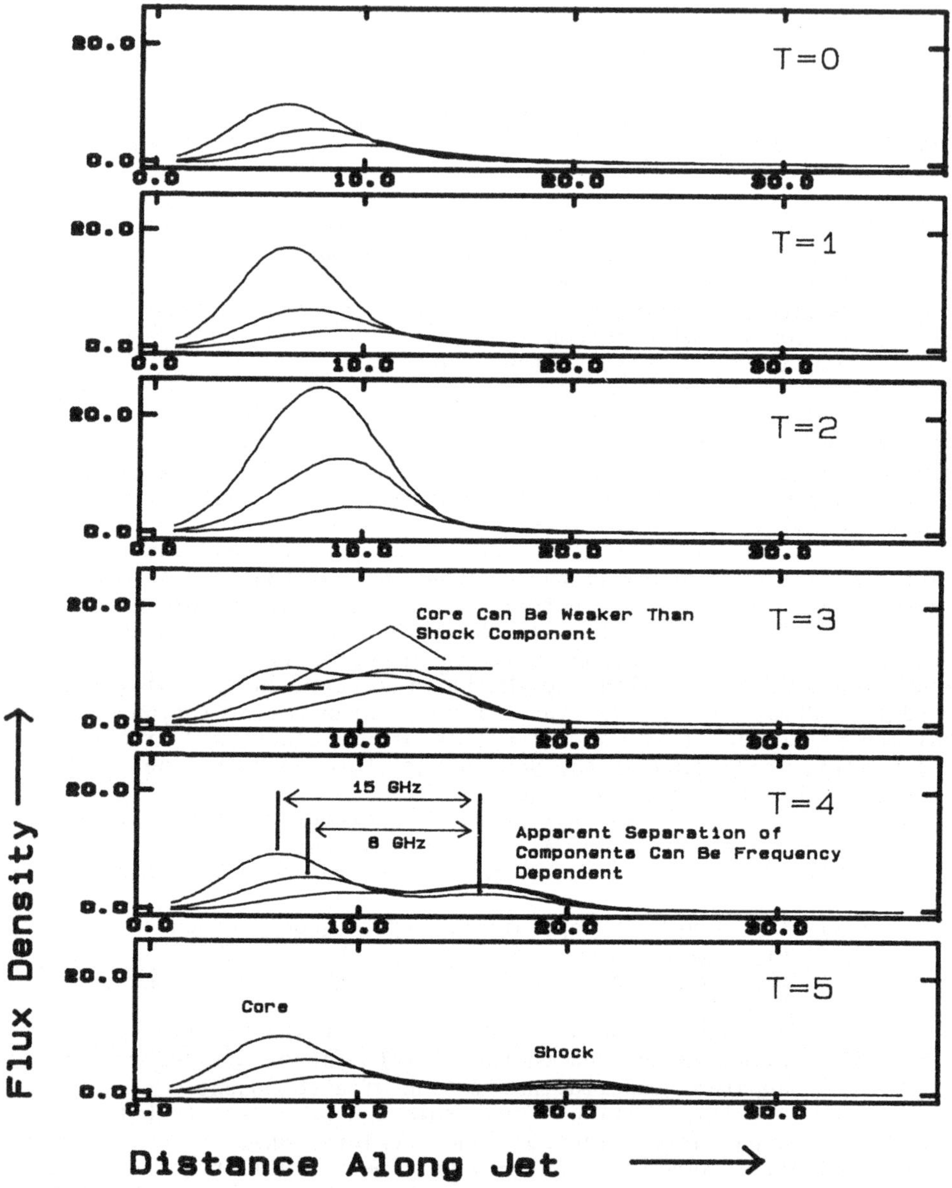

Figure 1. The brightness profiles for a shocked jet showing that the separation between core and shock component is frequency dependent and that the core can be weaker than the shock component. The top panel shows the quiescent jet, and the shock passes through the $\tau = 1$ surface at $T \approx 2$. The model has been convolved with an 'instrumental beam' that is three units HPBW.

ACTIVE EXTRAGALACTIC RADIO SOURCES WHICH SHOW SIGNATURES OF SHOCKS IN COMPACT JET STRUCTURES

M. F. Aller, H. D. Aller, and P. A. Hughes
Radio Astronomy Observatory
Dennsion Building
University of Michigan
Ann Arbor, Michigan 48109-1090

We have analyzed the cm-wavelength flux and polarization observations of four active objects (3C 279, OT 081, BL Lacertae, and 3C 446) for which the data exhibit signatures of shocks. Common characteristics of these sources are: high degrees of polarization, large changes in the polarized flux during outbursts; and stable position angles of the electric vector over long periods.

In 3C 279, OT 081 and BL Lac shock regions in the jet flow apparently dominate the integrated polarization from the source. Observational evidence for this are: 1) during highly polarized outbursts the position angles are parallel to the observed VLBI structures (evidence for an axial compression in the jet flow), and 2) the sources exhibit large, rapid fractional changes in polarized flux. During quiescent phases the polarization electric vector is often perpendicular to the VLBI structure (90 degrees from the orientation during bursts).

In 3C 446 the position angle has remained nearly constant, perpendicular to the orientation of the VLBI structure (Cohen et al., unpublished 1986 map), showing that the magnetic field component parallel to the jet dominates the integrated polarization of the source. We interpret the large downward excursions in polarization during outbursts in this source as being produced by shocked components in the jet.

Figure 1 shows a fit to the highly polarized bursts in BL Lac using a new model code which solves the radiative transfer equations (1977, Ap.J., 215, 236) following Jones & O'Dell, for a shocked adiabatically expanding jet with cellular magnetic field domains (see Hughes, Aller, & Aller, this meeting for model details). The model parameters required to fit both the total flux density and polarization evolution are consistent with those invoked to account for the VLBI structure evolution in this object (Phillips and Mutel this meeting). Figure 2 shows a fit of this new model to our OT 081 data using four shock events. This source is considerably more complex than BL Lac both because of a higher opacity in the emitting region and because of the overlap in time of the bursts. Both the data and the model predictions indicate that the polarization variations in this object are very rapid.

This work was supported in part by NSF Grant AST-8501093.

M. J. Reid and J. M. Moran (eds.), The Impact of VLBI on Astrophysics and Geophysics, 83–84.

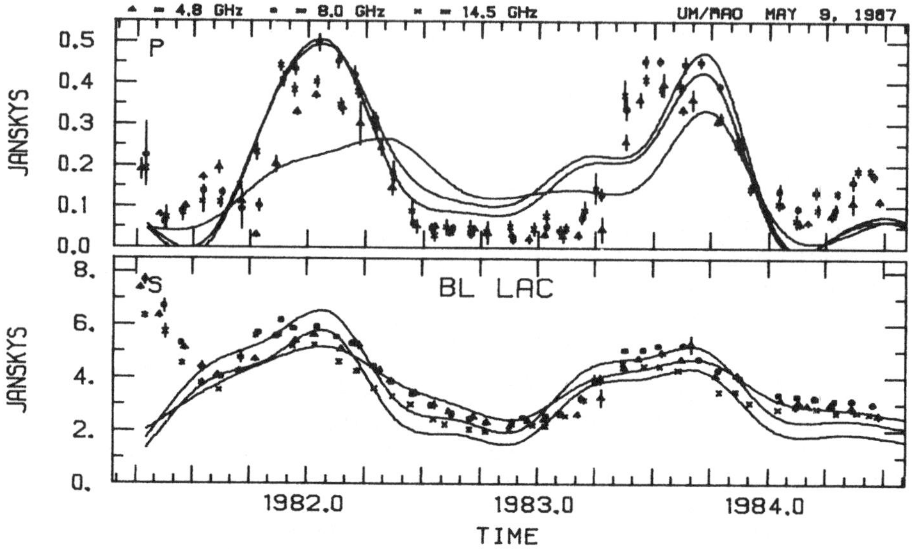

BL Lac. Numerical Simulation Model Of Shocks Forming In A Preexisting Jet Flow

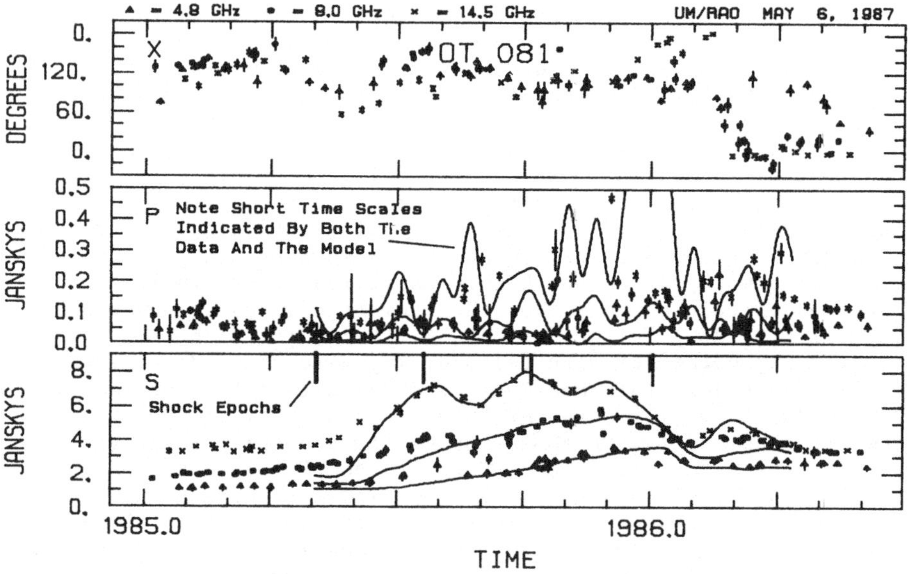

OT 081. Fit Of Shocked-Jet Model To Daily Observations

SHOCKWAVES IN EXTENDED NEAR-RELATIVISTIC JETS

Peter L. Biermann[1] and Peter A. Strittmatter[2]
[1]Max-Planck-Institut für Radioastronomie
Auf dem Hügel 69, D-5300 Bonn 1, W. Germany
[2]Steward Observatory, University of Arizona,
Tucson, AZ 85721, USA

ABSTRACT. The origin of the sharp near infrared cutoff in the continuous energy distribution of many compact non-thermal sources (active nuclei or knots in jets) is considered under the assumption that particle acceleration takes place in shockwaves. Energy losses due to synchrotron emission and Compton interactions set upper limits to both electron and proton energies. In this case the upstream disturbance of the flow is dominated by the most energetic protons which are postulated, by analogy with the solar wind, to excite a turbulent wave spectrum of Kolmogorov type in this region. We predict for near relativistic flows a spectral cutoff near 3 10^{14} Hz independent of magnetic field. The observation of a sharp spectral cutoff near 3 10^{14} Hz is thus independent evidence for near-relativistic flows in jets.

1. INTRODUCTION

Emission regions in certain jets as well as compact sources (BL Lac's, OVV's and red quasars) quite frequently show evidence of a sharp cutoff in the continuous spectrum near 3 10^{14} Hz (e.g. Meisenheimer and Röser 1986). We interpret this cutoff spectrum as synchrotron emission from an electron population with a sharp upper limit to the electron energies.

Within the framework of diffusive shock acceleration (e.g. Drury 1983) we discuss here the acceleration of both protons and electrons subject to synchrotron losses. Due to their lower losses the protons reach much higher energies, and thus go much further upstream than the electrons. We postulate that they excite the dominant turbulence spectrum which determines the resonant scattering of particles at lower than the maximal energy. By analogy with the solar wind plasma (e.g. Goldstein et al 1984) we assume that the spectrum of this turbulence is of Kolmogorov type.

In section 2 we derive the spectral cutoff frequency and in section 3 some consequences. A fuller account, including Compton interactions, neutrino emission and variability is being published elsewhere (Biermann and Strittmatter 1987).

M. J. Reid and J. M. Moran (eds.), The Impact of VLBI on Astrophysics and Geophysics, 85–86.

2. THE MODEL

In order to determine the energy and spatial scale we first derive the maximal particle energy for protons by setting acceleration and loss time equal to each other. This yields a maximal proton energy $E_{p,max}$ of

$$E_{p,max} = 5.6\ 10^{10}\ (b/B)^{1/2}\ U/c \quad GeV$$

where U is the upstream velocity, B the magnetic field strength and c the speed of light. b is the fraction of magnetic energy residing in the turbulence. Thus given the scale the analysis can be repeated for electrons from which we derive an upper limit so the synchrotron emission frequency ν^* of

$$\nu^* = 3\ 10^{14}\ (3b\ (U/c)^2)\ Hz.$$

On dynamical grounds it appears likely that $b \leq 1$; particle acceleration becomes inefficient for $(U/c)^2 \gtrsim 1/3$ (Webb 1985). Hence, $3\ 10^{14}$ Hz appears to be a strong upper limit. A generalized version of this model is discussed in Biermann and Strittmatter (1987).

3. CONSEQUENCES

Observations of jets (Eckart et al 1986, Johnston et al 1987, Biermann et al 1987), compact and extended if measured in kpc, demonstrate that a large fraction of jets have near-relativistic or relativistic velocities. Associated shockwaves should also be near-relativistic (or relativistic). Our model predicts that the emission knots in jets, interpreted as shock waves, should show a sharp spectral cutoff near $3\ 10^{14}$ Hz. We thus have an aspect-independent means to find relativistic motion.

For the outermost knot in a jet, which has near relativistic motion, magnetic fields of order 10^{-4} Gauss are inferred. Our model then predicts maximal proton energies of order 10^{12} GeV in accord with the highest energies observed in cosmic rays. We note that escaping high energy particles suffer very much less adiabatic losses than during an escape from a compact region near to an active nucleus itself. We thus may have identified with observations such as given by Meisenheimer and Röser (1986) the sources of very high energy cosmic rays.

REFERENCES:

Biermann P.L., Kühr, H., Snyder, W.A., Zensus, J.A.: 1987 Astron. & Astroph. (In press)
Biermann, P.L., Strittmatter, P.A.: 1987 Ap. J. (submitted)
Drury, L. O'C.: 1983, Rep. Prog. Phys. 46, 973
Eckart A., Witzel, A., Biermann, P., Johnston, K.J., Simon, R., Schalinski, C., Kühr, H.: 1986, Astron. & Astroph. 168, 17
Goldstein, M.L., Burlaga, L.F., Matthaeus, W.H.: 1984, J.G.R. 89, 3747
Johnston, K.J., Simon, R., Eckart, A., Biermann, P., Schalinski, C., Witzel, A., Strom, R.G.: 1987 Ap. J. 313, L 85
Meisenheimer, K., Röser, H.-J.: 1986, Nature 319, 459
Webb, G.M.: 1985 19th International Cosmic Ray Conference, Proc., vol. 3

Superluminal Accelerations Along a Helically Twisted Jet

Philip E. Hardee
Department of Physics and Astronomy
University of Alabama
Tuscaloosa, AL 35487

ABSTRACT. The relativistic dispersion relation describing the spatial growth of a helical perturbation to a relativistic supersonic jet is applied to observations of the nonlinear superluminal acceleration of component C4 along the inner jet in 3C345.

A HELICALLY TWISTED JET

The radio source 3C345 has a curved, inner-jet structure. Component C4 appeared at a position angle $\simeq$ - 135 deg and has exhibited nonlinear motion and apparent acceleration from $\beta \simeq 4$ to 12 ($H_o = 55$ Km/s/Mpc, $q_o = 0.05$) over a three year period with motion from 0.3 to 0.8 mas separation from the central component along a mean direction $\lesssim$ -75 deg and expansion indicating an opening angle of $\simeq$ 27 deg. (Biretta et al. 1983 - BCU ; Moore et al. 1983 - MRB; Biretta et al. 1986 - BMC). It can be shown that (Hardee 1987) the fastest growing helical wavelength along a constantly expanding jet in static pressure balance with an isothermal external medium can be written in the form

$$\lambda(z) = \lambda_1 (R/R_1)^c, \qquad (1)$$

where $\lambda_1 = \lambda(z_1)$, $R_1 = R(z_1)$ and $1 \leq c \leq 1.5$ with the lower limit set by isothermal jet expansion and the upper limit set by adiabatic expansion of a light jet, i.e., $\gamma\sqrt{\eta} << 1$ where $\eta \equiv n_j/n_{ex}$, and that following rapid initial growth in amplitude that the amplitude $A \simeq \alpha R$. Figure 1 shows the positions of C4 from 1981 to 1984, and an isothermal and adiabatic helical trajectory with wavelength varying according to equation (1) and helical amplitude $A = \alpha R$. The axis of the trajectories is at an angle of 2 deg with respect to the line of sight. At $z_1 = 10^{19}$ cm (0.0143 mas projected on the plane of the sky) $R_1 = 1.04 \times 10^{17}$cm and the intrinsic opening angle is $\simeq$ 0.6 deg, $1.6 \times 10^{20}\text{cm} > \lambda_1 > 2.5 \times 10^{19}$cm, $12.3 > M_1 > 1.9$, $3.8 > \alpha > 3.2$, and the upper and lower limits are given by the isothermal and adiabatic trajectories, respectively. The most likely case is initial adiabatic expansion with $\eta << 0.004$ so that $\gamma\sqrt{\eta} << 1$ at z_1, and with jet heating becoming important by 0.5 mas so that $1.7 < M < 12$ between 0.01 and 0.8 mas. C4 must be traveling with $\beta_o = 0.9980$, and $\gamma = 15.8$. A minimum apparent velocity occurs at position p*. Subsequently, we find an

M. J. Reid and J. M. Moran (eds.), The Impact of VLBI on Astrophysics and Geophysics, 87–88.

apparent motion $\langle\beta\rangle \simeq 5$ during 1981, followed by acceleration to $\langle\beta\rangle \simeq 10$ during 1982, and $\langle\beta\rangle \simeq 11.5$ during 1983. Values of the length of time that a feature takes to reach a position, of the angle ε that the velocity vector makes with the line of sight, of the angle θ that a shock normal would appear to make with respect to the line of sight, of the apparent velocity, β, and of the Doppler factor, $\delta \equiv 1/[\gamma(1 - \beta_o \cos \varepsilon)]$, are given in Table I.

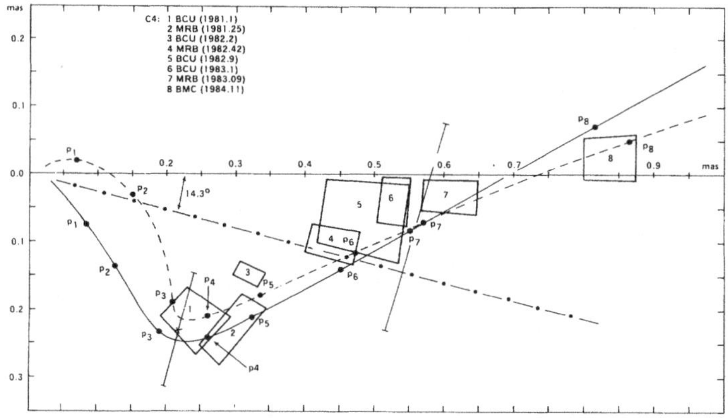

Figure 1 - Positions (error boxes) of component C4 observed by BCU, MRB, and BMC, and the adiabatic (dashed line), and isothermal (solid line) trajectories around the helical axis (dashed and dotted line). The jet's apparent cross section is indicated by the bars perpendicular to the helical axis.

Table I

Timescale, orientations, apparent speed and Doppler factor along the (a) Isothermal Trajectory ($\lambda \propto R$); (b) Adiabatic Trajectory ($\lambda \propto R^{1.5}$)

	a: t(yr)	ε^o	θ^o	β	δ	b: t(yr)	ε^o	θ^o	β	δ
p1	0.25	2.08	59.9	13.7	23.7	0.25	5.13	109.4	14.9	10.5
p2	0.50	1.36	41.1	10.5	22.7	0.50	3.95	95.1	15.7	14.5
p3	1.08	0.59	18.4	5.1	30.8	1.00	0.92	28.5	7.7	29.7
p*	1.25	0.56	17.6	4.8	30.9	1.29	0.40	12.7	3.4	31.2
p4	1.62	0.69	21.7	5.7	30.5	1.60	0.73	22.8	6.0	30.4
p5	2.03	0.91	28.1	7.3	29.7	2.00	1.06	32.7	8.3	29.1
p6	2.86	1.21	36.8	9.3	28.5	2.63	1.35	40.9	10.3	27.7
p7	3.15	1.38	41.8	10.4	27.6	3.04	1.47	44.2	10.9	27.1
p8	4.24	1.69	50.0	12.0	26.0	4.14	1.66	49.3	11.9	26.1

REFERENCES

Biretta, J.A., Cohen, M.H., Unwin, S.C., and Pauliny-Toth, I.I.K. 1983, *Nature*, 306, 42.

Biretta, J.A., Moore, R.L., and Cohen, M.H. 1986, *Ap.J.*, 308, 93.

Hardee, P.E. 1987, *Ap.J.*, 318, 78.

Moore, R.L., Readhead, A.C.S., and Baath, L. 1983, *Nature*, 306, 44.

DUAL RELATIVISTIC EFFECTS IN COMPACT RADIO SOURCES

S. J. Qian
Beijing Astronomical Observatory

ABSTRACT. Dual relativistic effects in compact radio sources are discussed, and as an illustrative example we try to show that the observed features of the knot C_4 in 3C345 (superluminal motion, superluminal expansion, apparent diameter and flux variation etc.) can be interpreted simultaneously.

Recent VLBI observations have shown that in the superluminal source 3C345 knots C_4 and C_2 are moving and expanding superluminally (Readhead(1984), Biretta(1985)). This dual apparently superluminal phenomenon may essentially be an evidence for the dual relativistic effect which we have introduced in the study of compact radio sources (Qian(1983), Qian et al.(1985)). We have shown with plasmon models that the dual relativistic effect is very promising in explaining the superluminal flux variations with time scales of 1-10 days at short centimeter wavelengths observed in certain compact sources. In favourable conditions it can increase significantly the probability of seeing a superluminal motion. The Lorentz factors required are moderate and in the range of the observed ones. Here we try to interpret the observed features of knot C_4 (as an illustrative example) in terms of dual relativistic plasmon model. The following observational facts, arguments and simplifications are taken into account in order to choose a reasonable set of parameters: (1) Although knot C_4 may presumably have a curved path very close to the core, its positions since 1980.5 (the epoch of first VLBI observations at 10.7GHz) can be fitted to straight line motion quite well; (2) For simplicity, we assume that the motion of C_4 is uniform during the period of 1980-1984, its apparent velocity V_a is taken to be 5.5c; (3) However, we suppose that knot C_4 would accelerate to an apparent velocity of ~10c similar to that of C_2 when it reaches distances of 5-6 mas to the core. A favourable evidence for this may be the similarity between the responsible outbursts at 10.7GHz, if their dynamic properties are considered to be dependent on their energetics. Thus we argue for its viewing angle θ being less than 12°, and we assume

M. J. Reid and J. M. Moran (eds.), The Impact of VLBI on Astrophysics and Geophysics, 89–90.

$\theta = 8°$;(4)According to the VLBI observations,the apparent expanding velocity V_d is taken to be 1.4c;(5)From inverse compton argument and X-ray observations the Doppler factor δ_x of C4 should be greater than 7.0;(6)Knot C4 is optically thin at 10.7GHz and 22.2GHz;(7)The energy spectral index is taken to be 1.0,i.e. in the reference system fixed to the centre of plasmon the electrons have a power law energy spectrum $N(E) = N_0E^{-1}$.

From points (1)-(4),we would assume that the bulk-motion velocity $V_b = \beta c$, $\beta = 0.9847$ ($\gamma_\beta = 5.74$ and $\delta_\beta = (\gamma_\beta(1-\beta\cos\theta))^{-1} = 7$ agreeing with point (5));the expansion velocity in the centre-fixed system is taken to be $V_e = \alpha c$, $\alpha = 0.20$ because of $V_d = \alpha\gamma_\alpha\delta_\beta$. Considering that the outburst started between 1979 and 1980,we derive the apparent diameter in 1983-1984 to be ~1.7pc which agrees with the observed ones.From point (6),particle acceleration and field amplification should be taken into account in order to explain the rising phase of the outburst.For simplicity we assume that in the centre-fixed reference system $N_0 = N_{00}t^{-n}$ and $H_0 = H_{00}t^{-m}$ with $n > 0, m < 0$ (in the rising phase $t < t_*$) and $m > 0$ (in the declining phase $t \geq t_*$);see Pacholczyk and Scott(1976).We choose $n = 3.0, m = -1.7$ ($t < t_*$) and $m = 1.7$ ($t \geq t_*$),$N_{00} = 0.25, H_{00} = 3 \times 10^{-6}$G and $t_* = 25$yr.The flux variation calculated for C4 is shown in Fig.1 which is consistent with the observations quite well. A characteristic of the model is that at the start of the outburst the electron energy is far greater than the field energy, but when $t \sim t_*$ it reaches towards the energy equipartition.

It can be seen from above, our model can simultaneously explain most of the observed features of knot C4:superluminal motion(5.5c),superluminal expansion(1.4c),apparent diameter(1.7pc),large angle subtended at the core(28°),flux variation time scales and spectral evolution.Only the accelerated motion and its effects remain to be studied.

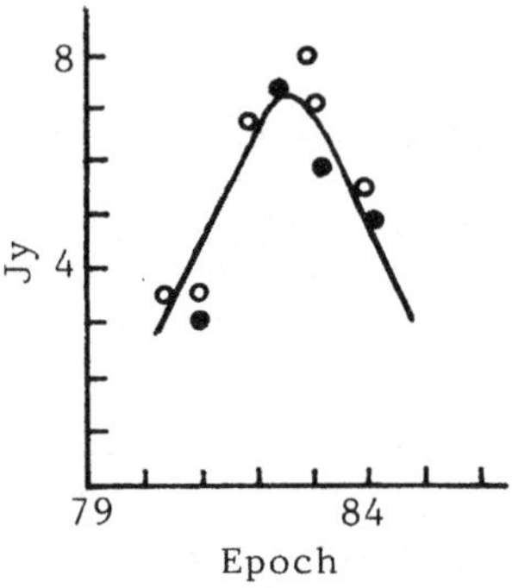

Fig.1. Flux variation of knot C4(• 22.2GHz, ○ 10.7GHz).

REFERENCES
Readhead,A.C.S.,1984,Proc. of Workshop on Quasat,pp139-145.
Biretta,J.A.,1985,Ph.D. Thesis,Caltech.
Qian Shan-jie,1983,Acta Astronomica Sinica,24,pp388-394.
Qian Shan-jie,Zhao Ren-yang,Yu Ning,1985,ibid.,26,pp197-206.
Pacholczyk,A.G. and Scott,J.S.,1976,Astrophys.J.,210,pp311-320.

KINEMATIC PROPERTIES OF THE EJECTED MATTER IN NGC 1275

J. M. Marr, D. C. Backer, M. C. H. Wright
Radio Astronomy Lab and Astronomy Department
University of California, Berkeley

A. C. S. Readhead, R. Moore
Astronomy Department
California Institute of Technology

ABSTRACT

We have mapped the nearby (z=0.018), active galaxy NGC 1275 (3C84) at 6 different epochs from 1981 to 1986 at 1.3 cm (22.3 GHz) with a global VLBI array of seven telescopes. We find a long-lived knot of emission separating from the brightest radio component with a projected velocity = 0.46±0.12 h^{-1} c. This knot moves through diffuse emission that also moves away from the main component with a slower projected velocity of 0.33±0.12 h^{-1} c. We show that the knot and diffuse emission result from two separate events that occurred around 1959 and 1968.

RESULTS

We have observed NGC 1275 at six different epochs from 1981 to 1986 at 1.3 cm (22 GHz) with a VLBI array. Our resultant maps (Fig. 1) display structure similar to that found in previous high frequency VLBI studies of this source (Romney et al. 1982, Readhead et al. 1983). The flux density distribution is dominated by a distinct component to the north which varies in structure and intensity on short time scales. Diffuse emission spreads out to the south with a relatively bright knot persisting through all epochs. We label this knot component K and focus this talk on its motion relative to the northern peak.

The overall size of the source expanded appreciably over the five year period at a rate in rough agreement with that estimated by Romney et al. (1982). We fit an expansion rate of 0.39±0.15 mas/yr (0.33 h^{-1} c) and an origin time of 1959. This time of expansion origin coincides with the start of activity in radio flux density (O'Dea, Dent, and Balonek 1984). Following Romney et al. (1984) we assume a core-jet model with the core located in the northern component. The diffuse emission to the south can then be interpretted as the result of particles ejected from the core in 1959.

The position of component K relative to the peak of the core fits a velocity of 0.55±0.15 mas/yr (0.46 h^{-1} c). This motion implies the

M. J. Reid and J. M. Moran (eds.), The Impact of VLBI on Astrophysics and Geophysics, 91–92.

knot emerged from the core in 1968±4. The flux density curves of O'Dea, Dent and Balonek (1984) display a radio flare that starts around 1970 at 90 GHz and moves to lower frequencies with time. We suggest that component K was ejected in this later event and is moving away from the nucleus independent of the diffuse emission.

Extrapolation of component K's motion backward in time suggest that the knot coincides with the central componet in the 2.8-cm images of Romney et al. (1982) between 1972 and 1977. The Romney et al. maps of 1979 and 1981 do not show this dominant central component. We suggest that the central component in the earlier maps is the same as our component K and that it has moved into the southern component by 1979.

Component K coincides in position with a region of flat or inverted spectra delineated by the spectral index map of Unwin et al. (1982). Following Miley (1980), we estimate from minimum energy arguments the magnetic field in K to be about 0.15 gauss, giving a radiative lifetime of under two years. Either the knot's internal energy is particle dominated or particle re-acceleration is needed.

REFERENCES

Miley, G., 1980, Ann. Rev., 18, 165
O'Dea, C. P., Dent, W. A., and Balonek, T. J., 1984, Ap. J., 278, 89
Readhead, Hough, Ewing, Walker, and Romney, 1983, Ap. J., 265, 167
Romney, Alef, Pauliny-Toth, Preuss, and Kellerman, 1982, IAU Symposium, 97, 291
Romney, Alef, Pauliny-Toth, Preuss, and Kellerman, 1984, IAU Symposium, 110, 137
Unwin, Mutel, Phillips, and Linfield, 1982, Ap. J., 256, 83

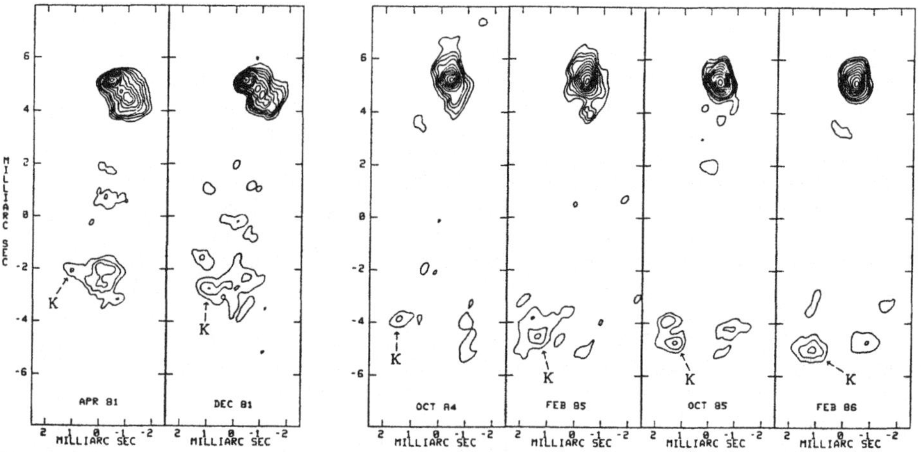

Figure 1

VLBI STRUCTURE OF NGC 1275 AT 3 MILLIMETERS

Melvyn Wright, Radio Astronomy Laboratory, UC Berkeley

We have made λ 3mm VLBI observations of 3C84 at 6 epochs between 1981 and 1987, using 6m antennas at the Hat Creek Radio Astronomy Observatory (HCRK), 10m antennas at the Owens Valley Radio Observatory (OVRO), the 12m Kitt Peak telescope (KTPK), and the 14m telescope at the Five College Radio Astronomy Observatory (QBBN). Model fitting suggests a core-halo structure, which we associate with the active nucleus of NGC 1275. The unresolved core, which is smaller than 0.1 pc (0.2 mas; z=0.018; H_0=50 km s^{-1} Mpc^{-1}) shows a slow decay corresponding to the decrease in total flux density since the flare in 1980. In 1985 we resolved the core component on a transcontinental baseline ($10^9\lambda$) into a compact nucleus, < 0.1 mas, and a "jet" extended at a position angle 205° and unresolved in width, < 0.1 mas. There is also evidence for a compact emission region located 10 mas from the nucleus at a position angle 170°. This structure corresponds closely with that observed at the same epoch in VLBI maps at 22 GHz[1,2].

The ~ 100μarcsec resolution provided by VLBI observations at λ 3mm on transcontinental baselines can probe the broad line emission and jet forming regions of quasars and the scale of an accretion disk around massive black holes in nearby active galaxies. Millimeter observations are required to probe the optically thick synchrotron components seen at centimeter wavelengths. For Sgr A* the measured size scales as λ^2; observations at λ 3mm provide a resolution ~ 1 AU. The small size of millimeter telescopes (6-14m), relatively high system temperatures (300-500K) and tropospheric turbulence limit on integration times (~ 100s) result in poor sensitivities. The limited uv coverage currently restricts our analysis to model fitting. Table 1 summarizes the experiments to date.

The peak flux density of the 3C84 flare was 45 Jy in early 1980[3]. The brightness temperature must be less than ~ 10^{12} K, corresponding to the inverse Compton limit; thus the minimum size in 1980 ~ 0.1 mas. The magnetic field, estimated from optically thick sychrotron emission, ~ 1 Gauss. The equipartition size is also ~ 0.1 mas. The core in our model-fit is smaller than 0.2 mas. If this core were the site of the mm flare then the emitting region did not grow in size by more than 0.15 ly in five years, as the flux density decayed. We suggest that the 1980 flare came from a wind or jet whose size at unity optical depth is proportional to the observed intensity which, in turn, is a measure of the relativistic particle injection rate. Future millimeter VLBI observations during a flare are needed to confirm this hypothesis. The observed decay of emission is not consistent with a single outburst in a magnetic field >~ 0.1 Gauss.

M. J. Reid and J. M. Moran (eds.), The Impact of VLBI on Astrophysics and Geophysics, 93–94.

There was a 20% increase in total flux density in 1984; this increase was not reflected in any of the compact components detected in the VLBI experiments. Our measurements of 3C345, detected in 1982, 1984 and 1985 supports our flux density calibration[4]. The data for 3C84 imply that the increase in total flux density in 1984 was due to a component larger than ~ 0.5 pc. This flare may result from secondary particle generation in a shock. The primary event may well be the flare in 1980; no other precursors are obvious in the data. Note that, if the details in the total flux density variations correspond to secondary particle generation, then we would not necessarily expect these details to be reflected in the evolution of the compact VLBI components.

We can look forward to improved sensitivity and uv coverage in the near future. Experiments phasing the antennas at HCRK and OVRO have been successful and use of a lens at HCRK has increased the aperture efficiency to 80% . A greatly improved uv coverage can be obtained by observing at 3 frequencies, 80 to 110 GHz, with existing antennas, and we can look forward to more antennas and increased bandwidth with the VLBA. Millimeter VLBI has a bright future.

Table I - Summary of λ 3mm VLBI Experiments

Epoch	Stations used	Sources detected
OCT 1981	HCRK-OVRO	3C84
MAY 1982	HCRK-OVRO-KTPK	3C84, 3C273, 3C345
APR 1983	HCRK-OVRO-KTPK	3C84, 3C273, OJ287, 1921-293
APR 1984	HCRK-OVRO-QBBN	3C84, 3C273, 3C345, 2145+06
MAR 1985	HCRK-OVRO-QBBN	3C84, OJ287, 3C273, 3C279, 1921-293, 3C345, 2145+067
MAR 1987	HCRK-OVRO-QBBN-KTPK	3C84, OJ287, 3C273, 3C279, 3C345, NRA0530

REFERENCES

1. A.C.S.Readhead, D.H.Hough, M.S.Ewing, R.C.Walker and J.D.Romney, 1983, Ap.J., 265, 107.

2. J.M.Marr, D.C.Backer, M.C.H.Wright, and A.C.S.Readhead, These proceedings.

3. W.A.Dent, C.P.O'Dea, T.J.Balonek, R.W.Hobbs, and R.J.Howard, 1983, Nature, 306, 41.

4. D.C.Backer, M.C.H.Wright, R.L.Plambeck, J.E.Carlstrom, C.R.Masson, A.T.Moffet, A.C.S.Readhead, D.Woody, A.E.E.Rogers, J.M.Moran, C.R.Predmore, and R.L.Dickman, 1987, Ap.J., 322, 74.

Structural Variability in the Core of 3C147

W. Alef, E. Preuss
Max-Planck-Institut für Radioastronomie, Bonn, FRG
K.I. Kellermann
National Radio Astronomy Observatory, Charlottesville, Va., USA
N. Whyborn, P.N. Wilkinson
Nuffield Radio Astronomy Laboratory, Jodrell Bank, Cheshire, UK

ABSTRACT. We have observed the core of the steep spectrum compact quasar 3C147 at a wavelength of 6 cm with an intercontinental VLBI array. The three observing epochs cover a period from 1981.25 to 1984.8. The core of 3C147 exhibits a complex, nonlinear, and asymmetric structure. It shows structural variability as well as variations of individual components. The observed changes in separation corespond to an apparent velocity of 2.6 ±1.0 c (H_0 = 50 km s^{-1} Mpc^{-1}; q_0 = 0.5). In addition a map from the 3-station observation in 1978 is presented.

1. Introduction

3C147 is a compact (≲ 1"), steep spectrum radio source identified with a quasar at z = 0.545. The arcsecond radio structure is two-sided and asymmetric with a bright 'core' ≲0.008" (60 pc) at the end of a jet ~0.2" in length pointing to the South West (p.a. ~ -130°). A larger elongated feature extends ~0.5" to the North East of the bright core in p.a. ~25° (e.g. Wilkinson et al. 1984).

2. VLBI Observations of the Core of 3C147 at 6 cm

We made high resolution VLBI observations at 6 cm wavelength in 1978.19 (3 stations), 1981.25 (5 stations, Preuss et al. 1982), 1982.95 (9 stations, Preuss et al. 1984), and 1984.80 (7 stations). The data from all four epochs were processed and reprocessed using the MPIfR Global Fringe Fitting procedure (Alef and Porcas 1986). As a result the phase quality was considerably improved and the signal could be detected on the weaker baselines in the broad and deep minima down to ~1σ. This enabled us to obtain maps of higher reliability than before.

The core structure of 3C147 is complex and non-linear with considerable "off-axis" structure (see fig.). The brightest component A is at the NE end of the core region. The structure is variable along the curved line connecting components A, B, and C. The spacing between A and B appears to have increased by about one-third of a beamwidth during the 3.55 year interval (1981 - 1984), corresponding to an apparent velocity of ~2.4 c. It should be noted that superluminal motion was predicted by Simon et al. (1983) on the basis of the observed low frequency variability and the low

M. J. Reid and J. M. Moran (eds.), The Impact of VLBI on Astrophysics and Geophysics, 95–96.

ratio of measured to expected (SSC) X-ray emission. The map from the first epoch in 1978 which was made from observations of three stations only and is therefore not very reliable does not confirm this picture. This map, however, agrees with the trend seen in the three later epochs: the overall appearance of the structure changes from linear with a sideways distortion to a strongly bent morphology. The separation of the components A and D remains constant. The resolution achieved by our 6 cm observations is not sufficient to allow a detailed analysis of the complicated morphology and kinematics of 3C147. Observations at higher frequencies are certainly needed to understand this unique source.

REFERENCES

Alef W. and Porcas R.W.: 1986, Astron. + Astrophys., **168**, 365.
Preuss, E. et al.: 1982, IAU Symposium No. 97, p. 289.
Preuss, E. et al.: 1984, IAU Symposium No. 110, p. 29.
Simon, R. et al.: 1983, Nature, **302**, 487.
Wilkinson P.N.W. et al.: 1984, IAU Symposium No. 110, p. 25.

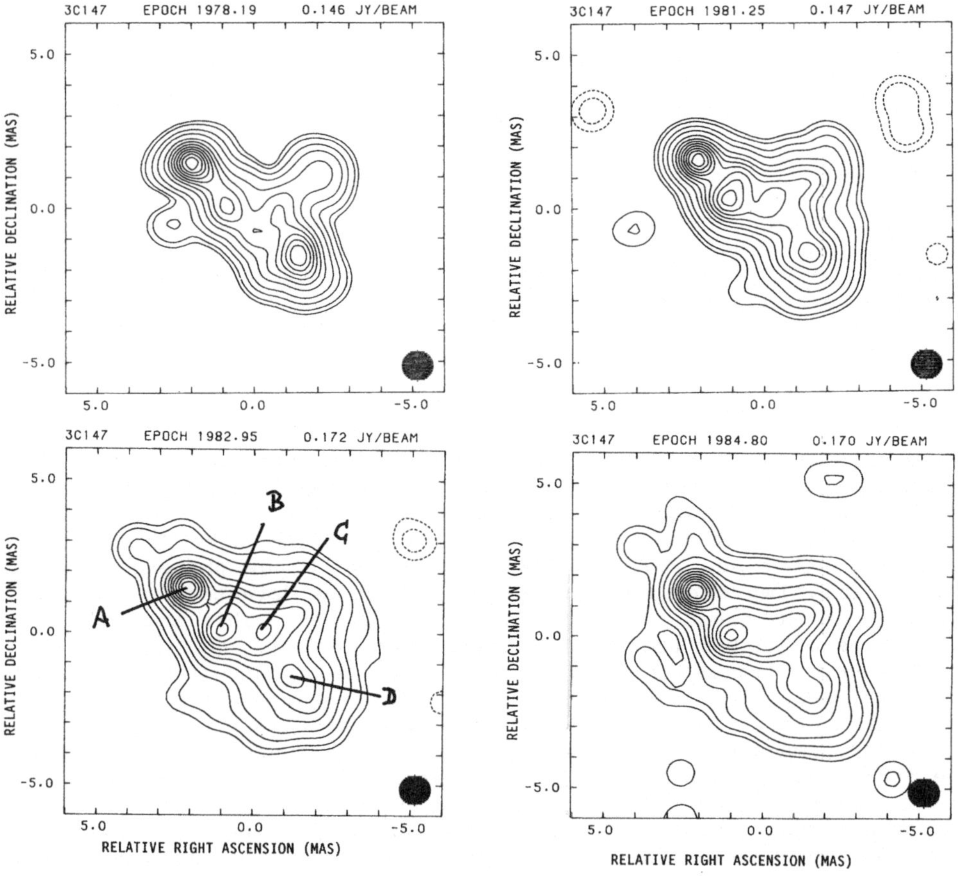

WSRT Observations of hourly flux variations in OJ287 at 6cm wavelength

A.G. de Bruyn
Netherlands Foundation for Radio Astronomy
Dwingeloo

Observations of short term flux variations provide one of the few means of determining the smallest scale sizes in compact radio sources. The fastest variations are generally found in BL Lac-type objects, OJ287 being one of the best studied. In 1985 Valtaoja et al. (Nature 314, 148) found evidence for variations at a level of (typically) a few % with a period of $15^m.7$ in OJ287 at frequencies of 22 and 37 GHz. [Such periodic variability was, however, not detected at 5 GHz during one of their observations by Dreher et al. (Nature 320, 239, 1986)].

In order to confirm and possibly improve on the reported results we have made two series of observations of OJ287, in the fall of 1985 and the fall of 1986, using the WSRT at 4.874 GHz. The September 1985 observations revealed a very dramatic drop in brightness and the 1986 observations were set up with sufficient calibration to ascertain the limiting factors of the WSRT for flux monitoring work. Contacts with D. Roberts (Brandeis) and J. Dreher (MIT) led to simultaneous coverage at the VLA for as much overlap as was possible.

In 1985 OJ287 was observed for three periods of 12^h on September 30 and November 10 and 16. The 1986 observations of OJ287 formed part of an $8x12^h$ project involving the sources 3C84, OJ287, CTD93 and BL Lac. Each source was observed twice with the expectation that, all things being equal, a division of successive 12^h lightcurves would eliminate all systematic, stable, system imperfections. The observations on 3C84, a supposedly nonvariable source, shows that this is indeed the case to an rms accuracy of 0.2-0.3% on scales of 1^m-12^h.

A total band of 80 MHz wide, in 8x10 MHz sub-bands, centered at 4874 MHz was recorded with a time resolution of 20^{sec}. The final resolution used was 60^{sec}. All combinations between two dipoles oriented at 0° and 90° position angle, were correlated. System temperature and gain calibration was done by firing a noise source every 10^{sec}. The theoretical noise level, for 1^m integration, was about 2 mJy, or 0.04% of the source flux density. The actual accuracy achieved was about 5 times worse for reasons still not understood.

We wished to obtain a continuous record of data on OJ287, and split-array observing being not (yet) possible, we could not switch to a flux reference at rapid intervals. [Because suitable reference sources

M. J. Reid and J. M. Moran (eds.), The Impact of VLBI on Astrophysics and Geophysics, 97–98.

are also much weaker than OJ287 the S/N ratio would have decreased had we made use of one]. We therefore chose to rely on the excellent stability of the WSRT and the diagnostic possibilities offered by the array redundancy.

The fastest variations seen are a 7% dip in a timespan of just over one hour in September 1985. Somewhat less dramatic, but more typical, variations are seen in the 1986 October data with $\Delta S/S \sim 1\%$ within one hour. The variations are clearly broadband (>80 MHz) and the suggestion of changes in the polarization state point to an intrinsic variation in the source. But this still needs to be further analysed.

Assuming a distance of about 1000 Mpc (z=0.3(?) and H_0=75 km/s/Mpc) the inferred brightness temperature approaches $\sim 10^{20}$K depending somewhat on whether the whole source varies or whether a local hotspot comes and goes.

If one wishes to explain these variations within the context of the relativistically beamed radio source models, for which OJ287 certainly is a prime candidate, there is the restriction that the Lorentz factor γ of the bulk motion cannot be greater than about 10. Otherwise the extended structure (about 0.2% of the core) would be too bright relative to the unbeamed core flux. Possible ways to explain these results, even for a $\gamma \sim 10$, are discussed in more detail in a paper soon to be submitted.

Preliminary analysis of the data reveals no periodic variations at any timescale. There is some evidence for a $\sim 15^m$ periodicity for about $1\frac{1}{2}$ hour around HA-50°→-40° on October 16, with an amplitude of about 0.3%. It is present in both polarisations but too close to the general noise level to be accepted as real. If such quasi-periodic variations are real large amounts of time are needed to confirm them.

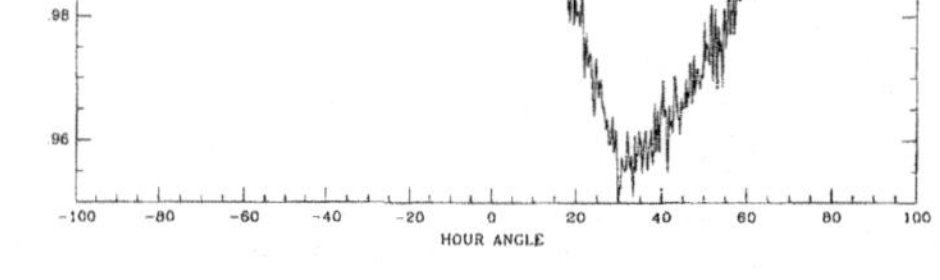

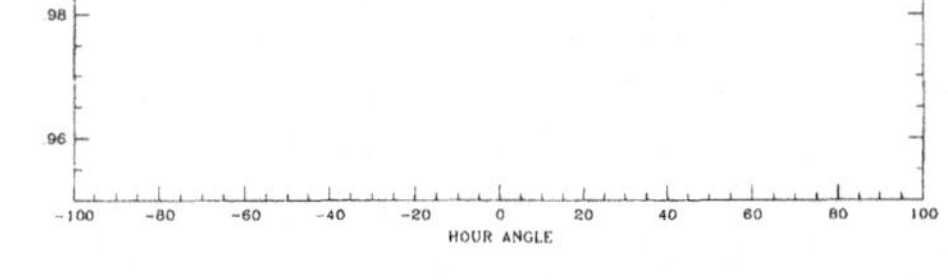

Figure 1. Two 10-hour lightcurves of OJ287 at 6 cm wavelength. The relative flux density (normalized to 4.6 Jy in 1986 and 4.0 Jy in 1985) is plotted versus the WSRT hour angle in degrees.

Relativistic Beaming and the Nuclei of Double-lobed Radio Quasars

D. H. Hough
Jet Propulsion Laboratory, 238-700, Pasadena, CA 91109

A. C. S. Readhead
California Institute of Technology, 105-24
Pasadena, CA 91125

ABSTRACT

We have defined a complete sample of double-lobed 3CR quasars with minimal orientation bias. Properties of these objects on the >kpc scale are consistent with the simple beaming model for their nuclei. VLBI maps of six nuclei reveal "core-jet" structures. Both nuclei mapped at multiple epochs are modestly superluminal (3C245, 3.1c; 3C263, 1.3c; H_0=100 km/s/Mpc, q_0=0.5). The distribution of apparent velocities in this sample will provide a stringent test of beaming.

The simple relativistic beaming model (Scheuer and Readhead 1979; Blandford and Königl 1979) predicts a strong dependence of source properties on the orientation of the beam axis to the line of sight, θ. In order to test such physical theories, we require a sample for which the distribution in θ is known. Thus we have defined a complete, flux density-limited sample of 26 steep-spectrum, double-lobed 3CR quasars (taken from Laing, Riley, and Longair 1983), which should be randomly oriented. For details, see Hough (1986) and Hough and Readhead (1987).

The statistics of >kpc-scale properties of these quasars are consistent with the assumption of random orientations and simple beaming. We find anticorrelations of projected linear size L with both nuclear strength R (=ratio of nuclear to extended emission at 5 GHz, emitted) and outer lobe misalignment angle C; there is a correlation between R and C. Others have found similar correlations (e.g., Kapahi and Saikia 1982), which could also be explained by interactions with the surrounding medium. Even if there is some bias in θ by our selecting quasars, the sample clearly covers a considerable range in θ. Further, even if Doppler-boosting effects are small and these nuclei are intrinsically far weaker than dominant cores, we can study the dependence of v on intrinsic luminosity.

We have mapped the nuclei of six objects at 10.7 GHz using the Mark III system. The sources are 3C207, 3C212, 3C245, 3C249.1, 3C263, and

M. J. Reid and J. M. Moran (eds.), The Impact of VLBI on Astrophysics and Geophysics, 99–100.

3C334. Each object exhibits a double or extended structure, which may be interpreted as a "core-jet". The pc-scale jets are aligned with >kpc-scale, one-sided jets, and the small- and large-scale jets lie on the same side of the compact core.

Thus far, two nuclei have been mapped at multiple epochs. We find superluminal expansion in 3C245 (Hough and Readhead 1987) with an apparent velocity v=3.1c (z=1.029) and in 3C263 (Zensus, Hough, and Porcas 1987) with a smaller v=1.3c (z=0.656). Since 3C245 has relatively large R and small L, it may be aligned at smaller θ than 3C263, which has moderate R and large L. Given a difference in orientation, the different v's are consistent with simple beaming using a bulk Lorentz factor $\gamma \geq 3$.

Porcas (1981) has argued that, based on superluminal motion in 3C179, simple beaming requires a typical $\gamma \geq 5$ for the nuclei of double-lobed quasars, not $\gamma \leq 2$ (Scheuer and Readhead 1979). Our observations of 3C245 further support the case for $\gamma \geq 2$. The three objects in this class (3C179, 3C245, 3C263) have an average $v \sim 3$c. The beaming interpretation for superluminal sources in general is supported by the larger average $v \sim 5$c for nine of the better-studied strongly core-dominated objects (see references in Hough and Readhead 1987). However, Barthel *et al.* (1986) point out that some of the core-dominated superluminals have quite large deprojected linear sizes.

The most direct test of beaming is the measurement of the v distribution in these nuclei (e.g., Scheuer 1984), and at least 85% of our sample is within reach of current VLBI sensitivity.

We are grateful for support at the U. S. and European observatories and processors. This work was partly supported by an NSF grant to Caltech. DHH is a National Research Council-NASA Research Associate.

REFERENCES

Barthel, P. D., Pearson, T. J., Readhead, A. C. S., and Canzian, B. J. 1986, *Ap.J.(Letters)*, **310**, L7.
Blandford, R. D., and Königl, A. 1979, *Ap.J.*, **232**, 34.
Hough, D. H. 1986, Ph.D. thesis, California Institute of Technology.
Hough, D. H., and Readhead, A. C. S. 1987, in *Superluminal Radio Sources*, ed. J. A. Zensus and T. J. Pearson (Cambridge:Cambridge University), p. 114.
Kapahi, V. K., and Saikia, D. J. 1982, *J.Astr.Ap.*, **3**, 465.
Laing, R. A., Riley, J. M., and Longair, M. S. 1983, *M.N.R.A.S.*, **204**, 151.
Porcas, R. W. 1981, *Nature*, **294**, 47.
Scheuer, P. A. G. 1984, in *QUASAT-A VLBI Observatory in Space*, ed. W. R. Burke (Noordwijk:European Space Agency), p. 149.
Scheuer, P. A. G., and Readhead, A. C. S. 1979, *Nature*, **277**, 182.
Zensus, J. A., Hough, D. H., and Porcas, R. W. 1987, *Nature*, **325**, 36.

SECOND EPOCH OF SIMULTANEOUS λ3.6 AND λ13 CM OBSERVATIONS OF THE PAIR OF QUASARS 1038+528 A,B

J.M. Marcaide[1], N. Bartel[2], R.J. Bonometti[3], B.E. Corey[4], W.D. Cotton[5], M.V. Gorenstein[2], R.A. Preston[6], M.I. Ratner[2], A.E.E. Rogers[4], and I.I. Shapiro[2]

[1]Instituto de Astrofisica de Andalucia, Granada, Spain
[2]Harvard-Smithsonian Center for Astrophysics, Cambridge, MA, USA
[3]US Military Academy, West Point, NY, USA
[4]Haystack Observatory, Westford, MA, USA
[5]National Radio Astronomy Observatory, Charlottesville, VA, USA
[6]Jet Propulsion Laboratory, Pasadena, CA, USA

1. INTRODUCTION

On 1983 May 10-11 we undertook simultaneous λ3.6 and λ13 cm Mark III VLBI observations of the quasars 1038+528 A,B. Our experimental conditions (i.e., synthesized band, uv-coverage, etc.) were almost identical to those we used on 1981 March 17-18. Thus, we could make a direct comparison of the results from both epochs. We were especially interested in:

a) Looking for structural changes in quasars A and B (Marcaide et al. 1985);
b) Determining accurately the plane-of-the-sky components of the ~33 arcsecond separation between quasar A and quasar B (Marcaide and Shapiro 1983); and
c) Confirming our previous discovery of a wavelength dependence of the location of the peak of brightness of the core of quasar A (Marcaide and Shapiro 1984).

2. RESULTS

a) We found no changes in the quasar 1038+528 B. The model obtained for this quasar from the 1981 (epoch 1) data (Marcaide et al. 1985) fits the 1983 (epoch 2) data to within the latter's uncertainties.

b) The quasar 1038+528 A showed the predicted evolution (Marcaide et al. 1985): A component ejected from the core circa 1980 was seen moving in the NE direction at an apparent speed more than three times the speed of light ($z=0.678$, $H_0=100$ km s^{-1} Mpc^{-1}, $q_0=0.5$). The

M. J. Reid and J. M. Moran (eds.), The Impact of VLBI on Astrophysics and Geophysics, 101–102.

position angle of this component with respect to the core was larger than in epoch 1 and was approaching the position angle of the large-scale structure. The trajectory suffered a bend of about 10 deg. (See Figure 1.)

c) We confirmed the wavelength dependence of the location of the peak of brightness of the core of quasar A.

d) At λ3.6 cm, the sky position of the peak of brightness of the core of quasar A with respect to a position on quasar B (the same position for both epochs) appears to have moved northwards by about 70 μas along PA -14 deg from epoch 1 to epoch 2. At both epochs the precision is better than 4 μas. We are carrying out extensive tests to determine whether this difference could arise from some effect not yet accounted for. If this 70 μas difference is not an "artifact," then it would suggest that the point where the jet becomes optically thin has moved by that amount in two years. It also appears that the jet curves by about 35 deg from its origin to its outer structure.

e) Combining the above result with astrometric results obtained since 1979, we conclude that the core of quasar A cannot be moving with respect to quasar B faster than about 40 μas/yr.

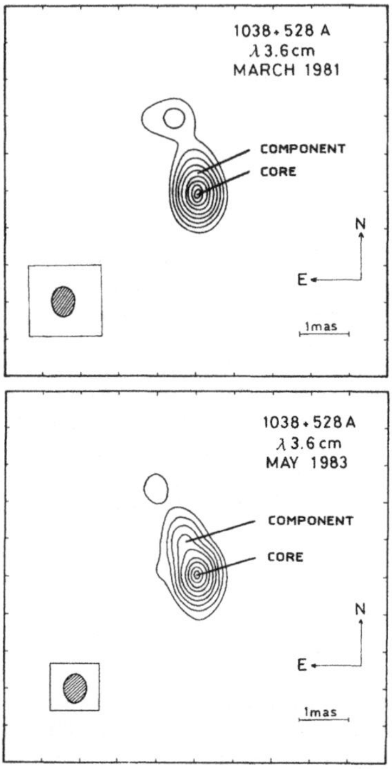

Figure. 1. Evolution of the structure of the quasar 1038+528 A. A component is moving out of the core. The core itself appears (nearly) stationary with respect to the quasar 1038+528 B. Contours in both maps correspond to 2, 5, 10, 18, 30, 50, 65, 85, and 95% of the peak of brightness.

3. CONCLUSIONS

The quasar 1038+528 B (z=2.296), not detectably changed in two years, is a good reference for studying the internal motions in quasar 1038+528 A. The apparent superluminal motion of a feature in quasar A takes place with respect to a (nearly) stationary core; its proper motion is bounded from above by about 40 μas/yr. The wavelength dependence of the location of the peak of brightness of the core of quasar A has been confirmed.

4. REFERENCES

Marcaide, J.M. and Shapiro, I.I. 1983 A.J. **88**, 1133.
Marcaide, J.M. and Shapiro, I.I. 1984 Ap. J. **276**, 56.
Marcaide, J.M. et al. 1985 A.A. **142**, 71.

BL Lacertae's Compact Variations, 1980-86: A Polarized, Superluminal Challenge.

Robert B. Phillips
Haystack Observatory, Westford MA 01886

Robert L. Mutel
The University of Iowa, Iowa City, IA 52242

BL Lac shows tightly correlated events in flux density and polarization. When the time scale slows enough for VLBI observations to track them, moving features which mimic piston-driven shocks seem to carry the disturbances. Evidence for BL Lac's shocks may ameliorate the difficulties posed by the excessive incidence of superluminal motion among core objects; superluminal motion may arise from any number of conditions combining shocks and optical depth, not just real jet or real component motions.

Circa 1980, the most popular world view for explaining bright, variable compact sources asserted that Little Sources were carefully aimed Big Sources, such that real bulk motions enhanced by many times the emission and shortened the time scale for variations, compared to those seen by a co-mover. Superluminality was seen as vindication by the view's boosters; the old light-travel-time canard was relegated to predicting mimimum source sizes following outbursts.

BL Lac's new era of powerful variations began in 1980, characterized particularly by its wild, intractable polarization rotations. Our earliest results (Mutel, Aller, Phillips 1980) indicated a source size larger than ct, and Phillips and Mutel (1982) substantiated components that implied a speed of 5c.

It was superluminal, but just what did that mean any longer? By 1983, VLBI investigators were beginning to pile up a surfeit of superluminal sources. BL Lac had begun a second career by showing two nearly identical polarization events - events which seemed to be carried by the moving (5c) features that faded before reaching a dozen or so light years from their spot of origin. Both events showed a de-rotated p.a. for fractional linear polarization P (E-vector) near the VLBI channel of motion (25 and 20 degrees versus 18 for the channel); the second P-event showed convincing evidence for apparent slowing and stagnation of the moving component. Let us **review** why we think the moving features are the carriers of P. First, the northern bright spot is always a feature in any map, has a flat spectrum, and can even be modeled as constant in strength (except when a new eruption is confused with it.) It always remains when BL Lac returns to P-quiescence. Finally, the fading of the moving VLBI features corresponds precisely with the fading of P-emission.

M. J. Reid and J. M. Moran (eds.), The Impact of VLBI on Astrophysics and Geophysics, 103–104.

The distinctive first P-event with B-field transverse to motion was sufficient to inspire Hughes, Aller, and Aller (1985) to investigate piston-shock models for intensifying B-fields and emission. One remarkable solution to these models shows a shock beginning to exhibit polarization after about 0.15 y; the scale for the driving piston's life is such that emission peaks at about 0.4 y and fades quickly (all in our frame.) This solution is a replica of the two P-events; small changes in optical depth up and down the jet might easily cause the perceived core position to move, leading to apparent decelerations in feature motions as we saw in 1984. Hughes (elsewhere in this volume) has synthesized brightness distributions for the 1982-84 era which show the general features of our VLBI maps, as well.

What is it that BL Lac offers us? Two lessons: one, superluminal features don't always spring full-blown like cannonballs from a core, but can be more ephemeral things that reach prominence well away from their apparent origin (the source core); two, all the bright features in variable compact sources should be viewed as places of localized high optical depth, whose location may wander not entirely in step with bulk flow. Future observations truly need polarization content and astrometric referencing to resolve the question of feature deceleration; BL Lac has already dropped a strong hint that superluminality is not always composed of cannonballs.

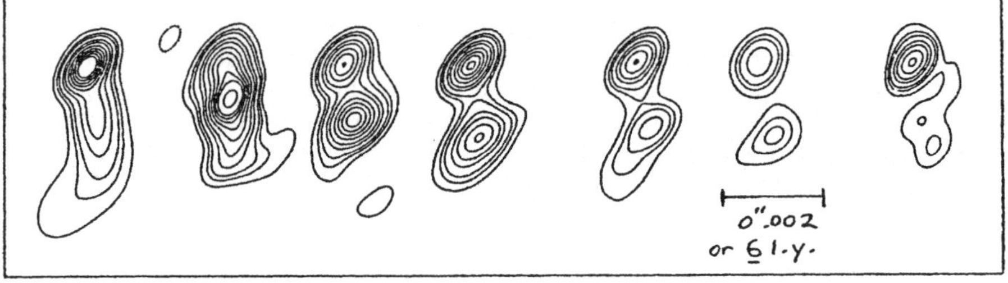

Fig. 1. Montage of BL Lac at 10.7 GHz spanning 1983.1-1985.3. This is the second of two P-events, and shows the apparent stagnation of the southern feature. Time scale is linear.

REFERENCES

Hughes, P.A., Aller, H.D., and Aller, M.F., 1985, Ap.J. **298**, 301.
Mutel, R.L., Aller, H.D., and Phillips, R.B., 1981, Nature **249**, 236.
Phillips, R.B., and Mutel, R.L., 1982, Ap.J. **257**, L19.

Superluminal Behaviour of the Double-Lobed Radio Galaxy 3C111

Eugen Preuss and Walter Alef
MPI für Radioastronomie, Auf dem Hügel 69, 5300 Bonn 1, FRG

Kenneth I. Kellermann
NRAO, Edgemont Road, Charlottesville, Va. 22903,USA

Due to sensitivity problems only about a dozen of the powerful double-lobed radio galaxies have so far been mapped with VLBI. Even less is known about the time dependence of the small scale structure in these objects (Preuss and Alef, 1987). We have recently reported the first results of our monitoring program of classical double sources. We observed strong changes of the pc-scale structure in 3C111 (Götz et al., 1987) and 3C390.3 (Alef et al., 1987) implying "superluminal behaviour" in both sources. This is the first time that such a phenomenon has been found in lobe-dominated radio galaxies. Both objects are Broad Line Radio Galaxies of type N, and their radio emission from m to cm wavelengths is dominated by their outer lobes which are ~300 kpc apart (H_0= 50 km/s/Mpc).

In this paper we give a brief interim report about another, more recent VLBI observation of 3C111 at 6cm wavelength. The central component of 3C111 is relatively strong (~1.5 Jy at 6 cm) and emits about 1/3 of the total flux density at 6 cm. We reobserved the source 1986.9 with an 8-station global array using the MKII (2 MHz) VLBI system. As a result, we obtained the map shown in Fig. 1 along with maps of previous observing epochs. As an example of the drastic changes in the measured data we also show in Fig. 1 the visibility curves for the baseline Effelsberg - Green Bank (fringe spacing ~1.9 milliarcs).

The overall structure of the maps 1980.4 and 1986.9 is elongated in P.A. ~63°. The bright core is slightly elongated in the same direction taking into account the ellipticity of the convolving beam. The direction and orientation of the milliarcs structure are the same as that of the one-sided large-scale jet ~100" in extent (Linfield and Perley, 1984). We note a certain similarity in the general morphology (a bright core at the SW end of an elongated "knotty" feature) of our 1980.4 and 1986.9 maps and Linfield's 1982.6 6 cm map (Linfield, 1987). But given the strong structural changes from epoch to epoch, unambiguous identification of subcomponents from different maps is not possible. More frequent sampling is needed for reconstructing the evolution of the source morphology over the time span of 6.5 years. But if interpreted as propagation effects, the rapid changes clearly imply a superluminal phenomenon. That is, the emergence of the secondary component visible in 1986.9 out of the bright core in a time span <1.5 y means apparent motion at a velocity >9.6 c.

M. J. Reid and J. M. Moran (eds.), The Impact of VLBI on Astrophysics and Geophysics, 105–106.

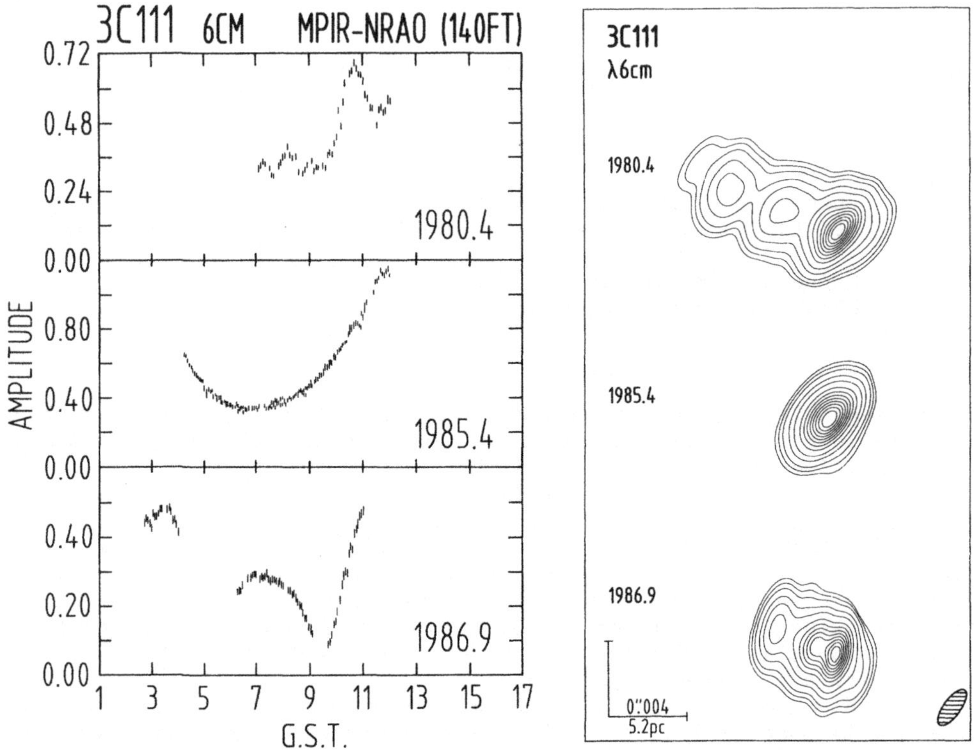

Fig.1: a) (left hand) Amplitudes of the correlated flux density vs. Greenwich Sidereal Time for the baseline Effelsberg- Green Bank for the 3 given observing epochs. b) (right hand) VLBI maps of 3C111 for 3 epochs. The contour levels are 0.5, 1, 2, 5, 10, 20, ..., 90% of the maximum brightness (~3 x 10^{11}K all epochs).

References

Alef, W., Götz, M.M.A., Preuss, E., Kellermann, K.I.: 1987, submitted to Astron. Astrophys.

Götz, M.M.A., Alef, W., Preuss, E., Kellermann, K.I.: 1987, Astron. Astrophys. **176**,171

Linfield,R., Perley,R.: 1984, Astrophys. J. **279**,60

Linfield,R.:1987, Astrophys. J., June

Preuss,E., Alef, W.: 1987, Proc. IAU Symp. No. 124, (eds) Hewitt et al., Reidel, Dordrecht.

RADIO GALAXIES AND RELATIVISTIC BEAMING

D. J. Saikia
University of Manchester
Cheshire, England

There have been recent suggestions and suspicions that the earlier unified scheme, which largely involved quasars, could be enlarged upon to include galaxies as well. In this scheme, the extended radio galaxies are believed to be close to the plane of the sky while the lobe and core-dominated quasars are seen at progressively smaller angles to the line-of-sight. In this paper, we study the correlations of the degree of core prominence with optical properties as well as radio structures on VLBI and larger scales for the both radio galaxies and quasars, in order to investigate whether they are consistent with the enlarged unified scheme. From earlier statistical studies of quasar samples, the degree of core prominence has been shown to be a reasonable statistical indicator of source orientation (cf., Saikia 1985 and references therein). Saikia (1984) also showed that jets in quasars are more easily detected in the ones with more prominent cores, consistent with the unified scheme. In the present paper, we present the results of our observations of 30 low-luminosity radio galaxies and investigate whether such a result holds for galaxies as well.

4. REFERENCES

Saikia, D. J., 1984, *M. N. R. A. S.*, **208**, 231.

Saikia, D. J., 1985, *Extragalactic Energetic Sources*, ed. V. K. Kapahi (Bangalore: Indian Academy of Sciences).

M. J. Reid and J. M. Moran (eds.), The Impact of VLBI on Astrophysics and Geophysics, 107.

X-RAY PROPERTIES OF COMPACT EXTRAGALACTIC RADIO SOURCES

D.M. Worrall
Harvard-Smithsonian Center for Astrophysics, Cambridge, MA, U.S.A.

I. OBSERVATIONS

A good correlation is known to exist between the X-ray and radio luminosity of flat-spectrum, core-dominated radio sources (*e.g.*, Owen, Helfand and Spangler 1981). Worrall (1987) presents a logarithmic plot of spectral luminosity in the source frame at 2 keV versus that at 5 GHz for a variety of QSOs, Highly Polarized QSOs (HPQs), and BL Lac Objects. Friedmann cosmology with $H_o = 100h$ km s^{-1} Mpc^{-1}, $q_o = 0$ is assumed. Exclusion of objects which are optically or X-ray selected, or in which the radio emission is not dominated by a flat-spectrum compact core, gives a sub-sample consisting of 50 QSOs, 20 HPQs, and 10 BL Lacs, of which 5,4,3, respectively, are known superluminals. The dispersion of these data about the log-log correlation (assuming a Gaussian distribution), is $\sigma_{obs} = 0.44 \pm 0.06$ (90% confidence errors for one interesting parameter).

II. COMPARISON WITH MODEL RESULTS

It is assumed that compact flat-spectrum radio sources are selected by preferential beaming towards the observer from a population undergoing relativistic motion of Lorentz factor γ. The Doppler factor is $\delta = [\gamma(1-\beta\cos\theta)]^{-1}$, where θ is the angle of motion to the line of sight. The enhancement of the radio luminosity relative to isotropically-emitted luminosity will be $\delta^{p+\alpha_r}$, where α_r is the spectral index of the radio emission, and $p = 3$ or 2 depending on whether or not emission is from discrete blobs (see Phinney 1985). Calculations below assume $p + \alpha_r = 2.5$. If the dispersion of $\sigma_{obs} = 0.44$ is less than the dispersion implied by projection effects of $\delta^{p+\alpha_r}$, then we must conclude that the X-ray emission is fully or partly beamed like the radio.

The dispersion in δ will depend on the following:

(a). The distribution of γ. A clue to this is available from the 12 superluminals in the sample. The transverse velocities measured for these objects suggest a likely range of γ between $4h^{-1}$ and $16h^{-1}$.

(b). The θ-distribution of the sources. Because the sources are radio selected and core dominated, they are presumably oriented between θ_{min} and θ_{max} such that $\theta_{max} < 180°$. Impey(1987) presents evidence that the optical radiation of compact flat-spectrum QSOs, HPQs and BL Lacs is a composite of beamed and isotropic radiation. Low-polarization QSOs are in the majority in the sample used here, suggesting $\theta_{max} \gtrsim 20°$, and the presence of BL Lac objects suggests $\theta_{min} \simeq 0°$.

Assuming that the X-ray emission is unbeamed and the radio emission is beamed, an expected value for σ can be calculated. If it exceeds σ_{obs}, the assumption of unbeamed X-ray emission is proved to be invalid.

M. J. Reid and J. M. Moran (eds.), The Impact of VLBI on Astrophysics and Geophysics, 109–110.

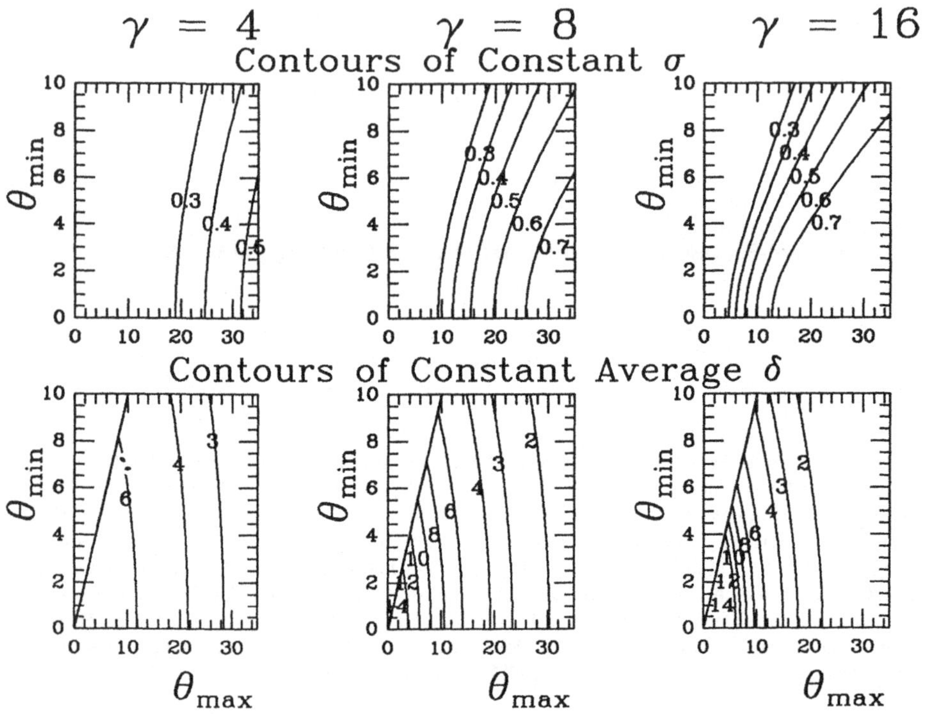

Figure 1. Calculations of σ and average value of δ, for three values of γ, as a function of the minimum and maximum angles of orientation of sources with respect to the line of sight, θ_{min} and θ_{max}. The model assumes that the radio emission is beamed and the X-ray emission is unbeamed, and that sources are oriented at random between θ_{min} and θ_{max}.

Likely parameters for the present sample (for which $\sigma_{obs} = 0.44$) are $4 \geq \gamma \leq 16$, $\theta_{min} = 0°$, and $\theta_{max} \gtrsim 20°$ [see (a) and (b) above]. For such parameter values, Figure 1 shows model predictions which prefer $\sigma > 0.44$. The discrepancy between the predicted σ and σ_{obs} becomes larger under the assumption that some of σ_{obs} is intrinsic or due to variability.

Results are necessarily weak due to the incompleteness of the sample and the uncertainty in the distributions for γ and θ. The best conclusion is that there is a *suggestion* that at least some of the X-ray emission must be beamed.

This work was supported by NASA Contract NAS8-30751 and Smithsonian Institutional funds.

REFERENCES

Impey, C. 1987, in *Superluminal Radio Sources*, ed. J.A. Zensus, and T.J. Pearson (Cambridge: Cambridge University Press), p. 233.

Owen, F.N., Helfand, D.J., and Spangler, S.R. 1981, *Ap.J. (Letters)*, **250**, L55.

Phinney, E.S. 1985, in *Astrophysics of Active Galaxies and Quasi-Stellar Objects*, ed. J.S. Miller (Mill Valley: University Science Books), p. 453.

Worrall, D.M. 1987, in *Superluminal Radio Sources*, ed. J.A. Zensus, and T.J. Pearson (Cambridge: Cambridge University Press), p. 251.

THE RADIO STRUCTURE OF COMPACT STEEP SPECTRUM RADIO SOURCES

C.Fanti[1,2], R.Fanti[1,2], P.Parma[2], R.Nan[3,6], R.T.Schilizzi[3], R.E.Spencer[4], W.J.M van Breugel[5], T.Venturi[1]

1) Dipartimento di Astronomia - Universita' di Bologna - ITALY
2) Istituto di Radioastronomia del CNR - Bologna - ITALY
3) Radiosterrenwacht - Dwingeloo - The Netherlands
4) The Nuffield Radio Astronomy Laboratories - Jodrell Bank - UK
5) Radio Astronomy Laboratory, University of California - Berkeley - USA
6) Beijing Astronomical Observatory - Beijing - CHINA

1. Introduction

Compact Steep Spectrum (CSS) radio sources are defined as having straight steep spectra typical of extended double-lobed radio sources and structures unresolved by conventional interferometers (Peacock and Wall 1982; van Breugel et al., 1984; Fanti et al., 1985). Their spectra remain straight up to quite high frequencies, indicating that no dominant core is present. Sometimes the spectrum bends at frequencies lower than 1 GHz. The angular sizes imply projected linear sizes generally < 10 kpc.

The number of CSS radio sources is not negligible, and varies from ~15% in catalogues compiled at low frequencies (e.g. 3CR, Laing et al., 1983) to ~30% in high frequency samples (e.g. Peacock and Wall, 1982).

If we assume that the extended radio sources in the 3CR catalogue are randomly oriented we can evaluate statistically the number of objects which appear of small size because they are oriented close to the line of sight and hence are shortened by projection. It is found (Fanti and Fanti, 1987) that the expected number of small angular sizes is much smaller than actually observed implying that the vast majority of the CSS radio sources are then physically small objects of sub-galactic dimensions.

2. The source sample

For some years now we have been studying a sample of CSS radio sources from the 3CR catalogue based on the 5 GHz maps made with the Cambridge 5 km radiotelescope (Jenkins et al., 1977). We define a source as CSS if it has at least 80% of its flux density at 5 GHz in a steep spectrum (α >0.5 above ~400 MHz; $S \propto \nu^{-\alpha}$) component, optically identified, and of projected linear size $\leq$10 kpc (Ho=100) (Fanti et al., 1985). These criteria are necessarily somewhat arbitrary, but were adopted to compile a sample which emphasizes the most important

M. J. Reid and J. M. Moran (eds.), The Impact of VLBI on Astrophysics and Geophysics, 111–114.

characteristics of the CSS radio sources. We also limit our analysis to powerful objects (Log P(178) >26 W/Hz), since the weaker ones might be governed by different mechanisms.

The sample includes 26 radio sources (15 QSS, 11 galaxies). VLBI maps at 18 cm exist for ~80% of the sample and for about 30% at 6 cm. MERLIN 18 cm and VLA data at 6 cm and 2 cm are also available for most objects.

3. The radio structure

An accurate morphological classification is not easy since it depends on the resolving power, frequency and sensitivity used. At high frequencies a weak core may show up. High sensitivity is important to establish whether any low brightness emission exists associated with the main component and hence to determine to what extent these objects are indeed small. Optical positions with high accuracy would also be of great importance for the identification of the true core among the source components.

The classification reported here is essentially based on the 18 cm VLBI data. Maps of the sources used as examples can be found in Fanti et al., 1985, unless differently specified. We divide the objects into:

- -doubles : two well separated components, occasionally with a weak core in between (which we refer to as "small doubles");
- -triples : doubles with a strong core;
- -jet-like: straight or bent, sometimes with a well visible core;
- -complex : any object which does not fit in any of the previous classes.

Small doubles (e.g. 3C343.1, 3C237) and triples are similar to the larger doubles, several tens of kpc in extent. They often show high brightness regions inside the lobes, which we may identify with "hot spots", and occasionally weak cores visible at high frequencies with the VLA (van Breugel et al., in prep.). They are also similar to the "compact doubles" of Phillips and Mutel (these proc.) of which they could be an evolved stage. Occasionally the two components may be very asymmetric in brightness (e.g. 3C49, 3C67, 3C268.3) compared to the wide doubles. The ratio of component size to separation is typically 1 to 10, as for their larger cousins, indicating that they are very unlikely to be extended doubles shortened by projection effects. The two or three point spectra of individual components tend to be equal and steep.

The core-jet like sources have significant similarities with those found in flat spectrum sources. Sometimes they are straight (e.g. 3C138, Geldzhaler et al., 1984) but may also show extreme examples of curved structure (e.g. 3C119, Nan et al., these proc.).

The complex sources may be core-jets very much wound up or with considerably distorted low brightness features (e.g. 3C380, Wilkinson et al., 1984a,1986; 3C287, Nan et al. these proc.) or, possibly, seen end on. But there are also cases like 3C343 (Nan et al. these proc.) which are very difficult to interpret.

The subdivision of 3C sources into morphological classes is summarized as follows:

- 9/11 of galaxies are double with weak cores and no visible jets. The only exceptions are: 3C299 which could be core-jet like and 3C346 which has a rather bright core, and possibly a jet, embedded in a very diffuse low brightness region (van Breugel et al., in prep.);
- 11/15 of the QSS's are either core-jet or complex;
- 4/15 of the QSS's are triple, i.e. double with a strong core.

Such a morphological segregation is independent of radio luminosity since weak quasars in our sample are still complex and powerful radio galaxies are still double. This marked structural difference was reported by Wilkinson et al., 1984b and Fanti et al., 1985. Here we strengthen those results, with higher resolution data.

Not only double structure is present in galaxies on all scales, but also the distribution of sizes for the galaxies of the entire 3C sample shows good continuity between our "small doubles" and the larger doubles (fig.1). This points to a single class of object undergoing the same evolutionary pattern.

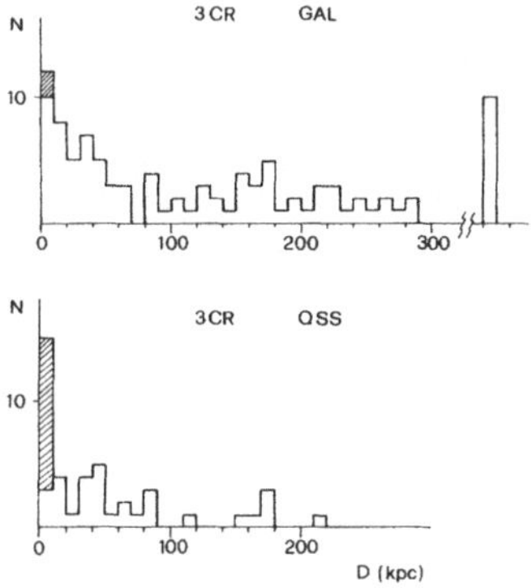

The relation between size and structure is more complex in quasars. With a few exceptions small quasars have complex morphologies with highly curved jet-like features. The distribution of quasar radio sizes in the total 3C sample shows an excess in the small size bin (fig.1) which is mainly due to compact distorted structures (dashed). The structure however seems to develop towards the double morphology as the overall size progressively increases. Quasars larger than 10 kpc are usually double.

Steep spectrum quasars have on average a much brighter core than radio galaxies. However, if we consider galaxies and QSS's together, after appropriate normalization, (Fanti and Fanti, 1987), we find that the distribution of the ratio of core-to-extended emission in CSS radio sources is consistent with that of the classical doubles in the 3CR sample. This tells us two things:

- it confirms that the large majority of CSS sources are randomly oriented, otherwise we would observe doppler-boosted radio cores;
- the radio cores in CSS radio sources have the same relative power as in the more extended objects.

4. Discussion

We have reported three arguments to support the idea that CSS radio sources are mostly intrinsically small objects: i) diameter statistics; ii) separation/component-size ratio; iii) core/extended emission ratio.

If CSS radio sources are mostly randomly oriented, we cannot ascribe the large observed distortions to enhancement by projection of small intrinsic bends. To explain their morphology it is tempting to invoke interactions with an ambient gas, as is known to occur in nearby

CSS radio galaxies (e.g. 3C305, Heckman et al., 1982). Ordered motions (e.g. 3C119, Fanti et al., 1986; Nan et al., in prep.), random motions and shocks (e.g. 3C380, Wilkinson et al., 1986) or a combination of these (van Breugel et al., these proc.) may play a role. In this scheme double and triple radio sources are either sources whose parent optical object has no turbulent medium or sources which have already escaped from such a medium.

The morphologies we find point to the first possibility for radio galaxies, since they have a double structure from the very beginning. The structure-size trend for quasars instead indicates the existence of a dense turbulent medium over scales of some kpc, strongly perturbing the radio structure. Only if (or when) the radio emitting material emerges from such a region the double structure can be formed. This is furthur discussed by van Breugel et al. (these proc.).

A number of objections can be brought against this simple scheme. For instance if optical boosting were present, as suggested by some authors, galaxies could be turned into quasars, and the diameter statistics we present would be wrong for such objects. It would however be difficult to reconcile this with the fact that we do not observe boosted radio cores (unless more ad hoc assumptions are introduced, e.g. weaker cores in CSS quasars, boosting occurring in both core and extended regions, etc.).

We conclude that the hypothesis that CSS radio sources are physically small objects, randomly oriented in the sky, is still the one requiring the minimum number of assumptions.

REFERENCES

Fanti,C., Fanti,R., Parma.P., Schilizzi,R.T., van Breugel.W.J.M: 1985 Astron.Astrophys, **143**, 292

Fanti,C., Fanti,R., Schilizzi,R.T., Spencer,R.E., van Breugel,W.J.M: 1986, Astron.Astrophys., **170**, 10

Fanti C., Fanti R., 1987, Superluminal Radio Sources, ed. J.A.Zensus and T.J.Pearson, Cambridge University Press, p. 174

Geldzahler,B., Fanti,C., Fanti,R., Schilizzi,R.T., Shaffer,D., Weiler,K.W.:1984, Astron.Astrophys., **131**, 232

Heckman,T.M., Miley,G.K., Balick,B., van Breugel,W.J.M., Butcher.H.: 1982,Astrophys.J., **262**, 529

Jenkins,C.J., Pooley,G.G., Riley,J.M.:1977,Mem.Roy.Astron.Soc.,**84**,61

Laing,R.A., Riley,J.M., Longair,M.S.: 1983, Monthly Notices Roy.Astron.Soc.. **204**,151

Peacock,J.A.. Wall,J.V.:1982, Monthly Notices Roy. Astron.Soc., **198**.843

van Breugel,W.J.M., Miley,G.K, Heckman,T.A.: 1984, Astron.J..**89**.5

Wilkinson.P.N., Booth,R.S., Cornwell,T.J., Clark,R.R., 1984a, Nature. **308**, 619

Wilkinson P.N., Spencer,R.E., 1984b, IAU Symp. 110. ed. R.Fanti et al., Reidel, p.25

Wilkinson.P.N., Kus,A.J., Pearson,T.J., Readhead,A.C.S., Cornwell,T.J., 1986, IAU Symp. 119, ed. G.Swarup and V.K.Kapahi, Reidel, p.165

SMALL STEEP-SPECTRUM SOURCES: JETS COLLIDING WITH AMBIENT GAS

Wil van Breugel, Radio Astronomy Laboratory, UC Berkeley, USA
Nan Ren-dong, Visiting Astr. at NFTA, Dwingeloo, The Netherlands
R. T. Schilizzi, NFRA, Dwingeloo, The Netherlands
C. and R. Fanti, Istituto di Radioastronomia, Bologna, Italy

A large fraction of sources in flux limited radio samples consists of compact steep spectrum (CSS) sources (Fanti *et al.* 1987). These objects have their dominant emission on kiloparsec scales and are found in a variety of active galaxies, including Seyfert II's, peculiar galaxies and quasars. Observations of very nearby CSS sources have shown good correlations between their complex radio structures and optical emission-line gas suggesting violent interactions between jets and dense gas in their ambient interstellar media (van Breugel *et al.* 1984). Similar jet/gas interactions may occur in more distant and more powerful CSS sources.

To investigate the radio morphologies of CSS *quasars* we have made global VLBI observations at 18 cm of 3C 119, 3C 287 and 3C 343. Using special imaging techniques high quality, noise limited maps were obtained (Nan Ren-dong *et al.* 1987). While the details of the low brightness features of these sources must be regarded with caution, it is clear from these data that the radio source structures are extremely complex, showing large curvature (3C 119, 3C 287) and seemingly detached regions of radio emission (3C 119, 3C 343). Similar radio morphologies have been found in the CSS quasars 3C 309.1 and 3C 380 (Wilkinson *et al.* 1985).

The extreme complexity of these CSS quasars suggests that the canonical "unified" scheme for quasars and galaxies, namely of relativistic jets seen end on, may be too simplistic. Other, statistical arguments which are difficult to accomodate in such a scheme are given by Fanti *et al* 1987. As before we suggest therefore that more realistic models include the "disturbing" effects on the (relativistic ?) jets by their environment.

Optical observations of powerful radio sources show increasingly that tidal interactions and merging of galaxies are *generally* (not rarely) important. Many powerful radio galaxies have nearby companions and exhibit peculiar morphologies indicative of merging activity (Heckman *et al.* 1986). Similarly, also many quasars have nearby companions (Heckman *et al.* 1984), and very suggestive evidence has been found from optical emission-line images that tidal interactions may occur in the *steep spectrum* (not flat spectrum !) quasars (Stockton and MacKenty, 1987).

Tidal interactions and merging of galaxies can affect radio source morphologies in important ways. The active nucleus might precess (temporarily) and/or exhibit orbital motion resulting in peculiarly twisted radio jets. If at least one of the interacting galaxies is gas rich, radio jets may collide with gaseous merging debris causing jet brightening (knots), deflection, decollimation and deceleration because of internal shocks, turbulence and possibly entrainment of ambient gas. In nearby radio galaxies both effects are well known to occur (Miley 1980; van Breugel 1985). The bright radio knots, large curvature and flaring of the jets at $\sim$ 100 pc from the nucleus in 3C 309.1 and 3C 380 (and possibly in 3C 119) indicates that this may also occur in CSS quasars.

M. J. Reid and J. M. Moran (eds.), The Impact of VLBI on Astrophysics and Geophysics, 115–116.

Precession models of (relativistic) jets in their simplest - ballistic - form can qualitatively account for many of the observed peculiar radio morphologies (Gower *et al.* 1982). It would seem worthwhile to perform more realistic numerical simulations in which the effects of a dense and inhomogeneous ambient environment would also be incorporated since large curvatures can in part also be attributed to jet/gas deflections and ram pressure effects.

An interesting consequence of the assumption that CSS sources are jets propagating through dense inhomogeneous interstellar media is that the VLBI observations then would seem to suggest that the complex CSS *quasars precess rapidly and may be intrinsically one-sided.* This is because twin jets, after violent collisions with dense gas, would both be expected to brighten and slow down so that both might also be expected to be observable. Radio observations suggest that most CSS quasar jets are one-sided, examples are 3C 119, 3C 309.1 and 3C 380, but the sample with good VLBI maps is still small.

By the same assumption, the CSS *galaxies do not precess and are intrinsically two-sided,* since they have simple double morphologies. If one were to insist that quasars and galaxies are related, one might suppose that (CSS) quasars are merging systems which could perhaps evolve into (CSS) galaxies when the merging process is completed: precession stops, ambient gas acquiesces, nuclear activity lessens and becomes more stationary, and steady twin jets develop fully.

We end with noting that the jet/gas interactions in CSS sources do not only have consequences for the jets, they may also have considerable impact on the interstellar media themselves, including heating of ambient gas and even inducing star formation. For example, a jet-induced starburst has been found in Minkowski's Object (van Breugel *et al.* 1985), and recent evidence suggests that this may commonly occur in powerful, distant ($z \gtrsim 0.5$) radio galaxies (McCarthy *et al.* 1987. CSS sources with evidence for ongoing star formation are for example NGC 1068, (Wilson 1986), 3C 459 (Ulvestad 1985) and 3C 48 (Neugebauer *et al.* 1985). High resolution (Space Telescope) optical imaging observations of these and other CSS sources would of course be needed to determine whether any enhanced star formation is indeed morphologically associated with the radio emission in these objects.

REFERENCES

Fanti *et al* 1987, these Proceedings.
Gower *et al.* 1982, Ap. J. 262, 478.
Heckman *et al.* 1986, Ap. J. 311, 526.
Heckman *et al.* 1984, Ap. J. 89, 958.
Miley 1980, Ann. Rev. Astr. Astroph. 18, 165.
McCarthy *et al.* 1987, preprint.
Nan Rendong *et al* 1987, these Proceedings.
Neugebauer *et al* 1985, Ap. J. 295, L27.
Stockton and MacKenty 1987, Ap. J. 316, 584.
Ulvestad 1985, Ap. J. 288, 514.
van Breugel *et al.* 1984, A.J. 89, 5.
van Breugel 1986, Can. J. Phys. 64, 392.
Wilkinson *et al.* 1985, in Proc. IAU Symposium 119, pg. 165.
Wilson 1986, in Proc. IAU Symposium 121, (in press)

HIGH RESOLUTION OBSERVATIONS OF LOW FREQUENCY VARIABLES AT 932 MHZ

Lars B. Bååth
Onsala Space Observatory
S-439 00 Onsala
Sweden

ABSTRACT. Fifteen sources known to be varying at low frequencies have been observed at six epochs during 1983-84 with a global VLBI array. Some of the sources show structural variations similar to the superluminals. Beaming effect may therefore play an important role at low frequencies as well as at higher.

A global VLBI array consisting of Kiruna(Sweden), Jodrell Bank(UK), Arecibo(Puerto Rico), and Green Bank and Owens Valley(USA), has been used to map fifteen low frequency variable radio sources at 932 MHz. The frequency was chosen to give maximum resolution while still being able to observe in the low frequency domain. The resolution achieved was 5 milliarcseconds (FWHM) with a typical noise of 30 milliJansky in the maps. Fifteen sources were observed at six epochs during 1983-84. The sources were selected for low frequency variablity in a declination range suitable for Arecibo. Eight of the sources have so far been mapped at three epochs. Work is progressing on the rest of the sample.

Some of the sources were nearly unresolved with a weak halo or a small extension to one side. Among these are OJ287 and DA406, both with small extensions. 0735+178 and 3C454.3 also had compact structures, but with extensions which are extrapolations from the structures at higher frequencies.

Other sources showed more complicated structure. CTA102 was the first in the sample to be mapped at several epochs. Two of the three components in the core separated from each other with an apparent speed of 0.65 ± 0.15 millliarcsec./year or $(18\pm4)h^{-1}c$ (Bååth 1987). Other sources also showed structure resembling that of the superluminals. 4C21.50 (Fig. 1 top) had a jetlike structure changing with time. 1117+14 (Fig.1 bottom) had at least three distinct components where two components seemed to move away from each other with time.

The flux variations in compact continuum radio sources at low frequencies can basically be intrinsic to the source, extrinsic, or a combination of both. Among the extrinsic models are noncosmological redshifts. Sholomitskii (1965) suggested in the original paper on flux variations that CTA102 was a Galactic object by comparing the angular size to the linear size implied by the time scale of the variations.

M. J. Reid and J. M. Moran (eds.), The Impact of VLBI on Astrophysics and Geophysics, 117–118.

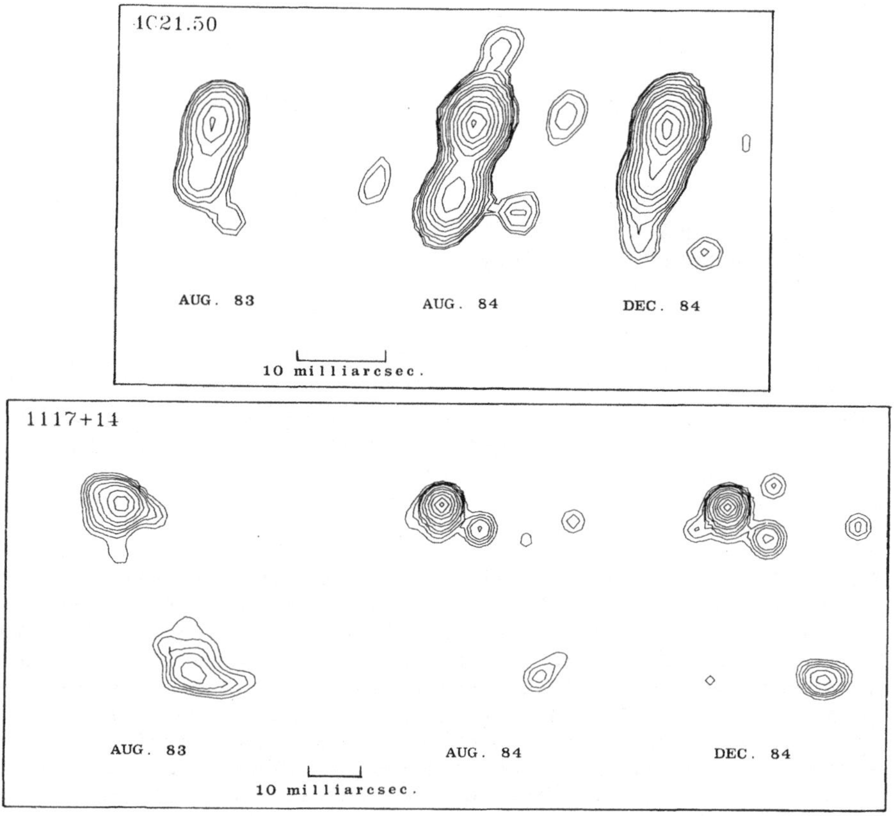

Figure 1. Hybrid maps of 4C21.50 (top) and 1117+14 (bottom). The contour levels are: 3*(7,8,10,12,15,20,25,30,40,50,60,70,80) mJy/beam area in the top panel and 2.5*(15,20,25,30,40,50,70,80,110,130,150, 170) mJy/beam area in the bottom panel.

Later years discussion has centered on slow, refractive scintillation caused by the interstellar medium as a possible extrinsic model for the flux variations (Ricket *et al.* 1984). Such scintillation may play the major role for the variations in the flux density of the very compact sources observed here (DA406,OJ287,3C454.3,0735+178). The superluminal motion in CTA102 and the structural changes in the more complex sources (4C21.50,1117+14 and others) indicate that intrinsic models as beaming effects play an important role at frequencies below the intermediate frequency gap (1 GHz) as well as at high frequencies.

REFERENCES:

Bååth,L.B. 1987, *Superluminal Radio Sources, ed. Zensus and Pearson,* Cambridge University Press, 206

Ricket,B.J., Coles,W.A., and Bourgois,G. 1984,*Astron.Astrophys.*,**134**,390

Sholomitskii,G.B. 1965,*Astron.Zh*,**42**,673

THE RADIO STRUCTURES OF THE COMPACT STEEP SPECTRUM SOURCES 3C119, 3C287 AND 3C343

Nan Ren-dong (Beijing Astronomical Observatory, China)
R.T. Schilizzi (NFRA, Dwingeloo, The Netherlands)
C. Fanti, R. Fanti (IRA, Bologna, Italy)
W.J.M. van Breugel (Radio Astronomy Laboratory, University of Berkeley, USA)
T.W. Muxlow (NRAL, Jodrell Bank, UK)

The compact steep spectrum radio sources, 3C119, 3C287 and 3C343, are identified with QSO'S of redshift 0.46, 1.055 and 0.988 and visual magnitude 20.0, 17.7 and 20.6 respectively. The overall physical extent of each source is ~500 pc (Fanti et al. 1985, 1986). Here we present the results of new global VLBI 18 cm observations which have been combined with MERLIN data to improve the U-V coverage. The data were mapped using the CALTECH and AIPS software packages and also using OLAF software system developed in Jodrell Bank for 3C119 and 3C343.

The resulting map of 3C119 is shown in Figure 1a. The RMS noise level at some distance from the source in the final cleaned map is 1 mJy/beam. The dynamic range in the map (peak value to RMS noise) is therefore about 3000 to 1. The four brightest components marked as A, B, C and D can also be found in the 6 cm map by Fanti et al., (1986). The brightness distributions at 6 cm and 18 cm are in good agreement when the two maps are superimposed. The spectrum of component A is inverted between 6 and 18 cm, suggesting that it is the true nucleus of 3C119. The high dynamic range and the good u-v coverage allows us to identify another three extended components with low surface brightness, E, F and G, amongst which only E is clear in the previous map at 18 cm by Fanti et al.

The resolution of the present map of 3C287 (Fig. 1b) is sufficient to see the structure in some detail. The bright peak can be fitted with a Gaussian model with HPW = 26 x 15 mas and position angle = 100°. There is a clear change in the source position angle going from the peak towards the north. The filamentary structures to the west and south west (S_{18} = 0.5 Jy) may be a continuation of the spiral. The shape of 3C287 suggests a curved "core-jet". An unpublished low quality EVN map of 3C287 at 6 cm shows a morphology similar to the present 18 cm map. A spectral index analysis (by T. Venturi, Bologna) indicates that the brightest peak in both maps has a steep spectrum.

The morphology of 3C343 is very complex (Fig. 1c). The brightness distribution is multi-peaked. If we consider the source to be composed

M. J. Reid and J. M. Moran (eds.), The Impact of VLBI on Astrophysics and Geophysics, 119–120.

of four parts A, B, C and D, their flux densities are roughly 2000, 400,500 and 1000 mJy. The three elongated components, A, B and C have position angles 40°, 90° and 180° and similar widths of ~30 mas; component D is a plateau of emission. "A" is obviously connected with the others by bridges of emission. Another isolated emission peak E with ~100 mJy appears well detached from the main emitting region. The map accounts for all the source flux density at 18 cm. The structure of 3C343 suggests a knotty jet which might be deflected, distorted and split by impact with the interstellar medium.

REFERENCES

Fanti, C., Fanti, R., Parma, P., Schilizzi, R.T., van Breugel, W.J.M., 1985, Astron. Astrophys. 113, 292.
Fanti, C., Fanti, R., Schilizzi, R.T., Spencer, R.E., van Breugel, W.J.M., 1986, Astron. Astrophys., 170, 10.

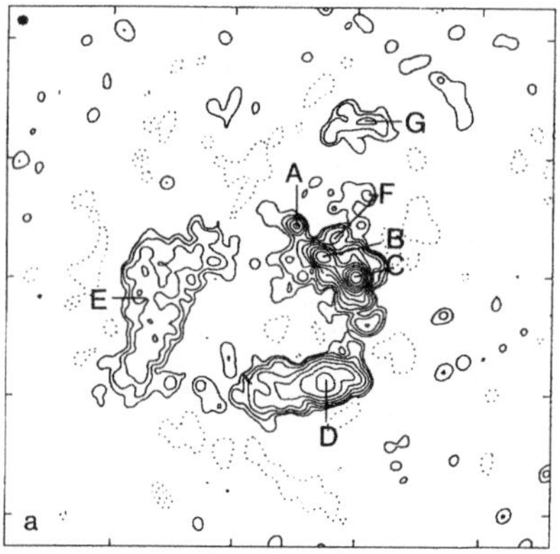

Figure 1:
a) Brightness distribution of 3C119 at 18 cm from AIPS/OLAF. Beam: 5 mas^2; tick separation: 60 mas; peak: 2.993 Jy/beam; contours (%): -0.10, -0.05, 0.05, 0.10, 0.20, 0.40, 0.80, 1.60, 3.20, 6.40, 12.80, 25.60,51.20.
b) Brightness distribution of 3C287 at 18 cm. Beam: 18 x 13 mas (p.a.=5°); tick separation 100 mas; peak: 1.480 Jy/beam; contour (%): -1, -0.5, 0.5, 1, 2, 5, 10, 20, 30, 40, 50, 60, 70, 80, 90.
c) Brightness distribution of 3C343 at 18 cm. Beam: 15 mas^2; peak: 341 mJy/beam; tick separation: 100 mas; contours (%): -1, 1, 2, 5, 10, 20, 30, 40, 50, 60, 70, 80, 90.

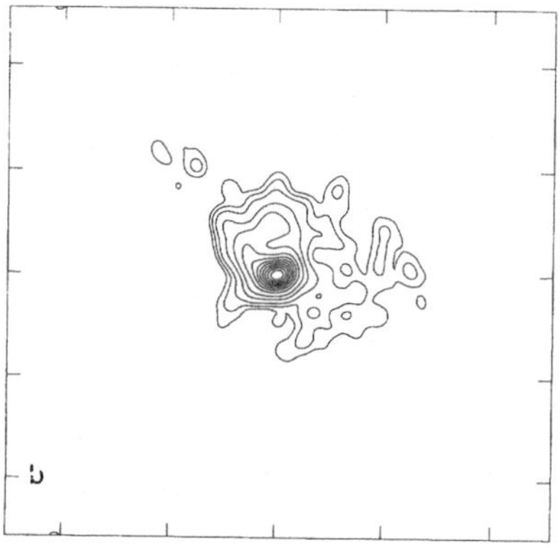

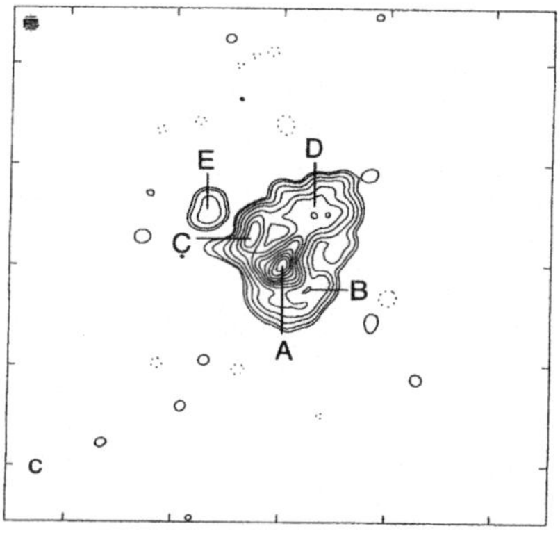

327 MHZ OBSERVATIONS OF 3C 84, 3C 120 AND 1148-001

V. K. Kulkarni and S. Ananthakrishnan
Radio Astronomy Centre (TIFR), P.O. Box 8,
Udhagamandalam -643001, India

We have made VLBI observations of a mixed sample of bright galaxies and quasars at 327 MHz using the Ooty Radio Telescope in India and the telescopes at Jodrell Bank, Westerbork, Torun, and Crimea in December 1983 and March 1986. MK II recording was used and the data were processed at MPI, Bonn.The preliminary results of the 1983 analysis (Ananthakrishnan and Kulkarni 1986) showed most of the nearby galaxies to be resolved over the longest baselines of 8 million wavelengths. The only seyfert galaxies for which good S/N ratio was available at the longest baselines were 3C84 and 3C120. Although UV coverage was poor we attempted to make maps of these objects. The radio source 1148-001 was used for gain calibration of the telescopes. These three sources are described below.

1148-001: This high redshift (z=1.982) flat spectrum quasar is known to be compact at low frequencies (Clarke et al., 1969, Galt et al., 1977 and Venugopal et al., 1985). Therefore it was used to calibrate the gains of the telescopes. The fringe amplitude and closure phase data for the source are consistent with a single component of size $\leq$12 mas. However, it is not clear whether the angular size observed is the intrinsic size of the source or could be due to the effect of interstellar scattering, which at 327 MHz is expected to be ~8 mas.

3C84: A self calibrated map of the source was made (Fig. 1) at the National Image Processing Facility at Ooty using the AIPS Package. A point source was used as a starting model and standard procedure was followed. The map was convolved with a restoring beam of 60x15 mas in PA 32°. A deconvolution of the map gives parameters shown in Table 1. Since the beam was elongated in approximately the same position angle as the source, model fitting was also done to ensure reliability (Table 1). The angular size value is consistent with the observations of Wilkinson et al. (1979) and the component may be the same as Comp. B of the map of Pauliny-Toth et al. (1976), but the flux density is slightly higher, suggesting a flattening of the spectrum. Whether this could be due to free free absorption rather than synchrotron self absorption needs to be investigated further.

3C120: UV coverage of the source was poor, the source being at a low declination and attempts to make a map did not succeed. The model fit gives two

M. J. Reid and J. M. Moran (eds.), The Impact of VLBI on Astrophysics and Geophysics, 121–122.

components (Table 1). The model resembles those of MERLIN maps made at 408 MHz (Browne et al. 1982), but on a much smaller scale.

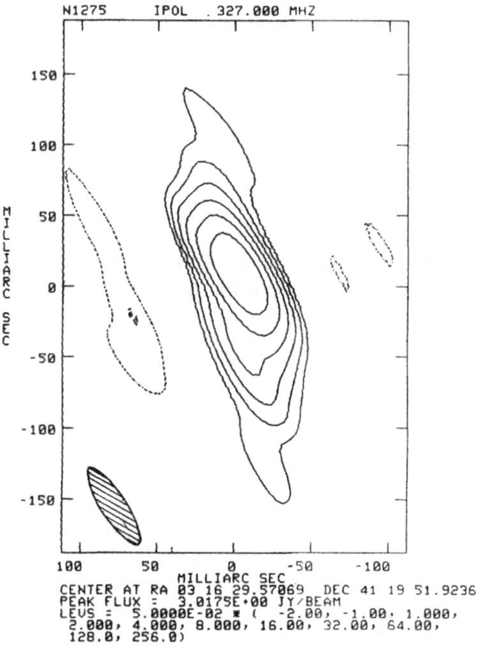

Fig. 1: Map of 3C 84

Table 1: Parameters for 3C84 and 3C120.

Source Name	Flux den. (Jy)	Rad. mas	Theta deg.	Maj. mas	Min. mas	PA deg.
3C84 (map fit)	6.9	0	0	31	20	18
(mod. fit)	6.9	0	0	36	20	18
3C120 (mod. fit)						
Comp. A	0.74	0	0	15	10	0
B	2.65	184	305	95	2*	119

* value uncertain

References:

Ananthakrishnan, S. and Kulkarni, V. K. (1986) Proc. of IAU Symp. **119** on 'Quasars', 169.
Browne, I. W. A. et al. (1982) Nature, **299**, 788.
Clarke, R. W. et al. (1969) Mon. Not. R. Astr. Soc., **146**, 381.
Galt, J. A. et al. (1977) Mon. Not. R. Astr. Soc., **178**, 301.
Pauliny-Toth et al. (1976) Nature, **259**, No. 5538, 17.
Venugopal, V. R. et al. (1985) Mon. Not. R. Astr. Soc., **215**, 685.
Wilkinson, P. N. et al. (1979) Astrophys. J., **232**, 365.

VLBI OBSERVATIONS OF COMPACT EXTRAGALACTIC RADIO SOURCES WITH STEEP SPECTRA

V. K. Kulkarni
Radio Astronomy Centre (TIFR)
P. O. Box 8, Ooty 643001, India

J. D. Romney
National Radio Astronomy Observatory
Edgemont Road, Charlottesville VA 22903-2475, U. S. A.

In recent years it has become clear that the common practice of dividing radio sources into two classes, compact flat-spectrum sources and extended steep-spectrum sources is too simplistic and that compact steep-spectrum sources (CSS) sources which seem to form a separate class in themselves, need to be taken into consideration. A majority of flat-spectrum sources are known to show asymmetric or D2 structures. In contrast to this many CSS sources when observed with subarcsecond resolution show symmetric structures typical of extended steep-spectrum sources. However, further high resolution observations of CSS objects are clearly needed to establish the structural differences, if any, between the two types of compact objects and to understand the interrelationship between the three classes of objects. In this connection we have undertaken a program to observe a sample of 18 CSS sources using the European VLBI network. The sample was selected from the compilation by Peacock and Wall (1982), according to the following criteria: (1) flux density > 1.8 Jy; (2) largest angular size < 1 arc sec; and (3) steep high frequency spectral index (> 0.5) and high degree of low frequency spectral curvature. Our aim was to find out the structures common to these sources and their relation to more commonly observed "centimetre excess" compact objects. Here we will present the preliminary results.

We observed each source in our sample using the European VLBI Network (Onsala, Effelsberg, Dwingeloo/Westerbork, Jodrell) at 18/21 cm, with a resolution of about 30 milliarcsec. The data were reduced at the Max Planck Institute for Radio Astronomy in Bonn. Based on consideration of aperture coverage and signal-to-noise ratio either a map was made or a model fitted to derive structural information about these sources. The maps are presented in Fig. 1.

The observations of our sample are not yet complete. Also we have data at only one frequency. Still based on our preliminary analysis we can draw some general conclusions. Our studies show that a large variety of structures is present. On the basis of our data

M. J. Reid and J. M. Moran (eds.), The Impact of VLBI on Astrophysics and Geophysics, 123–124.

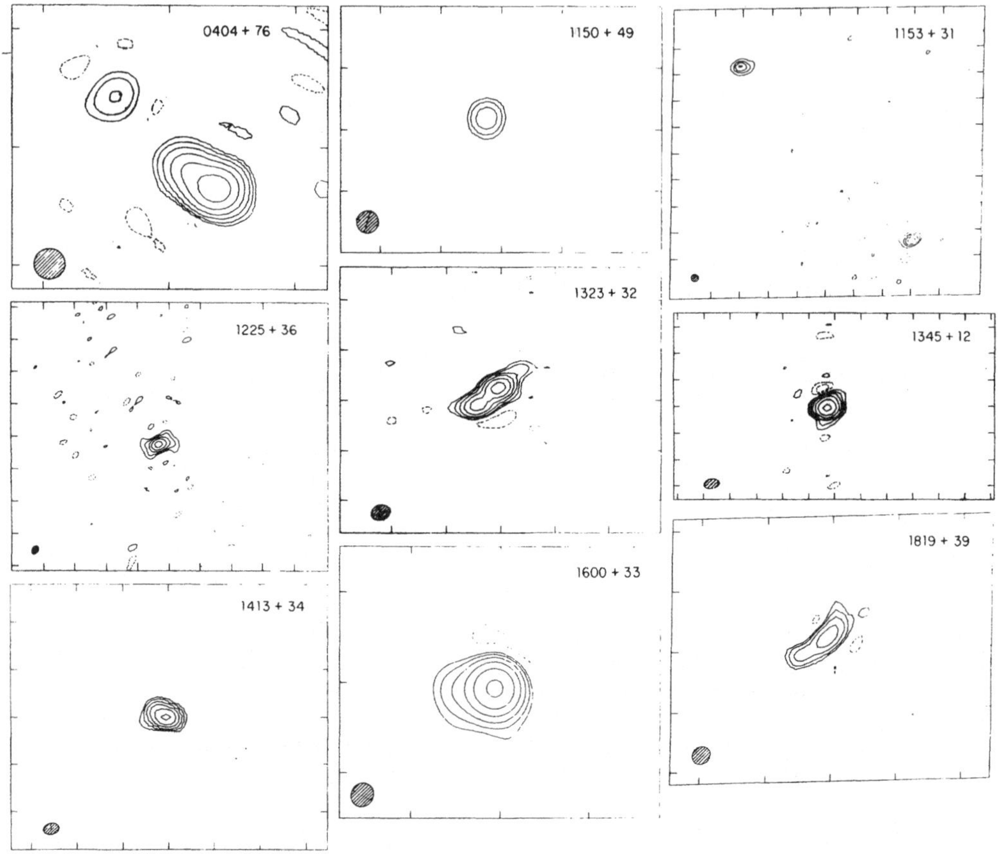

Fig. 1. VLBI maps of steep-spectrum compact sources. Tick marks are at intervals of 100 milliarcsec. Contours, in units of three times the r.m.s. noise in the residual map, are at levels of -2, -1, 1, 2, 4, 8, 16, 32, and 64.

we attempted a rough morphological classification of 14 objects (for which observations are complete) in the present sample. Five sources were found to be double. Six sources were found to consist of single components, in some cases with extensions. Complex structures were found in three sources. Thus compact steep-spectrum sources show a wide range in structures.

REFERENCES

Peacock, J. A., and Wall, J. V., 1982, MNRAS, **198**, 843.

MORPHOLOGY OF STEEP SPECTRUM RADIO SOURCES SHOWING VARIABILITY AT LOW FREQUENCY

Franco Mantovani[1], Tom Muxlow[2] and Lucia Padrielli[1]

1. Istituto di Radioastronomia del C.N.R. - Bologna, Italy
2. Nuffield Radio Astronomy Lab. - Jodrell Bank, England

INTRODUCTION

The observed variability at low frequency of the radio sources can be explained within the framework of the generally accepted models either extrinsic (refractive scintillation in the interstellar medium) or intrinsic (bulk relativistic motion along direction near the line of sight) for variability. Both explanations require a large fraction of the source flux density to be contained in a small high brightness component, of tens of m.a.s. in size. Radio sources with steep straight spectral index are usually tens of Kpc sized, with weak central components and they do not generally show low frequency variability.

Therefore, selecting low frequency variable sources with straight steep spectral index, it would mean to select compact steep spectrum sources characterized by: i) physical size of few Kpc; ii) absence of dominant flat spectrum radio core.

This contribution deals with 8 of these sources.

VARIABILITY

All but two sources show high frequency variability generally much lower than the low frequency one or even absent. This seems to be a point in favour of an extrinsic cause of the low frequency variability as suggested by Rickett (1986). He pointed out that a correlation exists between the variability parameters, galactic coordinates, interstellar medium parameters. Following Fanti et al. (1987) we have figured out $m\tau$ (m, fractional rms variability; τ, the short and long variability time scale), θ_s the scattering angular size, θ_o a parameter depending on the medium characteristics and L, the thickness of the scattering medium, (Tab. 1).

θ_i is the source angular size, as derived from VLBI observations mainly at frequencies lower than 1 GHz; b is the sources galactic latitude.

1524-136 shows two values of $m\tau$ which are very different indeed. This is not expected if the refractive scintillation is the mechanism responsible for the low frequency variability. Moreover, 1524-136 at 50 cm shows an extended structure with lack of compact component, (Fig. 1).

M. J. Reid and J. M. Moran (eds.), The Impact of VLBI on Astrophysics and Geophysics, 125–126.

Table 1.

Name		S(Jy) 408MHz	b deg	θi m.a.s.	m τ (days) s	l	θs m.a.s.	θo	L pc
0320+053	DA101	6.5	-41	< 25	10.5	16.3	1.5	2.6	876
0358+004	3C99	5.0	-37	< 30	4.4		1.4	1.5	237
0518+165	3C138	16.4	-11	< 45	17.5		1.0	1.0	713
0548+165	DA190	4.2	- 5	< 25	52.1		1.2	1.4	255
1422+202	4C20.33	5.1	67	< 10	3.3		0.4	0.7	1724
1524-136	OR-140	6.1	34	100	2.5	38.8	9	27	35
2033+181		3.3	-13	< 30	10.5		5.2	5.0	533
2147+145		6.5	-30	< 30	7.3		1.2	1.6	517

The values figured out for θo suggest that a better determination of this parameter, assumed 8 mas by Rickett, is needed.

COMPACT STEEP SPECTRUM SOURCES

Four sources in our list have measured red-shift. For the remaining sources we have assumed z=0.5.

All sources show Linear Diameters smaller than that of an optical galaxy. This suggests we are dealing with compact steep spectrum sources where the radio morphology might be determined by propagation of jets through the interstellar medium, (see Tab. 2).

Table 2.

Name		Fringe Vis. 430 MHz	1660	z	O.I.	Lin. Diam. Kpc
0320+053	DA101	0.55	0.4	0.5	G	< 0.1
0358+004	3C99		0.3	0.426	G	8
0518+165	3C138	0.1	0.9	0.76	Q	4
0548+165	DA190	0.51	0.6			
1422+202	4C20.33		0.3	0.87	Q	40
1524-136	OR-140		0.4	1.687	Q	35
2033+181			0.6	0.5	EF	1
2147+145		0.33	0.6	0.5	EF	1

Fig. 1

CONT.LEV.=5.0 10.0 20.0 30.0 40.0 50.0 70.0 90.0 %

DISTANCE BETWEEN TICKS 200.0 M.A.S.

BEAM = 200 X 60 (8)

MAX. 1032.9 MJY

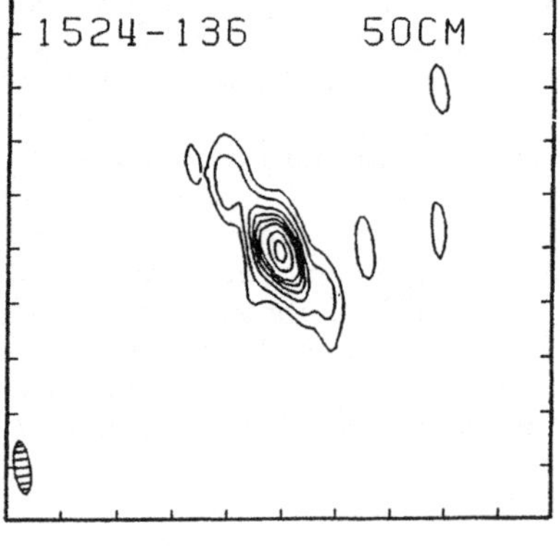

REFERENCES

Fanti R. et al., 1987, Superluminal R.S., ed. by J.A. Zensus, T.J. Pearson

Rickett B.J., 1986, Ap.J. 307,564

3C236 - THE GIANT RADIO GALAXY WITH A COMPACT STEEP SPECTRUM NUCLEUS AND A COMPACT TWO-SIDED JET

R.T. Schilizzi[1], E.D. Skillman[1], G.K. Miley[2], P.D. Barthel[3], J.M. Benson[4] and T.W.B. Muxlow[5]

1. Netherlands Foundation for Radio Astronomy, Dwingeloo, NL.
2. Space Telescope Science Institute, Baltimore, MD, USA.
3. California Institute of Technology, Pasadena, CA, USA.
4. National Radio Astronomy Observatory, Charlottesville, VA, USA.
5. Nuffield Radio Astronomy Laboratories, Jodrell Bank, UK.

3C236 is the largest radio source known. It has a 39 arcmin double structure (3.0 Mpc, Ho=75 kms^{-1} Mpc^{-1}) in which the SE lobe is narrow and edge-brightened and the NW lobe more diffuse and centre-brightened (Barthel et al., 1985). About half the flux density comes from a steep spectrum (α=-0.7) radio core of overall size 1.3 arcsec (2.2 kpc) located in the centre of a 17m elliptical galaxy with redshift 0.0988.

Figure 1 displays the λ18 cm radio structure of the core derived from a combined 16-station MK2 VLBI/6-station MERLIN observation made in April/May 1984. Figure 2 shows the structure of the south-east component in more detail.

We identify the nucleus with the brightest component in the centre of the structure in Figure 2. It is the most compact component and has an inverted spectrum (α_{18}^{6}). Low level emission can be seen to the north east and south west of the nucleus. The ridge-line of the south-east jet oscillates about PA 115° with a period of ~115 mas (~200 pc). It passes through a transvere structure ~55 mas (90 pc) in dimension at a distance of 135 mas (230 pc) from the nucleus. The north-west jet is not as well defined as its counterpart to the south-east. However, the cocoon or sheath emission can be seen clearly, as can the first stages of the jet close to the nucleus. 77 mas away from the nucleus in PA ~55° lies the second brightest feature in the whole south-east component. It lies to the northern edge of the cocoon and is resolved.

Figure 1 shows the north-west component is a quite relaxed structure whose steepest intensity gradient is to the north-east near the warmspot (15 mJy at 18 cm). It subtends an angle of ~25° at the nucleus, substantially larger than the north-west lobe of the large-scale structure (14°). If the nuclear jet enters the north-west component via the warmspot, one could interpret the extension of the component as due to diffusion. Alternatively if the north-west component defines a wide cone of relatively decollimated emission,

M. J. Reid and J. M. Moran (eds.), The Impact of VLBI on Astrophysics and Geophysics, 127–128.

recollimation will need to occur before the beam reaches the large-scale north-west lobe.

The identification of the nucleus with the brightest component in the SE structure leads to the conclusion that 3C236 has a compact two-sided jet, with approximately equal brightness per beam on each side. It is thus unlikely that superluminal motion (slm) will be observed in this source if slm is due to relativistic beaming close to the line of sight.

The oscillations in the SE and NW jets and in the NW large-scale lobe may be due to helical Kelvin-Helmholtz instability modes, or perhaps precession of the ejection axis in the nucleus.

The transverse structure seen in the SE jet may be evidence of a shock in the energy flow. Observations at 6 cm (and polarisation at 18 cm?) may help elucidate its nature.

The radio emission in the nucleus occurs on a scale commensurate with that expected for the narrow line region (few kpc). Thereafter no emission is seen for distances of the order of 1 Mpc until the large scale lobes are encountered. This suggests that the energy transport becomes more efficient (so that the jet is no longer visible) approximately 1 kpc from the nucleus. This may mark the outer boundary of the narrow-line region in this galaxy.

REFERENCE

Barthel, P.D. et al. (1985) A. and A. 148, 243.

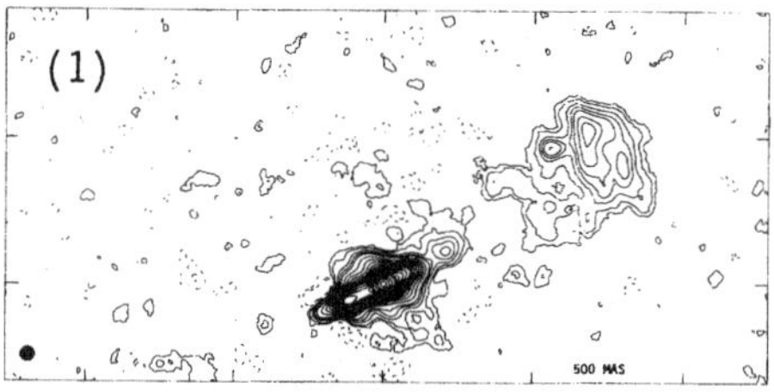

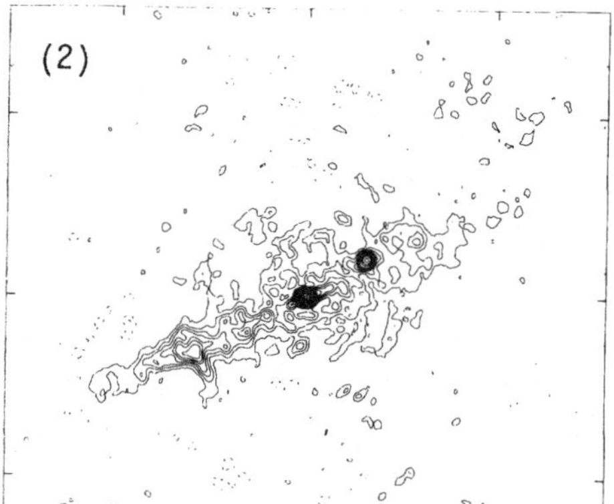

Figure 1: 3c236 at 18 cm; beam = 50 mas. Peak flux density is 372 mJy/beam, and the noise in the map is about 1.5 mJy/beam. Contours -5, -2.5, 2.5, 5, 10, 15, 20, 30, ..., 100, 120, ..., 200, 250, 300, 350 mJy/beam. The total extent of the central source is 2.2 kpc.
Figure 2: As for Figure 1 with a 10 mas beam and the NW component subtracted from the U-V data. Peak flux density is 218 mJy/beam, and the noise in the map is about 0.7 mJy/beam. Contours -4, -2, 2, 4, 6, 8, 10, 12, 15, 20, 25, ..., 50, 60, ..., 100, 125, 150, 200 mJy/beam.

VLBI STRUCTURE OF THE QSO 3C286

A. J. Kus, TRAO, Torun, Poland
A. Marecki, TRAO, Torun, Poland
S. Neff, NASA/GSFC, Washington, DC, USA
A. van Ardenne, Radiosterrenwacht, Dwingeloo, Netherlands
P. N. Wilkinson, NRAL, Jodrell Bank, United Kingdom

1. INTRODUCTION

The steep, small angular size QSO 3C286 ($z = 0.846$, $mv = 17^m$) (Burbidge and Burbidge 1969) is a bright object in the radio band, which has a steep spectrum ($\alpha = -0.65$, $S \sim f^{\alpha}$) from 1 GHz, peak in spectrum at about 100 MHz, and self-absorption below 50 MHz. 3C286 is a member of a well established class of radio sources (the CSSQSO) and is similar to 3C147, 3C309.1, and 3C380 in general.

VLBI hybrid maps of 3C286 were first made by Wilkinson *et al.* (1979) at 609 GHz, Pearson *et al.* (1980) at 1671 MHz, and Simon *et al.* (1980) at 327 MHz. The estimate of the overall angular size and the halo component contribution were suggested to be at least 100 mas and a third of the total flux at 609 MHz, respectively. The scintillation measurements (Readhead and Hewish 1974) suggest a halo component of the order of 0.6×0.3 arcsec (Wilkinson *et al.* 1979).

The VLBI maps mentioned earlier made with the U.S. or global array do not contain information about a structure larger than 50 mas. The ideally suited array for this purpose is the EVN at 609 MHz, which is supplementary to the MERLIN, the U.S., and global VLBI networks.

2. PRESENT OBSERVATIONS

The first test observation with the EVN at 609 MHz was conducted in August 1982 with the Mark I antenna at Jodrell Bank, one antenna at Westerbork, and, for the first time, the 15-m diameter telescope of the Torun Radio Astronomical Observatory (TRAO) at Torun, Poland. The purposes of the test were: (1) to measure the extended structure of 3C286, and (2) to check the performance of the network. The Mark IIc terminals were used. Jodrell Bank had a LHC polarized feed whereas Westerbork and Torun had linearly polarized feeds. The feed polarizations led to some calibration difficulties since 3C286 is highly polarized. Out of the 12 hours of observing time, we observed 3C273B as a calibrator for one hour, and four hours were lost due to TV interference at Torun. The tapes were correlated at MPIfR, and the hybrid mapping using the Jodrell Bank mapping package was done at TRAO, Torun. The 609 MHz map is shown in Fig. 1. For comparison, see the maps made at 327, 609, and 1671 MHz (Simon *et al.* 1980).

The 1420 MHz, 12-hour observing run with four EVN telescopes was performed in June 1984. In this project, the following antennas were used: the Mark I

M. J. Reid and J. M. Moran (eds.), The Impact of VLBI on Astrophysics and Geophysics, 129–130.

at Jodrell Bank, full array of the WSRT, Onsala and Torun. The data were processed at the MPIfR correlator in Bonn, and the hybrid mapping computation was initially started at Bonn but was finished in Torun. The resulting map is presented in Fig. 2. It shows core domination and a one-sided, curved jet about 150 mas long. The beam size is 23×28 mas^2. We account for all of the flux density observed at 1420 MHz.

3. GENERAL CONCLUSIONS

(1) The size of the source is 250×300 mas^2 (from the 609 MHz map), at 1420 MHz. The Gaussian component in the center dominates, the jet extends at least 150 mas (750 h^{-1} pc), the jet width is between 10 and 15 mas (50–75 h^{-1} pc). (2) Straight 80 mas (400 h^{-1} pc) jet at PA=222° curves rapidly within 70–90 mas (50–75 h^{-1} pc) from the core to the PA=260° toward the arcsec component (van Breugel *et al.* 1984). (3) No counter jet is observed ($T_{bjet}/T_{bcjet} \geq 200$). (4) Spectral index steepens gradually along the jet from 0.0 to -1.2. (5) Assuming domination of the central component (deconvolved size 30×12 mas^2), we find that the expected IC x-ray flux in the $0.1 - 3.0$ keV range is low ($\sim 10^{-16}$ erg cm^{-2} sec^{-1}), $B_{me} \approx 10$ mG, $P_{me} \approx 10^{58}$ ergs, and the surface brightness temperature is only 2×10^{10} K. 3C286 therefore seems to be not as energetic as the other 3C sources of the class.

4. REFERENCES

Burbidge, G. R., and Burbidge, E. M. 1969, *Nature*, **222**, 735.
Pearson, T. J., *et al.* 1980, *Ap. J.*, **236**, 714.
Readhead, A. C. S., and Hewish, A. 1974, *M. N. R. A. S.*, **78**, 1.
Simon, R. S., *et al.* 1980, *Ap. J.*, **236**, 707.
van Breugel, W. J. M., *et al.* 1984, *A. J.*, **89**, 5.
Wilkinson, P.N., *et al.* 1979, *Ap. J.*, **232**, 365.

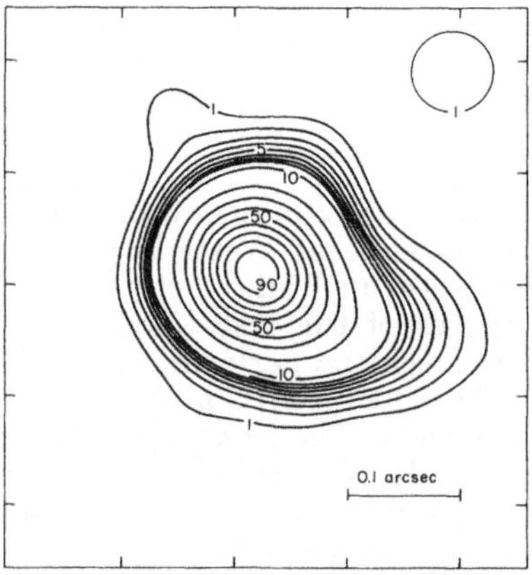

Figure 1. Hybrid map of 3C236 at 609 MHz (Jodrell–Westerbork–Torun). $S_{max} = 18.5$ Jy. $S_{TOTAL} = 27$ Jy. Beam = 90 mas.

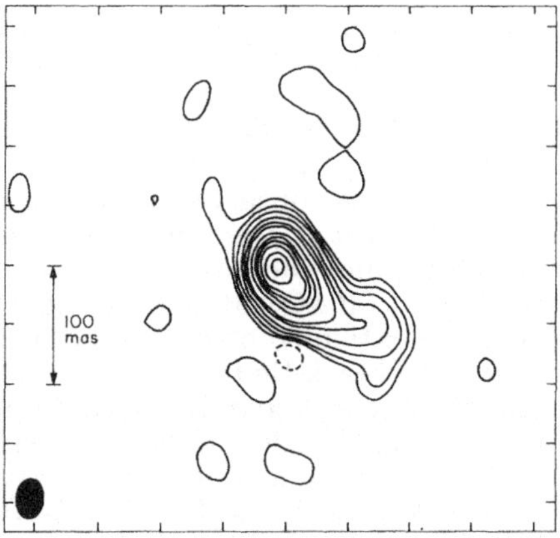

Figure 2. Map of 3C286 at 1420 MHz (Jodrell–Westerbork–Torun–Onsala). $S_{max} = 7.5$ Jy. $S_{TOTAL} = 15.7$ Jy.

HIGH DYNAMIC RANGE MAPPING OF A COMPLEX RADIO SOURCE - M87

T.W.B. Muxlow, W.Junor, R.E.Spencer, R.Simon, J.Benson & M. Reid
University of Manchester
Nuffield Radio Astronomy Laboratories
Jodrell Bank, Macclesfield, Cheshire, UK

The well known jet in M87 (Virgo A) has been extensively studied in the radio regime using conventional arrays with resultant resolutions in the range 0.22-1 arcsecond. We present here the results of a VLBI experiment to map the jet over its entire length (30 arcseconds) to high resolution. The observations were made in April 1984 at 1.67 GHz with a 'World Array' containing 18 VLBI telescopes. In addition, simultaneous observations were performed with the 6 station MERLIN array.

The spatial frequency coverage together with surface brightness constraints and computational limitations have restricted the use of the full resolution data set (5 mas) to the inner few arcseconds. The full ($\sim$30 arcsecond) jet can be studied at 75 mas resolution: and for this image we have used only the data from the MERLIN and European stations. The data were globally fringe-fitted and calibrated within the AIPS package. Subsequent telescope phase and gain corrections were performed with the difference-mapping program (1) of the Jodrell Bank OLAF system. This method provides a saving in computational resources by Fourier transforming only the difference between the current source model structure (the transform of the clean components) and the corrected UV data set. New clean components are obtained from this residual map and are then added into the current model. A full transform of all the clean components at each correction cycle is thus avoided. Telescope corrections are made in the usual way (2) every cycle. Interactive graphics facilitate user control at each major cycle; thus regulating the growth of the model structure where required. The corrected data were then mapped within the AIPS package to provide a new starting model for further difference-mapping cycles. Finally, the data were corrected for non-closing (baseline) errors. The overall mapping scheme is shown in Figure 1.

Figure 2 shows a contour plot of M87 at 1.67 GHz with a FWHM beam size of 75 mas. Detailed structure within the knots can be seen. The transverse ridge structure of knot A is clearly visible and is suggestive of an oblique shock. The deconvolved thickness of the leading edge is $\sim$160 mas, and that of the more complex trailing edge 300-400 mas; similar to those values found in the 15 GHz VLA image (3). Knot I, immediately preceding A, appears to be a precursor and is joined to A

M. J. Reid and J. M. Moran (eds.), The Impact of VLBI on Astrophysics and Geophysics, 131–132.

by a 'cone' of emission. Knot D is extended transverse to the jet with a deconvolved width of ∿200 mas, consistant with a constant jet opening angle between the core and knot A. Knot B, which follows Knot A, is complex in form although there is some suggestion of helical structure.

The features in the jet have in the past been explained in terms of shocks and Kelvin-Helmholtz instabilities. However, it is clear that the structure is so complicated that neither of these explanations can alone be responsible for the observed morphology.

(1) Cornwell, T.J. & Wilkinson, P.N., 1983 'Indirect Imaging', ed. J.A. Roberts, Cambridge University Press.

(2) Cornwell, T.J. & Wilkinson, P.N., 1981. Mon.Not.R.astr.Soc., 196, 1067.

(3) Biretta, J.A. Owen, F.O. & Hardee, P.E., 1983. Astrophys.J. 274, L27.

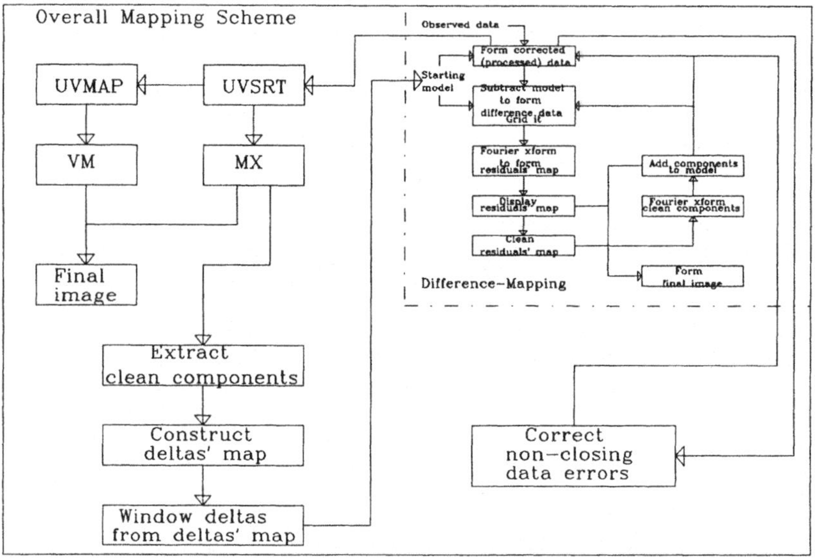

Figure 1 (left)

Figure 2 (below)

Virgo A (beam=75mas)

Peak=0.364 Jy/pixel

pixel spacing=25mas

Contour values:

1,2,3,4,5,6,7,8,9,10,11,12

13,14,15,16,17,18,19,20

50,100,200,500,800

x 200uJy/pixel

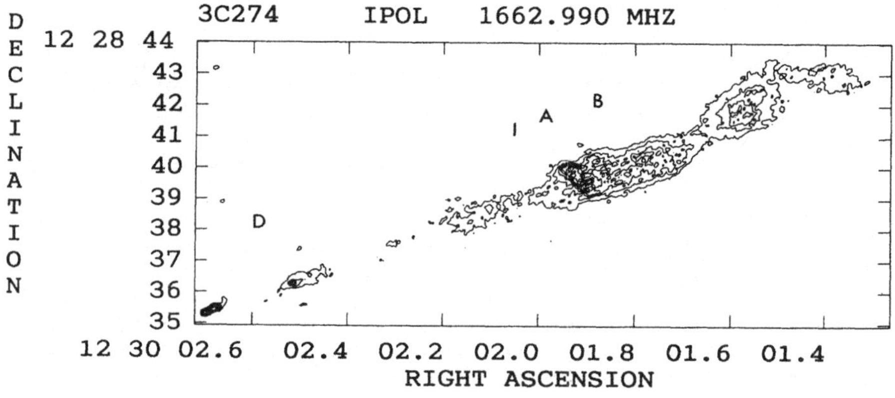

Observations of the cores of extended quasars

J.R. Hooimeyer Sterrewacht Leiden, the Netherlands
P.D. Barthel California Institute of Technology, Pasadena USA
R.T. Schilizzi NFRA, Dwingeloo, the Netherlands

A critical test of the simple relativistic beaming models is provided by the search for structural changes in the weak cores of lobe-dominated quasars. In these sources, which are expected to be close to the plane of the sky, superluminal motion should be comparatively rare. We are engaged in the study of a complete sample of 30 extended quasars (Barthel et al.1984), ranging in projected size from $40h^{-1}$ to $200h^{-1}$ kpc (H_o = 100h, q_o = 0.05). High resolution multi-epoch observations have been obtained for two objects, the quasars 1721+343 and 0742+318, at 5 GHz (1982.3, 1983.3, 1986.5) and 10.7 GHz (1986.2, 1986.5). Superluminal expansion with transverse velocities of $3.1h^{-1}$ c has been detected in the core of 1721+343, which is the largest quasar known (Barthel et al., these proceedings). No superluminal motion was found in 0742+318. The large and small scale structure of 1721+343 are shown in fig.1a,b. The spectra of the individual components point to component A as the core ($\alpha_5^{10.7} = -0.3$).
First epoch transatlantic 5 GHz observations have been obtained of the mas-structure of the quasars 1055+201, 1222+216, 1830+285, 2209+080 and 2251+134. The cores of 1222+216 and 1830+285 show the familiar asymmetric core-jet structure. The cores of 1055+201 and 2251+134 are Gaussian-shaped and elongated in the direction of the extended structure. The mas-structure of 2209+080 was too weak to be mapped with our observations.
The morphological characteristics of the quasars that have been studied can be summarized as follows:

- The core structures are unresolved or core-jet, similar to the morphologies observed in compact radio sources in general.
- In all sources the mas-scale emission joins smoothly with the outer lobe structure. There is no evidence for bending of the radio structure on pc-scales.
- Three of the four sources in the original sample for which multi-epoch VLBI observations, by ourselves and others, are available, are found to display superluminal expansion: 0850+581 (Barthel et al. 1986), 1137+660 (Zensus et al. 1986), and 1721+343 (Barthel et al., these proceedings). Superluminal motion is apparently not restricted to compact sources, and actually occurs surprisingly often in objects of very large size.

M. J. Reid and J. M. Moran (eds.), The Impact of VLBI on Astrophysics and Geophysics, 133–134.

The occurence of superluminal motion in quasars of large projected sizes presents a challenge to beaming models. Several mechanisms have been suggested to overcome the deprojection problem, such as changing the position angle of the ejection axis between core and extended structure, or postulating the jet to have a wide cone angle initially and be collimated further out. For 1721+343 however, with its high degree of linearity and narrow kpc-scale jet, neither mechanism seems very likely. Orientation-independent models decoupling the observed radiation from the bulk particle flow (e.g. screen models or oblique shock models (Lind and Blandford 1985)) may be able to account for the superluminal motion in very extended superluminals more easily; we feel that these models deserve more attention.

References:

Barthel,P.D., Miley,G.K., Schilizzi,R.T., Preuss,E. 1984 *Astr. Astroph.* **140**, 399

Barthel,P.D., Pearson,T.J., Readhead,A.C.S., Canzian,B.J. 1986 *Ap.J.Lett.* **310**, L7

Lind, K.R., Blandford, R.D. 1985 *Ap.J.* **295**, 358

Zensus, J.A., Porcas, R.W. 1986 *Proc. I.A.U.Symposium* **119** 167

Fig.1: kpc and pc-scale structure of 1721+343
(a) λ20 cm VLA map
(b) λ2.8 cm VLBI map

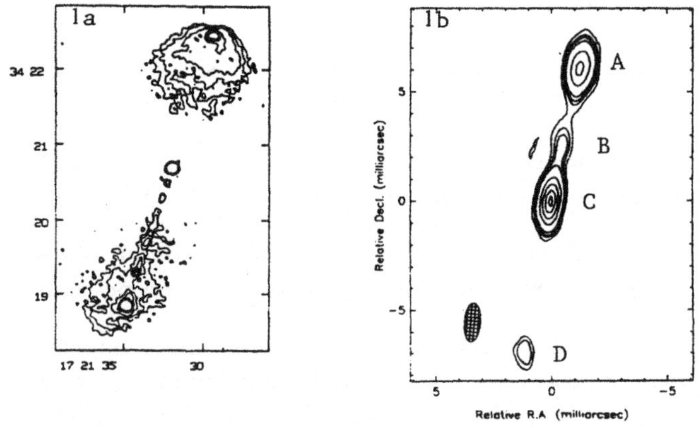

The Working Surface of a Powerful Radio Source: Global MkIII Observations of the Southern Lobe of 3C205.

Colin J. Lonsdale
Haystack Observatory, Westford MA 01886

Peter D. Barthel
California Institute of Technology, Pasadena, CA

The radio source 3C205 is identified with a quasar of redshift 1.53, and is, at first glance, a standard, powerful double radio source with hotspots. Upon closer examination (MERLIN, VLA A-array), the southern lobe is seen to consist of two distinct hotspots plus some more diffuse emission, some 50 kiloparsecs from the quasar ($H_o \sim 75$). The arcsecond-scale morphology and polarization structure of this hotspot complex strongly suggests that the larger and weaker off–axis secondary hotspot has been formed by outflow of material from the more compact primary hotspot (Lonsdale and Barthel 1984,6). This interpretation implies a collision between the presumed energy supply beam and something capable of deflecting the flow energy towards the secondary. Because the flow is so energetic, this obstacle would have to be both massive and dense. The main constraint on the density of the obstacle comes from measurements of the internal energy density of the compact primary hotspot coupled with model–dependent estimates for the maximum advance velocity of this feature, thus yielding a balance with the ram–pressure exerted by the obstacle medium.

Previous VLBI measurements of the primary hotspot were presented in Lonsdale and Barthel (1984), and consisted of a 4–station EVN Mark II 18cm hybrid map at 30 milliarcsecond resolution with approximately 10:1 dynamic range. The structure thus revealed consisted of a ridge of emission some 50 milliarcseconds in length and less than about 15 milliarcseconds wide, oriented roughly along the line joining the primary and secondary hotspots. There was also evidence for curvature of the ridgeline at one end towards the quasar, which gave the subjective impression of continuity from the quasar, through the primary hotspot to the secondary hotspot.

In order to probe the unresolved emission further, we conducted a global 7–station Mark III experiment at 18cm, wherein the longest baselines had a fringe spacing of 4 or 5 milliarcseconds. Fringes were detected on all the short baselines (inter–Europe, Haystack–Greenbank, OVRO–VLA), and intermittently on the most sensitive longer baselines (Greenbank–VLA and all transatlantic baselines except those involving Westerbork, Haystack and OVRO). A map made from the shorter baselines is shown, and has a restoring CLEAN beam of FWHM 15 milliarcseconds. On the longer baselines, the source is detected clearly over only

M. J. Reid and J. M. Moran (eds.), The Impact of VLBI on Astrophysics and Geophysics, 135–136.

a small range of hour angles, and during that time the visibility rises steeply to between 5 and 20 mJy. This behaviour has been fitted to a narrow ridge of length 10 milliarcseconds and axial ratio 5, in the same position angle as the main ridge in the map. What this amounts to is a faint but extremely narrow (probably still unresolved) linear feature pointing towards the secondary hotspot.

The preliminary results presented here were obtained without the benefit of sophisticated fringe–fitting and mapping techniques. We anticipate that the use of a global fringe-fitting algorithm and careful data editing will allow detection of fringes in a larger fraction of the long baseline data than we can presently claim, and that we will be able to produce a reliable hybrid map of this feature at 4 milliarcsecond resolution. Much of the data is clearly close to the detection threshold, as the closure phases appear to be non–random for substantial stretches.

The implications of these results for the interpretation of 3C205 include the need to postulate extremely dense clouds of matter more than 50 kiloparsecs from the parent galaxy. The nature of the deflection interaction should be constrained by the presence of the compact ridge, which some may choose to interpret as the long–hypothesized "working surface" where the jet impinges on the external medium (in this case a dense cloud). Finally, the detection of milliarcsecond structure in the outer lobe of a quasar raises the possibility of directly measuring the relative motion between the hotspot and core, one of the more important unknown quantities in radio source physics.

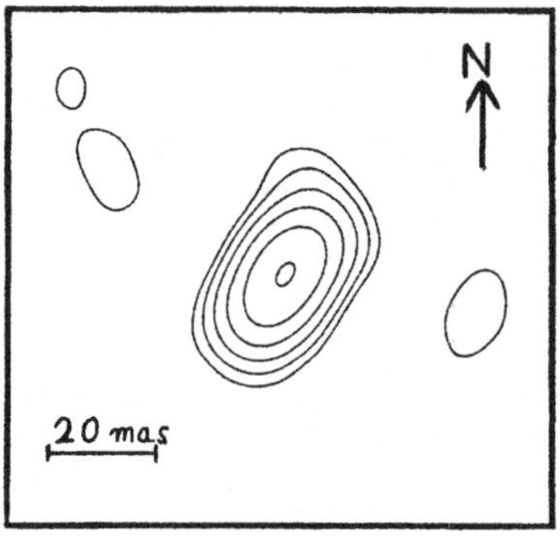

Fig. 1. MkIII VLBI map of 3C205 southern hotspot. The peak is about 50 mJy/beam, and the resolution is about 15 milliarcseconds. The bottom contour is 3% of the peak. This is a preliminary map utilizing only the shorter baselines (see text).

REFERENCES

Lonsdale C.J. and Barthel P.D., 1984, Astron. Astrophys. **135**, 45.
Lonsdale C.J. and Barthel P.D., 1986, Astron J. **92**, 1.

THE HIGH RESOLUTION STRUCTURE OF THE CENTAURUS A NUCLEUS AT 2.29 GHZ AND 8.42 GHZ

D.L. Meier[1], D.L. Jauncey[2], R.A. Preston[1], A.K. Tzioumis[3], A.E. Wehrle[4], R.A. Batchelor[2], J. Falkner[5], P.A. Hamilton[6], B.R. Harvey[7], R.F. Haynes[2], B. Johnston[8], A.P. Louie[1], P. McCulloch[6], G. Moorey[2], D.D. Morabito[1], G.D. Nicolson[9], A.E. Niell[10], J.A. Roberts[2], J.G. Robertson[11], G.W.R. Royle[6], L.J. Skjerve[1], M.A. Slade[1], O.B. Slee[2], A. Watkinson[12], and A. E. Wright[2]

[1]Jet Propulsion Laboratory, California Institute of Technology, Pasadena, California, 91109
[2]Division of Radiophysics, CSIRO, Epping, Australia
[3]School of Physics, University of Sydney, N.S.W., 2006, Australia
[4]Department of Astronomy, University of California, Los Angeles, California, 90024
[5]Department of Astronomy, University of Southern California, Los Angeles, 90007
[6]University of Tasmania, Hobart, Tasmania 7001, Australia
[7]School of Surveying, University of N.S.W. Kensington, N.S.W. 2033, Australia
[8]Ford Aerospace and Communication Corporation, Goldstone, California
[9]CSIR, National Institute for Telecommunications Research, Johannesburg, South Africa
[10]Haystack Observatory, NEROC, Westford, MA 01886
[11]Anglo-Australian Observatory, Epping Laboratory, Epping, NSW 2121, Australia
[12]School of Electrical Engineering, University of Sydney, N.S.W., 2006, Australia

ABSTRACT: VLBI observations of the nucleus of Centaurus A were made in April, 1982 at two frequencies with an array of five Australian radio antennas as part of the Southern Hemisphere VLBI Experiment (SHEVE). Observations were undertaken at 2.29 GHz with all five antennas, while only two were operational at 8.42 GHz. The 2.29 GHz data yielded significant information on the structure of the nuclear jet. At 8.42 GHz a compact unresolved core was detected as well.

We find that the source consists of the compact self-absorbed core and a 100 milliarcsecond jet containing a set of three knots with a very long, narrow component elongated along the same position angle as the knots. The position angle of the jet is in excellent agreement with that of the radio and X-ray structure on arcsecond and arcminute scales. The jet has brightened at 2.29 GHz by a factor of nearly three since the early 1970s, 30% of which occurred in the 1980-82 period. The present data do not distinguish between superluminal and subluminal motion in the core.

M. J. Reid and J. M. Moran (eds.), The Impact of VLBI on Astrophysics and Geophysics, 137–138.

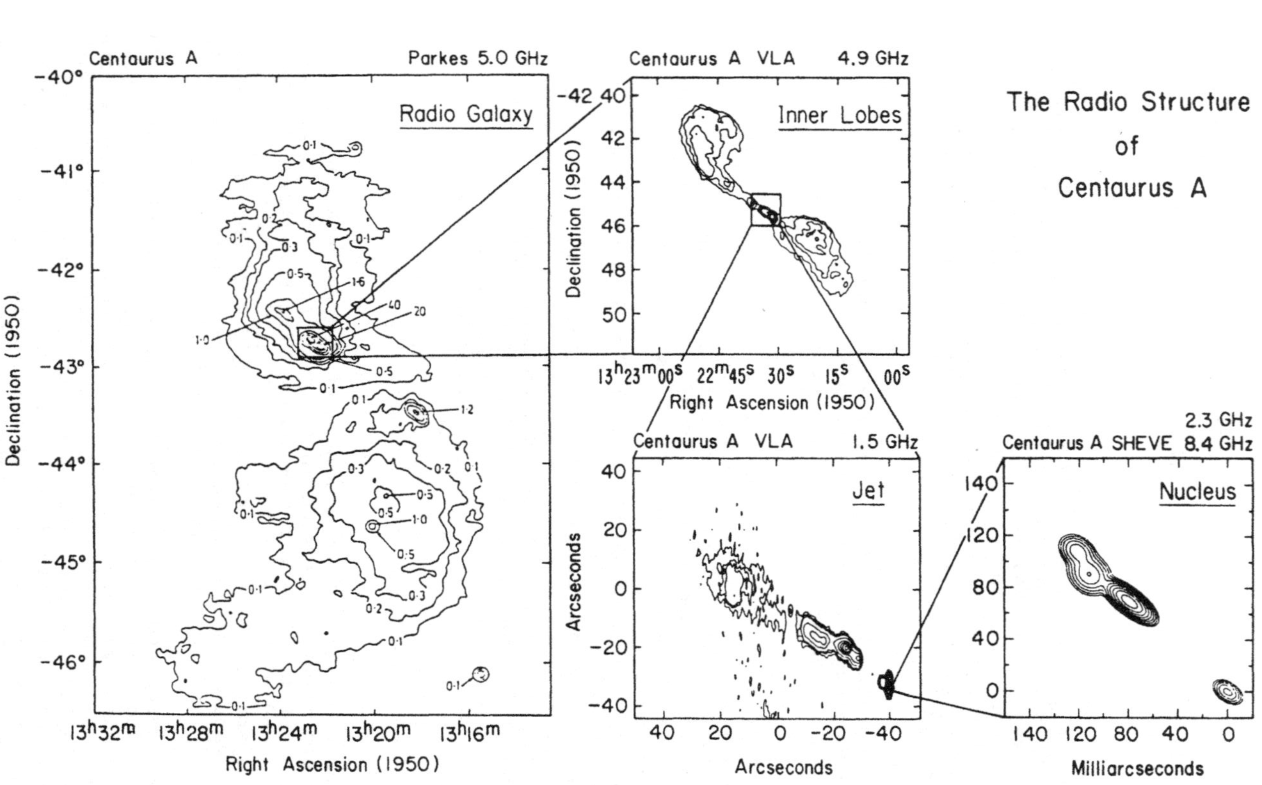

FIGURE 1. Montage of Centaurus A Radio Observations. Scales very over a factor of 10^6. First panel displays 5.0 GHz data from Haynes, Cannon, and Ekers (1983, *Proc. Astron. Soc. Australia,* **5**, 241). The second and third panels show VLA data from Burns, Feigelson, and Schreier (1983, *Astrophys. J.*, **273**, 128). The fourth panel shows our *dual frequency model* which was derived by fitting both 2.29 GHz and 8.42 GHz data simultaneously. The registration of the 2.29 GHz knots with the 8.42 GHz core is uncertain by about ±4 milliarcseconds along the jet. We do not show the long, slender jet component as its registration is uncertain.

THE SOUTHERN HEMISPHERE VLBI EXPERIMENT (SHEVE)

R.A. Preston[1], D.L. Jauncey[2], D.L. Meier[1], A.K. Tzioumis[3], J. Ables[2], R.A. Batchelor[2], J. Faulkner[4], B. Greene[5], P.A. Hamilton[6], B.R. Harvey[7], R.F. Haynes[2], B. Johnson[8], K. Lambeck[9], A.P. Louie[1], P. McCulloch[6], G. Moorey[2], D.D. Morabito[1], G.D. Nicolson[10], A.E. Niell[11], J.A. Roberts[2], J.G. Robertson[12], G.W.R. Royle[5], L.J. Skjerve[1], M.A. Slade[1], O.B. Slee[2], A. Stolz[7], A. Watkinson[3], A.E. Wehrle[13], and A.E. Wright[2].

[1]Jet Propulsion Laboratory, Pasadena, California
[2]CSIRO, Sydney, Australia
[3]School of Physics, Sydney University, Sydney, Australia
[4]Department of Astronomy, University of Southern California, Los Angeles, California
[5]Division of National Mapping, Canberra, Australia
[6]University of Tasmania, Hobart, Australia
[7]School of Surveying, University of New South Wales
[8]Ford Aerospace and Communications Corporation, Goldstone, California
[9]Research School of Earth Sciences, Australian National University, Canberra, Australia
[10]Hartebeesthoek Radio Astronomy Observatory, Johannesburg, South Africa
[11]Haystack Observatory, Westford, Massachusetts
[12]Anglo-Australian Observatory, Sydney, Australia
[13]Department of Astronomy, University of California, Los Angeles, California

ABSTRACT

Six radio telescopes were operated as the first southern hemisphere VLBI array in April and May 1982. Observations were made at 2.3 and 8.4 Ghz. This array produced VLBI images of 28 southern hemisphere radio sources, high accuracy VLBI geodesy between southern hemisphere sites, and sub-arcsecond radio astrometry of celestial sources south of declination -45 degrees. This paper discusses only the astrophysical aspects of the experiment.

Observations were performed for a two week period between 20 April and 3 May 1982. Three of the six sites -- Tidbinbilla (NASA Deep Space Station 43), Parkes, and Fleurs -- form an approximately equilateral triangle with antenna separations of about 250 km. The other two Australian antennas, Hobart and Alice Springs, are separated from the small triangle by about 1000 km and 2000 km, respectively. The sixth antenna, Hartebeesthoek, is located in South Africa about 9700 km from the small triangle.

All six antennas participated in the fine-scale mapping of celestial radio sources. The data from the five Australian antennas were used to

M. J. Reid and J. M. Moran (eds.), The Impact of VLBI on Astrophysics and Geophysics, 139–140.

produce hybrid maps and models of sources with minimum fringe spacings of about 12 milliarcseconds and an angular resolution of about 5 milliarcseconds, an improvement of almost four orders of magnitude over existing southern hemisphere radio mapping instruments.

The baselines from the five Australian antennas to the South African antenna yielded fringe spacings of about 3 milliarcseconds. These baselines provided information on the extremely fine scale structure of each source. These data were included directly in the modelling of the sources but not in the hybrid mapping due to the disparity in length between the baselines to South Africa and those within Australia, as well as due to the limited u-v coverage obtained with the former.

The compact structure of 27 southern hemisphere sources was investigated at 2.3 GHz. These sources included Centaurus A (the nearest active galaxy; see Meier et al., this volume), the flaring binary star system Circinus X-1, the Vela pulsar, and the most distant known radio quasar (2000-330). In addition, Sagittarius A, the center of our galaxy, and Centaurus A were observed at 8.4 GHz on the Parkes-Tidbinbilla baseline.

Each source was observed for periods totalling 3 to 15 hours, with the observing periods usually being shared with alternating observations of another source. The final schedule was a compromise between the need to produce a general idea of the angular structure of a large number of new (to VLBI) southern sources, and the desire to provide detailed maps of a small number of selected sources. Consequently detailed maps are available only for four sources, while general models have been derived for a further 25.

FIGURE 1. Examples of Source Models

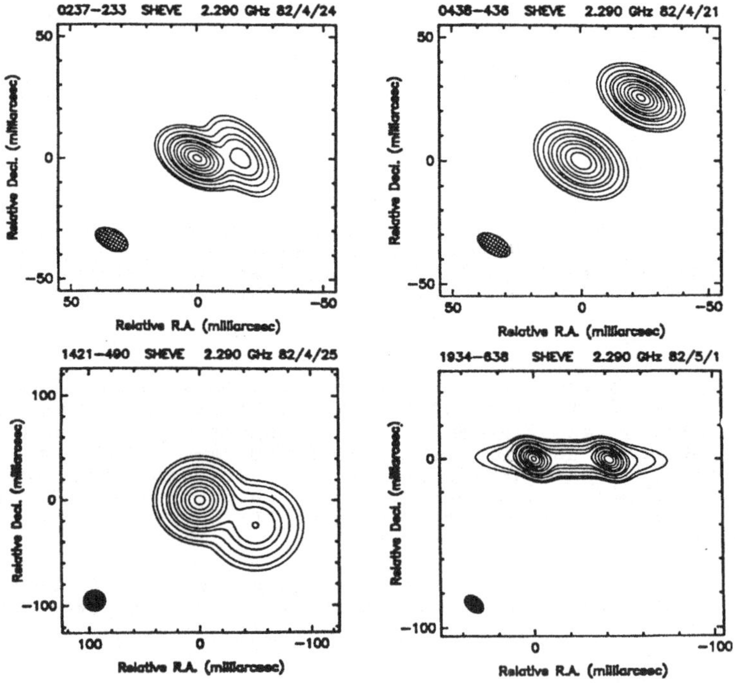

18 CM HIGH RESOLUTION OBSERVATIONS OF 15 EXTRAGALACTIC RADIO SOURCES

J. A. Waak, J. H. Spencer, R. S. Simon, and K. J. Johnston
E. O. Hulburt Center for Space Research
Naval Research Laboratory, Washington, D.C. 20375-5000

On 8-12 October 1981 a 5-station MkII VLBI experiment at 1660 MHz was performed using antennas at Westford, Maryland Point, Green Bank, Fort Davis, and Owens Valley. Fifteen extragalactic sources (Table 1) were observed with an average synthesized beam of roughly 10 mas. The data were processed using the NRAO correlator, globally fringe-fit, and mapped with the AIPS package. In addition to the usual self-calibration techiques, corrections were made to eliminate baseline-dependent calibration errors. Some of the sources, for example 1641+399 (3C345), display extended components not before observed at the dynamic range of the maps, which on the average is roughly 200:1.

The contour levels in the maps are normally at +/- 0.5, 1, 2, 5, 10, 20, 30, 50, 70, and 90% of the peak flux. The actual lowest contour level and other map parameters for each source are given in Table 1. One negative contour is drawn for each source at the level corresponding to the lowest positive contour for that source.

TABLE 1: MAP PARAMETERS

SOURCE	RESTORING BEAM (mas)	LOWEST CONTOUR (%)	PEAK BRIGHTNESS (mJy/beam)	RMS NOISE LEVEL (mJy/beam)	REMARKS
0428+000	16.4x5.8 @ -11.7°	2	1751	6.4	Component 35 mas to NW, P.A. -25°
0500+019	27.6x5.6 @ -6.8°	2	2284	12.1	Roughly unresolved
0711+356	28.4x6.1 @ -18.8°	-	1986	9.3	Roughly unresolved
0742+103	22.0x6.5 @ -16.0°	-	3729	7.1	Poor beam shape; extension to NW?
0828+493	10.9x5.5 @ -5.7°	2	1299	6.0	Roughly unresolved
0831+557	10.9x5.5 @ -5.7°	1	3431	13.6	*
0859+470	11.6x5.6 @ 0.0°	2	1535	9.1	Nearly unresolved
1311+678	11.2x5.1 @ -40.3°	-	638	8.5	**
1358+624	10.4x5.1 @ -41.0°	-	1110	13.3	***
1611+343	13.2x5.5 @ -1.0°	0.5	3157	7.5	Unresolved
1641+399	11.4x5.6 @ 0.0°	0.5	5234	6.1	Emission 15, 32, 55 mas from core
1803+784	8.7x5.5 @ -29.8°	2	1485	7.1	Extended core; comp. at P.A. -100°
1823+568	9.7x5.5 @ -18.4°	2	771	4.7	Slightly resolved, roughly N-S
2021+614	9.7x5.6 @ -37.6°	-	1287	9.5	****
2200+420	10.5x5.8 @ -31.4°	0.5	4470	7.3	Slightly resolved along P.A. -160°

NOTES TO TABLE 1:

* : Very extended along P.A. -35° ; apparently core-jet to approximately 100 mas; may be larger scale structure not on map due to lack of short baselines.
** : Complex structure out to approximately 40 mas; insufficient UV coverage to map reliably.
*** : Extended along P.A. -55° ; size >50 mas; too complex to map by our data.
****: Limited UV data; resolved on a 15 mas scale along P.A. 35°.

M. J. Reid and J. M. Moran (eds.), The Impact of VLBI on Astrophysics and Geophysics, 141–142.

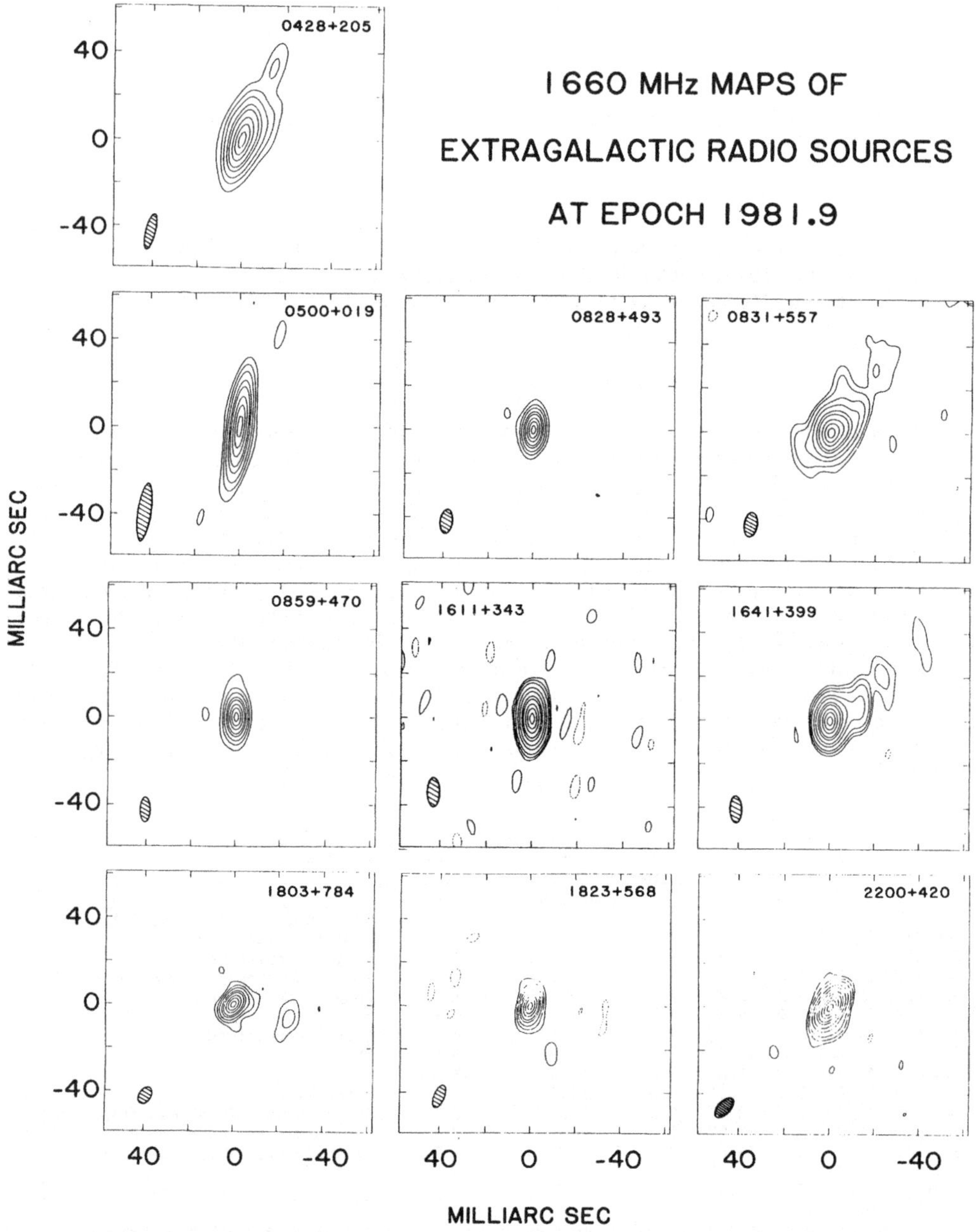

1660 MHz MAPS OF EXTRAGALACTIC RADIO SOURCES AT EPOCH 1981.9

THE LINEAR POLARIZATION OF EXTRAGALACTIC RADIO SOURCES AT MILLIARCSECOND RESOLUTION

J. F. C. Wardle and D. H. Roberts
Department of Physics
Brandeis University
Waltham, MA 02254 USA

ABSTRACT. We review the polarization sensitive VLBI observations of compact extragalactic radio sources that have been made to date, and discuss some preliminary physical results suggested by the data.

1. INTRODUCTION

Maps of the distribution of linear polarization in non–thermal radio sources can give a wealth of information that is not available from total intensity maps. This includes the topology of the magnetic field, the distribution of Faraday rotating material both inside and surrounding the radio emitting region, and the low–energy end of the relativistic electron spectrum. Also, relativistic and kinematic effects can impose distinctive signatures on the observed polarization. This information can allow important physical parameters to be measured and can help discriminate among theoretical models of radio sources.

The first map of the distribution of linear polarization in a compact radio source at VLBI resolution was of the quasar 3C454.3 at λ11.1 cm, published by Cotton *et al.* (1984). Since then, the group at Brandeis University has observed over 40 sources, some at as many as six epochs (Roberts *et al.* 1984; Wardle *et al.* 1986; Wardle and Roberts 1986a,b; Roberts and Wardle 1986, 1987; Roberts, Gabuzda and Wardle 1987; several papers, this symposium). Similar observational programs have also been initiated by several groups using the European VLBI network, but no results are available yet. Making polarization maps at VLBI resolution is not without its technical problems, but we shall not discuss them here (see Roberts *et al.* 1984, and in preparation).

In section 2, we summarize the results obtained to date. Eight sources are discussed in more detail in section 3. In sections 4 and 5 we attempt to draw some overall conclusions from these data; section 4 is concerned with the configuration of the magnetic field inferred from the radio (and optical) polarization data, while section 5 is concerned with the distribution of Faraday rotating material inside and surrounding the radio sources.

2. SUMMARY OF OBSERVATIONAL RESULTS

Table I summarizes twelve VLBI polarization observations on eight sources, and includes all results available to date. Column 1 gives the common name of each source, column 2 lists the epochs of observation, and column 3 gives the optical identification ('B'= BL Lac Object, 'Q'= quasar, 'G'= galaxy). Columns 4, 5 and 6 list the flux density, degree of

M. J. Reid and J. M. Moran (eds.), The Impact of VLBI on Astrophysics and Geophysics, 143–152.

polarization, and position angle (of the electric vector) for the 'core' component. Columns 7, 8 and 9 give the same quantities for the 'jet' component. Where the jet consists of more than one component, parameters are given for the component with the largest polarized flux density. Column 10 gives the position angle of elongation of the VLBI structure. The polarization position angles are uncorrected for Faraday rotation. This has not yet been measured directly at VLBI resolution, but for the 'jet' components it is probably small in most cases (*e.g.*, Rudnick and Jones 1983), though BL Lac itself is certainly an exception (RM=–198 rad m^{-2}, Rudnick *et al.* 1984). All observations were made at a wavelength of 6 cm, except 3C454.3 (λ11.1 cm).

Table I: Summary of VLBI Polarization Observations

Source	Epoch	ID	......Core......			......Jet......			VLBI PA
			S(Jy)	m(%)	χ(deg)	S(Jy)	m(%)	χ(deg)	θ(deg)
0235+164	1981.9	B	2.1	1.4	39	–	–	–	–
	1982.9		1.7	1.2	110	–	–	–	
0735+178	1982.9	B	1.3	4.1	5	0.2	8.3	46	72
OJ287	1981.9	B	2.9	3.4	160	0.3	64.	83	–116
	1982.9		2.8	4.6	99	0.4	23.	78	
1749+096	1981.9	B	1.8	6.5	39	–	> 40	–68	13
BL Lac	1984.2	B	0.8	2.8	–68	1.2	7.4	–35[a]	10
3C120	1982.9	G	1.0	< 0.7	–	0.2	11.	22	–96
3C345	1981.9	Q	6.4	< 1.0	–	1.8	11.	22	–82
	1983.9		9.0	1.6	–2	1.2	11.	30	
	1984.2		8.6	2.0	–27	1.0	12.	28	
3C454.3[b]	1980.6	Q	–	–	–	–	20.	–	12.

[a] RM=–198 rad m^{-2}, $\chi_{corr} = 5°$.
[b] λ11.1 cm.

3. DESCRIPTION OF INDIVIDUAL SOURCES

0235+164: This is a well known violently variable BL Lac object, that exhibits two absorption line redshifts ($z_a = 0.524, 0.852$), and shows variable HI absorption at the lower redshift (Wolfe *et al.* 1982). Cohen *et al.* (1986) have recently determined an emission redshift ($z_e = 0.94$). The radio polarization exhibited a position angle that rotated linearly with time during a large outburst in 1975 (Ledden and Aller 1978); this was interpreted by Blandford and Königl (1979) as changing relativistic aberration. Its VLBI structure is very compact at all wavelengths. Bååth (1984) found faint extensions to the north-east at three epochs, but at position angles that varied between 0° and 45°. We also find indications of structure to the north-east, but our dynamic range and resolution are insufficient to determine its nature.

0735+178: This BL Lac object has been called a 'cosmic conspirator' because several self-absorbed radio components 'conspire' to form a remarkably flat overall spectrum

(Cotton *et al.* 1980). Its optical spectrum shows a single absorption redshift ($z_a = 0.424$; Carswell *et al.* 1974). Bååth (1984) found the VLBI structure to consist of a bright core and a knot approximately 5 mas to the north-east, with evidence of a weaker superluminal component between these. Our observations (Wardle and Roberts 1986; Gabuzda *et al.* this symposium) show the two stronger components, together with a component located midway between them; it is unclear whether the latter is the superluminal component seen by Bååth. The inner knot is weakly polarized ($m \simeq 3\%$ at $\chi \simeq 20°$). The polarization of the outer knot is $m \simeq 8\%$ at position angle $\chi \simeq 45°$. The rotation measure for 0735+178 is given as -304 rad m^{-2} by Simard-Normandin *et al.* (1981). This value may have been affected by the source variability, since long-term monitoring programs indicate a much more modest value (Altschuler and Wardle 1976; Aller *et al.* 1985). If Faraday rotation is not important, then the magnetic field in the outer knot is roughly perpendicular to the jet.

OJ287: This is an extremely active BL Lac object at $z_e = 0.306$ (Sitko and Junkkarinen 1985). The details of the remarkable VLBI polarization structure are given in Roberts *et al.* (1987, and this symposium). At each epoch, the structure can be modeled as a core and two knots forming a jet at position angle $\theta \simeq -115°$. Over a year, both knots increased their separation from the core by about 0.3 mas, corresponding to an apparent superluminal velocity of about $3.3h^{-1}$c for each knot, assuming $q_0 = 0.5$ and a Hubble constant of $100h$ km/s·Mpc. The polarizations of the core and the (brighter) inner knot also changed markedly between the two epochs—the position angle of the core rotated by 61°, and the degree of polarization of the inner knot dropped from 64% to 23%. The rotation measure for OJ287 is quite uncertain because the source is so variable. If Faraday rotation is not important, then the magnetic field in the knots is roughly perpendicular to the direction of the jet.

1749+096: This a BL Lac object for which no redshift is available. The polarized structure consists of a comparatively strongly polarized component coincident with the bright core, and a 'knot' separated by about 0.8 mas at position angle $\theta \simeq 13°$ (Gabuzda *et al.* this symposium). Our observations were of insufficient dynamic range to detect this 'knot' in total intensity, but the degree of polarization of the knot must be in excess of 40% ($\chi \simeq 112°$). The integrated rotation measure of this source is given by Rudnick and Jones (1983) as 105 rad m^{-2}. If this applies to the knot, the magnetic field is roughly parallel to the direction of the 'jet'.

BL Lac: Our results for the prototype BL Lac object itself ($z_{gal} = 0.0688$; Miller *et al.* 1978) are preliminary (see Gabuzda *et al.* this symposium). Total intensity maps at λ2.8 cm by Mutel and Phillips (1984, and this symposium) show a complex, elongated, and curved structure, at position angle $\theta \simeq -170°$. The core is probably comparatively weak and located north of the brightest knot. The knots in the jet exhibit superluminal motion at $2.4h^{-1}$c. The bright northern knot is unresolved on transatlantic baselines (in both cross-hand and parallel hand fringes) and appears to be the component associated with the strongly polarized 1983 outburst (Aller, Aller, and Hughes 1985). Hughes, Aller, and Aller (1985) have convincingly modeled this outburst as a region of compression driven by a hot 'piston' traveling along the jet. Our observations were made at the end of this outburst. The rotation measure of BL Lac is large (-198 rad m^{-2}; Rudnick *et al.* 1984); assuming this applies to the strongest polarized component, then the magnetic field is transverse to the jet in that knot, as required by the model of Hughes *et al.*.

3C120: This galaxy provides the nearest ($z_{gal} = 0.033$) example of superluminal motion. The source contains a complex jet which exhibits structure on scales from milliarcseconds to arcminutes (Walker *et al.* 1987). The VLBI polarization map (Roberts and Wardle 1986) detected only the innermost knot of the jet, and most of the polarized flux was heavily resolved. No polarized emission was detected from the bright core, and its degree of polarization is certainly less than 1%. The integrated rotation measure is -3 rad m^{-2} (Simard-Normandin *et al.* 1981). If this applies to the detected knot, then its magnetic field is roughly perpendicular to the direction of the jet.

3C345: This well known superluminal quasar (z_e = 0.595) has been extensively observed with total intensity VLBI by the Caltech group (*e.g.*, Cohen *et al.* 1983; Biretta *et al.* 1986), and consists of a core and several bright knots in a curved jet with an average position angle $-75°$. The details of the polarization structure are given in Wardle *et al.* (1986, and this symposium). Each knot is detected in polarized emission as it emerges from the core, while the core itself appears to remain essentially unpolarized. The position angles of polarization of the four detected knots are all different, though the knot with the largest polarized flux density (C3) is polarized nearly perpendicular to the jet. If the integrated rotation measure (23 rad m^{-2}; Simard-Normandin *et al.* 1981) applies to this component, then the magnetic field in C3 is nearly parallel to the jet. In the other components, either the field is not parallel to the jet or there is a large gradient of rotation measure along the jet.

3C454.3: This quasar (z_e = 0.859) has been less extensively observed at VLBI resolution than 3C345, but certainly has a core–jet structure with evidence of superluminal motion at $v \sim 10h^{-1}$c (Pauliny–Toth *et al.* 1984). Great technical problems had to be overcome in order for Cotton *et al.* (1984) to make their polarization map at λ11.1 cm. They found that the position angle of polarization varied continuously down the jet. This, together with a rotation measure of -57 rad m^{-2} (Simard-Normandin *et al.* 1981), makes it difficult to determine the orientation of the magnetic field. Long-term monitoring of the integrated polarization of this source (Altschuler and Wardle 1976; Aller *et al.* 1985) show a preferred (intrinsic) position angle of polarization of 0 to 10°. This would imply a magnetic field aligned roughly along the jet, but it is not clear if this refers to the core or to the jet itself.

4. THE STRUCTURE OF THE MAGNETIC FIELD

The first point to be made from the data presented is that although the *integrated* polarizations of compact radio sources are typically quite modest (usually $\leq 5\%$, *e.g.* Altschuler and Wardle 1977; Aller *et al.* 1985), individual VLBI components can be very strongly polarized indeed. In OJ287, the degree of polarization approaches the theoretical limit for synchrotron radiation from a power-law electron energy distribution, implying an extremely well-ordered magnetic field. In general, much of the polarized flux comes from the optically thin 'jet' components, and the *degree* of polarization of the cores is always lower than that of the jets, and is sometimes $< 1\%$ (3C120, 3C345). Because of this, polarization maps can reveal extended structure that is not visible on total intensity maps (because of their finite dynamic range). The best example of this is OJ287 (Roberts *et al.* 1987, and this symposium). Total intensity observations show an extremely compact structure, and the source is frequently used as a calibrator. In our first epoch observations, the *polarized* structure is that of a well-resolved, almost equal double. Moreover, the polarized components appear to be separating at $3.3h^{-1}$c. None of this detail is apparent in total intensity maps made so far.

From Table I, we can compare the orientation of the projected magnetic field in the strongest jet component to the orientation of the VLBI jet itself. It is necessary to assume either that the Faraday rotation is not large, or that the integrated rotation measure applies to the VLBI component under consideration. While these assumptions usually are true (Rudnick and Jones 1983), we shall argue below that they are not always so. With this caveat in mind, four sources have a projected magnetic field roughly perpendicular to the direction of their jet (0735+178, OJ287, BL Lac, 3C120), and three sources have a projected magnetic field roughly parallel to the direction of their jet (1749+096, 3C345, and possibly 3C454.3).

These alignments are very rough, however, and many more observations are required to determine the true magnetic field topology and how (and whether) it changes with time. Nevertheless, this mild preference for magnetic fields to be aligned roughly parallel or perpendicular to the orientation of VLBI structure is in accord with the results of Aller *et al.* (this symposium), Wrobel *et al.* (this symposium), Rusk and Seaquist (1985), and Rusk (this symposium), based on *integrated* polarization measurements on a much larger number of sources. (It should be noted that the magnetic field directions inferred from such measurements may not be strictly comparable to those determined from VLBI polarization data, since the former may refer to the VLBI core component, or to structure on a larger scale that is resolved by VLBI). Two of these results seem to be particularly interesting. Aller *et al.* find several sources (of which BL Lac is one) for which the inferred magnetic field is parallel to the jet during quiescent periods, and perpendicular to the jet during strongly polarized outbursts. They ascribe this behavior to weak shocks in the jet that compress a nearly disordered magnetic field, thus greatly enhancing the transverse field components (see also Hughes, Aller and Aller 1985, and this symposium). Rusk has examined the magnetic field alignment as a function of optical identification. He finds that for galaxies and quasars the inferred magnetic field direction tends to be parallel to the jet, while for BL Lac objects, the distribution of alignments is bimodal, but with the majority of sources exhibiting a magnetic field transverse to the jet. This is an intriguing result that seems to have no obvious explanation within the standard theoretical models for compact radio sources. It might suggest that there is more than one way to form bright emission knots in a jet, or that the knots evolve in different ways, perhaps depending on the local environment (*e.g.*, the presence of gas?). Certainly, future studies of this type should pay attention to whether the radio source is in an active or a quiescent phase, and to any alignment changes during outbursts.

Optical polarization data also suggest certain correlations with the orientation of the VLBI structure which can give information on the magnetic field structure on scales smaller than those accessible to VLBI. 'Low polarization quasars' ($m_{opt} < 3\%$) tend to have the optical polarization position angle parallel to the radio structure (Stockman, Angel, and Miley 1979). In these sources, it is not clear that there is any non-thermal component to the optical continuum emission, and the polarization may be induced by scattering from an accretion disc or an oblate cloud of material aligned perpendicular to the radio axis. 'High polarization quasars' ($m_{opt} > 3\%$) and BL Lac objects, which are often very strongly polarized (sometimes as high as 40%), exhibit more complex behavior (Angel and Stockman 1980). These sources are believed to contain a strong component of non-thermal emission. The most variable sources exhibit no preferred position angle for the optical polarization. However, in sources where the range of optical position angles is not too extreme, the direction of the VLBI structure tends to fall within this range (Ulrich 1984; Rusk and Seaquist 1985; Sitko *et al.* 1985; Smith *et al.* 1987). Two sources that show this behavior are 3C345 and BL Lac. If the polarized optical emission is synchrotron radiation, then it comes from regions where the magnetic field has a tendency to be transverse to the direction of the jet (assuming that the emission region is optically thin at optical wavelengths). In some sources (*e.g.*, BL Lac; Moore *et al.* 1982, 1987), the optical polarization behavior is consistent with a random walk or 'Christmas tree' model, where the optical emission comes from many short-lived independent regions (size $\gtrsim 1$ lt-hr), each of high intrinsic polarization. By analogy with the variable radio emission from BL Lac, these too might be shocks, where intense particle acceleration is also taking place. Jones *et al.* (1985) have discussed a similar picture to explain the integrated radio polarization data.

The optical observations cannot, of course, determine the exact location of the emission, though it is usually assumed that the non-thermal optical emission originates deep inside the radio 'core', close to the central engine. For some sources, the radio and optical polarizations are sometimes well correlated (*e.g.*, 3C345 and BL Lac; Sitko *et al.* 1985). For both of these sources, we find that most of the polarized radio emission comes from components in the jet. It is therefore possible that part of the polarized optical emission is also associated with these jet components. If *changes* in the optical polarization are found that correlate with with *changes* in the radio polarization of a VLBI jet component, then this possibility would have to be taken seriously. Such a result would place extraordinary requirements on the particle acceleration mechanism in the jet components. To explore this possibility, M. Sitko is making optical polarization measurements simultaneously with our VLBI observing runs.

There are, of course, many ways of producing either predominantly parallel or transverse magnetic fields (Begelman, Blandford, and Rees 1984). In an expansion that conserves magnetic flux, $\mathbf{B}_{\perp} \sim r^{-1}$, and $\mathbf{B}_{\|} \sim r^{-2}$, where r is the transverse dimension of the jet. Thus the transverse field components are enhanced relative to the parallel component. This effect has certainly been seen in arcsecond scale jets (Bridle and Perley 1984). A transverse compression or shock has a similar effect: $\mathbf{B}_{\perp} \sim k^{-1}$, and $\mathbf{B}_{\|}$ is unchanged, where k is the degree of compression (unit length is compressed to length k; Hughes *et al.* 1985). Mild degrees of compression can produce quite high degrees of polarization when viewed from the correct angle (*i.e.*, nearly transverse to the plane of compression; see also Laing 1980). Note that the 'viewing angle' is determined in the rest frame of the emitting material. For a relativistic jet with a Lorentz factor Γ, the polarization is a maximum if the jet makes an angle $\sim 1/\Gamma$ to the line of sight, the same angle that maximizes apparent superluminal motion. Other effects of aberration on the observed polarization are discussed by Blandford and Königl (1979), Björnsson (1982), and Wardle and Roberts (1986a).

A predominantly longitudinal field is inevitably produced by shear due to any velocity gradient between the boundary and the center of the jet. If there is any interaction between the jet and the ambient medium then there will be boundary layer mixing and mass entrainment (De Young 1986), possibly observable as a center-brightened polarization distribution, and eventual deceleration due to momentum transfer. More complicated magnetic field structures have also been proposed. Two of these are the 'transverse self-similar' model of Chan and Henriksen (1980), which pays particular attention to the confinement of the jet, and the 'twisted rope' model of Königl and Choudhari (1985), which is a minimum energy configuration that predicts regular changes in polarization along the jet.

It is possible that all of these mechanisms could operate within the same jet. As the dynamic range of the polarization observations improves, we expect to see a rich variety of magnetic field topologies, and to be able to follow the dynamical evolution of the magnetic field in individual jet components.

5. THE DISTRIBUTION OF FARADAY ROTATING MATERIAL

In this section we will discuss how the polarization observations can be used to probe magneto-active plasma inside and in the vicinity of the radio emitting regions. We are particularly interested in why the core components in some sources are so weakly polarized.

The importance of Faraday rotation is measured by the 'Faraday depth', $\xi = 8.1 \times 10^5 \lambda^2 \int n_e \mathbf{B} \cdot d\mathbf{l}$, where λ(m) is the observing wavelength in the rest frame of the Faraday rotating plasma; n_e(cm^{-3}) is the density of thermal electrons; $\mathbf{B}$(gauss) is the magnetic field; and dl(pc) is an element of length along the line of sight. In general the integration extends from the back of the source to the observer. If there are reversals of the magnetic field along the line of sight, then the Faraday depth is, of course, reduced. Even if no thermal electrons are present at all, we may see Faraday rotation by the lowest energy relativistic electrons. The effective electron density to be inserted into Faraday depth calculations is $n_{eff} \sim n_{rel}/\gamma_l^2$, where γ_l is the low-energy cutoff in the electron energy spectrum (Wardle 1977; Jones and O'Dell 1977). If the synchrotron radiation is produced by equal numbers of electrons and positrons, then $n_{eff} = 0$.

Appreciable Faraday depth changes both the position angle and the degree of polarization (*e.g.*, Burn 1966). Only in the case of a Faraday 'screen' *external to and uniform over* the angular scale of the source do we expect the familiar λ^2 rotation of position angle over more than a radian or so (Laing 1984). This gives external Faraday rotation a comparatively unambiguous signature. Internal Faraday rotation can exhibit more complex behavior, and for an inhomogeneous source it may not give a λ^2 rotation of position angle at all (Blandford and Königl 1979). If there is a *spread* of Faraday depths of more than a radian or so within an element of resolution, then the degree of polarization will be appreciably reduced ('front-back' depolarization in the case of internal Faraday rotation, or 'side-side' depolarization in either case; see, *e.g.*, Conway *et al.* 1974).

It seems likely that there is not much Faraday rotation in most VLBI jets, although we wish to stress that this must be checked directly by simultaneous VLBI polarimetry at (at least) two wavelengths. Rudnick and Jones (1983) determined the rotation measures of 31 compact variable radio sources, using the VLA to separate extended (> arcsecond) structure. They found the resulting distribution of rotation measures to be indistinguishable from the distribution for extended sources, indicating that there is no additional nuclear component of rotation measure for these sources. Similar results have been found by Rusk (this symposium) and Wrobel *et al.* (this symposium). These measurements all include (or are made within) the λ18–20 cm band of the VLA. Therefore much of the polarized flux they detect is likely to come from components in the jet that either are further from the core than the components we observe or are resolved by us. Certainly, their rotation measures do not usually refer to the VLBI core itself, since this is often weakly polarized (see above). There is at least one source, 3C371 (Wrobel 1987), for which the 'VLA core' appears to have a much larger rotation measure (380 rad m^{-2}) than a bright component near the core (15 rad m^{-2}). We suspect there may be other cases of large core rotation measures which are discriminated against by the relatively long wavelength and low resolution of the VLA measurements.

Other evidence that the jet components usually have modest Faraday depths include the extremely high degree of polarization observed in OJ287 (this rules out significant *internal* Faraday rotation), and the rough correlations found between the direction of the magnetic fields and the direction of the jets. These would not exist if these components had large excess rotation measures. Rudnick and Jones (1983) used these results to place interesting constraints on the nature of any plasma surrounding the jet.

The situation may be quite different for the VLBI cores. First, the integrated rotation measures discussed above may not be relevant. Second, while the jets coexist with the narrow-line emitting region (at least in galaxies and quasars), the cores are much closer to the broad-line emitting region, where densities are much higher (*e.g.*, Ulrich 1984). Third, it is necessary to account for the very low degree of polarization observed in some of the cores ($m_{core} < 1\%$ for 3C120 and 3C345).

Optically thick synchrotron emission in a uniform magnetic field is about 10% polarized, so additional effects must be invoked. If the emitting region consists of N cells, each with a random magnetic field direction, then the degree of polarization is reduced by a factor $\sim N^{-1/2}$. For an optically thick source, we would require $N \geq 100$. For an inhomogeneous source the requirements are more severe. A Blandford–Königl type jet is much more strongly polarized if the magnetic field is ordered ($m \sim 50\%$), because most of the jet is always optically thin (Cobb *et al.*, this symposium). In this case we would require $N \geq 2500$. While this is possible, we consider it unlikely. The jets clearly contain regions of ordered magnetic field, and the high degrees of optical polarization imply ordered magnetic fields on a scale smaller than that of the VLBI core.

We therefore consider the possibility that the cores are depolarized by Faraday dispersion. This might be due to Faraday rotating material either inside the jet or surrounding the jet. Internal Faraday rotation is very inefficient at depolarizing a Blandford–Königl type jet, because the electron density and the magnetic field strength both decrease with distance down the jet, so the jet is always Faraday thin downstream. In terms of the Faraday depth at the $\tau = 1$ surface, $\xi_{\tau=1}$, it can be shown that $m \propto \xi_{\tau=1}^{-1/4}$ (Cobb *et al.* this symposium). Thus enormous values for the Faraday depth are required to produce a *net* polarization as low as 1%, and this is not compatible with the small Faraday depths observed in the jet components.

It therefore seems likely that the depolarization may occur outside the VLBI core, in the denser inner regions of the narrow-line emitting region. Physical parameters here are not well known, but if the magnetic field strength is not far below equipartition values, electron densities in the range $n_e \simeq 10^5$–10^7 cm^{-3} will provide the required Faraday depth. These densities are somewhat higher than are usually quoted for the narrow-line region in quasars and Seyfert I galaxies (*e.g*, Ferland 1981), but are consistent with more recent values found for the inner parts of this region (Fillipenko and Halpern 1984; De Robertis and Osterbrock 1984; Krolik and Vrtilek 1984).

Two pieces of circumstantial evidence are consistent with this picture. All the BL Lac objects that we have observed have appreciably more strongly polarized cores than 3C120 and 3C345. This is consistent with lack of a detectable emission line region in these objects. If observations of additional quasars and emission line galaxies show very weakly polarized VLBI cores, this would offer considerable support to this picture. In 3C345 we have observed new jet components, C4 and probably C5, emerging from the core (Wardle *et al.* this symposium). Their degrees of polarization are small, but appear to be increasing as their distance from the core increases, and their position angles of polarization differ from that of the bright knot C3 in the jet. This is also consistent with the picture of a Faraday thick screen surrounding the core. Indeed, it will be possible with VLBI polarimetry to *map* the radial distribution of Faraday depth in the nuclei of quasars and emission line galaxies by watching new components as they move away from the core. This will yield entirely new information on the narrow-line regions in these objects.

Financial support for this work was provided by the NSF under grants AST-82-13991 and AST-84-18636 (JFCW) and AST-83-15945 and AST-85-19529 (DHR).

REFERENCES

Aller, H. D., Aller, M. F., and Hughes, P. A. 1985, *Ap. J.*, **298**, 296.
Aller, H. D., Aller, M. F., Latimer, G. E., and Hodge, P. E. 1985, *Ap. J. Suppl.*, **59**, 513.
Altschuler D. R., and Wardle, J. F. C. 1976, *Mem. R. A. S.*, **82**, 1.
Altschuler D. R., and Wardle, J. F. C. 1977, *M.N.R.A.S.*, **179**, 153.

Angel, J. R. P., and Stockman, H. S. 1980, *Ann. Rev. Astr. Ap.*, **18**, 321.
Bååth, L. B. 1984, in *VLBI and Compact Radio Sources*, ed. R. Fanti, K. Kellermann, and G. Setti (Dordrecht: Reidel), 127.
Begelman, M. C., Blandford, R. D., and Rees, M. J. 1984, *Rev. Mod. Phys*, **56**, 255.
Biretta, J. A., Moore, R. L., and Cohen, M. H. 1986, *Ap. J.*, **308**, 93.
Björnsson, C.-I. 1982, *Ap. J.*, **260**, 855.
Blandford, R. D., and Königl, A. 1979, *Ap. J.*, **232**, 34.
Bridle, A. H., and Perley R. A. 1984, *Ann. Rev. Astr. Ap.*, **22**, 319.
Burn, B. J. 1966, *M.N.R.A.S*, **133**, 67.
Carswell, R. F., Strittmatter, P. A., Williams, R. E., Kinman, T. D., and Serkowski, K. 1974, *Ap. J. (Letters)*, **190**, L101.
Chan, K. L., and Henriksen, R. N. 1980, *Ap. J.*, **241**, 534.
Cohen, M. H., Unwin, S. C., Pearson, T. J., Seielstad, G. A., Simon, R. S., Linfield, R. P., and Walker, R. C. 1983, *Ap. J. (Letters)*, **269**, L1.
Cohen, R. D., Smith, H. E., and Burbidge, E. M. 1986, *B.A.A.S.*, **18**, 674.
Conway, R. G., Haves, P., Kronberg, P. P., Stannard, D., and Wardle, J. F. C. 1974, *M.N.R.A.S.*,**168**, 137.
Cotton, W. D. *et al.* 1980, *Ap. J. (Letters)*, **238**, L123.
Cotton, W. D., Geldzahler, B. J., Marcaide, J. M., Shapiro, I. I., Sanromá, M., and Rius, A. 1984, *Ap. J.*, **286**, 503.
De Robertis, M. M., and Osterbrock, D. E. 1984, *Ap.J.*, **286**, 171.
De Young, D. S. 1986, *Ap. J.*, **307**, 62.
Ferland, G. J. 1981, *Ap.J.*, **249**, 17.
Filippenko, A. V., and Halpern, J. P. 1984, *Ap.J.*, **285**, 458.
Hughes, P. A., Aller, H. D., and Aller, M. F. 1985, *Ap. J.*, **298**, 301.
Jones, T. W., and O'Dell, S. L. 1977, *Astron. Astrophys*, **61**, 291.
Jones, T. W., Rudnick, L., Aller, H. D., Aller, M. F., Hodge, P. E., and Fiedler, R. L. 1985, *Ap.J.*, **290**, 627.
Königl, A., and Choudhari, A. R. 1985, *Ap.J.*, **289**, 173.
Krolik, J. H., and Vrtilek, J. M. 1984, *Ap.J.*, **279**, 521.
Laing, R. A. 1980, *M.N.R.A.S*, **193**, 427.
Laing, R. A. 1984, in *Physics of Energy Transport in Extragalactic Radio Sources.*, NRAO Workshop No. 9, ed. A. H. Bridle and J. A. Eilek (Greenbank, WV: National Radio Astronomy Observatory), 90.
Ledden, J. E., and Aller, H. D. 1978, in *Pittsburgh Conference on BL Lac Objects*, ed. A. M. Wolfe (Pittsburgh: University of Pittsburgh), 60.
Miller, J. S., French, H. B., and Hawley, S. A. 1978, in *Pittsburgh Conference on BL Lac Objects*, ed. A. M. Wolfe (Pittsburgh: University of Pittsburgh), 176.
Moore, R. L., *et al.* 1982, *Ap. J.*, **260**, 415.
Moore, R. L., Schmidt, G. D., and West, S. C. 1987, *Ap. J.*, **314**, 176.
Mutel, R. L., and Phillips, R. B. 1984, in *VLBI and Compact Radio Sources*, ed. R. Fanti, K. Kellermann, and G. Setti (Dordrecht: Reidel), 117.
Pauliny-Toth, I. I. K., Porcas, R. W., Zensus, A., and Kellermann, K. I. 1984, in *VLBI and Compact Radio Sources*, ed. R. Fanti, K. Kellermann, and G. Setti (Dordrecht: Reidel), 149.
Roberts, D. H., Gabuzda, D. C., and Wardle, J. F. C. 1987, *Ap. J.*, submitted.
Roberts, D. H., Potash, R. I., Wardle, J. F. C., Rogers, A. E. E., and Burke, B. F. 1984, in *VLBI and Compact Radio Sources*, ed. R. Fanti, K. Kellermann, and G. Setti (Dordrecht: Reidel), 35.
Roberts, D. H., and Wardle, J. F. C. 1986, in *Quasars*, ed. G. Swarup and V. K. Kapahi,

(Dordrecht: Reidel), 141.
Roberts, D. H., and Wardle, J. F. C. 1987, in *Superluminal Radio Sources*, ed. J. A. Zensus and T. J. Pearson, (Cambridge: Cambridge University Press), 193.
Rudnick, L., and Jones, T. W. 1983, *A. J.*, **88**, 518.
Rudnick, L., Jones, T. W., Edgar, B. K., and Pedalty, J. A. 1984, *A. J.*, **89**, 316.
Rusk, R., and Seaquist, E. R. 1985, *A. J.*, **90**, 30.
Simard-Normandin, M., Kronberg, P. P., and Button, S. 1981, *Ap. J. Suppl.*, **45**, 97.
Sitko, M. L., Schmidt, G. D., and Stein, W. A. 1985, *Ap. J. Suppl.*, **59**, 323.
Sitko, M. L., and Junkkarinen, V. T. 1985, *P.A.S.P.*, **97**, 1158.
Smith, P. S., Balonek, T. J., Elston, R., and Heckert, P. A. 1987, *Ap. J. Suppl.*, **64**, 459.
Stockman, H. S., Angel, J. R. P., and Miley, G. K. 1979, *Ap. J. (Letts.)*, **227**, L55.
Ulrich, M.-H. 1984, in *VLBI and Compact Radio Sources*, ed. R. Fanti, K. Kellermann, and G. Setti (Dordrecht: Reidel), 73.
Walker, R. C., Benson, J. M., and Unwin, S. C. 1987, *Ap. J.*, **316**, 546.
Wardle, J. F. C. 1977, *Nature* (London), **269**, 563.
Wardle, J. F. C., Roberts, D. H., Potash, R. I., and Rogers, A. E. E. 1986, *Ap. J. (Letters)*, **304**, L1.
Wardle, J. F. C., and Roberts, D. H. 1986a, *Can. J. Phys.*, **64**, 434.
Wardle, J. F. C., and Roberts, D. H. 1986b, in *Continuum Emission in Active Galactic Nuclei*, Proceedings of KPNO Workshop, ed. M. L. Sitko, (Tucson: National Optical Astronomy Observatories), 190.
Wolfe, A. M., Davis, M. M., and Briggs, F. H. 1982, *Ap. J.*, **259**, 495.
Wrobel, J. M. 1987, in *Superluminal Radio Sources*, ed. J. A. Zensus and T. J. Pearson, (Cambridge: Cambridge University Press), 186.

CALCULATIONS OF THE LINEAR POLARIZATION OF INHOMOGENEOUS RELATIVISTIC JETS

W. K. Cobb, J. F. C. Wardle, and D. H. Roberts
Department of Physics
Brandeis University
Waltham, MA 02254 USA

ABSTRACT. We have performed exact numerical calculations of the Stokes parameters I, Q, and U for optically and Faraday thick inhomogeneous relativistic jets. We consider how changes in Faraday depth, as well as in the observer-jet geometry, change the resolved and unresolved properties of the jet.

1. INTRODUCTION

The standard model for relativistic jets in compact radio sources (Blandford and Königl 1979; Königl 1981) portrays jets as being conical regions of small semi-angle carrying a convected magnetic field in a strongly relativistic bulk flow of synchrotron emitting particles. The direction of motion makes a small angle to the line of sight.

We have generalized their model to include *radial* bulk motion (rather than motion strictly *parallel* to the jet axis) and internal Faraday rotation. The effect of the former is to give different Doppler factors for different parts of the jet. The Faraday rotation partially depolarizes the jet, as suggested by recent observations of very low degrees of polarization in some cores (Wardle *et al.* 1986). We also calculate lines of sight making small angles to the jet axis, and even those which lie inside of the jet cone itself (this may be necessary to explain observations of BL Lac objects). We are also able to treat arbitrary magnetic field models, and velocity gradients between the axis and boundary of the jet.

2. RESULTS

In the present calculations we assume $B(r) \sim r^{-2}$, $K(r) \sim r^{-2}$, $N(\gamma_e) \sim \gamma_e^{-2}$, and $\gamma_j = 5$ (in the notation of Königl 1981). For simplicity we take the magnetic field to be everywhere parallel to the bulk velocity.

Figure 1 shows the behavior of the integrated fractional polarization, m, as a function of the observer inclination angle θ. The jets here are *Faraday thin.* The three curves correspond (from top to bottom) to jet semi-angles ϕ of 2°, 5°, and 8° respectively. Note that the polarization in all three cases is largely independent of inclination angle when the observer is *outside* of the jet, but strongly dependent on inclination angle when the observer is *inside* the jet. This is just due to the projection onto the sky of the radial field lines.

Figure 2 shows the integrated fractional polarization, m, as a function of the Faraday depth ξ_1 at the $\tau = 1$ surface for the case $\phi = 3°$ and $\theta = 10°$. This was generated simply by varying the lower energy bound of the relativistic electron spectrum (Jones and O'Dell

M. J. Reid and J. M. Moran (eds.), The Impact of VLBI on Astrophysics and Geophysics, 153–154.

1977). The end points on the curve correspond to $\gamma_{e,lower}$ of 100 and 10 respectively. Even in the Faraday thick case the *integrated* polarization is still almost 12%. To Faraday depolarize a jet which would otherwise be highly polarized thus requires an *extremely* small value of $\gamma_{e,lower}$ (or an extremely high density of thermal electrons). This is because the jets always become Faraday thin downstream.

Figure 3 shows how the separation (here called $\Delta_{I,P}$) of the peaks of the I and P strip distributions (normalized by the distance of the $\tau = 1$ surface from the apex) varies as a function of the Faraday depth at the $\tau = 1$ surface. At low Faraday depths $\Delta_{I,P}$ is constant at about 0.3, but it rises sharply as the inner parts of the jet are depolarized. Depending on the jet geometry and model, this may provide a new observational means of putting bounds on the Faraday depth of a source.

Figure 1.

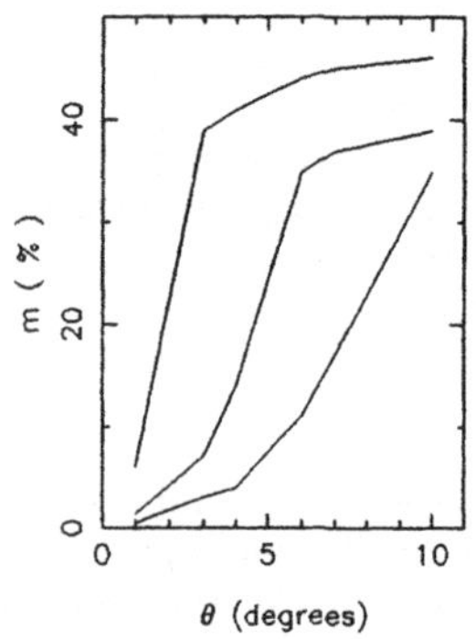

Figure 2.

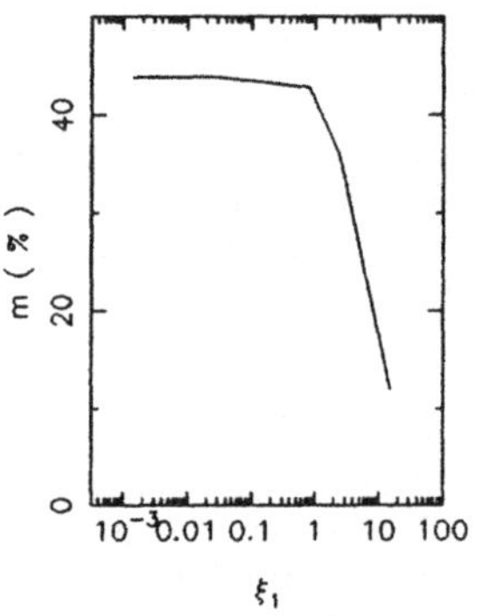

Figure 3.

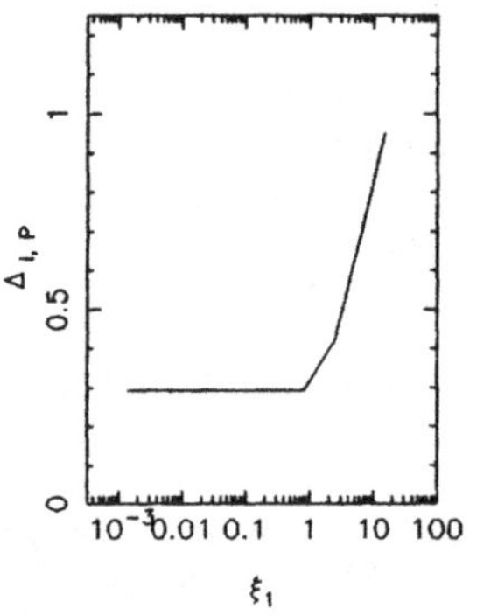

3. DISCUSSION

We wish to emphasize the preliminary nature of these results. We have just begun to investigate the enormous variety of models with which our code is capable of dealing. Nevertheless, we show that polarization observations can give useful constraints on the jet-observer geometry as well as on Faraday depth. In particular, internal Faraday rotation is not very efficient at depolarizing these jets because they always become Faraday thin further downstream.

ACKNOWLEDGEMENTS

Financial support was provided by the NSF under grants AST-82-13991 and AST-84-18636 (JFCW) and AST-83-15945 and AST-85-19529 (DHR).

REFERENCES

Blandford, R. D. and Königl, A. 1979, *Ap.J.*, **232**, 34

Königl, A. 1981, *Ap.J.*, **243**, 700

Wardle, J. F. C., Roberts, D. H., Potash, R. I.,and Rogers, A. E. E. 1986, *Ap.J.(Lett.)*, **304**, L1–L4

Jones, T. W., and O'Dell, S. L. 1977, *Ap.J.*, **215**, 236

GLOBAL FRINGE FITTING FOR POLARIZATION VLBI

L. F. Brown and D. H. Roberts
Department of Physics
Brandeis University
Waltham, MA 02254 USA

ABSTRACT. We present a global fringe fitting technique for the polarized fringes in VLBI. The standard search method imposes a signal-to-noise (*SNR*) limit on usable data. In our method the search procedure is circumvented and the *SNR* limitation removed.

1. INTRODUCTION

Standard VLBI data processing methods search a range of delays τ and delay rates $\dot{\tau}$ to "find the fringes;" this requires a large number of trials, typically $N=10^4$–10^5. The probability of false fringe detection depends on both N and the *SNR* of the fringe which is identified (Thompson *et al.* 1986). This places restrictions on the minimum *SNR* that can used if one wants to be confident of having found a true fringe; in Mark III VLBI this limit is 7. Compact radio sources are typically 1–10% linearly polarized, so the linearly polarized intensity (cross-hand, "XH") correlation coefficients are a tenth to a hundredth of the total intensity (parallel-hand, "PH") coefficients. The *SNR* restriction and reduced flux thus put a significant amount of VLBI polarization data at or below the limit of reliable detectability by the standard fringe search technique. Our technique, a simplified form of global fringe fitting (Schwab and Cotton 1983), was developed to circumvent the restrictions imposed by fringe searching and hence reclaim low *SNR* data.

2. THE R-L DELAY DIFFERENCE METHOD

Each VLBI station has different instrumental delays for right and left circular polarizations, so the PH (RR and LL) and XH (RL and LR) fringes are located at different residual delays. Our procedure predicts the location in $(\tau, \dot{\tau})$ space of the XH fringes by combining the observed τ and $\dot{\tau}$ of the PH fringes with the right-minus-left (R-L) station differences $\delta\tau$ and $\delta\dot{\tau}$. Using reliable ($SNR \geq 7$) PH and XH fringes, we determine $\delta\tau$ and $\delta\dot{\tau}$ for each station as a function of time throughout an experiment. For each baseline there are two independent estimates of the single- and multi-band $\delta\tau$ for each end of the baseline; for stations x and y on baseline xy,

$$\delta\tau_x \equiv \tau_x^R - \tau_x^L = (\tau_x^R - \tau_y^L) - (\tau_x^L - \tau_y^L) = \tau_{xy}^{RL} - \tau_{xy}^{LL} ,$$

$$\delta\tau_y \equiv \tau_y^R - \tau_y^L = (\tau_x^R - \tau_y^L) - (\tau_x^R - \tau_y^R) = \tau_{xy}^{RL} - \tau_{xy}^{RR} ,$$

plus two similar equations for reversed baseline order. The computation of $\delta\dot{\tau}$ mimicks that of $\delta\tau$. An example of the PH and XH multiband delays on one baseline is shown in

M. J. Reid and J. M. Moran (eds.), The Impact of VLBI on Astrophysics and Geophysics, 155–156.

Figure 1. Figure 2 shows the results of determining the R-L delay differences for one station. The individual station R-L differences are found to be fairly stable over the duration of an experiment (typically 24 hours), so mean $\delta\tau$ is adopted for each station. The average $\delta\dot{\tau}$s were all found to be zero, so the delay-rates for the XH fringes on a given scan are taken to be the same as that of the associated parallel-hand fringes. We predict the XH τ from the PH τ using the R-L differences; take an RL scan as an example:

$$\tau_{xy}^{RL}|_{\text{pred}} = \tau_{xy}^{LL}|_{\text{obs}} + \overline{\delta\tau_x} \quad \text{or} \quad \tau_{xy}^{RL}|_{\text{pred}} = \tau_{xy}^{RR}|_{\text{obs}} + \overline{\delta\tau_y} \ .$$

This procedure is followed for both the single- and multi-band delays, and the results are passed to the Mark III fringe parameter estimation program, which finds the amplitude and phase of the cross fringes without having to search $(\tau, \dot{\tau})$ space. Since the number of trials is reduced to one, the *SNR* restriction is bypassed.

3. RESULTS

The results shown here were derived from data taken in March of 1984 in a run using six stations: Bonn, Haystack, NRAO, Ft. Davis, VLA, and OVRO. The refringed to standard ratios should behave in the following manner: in the *SNR* range 0-5 the ratios should be essentially random as the standard fringes below *SNR*=7 have a high probability of being false detections; in the range 6-9 the scans should agree fairly well in amplitude and phase. In the high *SNR* range (> 10) there should be little disagreement between the two methods. This behavior is seen in the fitted results; for example, Figure 3 shows the complex ratio of the the R-L delay difference refringed scans to the standard processing scans in the refringed *SNR* range 6-9. Preliminary applications show the R-L delay difference technique to work well. There is a significant increase in the amount of usable data for a not undue amount of additional computation time. There is, however, some evidence for short term $\delta\tau$ drifts at some stations. Present work involves applying the R-L differences on a scan-by-scan basis, which can be done if there is at least one good set of XH fringes to each station for each scan.

REFERENCES

Thompson, R. A., Moran, J. M., and Swenson, G. W. 1986, *Interferometry and Synthesis in Radio Astronomy*, (New York: Wiley), p. 264.

Schwab, F. R., and Cotton, W. D. 1983, *A.J.*, **88**, 688.

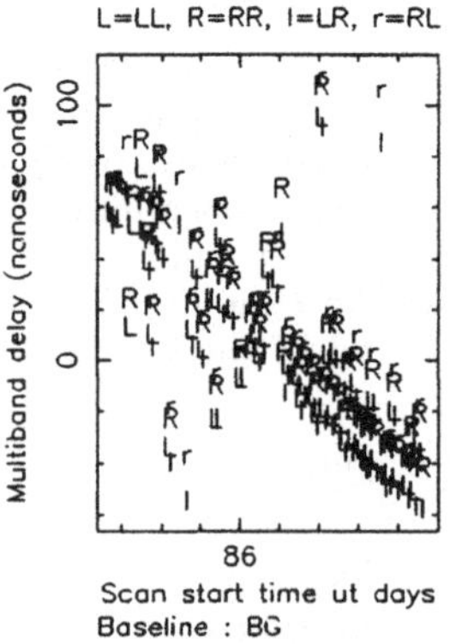

Figure 1

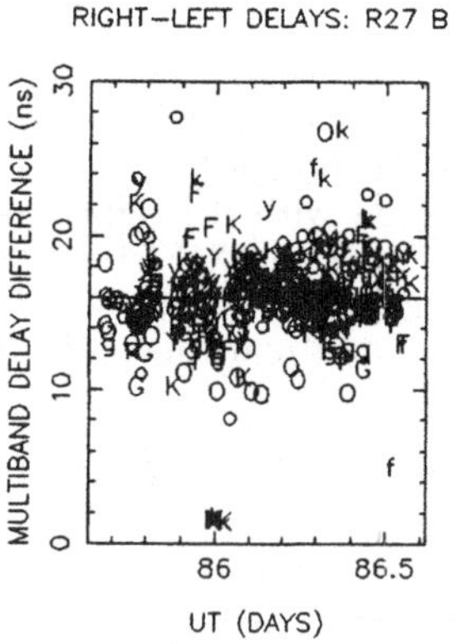

Figure 2

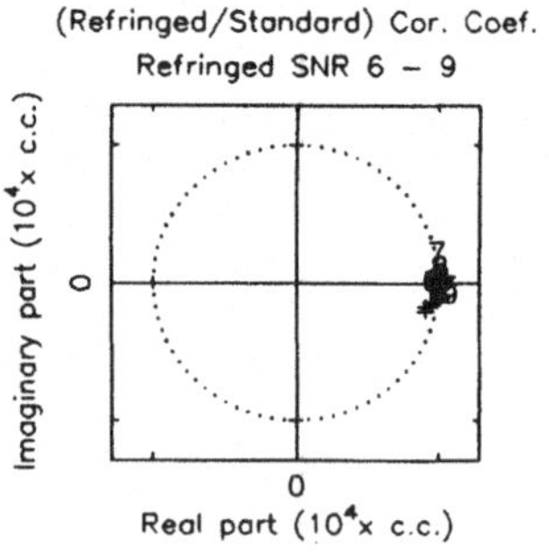

Figure 3

THE PHYSICAL NATURE OF SPURIOUS POLARIZATION TRANSFORMED FROM THE BACKGROUND NONPOLARIZED RADIATION OF COSMIC EXTENDED RADIO SOURCES

Chu Han-Shu
Purple Mountain Observatory
Academia Sinica
Nanjing, People's Republic of China

Polarized radiation of cosmic radio sources carries the important information of the radiation mechanism and magnetic field structure of the sources. The background nonpolarized radiation of extended radio sources may also be partly transformed into spurious polarization. No comprehensive theory concerning the whole physical process of radiation transfer has been developed.

Starting with the Hertz vector potential, we treated the polarization characteristics of three relevant aspects of the problem in a unified way, i.e., the polarization characteristics of the radio sources; the polarization characteristics of the exciting field of the antenna feed system; and the polarization characteristics of the antenna radiation field. A rather clear physical picture of the whole process with analytical formulas has been obtained. The spurious polarization due to nonpolarized radiation of extended radio are calculated from the aperture field excited thereby in various kinds of antenna feed systems. The results are consistent with existing observational facts.

In relation to practical observations, stress is given to the case of using a dual orthogonally polarized horn feed (or horn feed of circular aperture with dual orthogonal outlets). The spurious polarization derived is:

$$P_{sp}(\Phi) = \frac{\int\int\int\ 2I(\theta,\varsigma,\nu)\ U(\theta)\ V(\theta)\ \cos 2(\Phi-\varsigma)\ d\varsigma d\theta d\nu}{\int\int\int\ I(\theta,\varsigma,\nu)\ [U^2(\theta)+V^2(\theta)]\ d\varsigma d\theta d\nu} ,$$

where Φ is the position angle of the dual feed, I is the brightness distribution function of the extended source, θ and ς are two orthogonal coordinates of a point within the source with respect to the antenna beam axis, $U(\theta)$ and $V(\theta)$ are distribution functions of antenna main field pattern and cross polarization field pattern, respectively, and ν is the observational frequency.

As the formula shows, the spurious polarization changes in a cosine fashion as a function of the position angle of the dual feed. Thus, one can differentiate the spurious polarization from the intrinsic one by changing the position angle of the feed and measuring the amplitude of the cosine curve (see Figure 1).

Also, the spurious polarization can be minimized by minimizing $V(\theta)$.

M. J. Reid and J. M. Moran (eds.), The Impact of VLBI on Astrophysics and Geophysics, 157–158.

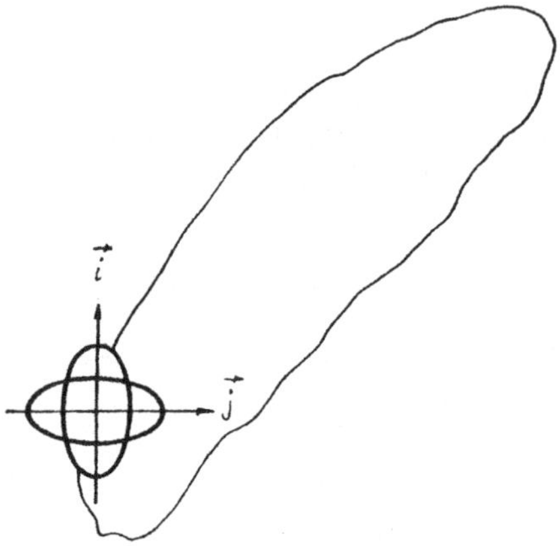

Figure 1. Spurious polarization can be differentiated from the intrinsic one by changing successively the position angle of the dual polarization feed and marking the cosine curve.

The derived formula might serve as a guide to eliminating the spurious polarization due to radiation of an extended source. Also, the systematic method of approach and the understanding obtained may be of use to the analysis of spurious polarization in VLBI polarization measurements.

The derived formula may also open open possibilities of improving and further extending polarization observations to very short wavelengths (i.e., in short mm and submm waves higher than 150–200 GHz). At such wavelengths, the realization of a corrugated horn feed or multimode feed with $V(\theta) = 0$ is quite difficult. Moreover, the losses at such frequencies are excessive. With the method mentioned above, the polarization observation can be extended to very short wavelengths.

BIBLIOGRAPHY

Chandrasekhar, S. 1950, *Radiative Transfer* (Clarendon Press, Oxford), p. 25.

Kellermann, K. I., and Pauliny-Toth, I. I. K. 1981, *Ann. Rev. Astr. Ap.*, **19**, 373.

Moran, J. 1976, in *Methods of Experimental Physics*, Vol. 12, Part C, ed. M. L. Meeks (Academic Press, New York), p. 174.

Silver, S. Microwave Antenna. MIT Radiation Series, Vol. 12 (McGraw-Hill, New York).

Wannier, P. G. 1983, *Ap. J.*, **267**, 126.

EVOLUTION OF THE MILLIARCSECOND POLARIZATION STRUCTURE OF THE BL LACERTAE OBJECT OJ287

D. H. Roberts, D. C. Gabuzda, and J. F. C. Wardle
Department of Physics
Brandeis University
Waltham, MA 02254 USA

ABSTRACT. The $\lambda 6$ cm total intensity and polarization structures of OJ287 have been determined at two epochs a year apart. At each epoch the source can be modeled as a linear structure consisting of a core and two knots at position angle $PA \simeq -115°$. This structure expanded slightly over a year, each knot increasing its separation from the core by $\simeq 0.3$ mas, suggesting an apparent superluminal velocity of $\simeq 3.3h^{-1}c$ for each.

1. INTRODUCTION

The BL Lacertae object OJ287 (0851+202) is an extremely active galactic nucleus at $z = 0.306$ (Miller, French, and Hawley 1978; Sitko and Junkkarinen 1985). At both optical and radio wavelengths it is violently-variable, both in intensity and polarization. Here we summarize the results of polarization-sensitive VLBI observations made at two epochs a year apart. Preliminary reports were given by Roberts and Wardle (1986, 1987); complete details may be found in Roberts, Gabuzda, and Wardle (1987).

2. OBSERVATIONS AND DATA PROCESSING

Polarization-sensitive $\lambda 6$ cm observations were carried out in December 1981 and December 1982 with the Haystack, Green Bank, and Owens Valley antennas and the phased-up VLA, using the Mark III data recording system. Maps of the total intensity were made by standard hybrid mapping techniques. After correction for instrumental contamination (Roberts *et al.* 1984), maps of the distribution of polarized intensity ($P = pe^{2i\chi} = mIe^{2i\chi}$, where $p = mI$ is the polarized intensity, m the fractional linear polarization, and χ is the position angle of the electric vector on the sky) were made by referencing the cross-hand fringes to the parallel-hand fringes using the antenna gains determined in hybrid mapping, Fourier transforming the result, and performing a complex CLEAN.

3. RESULTS

Maps of the I and P distributions at the two epochs are presented by Roberts and Wardle (1986, 1987). We fit simple models (Table 1) to the visibility data produced by the mapping process. Requiring that the models account for the I and P data simultaneously led to three components, a core C and two knots K1 and K2. The core C was weakly polarized ($\simeq 4\%$) at each epoch; its polarization position angle changed by at least $\simeq 60°$ between epochs.

M. J. Reid and J. M. Moran (eds.), The Impact of VLBI on Astrophysics and Geophysics, 159–160.

The inner knot K2 was strongly polarized at each epoch, and decreased its polarization from $\simeq$ 64% to $\simeq$ 23% between epochs. The outer knot K1 was only weakly polarized at each epoch, causing the predominant polarization structure to be double. The polarization position angle of each knot was roughly constant between epochs. The polarization of K2 in December 1981 is close to the maximum possible for optically-thin synchrotron radiation from electrons with a random pitch angle distribution in a homogeneous magnetic field.

Using a Hubble constant of $100h$ km/s·Mpc and assuming $q_0 = 0.5$, the radial motions of K1 and K2 each correspond to an apparent transverse velocity $v_\perp \simeq 3.3h^{-1}c$. While "two epochs do not a superluminal make," our observations are consistent with inclusion of OJ287 in the class of superluminal objects. It is interesting to note that the only other BL Lacertae object currently known to display superluminal motion, BL Lac itself, has an apparent expansion velocity of $v_\perp \simeq 2.4h^{-1}c$ (Mutel and Phillips 1984). If such modest superluminal velocities are representative of BL Lacertae objects in general, this favors the view that the jets in these objects are aligned particularly closely to the observer's line of sight. The maximum of the distribution of $v_\perp$ with angle θ between the line of sight and the jet velocity vector occurs at $\theta = 1/\gamma$. Modest superluminal velocities, such as those observed for OJ287 and BL Lac, would occur if the BL Lacertae objects constitute a population of sources in which $\theta < 1/\gamma$. This would also be consistent with the extreme "blazar" properties of these objects (Angel and Stockman 1980).

This work would not have been possible without the assistance of Dr. Alan Rogers. We also thank the staffs of the US VLBI Network observatories and of the Haystack Mark III Processor for their help. Financial support was provided by the NSF under grants AST-82-13991 and AST-84-18636 (JFCW) and AST-83-15945 and AST-85-19529 (DHR).

Table 1: Three-Component Models for OJ287

Component	α(mas)	δ(mas)	I(mJy)	p(mJy)	m(%)	χ(deg)	r(mas)[a]	θ(deg)[a]
				December 1981				
C	...	...	2910.	98.	3.4	160.	...	...
K2	−0.88	−0.43	284.	180.	64.	83.	0.98	−116
K1	−2.08	−1.20	243.	6.6	2.7	65.	2.40	−120
				December 1982				
C	...	...	2830.	129.	4.6	99.	...	...
K2	−1.15	−0.48	415.	95.	23.	78.	1.25	−113
K1	−2.55	−0.82	116.	4.0	3.4	72.	2.68	−108

[a] Separation and position angle with respect to the core.

REFERENCES

Angel, J. R. P., and Stockman, H. S. 1980, *Ann. Rev. Astr. Ap.*, **8**, 321–361.

Miller, J. S., French, H. B., and Hawley, S. A. 1978, in *Pittsburgh Conference on BL Lac Objects*, ed. A. M. Wolfe (Pittsburgh: University of Pittsburgh), 176–191.

Mutel, R. L., and Phillips, R. B. 1984, in *VLBI and Compact Radio Sources*, ed. R. Fanti, K. Kellermann, and G. Setti (Dordrecht: Reidel), 117–118.

Roberts, D. H., Gabuzda, D. C., and Wardle, J. F. C. 1987, *Ap. J.*, submitted.

Roberts, D. H., *et al.* 1984, in *VLBI and Compact Radio Sources*, ed. R. Fanti, K. Kellermann, and G. Setti (Dordrecht: Reidel), 35–38.

Roberts, D. H., and Wardle, J. F. C. 1986, in *Quasars*, ed. G. Swarup and V. K. Kapahi, (Dordrecht: Reidel), 141–147.

Roberts, D. H., and Wardle, J. F. C. 1987, in *Superluminal Radio Sources*, ed. J. A. Zensus and T. J. Pearson, (Cambridge: Cambridge University Press), 193–199.

Sitko, M. L., and Junkkarinen, V. T. 1985, *P. A. S. P.*, **97**, 1158–1162.

The Brightness and Polarization Structure of Compact Radio Sources

Raymond Rusk
Dept. of Astronomy
Univ. of Toronto
Toronto, Ontario, M5S 1A1

New VLA polarization data for 140 radio sources have been used to study the correlation between VLBI structural position angle and radio and optical polarization position angles reported in Rusk and Seaquist (1985). We confirm that there is a strong tendency, in active galaxies and quasars, for the radio **E** vector (of the core component) to lie *normal* to the VLBI structural axis. However, in BL Lac objects we find a tendency for the core radio **E** vector to be aligned *parallel* to the VLBI structural axis.

1. THE SURVEY

Many compact sources when observed at VLBI resolution display linear asymmetric structures consisting of an (apparently) optically thick core and an (apparently) optically thin jet. If the low-frequency turnover seen in the spectra of these sources is due to synchrotron self-absorption, then the optically thick core component should have a low degree of polarization. Most of the observed polarization must arise in the optically thin jet component which suggests that the direction of the radio **E** vector will be correlated with the direction of the VLBI asymmetry.

From a comprehensive literature search (complete to January 1985) we created a list of 156 sources, including 3 sources which we had mapped in a VLBI survey of highly polarized sources, which had VLBI structural position angles given by either hybrid maps or models. VLA snap-shots for 140 of these sources were made (at 2, 6, 18 and 22 cms) and VLA polarization measurements for an additional 14 sources were taken from the literature.

All of our data were self-calibrated, mapped and CLEANED using the NRAO AIPS package. Dynamic ranges of several-thousand-to-one were obtained in most cases. Off-core emission was mapped out so that core polarization values would reflect polarization from sub-arcsecond structures. For extended sources observed in both A- and B-arrays, scaled-array observations were used to form rotation measure and spectral index maps. The polarization data were corrected for positive bias, and the integrated rotation measure (RMc) and intrinsic polarization angle (IPAc) values for the core components were estimated.

2. DISCUSSION OF RESULTS

The left half of Figure 1 shows the misalignment angle distribution between VLBI structural and radio polarization position angles for 102 active galaxies and quasars. There is a sharp peak near 90° indicating that the sub-arcsecond magnetic field structure is aligned with the VLBI axis in most sources. The probable error in the misalignment value for any given source is roughly the bin size, so much of the scatter away from 90° is accounted for by error. However, some sources show large curvatures in their milliarcsecond jets or

M. J. Reid and J. M. Moran (eds.), The Impact of VLBI on Astrophysics and Geophysics, 161–162.

have continuously curved milliarcsecond to arcsecond scale structures. It is to be expected then that some large misalignments are real.

In light of the above discussion, it appears that for the great majority of sources, either their sub-arcsecond jets are fairly straight in the region producing the polarized emission and align with the observed milliarcsecond structure, or the observed polarization comes from a region near the observed VLBI jets. For about 30% of the sources, no arcsecond scale structure was found in the maps and the core was unresolved at sub-arcsecond resolution. For these sources at least, it is likely that nearly all the polarized flux comes from the observed VLBI structures.

The right half of Figure 1 shows the misalignment angle distribution between core radio polarization and VLBI position angles for 22 BL Lac objects. The slight bimodality apparent in the active galaxy and quasar distribution is very evident for the population of BL Lac objects. Some BL Lac objects have highly variable radio polarization making RMc and IPAc values uncertain (for these sources). However, variability alone is unlikely to produce this markedly bimodal distribution.

The tendency in BL Lac objects for the integrated polarization position angle to lie along the VLBI jet agrees with observations of BL Lacertae itself (Aller and Aller, 1984 and Mutel and Phillips, 1984) during times of outburst. (BL Lacertae has not been included in Figure 1.) Axial compression of the magnetic field in a shock could produce a magnetic field nearly perpendicular to the jet, as they discuss, and the strong outbursts of radio emission might result from electron enhancements within the shock.

While our analysis is still preliminary, and has not included a detailed study of the misalignment angle distribution for individual VLBI morphological classes, we can conclude that in most quasars and active galaxies there is a preferred structural axis defined by the VLBI morphology and that the sub-arcsecond magnetic field runs along this axis. The misalignment angle distribution for BL Lac objects is apparently distinct from that of quasars, in that the radio **E** vector is frequently aligned parallel, not normal, to the VLBI axis in these objects.

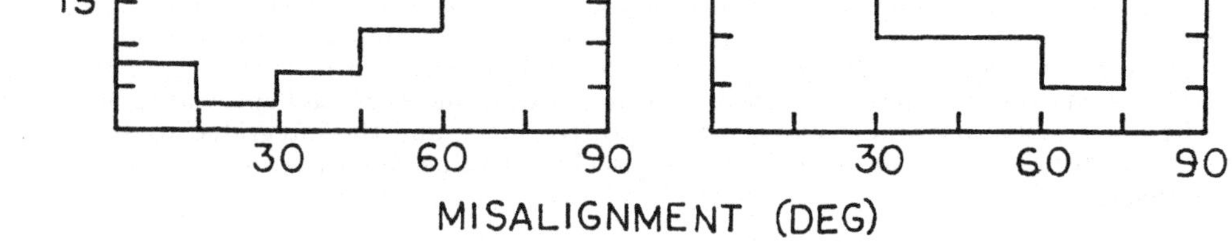

Fig. 1. The distribution of the acute angle between VLBI structural axis and intrinsic (core) radio polarization **E** vector for active galaxies and quasars (Left), and BL Lac objects (Right).

REFERENCES

Aller, H.D., and Aller, M.F.: 1984, IAU Symp. 110, p. 119.
Mutel, R.L., and Phillips, R.B.: 1984, IAU Symp. 110, p. 117.
Rusk, R. and Seaquist, E.R.: 1985, *Astron. J.* **90**, 30.

EVOLUTION OF THE MILLIARCSECOND POLARIZATION STRUCTURE OF THE SUPERLUMINAL QUASAR 3C345

J. F. C. Wardle, D. H. Roberts, L. F. Brown, and D. C. Gabuzda
Department of Physics
Brandeis University
Waltham, MA 02254 USA

ABSTRACT. The $\lambda 6$ cm milliarcsecond polarization structure of 3C345 has been determined at three epochs between December 1981 and March 1984. The knots C2, C3, and C4 all showed changes as they moved away from the core, which remained virtually unpolarized.

1. INTRODUCTION

We have embarked on a program to study the the structure and evolution of the polarization of 3C345 (and other sources) at milliarcsecond resolution. A first epoch (1981.92) map of 3C345 was presented by Wardle *et al.* 1986. Here we show *preliminary* maps made from data taken in December 1983 and March 1984, and give the first report of the evolution of the milliarcsecond polarization structure.

2. OBSERVATIONS AND DATA PROCESSING

Polarization-sensitive $\lambda 6$ cm observations were carried out in December 1983 and March 1984 with the Bonn (1984 only), Haystack, Green Bank, Fort Davis, and Owens Valley antennas and the phased-up VLA, using the Mark III data recording system. Maps of the total and polarized intensity were made as described by Roberts *et al.* (1984) and Wardle *et al.* (1986).

3. RESULTS

Maps of the I and P distributions of 3C345 in December 1983 and March 1984 are shown in Figures 1 and 2. The total intensity maps are in good agreement with those of Biretta, Moore, and Cohen (1986). The evolution of the polarization of the core and knots of 3C345, derived from summing CLEAN components or by model fitting, is summarized in Table 1. The polarization of the core of 3C345 is very small ($\lesssim 1\%$) at each epoch, requiring a very efficient depolarization mechanism (if the radiation is from the synchrotron process in an ordered magnetic field). The knots are moderately polarized, and as they move away from the core their electric vectors are seen to change. This may be due to changing orientation of the magnetic field, a changing Faraday rotation, or changing relativistic aberration. Additional observations of 3C345 at $\lambda\lambda$ 2.8 and 6 cm, which can decide among these possibilities, are being reduced.

M. J. Reid and J. M. Moran (eds.), The Impact of VLBI on Astrophysics and Geophysics, 163–164.

We thank the staffs of the US VLBI Network observatories and of the Haystack Mark III Processor for their help. Financial support was provided by the NSF under grants AST-82-13991 and AST-84-18636 (JFCW) and AST-83-15945 and AST-85-19529 (DHR).

REFERENCES

Biretta, J. A., Moore, R. L., and Cohen, M. H. 1986, *Astrophys. J.*, **308**, 93–109.

Roberts, D. H., *et al.*, 1984, in *VLBI and Compact Radio Sources*, ed. R. Fanti, K. Kellermann, and G. Setti (Dordrecht: Reidel), 35–38.

Wardle, J. F. C., Roberts, D. H., Potash, R. I., and Rogers, A. E. E. 1986, *Astrophys. J. (Letters)*, **304**, L1–L4.

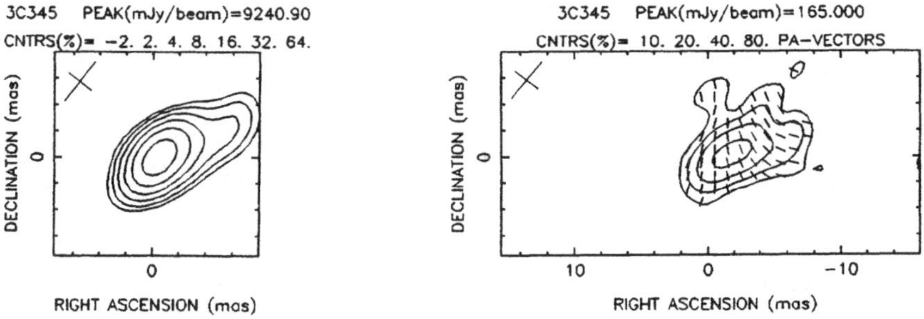

Figure 1: 3C345, epoch 1983.93 (one tick mark is 2 mas). (a) Total intensity. (b) Polarized intensity, with contours of polarized flux and electric vector position angles indicated by the line segments.

Figure 2: 3C345, epoch 1984.23. Same parts as Figure 1.

Table 1: Evolution of the Polarization of the Components of 3C345

	1981.92		1983.93		1984.23	
Component	$m(\%)$	χ(deg)	$m(\%)$	χ(deg)	$m(\%)$	χ(deg)
D	< 1	...	$\sim 1^a$	??	$\sim 1^b$	-72
C4	...	...	...	...	4.	-19
C3	11.	+22.	13.	+34.	14.	+28.
C2	6.	+83.	9.	+74.	6.	+61.

[a] Due in part to C4.

[b] Due in part to C5.

VLA POLARIMETRY OF VLBI CORE–JET SOURCES

J.M. Wrobel[1], T.J. Pearson[2], M.H. Cohen[2], and A.C.S. Readhead[2]

[1] Physics Department, New Mexico Institute of Mining and Technology
[2] Owens Valley Radio Observatory, California Institute of Technology

Radio imaging of VLBI core–jet sources can be used to examine the case for continuity of jet–like features between parsec and circumgalactic scales. Futhermore, polarimetry of such sources allows investigation of the dominant magnetic field topologies as a function of linear offset from the central engine. Examination of these continuity and field topology issues is essential for an understanding of how energy is channeled from the nuclear regions to the circumgalactic environment.

Linear polarization position angles (χ_λ) in radio images, especially those at wavelengths $\geq$ 18 cm, must be corrected for Faraday rotation before intrinsic magnetic field configurations can be inferred. Rotation measures of VLBI core–jet sources are difficult to determine from broad-band (e.g., 2, 6, and 20 cm) polarimetry, for two reasons (Rudnick and Jones 1983). First, the wavelength–dependent opacity alters the electron population emitting at different wavelengths. Second, structural complexities and time variations which do not appear simultaneously at all wavelengths make polarization position angle comparisons dubious. Rudnick and Jones (1983) demonstrated that these problems can be overcome by conducting simultaneous, multi–wavelength polarimetry within the 20–cm window of the NRAO VLA.

This technique was applied to 7 VLBI core–jet sources from the survey of Pearson and Readhead (1984). Some properties of these 7 are given in Table I. Snapshots of each source were taken at wavelengths of 21.6, 21.2, 20.5, 19.8, 18.3, and 18.0 cm. The resolution was $\sim 1''$. χ_λ is plotted in Figure 1 as a function of λ^2. The data for each source involve measurements either of its unresolved VLA core (size $< 1''$ or ≤ 4 kpc – see col. 4), or of its kpc–scale emission adjacent to the VLA core. For each source, the data were fit to a straight line by the method of least squares. Each fit yielded a zero–wavelength intercept (χ_0, col. 5) and a rotation measure (RM, col. 6).

The rotation measures resemble those of nine 3C sources with similar ranges of Galactic longitude and latitude (l^{II} = 75–115^0 and b^{II} = 20–50^0; Simard–Normandin, Kronberg, and Button 1981). This is consistent with a primarily Galactic origin for the rotation measures of both the (presumably steep–spectrum) 3C sources and the VLBI core–jet sources. If the Galactic disk behaves like an unresolved foreground screen, then it should cause Faraday rotation but no depolarization (Laing 1984). Such behavior is shown by 1803+784, the core–jet source with the largest rotation, for which the observed percent polarization p_λ is remarkably constant at 3.9 percent between 21.6 and 18 cm. For comparison, if the RM was due to mixed thermal and emitting plasma, then a simple uniform slab model would predict a factor of 5 drop in p_λ between 21.6 and 18 cm.

For 5 sources χ_0 refers to the unresolved VLA core and can be compared to the VLBI structural position angle χ_v at 6 cm for 1642+690 (Pearson *et al.* 1986) and at 18 cm for 0836+710, 1749+701, 1803+784, and 1928+738 (Eckart *et al.* 1987). Col. 7 gives the value of the acute angle A between χ_0 and χ_v. For sources with multiple VLBI components, χ_v refers to the outermost one. If the linearly polarized emission is due to optically thin synchrotron emission, then A $\geq 75^0$ (0836+710, 1642+690, 1928+738) indicates a projected magnetic field fairly well aligned with the VLBI jet. Bridle (1984) finds a similar trend among kpc–scale jets that, like the VLBI jets, are one–sided. The small A values found for 1749+701 and 1803+784 could be caused by, for example, the polarized emission originating in regions whose structural angles differ from those of the VLBI components; or else in regions with similar structural angles but whose magnetic fields are associated with shocks in the jet.

M. J. Reid and J. M. Moran (eds.), The Impact of VLBI on Astrophysics and Geophysics, 165–166.

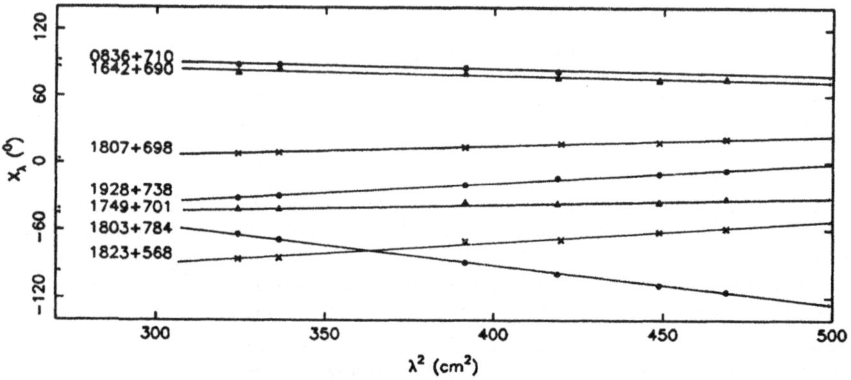

Fig. 1. – Linear polarization position angle χ_λ as a function of λ^2. Symbols distinguish between polarimetry of an unresolved VLA core (< 1″, filled symbols) and of adjacent kpc–scale emission (crosses). Straight lines are the results of least–squares fits.

References

Bridle, A.H.: 1984, *Astr. J.* **89**, 979–986.

Eckart, A., *et al.*: 1987, *Astr. Ap. Suppl.* **67**, 121–146.

Laing, R.A.: 1984, in *Physics of Energy Transport in Extragalactic Radio Sources*, ed. A.H. Bridle and J.A. Eilek (NRAO: Greenbank), pp. 90–96.

Lawrence, C.R., *et al.*: 1986, *Astr. J.* **91**, 494–495.

Pearson, T.J., and Readhead, A.C.S.: 1984, in *IAU Symposium 110, VLBI and Compact Radio Sources*, ed. R. Fanti, K. Kellermann, and G. Setti (Dordrecht: Reidel), pp. 15–24.

Pearson, T.J., *et al.*: 1986, *Ap. J. Letters* **300**, L25–L29.

Rudnick, L. and Jones, T.W.: 1983, *Astr. J.* **88**, 518–526.

Simard–Normandin, M., Kronberg, P.P., and Button, S.: 1981, *Astroph. J. Suppl.* **45**, 97–111.

Wrobel, J.M.: 1987, in *Superluminal Radio Sources*, ed. J.A. Zensus and T.J. Pearson (Cambridge Univ. Press: Cambridge), pp. 186–192.

Table I. Seven VLBI Core–Jet Sources

IAU	Common	Id[a]	Scale [b] (kpc/″)	χ_0 (°)	RM (rad m^{-2})	A (°)
0836+710	4C71.07	Q	4.0	+109±4	-11±2	75
1642+690	4C69.21	Q	4.1	+103±5	-11±2	88
1749+701	...	B	4.1	- 61±4	+10±2	16
1803+784	...	B	4.0	+ 51±2	-62±1	30
1807+698	3C371	G	0.7	- 20±2	+15±1[c]	...
1823+568	4C56.27	B	3.9	-148±6	+33±2	...
1928+738	4C73.18	Q	2.7	- 89±2	+31±1	83

Notes –

[a] Q = quasar, B = BL Lac object, G = galaxy.

[b] H_0 = 100 km s^{-1} Mpc^{-1} and q_0 = 0.5. Redshifts are from Pearson and Readhead 1984, Lawrence *et al.* 1986, Eckart *et al.* 1987, and Lawrence, private communication.

[c] Preliminary RM given by Wrobel 1987.

MILLIARCSECOND POLARIZATION PROPERTIES OF SEVERAL BL LACERTAE OBJECTS

D. C. Gabuzda, D. H. Roberts, J. F. C. Wardle, and L. F. Brown
Department of Physics
Brandeis University
Waltham, MA 02254 USA

ABSTRACT. We discuss the λ6 cm total intensity and polarization structures of a number of BL Lacertae objects at milliarcsecond resolution. 0235+164 was unresolved and weakly polarized at each of two epochs a year apart; each of the other objects displays structure in polarized flux. 0735+178 and 1749+096 can be adequately modeled by two or three point components—a "core" plus one or two "knots." The core components were moderately polarized ($\simeq$ 5%), while "knots" may be polarized at 8% or more, consistent with these components being optically thin. Preliminary results for BL Lac indicate that the total intensity structure can be modeled well by a set of four gaussian components; the polarization structure is complex, but is dominated by the northernmost knot in the jet.

1. INTRODUCTION

Here we report results of polarization-sensitive VLBI observations of the BL Lacertae objects 0235+164, 0735+178, 1749+096, and BL Lac. Polarization-sensitive λ6 cm observations were carried out in December 1981 and December 1982 with the Haystack, Green Bank, and Owens Valley antennas and the phased-up VLA, and in March 1984 with the Bonn and Fort Davis antennas in addition to these four. In all cases the Mark III data recording system was used. Maps of the total intensity and of the distribution of polarized intensity were made as described by Roberts, Gabuzda, and Wardle (these proceedings).

2. RESULTS

We fit simple models to the visibilities produced by the hybrid mapping process and, when dynamic range permitted, required that the models account for the I and P data simultaneously. Maps of the I and P distributions of BL Lac are presented in Figure 1; maps of 0735+178 can be found in Roberts and Wardle (1986). A more detailed discussion will be presented in Gabuzda, Roberts, and Wardle (1987).

Data for 0235+164 were taken at the earlier two epochs. At the first epoch, the object is unresolved and polarized $\simeq$ 1.4% at position angle 39°. At the second epoch, model fitting renders some evidence for a secondary component at a separation of $\simeq$ 1.8 mas from the core at position angle 45°, but the polarized fringes are too weak to determine the parameters of this component with any certainty. The core is $\simeq$1.2% polarized at position angle $\chi \simeq 110°$, indicating that χ for the core rotated by at least $\simeq$ 70° during the year between observations.

M. J. Reid and J. M. Moran (eds.), The Impact of VLBI on Astrophysics and Geophysics, 167–168.

Data for 0735+178 were taken in December 1982. At this epoch, this object can be represented by an unresolved core and a jet-like extension at structural position angle $\simeq 69°$ consisting of two knots, one located $\simeq 2.2$ mas from the core and one located $\simeq 4.7$ mas from the core. The inner knot has flux density $\simeq 200$ mJy and is $\simeq 3\%$ polarized at position angle $\chi \simeq 20°$; the outer knot has flux density $\simeq 160$ mJy and is $\simeq 8\%$ polarized at $\chi \simeq 45°$. The core is $\simeq 4\%$ polarized at $\chi \simeq 4°$. If the rotation measure for 0735+178 is small, as is indicated in the data of Altschuler and Wardle (1976) and Aller *et al.* (1985), these polarization position angles indicate a **B** field which is predominantly parallel to the jet at the position of the core and rotates further down the jet until it is predominantly perpendicular to the jet at the position of the outer knot.

Data for 1749+096 were taken in December 1981. The polarization structure can be represented by a $\simeq$ 2:1 double with a separation of $\simeq 0.8$ mas at position angle $\simeq 13°$. Our dynamic range was insufficient to identify an I counterpart to the northern P component. The core is $\simeq 6.5\%$ polarized at $\chi \simeq 40$; χ for the northern component is $\simeq 112°$. If the rotation measure for this object is small, as is again indicated by the data of Altschuler and Wardle and of Aller *et al.*, the **B** field direction inferred in the northern component is predominantly parallel to the position angle of the structure.

Data for BL Lac were taken in March 1984. *Preliminary* results indicate that the total intensity structure can be modeled by a set of four gaussian components—what may be a compact core plus three resolved knots in a jet which extends southward. The polarization structure has proved more difficult to model fit, but appears to be dominated by the northernmost knot.

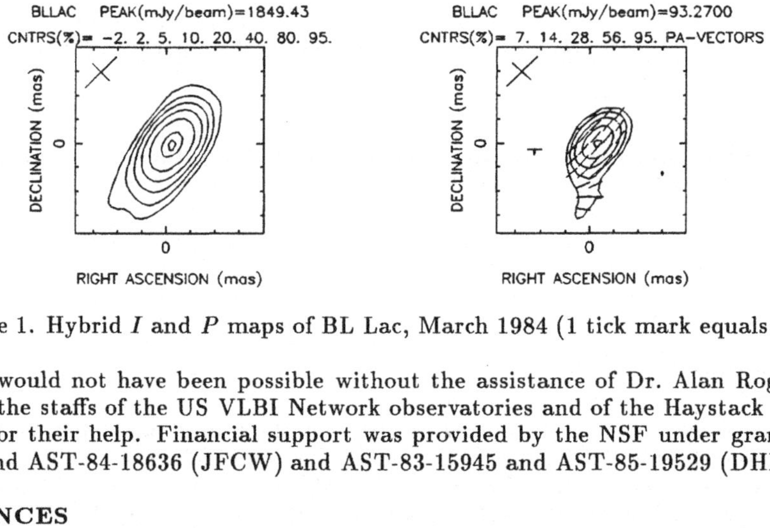

Figure 1. Hybrid I and P maps of BL Lac, March 1984 (1 tick mark equals 2 mas).

This work would not have been possible without the assistance of Dr. Alan Rogers. We also thank the staffs of the US VLBI Network observatories and of the Haystack Mark III Processor for their help. Financial support was provided by the NSF under grants AST-82-13991 and AST-84-18636 (JFCW) and AST-83-15945 and AST-85-19529 (DHR).

REFERENCES

Aller, H. D., Aller, M. F., Latimer, G. E., and Hodge, P. E. 1985, *Ap. J. Supp.* **59**, 513.

Altschuler, D. R., and Wardle, J. F. C. 1976, *Mem. R. astr. Soc.* **82**, 1.

Gabuzda, D. C., Roberts, D. H., and Wardle, J. F. C. 1987, in preparation.

Roberts, D. H., *et al.* 1984, in *VLBI and Compact Radio Sources*, ed. R. Fanti, K. Kellermann, and G. Setti (Dordrecht: Reidel), 35.

Roberts, D. H., and Wardle, J. F. C. 1986, in *Quasars*, ed. G. Swarup and V. K. Kapahi, (Dordrecht: Reidel), 141.

MASERS AND THE COSMIC DISTANCE SCALE

Mark J. Reid, James M. Moran, and Carl R. Gwinn
Harvard–Smithsonian Center for Astrophysics
60 Garden Street
Cambridge, Massachusetts 02138

ABSTRACT. Studies of H_2O masers have demonstrated the power of VLBI techniques to measure relative positions with sufficient accuracy ($\sim 10\ \mu$as) to determine proper motions and to estimate distances to maser sources throughout the Galaxy. The distance to four H_2O masers have been determined, and the distance to the center of the Galaxy has been determined to be 7.1 ± 1.5 kpc from observations of the H_2O masers in Sgr-B2. Proper motion distances for other H_2O masers, and possibly for OH masers, may allow the determination of the fundamental parameters describing the size (R_o) and rotation rate (Θ_o) of the Galaxy with accuracies of better than 10%. Finally, the measurement of the proper motions of H_2O masers in nearby galaxies (< 10 Mpc) is feasible and offers the possibility of direct calibration of the extragalactic distance scale.

1. INTRODUCTION

The combination of proper motion data and information about motion along the line-of-sight can be used to determine the distance to a source, as well as to understand the dynamics of the emitting material. The proper motion of a component, $\dot{\theta}_{app}$, is given by

$$\dot{\theta}_{app} = \frac{1}{d}\left(\frac{v \sin\theta}{1 - \beta\cos\theta}\right) \quad , \tag{1a}$$

or in the nonrelativistic limit by

$$\dot{\theta}_{app} = \frac{v \sin\theta}{d} \quad , \tag{1b}$$

where v is the speed of the emitting material moving at an angle θ to the line-of-sight, $\beta = v/c$, c is the speed of light, and d is the distance to the source. Line-of-sight velocities can be determined indirectly from intensity information, provided bulk relativistic motions are involved, via the equation

$$I_{app}(\nu_{app}) = D^3 I_{em}(\nu_{em}) \quad , \tag{2a}$$

M. J. Reid and J. M. Moran (eds.), The Impact of VLBI on Astrophysics and Geophysics, 169–174.

where $D = \sqrt{1-\beta^2}/(1-\beta\cos\theta)$, or directly from Doppler (or radial) velocities, v_R, of spectral lines via

$$v_R \approx v\cos\theta \quad , \tag{2b}$$

in the nonrelativistic limit.

Equations (1) and (2) relate measurable quantities to three physical parameters: the distance to the source (d) and the speed (v) and direction (θ) of the emitting material. VLBI observations of quasars and active galactic nuclei are often interpreted in the context of equations (1a) and (2a), and, assuming d from the source redshift, one can, in principle, determine the "jet" flow speed and direction. Alternatively, for molecular maser sources, equations (1b) and (2b) are appropriate. If proper motions and Doppler velocities are measured for a number of maser spots in one source, one can solve for the parameters by modelling the source dynamics or by assuming an isotropy constraint (e.g., $\langle\sin^2\theta\rangle = \langle\cos^2\theta\rangle$ as for a statistical parallax). Both techniques have been used to determine distances to H_2O maser sources.

2. MEASURED DISTANCES TO H_2O MASERS

Maser emission occurs in the envelopes of newly formed massive stars (interstellar masers) and evolved Red Giant stars (stellar masers). H_2O and OH molecules are trace constitutent of these envelopes, and population inversion of the molecular energy levels leads to the appearance of "spots" of emission $\sim 10^{13}$ cm is size with brightness temperatures as high as 10^{15} K. Because of the small size and high brightness of interstellar maser spots, they can be used to make precise astrometric measurements that allow proper motions to be determined. With VLBI techniques, relative positions across fields of $\sim 3''$ can be measured with an accuracy of ~ 10 microarcsec (μas). This accuracy is sufficient to determine proper motions and estimate distances throughout the Galaxy.

Genzel *et al.* (1981a) measured proper motions of H_2O maser spots in the Orion–KL source, and found that the spots are expanding about a central point near the massive young star IRc2. Using the two dimensions of spatial information (i.e., relative positions), and the three dimensions of velocity information (i.e., radial and proper motions) available for the maser spots, they constructed a three dimensional kinematic model of the source. The distance of the entire source, which is needed to relate the angular (proper) motions to true velocities, was then estimated by least-squares fitting techniques to be 480 ± 80 pc, in accord with optical luminosity distances of associated O-stars.

Genzel *et al.* (1981b) and Schneps *et al.* (1981) measured proper motions of H_2O masers in W51(M) and W51(N), respectively. The motions in these sources appeared largely random, and the technique of statistical parallax was employed to determine distances to the sources. Distance estimates are given by the ratio of the dispersion in Doppler velocities to the dispersion in (each dimension) of the proper motions. The distance to the W51 sources was estimated to be 7 ± 1.5 kpc. This value is consistent with the far kinematic distance to the W51 region.

Recently completed studies of the water masers in the galactic center region, Sgr-B2 (Reid *et al.* 1988), indicate a phenomenon similar to the Orion case. The H_2O maser spectrum of Sgr-B2(North) is composed of $\sim$100 maser spots with an LSR velocity range from 15 to 115 km s^{-1}. Spectral-line VLBI maps made from observations between 1980.9 to 1982.5 clearly show up to 50 spots stronger than

1 Jy at each epoch. With four telescopes spanning the United States, a synthesized beam of 2.4×0.4 milliarcseconds (mas) is obtained.

The maser spots in Sgr-B2(North) were slightly resolved and the location of peak brightness was estimated by fitting an elliptical Gaussian model. Formally, the center of a Gaussian brightness distribution can be estimated with a one-sigma precision given by $\sigma_\theta \approx 0.5\theta_{app}/SNR$, where θ_{app} is the apparent size (FWHM) of the maser spot (dominated by the synthesized beam) and SNR is the peak signal-to-noise ratio for the spot in the map. Plots of position versus time for features at the same line-of-sight velocity were made. These plots indicated that the maser spots typically appeared to move in straight lines on the sky to within a factor of two of the expected (theoretical) position uncertainties.

The dominant motion of the maser spots in Sgr-B2(North) appeared to be expansion. A least-squares fit of a uniformly expanding spherical source resulted in an estimate of the distance to the source of 7.1 ± 1.5 kpc. The estimated expansion velocity is about 45 km $s^{-1}$and the velocity residuals were about 15 km s^{-1}. These residuals are much greater than the measurement errors, suggesting that the source contains "turbulent" motions of about 15 km s^{-1}.

Sgr-B2 is almost certainly within 0.3 kpc of the galactic center. Thus, the Sgr-B2 distance estimate can be used directly as an estimate of R_o, the distance between the Sun and the galactic center. The H_2O maser distance estimate of 7.1 kpc to the center of the Galaxy is significantly lower than the previous standard value of 10 kpc and in better agreement with the newly revised IAU value of 8.5 kpc (Kerr and Lynden-Bell 1986).

A decrease in the value of R_o has widespread impact on astrophysics. The following is a brief list of items that would be affected by reducing R_o:

1) decrease all kinematic distances;
2) reduce the mass of the Galaxy and of the galactic center;
3) decrease some extragalactic distances (i.e., increase H_0) by a) favoring recent revisions of the absolute magnitudes of RR Lyrae variables (Frenk and White 1982), which in turn affects distances in the Local Cluster, and b) bringing spiral galaxies closer to match Milky Way size (e.g., DeVaucouleurs 1983);
4) decrease the luminous mass in the Milky Way and increase the "dark matter" in the Local Group (see Trimble 1986); and
5) reduce the total luminosity of X-ray bursts, some of which appear super-Eddington for $R_o = 10$ kpc (see Ebisuzaki *et al.* 1984).

3. FUTURE PROSPECTS FOR GALACTIC DISTANCE MEASUREMENTS

Analysis of existing data as well as future observations hold the promise of greatly increasing the accuracy of estimates not only of R_o, but also of the Galactic rotation parameter, Θ_o. If one assumes circular rotation in the Galaxy, then the observed radial velocity, v_R, and absolute source proper motion, μ_ℓ, are given by

$$v_R = R_0 \sin \ell \left(\frac{\Theta}{R} - \frac{\Theta_0}{R_0} \right) \quad , \tag{3}$$

and

$$\mu_\ell = \frac{\Theta}{d} \left[\left(\frac{R_0}{R} - 1 \right) \cos \ell - \frac{d}{R} \right] \quad , \tag{4}$$

where R and Θ are the distance and rotation speed of the source with respect to the galactic center, d is the distance of the source from the Sun, and ℓ is the galactic longitude of the source. After some algebraic manipulation (and with the assumption of a flat rotation curve), one can use measurements of d, v_R, and μ_ℓ to give estimates of R_0 and Θ_0 as follows:

$$R_0 = \frac{\rho^2 - d^2}{2(\rho - d\cos\ell)} \quad , \tag{5}$$

and

$$\Theta_0 = -\frac{v_R}{\sin\ell}\left[\frac{\rho^2 - 2\rho d\cos\ell + d^2}{2\rho(\rho - d\cos\ell)}\right] \quad , \tag{6}$$

where

$$\rho = R_0 - R = \left(\frac{\cos\ell}{d} - \frac{\mu_\ell \sin\ell}{v_R}\right)^{-1} \quad . \tag{7}$$

From these equations one can see that, in general, measurement of the absolute proper motion as well as the source distance are needed to determine R_0 and Θ_0 with data from a single source. A proper motion experiment to measure the motion of a source against the extragalactic background, such as that of Backer and Sramek (1982) for Sgr A*, is possible using H_2O and/or OH masers. Analysis of equations (5) and (6), however, indicates that only sources near the solar circle (and not too close to the Sun) or sources near the galactic center can be used if deviations from circular motions (~ 10 km s^{-1}) and errors in distances ($\approx 10\%$) are not to affect the estimates of R_o and Θ_o too seriously.

Note that for a source on the solar circle, such as W49, the results just derived have relavence for distances estimated from internal motions, even if the absolute proper motion of the source is not measured. Near the solar circle, the terms involving ρ ($= R_0 - R \approx 0$) in equations (5) and (6) can be much smaller than d. This results in an estimate of R_0 dependent only on the measurement of d. In this case, equation (5) becomes

$$R_0 \approx \frac{d}{2\cos\ell} \quad , \tag{8}$$

independent of the measurement of the absolute proper motion μ_ℓ. The formal uncertainty in estimating R_o, σ_{R_0}, is given by

$$\frac{\sigma_{R_0}}{R_0} \approx \frac{\sigma_d}{d} \quad . \tag{9}$$

Unfortunately, an estimate of Θ_0 depends on both μ_ℓ and d, and near the solar circle, equation (6) reduces to

$$\Theta_0 \approx -\frac{d\,\mu_\ell}{2\cos\ell} \quad . \tag{10}$$

The most luminous maser in the Galaxy, W49, is known to be on the far side of the galactic center (from hydrogen absorption studies). It has a low radial velocity and, therefore, is nearly on the solar circle. Thus, R_o can be accurately determined from the distance to W49, even assuming a crude value for Θ_o. For $\ell_{W49} = 43°$ and

$d_{W49} = 12$ kpc, circular rotation (with $R_0 = 8.5$ kpc and $\Theta_0 = 220$ km s^{-1} pc^{-1}) gives $v_R = 6$ km s^{-1}and $\mu_\ell = -5.4$ mas yr^{-1}. By measuring v_R and μ_ℓ *exactly* and d with an error of 1.2 kpc (10%), R_0 and Θ_0 are estimated with uncertainties of 0.8 kpc and and 20 km s^{-1}, respectively. If we add "noise" to the motion of the source in the amount of 10 km s^{-1} to simulate deviations from circular motion or measurement error in μ_ℓ, the uncertainties for R_o and Θ_o are increased only slightly to 1.0 kpc and 23 km s^{-1}, respectively.

4. EXTRAGALACTIC DISTANCES

H_2O masers have recently been discovered in about a dozen galaxies. For example, several sites of maser activity are known in M33, a galaxy about 0.7 Mpc from the Sun. The strongest H_2O maser in M33 has a flux density of about 5 Jy. The most distant extragalactic H_2O maser known is in the nucleus of the active galaxy NGC 3079. This maser reaches a peak flux density of about 10 Jy; at the distance of 16 Mpc this maser is emitting $500L_\odot$ in the H_2O line, assuming istropic emission. Detections of VLBI fringes have been reported from the galaxies NGC 3079 by Haschick *et al.* (1987) and from NGC 4258 by Claussen *et al.* (1987).

Two types of proper motion studies can be done to estimate distances to galaxies: 1) measurement of the internal motions within one source (as done for example for Orion, W51 and Sgr-B2), and 2) measurement of the orbital motion of an entire source around the galaxy. Regarding the first type of study, galactic H_2O maser sources have internal motions with a dispersion of about 30 km s^{-1}. At a distance of 1 Mpc, such motions correspond to transverse angular velocities of 6 μas yr^{-1}. The second type of proper motion study involves measuring orbital motions due to galactic rotation of H_2O masers with respect to a "reference" source, e.g., an H_2O maser or continuum source located elsewhere in the galaxy (perhaps at the center). In this latter case, the relative velocities are $\sim$ 300 km s^{-1}– an order of magnitude greater than internal source motions. If the orbital speed (rotation curve) and inclination of the galaxy are known, then measurement of the proper motion directly yields the galaxy's distance. A maser located away from the center of a face-on galaxy 1 Mpc from the Sun moves 60 μas yr^{-1}, a motion that over one year is a significant fraction of the synthesized beam of intercontinental VLBI arrays and comparable to the synthesized beam of proposed space-based VLBI observations.

The most crucial aspect in the measurement of proper motions of H_2O masers is to locate reliably each maser spot at several epochs. Maser spots often cluster together. For example, the strongest maser source in the Milky Way is in W49 ($d \approx 12$ kpc). This source contains about 100 maser spots stronger than 100 Jy, with large numbers of these maser spots separated by less than 0.03 arcsec. At a distance of 1 Mpc, this separation would be about 0.3 mas, comparable to the best resolution obtainable by an Earth-based array. Once two or more maser spots with similar radial velocities fall within the interferometer beam, intensity variations mimic position shifts. This leads to a "confusion" limit. For example, an Earth-based array observing a source like W49 in a galaxy 1 Mpc distant would not be able to determine the proper motions of the clustered spots. The higher angular resolution provided by future space-based VLBI observations may prove crucial in resolving the structure of H_2O maser clusters, thereby removing the "confusion" problem (as well as allowing motions to be determined over a shorter time base).

Even now, measurements of extragalactic proper motions are feasible for known sources with the largest existing radio telescopes and reasonable integration times under good weather conditions. However, we note that no nearby galaxy has yet been searched *completely* for H_2O masers. A major program to find the best candidate galaxies for proper motion studies still needs to be done. Since H_2O proper motions yield a direct estimate of the distance to a source, independent of absolute luminosities and extinctions, they share none of the sources of systematic error usually associated with distance estimates. Thus, obtaining a distance to even one galaxy would provide a crucial check on the calibration of Cepheids and of other distance indicators, and hence could remove significant sources of error in the determination of the Hubble constant.

5. REFERENCES

Backer, D. C., and Sramek, R. A. 1982, *Ap. J.*, **260**, 512.
Claussen, M., *et al.* 1987, this volume.
DeVaucouleurs, G. 1983, *Ap. J.*, **268**, 451.
Ebisuzaki, T., Hanawa, T., and Sugimoto, D. 1984, *P.A.S. Japan*, **36**, 551.
Frenk, C. S., and White, S. D. M. 1982, *M. N. R. A. S.*, **198**, 173.
Genzel, R., Reid, M. J., Moran, J. M., and Downes, D. 1981a, *Ap. J.*, **244**, 884.
Genzel, R., *et al.* 1981b, *Ap. J.*, **247**, 1039.
Haschick, A. D., *et al.* 1987, this volume.
Kerr, F. J., and Lynden-Bell, D. 1986, *M. N. R. A. S.*, **221**, 1023.
Reid, M. J., Schneps, M. H., Moran, J. M., Gwinn, C. R., Genzel, R., Downes, D., and Ronnang, B. 1988, *Ap. J.*, in press.
Schneps, M. H., Lane, A. P., Downes, D., Moran, J. M., Genzel, R., and Reid, M. J. 1981, *Ap. J.*, **249**, 124.
Trimble 1986, *Comments Astrophys.*, **11(5)**, p. 257.

DETERMINATIONS OF DISTANCES TO RADIO SOURCES WITH VLBI

Norbert Bartel
Harvard-Smithsonian Center for Astrophysics

ABSTRACT. The most direct method of determining distances to galactic sources outside the Solar System is the measurement of annual parallaxes. Using as an example astrometric observations of the quasars 3C345 and NRAO512 which are separated by ~ 0°.5 at the sky, we show that under certain conditions, annual parallaxes of galactic radio sources can be determined with VLBI with an accuracy of 50 μas (standard error). This accuracy is about tenfold higher than any annual parallax accuracy ever obtained with optical astrometry.

The, so far, most direct method of determining distances to extragalactic sources combines VLBI measurements of a supernova's angular expansion velocity with optical spectroscopic measurements of the supernova's radial expansion velocity. We have estimated the distance to SN1979C in M100 in the Virgo cluster to be 20^{+7}_{-6} Mpc, a lower bound on the distance to SN1980K in NGC6946 to be 5 Mpc, and H_0 to be 65^{+25}_{-20} km s^{-1} Mpc^{-1}, with the uncertainties intended to represent standard errors.

1. INTRODUCTION

Determinations of distances to celestial objects have been made through the decades usually from data obtained with optical telescopes. However under certain conditions significantly more accurate determinations of such distances can be made with radio telescopes and the technique of very-long-baseline-interferometry (VLBI).

The most direct determinations of distances to celestial objects located outside the solar system have been those derived from annual parallax measurements. Optical observations of stars (Monet *et al.* 1987) and VLBI observations of pulsars (Gwinn *et al.* 1986) have yielded determinations of annual parallaxes with a standard deviation as low as ~ 0.5 mas. A tenfold more accurate determination of an upper bound on an annual parallax has been obtained by Bartel *et al.* (1986) for the quasar 3C345 relative to the quasar NRAO512. The technique used allows under certain conditions measurements of annual-parallax distances to galactic sources with unprecedented accuracy. See Reid (1988; this volume) for measurements of statistical-parallax distances.

VLBI determinations of distances, or lower bounds on them, to extragalactic sources have been obtained through supernova observations. By combining

M. J. Reid and J. M. Moran (eds.), The Impact of VLBI on Astrophysics and Geophysics, 175–184.

VLBI measurements of the supernova's angular expansion velocity with optical spectroscopic measurements of the supernova's radial expansion velocity, we have estimated the distance to SN1979C in M100 in the Virgo cluster, a lower bound on the distance to SN1980K in NGC6946, and H_0. We also determined the angular diameter of SN1986J in NGC891. In the remainder I discuss our measurements in more detail.

2. GALACTIC DISTANCE DETERMINATIONS WITH ANNUAL-PARALLAX MEASUREMENTS

2.1 Observations of a quasar pair as a test case

VLBI observations at nine epochs between 1971 and 1983 have been used to determine the positions of the "core" and the jet components of the quasar 3C345 ($1641 + 399$, $z = 0.595$) relative to the more distant compact quasar NRAO512 ($1638+398$, $z = 1.67$) which is separated from the former by $\sim 0.^\circ5$ at the sky. The observations of the 1970's and their reduction are detailed by Wittels *et al.* (1976) and Shapiro *et al.* (1979), and the observations of the 1980's and a description of the reduction of all data are given by Bartel *et al.* (1986). The observations had the primary result that the core of the superluminal quasar 3C345 is stationary with respect to the quasar NRAO512, whereas the jet components of 3C345 are moving (Figure 1). Here we want to ignore the determinations of the positions of the jets and focus on those of the core only. The nine determinations of the core positions relative to those of NRAO512 are shown in Figure 2. The uncertainties (σ: standard deviation) in relative right ascension, $\delta\Delta\alpha$, and relative declination, $\delta\Delta\delta$, are the root-sum-squares of the corresponding standard deviations due to:

a) the ionospheric effects (only for sessions #1-#4 and #7 with single-band observations)
b) the core identification uncertainties caused by the finite angular resolution and dynamic range of the quasar maps,
c) the statistical uncertainties after the latter were (uniformly) scaled for each session so that the χ^2 per degree of freedom of the postfit residuals of the different phase delays was unity,
d) the uncertainties of the atmospheric delay,
e) the uncertainties of UT1, and
f) the uncertainties of the coordinates of the pole position, antenna sites, NRAO512 and of the values for the earth tides.

These contributions are listed for each of our observations in the 1980's in Table 1. The smallest total standard deviation was obtained for the last session, 39 μas in $\Delta\alpha$ and 49 μas in $\Delta\delta$, with the uncertainty of locating the core in the brightness distribution of 3C345 dominating the error budget. This uncertainty was 35 μas in both coordinates which is equal to $\sim$ 7% of the FWHM of the Gaussian restoring beam for our high-frequency map. Any further improvement in the accuracy of position determination of any source could most likely only come from a decrease of this dominant error, either by improving the dynamic range of the map or, more effectively, by increasing the angular resolution of the VLBI array.

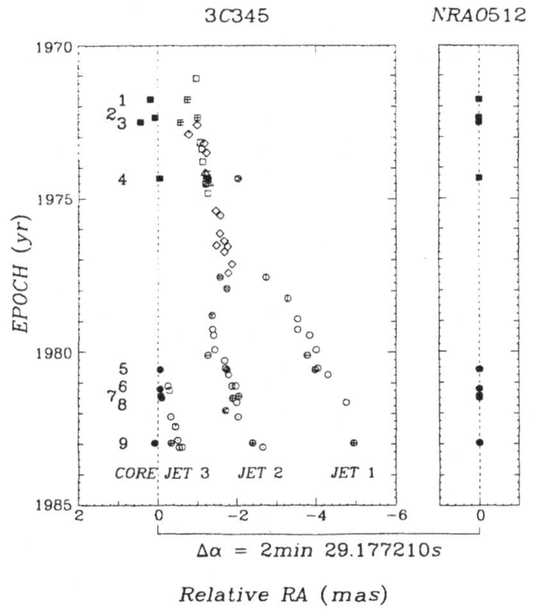

Figure 1. The differences in right ascension (J 2000) between the core component of 3C345 and the single compact component of NRAO512 for the less accurate observations in the 1970's (■, #1-#4) and for the more accurate observations in the 1980's (●, #5-#9). The dashed line marks the weighted mean of these differences. Other symbols indicate the positions of the jet components of 3C345.

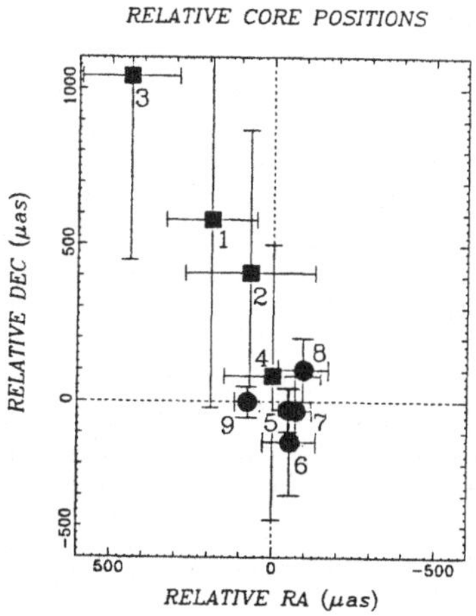

Figure 2. The positions of the core component of 3C345 relative to the position of the single component of NRAO512. The origin of the diagram lies at the weighted mean of the relative position estimates.

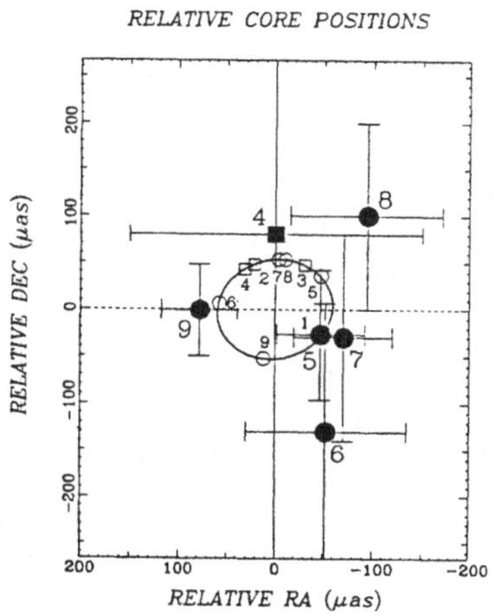

Figure 3. Enlarged inner portion of Figure 2. The ellipse indicates the predicted apparent motion of the core of 3C345, and the light symbols the core's predicted apparent positions at epochs, for the upper bound on any annual parallax of 60 μas.

We estimated the parallax, π, along with the relative right ascension, $\Delta\alpha$, and declination, $\Delta\delta$, at the epoch of our first observations, and the proper motion components, μ_α, μ_δ, by a weighted-least-squares fit to our position determinations.

Table 1
Error Budget for Observations in the 1980's of 3C345 and NRAO512

#	Epoch	Observ. freq. (GHz)	Ionosph. effect at 8.4 GHz[1]	Contributions due to:[2] a)[3]	b)	c)	d)[4]	e)[5]	Total (a-f)
			$\mid\delta\Delta\alpha\mid$, $\mid\delta\Delta\delta\mid$ (μas), (μas)						
5	1980.57	2.3/8.4	6,50	—	35,35	28,53	7, 18	3,21	45, 69
6	1981.20	2.3/8.4	12,10	—	70,70	16,34	38,121	10,82	82,170
7	1981.43	10.7	—	30,30	35,35	8,17	14, 43	12,82	50,110
8	1981.45	23./8.4	67,68	—	70,70	35,66	8, 25	1,17	78,100
9	1982.96	2.3/8.4	83,49	—	35,35	15,22	8, 19	3,18	39, 49

[1] The magnitudes of the differences between the values for the positions determined from the 8.4 GHz data only and the (largely ionosphere free) 2.3/8.4 GHz data. The magnitudes indicate the effect of the ionosphere at 8.4 GHz and were used to estimate the ionospheric effect a) on our astrometric result obtained for epoch 1981.43 from single-band data only.

[2] Contributions due to f) are smaller than 3 μas in $\Delta\alpha$ and 10 μas in $\Delta\delta$.

[3] Dashes indicate that 2.3/8.4 GHz data were used for the analysis. These data are largely free from ionospheric effects. We, however, caution that any mismatch of the fiducial points in the 2.3 GHz and 8.4 GHz maps is carried over to the final position determination at the 7% level.

[4] The uncertainties of the atmospheric zenith delays were assumed to be 0.3μs, except for stations where meteorological data were recorded, or where such delays could be estimated from geodetic VLBI observations immediately preceding or following our astrometric observations. In the latter cases we assumed uncertainties of 0.1 μs.

[5] The uncertainties of UT1 were assumed to be 0.5 ms, except for sessions #5, #8, and #9 for which UT1 could be estimated from above mentioned geodetic VLBI observations. For these sessions we assumed uncertainties of 0.1 ms.

The result is $\pi = 10 \pm 50$ μas, with the uncertainty denoting 1σ, scaled so that it corresponds to χ^2 per degree of freedom of unity. Figure 3 shows an enlarged portion of the inner part of Figure 2 with the ellipse describing the expected apparent motion of the core of 3C345 due to an annual parallax of 60 μas (upper bound of above determination).

From the upper bound on the relative annual parallax of 3C345 and NRAO512, we can place a lower bound on the distance to 3C345 of 15 kpc (85%

confidence limit), if NRAO512 is at a sufficiently greater distance from Earth, i.e. if NRAO512 can be regarded as a background source. This bound is, by far, the largest yet obtained for an annual-parallax distance and with the above assumption, would place the quasar outside the Galaxy.

2.2 Possible observations of pairs of galactic sources and quasars

Instead of promoting the usefulness of determinations of annual parallaxes, or upper bounds on them, for more extragalactic sources, we use our result as an example for the accuracy that could be obtained for the determination of annual parallaxes of galactic sources. Prime candidates for annual-parallax determinations are radio pulsars and radio stars, the latter being, e.g., RS CVn systems, x-ray emitters, and Mira variables (see Lestrade 1988 for a description of radio stars). If, e.g., for a galactic radio source six position determinations over the course of a year are made with only 1/4 of the accuracy we obtained for the last determination of the position of 3C345 relative to that of NRAO512, then annual-parallax distances can be determined with an accuracy of $\sim$ 10% for sources up to 1 kpc away from Earth.

3. EXTRAGALACTIC DISTANCE DETERMINATIONS WITH SUPERNOVA OBSERVATIONS

3.1 SN1979C

The supernova SN1979C in the galaxy M100 [NGC4321, distance= 11.8 Mpc (de Vaucouleurs 1982) and 22.5 Mpc (Sandage and Tammann 1985)] in the Virgo cluster has been emitting radio radiation; at 5 GHz the flux-density maximum of $\sim$ 7 mJy occurred about one year after the explosion which we assume to have happened on 1979 April 1. The flux density has decreased since the time of the maximum by $\sim$ 50% (Weiler *et al.* 1986). Figure 4 displays a radio map of the galaxy with the supernova located in one of the galaxy's spiral arms.

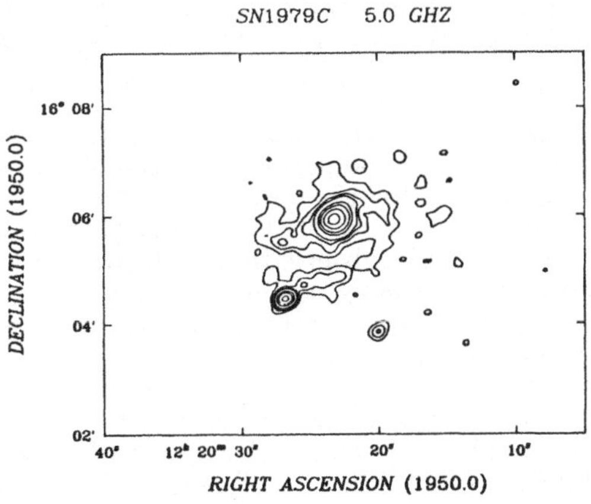

Figure 4. Radio map of SN1979C, located at the southern edge of a spiral arm of the galaxy, M100. The map was made with the VLA on 1982 Dec 8.

A VLBI determination, at time t after the explosion, of the angular size, Θ, of a supernova's radiosphere, coupled with an optical-spectroscopic determination of the radial expansion velocity v_{ph}, at time t_{ph} after the explosion, of the supernova's photosphere allows estimates to be made of the distance, D, to the supernova's host galaxy and of Hubble's constant (Bartel 1985, Bartel *et al.* 1985, see also Bartel 1986, for a comparison of this method with other methods that use supernovae as distance indicators):

$$D = \frac{2\mu\ \eta\ v_{ph} t}{\kappa \Theta\ m} \left[\frac{t}{t_{ph}}\right]^{m-1} \tag{1}$$

$$H_0 = \frac{v_{redshift} + v_{infall}}{D}\ , \tag{2}$$

The parameter μ describes the effect of any deviation from spherical symmetry on the distance determination, η the ratio between the expansion velocity of the radiosphere and that of the photosphere, m a measure of the deceleration (or acceleration) of Θ, with $\Theta \alpha t^m$, and κ the model dependence of Θ. We define the parameter κ so that $\kappa = 1.0$ for a supernova having a shell-like brightness distribution with a shell thickness of $\sim$ 15%. The parameter κ ranges between $\sim$ 1.3 for the brightness distribution being a uniform sphere and $\sim$ 0.8 for the distribution being a ring (see, e.g., Marscher 1985). For comparison, $\kappa = 0.7$, for the FWHM of a Gaussian. The parameter $v_{redshift}$ denotes the redshift velocity of the host galaxy, and v_{infall} its correction due to local peculiar velocities. For a uniformly expanding, spherically symmetric supernova shell equation (1) becomes simply: $D = \frac{2\eta v_{ph} t}{\Theta}$.

The goal is to determine observationally each of the parameters in eq. (1) and thereby to estimate a host galaxy's distance and H_0. We made VLBI observations of SN1979C at several epochs with an array of usually four telescopes, three located in the US (NRAO, OVRO, VLA) and one in Europe (Bonn). The measured visibility amplitudes at an observing frequency of 5 GHz and the predictions from a fit of a uniform sphere model to the amplitudes are given in Figure 5. The corresponding angular-diameter determinations for $\kappa = 1.0$ are displayed in Figure 6, with each of the uncertainties denoting the root-sum-square of the statistical standard error and a contribution due to antenna gain uncertainties. This contribution was taken as the standard error of the diameter estimates obtained by changing, in turn, the sensitivities of each antenna of the array by $\pm$10% while keeping the sensitivities of the other antennas fixed.

From the six dependent parameters in eq. (1), two have so far been determined with VLBI:Θ and m. From a weighted least-squares fit to the diameter determinations we estimated: $\Theta = 0.83 \pm 0.05$ mas for the time of our first VLBI observations 3.69 yr after the supernova explosion, and $m = 1.03 \pm 0.15$. One other parameter, v_{ph}, is known from optical spectroscopic observations (Branch *et al.* 1981): $v_{ph} \sim 10000$ km s^{-1}, for $t_{ph} = 0.6$ yr, after the explosion.

Since the remaining three parameters have not yet been determined, we assume reasonable values and standard errors for them.

Based on the assumption that the radio emission emanates from the shock front region that expands with the largest velocity, and the observation that the largest expansion velocity of SN1979C was $\sim$ 12000 km s^{-1} (Branch *et al.* 1981), we adopt a value for η, so that $\eta v_{ph} = 11000 \pm 1000$ km s^{-1}. The assumed value

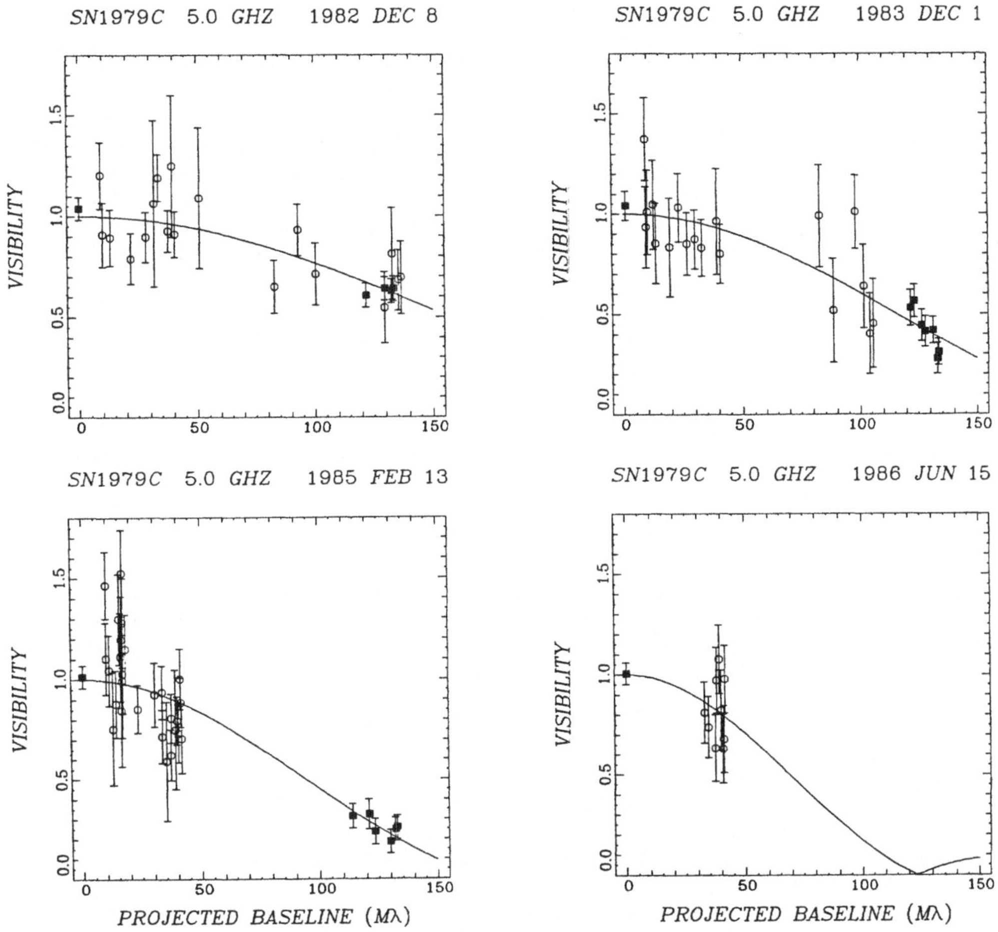

Figure 5. Measured visibility amplitudes at 5.0 GHz and predictions from the fit of a uniform sphere model. The filled squares show the most significant data points, obtained with the VLA alone and with the Bonn-VLA interferometer.

for η is consistent with the results obtained from VLBI observations of SN1987A in the Large Magellanic Cloud (Shapiro *et al.* 1988). Based on observations that many SNRs are approximately circularly shaped (Green 1984), we adopt $\mu = 1.0 \pm 0.15$. Based on observations that $\sim 90\%$ of all known SNRs with discernible morphology have a shell-like structure (Green 1984), we adopt $\kappa = 1.0 \pm 0.15$.

With these values and uncertainties we get:

$$D = 20^{+7}_{-6} \text{ Mpc} \quad .$$

If we further take $v_0 + v_{\text{infall}} = 1250 \pm 150 \ \text{km s}^{-1}$ for the Virgo cluster center (see 12 references in Bartel *et al.* 1985) and assume that the distance to M100 is within

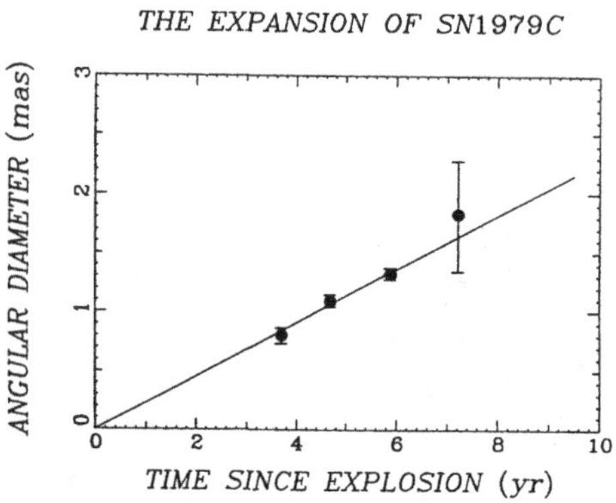

Figure 6. The angular diameter determinations for a shell model. For the extreme models of a ring and a uniform sphere, the ordinate scale has to be multiplied by 0.8 and 1.3, respectively. The solid line represents uniform expansion ($m = 1$), which is consistent with our weighted least-squares solution for m.

10% of the distance to this center, since M100 is within 5° of the center and has a redshift consistent with this location, we get

$$H_0 = 65^{+25}_{-20} \text{ km s}^{-1} \text{ Mpc} \quad ,$$

with the uncertainties of both estimates intended to represent standard errors. A more complete error analysis is pending. The accuracy of these updated estimates is somewhat higher than the accuracy of our earlier estimates (Bartel *et al.* 1985) mostly because more data have been obtained and analyzed since then.

3.2 SN1980K

The supernova SN1980K in the galaxy NGC6946 [distance = 7.1 Mpc (de Vaucouleurs 1979) and 10.5 Mpc (Sandage and Tammann 1974)] emitted radio radiation of $\sim$ 2.5 mJy, at 5 GHz 100 days after the explosion, and has since then also weakened considerably.

We conducted VLBI observations of SN1980K at 2.3 GHz on 1983 May 7 with the sensitive NASA 64-m antennas at Goldstone, CA and Madrid, Spain. The supernova was detected in two out of three adjacent 13-min scans. The visibility amplitudes determined in the two scans, and an upper bound determined in the third scan, are shown in Figure 7. Since the visibility amplitudes of the three segments are expected to be approximately Gaussian distributed, we can compute their mean value and standard error and infer that SN1980K was unresolved at the epoch of our VLBI observations, 2.54 yr after the assumed date of explosion on 1980 Oct. 17. The formal value for the diameter of a shell model is: $\Theta = 1 \pm 1$ mas, equivalent to $\Theta < 2$ mas. The prediction from this model is also shown in Figure 7.

We computed a lower bound on the distance to SN1980K and its host galaxy by again using eq. (1). The last optical spectroscopic observation made 1.82 yr after the explosion of the supernova, indicates $v_{\rm ph} \sim 7900$ km s^{-1} (Uomoto and Kirshner 1986). We take $\eta v_{\rm ph} = 9500 \pm 1000$ km s^{-1}, consistent with the supernova's maximum expansion velocity of ~ 9500 km s^{-1} about one month after the explosion. Assuming the remaining parameters in eq. (1) to

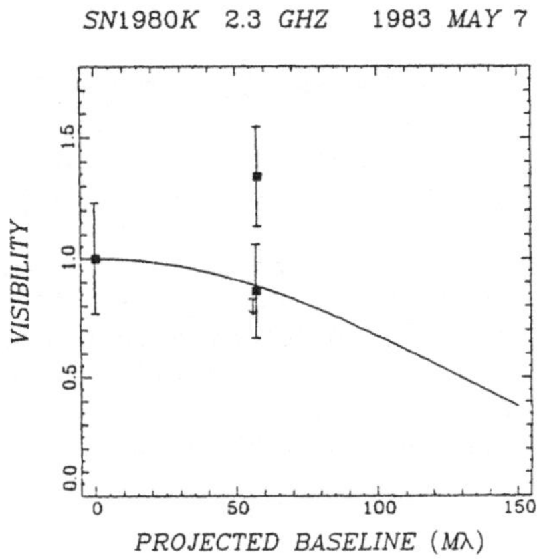

Figure 7. Visibility amplitudes and the prediction from a uniform sphere model. The visibility amplitude at zero spacing was obtained from interpolating between several measurements made with the VLA by Weiler *et al.* (1986) at 1.4 and 5 GHz two days to seven weeks prior to our VLBI observations.

be: $m = 1.0 \pm 0.15$, $\mu = 1.0 \pm 0.15$ and $\kappa = 1.0 \pm 0.15$, we formally get: $D = 10^{+\infty}_{-5}$ Mpc, equivalent to

$$D > 5 \text{ Mpc} \quad ,$$

with the uncertainties being again intended to represent standard errors.

3.3 SN1986J

The supernova SN1986J in the galaxy NGC891, at 0.56 the distance to the center of Virgo (Aaronson *et al.* 1982), has been by far the strongest radio supernova detected to date. The flux density reached 128 mJy at 5 GHz on 1986 May 1 (Rupen *et al.* 1987). Optical spectroscopy revealed an extremely small expansion velocity of $v_{\rm ph} \sim 1000$ km s^{-1} (Rupen *et al.* 1987).

We conducted VLBI observations of SN1986J at 1.7 GHz on 1986 Sept. 29 (Bartel *et al.* 1987) and at 10.7 GHz on 1987 Feb. 23, and determined the supernova's angular sizes for $\kappa = 1$: 2.0 ± 0.3 mas and 1.4 ± 0.1 mas, respectively, with the former diameter at 1.7 GHz being most likely broadened appreciably by scattering in the interstellar medium.

Unfortunately for our ability to estimate the supernova's distance, the supernova's date of explosion has not yet been well determined. Since the supernova remained undetected till at least 1975 (Allen *et al.* 1978) and was first detected in 1984 (van der Hulst *et al.* private communication with Rupen *et al.* 1987), the supernova probably exploded sometime in between. Further, optical spectroscopic observations commenced only after the discovery of the supernova at radio wavelengths in 1986. Crucial data from the early expansion of the supernova were therefore never obtained. However, if we assume for SN1986J the same values and uncertainties for the parameters m, η, μ and κ, we determined or adopted for SN1979C, then our value for the ratio between the distances to SN1986J and Virgo is at least 40% smaller than that of Aaronson *et al.* (1982). More VLBI and optical spectroscopic observations are needed to determine accurately the supernova's expansion velocity and morphology and to shed more light on this rather unusual source.

4. CONCLUSIONS

Astrometry with VLBI has enabled us to determine an upper bound on the annual parallax of the quasar 3C345 relative to the quasar NRAO512 with a standard error of 50 μas. Such accuracy is sufficient for the determination of annual-parallax distances to galactic compact radio sources up to $\sim$ 1 kpc away from Earth with an uncertainty smaller than 10%.

Distances to extragalactic sources and Hubble's constant can be determined by using a combination of VLBI and optical spectroscopy of supernovae. This combination represents the, so far, most direct method of such determinations and should allow, under certain conditions, measurements of distances to galaxies up to 40 Mpc away from Earth with an uncertainty smaller than 20%.

This work was supported by the NSF under grant no. AST 85-19763.

5. REFERENCES

Aaronson, M. *et al.* 1982, *Ap. J. Suppl.*, **50**, 241.

Allen, R. J., Baldwin, J. E. and Sancisi, R. 1978, *Astr. Ap.*, **62**, 397.

Bartel, N. 1985, in *Supernovae as Distance Indicators, Lecture Notes in Physics*, ed. N. Bartel (Springer-Verlag, Berlin), **224**, 107.

Bartel, N. 1986, in *Highlights of Astronomy*, ed. J. P. Swings, **7**, 655.

Bartel, N., Herring, T. A., Ratner, M. I., Shapiro, I. I., and Corey, B. E. 1986, *Nature*, **319**, 733.

Bartel, N., Rogers, A. E. E., Shapiro, I. I., Gorenstein, M. V., Gwinn, C. R., Marcaide, J. M., and Weiler, R. W. 1985, *Nature*, **318**, 25.

Bartel, N., Rupen, M., and Shapiro, I. 1987, *IAU Circ.* 4292.

Branch, D., Falk, S. W., McCall, M. L., Rybski, P., Uomoto, A., and Wills, B. J. 1981, *Ap. J.*, **244**, 780.

de Vaucouleurs, G. 1979, *Ap. J.*, **227**, 380.

de Vaucouleurs, G. 1982, *Ap. J.*, **253**, 520.

Green, D. A. 1984, *M. N. R. A. S.*, **209**, 449.

Gwinn, C. R., Taylor, J. H., Weisberg, J. M., and Rawley, L. A. 1986, *A. J.*, **91**, 338.

Lestrade, J.- F. 1988, this volume.

Marscher, A. P. 1985, in *Supernovae as Distance Indicators, Lecture Notes in Physics*, ed. N. Bartel (Springer-Verlag, Berlin), **224**, 130.

Monet, D. G., Dahn, C. C., Harris, H. C., Ables, H. D., Luginbuhl, C. B., Pier, J. R., and Vrba, F. J. 1987, *Bull. AAS*, **19**, 79.09.

Reid, M. J. 1988, this volume.

Rupen, M. P., van Gorkom, J. H., Knapp, G. R., Gunn, J. E., and Schneider, D. P. 1987 *A. J.*, **94**, 61.

Sandage, A., and Tammann, G. H. 1974, *Ap. J.*, **194**, 559.

Sandage, A., and Tammann, G. H. 1985, in *Supernovae as Distance Indicators, Lecture Notes in Physics*, ed. N. Bartel (Springer-Verlag, Berlin), **224**, 1.

Shapiro, I. I. *et al.* 1979, *A. J.*, **84**, 1459.

Shapiro, I. I. *et al.* 1988, this volume.

Uomoto, A., and Kirshner, R. P. 1986, *Ap. J.*, **308**, 685.

Weiler, K. W., Sramek, R. A., Panagia, N., van der Hulst, J. M., and Salvati, M. 1986, *Ap. J.*, **301**, 790.

Wittels, J. J. *et al.* 1976, *A. J.*, **81**, 933.

VLBI OBSERVATIONS OF SUPERNOVA 1987A

I. Shapiro[1], N. Bartel[1], R. Preston[2], D. Jones[2], A. Kemball[3], G. Nicolson[3], D. Jauncey[4], J. Reynolds[5], A. Whitney[6], A. Rogers[6], R. Phillips[6], T. Clark[7], D. Robertson[8],

[1] Harvard-Smithsonian Center for Astrophysics, U.S.A.
[2] Jet Propulsion Laboratory, California Inst. of Technology, U.S.A.
[3] Hartebeesthoek Radio Astronomy Observatory, South Africa
[4] CSIRO, Australia
[5] Mt. Stromlo and Siding Springs Observatory, Australia
[6] NEROC Haystack Observatory, U.S.A.
[7] Goddard Space Flight Center, U.S.A.
[8] National Geodetic Survey, U.S.A.

ABSTRACT. VLBI observations at 2.3 GHz of SN1987A on 28 February 1987 yielded no fringes, implying, for an optically thin shell, a lower bound on the (outer) diameter of 1.9 mas. From the comparison of the VLBI and optical results, we infer that the radiosphere of SN1987A was either about equal to, or larger than, the photosphere of the supernova five days after the explosion.

1. OBSERVATIONS

Following the detection of radio emission from Supernova 1987A (Turtle *et al.* 1987), we made VLBI observations on each of the days 28 February and 1,2 March 1987 with an interferometer composed of the 26-meter-diameter antenna in Hartebeesthoek, South Africa, and a 34-meter-diameter antenna (DSS 42) in Tidbinbilla, Australia. The observations of SN 1987A were alternated with those of calibrator sources, mostly of 0537-441 but sometimes also of 0521-365, and lasted for 2.5 to 3.5 hours on each day. The fringe spacing was about 2.9 mas. A bandwidth of 56 MHz, centered at 2280 MHz, was recorded at each site (with an offset of 2 MHz having been inadvertently introduced between the band recorded at one site and that at the other). The data were correlated with the Mark III processor at the Haystack Observatory.

2. RESULTS

No fringes have yet been found from the supernova's radio emissions. Fringes were obtained for both (time-variable) calibrator sources with estimated standard errors of about 10% for the correlated flux densities (2.3 Jy and 0.44 Jy for 0537-441 and 0521-365, respectively) for each day of observations. The results agreed, perhaps coincidently, to within these tolerances with VLBI measurements made at the same resolution, but five years earlier (Preston *et al.* 1985; Nicolson 1987, private communication). The peaks observed in the correlation coefficients from the supernova data for 28 February correspond to correlated flux densities between 20

M. J. Reid and J. M. Moran (eds.), The Impact of VLBI on Astrophysics and Geophysics, 185–186.

and 25 mJy and to positions on the sky widely scattered over our search area of about 7$''$ in right ascension and 3$''$ in declination, centered at $\alpha(1950.0) = 05^h35^m49^s.942$ and $\delta(1950.0) = 69°17'57''.6$. Each of these coordinates is within the uncertainty of about 0$''$.3 of White and Malin's (1987) optical coordinates of the supernova. Thus, we conclude that there likely was no signal from the supernova on 28 February with a correlated flux density exceeding 25 mJy. The observations from the other two days have not yet been processed so thoroughly. A somewhat better bound (about 30% lower), or a detection, is expected when the supernova data are processed in an optimum manner (nonstandard processing in the correlation is required due to the 2 MHz offset, mentioned above).

We compared our upper bound on the supernova's correlated flux density with its total flux density of 69 ± 10 mJy (Turtle *et al.* 1987), determined with the Parkes-Tidbinbilla interferometer at 2.3 GHz a few hours before our VLBI observations on 28 February. Measurements with this short-baseline interferometer (baseline length: 275 km) yielded only upper bounds of about 40 mJy at 2.3 GHz for the flux density from the supernova on 1 and 2 March. Hence, no correlated flux density should be expected from the long baseline on these dates.

By matching an optically thin uniform sphere to the upper bound on the correlated flux density, we estimated a lower bound on the sphere's diameter of 2.5 mas for 28 February, *i.e.* about 120 AU, given a distance to SN1987A of 50 kpc. For an optically thin shell with a shell thickness of about 15% of the shell radius, we obtained a lower bound on the corresponding (outer) diameter of about 1.9 mas. These bounds on the diameter, of course, scale directly with the (assumed) distance to SN1987A.

If we assume that the radiosphere has been expanding linearly from a zero size at the epoch of the neutrino detections (Koshiba 1987), our lower bound on the diameter for the assumption of an optically thin shell corresponds to a lower bound on the expansion velocity of the radiosphere of $\sim 1.6 \times 10^4$ km s^{-1}. Such a bound is comparable to the largest value for the expansion velocity of the photosphere of 1.8×10^4 km s^{-1} inferred from the blueshifted minimum of the absorption trough of the P-Cygni profile of the Hα line on 25 February 1987 (Blanco *et al.* 1987; Hanuschik and Dachs 1987) and thereafter.

3. ACKNOWLEDGEMENTS

We thank Turtle *et al.* (1987) for making their first results available to us prior to publication and Tom Reid, Paul Hutchinson and the staff of the Deep Space Communications Complex, Tidbinbilla, for their support. Partial support for this research was provided by the NSF under Grant AST-8519763.

4. REFERENCES

Blanco, V. M. *et al.* 1987, preprint.

Hanuschik, R. W. and Dachs, J. 1987, *Astr. Ap.* (Letters), to be published.

Koshiba, M. 1987, *IAU circ.*, **5338**.

Preston, R.A., Morabito, D.D., Williams, J.G., Faulkner J., Jauncey, D.L., and Nicolson, G.D. 1985, *Astr. J.*, **90**, 1599.

Turtle, A.J., Campbell-Wilson, D., Bunton, J.D., Jauncey, D.L., Kesteven, M.J., Manchester, R.N., Norris, R.P., Storey, M.C.,and Reynolds, J.E. 1987, *Nature*, **327**, 38.

White, G.L. and Malin, D.F. 1987, *Nature*, **327**, 36.

SHELL-LIKE STRUCTURE IN 41.9+58, A POWERFUL SNR IN M82

P.N. Wilkinson, NRAL, Jodrell Bank, U.K.
A.G. de Bruyn, NFRA, Dwingeloo, The Netherlands.

We have studied the brightest of the many compact radio components, designated 41.9+58, in the central region of M82 (Unger et al. 1984; Kronberg et al. 1985) with the EVN at frequencies of 1.6 and 5 GHz (Wilkinson and de Bruyn, 1984; 1987).

The radio image (resolution 5 m.a.s.) of 41.9+58 reveals an elongated shell-like structure (Figure 1) which at the 5% intensity level measures 33x16 m.a.s. At the distance of M82 (D=3.3 Mpc) this corresponds to a linear size of 0.53x0.26 pc. The total flux density of 41.9+58 at the epoch of our observations, October 1985, was about 60 mJy which corresponds to a 5 GHz luminosity some 80 times that of Cas A at its present age.

Our EVN image, and the steady overall flux decay, leads us to favour the hypothesis that 41.9+58 is indeed a SNR as was first suggested by Kronberg and Wilkinson (1974). The flux density of the source has been decaying steadily for the last 20 years, the decay rate being about 8.5%/year at present at both 1.6 and 5 GHz (Kronberg and Sramek, 1985; Unger et al., in preparation). Since the earliest detection of the source by Bash (1968) the source has decreased in flux density by about a factor 3.5. At 1.4 GHz the reduction in the total flux of M82 (3C231) due to the decay of 41.9+58 alone is 0.7±0.2 Jy in a 25 year timespan and this can be detected in the catalogued and recently measured total flux densities.

We can set a (conservative) upper limit to the angular expansion rate along the major axis at 5 GHz between 1980.0 and 1985.8 of 1 m.a.s./year using our EVN data only. (The formal fits yield an expansion rate of about 0.7 m.a.s./year). A careful analysis of the many fragmentary observations of 41.9+58 which have been made since the mid 1970s shows them to be consistent with this estimate and, in fact, all point to a source diameter greater than 15 m.a.s. We have therefore been forced to discount Geldzahler et al.'s (1977) report that 41.9+58 had a source diameter of 1.5 m.a.s. in 1974. The upper limit of 1 m.a.s./year on the angular expension rate translates to a limit on the radial expansion velocity of 7800 km/s along the major axis. The kinematic age, for undecelerated expansion, is therefore ≥35 years.

M. J. Reid and J. M. Moran (eds.), The Impact of VLBI on Astrophysics and Geophysics, 187–188.

If we fit the observed flux density decay of 41.9+58 by a powerlaw of the form $S \propto t^{-\beta}$ then $dS/Sdt = -\beta/t$. The present rate dS/Sdt equals -8.5%/year hence $\beta/T = (12\ \mathrm{years})^{-1}$, where T is the age of the object. Within the context of Shklovsky's (1968) adiabatically expanding, magnetic flux conserving model, which predicts $S \propto r^{-2\gamma}$, where $\gamma = 1-2\alpha$ and $\alpha \sim -0.75$, the value of $\beta = 5$ (for dr/dt=constant); the decay age T would then be 60 years. The "Shklovsky' method, when applied to Cas A, yields an age of about 500 years which is about 1.6 times larger than the canonical value of about 300 years. Applying, naïvely, this factor of 1.6 to 41.9+58 would bring its decay age remarkably close to the limit on the kinematic age. We therefore suggest an age of 40 years, i.e. an explosion date around 1945, for the SN that gave birth to 41.9+58.

The extra ordinary radiopower of 41.9+58 puts in in the same class at that of SN1979c, the SNR in NGC4449 (Seaquist and Bignell, 1978; de Bruyn, 1983), and the recently discovered SN1986j in NGC891. We speculate that these SNR's derive from a separate class of type II SNe (radio hyper-novae!) which become about 25-50 times as bright as most (optically normal) type II SNe like SN1970g, SN1980k and SN1981k (Weiler et al., 1986). Cas A might also belong to the class of radio hypernovae. Evidence in support of this contention is presented, along with a more extensive account of the material presented here, in Wilkinson and de Bruyn (1987).

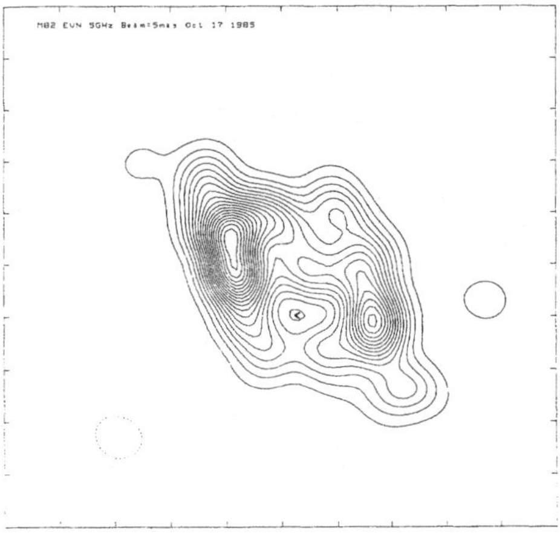

Fig. 1: 5 GHz EVN 5-station hybrid map of the brightest radio source in M82. The total flux density of the source was 60 mJy. Contours are drawn at levels of -5, 5 (5) 95% of the peak flux density of 8.9 mJy. The circular restoring beam measures 5 m.a.s. at half power. Tickmarks are 6 m.a.s. apart, north is up and east to the left.

References:

Bash, F.N., Astrophys. J. Suppl. 16, 373 (1968).
de Bruyn, A.G., Astron. Astrophys. 119, 301 (1983).
Geldzahler, B.J. et al., Ap. J. (letters), 215, 15 (1977).
Kronberg, P.P. and Wilkinson, P.N., Ap. J. 200, 430 (1974).
Kronberg, P.P. et al., Ap. J. 291, 693 (1985).
Kronberg, P.P. and Sramek, R., Science, 227, 28 (1985).
Seaquist, E.R. and Bignell, R.C., Ap.J. 226, L5 (1978).
Shklovsky, I. in "Supernovae", Wiley-Interscience, London (1968).
Unger, S.W., et al. M.N.R.A.S., 211, 593 (1984).
Weiler, K.W. et al., Ap.J. 301, 790 (1986).
Wilkinson, P.N. and De Bruyn, A.G., M.N.R.A.S., 211, 593 (1984).
Wilkinson, P.N. and De Bruyn, A.G., M.N.R.A.S., 1987 (in press).

INITIAL RADIO OBSERVATIONS OF SN1987a IN THE LARGE MAGELLANIC CLOUD

A. J. Turtle and D. Campbell-Wilson
School of Physics, Sydney University, Sydney 2006
J. D. Bunton
Department of Electrical Engineering, Sydney University, Sydney 2006
D. L. Jauncey, M. J. Kesteven, R. N. Manchester, R. P. Norris,
M. C. Storey, and G. L. White
Division of Radiophysics, CSIRO, P.O. Box 76, Epping, 2121
J. E. Reynolds
M.S.S.S.O., Private Bag, Woden, ACT, 2606
D. F. Malin
Anglo-Australian Observatory, P.O. Box 296, Epping, 2121

A prompt radio burst has been observed from the supernova 1987a in the Large Magellanic Cloud. Observations were made at 0.843, 1.415, 2.29, and 8.41 GHz. At frequencies around 1 GHz, the peak flux density reached about 150 mJy and occurred within four days of the supernova. This event may be a weak precursor to a major radio outburst of the type previously observed in other extragalactic supernovae. Radio monitoring of the supernova is continuing at each of the above frequencies, and coordination is underway of a southern hemisphere VLBI array to map the radio outburst region as it expands. Differential astrometry carried out on prime-focus plates taken with the Anglo-Australian telescope indicates that the component, star 1, of Sanduleak's star SK-69202 is within 0.05 ± 0.13 arcsec of the supernova.

A complete discussion of our observations can be found in *Nature*, **327**, 38-40 (1987) and *Nature*, **327**, 36-38 (1987).

M. J. Reid and J. M. Moran (eds.), The Impact of VLBI on Astrophysics and Geophysics, 189.

VLBI OBSERVATIONS OF GRAVITATIONAL LENS SYSTEMS

Edwin L. Turner

Princeton University Observatory
Princeton, NJ 08544 USA

ABSTRACT

Four specific and particularly powerful types of possible VLBI lens studies are discussed. First, comparison of mas scale structure in putative pairs of images separated by arc seconds can provide a powerful additional test of the lens hypothesis in specific candidate systems. Second, VLBI searches for lens systems with image separations too small for resolution by optical or VLA searches will limit (or even determine!) the cosmological density of condensed objects with individual masses $\sim 10^6 M_\odot$. Third, study of multiply imaged superluminal expansion events will allow a determination of the light travel time delay between different images in a lens system, a quantity which is quite difficult to measure by other means but which would allow profound cosmological tests. Fourth, VLBI data can be used to determine relative image parities and even the full magnification matrix of various images in a lens system, thus providing powerful additional constraints on detailed lens models. Finally, the speculative possibility of detecting Galactic stellar lensing events using VLBI techniques is discussed.

1. INTRODUCTION

Since I have participated only very marginally in VLBI observations of gravitational lens systems and since nearly all of those who have made the major contributions to this fledgling subfield are in attendance and will be reporting many of their results in other papers, no review of the current state of the subject will be given. In addition, although the customary brief summaries of the history and basic ideas of gravitational lens theory were included in the presentation of this paper, they have been stated so well and so often elsewhere (see for example, Canizares 1987, Gott 1987, Turner 1987a, Peacock 1983, Walsh 1983, and Gunn 1981) that they will not be included in the written version. Instead I

M. J. Reid and J. M. Moran (eds.), The Impact of VLBI on Astrophysics and Geophysics, 191–199.

shall here concentrate on the special potential of VLBI observations for helping to realize some of the many possible astrophysical and cosmological applications of gravitational lens studies. In effect this paper is an advertisement intended to promote VLBI observations of gravitational lenses.

Before narrowing the focus to possible VLBI observations, it may be well to briefly state the fundamental scientific justification for studying gravitational lensing. The primary reason for the current level of interest in lenses is that they may provide a fundamentally new tool for attacking many difficult problems in cosmology and extragalactic astronomy. In principle, lens observations could determine all three of the standard cosmological parameters (H_0, Ω_0, Λ), allow an independent measurement of galaxy masses, measure or constrain many of the properties of dark matter, facilitate studies of the spatial structure of the IGM, give some indication of the structure of quasars on very small angular scales, help explain apparent clustering on the sky of objects of very different redshifts, account for rare but extreme objects and phenomena (e.g., BL Lacs), and so forth. The list is being steadily expanded by theoretical work. The underlying reasons for this remarkable potential are that lensing involves optical geometry on scales of order the horizon size and that any material which affects the metric, globally or locally, can influence lensing events independent of its emission and absorption properties. It remains to be seen how well many of these potential applications can be achieved in hard observational reality. Nevertheless, these problems are so important and so difficult to address by other available means that enthusiasm for gravitational lens studies is undiminished.

2. TESTING THE LENSING HYPOTHESIS

Currently the most vexing observational problem associated with gravitational lens studies is that of deciding whether or not particular observed systems are actually due to lensing (Djorgovski et. al. 1987, Shaver, Wampler, and Cristiani 1987; Turner 1987b). The alternative hypothesis is usually that we are simply observing two or more separate but physically associated (i.e., clustered) quasars with similar properties. The situation is logically difficult since no degree of similarity can absolutely exclude separate twin objects. Oddly, neither can observed differences, particularly small or moderate ones, really exclude the lensing hypothesis since they can quite plausibly be ascribed to source variability plus differential image time delay or to different microlensing structure in the images (Paczynski 1986; Kayser, Refsdal, and Stabell 1986). At present, no clear cut and generally accepted critical test of the lensing hypothesis exists; each case must be judged on an individual basis.

VLBI observations can provide a very powerful additional test of the lensing hypothesis in systems which are sufficiently strong radio sources by allowing a comparison of the mas scale structure in the separate images. The first order test is just whether or not both images have compact components carrying similar fractions of the total flux. A second order test is possible if high quality VLBI observations yield a phase closure map of the compact components in each image. In this case it should be possible to match these maps to each other after the application of a 2x2 magnification matrix. This test can only be applied to structures with angular sizes much smaller than the image separation (typically a few arc seconds) otherwise much more complex distortions due to highly variable magnification must be considered. Thus, for currently available observing techniques and known candidate lens systems, it is unique to VLBI. Of course, since many sources show great intrinsic variability in flux and structure on mas scales, differential image time delays can confuse these tests. Comparison of source structure at the same emission epoch may be crucial in some cases (see section 4 below).

VLBI observations of the sort described above have lent important support to the best documented case of gravitational lensing 0957+561 (Gorenstein et. al. 1988, Gorenstein et. al. 1984, Porcas et. al. 1981), have yielded preliminary results for another well studied systems 2016+112 (Heflin et. al. 1988), and are being pursued for two other lens candidates 0023+171 and 1042+178 (Hewitt 1987).

3. THE COSMOLOGICAL DENSITY OF LOW MASS COMPACT OBJECTS

The probability of gravitational lensing events for distant sources is proportional to the mass density of the lensing objects in the intervening volume and, for high redshift sources, is of order the lensing objects' contribution to Ω_o (Turner, Ostriker, and Gott 1984; Press and Gunn 1973). Thus, a determination of the frequency of lensing events (or a limit there on) in a sample of distant objects is of fundamental cosmological significance. In order to detect a lensing event it is generally necessary to resolve two or more of the multiple images of the distant source. For fixed source and lens distances the characteristic image separation is proportional to $M_{lens}^{1/2}$ for compact (i.e., point like) lensing objects. Thus, good angular resolution is necessary to detect lensing events due to low mass objects.

The particular relevance of VLBI observations for this problem arises from the fact that the image separation produced by a lens at a typical redshift ($\sim$0.5) for a high redshift ($\sim$2) quasar is $\gtrsim 2(M_{lens}/10^6 M_{\odot})^{1/2}$ mas. This means that only VLBI observations can produce a lensing limit on (or detection of!) a population of compact objects with masses in the particularly theoretically interesting range near $10^6 M_\odot$. This mass range corresponds to the Jeans mass in the cosmic plasma just after recombination (Peebles and Dicke 1968) when the development of significant large scale structure in the Universe probably began, agrees with the mass of dark halo objects needed to heat the old Galactic disk population by tidal shocks (Lacey and Ostriker 1985), and is the mass scale for candidate macroscopic forms of dark matter least constrained by other direct and indirect arguments (Carr 1978). Unfortunately the high frequency of intrinsic structure on mas scales in high redshift radio sources renders this otherwise straightforward cosmological test difficult; structure due to lensing must be distinguished from real source structure in some way. Nevertheless, this cosmological experiment is so powerful, so unique, and so important, a serious attempt to carry it out is clearly indicated in my view.

Despite having been first proposed nearly 15 years ago (Press and Gunn 1973), no serious attempt to use VLBI data for the test described above has been carried out to my knowledge. Recently Hewitt (1986) carried out an analogous experiment using VLA data and derived a limit on the cosmic mass density of compact objects with masses near $10^{11} M_\odot$.

4. DETERMINING DIFFERENTIAL IMAGE TIME DELAYS

The light travel times along the various paths to the source in a multiple image gravitational lens system are not generally equal both because they are of different physical lengths and because they pass through varying depths of the lens's potential field producing varying gravitational time dilations. Thus the source is seen at slightly different look back times in each image. The differences in these light travel times are differential image time delays Δt, and we may write $c\,\Delta t \sim (\Delta\theta)^2 D_{lens}$ where $\Delta\theta$ is the image separation (an observable!) and D_{lens} is the distance to the lens. It should be emphasized that the dimensionless factor required to write this relation as an equality is sensitive to the details of the mass distribution in the lensing object. Thus, a measurement of the differential time delay in a lens systems provides information on a combination of the distance to the lens and its mass distribution. Both are of great interest. A detailed model of the lens is required to realize many of the suggested lens applications, particularly the study of dark matter. If the lens mass distribution can be modeled satisfactorily, the lens distance falls out, and if the lens redshift can be measured, a value of H_0 follows. This possibility of a *direct physical* determination of Hubble's constant on a very large scale, a sort of cosmological Holy Grail, is probably the best known and most assiduously pursued potential application of gravitational lenses.

Since most quasars vary in brightness, the differential image time delays can be measured in principle just by recording the light curves of each image and cross correlating. In practice, a wide variety of intrinsic, technical, practical, and organizational problems make the accumulation of adequate light curve data quite difficult. No unambiguous time delay has yet been determined by this method despite substantial efforts. A possible alternate approach to the problem could be made if VLBI observations of a lens system revealed typical superluminal jet activity. In many superluminal expansion systems two components are observed to be separating with a linear proper motion so that measurement of the separation at any pair of epochs allows extrapolation back to a unique time of zero separation. In the various lens images of a single such superluminal source the zero separation times would differ by the differential time delays which could be determined therefore by a single pair of observational epochs. In reality one would probably want several epochs to check the linearity of the proper motions in each image and to make sure one was viewing the same pair of components in each image and not some previous or successive generation.

Unfortunately, no report of superluminal expansion in a lens system has yet appeared. Continued monitoring of 0957+561 which has a compact core-jet structure (Porcas et. al. 1981, Gorenstein et. al. 1984) and any other lens system discovered to have suggestive compact structure is clearly warranted.

5. MEASURING THE MAGNIFICATION MATRIX

A detailed model of a particular gravitational lens system predicts a mapping of each point in the source plane onto one or more points on the image plane. In any small region of the image plane this mapping can be characterized by a 2x2 magnification matrix applied to some corresponding region of the source plane (Falco, Gorenstein, and Shapiro 1985). Unfortunately, only a tiny fraction of this rich and complex set of predictions can be tested observationally to determine the validity of the model. Naturally, only those regions of the source plane which are emitting detectably produce any signature in the image plane. Furthermore, if the source cannot be resolved as it appears in the image plane, only ratios of the total area magnification (determinant of the 2x2 matrix) can be measured. Thus, in typical situations, a variety of detailed lens models can predict degenerate observational consequences.

As mentioned in section 2, VLBI offers the possibility of partially alleviating this problem. Comparison of resolved mas scale structure in lens images separated by arc seconds can lead to a direct determination of the 2x2 matrix at each image position. Such information can provide powerful and unique constraints on lens models which often must form the basis of other conclusions derived from a particular lens system.

VLBI studies of the gravitational lens 0957+561 have achieved this goal and provided crucial inputs to our best detailed models of that system (Falco et. al. 1985).

6. DETECTING GALACTIC STELLAR GRAVITATIONAL LENSES

Einstein (1936) published the first detailed analysis of gravitational lensing. He considered lensing of stars in the Galaxy by other stars and, after outlining the basic theory, concluded that no such effect was likely to be observed in practice. His grounds for pessimism were 1) that the multiple images were unlikely to be resolvable, and 2) that such alignments are extremely unlikely. In the next year Zwicky (1937) suggested that extragalactic lensing events would be more observationally accessible, and the subject was thereafter relegated to cosmologists. Here I would like to point out that VLBI observations could contradict these expectations by detecting Galactic stellar lensing events.

First the resolution issue is easily addressed by VLBI. The image separation will be $\Delta\theta \gtrsim 2(M/M_{\odot})^{1/2}(D_{lens}/8\text{ kpc})^{-1/2}$ mas for an infinitely distant source and half as great for a source at twice the distance of the lens. Thus stars all over the Galaxy will produce detectable image separations at the best achievable VLBI resolutions.

The low probability of lensing events is a more challenging obstacle. A random line of sight through the Galaxy has a probability $\sim 10^{-6}$ of passing within the standard critical impact parameter of a star required to produce a strong lensing event. Of course, we need not choose random lines of sight; directions very close to the Galactic Plane and even toward the Galactic Center may be chosen. This is sufficient to raise the lensing probability to $\sim 10^{-4}$ or even $\sim 10^{-3}$ in the most favorable directions. Further improvements are possible if one considers not just strong lensing events but also the more probable large impact parameter cases in which one greatly demagnified and one nearly unperturbed image occur (Hewitt 1988, 1986; Press and Gunn 1973). If VLBI maps with dynamic ranges of order 10^3 are achieved, the chances of detecting an event will be increased by a factor of about 30 so that total probabilities can reach levels of 0.003 up to 0.03. At this point the detectable lensing frequency produced by Galactic stars is comparable to that due to external galaxies for high redshift quasars (Turner et. al. 1984) and thus not impossibly rare.

What sorts of VLBI sources should be examined for Galactic stellar lensing? Fortuitously positioned extragalactic sources, masers, and perhaps pulsars come to mind. The masers seem most promising to me since they are bright (allowing large dynamic ranges), numerous, concentrated to the Galactic Plane, and merit much observational attention for other reasons in any case.

How would Galactic stellar lenses be distinguishable from intrinsic source structure? The simplest test would be spectral similarity, in particular identical radial velocities and line shapes. Individual maser components often show complex and individualistic line profiles which should make excellent signatures. A more powerful test could be based on proper motion effects. Galactic rotation will produce relative proper motions of the lens and source of order 6 mas/yr and even parallax effects are not entirely negligible ($\gtrsim$ 0.25 mas), especially for nearby lensing stars. Of course, the masers themselves also have proper motions. Moreover, since the lensing objects are effectively simple point masses (or at worst, binary systems) and the differential time delays are tiny, there will be little ambiguity in fitting a lens model to observations at a single epoch. Observations at multiple epochs must be consistent with this model and some reasonable relative proper motion of the source and lens. A few epochs of observation should provide an entirely conclusive test of the lens hypothesis.

Finally, other than the minor triumph of proving Einstein wrong (albeit, in a trivial way), what value would observations of Galactic stellar lensing have? The most obvious answer would be that information on stellar masses and distances (and thus on stellar interiors theory and Galactic structure) of a new kind would be forthcoming. One might also learn something about masers (or whatever source), particularly about their very small scale angular structure. A more speculative possibility is that we might learn something of the nature of the "Oort dark matter" which seems to make up roughly half of the material in the Galactic Plane (Bahcall 1984).

7. CONCLUSIONS

The two following conclusions suggest themselves: VLBI data will probably prove crucial to achieving many of the most important potential gravitational lens experiments. Gravitational lens observations may well provide some of the most scientifically important applications of VLBI techniques. These logically independent conclusions may be put in a predictive form: Future symposia on gravitational lens will someday feature VLBI observations as prominently as they now feature (for instance) optical spectroscopy, and future symposia on VLBI will eventually devote roughly as much attention to lens results as they now do to (for instance) masers or superluminal sources.

Conversations with numerous colleagues, including particularly B. Burke, M. Gorenstein, J. Hewitt, J. Ostriker, and I. Shapiro, provided the background for this discussion. A. Rogers pointed out the importance of Galactic rotation for the discussion in section 6. This work was supported in part by NASA grant

NAGW-765 and NSF grant AST84-20352.

REFERENCES

Bahcall, J. N. 1984, *Ap. J., 276,* 169.

Canizares, C. 1987, in *Observational Cosmology: Proceedings of IAU Symposium No. 124,* eds. A. Hewitt and G. Burbidge (Boston: D. Reidel Publishing), in press.

Carr, B. J. 1978, *Comm. Ap., 7,* 161.

Djorgovski, S., Perley, R., Meylan, G., and McCarthy, P. 1987, preprint.

Einstein, A. 1936, *Science, 84,* 506.

Falco, E. E., Gorenstein, M. V., and Shapiro, I. I. 1985, *Ap. J. Lett., 289,* L1.

Gorenstein, M. V., Bonometti, R. J., Cohen, N. L., Falco, E. E., Shapiro, I. I., Bartel, N., Rogers, A. E. E., and Marcaide, J. M. 1988, in this volume.

Gorenstein, M. V., Shapiro, I. I., Rogers, A. E. E., Cohen, N. L., Corey, B. E., Porcas, R. W., Falco, E. E., Bonometti, R. J., Preston, R. A., Rius, A., and Whitney, A. R. 1984, *Ap. J., 287,* 538.

Gott, J. R. 1987, in *Dark Matter in the Universe: Proceedings of IAU Symposium No. 117,* eds. J. Kormendy and G. R. Knapp (Boston: D. Reidel Publishing), 219.

Gunn, J. E. 1981, in *Proceedings of the Tenth Texas Symposium on Relativistic Astrophysics,* eds. R. Ramaty and F. C. Jones (New York: New York Academy of Sciences), 287.

Heflin, M. B., Gorenstein, M. V., Falco, E. E., Shapiro, I. I., Burke, B. F., Hewitt, J. N., Rogers, A. E. E., and Lawrence, C. 1988, in this volume.

Hewitt, J. N. 1986, MIT Ph. D. Thesis Dissertation.

Hewitt, J. N. 1987, private communication.

Hewitt, J. N. 1988, in preparation.

Kayser, R., Refsdal, S., and Stabell, R. 1986, *Astron. Ap., 166,* 36.

Lacey, C. G., and Ostriker, J. P. 1985, *Ap. J., 199,* 633.

Paczynski, B. 1986, *Ap. J., 301,* 76.

Peacock, J. A. 1983, in *Quasars and Gravitational Lenses: Proceedings of the 24th Liege International Astrophysical Colloquium,* ed. J. P. Swings (Liege: Universite de Liege), 86.

Peebles, P. J. E., and Dicke, R. H. 1968, *Ap. J., 154,* 891.

Porcas, R. W., Booth, R. S., Browne, I. W. A., Walsh, D., and Wildenson, P. N. 1981, *Nature, 289,* 758.

Press, W. H., and Gunn, J. E. 1973, *Ap. J., 185,* 397.

Shaver, P. A., Wampler, E. J., and Cristiani, S. 1987, *Nature, 327,* 40.

Turner, E. L. 1987a, in *Dark Matter in the Universe: Proceedings of IAU Symposium No. 117,* eds. J. Kormendy and G. R. Knapp (Boston: D. Reidel Publishing), 227.

Turner, E. L. 1987b, in *Observational Cosmology: Proceedings of IAU Symposium No. 124,* eds. A. Hewitt and G. Burbidge (Boston: D. Reidel Publishing), in press.

Turner, E. L., Ostriker, J. P., and Gott, J. R. 1984, *Ap. J., 284,* 1.

Walsh, D. 1983, in *Quasars and Gravitational Lenses: Proceedings of the 24th Liege International Astrophysical Colloquium,* ed. J. P. Swings (Liege: Universite de Liege), 106.

Zwicky, F. 1937, *Phys. Rev, 51,* 290.

VLBI OBSERVATIONS OF THE 0957+561 GRAVITATIONAL LENS SYSTEM

M.V. Gorenstein (CfA), R.J. Bonometti (MIT), N.L. Cohen (Cornell),
E.E. Falco, I.I. Shapiro, N. Bartel (CfA), A.E.E. Rogers (Haystack),
J.M. Marcaide (IAA), T.A. Clark (GSFC)

ABSTRACT. A series of VLBI observations of the gravitational lens system 0957+561 at λ13 cm has yielded the positions of the A and B images, the relative magnification of their largest discernible radio structures, and the time variability of their smallest discernible radio structures. These observations have also allowed upper limits to be placed on the flux density of an expected third image. The positions and relative magnification of the A and B images provide new information with which to constrain models of the lens that forms the images. The detection of variations in the flux densities of the cores of A and B suggests that observations at shorter wavelengths may reveal superluminal motion, which may in turn provide a means to measure the relative time delay.

1. INTRODUCTION

The 0957+561 gravitational lens system (Walsh, Carswell, and Weymann 1979) is the first known and best studied example of the gravitational lens effect. VLBI observations can contribute to our understanding of this, and other, lens systems through the determination of the observable properties of the images. For two images of a common source, VLBI observations can determine their (i) relative positions to sub-milliarcsecond accuracy from the difference of their interferometer phase delays; (ii) relative magnification from their partially-resolved time-independent brightness distributions; and (iii) relative time delay, $\Delta\tau_{BA}$, from correlations in the evolution of their partially-resolved time-variable brightness distributions. These results can check the basic lensing hypothesis, add new information with which to constrain lens models, and, perhaps, provide a new means with which to estimate the Hubble constant.

In 1981, at λ13 cm (Cohen 1985), and in 1983, at λ13 cm, λ6 cm, and λ3.6 cm (Bonometti 1985), we observed with VLBI the A and B gravitational images of the quasar Q0957+561. Here, we discuss only the λ13 cm data taken at these two epochs. We employed the Mark-III VLBI data acquisition system (Rogers *et al.* 1983), recording a 56-MHz band (Mode-A), at participating stations located at Effelsberg, FGR (100-m diameter antenna); Onsala, Sweden (25-m); Madrid, Spain (64-m); Green Bank, WV (43-m); Owens Valley, CA (40-m); and Goldstone, CA (64-m). The minimum fringe spacing, obtained on the Effelsberg–Goldstone baseline, was 3.5 milliarcseconds (mas).

In §II we describe models for the brightness distributions of the A and B images obtained from our VLBI data, and the properties of the images that follow from these models. In §III we discuss the detection of the radio core of the main lensing galaxy G1 in the 0957+561 system and the limits on the flux density of a third image of Q0957+561. The following paper in the volume (Falco, Gorenstein, and Shapiro 1987a) presents a model of the mass distribution of the lens constrained, in part, by the data presented here.

M. J. Reid and J. M. Moran (eds.), The Impact of VLBI on Astrophysics and Geophysics, 201–206.

2. OBSERVABLE PROPERTIES OF THE A AND B IMAGES

In §2.1, we present models of the brightness distributions of A and B and demonstrate that they are images of a common source and have opposite parity, as predicted from models for the lens' mass distribution. In §2.2, we give our result for the relative position of the images. In §2.3, we discuss evidence for the time variability of the cores' flux density.

2.1 VLBI Structure of the Images — Confirmation of the Lensing Hypothesis

We present in Figure 1 a model ("Model I") for the brightness distribution of each image obtained from the 1981 data (Cohen 1985; Cohen *et al.* 1987). We modeled each image with four elliptical Gaussian components: two compact components (Core and Jet 1) and two extended components (Jet 2 and Jet 3). The model parameters were constrained by correlated flux densities, closure phases, and difference phases (for definitions of difference phases, see §2.2). The χ^2 contribution per degree of freedom (DOF) for the (combined) models is 2.2. For these preliminary results, the errors in the magnification parameters and in the A−B difference position presented in Table 1, and in the data presented in Figure 2, are all a factor of two larger than the statistical errors. Final values for these errors will be presented in Cohen *et al.* (1987).

The approximate correspondence in relative position, flux density, size, shape, and orientation of the respective components in A and B suggests that they are images of the same source. We test these correspondences quantitatively as follows: First, we note that the angular extent of each image ($\sim$ 60 mas) is small compared to the arcsecond extent of the lens, suggesting that over the image's brightness distribution, the deflection caused by the lens changes linearly with position. Geometric optics implies then that the brightness distribution of an image should be related to the source's distribution by a three-parameter transformation that corresponds to a magnification and an astigmatism. Further, the brightness distribution of one image can be related to that of the other image by a four-parameter transformation, described by the four elements of a 2×2 "magnification matrix" [see Gorenstein *et al.* (1984) for details]. This magnification matrix can be estimated from corresponding components of images observed at one epoch, provided the time evolution of those components on the time scale of the relative time delay can be neglected.

For the 1981 data, we also used a second model ("Model II") that had parameters for the structures of the images and for the magnification matrix. The magnification parameters were used to relate only the flux densities, shapes, and orientations of Jet 2 and Jet 3 in the A and B images, and their positions relative to the centers of the respective Core components. As Jet 2 and Jet 3 are at least a few hundred light-years in extent, we concluded that we could neglect any evolution in their structure.

For Model II, the parameters describing the flux densities, shapes, and relative positions of the Core and Jet 1 components were estimated independently for both images, since, because of possible evolution, these parameters may not be related accurately by the magnification matrix. As these components are only some tens of light years in extent, or less, their brightness distribution could evolve significantly over a time delay of order 1 yr.

Inspection of Figure 1 suggests that the B image, aside from parts affected by evolution, can be obtained from the A image by a magnification of about 1.2 applied along the jet axis. Model I shows that the ratio of flux densities of Jet 2 and Jet 3 is $B/A \sim 0.6$. This result suggests a demagnification along an axis perpendicular to the jet axis of about $0.6/1.2 = 0.5$. Estimates of the magnification parameters conform to these results, as shown in Table 1. The total relative magnification, B/A, is 0.64 ± 0.05. The χ^2 per DOF for Model II is 2.2; χ^2 for Model II is 7 larger than for Model I for an increase of 8 degrees of freedom, which shows that the models are statistically indistinguishable. The estimated values for these parameters are independent of possible systematic errors in the overall calibration, as the magnification depends only on the *ratios* of flux densities. However, we have not yet checked the sensitivity of the magnification parameters to systematic calibration errors that may be produced at each station.

Models for the lens in the 0957+561 system predict that the A and B images have opposite parity (Young *et al.* 1981). Model I, presented in Figure 1, confirms this prediction, albeit only at the lowest, least significant, contour level: for example, Jet 2 appears to the left of Jet 3 in A, but to the right of Jet 3 in B. To test the significance, we analyzed a third model in which the relative parity of the images was constrained to be positive; χ^2 for this third model was larger than for Model I, but by an amount too small to rule out the alternative hypothesis with high confidence.

The apparent existence of the same number of components in each image, their satisfaction of the magnification relationship, including parity reversal, all are consistent with, and therefore serve to confirm, the lens hypothesis.

2.2 Difference Position

Because of their close $\sim 6''$ separation, both images were observed simultaneously by all antennas during the observations. For each 13 min 'scan', we obtained, for each interferometer, a phase delay for each of the images; we employed the difference of these phases in our astrometric analysis. The difference of these phases eliminates common contributions to both that arise from the atmosphere, instrumentation, and clock behaviors; the difference phase equals the sum of the difference phases due to the images' different sky locations and brightness structures.

We employed these difference phases, together with the visibility amplitudes and closure phases for each source, to obtain a difference position for the Core components in A and B (Cohen *et al.* 1987), presented in Table 1.

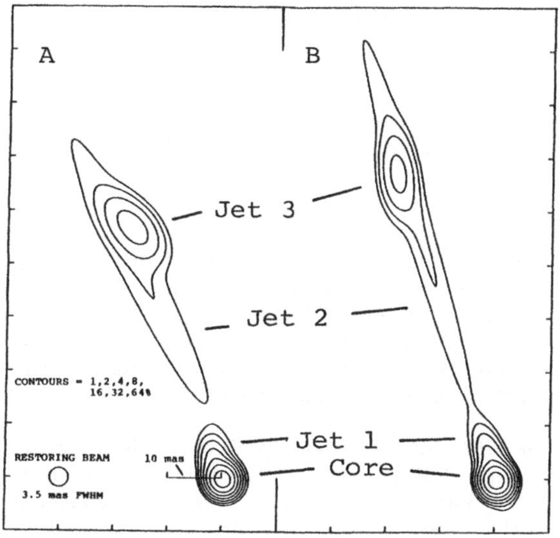

Figure 1 — The brightness distributions of 0957+561A and B. Four elliptical Gaussian components account for the data obtained from the 1981 λ13 cm VLBI observations for A and for B (Model I—see text). The correspondences in the number and in the properties of the components of the respective models, and the evidence for a change in parity from one image to the other, all support the hypothesis that A and B are images of a single object.

Redshifts

Images[a]	G1[b]
1.4136 ± 0.0002	0.36 ± 0.01

Image and Lens Positions

	A–B[c] (1950.0)	G1–B[d] (1950.0)
$\Delta\alpha$	$-0^{s}.149\,866 \pm 0^{s}.000\,002$	$0''.181 \pm 0''.001$
$\Delta\delta$	$6''.046\,63 \pm 0''.000\,02$	$1''.029 \pm 0''.001$

Relative Magnification Matrix[c] (B/A)

Linear Magnification (Eigenvalues)	Direction of Magnification Axis (Pos. Ang. of Eigenvectors)
+1.21 ± 0.01	$+18^{\circ}.9 \pm 0^{\circ}.2$
−0.53 ± 0.04	$-70^{\circ} \pm 10^{\circ}$

Relative Magnification (Product of Eigenvalues)
−0.64 ± 0.05

[a] From optical observations, Weymann *et al.* (1979).
[b] From optical observations, Young *et al.* (1981).
[c] From 1981 λ13 cm VLBI observations, Cohen *et al.* (1987).
[d] From 1981 λ13 cm VLBI observations, Gorenstein *et al.* (1983).

Table 1 — The observable properties of the images and of the lens of the 0957+561 system. We employ these properties to constrain a model for the mass distribution of the 0957+561 system (Falco, Gorenstein, and Shapiro 1987a).

2.3 Time Variability of Images.

The relative time delay, $\Delta\tau_{BA}$, for the images results from the differing lengths and gravitational potentials associated with the respective ray paths. A measurement of $\Delta\tau_{BA}$ is crucial because, first, a detection of definitive correlated variations between the images would provide a final proof of the reality of the gravitational lens effect, and, second, the value for $\Delta\tau_{BA}$ is inversely proportional to the Hubble constant (Falco, Gorenstein, and Shapiro [1985] and references therein). The value for $\Delta\tau_{BA}$, together with a model for the mass distribution of the lens, can yield a value for H_0 and for the mass of the lens.

In 1983 we re-observed the 0957+561 system at λ13 cm to search for possible time evolution of the Core and Jet 1 in each image. Such evolution, if detected, could lead to a measurement of $\Delta\tau_{BA}$. The models for the Jet 2 and Jet 3 components in the A and B images obtained from the 1983 data were consistent with the corresponding models obtained from the 1981 data. However, the flux densities of the Core and Jet 1 in both images changed. In Figure 2, we plot the sum of the Core+Jet 1 flux densities for A and for B for both epochs. (We consider only the sum of these flux densities because the components were barely resolved with the 3 mas resolution of the longest-baseline available, causing the estimates of the individual flux densities to be highly correlated for each image for each epoch.) The Core+Jet 1 flux densities in 1983 can be seen to be significantly lower than in 1981 for both images. Further, the difference between the (scaled) flux densities for A and B obtained at the *same* epoch gives evidence for the time evolution of the flux density of the Core+Jet 1 for the source on the time scale of the time delay: If there were no time evolution, the Core+Jet 1 flux density for B, when scaled for the magnification as determined by the outer jets, would equal, within the measurement errors, the Core+Jet 1 flux density for A.

A reliable value for $\Delta\tau_{BA}$ cannot be extracted from these data; however, we can obtain a model-dependent value. For example, if we *assume* the decay to be linear in time, we obtain $\Delta\tau_{BA} = 4 \pm 1$ yr (A varies first); in Figure 2, the open triangles represent the B data translated by this amount, with the corresponding light curve being indicated by the dashed line. The data do not distinguish this light curve from other possible light curves. Although our data do demonstrate that the flux density of the source varies on a time scale comparable to $\Delta\tau_{BA}$, they do not provide a definitive demonstration of correlated evolution of the images, nor do they provide a reliable estimate for the time delay.

Motion of multiple components in a radio source can provide another means with which to determine $\Delta\tau_{BA}$, as pointed out by Vanderriest (1982) and Gorenstein *et al.* (1984). For example, a core-jet separation could be determined for each image at each of two epochs; then, if the deduced separation velocities were constant, but non-zero and detectable, we could determine for each image the epoch of zero apparent separation. The difference in these epochs yields $\Delta\tau_{BA}$. Our original goal in performing a second epoch observation at λ13 cm was to detect a change in the separation of Jet 1 from the Core in each image. Our 1983 data showed that these structures were still barely resolved, and no statistically significant change in their separation from 1981 to 1983 was evident.

The changes in the flux density of the images over two years demonstrates that the brightness distribution of the source is evolving on a useful time scale. This evolution gives us hope that similar evolution may occur at shorter wavelengths, where we may be able to detect observable changes in the images' brightness distributions. We are planning observations at λ6 cm and, ultimately, at λ3.6 cm, in order to search for evidence for such structural evolution.

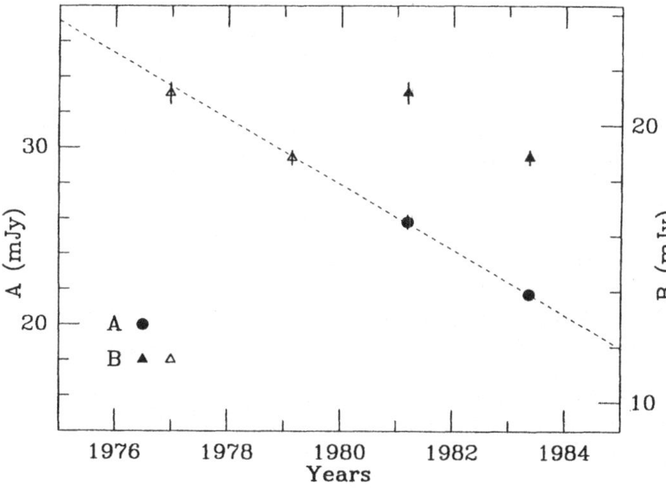

Figure 2 — **The flux densities of the Core+Jet 1 components in A and in B at λ13 cm in 1981 and 1983. The A (B) flux densities are plotted with respect to the left- (right-) hand vertical axes; the scale of the latter axis reflects the ratio of the image magnification ($B/A = 0.64$). These data are consistent with $\Delta\tau_{BA} = 4 \pm 1$ years, if the flux density of the Core+Jet 1 is assumed to decay linearly (see text).**

3. WHITHER THE THIRD IMAGE?

Theories of lensing predict that every transparent lens should produce an odd number of images, except for "singular" relative orientations of source, lens, and observer (Burke 1981). Our search for a third image in the 0957+561 system using our 1981 data led to the discovery of a weak (0.6 ± 0.1 mJy), compact (< 2 mas) radio source, G′, located on the sky at or near the center of the foreground galaxy G1 (Gorenstein *et al.* 1983). The size, flux density, and location of G′ were consistent with it being either a third image of Q0957+561, radiation from G1, or, perhaps, some combination of the two.

Our 1983 data revealed that the correlated flux density of G′ increased from its 1981 value by nearly 100% to 1.0 ± 0.1 mJy. If G′ were an image, then this increase could be due to a corresponding brightening of the source, or microlensing by a compact mass in G1 traversing the ray path. However, the flux density for the nearby B image decreased by about 5% during this period, while our models for the lens predict that the propagation times for B and the possible third image G′ should differ by, at most, a few weeks (Falco, Gorenstein, and Shapiro 1987b). Further, the apparent angular size of such a demagnified image should be no less than ~ 0.1 mas (1 ly for $H_0 = 100$ km sec^{-1} Mpc^{-1}); so a suitable microlens must have a speed near or greater than c in order to traverse this distance in 2 yr. A straightforward interpretation of these results implies that G′ is not the third image, but rather the radio core of G1. The variations in, and the values of, the flux density of G′ are within the range expected from observations of the cores of other elliptical galaxies (e.g., Sramek 1975).

We combined the most sensitive data from both the 1981 and 1983 observations and searched for evidence for an additional image ("C") over a field of width $1''\!.0$ in α and $1''\!.6$ in δ that was centered $1''$ due north of B. Our search yielded no evidence for C, setting a (5σ) limit of 0.2 mJy, or one part in 80 of the flux density of the compact Core in the B image. A magnification ratio (C/B) of less than one part in 80, implies, according to one simple mass model for the G1 lens galaxy, that $\Delta\tau_{CB}$ is less than 3 weeks, the separation of C from the center of G1 is less than $0''\!.2$, and the surface mass density at the center of G1 is greater than eight times the critical surface mass density (Falco, Gorenstein, and Shapiro 1987b). The critical surface mass density for a given lens system is determined only by the redshifts of the lens and the source and by the cosmology (see Falco, Gorenstein, and Shapiro 1985). For the 0957+561 system and for $H_0 = 100h$ km sec^{-1} Mpc^{-1} and $q_0 = 0$, the critical surface mass density is $4.4 \times 10^{10} h^{-1} M_\odot$ arcsec^{-2}.

We gratefully acknowledge support from NSF grants AST8519763 and PHY8409671. One of us (RJB) acknowledges fellowship support from the Hertz Foundation.

Bonometti, R. J. 1985, *Thesis, Mass. Inst. of Tech.*

Burke, W. L. 1981, *Ap. J. (Lett.)*, **244**, L1.

Cohen, N. L. 1985, *Thesis, Cornell University.*

Cohen, N. L., Gorenstein, M. V., Shapiro, I. I., Rogers, A. E. E., Bonometti, R. J., Falco, E. E., Bartel, N., and Marcaide, J. M., 1987, in preparation.

Falco, E. E., Gorenstein, M. V., and Shapiro, I. I. 1985, *Ap. J. (Lett.)*, **289**, L1.

Falco, E. E., Gorenstein, M. V., and Shapiro, I. I. 1987a, in this volume.

Falco, E. E., Gorenstein, M. V., and Shapiro, I. I. 1987b, in preparation.

Gorenstein, M. V. *et al.* 1983, *Science*, **219**, 54.

Gorenstein, M. V. *et al.* 1984, *Ap. J.*, **287**, 538.

Rogers *et al.* 1983 *Science*, **219**, 51.

Sramek, R. 1975 *Ap. J. (Lett.)*, **198**, L13.

Vanderriest, C. 1982, *Astron. Astrop.*, **106**, L1.

Walsh, D., Carswell, R. F., and Weymann, R. J. 1979, *Nature*, **279**, 381.

Weymann, R. J. *et al.* 1979, *Nature*, **285**, 641.

Young, P. *et al.* 1981, *Ap. J.*, **244**, 736.

CAN WE MEASURE H_0 WITH VLBI OBSERVATIONS OF GRAVITATIONAL IMAGES?

E. E. Falco, M. V. Gorenstein, and I. I. Shapiro
Harvard-Smithsonian Center for Astrophysics
60 Garden Street
Cambridge, Massachusetts 02138
U.S.A.

ABSTRACT. We have used the relative positions and magnifications of the A and B images in the gravitational lens system 0957+561, obtained from VLBI observations, to constrain a model for the surface mass distribution of the lens. With measurements of the difference $\Delta\tau_{BA}$ in propagation times associated with A and B (the "relative time delay") and of the velocity dispersion of the main lensing galaxy, both to be obtained, our model will yield a value for H_0 with an uncertainty of $\sim 20\%$ due mainly to uncertainties in our assumptions.

1. INTRODUCTION

The images A and B of the quasar Q0957+561 (Walsh, Carswell and Weymann 1979) appear to be formed by the gravitational lens action of an intervening cluster of galaxies (Young *et al.* 1981; hereafter YI). The main light deflector, G1, is a cD giant elliptical galaxy, the brightest in that cluster. Models that describe the mass distribution of the lens have been proposed by different authors [YI; Greenfield *et al.* 1985; Subramanian, Chitre and Narasimha 1985; Falco, Gorenstein and Shapiro 1985 (hereafter Paper I)]. We present here a new model, based on available redshift measurements and on our own VLBI observations, that accounts for the relative positions and magnifications for the two images. We discuss the image and lens observables in §2. In §3, we present and discuss our new model. The properties of similar lens models are described in Paper II (Gorenstein, Falco and Shapiro 1987); the results presented here are discussed in detail in Paper III (Falco, Gorenstein and Shapiro 1987).

2. OBSERVABLES

The relative magnification matrix for the two images (Gorenstein *et al.* 1984) and the other information we used to constrain our model are presented elsewhere in these proceedings (Table 1, Gorenstein *et al.* 1987). Since the matrix was obtained from the jets that emanate from the cores of A and B and extend over hundreds of parsecs (Cohen *et al.* 1987), our values should be insensitive to fluctuations in the quasar's flux density on time scales of the order of $\Delta\tau_{BA}$. We also utilized the optical observations described in YI, which indicate that the cluster is spread over an area of the sky much larger than the separation between A and B.

3. MODEL

We describe G1 as the superposition of: (1) a King surface mass density (in the approximation described in YI), parametrized by its velocity dispersion σ_v and its structural length; and

M. J. Reid and J. M. Moran (eds.), The Impact of VLBI on Astrophysics and Geophysics, 207–208.

(2) a point-like object at the center of the King component, parametrized by its mass. We approximate the cluster as a "linear" deflector (Paper II), parametrized, as in Paper I, by a focusing amplitude Q_S, an astigmatic amplitude Q_A and a position angle γ. Application of the "magnification" transformation (Paper II) shows that the observables constrain only a particular combination of σ_v and Q_S. Table 1 shows the estimates of the parameters for this model; the resultant χ^2 per degree of freedom is ~1.3. The uncertainties shown were obtained by propagating the measurement standard errors through the model (Gorenstein *et al.* 1987). Since we lack values for $\Delta\tau_{BA}$ and σ_v, we cast the result for H_0 as:

$$H_0 = (97 \pm 20)\left(\frac{\Delta\tau_{BA}}{1\,\mathrm{yr}}\right)^{-1}\left(\frac{\sigma_v}{390\,\mathrm{km\,s^{-1}}}\right)^2\ \mathrm{km\,s^{-1}\,Mpc^{-1}}.$$

The uncertainty shown here arises primarily from the uncertainties in the model, in particular the indeterminacy of the surface mass density of G1 between the radii r_A and r_B, and the distances, respectively, of A and B from the center of G1. Allowance was also made for the value of the cosmological matter density being between zero and one, and for density inhomogeneities being up to ~50% of the mean along the line of sight (Paper III). Because of the positivity of mass, we must have $Q_S \leq 0$ (Paper I), so the value given in Table 1 sets an upper limit on σ_v of 390 km s^{-1} and, therefore, on the product $\Delta\tau_{BA}H_0$. The point-like component, which contributes nearly 50% of the total enclosed mass within r_B (~1″), complicates the interpretation of measurements of the velocity dispersion of G1, as we discuss further in Paper III.

We conclude that point estimates of H_0 may be made if VLBI observations of gravitational images are complemented by further observations to determine quantities such as $\Delta\tau_{BA}$ and σ_v. We gratefully acknowledge support from NSF grants AST8519763 and PHY8409671.

Table 1: Model for 0957+561

Parameter	*Estimated value*
Mass of G1's point-like component	$(1.252 \pm 0.004)\times(\Delta\tau_{BA}/1\,\mathrm{yr})\,10^{11}M_\odot$
Structural length of G1	$0.''94 \pm 0.''05$
Q_S	$-(1-(\sigma_v/390\,\mathrm{km\,s^{-1}})^2)$
Q_A	$(0.29 \pm 0.02)\times(\sigma_v/390\,\mathrm{km\,s^{-1}})^2$
γ	$63^\circ \pm 1^\circ$

Cohen, N. L., Gorenstein, M. V., Shapiro, I. I., Rogers, A. E. E., Bonometti, R. J., Falco, E. E., Bartel, N., and Marcaide, J. M. 1987, in preparation.

Falco, E. E., Gorenstein, M. V., and Shapiro, I. I. 1985, *Ap. J. Letters*, **289**, L1.

Falco, E. E., Gorenstein, M. V., and Shapiro, I. I. 1987, to be submitted to *Ap. J.*

Gorenstein, M. V. *et al.* 1984, *Ap. J.*, **287**, 538.

Gorenstein, M. V., Falco, E. E., and Shapiro, I. I. 1987, to be submitted to *Ap. J.*

Gorenstein, M. V. *et al.* 1987, in these *Proceedings.*

Greenfield, P. E., Roberts, D. H., and Burke, B. F. 1985, *Ap. J.*, **293**, 370.

Subramanian, K., Chitre, S. M., and Narasimha, D. 1985, *M.N.R.A.S*, **210**, 79.

Walsh, D., Carswell, R. F., and Weymann, R. J. 1979, *Nature*, **279**, 381.

Young, P., Gunn, J. E., Kristian, J., Oke, J. B., and Westphal, J. A. 1981, *Ap. J.*, **244**, 736.

VLBI OBSERVATIONS OF THE GRAVITATIONAL LENS SYSTEM 2016+112

M. B. Heflin (MIT), M. V. Gorenstein, E. E. Falco, I. I. Shapiro (CfA)
B. F. Burke (MIT), J. N. Hewitt, A. E. E. Rogers (Haystack)
C. R. Lawrence (CIT)

ABSTRACT. On June 1, 1984 we conducted a seven station 18-cm VLBI observation of the 2016+112 gravitational lens system. Preliminary brightness distributions for A and B have been obtained via model fitting. Weak correlated flux density was detected in the C component region.

The 2016+112 gravitational lens system has three radio components, labelled A, B, and C respectively (1). A and B appear to be gravitationally lensed images of a single source (1). They are both unresolved by the VLA at 6-cm and have identical redshifts of 3.2733 within measurement errors. C is resolved by the VLA at 6-cm. Optical observations suggest that most of the light from the C component comes from a foreground elliptical galaxy at a redshift of roughly .8, possibly with some contribution from a weak third image (2).

On June 1, 1984 we conducted a seven station 18-cm VLBI observation of the system. The stations involved were Bonn (B), Green Bank (G), Fort Davis (F), Owens Valley (O), Haystack (K), the VLA (Y), and Arecibo (A). Mark-II recorders were present at all stations. Mark-III recorders were present at all stations except Arecibo. The following analysis involves only the Mark-III data.

Only the A and B components were unambiguously detected in fringe rate spectra corresponding to 13 minute integrations. A and B were not detected by the KF and KO baselines. There were detection and resolution problems with the F-Y-O triplet data. Our preliminary analysis involves the remaining baselines, which detected and resolved A and B.

Standard gain calibration was performed. Plots of correlated flux density suggested two-component circular-gaussian models for the A and B images. Residual difference phases were obtained by subtracting an estimate of the geometric difference phase from the observed difference phase. Correlated amplitude, closure phase, and residual difference phase data were used to constrain the A and B models. The results, which represent a stable chi-squared fit, are illustrated in Figure 1.

VLA observations give values of 56.0 and 61.7 mJy for the A and B 20-cm fluxes respectively (1). Our models yield $S_A = 36.1$ mJy and $S_B = 38.8$ mJy, suggesting that extended structures are being overresolved by the baselines included in our analysis. The major discrepancy between the models and the data is found in excess correlated flux of the B data on the K-G baseline near UT = 13. This excess suggests a jet with a position angle of -95 degrees. Later analysis, including

M. J. Reid and J. M. Moran (eds.), The Impact of VLBI on Astrophysics and Geophysics, 209–210.

the B jet and more short baseline data, may allow matching of A and B features. If such correspondence can be established, calculation of magnification parameters will follow (3). These parameters, which describe a linear mapping of one image into the other, help to constrain models of the lensing mass distribution.

The C component was too weak to be detected unambiguously in individual 13 minute fringe rate spectra. We did, however, detect weak correlated flux density in the C component region by coherently summing 403 minutes of data from the G-Y baseline, using B as a phase reference (4). The C component region has a peak signal to noise ratio of 61. The B component region has a peak signal to noise of 233, implying $S_B/S_C = 3.8$ at 18-cm.

Future work includes addition of a jet component to the B model, and incorporation of more short baseline data. Also, VLBI observations at 6-cm will be made in order to detect core-jet morphology at 1 mas resolution. If such structures are found, follow-up observations will search for time dependent source evolution in order to measure the time delay between the A and B images. A well constrained mass distribution for the lens of 2016+112, along with a time delay between the images, can lead to an estimate of the Hubble constant.

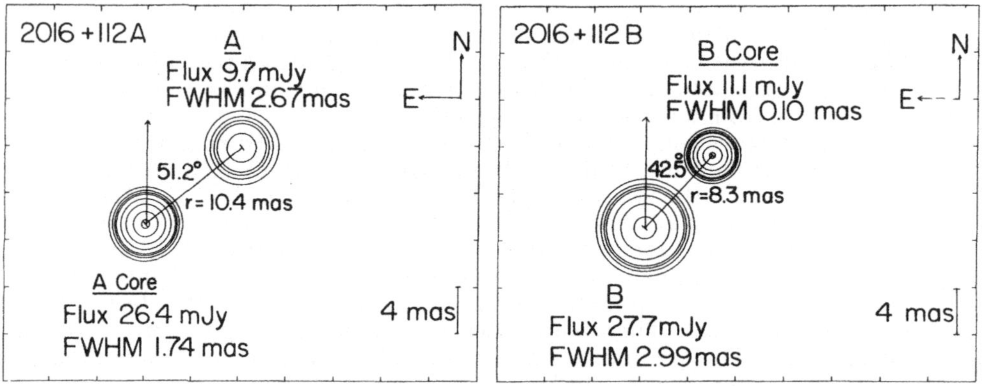

FIGURE 1
Two component models for 2016+112 A and B. All of the components are circular gaussians. The more compact component in each model is called the core. The separation A Core - B Core in coordinates of date is 3.02768 $''$ ± .00006 $''$ R.A. and 1.5014 $''$ ± .0003 $''$ Dec. The B Core FWHM has an uncertainty of 0.78 mas. All other parameter errors are roughly 5%. These models are based on Mark-III VLBI observations at 18-cm.

1. Lawrence, C. R., *et al.* 1984, *Science* **223**, 46.
2. Schneider, D. P., *et al.* 1985, *Ap. J.* **294**, 66.
3. Gorenstein, M. V., *et al.* 1984, *Ap. J.* **287**, 538.
4. Gorenstein, M. V., *et al.* 1983, *Science* **219**, 54.

VLBI STUDIES OF HIGH-REDSHIFT 21CM ABSORPTION LINES

F. H. Briggs
University of Pittsburgh
Pittsburgh, PA 15260 USA

VLBI experiments can contribute to our understanding of absorbing gas observed in the spectra of high-redshift quasars by measuring the spatial extent of the absorbers. An optical survey conducted by Wolfe, Turnshek, Smith, and Cohen (1986) has turned up a class of absorbers characterized by large HI column density and a rich spectrum of metal absorption lines. In these respects the absorbers resemble spiral galaxies, but their occurence is too frequent for interception probabilities based on galaxy cross sections at the present epoch. Recent radio observations to determine the nature of such an absorber at z=2.04 in the spectrum of PKS0458-020 include both continuum observations at 606 and 1590 MHz to define the structure of the background quasar as well as spectral line VLBI at 467 MHz. The observations reveal structure in the continuum source on a wide range of angular scales, making the source nearly ideal for spectral line VLBI on a range of baselines. Early spectroscopic results show that the absorber must have a spatial extent of galaxian size, consistent with the hypothesis that these absorbers are galaxies. The further implication is that galaxies must have been larger and richer in gas at the epoch around z=2 in order to provide the likelihood of interception that is observed optically. (A complete description of the result is in preparation by Briggs, Wolfe, Liszt, Davis and Turner.)

VLBI has also proved to be a useful technique for performing spectral line studies at frequencies that fall outside the bands protected for radio astronomy: The systems studied often have redshifts that places the absorption lines in heavily used communications bands. High-redshift, radio-loud quasars often have compact structure that remains unresolved on long baselines, permitting VLBI observations that are immune to moderate levels of interference which is local to the individual stations and does not cohere in the final cross-correlation spectrum.

REFERENCES

Wolfe, A.M., Turnshek, D.A., Smith, H.E., and Cohen, R.D. 1986, Ap. J. Suppl., 61, 249.

M. J. Reid and J. M. Moran (eds.), The Impact of VLBI on Astrophysics and Geophysics, 211.

OBSERVATIONS OF COSMIC MASERS

P.J.Diamond
National Radio Astronomy Observatory
Edgemont Road
Charlottesville VA 22903-2475

ABSTRACT.

A review is presented of recent results from interferometric observations of cosmic masers. Several topics will be discussed, including MERLIN/VLA/VLBI observations of megamasers; recent results on masers in star forming regions and around late-type stars and some new results on the effects of interstellar scattering on maser spot sizes.

INTRODUCTION

Astronomical masers were discovered in 1965 (Weaver *et al.*), but although they have been studied extensively in the intervening 22 years very little has been uncovered concerning the physics of their emission, or the conditions that give rise to it. Indeed, the study of masers *per se* reveals almost no information concerning the physical conditions in their immediate environment. This was summarized in a comment I once overheard in the corridors of a well-known European observatory : "Why observe masers, the damn things tell you nothing!". However this is not, of course, strictly true; observations of masers can provide detailed information on the dynamics of a region, be it a circumstellar shell, an outflow near an HII region or the central parts of a galaxy; they can also provide accurate estimates of the distance to the maser sources through observations of the proper motions of H_2O masers, or the method of phase-lags (Schultz *et al.* 1978); the circular polarization observed in OH and SiO masers can be used to determine the magnetic field strengths in the immediate vicinity of the masers; observation of the sizes of the masers provide extremely useful information on the distribution of scattering material in our galaxy. These are some of the many reasons why we observe masers and I shall discuss some of the more recent results obtained.

M. J. Reid and J. M. Moran (eds.), The Impact of VLBI on Astrophysics and Geophysics, 213–222.

EXTRAGALACTIC MASERS

Extragalactic masers were first discovered in 1973 in the galaxy NGC253 (Whiteoak and Gardner 1973). The OH masers observed, although one to two orders of magnitude stronger than Galactic masers, were not particularly accesible due to their large distance and thus low flux density. In 1982 a different class of extragalactic OH maser was discovered in the peculiar galaxy IC4553 (Baan *et al.* 1982). It was immediately dubbed a "megamaser" since its intrinsic strength was about 7×10^7 that of the W3(OH) masers in our Galaxy. Following the initial excitement of the discovery telescopes all over the world began searching for new megamasers until, at the time of writing, 15 are known (see Table 1).

In addition to OH two other molecules have been observed to be masing in other galaxies. In 1986 Baan *et al.* (1986) reported the detection of a powerful H_2CO maser in IC4553 (and possibly one in NGC 3079). H_2O masers were detected in external galaxies as long as 10 years ago (Churchwell *et al.* 1977; Lepine and Dos Santos 1977; Huchtmeier *et al.* 1978, 1980; Scalise and Braz, 1981). However these sources are comparable in luminosity to the strongest Galactic masers (≤ 1 L$\odot$). Much more luminous H_2O masers have since been found and to date 6 H_2O megamasers are known (see Table 2).

TABLE 1
KNOWN OH MEGAMASER GALAXIES

Galaxy	References
IC4553	Baan, Wood, and Haschick 1982
NGC 3690	Baan 1985
Mrk 231	Baan 1985; Kázes and Dickey 1985
Mrk 273	Baan, Haschick and Schmelz 1985
	Bottinelli *et al.* 1985
IIIZW 35	Chapman *et al.* 1986
IRAS 11506-3851	Norris *et al.* 1986
Arp 148	Bottinelli *et al.* 1986
IIZW 96	Bottinelli *et al.* 1986
	Baan, Henkel and Haschick 1987
IRAS 17208-0014	Bottinelli *et al.* 1986
ZW15107+0724	Baan, Henkel and Haschick 1987
IRAS 16399-0937	Staveley-Smith *et al.* 1986
ZW475.056	Mirabel and Sanders 1987
IRAS 1017+08	Mirabel and Sanders 1987
IRAS 1211+03	Kázes *et al.* 1987
NGC 4418	Bottinelli *et al.* 1987

TABLE 2
Known H_2O Megamaser Galaxies

Galaxy	References
NGC 4945	Dos Santos and Lepine 1979
Circinus	Gardner and Whiteoak 1982
NGC 1068	Claussen *et al.* 1984
NGC 4258	Claussen *et al.* 1984
NGC 3079	Henkel *et al.* 1984
IC 10	Henkel *et al.* 1986

The extragalactic masers so far discovered are all believed to be associated in some way with regions of star formation, whether they are physically associated with OB stars as might possibly be the case for extragalactic H_2O masers, or whether their emission is indirectly triggered by a burst of star formation. However Wood *et al.* (1986) and Whiteoak *et al.* (priv. comm.) have recently discovered several 1612 MHz OH masers associated with OH/IR stars in the Large Magellanic Cloud. These masers appear to be weaker than their Galactic counterparts but show otherwise similar characteristics.

Interferometric observations of extraglactic masers

IC4553 is the only extragalactic OH megamaser whose structure is at all well determined. Baan and Haschick (1984) used the VLA to map the spatial distribution of the OH emission. The source was only slightly resolved but two interesting facts emerged from the data; a) the nuclear continuum source and OH emission region appeared to be perfectly superposed, and b) the velocity structure of the OH emission was consistent with solid body rotation of a molecular disc. Norris *et al.* (1985) used MERLIN to produce much higher resolution maps of the OH and 18cm continuum emission. They found the nucleus to be a core-dominated triple similar to those seen in other active galactic nuclei and the OH masers appeared to amplify the triple structure (it now appears that the triple structure observed was caused by the symmetrization propeties of self-cal and that the true structure is a double, this does not however change any of the results). The OH emission from IC4553 has also been detected with VLBI (Diamond *et al.* 1987), the VLB image is seen to be a double , and the eastern component appears to breakup into a number of smaller masers. No continnuum emission was detected with VLBI. Four other OH megamasers have been observed with the VLA; Mrk 273 (Schmelz 1987); IIIZw35 (Chapman priv. comm.); IRAS 17208-0014 (Martin priv. comm.) and IRAS 11506-3851 (Norris priv. comm.). As expected the OH emission from each is superposed on the continuum emission but none of the sources were resolved.

Interferometric observations of extragalactic H_2O masers have been conducted recently by Claussen and Lo (1986) and Haschick *et al.* (1987). These observations have revealed the surprising result that the extragalactic H_2O masers in NGC 4258, 1068 and 3079 are confined to very small regions (≤ 4 pc) and appear to be coincident with compact sources of radio emission in the nuclei of the galaxies. In addition the masers have luminosities ≥ 120 L$\odot$ (~ 500 L$\odot$ for NGC3079), two orders of magnitude more powerful than the Galactic H_2O masers in W49. It is evident that standard pumping schemes (which have difficulty explaining the Galactic H_2O masers) cannot explain the extragalactic masers. Strelnitskij (1984) and Kylafis and Norman (1986) have proposed collisional pumping schemes involving hydrogen and electrons at different kinetic temperatures which may be able to explain these powerful masers since collisional quenching of the pump does not occur, thereby giving rise to more powerful masers.

A Model for Extragalactic Megamasers

Due to the relative paucity of observational data on extragalactic masers no detailed model explaining them has yet emerged. However some common links are evident. First, almost all megamasers appear to be superposed on continuum emission originating in the host galaxy, this would imply (at least in the case of OH megamasers) that the maser is a result of low gain amplification of the continuum by excited foreground molecular gas. Secondly, the galaxies with megamaser emission are usually very strong infrared sources, the observed IR emission is quite sufficient to excite the molecular gas containing the masers. Baan *et al.* (1987) have demonstrated that for the OH megamasers L_{OH} is proportional to L_{IR}^2. It is suggested that the cause of the continuum emission and high IR luminosity is that the megamaser galaxies recently underwent, or are undergoing, a burst of star formation. A third link common to the OH megamaser galaxies is that many (if not all) of them appear to have an edge-on molecular disc, this increases the probability of seeing inverted molecular clouds along the line of sight.

Baan *et al.* 1987 suggest that the OH megamaser galaxies are a very early stage of Seyfert 2 or HII nuclear activity in a dusty galaxy and that possibly all Seyfert and HII nucleii exhibited maser activity when they turned on.

The H_2O megamaser galaxies tend to have slightly different properties than the OH emitting galaxies. Their IR luminosities are less and the compact continuum sources also tend to be weaker, although the masers do coincide with continuum sources. Haschick *et al.* (1987) suggest that the coincidence of the masers and continuum sources may be causally related due to either a) the H_2O masers may be amplifying the background radiation; or b) the radio source is an indication of a massive cluster of OB stars that may be providing the maser pump power (although see above).

INTERSTELLAR MASERS

Interferometric observations of Galactic interstellar masers is not a subject which has been pursued with any great vigour over the last few years, however a few results do stand out. The proper motion studies of H_2O masers spots in HII regions have produced accurate distances to several sources; Genzel *et al.* (1981a) have determined a distance of 480 ± 80 pc for Orion-KL; Genzel *et al.* (1981b) and Schneps *et al.* (1981) determined the distance to the W51 molecular cloud to be 7 ± 1.5 kpc; more recently Reid *et al.* (1987) have determined the distance to Sgr-B2 to be 7.1 ± 1.5 kpc implying that the Galactic centre is significantly closer than the previous standard value of 10 kpc.

OH masers are also associated with regions of star formation and can provide the observer with a wealth of information on the dynamics of such objects. Gaume and Mutel (1987) have observed 11 star forming regions with the VLA and, for the first time, are able to produce some statistics concerning the relationship of OH masers and HII regions. They find that all four ground state OH transitions are observed but that the mainline transitions (at 1665 and 1667 MHz) are usually the strongest. They also observed that the different maser transitions are frequently found very close to each other ($\leq 0.5''$) suggesting that several transitions can be pumped together. Regarding the positional relationship between OH masers and the 15 GHz structure of the HII regions they find that the overall distribution of maser features is strongly peaked near the edge of the continuum sources, and that in those HII regions which have a cometary-like structure (suggesting motion between the HII region and the surrounding neutral material) the strongest OH masers are projected onto the upstream leading edge of the HII regions.

Baart and Cohen (1985) and Baart *et al.* (1986) have used MERLIN to observe the OH masers in G45.1+01 and W75N, both active regions of star formation. In both sources they find that the OH masers lie near the edges of the HII regions, similar to the result obtained by Gaume and Mutel (1987); they also suggest that there is a tendency for OH masers to lie at greater distances from more luminous stars. There is some discussion as to the nature of the OH/HII regions; Reid *et al.* (1980), in their definitive study of the OH masers in W3(OH), suggested that the masers were part of a remnant-accreting envelope which is still collapsing towards a newly formed star. This conclusion was arrived at since many of the masers detected by Reid *et al.* (1980) were projected against, and appeared to be red-shifted with respect to, the underlying HII region. However more sensitive observations of the W3(OH) masers by Norris *et al.* (1982), and observations of other HII region masers by Baart and Cohen (1985) and Gaume and Mutel (1987) suggest that in general OH masers are expanding away from their parent HII regions.

W3(OH) is probably the best studied of all OH/HII region masers sources and observations have continued with recent VLBI observations of excited state ($^2\pi_{1/2}, J = 1/2, \Delta F = 1 - 0$) OH masers at a frequency of 4765 MHz by Baudry *et al.* (1987). The VLBI data implied apparent brightness temperatures $\geq 9 \times 10^8$ K, and some spatial and velocity correlation of the 4765 MHz masers with the 1665/7 MHz and 6035 MHz masers is observed.

Another strong masing molecule seen in W3(OH) is CH_3OH. There are several transitions detected in various sources with the $2_0 \rightarrow 3_{-1}E$ transition recently observed to have fluxes between 100 - 1200 Jy in some sources (Batrla *et al.* 1987; Norris *et al.* 1987), but at the relatively inaccesible frequency (for most interferometers) of 12 GHz. However the $9_2 \rightarrow 10_1A^+$ transition at 23 GHz is observed in W3(OH) at a flux density of ~10 Jy. This transition was recently observed with both the VLA (Menten priv. comm.) and VLBI (Johnston priv. comm.). The VLBI experiment was unable to detect any line fringes implying spot sizes of between 5 and 40 mas, but the lack of VLBI fringes may be due to instrumental problems. The image produced with the VLA data is a great surprise since the CH_3OH masers are seen to correlate almost exactly with the gross spatial and velocity structure of the 1665 OH masers.

STELLAR MASERS

Molecular masers are also associated with oxygen-rich M-type giant and supergiant stars. Strong OH, H_2O and SiO masers are observed in such objects and it is now commonly accepted that the masers lie in an expanding circumstellar shell. Since the masing transitions of the three molecular maser species are excited at very different temperatures it is assumed that they lie at increasing distances from the stellar surface with the SiO being excited close to the star and the OH at typical distances $\geq 10^{16}$ cm from the star. The envelopes typically have expansion velocities of $5 \rightarrow 40$ km $s^{-1}$and mass loss rates of $\sim 10^{-6} \rightarrow 5 \times 10^{-4} M_\odot yr^{-1}$. In addition to the masers in oxygen-rich stars Guilloteau *et al.* (1987) have recently discovered a strong maser (~ 75 Jy) from a vibrationally excited transition of HCN in the envelope of the carbon-rich star CIT 6.

Thers is now a large database of interferometric observations of stellar masers. Booth *et al.* (1981) first demonstrated convincingly that OH masers lay in an expanding, spherical envelope with observations of the 1612 MHz OH masers in OH127.8-0.0. This model has since been confirmed for many sources by VLA and MERLIN observations (e.g. Bowers *et al.* 1983; Diamond *et al.* 1985a and references therein). Models of maser emission predict (as mentioned above) that the different molecular masers should lie at different distances from the central exciting star; it is only in the last year or so that detailed observations of the different molecular species have become available (Chapman and Cohen, 1986; Diamond *et al.* 1985b; Diamond *et al.* 1987) that show the models to be correct in their overall details.

VLBI observations of stellar OH masers have been useful to determine the relative locations and sizes of individual masers spots but, until recently, had only been performed on a few sources since it had been assumed that in general stellar masers were larger than the equivalent interstellar masers. However recent VLBI observations of OH masers have revealed the presence of very compact emission spots (Norris *et al.* 1984; Sivagnanam *et al.* 1987) in both distant OH/IR stars and Mira Variables. Such observations are important since it appears the masers are amplifying the thermal emission from the star and thus lie directly along the line of sight to it implying that with both radio and optical astrometric observations these sources could be used to relate the FK4 optical reference frame with the radio reference frame (see de Vegt *et al.* 1986).

Unlike interstellar OH masers, most stellar sources exhibit little or no polarization. There are some very notable exceptions such as the supergiant stars and the variables W Hya and U Ori. It is assumed that the Zeeman effect is responsible for the observed polarization and, given that, it should be possible to determine the magnetic field strength in the immediate vicinity of the masers and thus extapolate back to determine the field strength at the stellar surface. Early experiments (Reid *et al.* 1979) estimated the magnetic field at the surface of U Ori to be ~ 10 G and at the surface of IRC+10420 to be ~ 100 G. Recently Barvainis *et al.* (in prep.) have observed significant circular polarization in SiO masers associated with late-type stars which imply magnetic fields in the immediate vicinity of the masers $\sim 10 \rightarrow 100$ G, this is a much more direct measurement of the magnetic field. McIntosh (priv. comm.) has made VLBI maps of the SiO maser distribution in the envelopes of two late-type stars which he combined with single-dish polarimetric data to produce maps which may show the pattern of the magnetic field lines around the stars.

INTERSTELLAR SCATTERING

Previous VLBI observations of the spot sizes of masers (Bowers *et al.* 1980) suggested that there was a tendency for more distant masers to be larger. Since it is unlikely that this is an intrisic property of distant masers it was assumed that the spot sizes of these masers were broadened by scattering by intervening diffuse HII regions. To date very little in the way of observational data of the distribution of maser spot sizes within the Galaxy has been gathered, howvever Kemball *et al.* (1987) have observed 17 OH maser sources in the galactic plane from longitude $355°$ to $50°$, they only detect sources at longitude $\geq 35°$ implying that interstellar scattering increases at low Galactic longitudes.

ACKNOWLEDGEMENTS

I would like to thank Willem Baan, Jessica Chapman, Jim Cohen, Aubrey Haschick, Christian Henkel, Gordon Mcintosh, Ray Norris, John Whiteoak and Tom Wilson who willingly and quickly provided me with reams of preprints and yet to be published data without which this review could not have been written.

REFERENCES

Baan, W. A. 1985, *Nature*, **315**, 26.
Baan, W. A. and Haschick, A. D., 1984, *Ap. J.*, **279**, 541.
Baan, W. A., Wood, P. A. D. and Haschick, A. D., 1982 *Ap. J. (Letters)*, **260**, L49.
Baan, W. A., Haschick, A. D. and Schmelz, J. T., 1985 *Ap. J. (Letters)*, **298**, L51.
Baan, W. A., Güsten, R. and Haschick, A. D., 1986, *Ap. J.*, **305**, 830.
Baan, W. A., Henkel, C. and Haschick, A. D., 1987, *Ap. J.*, in press.
Baart, E. E. and Cohen, R. J., 1985, *M.N.R.A.S.*, **213**, 641.
Baart, E. E., Cohen, R. J., Davies, R. D., Norris, R. P. and Rowland, P. R., 1986, *M.N.R.A.S.*, **219**, 145.
Barvainis, R., McIntosh, G. and Predmore, C. R., 1987, preprint.
Batrla, W., Matthews, H. E., Menten, K. M. and Walmsley, C. M., 1987, *Nature*, **326**, 49.
Baudry, A., Diamond, P. J., Graham, D. A., Walmsley, C. M., Guilloteau, S. and Booth, R. S., 1987, in preparation.
Booth, R. S., Kus, A. J., Norris, R. P. and Porter, N. D., 1981, *Nature*, **291**, 382.
Bottinelli, L., Fraix-Burnet, D., Gougenheim, L., Kázes, I., Le Squeren, A. M., Patey, I., Rickard, L. J. and Turner, B. E., 1985 *Astr. Ap. (Letters)*, **151**, L7.
Bottinelli, L., Dennefeld, M., Gougenheim, L., Le Squeren, A. M., Martin, J. M. and Paturel, G., 1986, Activity in Stars and Galaxies, in press.
Bottinelli, L., Gougenheim, L., Le Squeren, A. M., Martin, J. M., Dennefeld, M. and Paturel, G., 1987, IAU Circ. 4379
Bowers, P. F., Johnston, K. J. and Spencer, J. H., 1983, *Ap. J.*, **274**, 733.
Bowers, P. F., Reid, M. J., Johnston, K. J., Spencer, J. H. and Moran, J. M., 1980, *Ap. J.*, **242**, 1088.
Chapman, J. M. and Cohen, R. J., 1986, *M.N.R.A.S.*, **220**, 513.
Chapman, J. M., Cohen, R. J., Pointon, L., Staveley-Smith, L. and Unger, S. W., 1986, IAU Circ. 4180
Churchwell, E., Witzel, A., Huchtmeier, W. K., Pauliny-Toth, I., Roland, J., Sieber, W., 1977, *Astr. Ap.*, **54**, 969.
Claussen, M. J., Heiligman, G. M. and Lo, K. -Y., 1984, *Nature*, **310**, 298.
Claussen, M. J. and Lo, K. -Y., 1986, *Ap. J.*, **308**, 592.
de Vegt, Ch., Kleine, Th., Johnston, K. J., Bowers, P. F. and Spencer, J. H., 1987, preprint.

Diamond, P. J., Norris, R. P., Rowland, P. R., Booth, R. S. and Nyman, L. -A., 1985a, M.N.R.A.S., **212**, 1.
Diamond, P. J., Norris, R. P. and Booth, R. S., 1985b, M.N.R.A.S., **216**, 1p.
Diamond, P. J., Johnston, K. J., Chapman, J. M., Lane, A. P., Bowers, P. F., Spencer, J. H. and Booth, R. S., 1987, Astr. Ap., **174**, 95.
Diamond, P. J., Norris, R. P., Baan, W. A. and Booth, R. S., 1987, in preparation.
Dos Santos, P. M. and Lepine, J. R. D., 1979, Nature, **278**, 34.
Gardner, F. F. and Whiteoak, J. B., 1982, M.N.R.A.S., **201**, 13p.
Gaume, R. A. and Mutel, R. L., 1987, preprint.
Genzel, R., Reid, M. J., Moran, J. M. and Downes, D., 1981a, Ap. J., **244**, 884.
Genzel, R., Downes, D., Schneps, M. H., Reid, M. J., Moran, J. M., Kogan, L. R., Kostenko, V. I., Matveyenko, L. I. and Rönang, B., 1981b, Ap. J., **247**, 1039.
Guilloteau, S., Omont, A. and Lucas, R., 1987, Astr. Ap. (Letters), **176**, L24.
Haschick, A. D., Baan, W. A., Schneps, M. H., Reid, M. J. and Moran, J. M., 1987, in preparation.
Henkel, C., Gusten, R., Downes, D., Thum, C., Wilson, T. L. and Biermann, P., 1984, Astr. Ap. (Letters), **141**, L1.
Henkel, C., Wouterloot, J. and Bally, B., 1986 Astr. Ap., **155**, 193.
Huchtmeier, W. K., Witzel, A., Kuhr, H., Pauliny-Toth, I. and Roland, J., 1978, Astr. Ap., **64**, L21.
Huchtmeier, W. K., Richter, O. -G., Witzel, A. and Pauliny-Toth, I., 1980, Astr. Ap., **91**, 259.
Kázes, I. and Dickey, J. M., 1985, Astr. Ap. (Letters), **229**, L91.
Kázes, I., Mirabel, I. F. and Sanders, D. B., 1987, IAU Circ. 4362.
Kemball, A., Diamond, P. J. and Mantovani, F., 1987, in preparation.
Kylafis, N. D. and Norman, C., 1986, Ap. J. (Letters), **300**, L73.
Lepine, J. R. D. and Dos Santos, P. M., 1977, Nature, **270**, 501.
Mirabel, I. F. and Sanders, D. B., 1987, preprint.
Norris, R. P., Booth, R. S. and Diamond, P. J., 1982, M.N.R.A.S., **201**, 209.
Norris, R. P., Booth, R. S., Diamond, P. J., Nyman, L-A, Graham, D. A. and Matveyenko, L. I., 1984, M.N.R.A.S., **208**, 435.
Norris, R. P., Baan, W. A., Haschick, A. D., Diamond, P. J. and Booth, R. S., 1985, M.N.R.A.S., **213**, 821.
Norris, R. P., Whiteoak, J. B., Gardner, F. F., Allen, D. A. and Roche, P. F., 1986, M.N.R.A.S., **221**, 51p.
Norris, R. P., Caswell, J. L., Gardner, F. F. and Wellington, K. J., 1987, preprint.
Reid, M. J., Moran, J. M., Leach, R. W., Ball, J. A., Johnston, K. J., Spencer, J. H. and Swenson, G. W., 1979, Ap. J. (Letters), **227**, L89.
Reid, M. J., Haschick, A. D., Burke, B. F., Moran, J. M., Johnston, K. J. and Swenson, G. W., 1980, Ap. J., **239**, 89.
Reid, M. J., Schneps, M. H., Moran, J. M., Gwinn, C. R., Genzel, R., Downes, D. and Rönang, B., 1987, IAU 115.
Scalise, E. and Braz, M. A., 1981, Nature, **290**, 36.
Schmelz, J. T., 1987, Ph.D Thesis

Schneps, M. H., Lane, A. P., Downes, D., Moran, J. M., Genzel, R. and Reid, M. J., 1981, *Ap. J.*, **249**, 124.
Schultz, G. V., Sherwood, W. A. and Winnberg, A., 1978, *Astr. Ap. (Letters)*, **63**, L5.
Sivagnanam, P., Diamond, P. J., Ls Squeren, A. M., Daigne, G., Biraud, F. and Ortega-molina, A., 1987, in preparation.
Staveley-Smith, L., Unger, S. W., Cohen, R. J., Chapman, J. M. and Pointon, L., 1986, IAU Circ. 4248
Strelnitskij, V. S., 1984, *M.N.R.A.S.*, **207**, 339.
Weaver, H., Williams, D. R. W., Dieter, N. H. and Lum, W. T., 1965, *Nature*, **208**, 29.
Whiteoak, J. B. and Gardner, F. F., 1973, *Ap. Letters*, **15**, 211.
Wood, P. R., Bessell, M. S. and Whiteoak, J. B., 1986, *Ap. J. (Letters)*, **306**, L81.

THEORY OF ASTRONOMICAL MASERS

Nikolaos D. Kylafis
University of Crete
Physics Department
P. O. Box 470
714 09 Iraklion, Crete
GREECE

The theory of astronomical masers is reviewed. As with laboratory masers, masing occurs when a transition between two energy levels of a molecule exhibits inverted populations. In order to present the basic concepts about masers, an idealized two-level system is used. The exact energy level structure is taken into account later on when the pumping of specific molecules is discussed. Unlike laboratory masers, where the radiation must be bounced between two mirrors to accumulate gain, the propagation of radiation in astronomical masers is a lot simpler. This is because astronomical masers are single-pass and broadband. Thus, the main theoretical effort has concentrated on inventing efficient mechanisms that produce population inversion. Specific pumping mechanisms for the three molecules (H_2O, SiO, and OH) that exhibit strong masing are presented and their ability to explain the observations is discussed.

1. BASIC CONCEPTS

I would like to start with a qualitative description of masers and then proceed to a discussion of the basic concepts regarding astronomical masers (for a more extensive discussion, see Elitzur 1982a). I will briefly discuss the differences between laboratory masers and astronomical masers, and I will devote the major part of this review to the specific pumping mechanisms proposed for the different molecules in various astronomical environments.

The name MASER comes from the capitalized initials of the following words: Microwave Amplification through Stimulated Emission of Radiation. Microwave radiation is emitted when a radiative transition occurs between two energy levels of a molecule. For the purposes of a qualitative discussion, we will idealize the energy level diagram of a molecule as follows. We will isolate the two energy levels, between which the maser transition occurs, and treat all the other levels as a reservoir. As a result of collisional and radiative transitions, the molecules circulate between the reservoir and the masing levels. If, as a consequence of this circulation, and when steady state is achieved, the population per magnetic sublevel of the upper masing

M. J. Reid and J. M. Moran (eds.), The Impact of VLBI on Astrophysics and Geophysics, 223–230.

level is larger than that of the lower masing level, then the maser phenomenon takes place. This is because a photon that was emitted by a spontaneous emission from the upper to the lower masing level is more likely as it propagates to cause a stimulated emission than to be absorbed. Under normal conditions, the optical depth of the medium, which is proportional to the population per magnetic sublevel of the lower level minus that of the upper level, is positive and causes attenuation of the radiation. However, under conditions of inverted populations, the optical depth is negative and causes amplification of the radiation. This amplification is exponential because each maser photon causes the emission in the same direction of an identical photon.

The question that naturally arises at this point is: Can the intensity of a maser grow exponentially forever, assuming that we have a large enough region? The answer is clearly no, because with each stimulated emission a molecule goes from the upper masing level to the lower one. Thus, stimulated emission tends to reduce the population inversion. We will say that a maser is *saturated* when the stimulated emission rate is comparable to or greater than the intrinsic pumping rate. After saturation sets in, the growth of the intensity is no longer exponential. In the case of a one-dimensional maser, the growth after saturation is linear.

It is often argued that astronomical masers are strongly beamed. In order to understand how strong beaming can occur, we consider a maser region that is not spherically symmetric but is elongated. The maser photons travel, of course, in all directions. Nevertheless, for simplicity we approximate the maser radiation field as consisting of two beams, one in the direction of elongation and the other perpendicular to it. From geometrical considerations alone, the beam propagating in the direction of elongation is stronger than the other. This anisotropy is further enhanced because at every point in the maser, the two beams compete for the available pump photons and the more intense beam wins. Thus, an anisotropy in the shape of the masing region results in strong beaming. This statement has been quantified by Alcock and Ross (1985a,b; 1986a). Observers usually quote *isotropic* luminosities for the various sources. That is, they make the assumption that the observed flux is the same in all directions. It should be kept in mind that this assumption may result in significant overestimates of the true luminosity.

Many maser sources exhibit significant amounts of polarization. In order for a masing region to emit polarized radiation, the molecules must not only exhibit *inverted* populations, but also *unequal* populations in the magnetic sublevels of a maser level. Magnetic fields in the masing region play a central role in determining the polarization of emitted radiation, as demonstrated in a classic paper by Goldreich, Keeley, and Kwan (1973). This work has been generalized and extended in recent years (Western and Watson 1983a,b,c; 1984; Deguchi and Watson 1986a,b; Deguchi, Watson and Western 1986). These authors have shown that anisotropic radiation fields in the maser region can also cause polarization. Thus, it is not always clear what is the cause of the observed polarization. Notable exceptions, however, do exist. For example, Moran *et al.* (1978) identified 12 Zeeman pairs of left and right circular polarization in the spectrum of W3(OH) and determined the magnetic field strength in the region to be 2–9 mG.

In closing the qualitative discussion about astronomical masers, I would like to mention the major differences between laboratory masers and astronomical masers. In laboratory masers, the maser region is typically ~ 1 m long. In order to have significant gain, the radiation is bounced between two mirrors, one of which is semi-transparent to allow the amplified radiation to escape. In contrast, astronomical masers are single-pass, broadband, and have linear dimensions of a few AU or larger. Because the radiation transfer in astronomical masers is relatively simple, the main theoretical effort has concentrated on inventing efficient pumping mechanisms. In what follows, I will discuss in detail the various pumping mechanisms that have been proposed. For each of the three molecules that exhibit the strongest maser emission, namely H_2O, SiO, and OH, I will discuss both the environments that host them and the specific pumping mechanism that is believed to be operating there.

2. PUMPING MECHANISMS

2.1. H_2O Masers

The H_2O molecule exhibits one known maser transition between levels 6_{16} and 5_{23}. An inspection of the energy level diagram of H_2O (see, e.g., de Jong 1973, his Fig. 1) shows that the above two levels accidently lie close to each other, thus giving rise to a transition of 22 GHz. It is not an exaggeration to say that if you abuse this molecule in some way, it will most likely mase. A common method of abuse in astronomical environments is to expose the H_2O molecules to collisions and/or radiation. Thus, as collisions and/or radiation move the H_2O molecules among the energy levels, the upper masing level always gets more than the lower one, and population inversion is established. This is only a fluke of the transition rates. As it will be stressed later on, this pumping mechanism has its limitations.

2.1.1. *H_2O masers in the envelopes of late-type stars.* This class of H_2O masers presents no problem in its theoretical understanding because the masers in the envelopes of late-type stars are not intense. Cooke and Elitzur (1985), using plausible parameters to describe the environment, have shown that collisional pumping more than accounts for the observed luminosities.

2.1.2. *H_2O masers in star-forming regions.* These masers are the most intense of all galactic masers. Contrary to previous beliefs, Strelnitskij (1984) demonstrated that all models, suggested up to that time, suffer difficulties in explaining the extremely strong sources such as W49, W51, and others. The basic problem with these models is that the maximum power they can produce is lower than what is observed, and, if the models are pushed to their limits, the H_2O molecules thermalize.

Strelnitskij (1980; 1984) proposed the elegant idea that collisional pumping of H_2O masers could avoid the difficulty of thermalization if the H_2O molecules collide with two kinds of particles (e.g., H_2 molecules and electrons) at different kinetic temperatures, but with comparable collision rates. In other words, if the H_2O molecules find themselves in contact with two heat baths, they do not know which of the two Maxwellian distributions to obey and in fact they obey neither. With the help of analytical and numerical model calculations, it has been demonstrated (Kylafis and Norman 1986; 1987) that collisional pumping of H_2O, in environments where the temperature of the neutral particles T_H is significantly less than that of the electrons T_e, *can* sustain inverted populations. Such environments exist in the magnetic precursors of MHD shocks (Draine 1980; 1981; Draine, Roberge and Dalgarno 1983).

To understand physically why inversion occurs when $T_H \ll T_e$, consider the artificial situation where the electron collisions are turned off. Then, the H_2O molecules are in thermal equilibrium with the neutral particles. This means that for low enough T_H, practically all H_2O molecules are in the lowest rotational state (level 1_{01}). If the electron collisions are now turned on, there is a flow of H_2O molecules upward in energy space. The electron collision rates for these upward transitions are such that there is a wide "pathway" leading to the upper masing level but a narrow one to the lower masing level. Since the downward collision rates by the neutral particles do not destroy this asymmetry, population inversion is established. For more details see Kylafis and Norman (1987). The two-temperature model appears to explain the strong galactic H_2O maser sources, but it needs testing.

2.1.3. *Extragalactic H_2O masers.* Extragalactic H_2O megamasers with luminosities roughly in the range of 100 to 1000 solar luminosities have been observed. For reviews, see Baan (1985) and Diamond (1987). As discussed at the beginning of this section, infrared radiation *can* cause inversion in the H_2O molecules. However, the observations suggest that the infrared luminosity is not enough to pump the masing regions. It is possible that two-temperature collisional pumping is operating there (Claussen and Lo 1986). Because of this, I would like to point out that hot electrons and cold neutrals can also coexist in ionization fronts (Kylafis and Norman 1987). It is therefore interesting to note the similarity between galactic star-forming regions and the region somewhere between the broad line (BLR) and narrow line regions (NLR) in active galactic nuclei. Both have densities $\sim 10^9 - 10^{11}$ cm^{-3} and characteristic lengths $\sim 10^{11} - 10^{13}$ cm. If the extragalactic H_2O masers are indeed located in the BLR/NLR, then VLBI observations can probe these regions and provide invaluable information.

2.2. SiO Masers

The SiO molecule exhibits the most masing transitions of all other molecules. Maser emission has been observed in rotational transitions as high as $J = 5$ to $J = 4$ and in vibrational transitions as high as $v = 3$. Therefore, the maser emission in SiO is a general feature of the molecule and does not depend on the properties of a particular transition and the associated energy levels. For a review of the observations and a diagram of the energy levels of SiO, see Lane (1982, her Fig. 1).

The pumping mechanism that is believed to cause population inversion in SiO molecules was first suggested by Kwan and Scoville (1974). The basic idea is that if the radiative transition from the vibrational state v to $v-1$ is optically thick, then the transition rate is inversely proportional to $2J+1$, where J is the rotational level in the vibrational level v. Thus, the smaller the value of J is, the faster the level is drained and inversion is established. The upward transitions can be caused either by collisions (Elitzur 1980) or infrared radiation (Deguchi and Iguchi 1976). In the case of infrared pumping, the upward transitions must be optically thin while the downward ones must be optically thick. In other words, the optical depth must be anisotropic and larger than unity when it is averaged over all directions.

2.2.1. *SiO masers in the envelopes of late-type stars.* Detailed model calculations, taking into account both collisions and radiation, of SiO maser emission in a smooth stellar wind around a late-type giant have been performed by Langer and Watson (1984). They have found that, contrary to previous beliefs, for a variety of system parameters the model fails to produce the observed power. VLBI observations (Lane 1982) indicate that the maser regions are blobs or shocks. No model calculations have as yet been performed for such maser regions (see, however, Alcock and Ross 1986b).

I would like to point out here that strong magnetic fields (of order 10–80 G) have been inferred in the maser regions of late-type giants and supergiants (McIntosh 1987). One arrives at a similar conclusion by scaling a magnetic field strength of ~ 1 mG at $R \sim 3 \times 10^{16}$ cm for supergiant OH–IR sources (Cohen *et al.* 1987) down to a few stellar radii. If that is the case, then the magnetic field dominates the dynamics in the SiO maser region. Detailed studies of maser polarization with VLBI would therefore be of great help to the theorists.

2.2.2. *SiO masers in star-forming regions.* Up to 1985, it was thought that only IRc2 in Orion exhibited SiO maser emission. A model by Elitzur (1982b) could be made to work if stringent requirements on the parameters were imposed. This led Elitzur to suggest that the model automatically explained the lack of SiO maser emission in other star-forming regions.

Hasegawa *et al.* (1985) reported observations of SiO masers in the star-forming regions W51 IRS2 and Sgr B2 MD5. These SiO masers coincide in position with strong H_2O masers within a positional uncertainty of 5 arcseconds, and their isotropic luminosities are comparable to that in Orion. Further theoretical work is therefore needed for this class of SiO masers.

SiO masers in other galaxies have not been observed yet.

2.3. OH Masers

The OH molecule exhibits maser emission in all four allowed transitions in the ground state and also in several higher rotational levels (Reid and Moran 1981, see their Table 1). For a diagram of the energy levels of OH, see, for example, Elitzur (1982a, his Fig. 1). Of all the observed maser transitions in OH, the best studied ones are the four transitions in the ground state. The F-conserving transitions $F = 2$ to $F = 2$ (1667 MHz) and $F = 1$ to $F = 1$ (1665 MHz) are called *main* lines. The F-changing ones $F = 1$ to $F = 2$ (1612 MHz) and $F = 2$ to $F = 1$ (1720 MHz) are called *satellite* lines. As with the other two molecules discussed above, the OH molecules are pumped by radiation and collisions from the ground state to the higher rotational levels and then cascade back. The route they follow on their cascade back to the ground state determines which of the two satellite lines gets inverted (for details, see Elitzur 1982a). Unlike the satellite lines, the inversion of the main lines requires the pump process to have a parity selectivity built in. Radiative pumping can have this selectivity if the intensity of the radiation field that causes the excitation is a monotonically increasing function of frequency (Elitzur 1982a).

2.3.1. *OH masers in the envelopes of late-type stars.* This class of OH masers is fairly well understood theoretically. The pumping is done radiatively (Elitzur 1982a), and the stellar radiation is sufficient to do it (Dickinson 1987). The masers lie more or less in a spherical shell (as an example, see Chapman and Cohen 1986), but observations and theory suggest that they are located in blobs in the shell (Alcock and Ross 1986c).

2.3.2. *OH in star-forming regions.* The OH masers in star-forming regions are associated with compact HII regions. There is currently a controversy as to where exactly the OH masers lie, but I will not enter into it. The interested reader can consult a review of the observations by Genzel (1986) and an assessment of the theoretical understanding by Elitzur (1986). I will restrict myself to pumping considerations. Unlike the OH masers in the envelopes of late-type stars, radiative pumping here can be excluded because the number of exciting photons in the UV, the near infrared, and the far infrared are smaller than the number of maser photons observed. Contrary to previous beliefs, collisions with neutral particles also fail to establish inversion (Andresen *et al.* 1984). Johnston (1967) showed that if the OH molecules are exposed to collisions with a stream of electrons, then the excitation rate is larger than the de-excitation rate for $\Delta m = 0$ transitions and population inversion is established. The problem with this mechanism is how to get these streams of electrons. In order to speak of a stream, the ordered velocity of the electrons must be larger than their thermal speed, which is difficult to achieve. Because of this, Elitzur (1979) suggested ion streams. Ions stream when their ordered velocity exceeds ~ 1 km/s, a condition which is easily satisfied in MHD shocks. However, the collision rates of ions with OH are not known, and quantitative calculations cannot be performed. It would be very interesting to carry out two-temperature calculations in MHD shocks, similar to the ones for H_2O, to see whether this pumping mechanism works also for OH.

2.3.3. *Extragalactic OH masers.* Extragalactic OH megamasers have been observed in active galaxies with high far infrared emission. Baan (1985) and Staveley-Smith *et al.* (1987) give a detailed account of the observations and report the results of systematic searches for such masers. Their luminosity is comparable to that of H_2O masers, i.e., roughly in the range of 100 to 1000 solar luminosities, and it comes from amplification of the background radio source. The pumping is done by infrared radiation, which is abundant in the host galaxies (Baan 1985).

One point that I would like to stress before closing this review is the fact that high velocity wings have been observed (Staveley-Smith *et al.* 1987) in the OH maser line profiles of the Seyfert galaxies Mrk 231 ($\sim$ 800 km/s) and Mrk 273 ($\sim$ 1100 km/s). These velocities are comparable to those in the forbidden line regions of Seyfert galaxies. VLBI observations are therefore recommended in order to locate the emission regions relative to the nuclei of the galaxies.

I would like to thank Colin Norman for useful discussions and suggestions and the Space Telescope Science Institute for hospitality and financial assistance. This work has been supported in part by NATO through travel grant RG 86/82.

3. REFERENCES

Alcock, C. and Ross, R. R. 1985a, *Ap. J.*, **290**, 433.
Alcock, C. and Ross, R. R. 1985b, *Ap. J.*, **299**, 763.
Alcock, C. and Ross, R. R. 1986a, *Ap. J.*, **306**, 649.
Alcock, C. and Ross, R. R. 1986b, *Ap. J.*, **310**, 838.
Alcock, C. and Ross, R. R. 1986c, *Ap. J.*, **305**, 837.
Andresen, P., Hausler, D., Lulf, H. W., and Kegel, W. H. 1984, *Astr. Ap.*, **138**, 617.
Baan, W. A. 1985, *Nature*, **315**, 26.
Chapman, J. M., and Cohen, R. J. 1986, *M.N.R.A.S.*, **220**, 513.
Claussen, M. J., and Lo, K. -Y. 1986, *Ap. J.*, **308**, 592.
Cohen, R. J., Downs, G., Emerson, R., Grimm, M., Gulkis, S., Stevens, G., and Tarter, J. 1987, *M.N.R.A.S.*, **225**, 491.
Cooke, B., and Elitzur, M. 1985, *Ap. J.*, **295**, 175.
Deguchi, S., and Iguchi, T. 1976, *Pub. Astr. Soc. Japan*, **28**, 307.
Deguchi, S., and Watson, W. D. 1986a, *Ap. J. (Letters)*, **300**, L15.
Deguchi, S., and Watson, W. D. 1986b, *Ap. J.*, **302**, 750.
Deguchi, S., Watson, W. D., and Western, L. R. 1986, *Ap. J.*, **302**, 108.
de Jong, T. 1973, *Astr. Ap.*, **26**, 297.
Diamond, P. J. 1987, these proceedings.
Dickinson, D. F. 1987, *Ap. J.*, **313**, 408.
Draine, B. T. 1980, *Ap. J.*, **241**, 1021.
Draine, B. T. 1981, *Ap. J.*, **246**, 1045.
Draine, B. T., Roberge, W. G., and Dalgarno, A. 1983, *Ap. J.*, **264**, 485.
Elitzur, M. 1979, *Astr. Ap.*, **73**, 322.
Elitzur, M. 1980, *Ap. J.*, **240**, 553.
Elitzur, M. 1982a, *Rev. Mod. Phys.*, **54**, 1225.
Elitzur, M. 1982b, *Ap. J.*, **262**, 189.
Elitzur, M. 1986, in *Masers, Molecules, and Mass Outflows in Star Forming Regions*, ed. A. D. Haschick (Haystack Observatory), p. 299.
Genzel, R. 1986, in *Masers, Molecules, and Mass Outflows in Star Forming Regions*, ed. A. D. Haschick (Haystack Observatory), p. 233.
Goldreich, P., Keeley, D. A., and Kwan, J. Y. 1973, *Ap. J.*, **179**, 111.
Hasegawa, T., Morita, K.-I., Okumura, S., Kaifu, N., Suzuki, H., Ohishi, M., Hayashi, M., and Ukita, N. 1986, in *Masers, Molecules, and Mass Outflows in Star Forming Regions*, ed. A. D. Haschick (Haystack Observatory), p. 275.
Johnston, I. D. 1967, *Ap. J.*, **150**, 33.
Kwan, J., and Scoville, N. 1974, *Ap. J. (Letters)*, **194**, L97.
Kylafis, N. D., and Norman, C. 1986, *Ap. J. (Letters)*, **300**, L73.
Kylafis, N. D., and Norman, C. 1987, *Ap. J.*, in press.
Lane, A. 1982, Ph.D. Thesis, University of Massachusetts.
Langer, S. H., and Watson, W. D. 1984, *Ap. J.*, **284**, 751.
McIntosh, G. 1987, Ph.D. Thesis, University of Massachusetts.
Moran, J. M., Reid, M. J., Lada, C. J., Yen, J. L., Johnston, K. J., and Spencer, J. H. 1978, *Ap. J. (Letters)*, **224**, L67.
Reid, M. J., and Moran, J. M. 1981, *Ann. Rev. Astr. Ap.*, **19**, 231.
Staveley-Smith, L., Cohen, R. J., Chapman, J. M., Pointon, L., and Unger, S. W. 1987, *M.N.R.A.S.*, in press.
Strelnitskij, V. S. 1980, in *IAU Symposium 87, Interstellar Molecules*, ed. B. H.

Andrew (Boston: Reidel) p. 591.
Strelnitskij, V. S. 1984, *M.N.R.A.S.*, **207**, 339.
Western, L. R., and Watson, W. D. 1983a, *Ap. J.*, **268**, 849.
Western, L. R., and Watson, W. D. 1983b, *Ap. J.*, **274**, 195.
Western, L. R., and Watson, W. D. 1983c, *Ap. J.*, **275**, 195.
Western, L. R., and Watson, W. D. 1984, *Ap. J.*, **285**, 158.

VERY HIGH RESOLUTION OBSERVATIONS OF THE LUMINOUS WATER MASERS IN NGC 4258

M. J. Claussen (FCRAO, Univ. of Massachusetts, Amherst, MA USA),
M. J. Reid, M. H. Schneps (CfA, Harvard Univ., Cambridge, MA USA),
K.-Y. Lo (Univ. Illinois, Urbana, IL USA),
J. M. Moran (CfA, Harvard Univ., Cambridge, MA USA), and
R. Gusten (MPI, Bonn, FRG)

Abstract

We summarize the results and interpretation of a four station transcontinental VLBI experiment of the luminous water masers in the nearby galaxy NGC 4258. At a distance of 5 Mpc, the longest baseline of the experiment provides spatial resolution of less than 10^{16} cm. The strongest maser emission was detected on all baselines, and was found to consist of at least two features separated by about 0.1 mas (10^{16} cm). Weaker features are possibly spread over a region up to 1 mas in size. These results provide evidence that supports the scenario described by Claussen and Lo (1986) which suggests that the very luminous water masers reside in molecular gas that immediately surrounds the central, active nucleus.

Introduction

Five nearby galaxies have been found to have very luminous water masers residing in or near the nucleus (see Claussen and Lo 1986 and references cited therein). The total isotropic luminosity of the masers in these galactic nuclei range from 30 to 500 $L_{\odot}$. High resolution observations (35 mas) using the Very Large Array of the NRAO have shown that, for NGC 1068, NGC 4258, and NGC 3079, all the masers are coincident and in the nucleus of their respective galaxy (Claussen and Lo 1986, Haschick *et al.* 1987). When compared with the size scales and luminosities of masers in the Galaxy, these values are quite remarkable in that a large number of maser sites are required in a rather small volume. It is noteworthy that all five galaxies show some evidence for nuclear activity.

Three competing suggestions have been made in order to explain the origin of such luminous masers: 1) the masers are associated with the active nucleus of each galaxy and may reside in a circumnuclear disk of dust and molecular gas (Lo and Claussen 1986); 2) the masers arise from foreground gas which amplifies background continuum radiation (Haschick and Baan 1985); or 3) the masers are the very high end of a luminosity function of masers similar to those found toward star forming regions in the Galaxy (Ho *et al.* 1987). In order to determine the spatial distribution of the masers in the nucleus, and to help to distinguish these suggestions, very high resolution observations are necessary.

M. J. Reid and J. M. Moran (eds.), The Impact of VLBI on Astrophysics and Geophysics, 231–232.

Observations, Analysis and Results

A VLBI spectral line experiment of the masers in NGC 4258 was performed on October 7, 1984, using a transcontinenal array with the following stations: the 40-m at OVRO, the phased VLA, the 43-m at NRAO-Green Bank, and the 100-m at Effelsberg, FRG. The longest baseline (OVRO-Eff) provides a fringe spacing at 22 GHz of about 0.2 mas. The Mark II recording system was used, providing an instantaneous bandwidth of 2 MHz (27 km sec^{-1}). Since this velocity coverage is insufficient to observe all the emission from NGC 4258, switching between video converters set at different center frequencies was used to cover 6 MHz. The switching was done at a rate of 1 Hz. The data were correlated at the NRAO spectral line processor in Charlotesville, VA, and standard software was used to perform necessary calibration. The data has been analyzed using both fringe-rate and fringe-phase techniques. The results of fringe-phase analysis from the longest baseline shows that the strongest feature in the spectrum is split into two components separated by about 0.1 mas, as shown in Figure 1. Fringe-rate analysis on weaker features in the spectrum show that they lie within 1 mas of the strong feature.

These results, taken in conjunction with similar results on NGC 3079 (Haschick *et al.* 1987), show that the maser emission in these nuclei is distributed in regions of circumstellar rather than interstellar sizes. It seems unlikely that a single star-forming region (typical Galactic size of $\sim 10^{17}$ cm) could give rise to such luminous masers. If the luminous masers are simply the high-luminosity tail of a smooth distribution then one might expect to see more galaxies with nuclear masers of a few $L_{\odot}$. Present survey statistics do not support such a hypothesis (see Claussen and Lo 1986, Haschick and Baan 1985, and Henkel *et al.* 1985). Systematic velocity/spatial offsets could determine the kinematic structure of the gas in this region. Aperture synthesis techniques on this data set are now being used in order to determine if such structures exist.

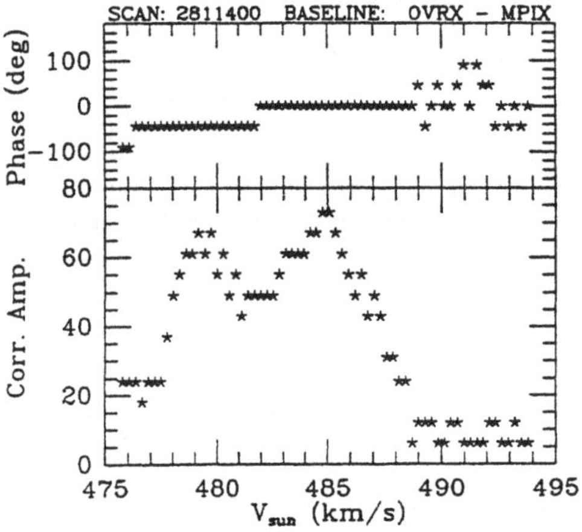

Figure 1 Cross-correlation spectrum of strong water maser feature in NGC 4258

References

1. Claussen, M. J. and Lo, K.-Y. 1986, *Ap. J*, **308**, 592.
2. Haschick, A. D., Schneps, M. H., Reid, M. J., Baan, W. A., and Moran, J. M., 1987, this volume.
3. Haschick, A. D. and Baan, W. A. 1985, *Nature*. **314**, 144.
4. Henkel, C., Gusten, R. Downes, D., Thum, C., Wilson, T. L., and Biermann, P. 1984, *Astr. Ap.*, **141**, L1.
5. Ho, P. T. P., Martin, R. N., Henkel, C. and Turner, J. L. 1987, preprint.

THE SUPER LUMINOUS MASER SOURCE IN THE NUCLEUS OF NGC 3079

Aubrey D. Haschick
Haystack Observatory, Westford, MA, USA
Willem A. Baan
Arecibo Observatory, Arecibo, PR, USA
Matthew H. Schneps, Mark J. Reid and James M. Moran
Center for Astrophysics, Cambridge, MA, USA

On 1984 October 6 we conducted a 3-station intercontinental Mark II VLBI experiment in order to study the very luminous water vapor maser source in the nucleus of the galaxy NGC 3079, which was detected first by Haschick and Baan (1985) using the Haystack Observatory 36.6 m antenna. The cross correlation spectrum for the longest Owens Valley to MPI baseline is presented in Figure 1 and shows the phase variation across the width of the brightest feature at 955.7 km s^{-1} to be less than 10 degrees of phase.

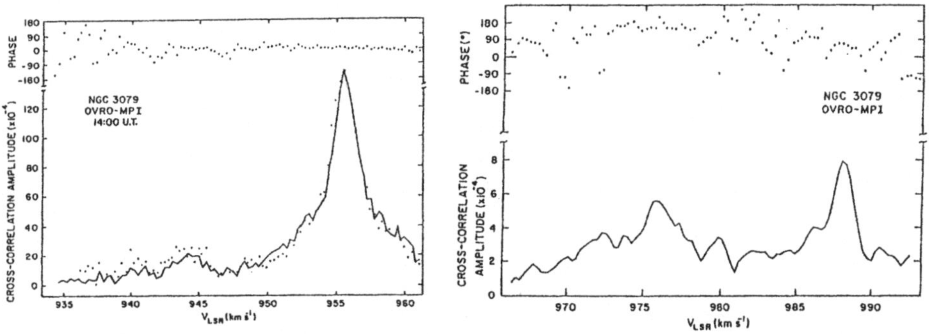

Figure 1. Cross correlation and autocorrelation spectra of NGC 3079 for two separate 2 MHz windows.

The phase difference between the brightest feature at 955.7 km s^{-1} and the two weak features at 944 and 940 km s^{-1} is less than 40° in Figure 1 and shows a slow variation by ~40° over the 7 hour period of the experiment. An autocorrelation spectrum for the MPI 100m telescope is plotted superimposed on the cross correlation in Figure 1a and shows identical features to the cross-correlation. In Figure 1b the cross-correlation spectrum for the 2 MHz window centered on 980 km s^{-1} is given; the cross-correlation amplitude spectrum is the scalar average of the amplitude of the seven 40 minute scans in the experiment. The phase spectrum is for a single scan and shows a clear separation of the phases of the features indicated in the amplitude spectrum. To

M. J. Reid and J. M. Moran (eds.), The Impact of VLBI on Astrophysics and Geophysics, 233–234.

illustrate the separation of the features further, the phases of four features at 976, 980, 986 and 988 km s^{-1} are plotted with respect to time in Figure 2 and show variations which are less than ~360°. The 976 km s^{-1} feature has a phase variation of less than 90° and this indicates its relative proximity to the reference feature.

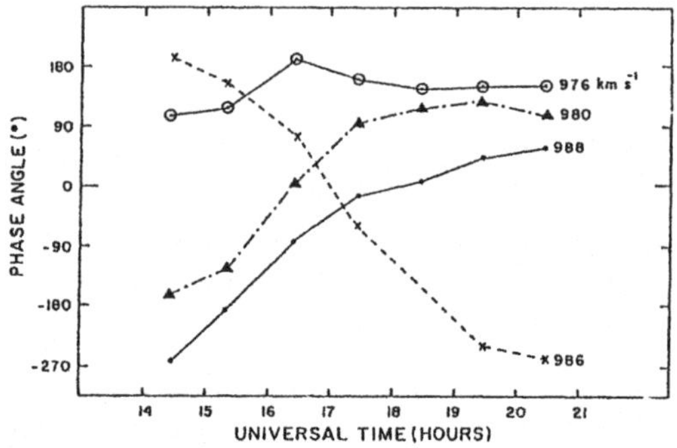

Figure 2. Plot of the variation of phase with time for the four velocity features in Figure 1b.

From the above data we may conclude that the water vapor maser features covering a velocity range from 940 to 988 km s^{-1} are separated on the sky by less than 0.3 milliarcsec ($7x10^{16}$ cm for a distance of 16.5 Mpc). The maser source in NGC 3079 has a luminosity which is 500 times higher than the brightest H_2O maser source in our Galaxy, W49N, and has a sizescale of at least a factor of two less than those of Galactic maser regions. In order for the maser source in NGC 3079 to be pumped by OB stars similar to W49N and Orion A in our galaxy it would require an unusually high density of OB stars. We favor a model in which the outflow from the galactic nucleus, as indicated by the 180 km s^{-1} blueshift of the maser features from the galaxy systemic velocity, interacts with a molecular shell or disk surrounding the nucleus, and creates high density ($>10^{10}$ cm^{-3}) clumps or filaments. The hot electron-cold neutrals model of Kylafis and Norman (1986) is capable of generating the observed high luminosity maser at densities of $10^{11}cm^{-3}$ without the occurrence of thermalization. The H_2O maser source is coincident with a 60 mJy nuclear source (Haschick & Baan 1986) and may be amplifying the background continuum. An unsaturated gain of only 4 is required to produce the observable emission.

REFERENCES

Haschick, A. D. and Baan, W. A., 1985, *Nature*, 310, 298.

Haschick, A. D. and Baan, W. A., 1986, in *Masers, Molecules and Mass Outflows in Star Forming Regions*, ed. A. D. Haschick (Haystack Observatory: Westford, MA), p. 249.

Kylafis, N. D. and Norman, C. 1987 in *Masers, Molecules and Mass Outflows in Star Forming Regions*, ed., A. D. Haschick (Haystack Observatory: Westford, MA), p. 341.

FIRST VLBI OBSERVATIONS OF 6.3 cm OH MASERS IN COMPACT HII REGIONS

A. Baudry [1], P.J. Diamond[2], D. Graham [3], M. Walmsley [3],
R. Booth [4], N. Brouillet [1], G. Daigne [1]
[1] Observatoire de Bordeaux, 33270 Floriac, France
[2] NRAO, Charlottesville, 2015 Ivy Rd, Charlottesville, VA 22903
[3] Max-Planck-Institut für Radioastronomie, Auf dem Hügel 69, D-5300 Bonn 1, FRG
[4] Onsala Space Observatory, S-43900 Onsala, Sweden

ABSTRACT. VLBI synthesis observations of the $^2\Pi_{1/2}$, J = 1/2 excited-state of OH at 6.3 cm (ΔF = 1-0 = 4765.562 MHz) have been made towards four compact HII regions. Detailed maps have been produced for W3(OH) where three groups of sources are distributed over a region $\sim$ 0.01 pc in size. The brightest sources are shown to be saturated masers with $T_B \gtrsim 10^9$ K. In ON1 a weak feature gives fringes with a $\sim$ 0.05" lobe spacing.

1. INTRODUCTION

Synthesis maps of the ground- and excited-state OH sources associated with compact HII regions provide, together with model calculations, a unique mean to understand the physical conditions prevailing in cosmic masers. Detailed maps also probe the spatial distribution and the excitation mechanisms of the OH gas.

The only VLBI observations of excited-states of OH that have been published so far concern the 6035 MHz line, 84 cm^{-1} above the ground-level (Moran et al., 1978) and the 4765 MHz line, 126 cm^{-1} above the ground-level (Baudry et al., 1987).

2. OBSERVATIONS

The $^2\Pi_{1/2}$, J = 1/2 line of OH at 4765.562 MHz was observed with 4 or 3 telescopes of the European VLBI Network (EVN) and with transatlantic baselines to Owens Valley, in October 1984 and 1985 and in November 1986. The lobe spacings with the EVN were $\sim$ 0.018 to 0.050" and the spectral resolution in the line maps was $\sim$ 0.2 km/s. The type I OH sources associated with compact HII regions which were searched for 6.3 cm OH emission are given in Table I. These regions are known to be sites of active stellar formation.

3. W3(OH)

Our VLBI maps are extensively discussed in Baudry et al. (1987). The 6.3 cm emission consists of 3 groups of sources distributed over a region $\sim$ 1" in size. They lie on the edge of the associated continuum

M. J. Reid and J. M. Moran (eds.), The Impact of VLBI on Astrophysics and Geophysics, 235–236.

TABLE I

Sources	RA (1950)	DEC	Detection	Year of observations (EVN)
W3	$02^h21^m53^s.35$	$61^\circ52'23''.50$	No(a)	1984
W3(OH)	02 23 16.32	61 38 57.78	Fringes	1984, 1985, 1986
ON1	20 08 09.80	31 22 41.00	Fringes	1986
DR21(OH)	20 37 24.70	42 12 09.00	(b)	1986

(a) W3 is very time variable. No fringes at the time of the observations.
(b) Data not yet processed.

source. Some sources are unresolved with the EVN and have sizes $\leqslant$ 15-25 milli arc sec implying $T_B \geqslant 10^9$ K. All sources are resolved with the Effelsberg-Owens Valley baselines. The strongest masers in W3(OH) appear to be saturated. In addition the cross-correlated spectra obtained in 1984, 1985 and 1986 with the Effelsberg-Westerbork baseline are fairly similar. The total power spectrum remains also unchanged. The observations show that there is a good spatial coincidence of the 4765 MHz masers with the 6035 MHz and the 1665-1667 MHz masers in comparable velocity ranges. This suggests similar pumping conditions for $^2\Pi_{1/2}$ and $^2\Pi_{3/2}$ OH although theory and experiment predict that collisions have opposite effects on the two OH energy ladders. However model calculations (Guilloteau et al., 1985) prove that the 4765 and 1665-1667 MHz transitions can be inverted simultaneously.

4. ON1 (OH 69.5 - 1.0)

ON1 is an isolated compact HII region with strong type I OH emission. Weak fringes were detected with the Effelsberg-Westerbork baseline near 14.9 km/s where 6035 MHz OH emission is also present (Rickard et al., 1975). Inspection of our cross-correlated spectra obtained with a lobe spacing of 0.05" suggests that the spatial structure across the spectral feature is complex. Detection of this feature with longer, less sensitive baselines is difficult because it is weak. Although final calibration of the data is not yet available we deduce, by comparison with the unresolved feature in W3(OH), that the peak flux is $\sim$ 0.45 Jy.

REFERENCES

Baudry, A., Diamond, P.J., Graham, D., Walmsley, M., Booth, R., 1987, *Astron. Astrophys.*, in preparation

Guilloteau, S., Baudry, A., Walmsley, M., 1987, *Astron. Astrophys.* **153**, 179

Moran, J.M., Reid, M.J., Lada, C.J., Yen, J.L., Johnston, K.J., Spencer, J.H., 1978, *Astrophys. J.* **224**, L67

Rickard, L.J., Zuckerman, B., Palmer, P., 1975, *Astrophys. J.* **200**, 6

CIRCULAR POLARIZATION OF ASTROPHYSICAL MASERS

S. Deguchi, G. Nedoluha and W. D. Watson
University of Illinois at Urbana-Champaign
1110 West Green Street, Urbana, IL 61801

ABSTRACT

Results of further calculations are presented to explore the non-linear, Zeeman overlap effect as the cause for the circular polarization of astrophysical masers. Emphasis is placed on the regime in which the Zeeman splitting is small and on the variation of the polarization with maser saturation.

Strong net circular polarization has long been recognized as a striking characteristic of astrophysical OH masers. It is especially prominent for OH masers in molecular clouds (star-forming regions) for which the measured, fractional circular polarization exceeds fifty percent in perhaps half of the masers (e.g., Reid and Moran 1981). Linear polarization is also common. More recently, high resolution studies of the other general class of OH masers--those associated with circumstellar environments--have found significant, net circular polarization in these as well (Mutel et al. 1979; Cohen et al. 1987). In a very few of both classes of OH masers, Zeeman pairs can be identified and tend to indicate magnetic fields of a few milligauss or so. Very recently, circular polarization (up to about ten percent, fractionally) has been detected in the radiation of the circumstellar SiO masers (Barvainis, McIntosh and Predmore 1987). Because of the much smaller magnetic moment of SiO (in comparison with OH), the occurence of this circular polarization is of special, potential significance.

Here, we emphasize some additional points and present some further calculations beyond our basic paper (Deguchi and Watson 1986) on the previously unrecognized non-linear, Zeeman overlap mechanism as a cause for the circular polarization of astrophysical masers. Our calculations treat the case of an angular momentum J=1-0 transition and utilize the Sobolev approximation. The velocity gradients are cylindrically symmetric with the gradient in the axial direction being much smaller than the gradient in the other directions so that the maser is essentially one-dimensional in the axial direction.

M. J. Reid and J. M. Moran (eds.), The Impact of VLBI on Astrophysics and Geophysics, 237–238.

Our previous results were directed toward the regime in which (Zeeman splitting/local Doppler breadth) > 1. Here, we show how the polarization characteristics vary with this ratio as it approaches zero. Ten percent circular polarization remains at a value of about 0.1 for this ratio. Note also that the linear polarization (unlike the circular) is non-zero (as it should be) when there is no Zeeman splitting. This is most likely to be the relevant regime for species such as SiO with small magnetic moments.

We also consider in more detail how the circular polarization varies with the degree of saturation (R_3/Γ) of the maser. It rapidly reaches to near its asymptotic value when (R_3/Γ) is only a few. For (Zeeman splitting/local Doppler breadth) $\gtrsim$ 1 and at least partly saturated masing the sign of the circular polarization tends to be that of the Zeeman transition of the highest frequency when the masing region is expanding (and vice-versa).

This research has been supported in part by the National Science Foundation.

REFERENCES

Barvainis, R., McIntosh, G., and Predmore, C. R. 1987, submitted.

Deguchi, S. and Watson, W. D. 1986, Ap. J. (Letters), 300, L15.

Cohen, R. J., Downs, G., Emerson, R., Grimm, M., Gulkis, S., Stevens, G., and Tarter, J. 1987, MNRAS, 225, 491.

Mutel, R. L., Fix, J. D., Benson, J. M. and Webber, J. C. 1979, Ap. J., 228, 771.

Reid, M. J. and Moran, J. M. 1981, Ann. Rev. Astr. Ap., 19, 231.

PHYSICS OF COSMIC MASERS

V. S. Strelnitskij
Astronomical Council, U.S.S.R. Academy of Sciences,
Moscow, U.S.S.R.

1. INTRODUCTION

Natural masers, which emit powerful narrow-band radiation in molecular microwave lines, were discovered about 20 years ago. They draw the astrophysicist's attention in two ways: 1) as a unique physical phenomenon and 2) as very informative probes of physical conditions in those astrophysical objects (very young and very old stars) with which they are intimately connected. In this paper, the first aspect of masers will be stressed, though some consideration will be given to the astrophysical aspect as well. The limited space here will not allow a complete discussion. For a more systematic account, refer to reviews on astrophysical masers by Litvak 1974; Strelnitskij 1974; Kegel 1975; Cook 1977; Moran 1976; Reid and Moran 1981; Elitzur 1982. I shall limit myself only to several topics of general character, which, to me personally, seem to be particularly important for understanding the fundamental principles of cosmic maser physics. I shall then try to illustrate these general principles by the example of H_20 masers in star formation regions. These masers, in my opinion, are the most intriguing of all the cosmic masers known at present. And (to make the subjectivity of the report complete!) I shall lay special stress on the results obtained by our group in the last few years.

2. SOME HISTORICAL REMARKS

Cosmic maser radiation was first observed in 1965 – in OH radio lines near 18-cm wavelength (Weaver *et al.* 1965). It would be wrong to consider this discovery as a *complete* surprise. For example, long before the discovery of not only astrophysical, but also laboratory, masers the astrophysicist Menzel (1937) had pointed out the possibility, in principle, of the "negative absorption" (i.e., amplification) of radiation in spectral lines in the absence of local thermodynamic equilibrium (LTE). A really unexpected result was the *strength* of the maser effect detected in the interstellar medium: amplification in many sources attains $10^{10} - 10^{12}$, so that radiation brightness temperatures reach 10^{16}K, and the luminosity in a narrow radio line is sometimes comparable to the total (bolometric) luminosity of a solar class star (assuming isotropic emission from the maser).

Soon after the discovery of the OH masers Shklovskij (1966) suggested that the maser radiation was generated in the upper layers of protostars. It then became clear that a unique observational method for probing the star-formation regions became available to astrophysicists.

M. J. Reid and J. M. Moran (eds.), The Impact of VLBI on Astrophysics and Geophysics, 239–247.

In 1969 the Berkeley group (Cheung *et al.* 1969) discovered the first interstellar H_20 maser at 1.35-cm wavelength ($6_{16} - 5_{23}$ rotational transition). It was soon found that some H_20 masers exhibit very broad spectra, up to several hundred km/s in radial velocity. Strelnitskij and Sunajev (1972) pointed out that the simplest, kinematical, interpretation of such wide spectra automatically implies dynamical instability of the corresponding system of maser condensations. Thus, they predicted the proper motions of the maser spots of the order of milliarcseconds per year, a prediction that was confirmed a decade later (Genzel *et al.* 1981). They also suggested that the systems of maser condensations are systems of *light* (planetary masses) gas condensations blown away from very young stars by strong stellar winds or other mechanisms. Many subsequent observations by different techniques confirmed extremely strong gas outflows (up to $\sim 10^{-2}\ M_\odot$/year) from very young stars. As a rule, these flows have a well-defined "bipolar" structure (e.g., Bally and Lada 1983).

The first *extragalactic* masers, probably also related to the star-formation activity in other galaxies, were discovered in the late seventies (Churchwell *et al.* 1977). More than a dozen "masing" galaxies (in OH, H_20, or H_2CO lines) are known now.

In 1968, the OH maser emission was first observed from the *old* red giant stars (Wilson and Barrett 1968). Later on, many of these stars were also found to have H_20 and Si0 maser radiation.

Weak masers in CH_3OH (methanol) lines at 1.2 cm wavelength have been discovered in Orion by Hills, Pankonin, and Landecker (1975). Stronger methanol masers at 2.5 cm wavelength appear to be common (Batrla, Matthews, and Menten 1987). Weak maser effects have also been observed in the radio lines of some other molecules such as H_2CO and CH.

In parallel with the observations of cosmic masers was the development of maser theory. The major relations for the simplest, one-dimensional, model of a cosmic maser amplifier were formulated by Litvak *et al.* (1966) soon after the discovery of the OH masers. In 1972–74, a series of papers by Goldreich and others (e.g., Goldreich and Keeley 1972; Goldreich and Kwan 1974a,b) appeared, which considerably extended our understanding of the physical processes in astrophysical masers.

The study of cosmic masers is closely connected with the development of VLBI. The first interferometric measurements of masers (Rogers *et al.* 1966; Cudaback, Read, and Rougoor 1966) showed that their brightness temperatures were in excess of 10^6 K. Shortly thereafter, VLBI observations (Moran *et al.* 1968) revealed the extremely high brightness temperatures of the OH sources (as high as 10^{13} K), making the maser interpretation of these sources inevitable.

3. KINETICS AND THERMODYNAMICS OF COSMIC MASERS

The relaxation time scales in strong cosmic masers ($\sim 10^{-1} - 10^4$ s) are much smaller than the time scales of their variability, and this condition is believed to be sufficient for the assumption of *stationarity*. The level populations of a stationary non-LTE system are described by the stationarity equations:

$$\sum_{k\neq i} n_k W_{ki} - n_i \sum_{k\neq i} W_{ik} = 0 \ , \qquad \sum_{k} n_k = N \ \ (k = 1, 2, ..., m) \ , \tag{1}$$

where n_k is the population of level k, W_{ik} is the total probability of the ik transition, m is the total number of the levels involved, and N is the total population. Transitions are produced by collisions of the "working" molecules with other particles, by their interactions with the radiation fields, and as a result of chemical processes. These external agents can be regarded as "thermostats" that tend to bring the level populations into equilibrium with their own temperatures. These temperatures need not be strict equilibrium temperatures; they can be defined as the inverse derivatives of the entropy with respect to the energy in nonequilibrium situations, or simply as convenient temperature-like parameters, which, together with another well-known temperature parameter, the excitation temperature of the transition, satisfy the usual rule of the direction of energy transfer between energy reservoirs (Strelnitskij 1983).

Assuming some physical model of the maser, one can, in principle, fix all the rate coefficients in the system (1), or indicate the procedure by which they can be found. Solution of this system allows us to say whether the signal transition will or will not be *inverted* in such a physical situation, a condition for a maser to function. In practice, however, it is impossible to take into consideration all the possible processes, and one tries to *estimate* the most important nonequilibrium process leading to inversion: the *pump mechanism*. A clear understanding of the thermodynamical properties of nonequilibrium quantum-mechanical systems can help substantially in this analysis.

In a stationary system, the transfer of populations can only be *cyclic*. It can be shown (e.g., Strelnitskij 1983) that each pump cycle leading to inversion of the signal transition works as a "heat engine." As such, the pump cycle must have at least one step corresponding to the receipt of energy from a high-temperature thermostat and at least one corresponding to the sink of energy to a low-temperature thermostat. The net population transfer realizing these steps can be of radiative (R), collisional (C) or chemical (X) nature. Thus, all the pump mechanisms can be divided into nine types, according to the combination of the kinds of energy source and sink:

$$\begin{matrix} RR & RC & RX \\ CR & CC & CX \\ XR & XC & XX \ . \end{matrix} \tag{2}$$

To describe precisely the type of the pump one must also indicate the kind of transitions by which the pump cycles are constructed: electronic (e), vibrational (v), or rotational (r). The matrix (2) is a more informative and heuristic classification than the previous division of all pump mechanisms into radiative, collisional, or chemical, solely based on the kind of energy source. The scheme described here serves as a relatively good guide for the search of the adequate pump mechanism by the preliminary comparative analysis of possible pump schemes and crude estimates of their pump rates (the "equivalent transition approximation," Strelnitskij 1983,

1984). When a potentially effective mechanism is chosen in this manner, it can be investigated in more detail by solving the system of equations (1). The effective procedures to identify the real pump cycles and to describe the type of the pump after solving equation (1) were worked out by Sobolev (1987) and references therein.

4. MASER SATURATION AND LINE WIDTH

Neglecting spontaneous emission, which is not substantial for strong masers, one can give the one-dimensional transfer equation of maser radiation the following form (Goldreich and Kwan 1974a):

$$\frac{dI_\nu}{dz} = k_\nu \, (I_\nu^{-1} + I_s^{-1})^{-1} \quad , \tag{3}$$

where

$$k_\nu = \frac{h\nu B}{4\pi} \frac{\Delta\lambda_\nu}{\Gamma} \tag{4}$$

is the amplification coefficient at frequency ν within the line, B is the Einstein B-coefficient, $\Delta\lambda_\nu$ is the difference of pump rates for the upper and lower maser levels (per unit volume, per unit frequency interval), Γ is the decay constant, describing the relaxation (collisional and/or radiative), which is assumed to be the same for both levels, h is Planck's constant, and

$$I_s = \frac{2\pi\Gamma}{B\Omega} \tag{5}$$

is the intensity at which saturation effects become important. What does saturation mean? The solution of equation (3) is:

$$\ln \frac{I_\nu}{I_\nu(0)} + \frac{I_\nu - I_\nu(0)}{I_s} = \int_0^z k_\nu \, dz \equiv \tau_\nu \quad , \tag{6}$$

where z is the distance along the line of sight through the maser, and τ_ν is the optical depth. It is readily seen that when $I_\nu \ll I_s$, the intensity depends exponentially on the optical depth:

$$I_\nu \propto e^{\tau_\nu} \quad . \tag{7}$$

However, when $I_\nu \gg I_s$ (i.e., when $2R \gg \Gamma$, $R \equiv I_\nu B\Omega/4\pi$, the stimulated emission rate, and Ω is the solid angle of the maser beam), I_ν depends only linearly on τ_ν:

$$I_\nu \propto \tau_\nu \quad . \tag{8}$$

It must be stressed that the decay constant Γ, which determines the saturation condition, accounts for *all* the relaxation processes, both collisional and radiative. Ignoring this fact leads to misunderstandings.

One potential misunderstanding concerns line broadening. It is well known that the exponential dependence of I_ν on τ_ν in the case of unsaturated maser amplification leads to line narrowing relative to the Doppler width $\Delta\nu_D$:

$$\frac{\Delta\nu}{\Delta\nu_D} \approx \tau_0^{-0.5} \quad , \tag{9}$$

where $\Delta\nu$ is the linewidth, and τ_0 is the optical depth at the line center (the proof can be found, for example, in Strelnitskij 1982a). Comparing equations (7) and (9), we find that

$$\Delta\nu \propto (\ln I_0)^{-0.5} \quad , \tag{10}$$

where I_0 is the intensity at line center. Hence, there is only a *very weak* inverse dependence of $\Delta\nu$ on I_0.

In the last few years many cases of much stronger $\Delta\nu(I_0)$ dependence have been reported for H_20 sources:

$$\Delta\nu \propto I_0^{-\alpha} \quad , \tag{11}$$

where α varies from ≈ 0.3 (Lada *et al.* 1981) to ≈ 1 (Rowland and Cohen 1986), with $\alpha \approx 0.5$ as the most typical value (Mattila *et al.* 1985; Rowland and Cohen 1986). In the last two papers, the $\Delta\nu \propto I_0^{-0.5}$ dependence is explained by the mechanism of radiative relaxation proposed by Goldreich and Kwan (1974a), but this explanation seems to be incorrect since it does not include the radiative relaxation in Γ (Strelnitskij 1986). Another explanation for the observed line width dependence is described in the next section.

5. PHYSICS OF H_20 MASERS IN REGIONS OF STAR FORMATION

H_2O maser sources are particularly interesting because: 1) they are the strongest of all known masers; 2) they are probably connected with the earliest stages of stellar evolution observed in maser emission; 3) their kinematics promise new and very attractive possibilities of the distances measurements in the Galaxy and beyond (Reid, Moran, and Gwinn 1986, 1987).

The radiative-radiative and the collisional-radiative pump schemes (*RRv*, *CRr*, *RCv*, and *CRv*) that have been proposed (Litvak 1969; Strelnitskij 1971; de Jong 1973; Goldreich and Kwan 1974b; Shmeld, Strelnitskij, and Musilev 1976; Norman and Silk 1979) are barely capable of explaining the extremely strong sources such as W49, Ori A or W51. The basic reason is the existence of an upper limit for the number density in the maser condensation (above which the radiative population transfer is hindered by collisions). With this limitation and the limitation on the dimensions of the source determined by VLBI measurements, these pump schemes cannot account for the observed maser emission power (Strelnitskij 1984).

The analysis of matrix (2) suggested to the author a new type of possible pump, *CC*, which is able to work, in principle, at *any* density (Strelnitskij 1980, 1984). For the *CCr* pump, two kinds of colliding particles with different kinetic temperatures are required, e.g., H_2 molecules and electrons. The hot particles provide the source, and the cool particles provide the sink of the quantum heat engine. Solutions of the stationarity equations obtained by Bolgova (1981) and by Bolgova, Makarov, and Sobolev (1987) have shown a strong tendency of the CCr pumping with $T_H > T_e$ to invert the signal transition $6_{16} - 5_{23}$ as well as other transitions. This inversion takes place over a large range of fractional ionization ($10^{-5} - 10^{-1}$) and kinetic temperature ($\sim 10^2 - 10^3$ K). In these calculations, the poorly known collision probabilities were varied over a wide range.

In the astrophysical model of the H_20 maser with the *CCr* pump proposed by the author (Strelnitskij 1984), the required condition $T_H > T_e$ arises behind the shock front in a circumstellar gas condensation impinged upon by a strong stellar

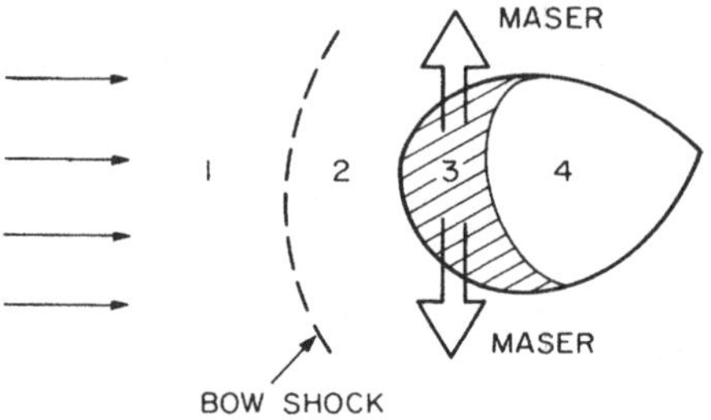

Figure 1. A stellar wind impinging on a maser condensation. 1—the unperturbed stellar wind; 2—the compressed gas behind the bow shock; 3—the gas of the condensation compressed by the internal shock; 4—the unperturbed gas of the condensation.

wind (regions 3 and 4 in Fig. 1). The required ionization can be produced by high-energy (MeV) protons accelerated by plasma instabilities behind the shock front (regions 1 and 2). The main advantage of this model, described in detail by Strelnitskij (1984), is the possibility to explain the observed extremely high flux densities from the H_2O sources. It can also explain many other properties of the sources. Let us see here how the model can explain the observed strong anticorrelation of $\Delta\nu$ and I_0 mentioned in the preceding section. The optical depth in the line center is inversely proportional to $\Delta\nu_D$ if the pump rate, as is commonly assumed, does not depend on the velocities of the molecules:

$$\tau_0 \propto k_0 \propto \frac{\Delta\Lambda}{\Delta\nu_D} \quad , \tag{12}$$

where k_0 is the amplification coefficient at line center, and $\Delta\Lambda$ is the *full* difference of the pump rates for the upper and lower maser levels. The previous unsuccessful (see §4) explanations of the observed variations of I_0 and $\Delta\nu$ were based on the assumption that these variations are due to changes of $\Delta\Lambda$, leaving $\Delta\nu_D$ constant. Our hypothesis is that I_0 and $\Delta\nu$ change because $\Delta\nu_D$ changes, e.g., due to changes of the kinetic temperature in the amplification region.

We note now that in *all* stages of maser saturation, the observed $\Delta\nu$ will be proportional to $\Delta\nu_D$, whereas τ_0 is inversely proportional to $\Delta\nu_D$. Suppose that the maser is partly saturated. Then we can write $I_0 \propto \tau_0^{\beta}$ (see Fig. 2). For a large part of the $I_0(\tau_0)$ curve, β is between ~ 1 and ~ 3, and one will have $\Delta\nu \propto I_0^{0.3-1}$, as observed.

In the model shown in Figure 1, the rise of kinetic temperature necessary to explain the rise of $\Delta\nu_D$ and the decrease of I_0 can be connected with perturbations in the stellar wind coming to the shock front 1–2. Any such perturbation will change the conditions behind the shock and in particular will change the number and energies of protons accelerated here and producing the ionization in the pump region 3.

At first, the wind condensation coming to the front 1–2 will increase the maser intensity due to these extra protons, which reach the maser region in less then 1 day due to their high velocities. The gas-dynamical perturbation, on the other hand, takes $\sim$ 10 days to cross into region 2 and reach the amplification layer 3. Thus, within this period, I_0 rises, but $\Delta\nu$ *remains almost unchanged.* Then the perturbation reaches regions 2–3, and the kinetic temperature within the

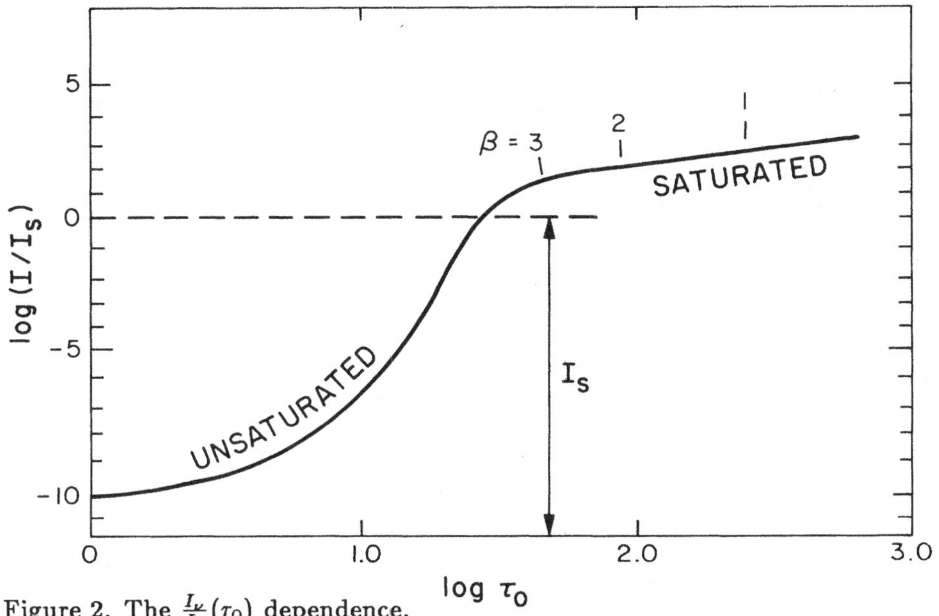

Figure 2. The $\frac{I_\nu}{I_s}(\tau_0)$ dependence.

amplification region begins to rise, increasing $\Delta\nu_D$ and decreasing I_0. The time scale for this phase is of the order of sound propagation time through region 3, which is approximately several months. This simple model satisfactorily explains the typical time scales in H_2O maser outbursts observed by Rowland and Cohen (1986). It also explains why, for a well-studied, flare-up of the -11 km/s feature in Cepheus A (Rowland and Cohen 1986; Mattila *et al.* 1985), the increase of I_0 was not accompanied by changes of $\Delta\nu$, whereas the decrease of I_0 was. Recently Kylafis and Norman (1986) proposed a variant of the CCr-pump model in which $T_H \ll T_e$. According to these authors, such a situation can arise in the magnetic precursor of a MHD J-type wave.

6. CONCLUDING REMARKS

I would like to conclude with three suggested observations for VLBI specialists (without regard to their technical difficulty):

1. Try to measure the spatial evolution of outbursts described in §§4–5.
2. Try to measure the proper motion of the flared-up 8 km/s feature in Orion in order to discriminate between the "stationary" (Matveenko 1981) and "outflowing" (Strelnitskij 1982a) models of this source.
3. Try to measure the time evolution of the H_2O maser distribution in Mira variables in order to discriminate between the "expanding shock front" model and "stationary base of the stellar wind" model of these sources (Strelnitskij 1982b).

Good luck!

REFERENCES

Bally, J., and Lada, C. J. 1983, *Ap. J.*, **265**, 824.
Batrla, W., Matthews, H. E., and Menten, K. M. 1987, *Nature*, **326**, 49.
Bolgova, G. T. 1981, *Nauchnye Inform. Astron. Sov. AN SSSR*, **47**, 9.
Bolgova, G. T., Makarov, S. V., and Sobolev, A. M., 1987, *Astrofisika*, submitted.
Cheung, A. C., Rank, D. M., Townes, C. H., Thornton, D. D, and Welch, W. J. 1969, *Nature*, **221**, 626.
Churchwell, E., Witzel, A., Huchtmeier, W., Pauliny-Toth, I., Roland, J., and Sieber, W. 1977, *Astr. Ap.*, **54**, 969.
Cook, H. N. 1977, *Celestial Masers* (Cambridge: Cambridge University Press).
Cudaback, D. D, Read, R. B., and Rougoor, G. W. 1966, *Phys. Rev. Lett.*, **17**, 452.
de Jong, T. 1983, *Astr. Ap.*, **26**, 297.
Elitzur, M. 1982, *Rev. Mod. Phys.*, **54**, 1225.
Genzel, R., Reid, M., Moran, J., and Downes, D. 1981, *Ap. J.*, **244**, 884.
Goldreich, P., and Keeley, D. A. 1972, *Ap. J.*, **174**, 517.
Goldreich, P., and Kwan, J. 1974a, *Ap. J.*, **190**, 27.
Goldreich, P., and Kwan, J. 1974b, *Ap. J.*, **191**, 93.
Hills, R., Pankonin, V., and Landecker, T. L. 1975, *Astr. Ap.*, **39**, 149.
Kegel, W. H. 1975, in *Problems in Stellar Atmospheres and Envelopes*, ed. B. Bascheck *et al.* (Berlin: Springer), p. 257.
Kylafis, N. D., and Norman, C. 1986, *Ap. J. (Letters)*, **300**, L73.
Lada, C. J., Blitz, L., Reid, M. J., Moran, and J. M. 1981, *Ap. J.*, **243**, 769.
Litvak, M. 1969, *Science*, **165**, 855.
Litvak, M. M., McWhorter, A. L., Meeks, M. L., and Zeiger, H. J. 1966, *Phys. Rev. Lett.*, **17**, 821.
Litvak, M. M. 1974, *Ann. Rev. Astr. Ap.*, **12**, 97.
Mattila, K., Holsti, N., Toriseva, M., Anttila, R., and Malkamäki, L. 1985, *Astr. Ap.*, **145**, 192.
Matveenko, L. I. 1981, *Sov. Ast. Lett.*, **7(1)**, 54.
Menzel, D. 1937, *Ap. J.*, **85**, 330.
Moran, J. M. 1976, in *Frontiers of Astrophysics*, ed. E. Avrett (Cambridge: Harvard University Press), p. 385.
Moran, J. M., Burke, B. F., Rogers, A. E. E., Carter, J. C., Ball, J. A., and Cudaback, D. D. 1968, *Ap. J. (Letters)*, **152**, L97.
Norman, C., and Silk, J. 1979, *Ap. J.*, **228**, 197.
Reid, M. J., and Moran, J. M. 1981, *Ann. Rev. Astr. Ap.*, **19**, 231.
Reid, M. J., Moran, J. M., and Gwinn, C. R. 1986, in *Radio Astronomy in Space*, Proceedings of Green Bank Workshop, ed. K. W. Weiler (Green Bank, WV: NRAO), p. 145.
Reid, M. J., Moran, J. M., and Gwinn, C. R. 1987, this volume.
Rogers, A. E. E., Moran, J. M., Crowther, P. P., Burke, B. F., Ball, J. A., Meeks, M. L., and Hyde, G. M. 1966, *Phys. Rev. Lett.*, **17**, 450.
Rowland, P. R., and Cohen, R. J. 1986, *M. N. R. A. S.*, **220**, 233.
Schmeld, I. K., Strelnitskij, V. S., and Musilev, V. V. 1976, *Astron. Zh.*, **53**, 728.
Shklovskij, I. S. 1966, *Astron. Tsirc.*, **372**, 1.
Sobolev, A. M. 1987, *Astron. Tsirc.*, submitted.
Strelnitskij, V. S. 1971, *Astron. Tsirc.*, **609**, 1.
________. 1974, *Usp. Fis. Nauk*, **113**, 463.

________. 1980, *Pis'ma v Astron. Zh.*, **6**, 354.
________. 1982a, *Pis'ma v Astron. Zh.*, **8**, 165.
________. 1982b, *Publ. Astrophys. Obs. Potsdam*, **110**, 121.
________. 1983, *Nauchnye Inform. Astron. Sov. AN SSSR*, **2**, 75.
________. 1984, *M. N. R. A. S.*, **207**, 339.
________. 1986, *Astron. Tsirc.*, **1465**, 1.
Strelnitskij, V. S., and Sunjaev, R. A. 1972, *Astron. Zh.*, **49**, 704 [*Sov. Ast. A. J.*, **16**, 579, 1973].
Weaver, H., Williams, D. R. W., Dieter, W. H., and Lum, W. T., 1965, *Nature*, **208**, 19.
Wilson, W. J., and Barrett, A. H. 1968, *Science*, **161**, 728.

W43A: A MAVERICK MASER SOURCE

P.J. Diamond
National Radio Astronomy Observatory, Charlottesville, Va.
L.-Å. Nyman
Center for Astrophysics, Cambridge, Ma.

We report here on some preliminary results from a multi-epoch monitoring project of the OH and H_2O masers in the source W43A. Our results suggest that W43A is a very young stellar object which has just entered a strong stellar wind stage.

Type I OH masers are normally associated with regions of star formation and are known to have a complex spatial structure, they are also associated with strong H_2O masers and $\Delta V_{H2O} \geq \Delta V_{OH}$. Type II OH masers are associated with late-type giant stars and tend to exhibit a definite shell structure indicative of an expanding circumstellar envelope. In many cases Type II masers are also accompanied by H_2O masers and $\Delta V_{H2O} \leq \Delta V_{OH}$. The maser source W43A is a dramatic exception to the standard classification. The two H_2O emission complexes are observed to be separated by $\sim 180 kms^{-1}$ while the two principal OH complexes are separated by $\sim 16 kms^{-1}$. Due to the unusual nature of these observations we have observed the maser emission with various interferometers in an attempt to deduce the nature of the source.

To date we have five epochs of interferometric observations of W43A: 1) March 1981: MERLIN observations of the OH 1612 MHz masers (see Figure 1a); 2) February 1985: VLA observations of the OH 1612 MHz and H_2O masers. The VLA OH observations confirmed the MERLIN data, the VLA H_2O map is shown in Figure 1b. 3) May 1986: VLBI frequency-switched observations of the H_2O masers searching for component proper motions (data being reduced); 4) September 1986: VLBI observations of the OH 1612 MHz masers, the map resulting from these data shows essentially the same picture as the MERLIN data from 5 years previously but in more detail; the maser distribution is still seen to be peculiar with no hint of the shell structure we have come to expect in Type II OH masers. 5) November 1986: second epoch VLBI observation of the H_2O masers.

The MERLIN map of the OH 1612 MHz masers revealed that the blue and red-shifted components are separated by 0.21 ± 0.03 arcsec and that each of them consists of a number of point sources located within the same region. The blue-shifted masers lie at a position angle of $\sim 285°$ with respect to the red-shifted. The VLA OH 1612 MHz data confirmed

M. J. Reid and J. M. Moran (eds.), The Impact of VLBI on Astrophysics and Geophysics, 249–250.

the MERLIN data but the VLA H_2O map was a great surprise. In Figure 1b it can be seen that the separation of the two H_2O components is ~ 0.5 arcsec, over twice that of the OH masers; normally H_2O masers lie closer to the central exciting source than OH masers. More surprisingly the position of the blue and red-shifted H_2O maser components is reversed compared with the OH masers, i.e. the velocity gradient for the H_2O masers is in the opposite direction across the source than that for the OH masers.

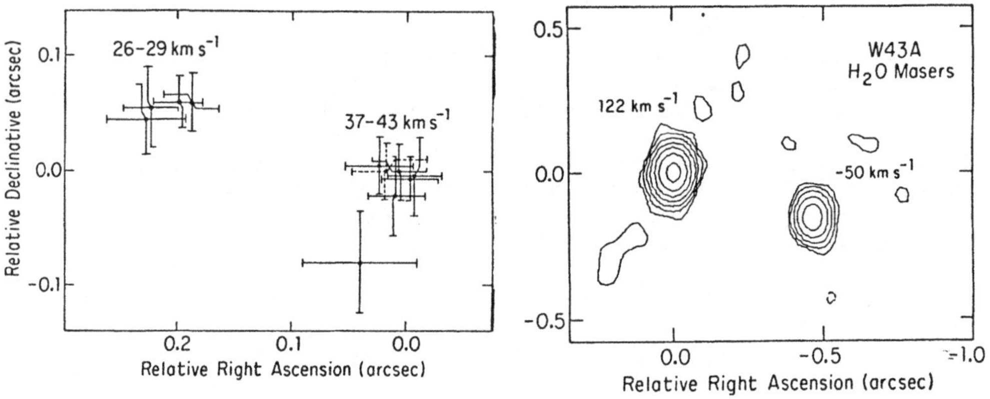

Figure 1:a) shows the MERLIN map of the OH 1612 MHz data, the velocities given are with respect to the lsr; b) the VLA map of the H_2O masers.

What is W43A? From CO data we have obtained and from IRAS observations which imply a dust temperature ≤ 30 K we believe W43A is in fact a young stellar object and not an evolved star. The OH and H_2O maser observations reveal the presence of a disc structure suggestive of either rotation or some mechanism confining the maser emission to a plane. The separation of the H_2O masers is about twice that of the OH masers suggesting either a very peculiar excitation mechanism and/or temperature gradients or a geometric effect. The large difference in ΔV_{OH} and ΔV_{H2O} suggest either strong acceleration or deceleration occurring in the W43A region, however, the opposite polarity of the two velocity gradients again suggests an odd geometric arrangement.

A whole series of models can be constructed in an attempt to explain our observations of this source but we favour one in which the central object is a new protostar which has just entered the stellar wind stage, this wind may be pushing on the inner regions of a disc of gas which contains the H_2O masers causing a rapid expansion. The OH masers are then presumed to lie in the less dense outer regions of the disc which are still falling towards the central object. A rotational component in the outer regions would produce the peculiar velocity distribution observed and could be supported by a central object with a mass of 1-20 $M_\odot$.

THE GROUND STATE HYDROXYL MASERS ASSOCIATED WITH OH 340.78−0.10

Ralph A. Gaume
Department of Astronomy
University of Michigan
Ann Arbor, MI 48109

The dynamics of the molecular envelopes surrounding ultracompact H II regions has been the subject of much debate. Since the envelopes often exhibit molecular line emission, line observations have been the primary method of probing these circumstellar shells. The existence of expansive motions in the molecular clouds surrounding massive newly formed stars has been firmly established. Proper motion studies of galactic water masers indicate general expansion, whereas high resolution CO studies often show collimated bipolar outflows centered on continuum H II regions. Maps of the OH maser emission, often associated with these regions, show the masers scattered, apparently randomly, in both position and velocity across an area a few arcseconds in size. Typical velocity ranges for the associated OH maser emission are on the order of 10 km s^{-1}. A number of models of the dynamical and physical conditions of the masing envelope have been developed to explain the properties of specific regions, e.g., collapsing remnant accretion envelopes, rotating disks or tori of neutral material, and expanding shells of shocked material. There is evidence from comparing the velocities of the OH masers to that of recombination lines associated with the underlying H II regions that in most sources the OH masers are formed in material still accreting toward the central object (Garay et al., 1985) (however see Welch and Marr, 1986). Mirabel et al., 1986, have reported the detection of (thermal and subthermal) high velocity OH outflow in several regions of star formation, demonstrating that OH molecules, at least in some instances, are also involved in collimated flows from central stellar objects.

OH 340.78−0.10 is a maser source whose main-line ground state spectrum covers a moderately large velocity range, and tends to resemble the double-peaked spectrum of OH/IR stellar sources (see Figure 1). We have mapped the OH masers and associated continuum emission toward OH 340.78−0.10 with the VLA of the NRAO. The observations were conducted 11 September 1986 in the B configuration. We observed the four ground state OH maser transitions in a 32 channel spectral line mode, with a velocity resolution of ≈ 1.1 km s^{-1}. Typical 1σ relative positional uncertainties for masers features of a given transition are on the order of 1.5 by 0.3″ (p.a. 0°), with larger uncertainties for the weaker features. The continuum observations were made at $\lambda = 6$ cm. The continuum source is unresolved by the 4.6″ by 0.9″ beam. The uncertainty in the position of the continuum source with respect to the position of the OH masers is ≈ 1.0 by 0.5″ (p.a. 0°).

M. J. Reid and J. M. Moran (eds.), The Impact of VLBI on Astrophysics and Geophysics, 251–252.

The spatial distribution of the OH maser emission is unusual for a region of star formation. Even though the emission covers a broad velocity range, all the features appear to be spatially coincident (within positional errors). Of the regions that have been mapped with high resolution techniques, only W49 N - G, where OH masers in a 64 km s^{-1} velocity range are found within a 0.5″ area (Gaume and Mutel, 1987), displays a similar (though larger) velocity dispersion for spatially coincident maser features. These sources apparently do not fit any of the standard dynamical models that have been proposed for other regions of OH maser emission. Instead, we suggest the masers in these regions are tracing dense OH clumps driven by mass outflow from a central star. The masers could be collisionally pumped as the individual OH clumps are decelerated via collisions with the surrounding molecular cloud. Alternatively, the masers could trace regions of unusually favorable velocity coherence in a region of general OH outflow, such as those OH outflows detected by Mirabel et al. (1986).

REFERENCES

Garay, G., Reid, M., and Moran, J. 1985, *Ap. J.*, **289**, 681.
Gaume, R., Mutel, R., 1987, *Ap. J. Supp.*, in press.
Mirabel, I., Ruiz, A., Rodríguez, L., Cantó, J., 1986, in *Masers, Molecules, and Mass Outflows in Star Forming Regions*, ed. A.D. Haschick, Haystack Obs., p189.
Welch, W., and Marr, J., 1986, B.A.A.S., **18**, 920.

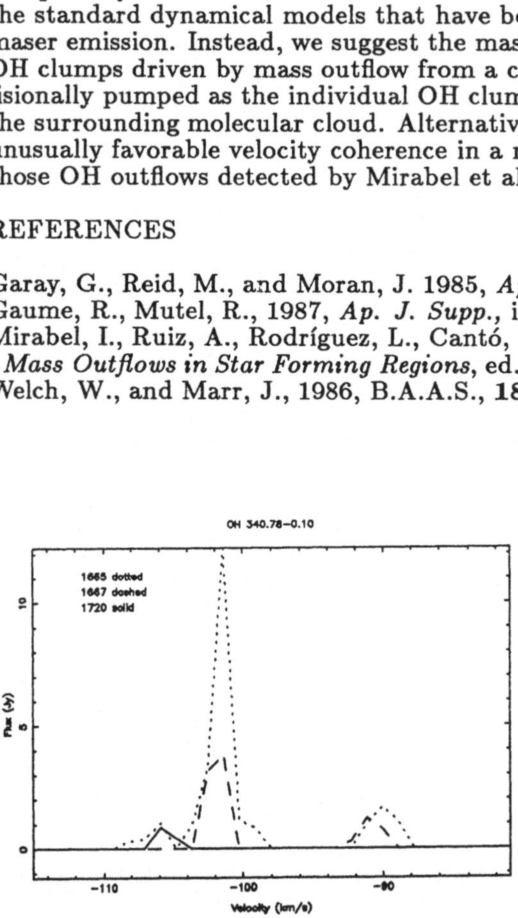

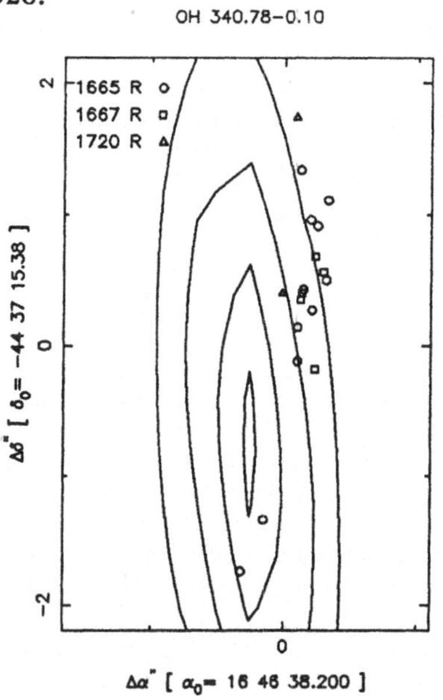

Figure 1. (left) Summed VLA spectra of the RCP OH maser emission associated with OH 340.78−0.10.
Figure 2. (right) Distribution of the OH 340.78−0.10 ground state OH maser emission superposed on 5 GHz continuum contours. Contour levels are 25, 50, 75, and 95% of the peak emission (8 mJy). The data's relative and absolute positional uncertainty is discussed in the text.

VLBI Mapping of 43 GHz SiO Emission Associated with IRc2 in Orion-KL

L. J. Greenhill, Center for Astrophysics, Cambridge, MA
J. M. Moran, Center for Astrophysics, Cambridge, MA
M. J. Reid, Center for Astrophysics, Cambridge, MA
C. R. Predmore, University of Massachusetts, Amherst, MA
G. C. McIntosh, University of Massachusetts, Amherst, MA
A. E. E. Rogers, Haystack Observatory, Westford, MA

1. INTRODUCTION

Information on the structure of the molecular flow within 1″of IRC-2, in Orion-KL, is sparse. Measurements of the continuum at 7.8μ and 12.5μ show a disk of size 1.″1 x 1.″9 and suggest that the center of the disk may be dust free (Lester *et al.* 1985). Aperture synthesis mapping of water maser shell features (Sylber 1986) has provided information on the 0.″1 scale. Smaller scales (0.″01) can be studied by mapping SiO maser emission. We observed the 43 GHz, v=1, J=1→0, transition of SiO using a 2 station interferometer with a 74 km baseline between Haystack Observatory, Westford, MA and Five College Radio Observatory, New Salem, MA. The fringe spacing was 20 milliarcseconds (mas) and the velocity resolution was 0.25 km-s^{-1}. Our results provide the highest resolution view to date of what is likely to be the inner 0.″1 of IRC-2.

2. RESULTS

The spectrum of the maser emission at the epoch of the experiment was double peaked, as it has been since at least 1974 (Snyder and Buhl 1974; Genzel *et al.* 1979; Lane 1982). Maps were obtained by fitting, for each velocity channel, the relative phase with respect to the reference feature. The map in Figure 1. shows the most significant groupings of features. The 1σ error bars are typically < 2 mas in R.A. and < 6 mas in declination, and include contributions from systematic errors. The systematic errors probably result from residual phase slopes across the bandpasses, which could not be entirely removed in the data reduction.

The map shows seven clumps of maser features, organized in three spatially distinct centers of maser activity in the radial velocity interval $V_{lsr} = -9.5$ to 21.5 km-s^{-1}. All the maser emission originates within an area of radius ≈ 80 mas (40 AU at a distance of 480 pc). The clustering of clumps of different radial velocities within the centers of activity is neither a consequence of the Zeeman Effect, for magnetic fields <1000 G, nor of hyperfine splitting.

The linear sizes of the maser regions were estimated from analysis of the normalized fringe visbilities of the individual maser features. Most maser features have sizes of ≈ 2 to 10 mas (≈ 1 to 5 AU). The implied brightness temperatures are ≈ 10^9 to 10^{10} K, and the masers are probably at least partially saturated.

3. DISCUSSION

A current model of IRC-2 (Genzel 1985) includes an expanding torus around a compact source. The apparent major axis of the torus is tilted ≈60° east of north, and the torus is inclined to the

M. J. Reid and J. M. Moran (eds.), The Impact of VLBI on Astrophysics and Geophysics, 253–254.

line of sight (Vogel *et al.* 1985). Inside the torus, expansion at $\approx$ 20 km-s^{-1} is the dominant motion (Plambeck *et al.* 1982). The H_2O masers that were mapped by Genzel *et al.* (1981) are in the outer edge of the expanding torus, located perhaps where it collides with the quiescent molecular cloud. The H_2O shell masers (Sylber 1986) are located toward the inner edge of the torus.

Our results may agree with this model of IRC-2, locating the SiO masers inside the torus, close to the central compact object. However, the absolute positions of the infrared continuum, radio continuum, and 86 GHz SiO masers are not well enough established to confirm the model (Lester *et al.* 1985; Moran *et al.* 1987; Wright and Plambeck 1983). If the SiO masers are closely associated with IRC-2, one expects the central object to lie between the red and blue shifted maser groups (stellar velocity $\approx$5 km-s^{-1}). The masers would lie as close as 40 AU to the central compact object. Conservation of mass and assumptions of constant outflow velocity (18 km-s^{-1}), uniform mass loss (10^{-3} $M_{\odot}$-yr^{-1}), and spherical symmetry, close into IRC-2, would imply a density, at the positions of the maser groups of 5×10^9 cm^{-3}. These physical conditions would quench the formation of a spherical HII region in IRc2. In order to explain both the observed HII region and large mass loss, it may be necessary to invoke an anisotropic model of the mass loss.

4. REFERENCES

Genzel, R. 1985 in *Masers, Molecules and Mass Outflows in Star Forming Regions*, ed. A. Haschick, 233.

Genzel, R., Reid, M. J., Moran, J. M., and Downes, D. 1981, *Ap. J.*, **244**, 884.

Genzel, R., Moran, J. M., Lane, A. P., Predmore, C. R., Ho, P. T. P., Hansen, S. S., and Reid, M. J. 1979, *Ap. J. (Letters)*, **231**, L73.

Lane, A. P. 1982, Ph.D. thesis, University of Massachusetts, Amherst.

Lester, D. F., Becklin, E. E., Genzel, R., and Wynn-Williams, C. G., 1985, *A. J.*, **90**, 2331.

Moran, J. M., Garay, G., Reid, M. J. 1987, in preparation.

Plambeck, R. L., Wright, M. C. H., Welch, W. J., Bieging, J. H., Baud, B., Ho, P. T. P., and Vogel, S. N. 1982, *Ap. J.*, **259**, 617.

Sylber, M. 1986, private communication.

Vogel, S. N., Bieging, J. H., Plambeck, R. L., Welch, W. J., and Wright, M. C. H. 1985, *Ap. J.*, **296**, 600.

Wright, M. C. H. and Plambeck, R. L. 1983, *Ap. J. (Letters)*, **267**, L115.

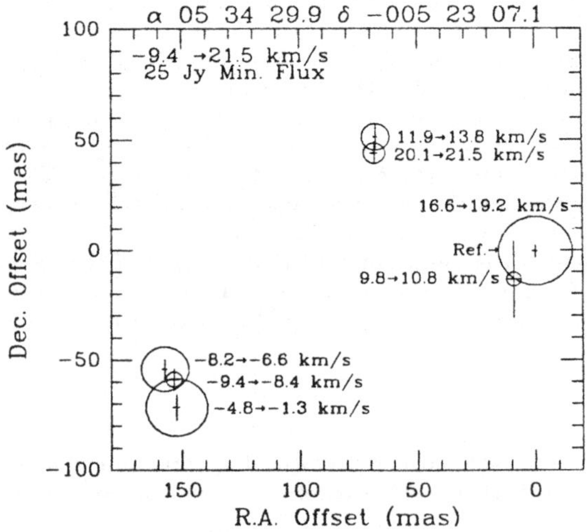

Figure 1. A simplified map of the SiO maser emission showing details of the 3 regions of maser activity. The areas of the circles are proportional to the integrated flux of the maser features. The error bars indicate the total uncertainty in position, due to formal and systematic errors.

A SEARCH FOR VERY COMPACT STRUCTURE IN 17 OH MASERS

A. J. Kemball[1], P. J. Diamond[2] and F. Mantovani[3]
[1] Hartebeesthoek Radio Astronomy Observatory
[2] Max-Planck Institut für Radioastronomie, Bonn
[3] Istituto di Radioastronomia CNR, Bologna, Italy

The apparent spot sizes of OH masers appear to be significantly broadened when seen through the inner galaxy or large extents of the galactic disk (Burke 1968). Bowers et al (1980) found evidence of small-scale structure ($\lesssim$50 mas) in OH sources at distances of less than 5 kpc but this was characteristically absent in very distant sources ($\gtrsim$8kpc) at galactic longitudes $l \lesssim 40°$. This result is typically explained in terms of interstellar scattering (ISS) by intervening diffuse HII regions.

Determinations of the galactic distribution of ISS have used both pulsars (Cordes 1983) and compact extragalactic radio sources (Dennison et al 1984) as probes of the interstellar medium. Surveys of OH maser spot sizes can provide useful complementary information on the low latitude galactic distribution of scattering material. In this context the intrinsic size of OH maser components is also of interest. Interferometric observations of interstellar hydroxyl masers in W3(OH) by Reid et al (1980) indicate characteristic sizes of 2×10^{14} cm with 25% of the components in the masing region exhibiting even more compact structure.

We present here the results of a pilot VLBI search for very compact structure in a sample of 17 strong 1665 MHz OH masers selected from the catalogues of Caswell and Haynes (1983a and 1983b). This was a single-baseline experiment involving Medicina (Bologna) and Hartebeesthoek (Johannesburg). The sources cover the range of galactic longitude from 355° to 50°.

The observations were carried out during three 8 hour sessions during 1986 April 17-19. All observations were made with left circular polarization and a bandwidth of 62.5 kHz. The data were taken on the MKII video tape recording system. Each source was observed for four 15 min scans distributed in hour angle. The long north-south interferometer geometry yielded limited u-v coverage about a projected baseline length of 40 Mλ with an interferometer fringe spacing of 5 mas.

M. J. Reid and J. M. Moran (eds.), The Impact of VLBI on Astrophysics and Geophysics, 255–256.

Standard techniques were used in data reduction (Reid et al 1980). The cross-power spectra were coherently averaged for 3 min yielding 3 Jy rms noise in cross-power amplitude. The spectra were Hanning weighted and have a resolution of 0.088 kms^{-1}.

Positive detections were obtained for three of the 17 sources. The apparent angular size of the most compact feature in each spectrum was calculated from the normalised visibility based on a circularly symmetric Gaussian brightness distribution model. Intrinsic spot sizes, for both the near and far kinematic distances, and the apparent brightness temperatures of the detected components are indicated in Table I. For the three sources where compact components were detected it is almost certain that the distance ambiguity is resolved (Downes et al 1980) but both distances are included for completeness. Less likely values are included in parentheses.

TABLE I. LIST OF COMPACT COMPONENTS DETECTED

Source name	Velocity (kms^{-1})	Flux (Jy)	Angular size (FWHM / mas)	Linear size (10^{14} cm)	T_b (10^{13}K)
OH34.26+0.15	55.9	39.5	1.4 ± 0.6	0.75 (2.2)	1.0
OH35.20-1.73	42.4	30.9	0.8 ± 0.8	0.30 (1.4)	2.5
OH45.47+0.13	59.3	26.4	2.5 ± 0.4	(1.5) 3.0	0.21

It is noted that all the detected sources are at galactic longitudes $l \gtrsim 35°$. Two of the sources, OH 34.26+0.15 and OH 35.20-1.73, are at probable distances of $\lesssim 4$ kpc. Clearly this is not statistically significant but the result is consistent with previous determinations of the interstellar broadening of compact low galactic latitude radio sources (Dennison et al 1984). The data also possibly indicate the existence of maser spots with intrinsic sizes somewhat smaller than the commonly accepted value of 2×10^{14} cm for 1665 MHz OH maser components. This in turn implies larger apparent brightness temperatures than previously reported. As a pilot survey the experiment was successful in the detection of very compact structure in interstellar OH masers.

REFERENCES

Bowers, P.F., at al.: 1980, Astrophys. J., **242** , 1088.
Burke, B.F., et al.: 1968, Astron. J., **73** , S168.
Caswell, J.L., and Haynes, R.F.: 1983a, Aust. J. Phys., **36** , 361.
Caswell, J.L., and Haynes, R.F.: 1983b, Aust. J. Phys., **36** , 417.
Cordes, J.M.: 1983 in IAU Symposium No 110, VLBI and Compact Radio Sources, ed. R. Fanti, K. Kellermann and G. Setti:(Dordrecht:Reidel),303
Dennison, B., et al.: 1984, Astron Astrophys., **135** ,199.
Downes, D., et al.: 1980, Astron. Astrophys. Suppl. Ser., **40** , 379.
Reid, M.J., et al.: 1980, Astrophys. J., **239** , 89.

VLBI OBSERVATIONS OF THE H_2O MASER OUTBURST IN ORION KL

L.I. Matveyenko
Space Research Institute, Moscow, USSR
D.A. Graham
Max-Planck Institut fur Radioastronomie, Bonn, F.R.G.
P.J. Diamond
National Radio Astronomy Observatory, Charlottesville, U.S.A.

We present preliminary results from a seven year VLBI study of the H_2O maser outburst in Orion-KL. Our results reveal changes in the profile, in the spatial structure and in the polarization characteristics of the H_2O masers. We interpret our data in terms of edge-on rings rotating about a protostellar object.

In August 1979 Abraham *et al.*(1981) observed an enormous outburst of the H_2O maser emission in Orion-KL. The peak flux density at a velocity of ~ 8 km s^{-1}rose from a $\sim 10^4$ Jy to $\sim 10^6$ Jy. The outburst emission was observed to have strong linear polarization ($P \sim 60\%$) and was concentrated inside a linewidth $\Delta f \approx 30$ kHz. The profile shape and flux density is changing with time. See Fig.1 for a representative spectrum and its linear polarization structure.

We have performed multi-epoch VLBI observations of the H_2O maser outburst in order to study the spatial structure as a function of time. Our results (so far obtained only by model-fitting) demonstrate that the structure is changing and complex, but in general consists of a few compact maser components distributed along a position angle $\phi \sim -80°$.

We briefly descibe a subset of our results.
25 September 1979. Observations with the Crimea-Puschino baseline revealed a simple core-halo structure, the core has $4-5$ compact components and a total size ~ 0.6 AU and a halo size ~ 1.2 AU. The brightness temperature T_B was observed to be $\geq 2 \times 10^{16}$ K. The compact emission region had a total linewidth ≤ 10 kHz implying $T_K \sim 100-200$ K if the maser is not saturated, this suggests an IR pumping mechanism. Observations with Crimea-Effelsberg and Greenbank-Haystack in Nov. 1979 showed a very similar structure.
1 December 1982. Observations with Crimea-Onsala-Effelsberg revealed 4 compact components each of size ≤ 0.2 AU implying $T_B \geq 10^{17}$ K. The masers were located along $\phi \sim -80°$ (see Fig.2a), a distinct velocity gradient was visible implying $\Delta V/\Delta L \approx 0.44$ km s^{-1}/AU. We suggest that this gradient implies the masers lie in a disc which is rotating around

M. J. Reid and J. M. Moran (eds.), The Impact of VLBI on Astrophysics and Geophysics, 257–258.

a protostar. The masers have a velocity ~ 11 km s^{-1}relative to the protostar (deduced from the velocity of the ambient molecular gas) and lies at a distance of 12 AU to the west of the masers.

10 October 1983. Observations with Crimea-Effelsberg revealed a structure similar to that seen in Dec 1982. Other observations from the Crimea-Puschino baseline combined with single dish data showed that the compact regions have $P \geq 85\%$ and that the polarization position angle was changing across the profile (see Fig.1).

26 May 1985. We were able to obtain a global observation of the outburst using Crimea-Effelsberg-Onsala-Greenbank-Haystack-VLA-OVRO. Our initial model-fitting showed that a new feature was now visible in the source (see Fig.2b) at a ~ 6.5 AU from the centroid of the previous structure.

Our observations have revealed the presence of maser components with $T_B \sim 10^{16-18}$ K. These are the highest T_B yet recorded and are probably caused by a high degree of beaming ($\Omega < 10^{-4}$). The beaming may be due to a) coherent emission, which we do not regard as physically likely; or b) the masers being filaments of gas which lie in edge-on rings rotating around the protostar. We suggest that the large degree of linear polarization is caused by an anisotropic pump engine due to IR radiation from an unidentified source (possibly the protostar). From the velocities and distances observed the motion could be supported if the protostar had a mass $\sim 1.8 M_\odot$. The observed change in polarization position angle across the maser profile and emission region could be explained if the proposed rotating rings also had an expansional velocity component and were permeated by a magnetic field.

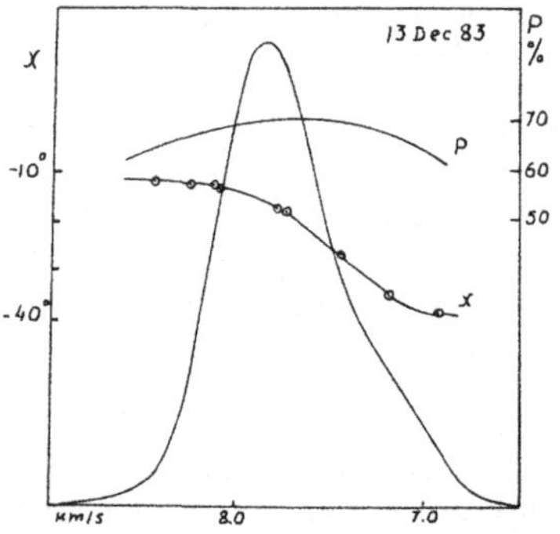

Figure 1. A single dish profile of the H_2O maser in Orion-KL, the solid line (P) is the percentage linear polarization, the line (χ) is the polarization position angle. Figure 2 shows maps of the maser distribution at different epochs.

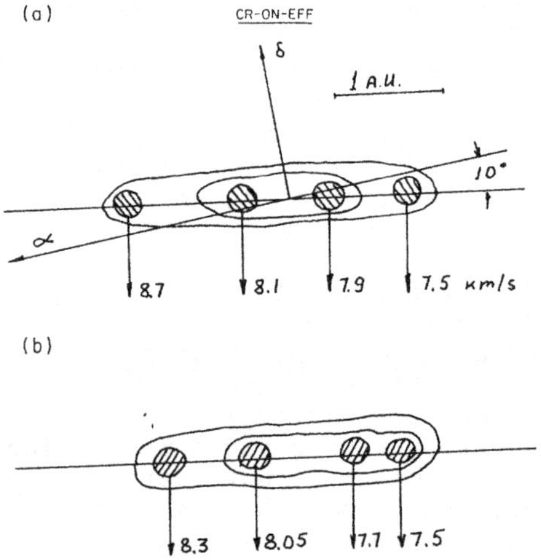

Abraham, Z., Cohen, N. L., Opher, R., Raffaeli, J. C. and Zisk, S. H., 1981, Astr. Ap., **100**, L10.

SiO MASER OBSERVATIONS - VLBI AND POLARIMETRY

G.C. McIntosh, C.R. Predmore
University of Massachusetts
J.M. Moran, L.J. Greenhill, M.J. Reid
Harvard-Smithsonian Center for Astrophysics
A.E.E. Rogers
Haystack Observatory
and R.E. Barvainis
National Radio Astronomy Observatory

ABSTRACT. The spatial distribution and polarization characteristics of the SiO (v=1, J=1-0) maser emission from several late type stars have been observed. The spatial distribution, derived from VLBI observations, generally shows a number of emitting regions but no clear velocity pattern or geometry. Some of these regions have well defined polarization characteristics. The results of high spatial resolution polarization measurements of RCas are similar to the lower spatial resolution polarimetry performed on this source.

1. The VLBI Experiment and Polarimetry Measurements

The VLBI experiment was performed on 25, 26 Mar 1985 using the Five College Radio Astronomy Observatory (FCRAO), New Salem, MA, and Haystack Observatory, Westford, MA. These observatories provide a 75 km baseline with a minimum fringe spacing of 20 milliarcseconds at 43 GHz. Each observatory was equipped with a Mark III VLBI station and a polarimeter. The VLBI observations were made at polarization position angles of 0, 45, and 90 degrees (measured east of north). The data were processed at Haystack Observatory and further reduced at the Center for Astrophysics. On 27 Mar 1985 the linear polarization characteristics of the sources were measured at FCRAO. A complete description of the the polarimetry system and VLBI experiment is available in McIntosh (1987).

2. RCas

For RCas, a long period variable star, the high resolution spatial map derived from the VLBI data showed four regions of maser emission. These regions occurred at V_{LSR}'s of approximately 21.5 km/s, 22.3 km/s, 24 km/s, and 32 km/s. Figure 1 shows the spatial offset positions with errors and the combined spatial/polarization map. The spatial/polarization map shows the fractional polarization and position angle of each velocity channel and was derived from the VLBI positions and the single dish polarization measurements.

M. J. Reid and J. M. Moran (eds.), The Impact of VLBI on Astrophysics and Geophysics, 259–260.

The 32 km/s feature had a fractional polarization of 40.5±9.9% at a position angle of -12.9±2.6 degrees derived from the VLBI fringe amplitude data. The polarization characteristics of this highly polarized feature were extracted from the VLBI data by analyzing the fringe amplitude as a function of the polarization position angle of the observation. For the single dish polarization measurements made at FCRAO the 32 km/s feature had a fractional polarization of 48.9±9.1% at a position angle of -5.0±12.9 degrees. These results are comparable within the errors and indicate that the increased spatial resolution available from the VLBI data does not greatly change the measured polarization characteristics of the maser feature.

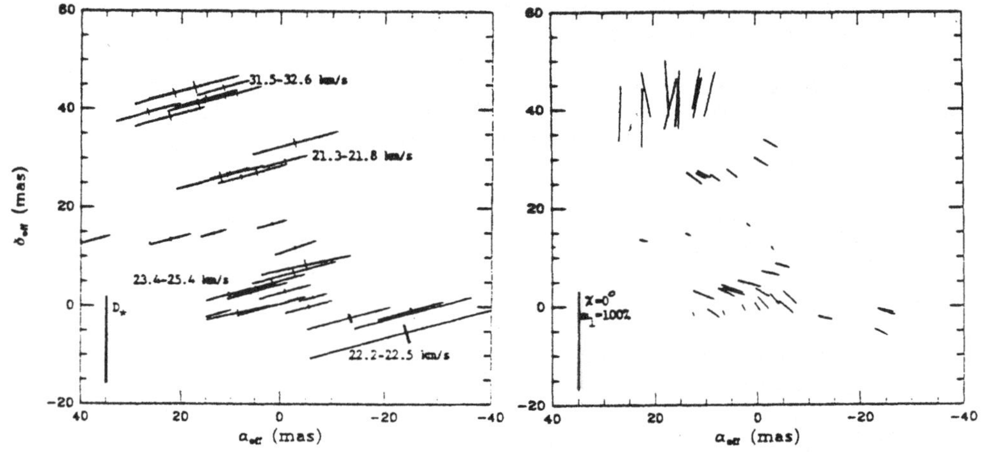

Figure 1. (Left) Map of the maser emission towards RCas. The stellar diameter is shown in the lower left. (Right) Polarization map. The bar length is proportional to the fractional polarization and the orientation to the polarization position angle.

3. VXSgr

VXSgr, a supergiant star, showed a number of maser features. The distance to this source (1500 pc), the large velocity range of emission, low flux density, and short UV track available makes it difficult to differentiate individual features on the high spatial resolution map. Some velocity ranges do show well defined polarization characteristics and spatial offsets indicating unique maser features.

FCRAO is operated with support of the National Science Foundation (grant AST85-12903) and with permission of the Metropolitan District Commission of the Commonwealth of Massachusetts. Radio Astronomy at the Haystack Observatory of the Northeast Radio Observatory Corporation is supported by the National Science Foundation (grant AST-8512598).

4. Reference

McIntosh, G.C. 1987, Ph. D. Dissertation, Univ. of Massachusetts.

OH MEGAMASERS AND ACTIVE GALACTIC NUCLEI

R.P.Norris
CSIRO Division of Radiophysics
PO Box 76, Epping, NSW 2121,
Australia

ABSTRACT. OH megamasers are believed to be active galaxies in which a substantial fraction of the OH gas in the disk of the galaxy is stimulated by the intense far-infrared flux from the active nucleus. The result is that the galactic disk acts as a maser amplifier, producing in the OH line an amplified image of the radio continuum source in the nucleus. Megamasers promise to be powerful tools for the study of active galaxies, provided we can determine what it is that turns an active galaxy into a megamaser. Here I examine the archetypal megamaser galaxy Arp220 and ask the question: what makes it different from other active galaxies?

1. OH MEGAMASERS.

The first megamaser was discovered when Baan et al. (1982) discovered intense OH emission from the peculiar galaxy Arp220 (IC4553). This galaxy was independently found by IRAS (Soifer et al. 1984) to have the highest infrared excess of any galaxy in the UGC catalogue, and it is now known to have a total luminosity which ranks it amongst quasars, but a level of extinction which gives it an undistinguished optical appearance.

Six other OH megamaser galaxies have since been found (Norris et al., 1986, and references therein), and each is distinguished by a high infrared luminosity (10^{11}-10^{12} L_o) and a remarkably uniform infrared colour temperature (55-71 K). Hundreds of other active galaxies have been searched by various groups with a spectacular lack of success which implies that megamasers require very special conditions. The colour temperature undoubtedly represents one of these conditions, since only a small fraction of active galaxies have a temperature in this range. This suggests that the population inversion is accomplished by a far-infrared pumping mechanism, perhaps operating through the 35μm or 80μm OH transitions.

A megamaser also requires a radio input, which in the case of Arp220 is provided by a central radio continuum source in the galaxy. Theory indicates that other megamaser galaxies operate in a similar manner. Thus OH megamasers seem to occur in galaxies which contain an active radio nucleus and a cool but infrared-bright disk. However, the large number of galaxies which satisfy these criteria and yet do

M. J. Reid and J. M. Moran (eds.), The Impact of VLBI on Astrophysics and Geophysics, 261–262.

not exhibit OH megamaser emission attests to the existence of some other condition - perhaps that of orientation or gas density.

2. ARP220

Arp220 contains at its nucleus a pair of compact radio sources. Their size at 1.3cm (Norris, 1987) is about 50 pc, whereas VLBI results (Diamond et al., in preparation) suggest an even smaller size. These and other results have been interpreted by Norris (1987) as evidence that Arp220 possesses two active nuclei separated by about 350 pc. This directly supports the hypothesis, advanced by Joseph et al. (1984) on the basis of molecular hydrogen observations, that Arp220 is the result of a merger. These and other observations suggest that Arp220 contains in its core a pair of compact accreting objects, but that these are surrounded by regions of starburst activity.

This conjunction of Seyfert and starburst activity is becomingly increasing attractive for all types of active galaxy. It should therfore be noted that the essence of the megamaser phenomenon is a requirement for both starburst activity (to accompany the dust and molecular column density) and Seyfert activity (to provide the central radio source). Thus megamaser galaxies perhaps represent some extreme example of the the generic active galaxy phenomenon.

3. CONCLUSION.

Megamasers pick out those galaxies which are disrupted by both starburst and Seyfert activity. In this respect they represent extreme cases of what makes an active galaxy, although there is in addition some unrecognised condition necessary for megamaser activity. Identifying this condition would permit us to wield the megamasers as powerful tools for the study of active nuclei.

4. REFERENCES.

Baan, W.A., Wood, P.A.D., & Haschick, A.D., 1982, *Astrophys. J.*, **260**, L49.

Joseph, R.D., Wright, G.S., & Wade, R., 1984, *Nature*, **311**, 132.

Norris, R.P., Baan, W.A., Haschick, A.D., Booth, R.S., & Diamond, P.J., 1985, *Mon. Not. R. astr. Soc.*, **213**, 821.

Norris, R.P., Whiteoak, J.B., Gardner, F.F., Allen, D.A., & Roche, P.F., 1986, *Mon. Not. R. astr. Soc.*, **221**, 51p.

Norris, R.P., 1987, *Mon. Not. R. astr. Soc.*, in press.

Soifer, B.T., Helou, G., Lonsdale, C.J., Neugebauer, G., Hacking,P., Houck, J.R., Low, F.J., Rice,W., & Rowan-Robinson,M., 1984, *Astrophys. J.*, **283**, L1.

CIRCULAR POLARIZATION AT 1665 AND 1667 MHz TOWARDS OH MASERS IN NGC 6334N, NGC 7538N, NGC 7538S, AND G45.07+0.13

X. W. Zheng,[1,2] J. M. Moran,[1] M. J. Reid,[1] M. H. Schneps,[1]
J. A. Garcia-Barreto,[3] G. Garay[4]

[1] Harvard–Smithsonian Center for Astrophysics
[2] Astronomy Department, Nanjing University
[3] Universidad Nacional Autónoma de México
[4] European Southern Observatory

VLBI OH maser observations simultaneously in right and left circular polarization at the 1665 and 1667 MHz transitions were conducted by taking advantage of the multichannel capability of the Mk III system. The OH maser maps of the right and left circular polarization at both transitions were obtained toward the OH maser sources NGC 6334N, NGC 7538N, NGC 7538S, and G45.07+0.13.

We found five Zeeman pairs in NGC 6334N, one in NGC 7538S, and one in G45.07+0.13. The inferred magnetic field strengths range from 2 to 7 mG, respectively. For NGC 6334N, the magnitude of magnetic field is roughly constant over the source, and no field reversals are seen. The magnetic field in this source may be relatively homogeneous. The shift among features in the RCP and LCP spectra for NGC 7538N is less than 0.2 km s^{-1} and may represent a single partially polarized σ–component.

All four OH masers are associated with H_2O maser sources. The OH maser emission may be located closer to the center of the compact H II region than H_2O maser emission, but it is difficult to be more specific given the large error bars on the H_2O positions. The OH maser emission regions in NGC 7538N, NGC 6334N, and G45.07+0.13 are losely associated with the compact H II regions. The majority of OH spots are projected against the compact H II region within the half-power diameter. No radio continuum counterparts have been found for the OH source in NGC 7538S. One possible reason is that small HII regions less than $0\rlap{.}''1$ in size have not yet or have just begun to form and are difficult to detect.

An interesting characteristic of our map of NGC 6334N (Figure 1) is that the majority of the higher velocity features lie in the core region. It appears that the velocities of the features systematically decrease from southwest to northeast (see Figure 2). The gradient is 8 km s^{-1} $arcsec^{-1}$. A rotating disk with an approximate mass 10 $M_\odot$ is inferred for this region. In most OH masers, there is little spatial correspondence between the features at two main line frequencies, even though there is velocity overlap in the spectra. Different transitions may come from the different layers surrounding the central star. However, in NGC 6334N, most of the maser spots in both transitions are coincident within 3×10^{15} cm. The turbulence or complicated movement of molecular clumps probably blend these layers.

M. J. Reid and J. M. Moran (eds.), The Impact of VLBI on Astrophysics and Geophysics, 263–264.

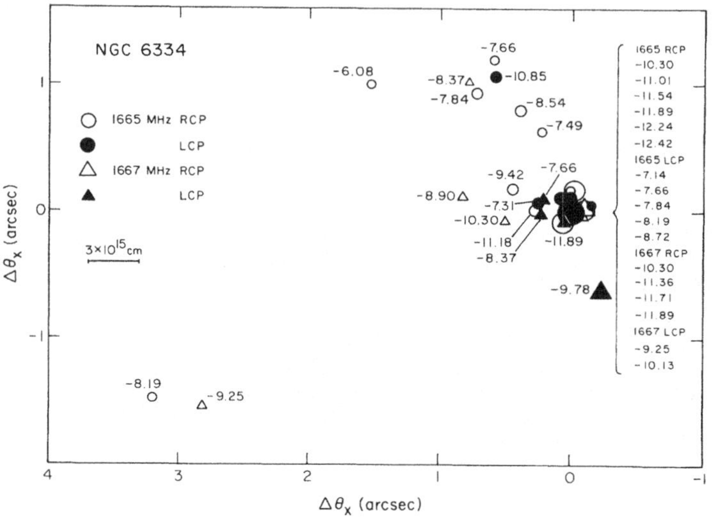

Figure 1. Map of OH masers at 1665 MHz towards NGC 6334N.

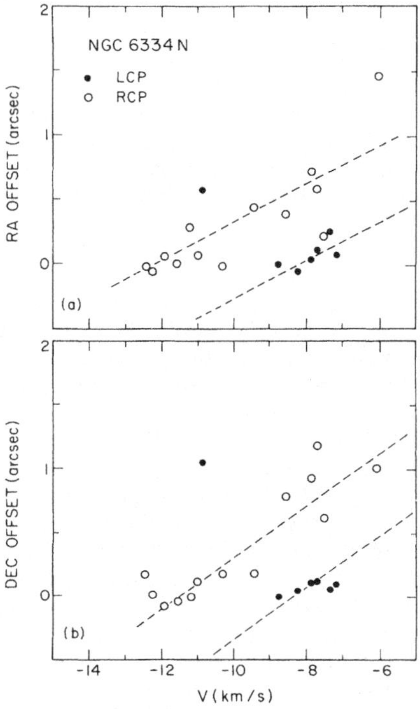

Figure 2. The RA and declination offsets versus velocity of maser features shown in Figure 1. The tendency of the features to cluster around the dashed lines suggests the presence of rotation and of a uniform magnetic field of strength ~ 5 mG.

VLBI OBSERVATIONS OF RADIO STARS

Jean-François Lestrade
Bureau des Longitudes
77 Avenue Denfert-Rochereau, F-75014, Paris, France
and Jet Propulsion Laboratory, Pasadena, CA

1. INTRODUCTION

In recent years, observations of stars across the whole electromagnetic spectrum have yielded a wealth of information on their atmospheres and borne out new insights on their internal structures. Indirect evidences for chromospheres, coronae and winds were discovered in stars other than the Sun. For instance, ultra-violet and optical emission lines from collisionally excited atoms are evidences for chromospheric activity with temperatures between 10^4 and 2×10^5 K. Soft X-ray emission with thermal spectrum is evidence for hot coronae with temperatures between 10^6 and 3×10^7 K. The spindown of stars with age is best interprated by loss of mass and of angular momentum via magneticly coupled stellar winds originating in hot stellar coronae.

It is common practice to divide the stars into early-type (earlier than spectral type F0) and late-type. The early-type stars have massive winds driven by their intense photospheric radiative flux. The discovery of X-ray emitting gas from these stars was a surprise. It is currently interprated as hot gas confined by small scale magnetic loops while the radiatively driven wind escapes through the "holes" between them. The late-type stars have atmospheres which are dominated by chromospheric and coronal activities and have relatively moderate winds. It is thought that magnetic fields shape the atmospheres and control the energetics of late-type stars. The corona is considered an assemblage of magnetic loops emerging from the stellar surface and confining a plasma which is too hot to be gravitationally bound to the star. The magnetic fields at the stellar surface are stressed by turbulent fluid motions and generate heating of the chromosphere and corona by dissipation of DC current or/and MHD waves. The moderate wind escapes through coronal holes which are regions of open magnetic lines between the loops.

At radio wavelengths, stellar *thermal* emission has been detected from early-type stars and is an important tool to determine their mass loss rates. The large scale winds and ionized circumstellar envelopes of these stars are best studied at the Very Large Array with arcsecond resolution. We shall not review these observations and refer the reader to Abbott (1985) and Kwok (1985). Stellar *non-thermal* continuum radio emission has been detected from single late-type stars and from close binaries having at least a late-type component. This non-thermal radio activity is sporadic and occurs on the Sun too where it is interprated with a variety of coherent and incoherent gyromagnetic emission mechanisms (Dulk 1985). We shall review the contributions of VLBI to a better understanding of the emission process(es) and of the physical properties of the non-thermal radio stars. We exclude pulsars, stellar molecular masers and super nova remnants from this discussion.

The RS CVn (RS Canum Venaticorum) close binaries are the most active non-thermal

M. J. Reid and J. M. Moran (eds.), The Impact of VLBI on Astrophysics and Geophysics, 265–274.

emitters and most VLBI observations of radio stars have been conducted on them. They are detectable at distances of less than 100 parsecs and their angular sizes are between 1 and 3 milliarcseconds matching the angular resolution of VLBI. The relatively successful VLBI observations of these close binaries in recent years have come with the implementation of the Mark III VLBI data acquisition system (Rogers et al 1983). This system provides the sensitivity required for these objects which are weak by VLBI standards (typically < 100 mJy).

VLBI observations of RS CVn are motivated both by astrophysic and astrometric goals. In astrophysics, direct measurements by VLBI of their radio source sizes and of their brightness temperatures as well as imaging of their milliarcsecond radio emitting regions are helping to understand their emission mechanism which produces at least 10^4 times more energy than the Sun at radio wavelengths. VLBI observations yield also the values of important model-dependent parameters, e.g. magnetic field strength and electron number density, and will possibly unveil the nature of the magnetic structure which might be the dominant energy source in late-type stars. In astrometry, the RS CVn binaries being both radio active and optically bright can be used as link objects to tie the radio and optical celestial reference frames together. The quality of such a tie depends on the degree of coincidence between the optical and radio components. This can be answered with a comprehensive physical model of these stars.

Five X-ray stars (Cyg X3, Cyg X1, Circ X1, LSI 61°303 and Sco X1) and the exotic objects SS 433 have also been studied by VLBI technique. These objects have uncommon properties and cannot be easily classified. Their energetics might be analogous, although on a different scale, to the mechanisms powering the extragalactic sources. Hence, they are an interesting connection with the physics of extragalatic radio sources.

2. RS CVn AND ALGOL CLOSE BINARIES

a. PHYSICAL PROPERTIES

RS CVn close binaries (Fig. 1a) have detached stellar components with mass ratio close to unity. Both components are late-type stars, typically GV for the primary and K0IV for the secondary, tidally locked with rotation-orbital periods generally less than 20 days. These close binaries were first noticed for the conspicuous wavelike distorsions of their light curves (Hall 1975) presumably caused by large dark starspots on the photosphere of the K-subgiant. The evolution of these starspots (in size and location) has been studied by photometry (Rodonò 1983) and recently by a Doppler imaging technique (Vogt and Penrod 1983). Algol (Fig. 1b) is as active in radio as RS CVn but is a semi-detached close binary (belonging to a multiple system) characterized by a late-type component filling its Roche lobe and spilling over gas on an early-type star through an accretion disc.

Radio surveys of RS CVn binaries were undertaken by Hjellming, Gibson, Spangler, Owens and Feldman with single dishes or short baseline interferometers in the 70's. Non-thermal outbursts with flux density ranging from a few tens to a few hundreds mJy, and exceptionally up to 1.2 Jy for HR1099, were discovered by these astronomers. The corresponding radio luminosity is between 2×10^{15} and 6×10^{17} ergs/s/Hz or 10^4 to 10^6 times larger than the Sun. A dozen of radio RS CVn were discovered during these early surveys. Later, the sensitivity of the VLA and its imaging capability to eliminate confusion sources led to the discovery of more than 30 active RS CVn compiled in Mutel and Lestrade (1985).

Pallavicini (1986) in his review of coronal magnetic fields stresses that early-type stars have no convective zone while late-type stars do have such a zone. This leads to think that the dominant energy source in late-type stars, as RS CVn and Algol-type, might be a dynamo mechanism involving the interaction of rotation and convection. Such interaction generates intense magnetic fields emerging at the stellar surface and penetrating through all layers of the stellar atmosphere. Stellar components of close binary have much higher rotation rates than single star with the same spectral type because rotation is maintained against spin-down by tidal torque. This rapid rotation increases the efficiency of the dynamo mechanism and might be responsible for the exceptional magnetic activity of RS CVn as evidenced by their starspots which might cover up to 30 % of their surface and should be associated with intense magnetic fields. The variable degree of circular polarization observed from RS CVn, usually between 0 and 30 % and exceptionally more, is another evidence for large magnetic structure. Mutel et al (1987) have noticed than the sense of circular polarization at a given frequency stays the same over several years. This is evidence for the existence of permanent large magnetic structures on these stars.

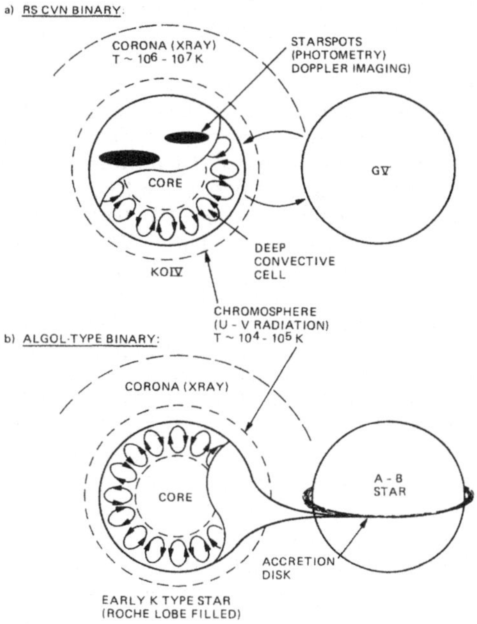

Fig. 1: sketches of RS CVn and Algol-type close binary systems.

The very hot X-ray coronae ($\leq 3 \times 10^7$ K) discovered in RS CVn have been estimated to be at least 3 stellar radii in extent for the RS CVn system AR Lac (Walter, Gibson and Basri 1983) and for Algol (White et al 1985) when X-ray eclipses failed to be detected during optical eclipses of these systems. This is unlike the solar magnetic structures which have scale heights much smaller than a solar radius ($< 0.01\ R_\odot$). Such linear sizes correspond to milliarcsecond extent at the distances of these RS CVn. The lack of spatial resolution of X-ray observations makes the topology of coronal magnetic fields confining hot plasma unknown in stars other than the Sun. VLBI observations can acheive milliarcsecond resolution in radio and therefore is presently the only technique capable of probing the magnetic structures of these non-thermal stars.

b. VLBI OBSERVATIONS OF CLOSE BINARIES

The first radio star observed by the VLBI technique was Algol by Clark B.G., Kellermann and Shaffer (1975) and Clark T.A. et al (1976) at 8.4 GHz on a single baseline during two exceptional outbursts (> 0.5 Jy). The fringe visibility functions they obtained provided the first direct measurements of the radio source size (5 mas and 2 mas) and of the brightness temperature ($T_B > 10^8$ K) for continuum radio emission from a star. The important result was that T_B was high enough to be clear evidence for a non-thermal emission mechanism. The data were also used to derive the VLBI position of Algol (Clark et al 1976) at epoch B1950 and were recently re-analysed by Ma et al (1986) in the J2000 frame.

In 1978, Mark II VLBI observations were successfully conducted on the two most active RS CVn; UX Ari by Backer and Srameck at 1.4 GHz between Arecibo and Greenbank (unpublished) and HR1099 by Resch at 8.4 GHz between Goldstone and Owens Valley (unpublished). The visibility functions were thought unreliable by these authors because of lack of proper flux scale calibration and so neither source sizes nor brightness temperatures were inferred from these data. In 1981, Cohen detected the RS CVn system HR5110 during Mark III VLBI observations at 5 GHz between Owens Valley and Greenbank and inferred a source size of 3.7 milliarcseconds (Little-Marenin et al 1986).

In December 1982, we initiated a series of VLBI observations with the Mark III recording system and the US VLB Network including stations in Europe and from the Deep Space Network. These VLBI observations were conducted at various centimetric wavelengths and sometimes at dual frequency and dual polarization. The first five experiments were designed to survey about twenty RS CVn and Algol-type stars as well as two X-ray stars. Later, we devoted several experiments to the most active stars. A recent experiment was also conducted to observe southern hemisphere RS CVn in order to increase the astrometric coverage of the celestial sphere. From all these experiments, we detected 9 RS CVn (UX Ari, HR1099, HR5110, σ CrB, AR Lac, II Peg, SZ Psc, HD26337, HD77137), as well as Algol and the two X-ray stars LSI 61°303 and Cyg X1.

Palagi et al (these proceedings) have also conducted Mark III VLBI observations of 4 RS CVn and Algol at 5 GHz with a multi-station VLB array in 1986.

One of the difficulties in observing RS CVn is that their flux densities are extremely variable on a time scale as short as a day. None of these stars is dependable at a flux density level of 10 mJy. However when an outburst occurs, it might last for several days. Depending on the stage of the outburst, the flux density might change on an hour time scale, limiting the hybrid mapping of these stars to snapshots.

c. RESULTS AND INTERPRETATION: CORE-HALO IN CLOSE BINARIES

Table 1 summarizes the most precise VLBI measurements made for 5 RS CVn and Algol.

Star	Flux density (mJy)	Measured angular size (mas)	Linear size 10^{11} cm	Measured Brightness Temp. (10^8 K)	Magnetic field (Gauss)	Refer.
HR5110	15 - 165	0.9 - 1.4	4 - 11	4 - 20	5 - 150	a,d
UX Ari	14 - 145	0.4 - 3.2	3 - 26	1 - 100	5 - 20	b,d
HR1099	12 - 400	0.3 - 2.1	4 - 12	1 - 300	5 - 200	b,c,d
Algol	5 - 220	0.5 - 2.2	2 - 8	3 - 300	4 - 200	d,e
σ CrB	14	0.6	2	30	5 - 90	d
SZ Psc	37	0.6	8	80	5 - 20	d

Table 1: Source sizes and brightness temperatures directly measured by VLBI for 5 RS CVn and Algol. References: a: Lestrade et al 1984a, b: Mutel et al 1984, c: Lestrade et al 1984b, d: Mutel et al 1985, e: Lestrade et al 1987.

The main result is the discovery that radio emission from these binaries exhibits variable core-halo structure. The brightness temperature of the core is typically $(1 - 3) \times 10^{10}$ K and the corresponding effective energy of the radiating electrons is $E_{eff} \sim 5$ MeV while the brightness temperature of the halo is typically $(5 - 10) \times 10^8$ K with $E_{eff} \sim 0.5$ MeV for the radiating electrons. For the most precise measurements made, the halo component is comparable to the overall size of the binary system while the core is smaller than the K-subgiant which is presumably the active star. These high measured brightness temperatures are incompatible with a thermal process and gyrosynchrotron emission must be invoked, i.e. synchrotron emission from mildly relativistic electrons (Lorentz factor ≤ 10) with a power-law energy distribution.

Figures 2 and 3 are typical visibility curves measured for the two RS CVn systems HR1099 and UX Ari observed during outbursts at 10.6 GHz and 5 GHz, respectively (see also Palagi et al in these proceedings for UX Ari). These curves show that the halo is strongly resolved over baselines longer than $\sim 60 \times 10^6 \lambda$ corresponding to source sizes of 2 - 3 milliarcseconds if the first part of the curve is fitted with a Gaussian brightness distribution. Then, there is a plateau which extends up to baselines as long as $220 \times 10^6 \lambda$ in Figure 3. This is evidence for a radio core smaller than 0.3 milliarcsecond in HR1099 or smaller than 25 % the diameter of its K-subgiant during these observations (Lestrade et al in preparation).

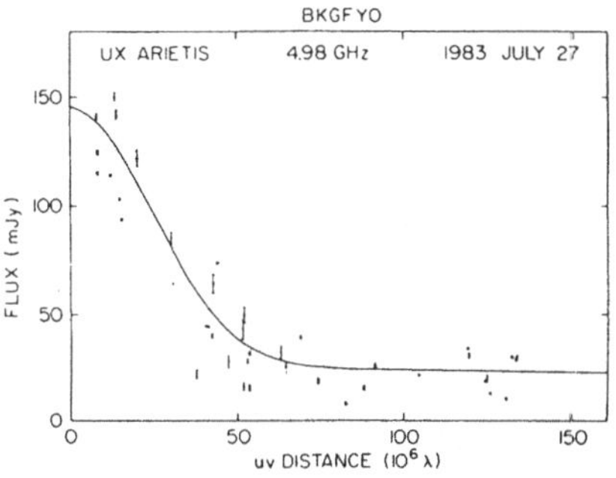

Fig. 2: VLBI visibility curve of UX Ari. Two Gaussian components are fitted to the data (source sizes: 3.2 and ≤ 0.4 mas)

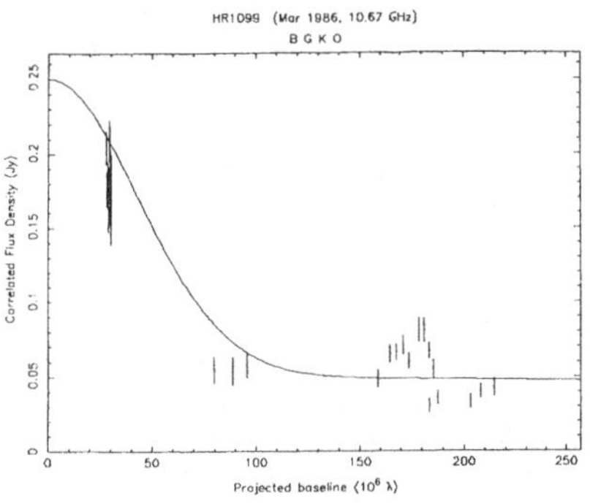

Fig. 3: VLBI visibility curve of HR1099. Two Gaussian components are fitted to the data (source sizes: 1.8 and ≤ 0.3 mas)

It is interesting to go back to the puzzling result of the experiment designed by Owens and Spangler (1977) to probe the radio source size of the eclipsing RS CVn system AR Lac. Based on the non-thermal and rapidly variable total flux densities of RS CVn, they expected the radio source size to be small. However, they failed to detect any simultaneous optical and radio eclipses and concluded, on the contrary, that the radio source was at least as large as the whole binary system. A compact active core and a large halo found by VLBI reconcile these two seemingly opposite views.

If the gyrosynchrotron process is responsible for the radio emission of RS CVn and Algol and if their brightness temperatures T_B are measured by VLBI, the magnetic field strength B of the emitting region can be inferred since $T_{eff} \propto (\nu/2.8B)^{0.5+0.5\delta}$ for this process. In this formula, ν is the observing frequency, δ is the energy spectral index of the electron energy distribution ($N(E) \propto NE^{-\delta}$) and is assumed to be between 2 and 7, the effective temperature T_{eff} is approximately the measured brightness temperature T_B if the optical depth of the source is assumed to be close to unity. The ranges of magnetic field strengths in Table 1 have been inferred with these assumptions. It is interesting to note that VLBI observations can provide indirect measurements of the magnetic field strengths in stars while this is not possible by any other technique for fields weaker than 1000 Gauss.

A simple physical model accounting for the observed angular size, polarization and spectrum is proposed in Mutel et al 1985 by analogy with some types of "Type IV" radio emission from the Sun. Some energy deposition mechanism (presently unknown) produces enough electrons with energy > 1 MeV in the magnetic structure of the active star to form a small self absorbed (optically thick) gyrosynchrotron source. This core is so small according to our measurements (e.g. < 0.3 milliarcsecond for HR1099) that the energy deposition can occur in a single magnetic loop of the active K-subgiant. As the outburst evolves, this coronal loop expands to a size comparable with the binary system itself. Alternatively, the high energy electrons suffer rapid radiation loss in the active core and are eventually injected with low energy (E_{eff} < 1 MeV) in a joint magnetosphere formed of loops connecting between the two stellar components as proposed by Uchida and Sakurai (1983). These low energy electrons can produce an optically thin and circularly polarized halo if the magnetospheric field is sufficiently strong. Such circularly polarized halo associated with an unpolarized core have actually been observed for Algol at 5 GHz (Mutel et al 1985). The radiative lifetime of the high energy electrons in the core is only a few hours while it is several days for the low energy electrons in the halo if the values for E_{eff} and the magnetic field strengths inferred above are used.

This model agrees with the most interesting data we obtained in July 83 with a six-station VLB array on UX Ari undergoing a strong outburst of 145 mJy at 5 GHz. All the 15 baselines detected the star and the resulting visibilities (including closure phases) were used to map the source with the hybrid mapping technique. This map in Fig. 4 shows a core-halo morphology with the brightness temperature twelve times higher in the core than in the halo. The map shows also that the centroids of the core and halo have a position offset about equal to the projected major axis of the binary at the time of observation. This finding led us to locate the optical components of UX Ari relatively to the radio map as shown in Figure 4 but this is merely a conjecture. The elongated radio halo of Fig. 4 favors the joint magnetosphere model hypothesized by Uchida and Sakurai (1983). The halo contains 80 % of the total flux density and has a size of 3.2 milliarcseconds (almost twice as large as the overall binary size), making it the most extended radio source from RS CVn systems detected so far. The core was unresolved on the longest baseline, indicating a size < 0.4 milliarcsecond, i.e. smaller than 75 % of the diameter of the

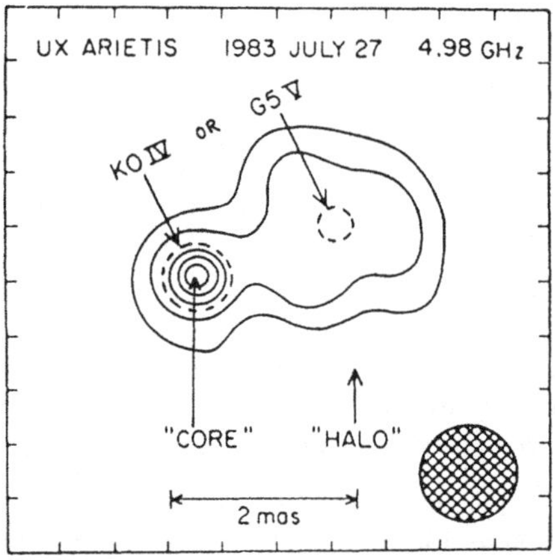

Fig. 4: hybrid map of UX Ari at 5 GHz. The contour levels are at 25, 35, 50, 70 and 90 per cent of the peak brightness. The location of the two optical stellar components relative to the radio map is conjectural (see text).

K-subgiant.

d. DUAL FREQUENCY AND DUAL POLARIZATION VLBI OBSERVATIONS OF ALGOL

An exceptional event occurs on Algol when we were conducting VLBI observations simultaneously at 2.3 and 8.4 GHz on 1983 May. The total flux density of Algol changes from 44 to 122 mJy at 2.3 GHz and from 42 to 240 mJy at 8.4 GHz in a 3-hour period. From the pre-outburst to outburst phase, the source size stays constant at 2.3 GHz (1.9 mas or $3R_{K-subgiant}$) while it decreases from 0.6 to 0.45 mas at 8.4 GHz. An enhanced total flux density could be due to the expansion of an optically thick radio source in Algol but this is clearly not what we observed in this early stage of the outburst. On the contrary, the source size has decreased at 8.4 GHz implying that it is the brightness temperature which has increased. This is a direct evidence for energy deposition and production of high energy electrons ($\sim$ 5 MeV) in a region as small as a single loop.

Klein and Chiuderi-Drago (1987) have simulated gyrosynchrotron emission from single loop and loops connecting the components of a typical RS CVn system and found the same frequency dependence of the source size as we observed.

Another exceptional event occurs on Algol when we were conducting VLBI/VLA observations at 1.6 GHz recording simultaneously both senses of circular polarization. Figure 5 shows the total flux density history from the VLA data during 3 scans spaning 6 hours. The first and third scans are only slightly circularly polarized while a highly left circularly polarized outburst ($\pi_c \sim 50\%$) occurs during the second scan with a high brightness temperature ($> 5 \times 10^9$ K) directly measured by VLBI. The combination of large fractional circular polarization $\pi_c \geq 50$ % and high brightness temperature ($> 10^9$ K) is difficult to produce by any incoherent gyrosynchrotron mechanism because high T_{eff} implies both high E_{eff} of the radiating electrons and low magnetic field while high degree of circular polarization requires the opposite conditions. Melrose and Dulk (1982) propose that such events, which are known on the Sun too, be coherent emission. They developed the theory of an electron cyclotron maser source with radiation escaping at the second harmonic and driven by a loss cone in the foot of a magnetic loop. A magnetic field of 300 Gauss in the maser region is inferred during our observations of Algol from this model.

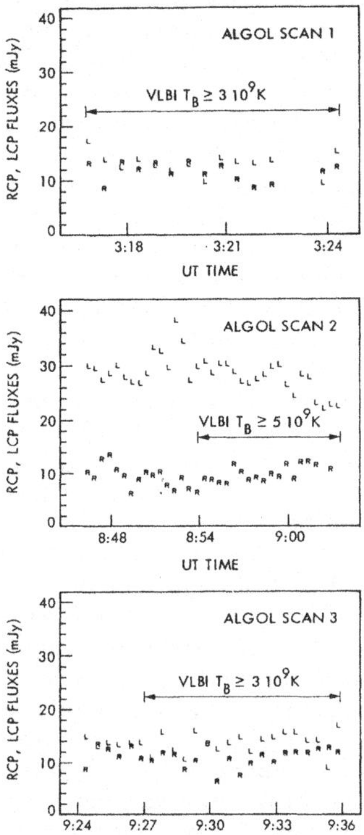

Fig. 5: high brightness temperature associated with a highly circularly polarized outburst of Algol at 1.6 GHz. A coherent emission process might be responsible for this event.

3. X-RAY BINARIES AND THE EXOTIC OBJECT SS433

Eleven major known radio emitting X-ray stars are listed in Hjellming and Johnston (1985). These objects are binary systems with X-ray emission from accretion discs. They exhibit relativistic expanding synchrotron sources which are common in extragalactic objects. The periodicities of the X-ray and radio emissions in three of these objects coincide with their orbital periods (LSI 61°303 (26.5 days), Cir X1 (16.59 days), Cyg X3 (4.95 hours). The formation and evolution of jets in these stellar objects can be studied on a time scale comparable to their orbital periods, i.e. considerably shorter than in extragalactic sources.

Cyg X3: Molnar et al (1984) studied the low-level radio emission which is characterized as a succession of synchrotron flares. The timing of these flares suggests that they are triggered by small variation in the binary separation. Molnar (1986) (see also Molnar in these proceedings) conducted Mark III VLBI observations during one flare of Cyg X3 at 1.3 cm. The VLBI data are consistent with a model composed of an elongated synchrotron source expanding with time at a rate of 0.41 mas/h. Molnar (1986) proposes that the expansion rate corresponds to the tranverse velocity of a double-sided jet.

Cyg X1: this black hole candidate binary was detected at 5 GHz on 1983 July 26. The 15 mJy source was unresolved and the upper limit of its source size is 0.5 milliarcsecond (Mutel et al 1985). The data were not sufficient to probe the morphology of the source.

LSI 61°303: this binary (Be star + neutron star ?) is a periodic non-thermal radio source. It has been detected on VLBI baselines at 1.6 GHz on 1983 October 16 during its quiescent period (25 mJy). A Gaussian brightness distribution was fitted to the VLBI visibility function yielding a source size of 4.1 mas (unpublished). At 2300 parsecs, this size is twelve times the major axis of the binary system. It is not clear if such a large source size is due to interstellar scattering or if radiating electrons in the environment of the binary are accelarated by Compton collision with γ-rays produced by the neutron star (Vestrand 1983).

Circ X1: this periodic radio emitting binary was observed on VLBI baselines at 2.3 GHz by Preston et al (1983) to measure its source size during the quiescent period (size > 0".2) and during flaring (size between 0".0015 and 0".015). As for LSI 61°303, it is not clear if interstellar scattering or relativistic expansion of a jet is responsible for such a large source.

Sco X1: Geldzahler and Fomalont (these proceedings). report on the first epoch Mark III VLBI observations of the variable radio source Sco X1 to determine the proper motions of the core and of the two surrounding lobes.

SS433: the optical emission lines of H I and He I in the SS433 star system (V1343 Aql) have twin Doppler shifts, corresponding to velocity of 80 000 km/s, with a period of 164 days. A precessing twin-jet model was proposed for these optical data but the two angles between the ejection axis and precession axis and between the precession axis and the line of sight cannot be unambiguously determined from optical data. Walker et al (1981) conducted VLBI observations of SS433 and found that the data were consistent with a model composed of a compact core and of collinear elongated jets of ~ 0".1 and 2" in length. They found that the extended components are approximately aligned with the outer bulges of the super nova remnant W50 on which SS433 is superposed, suggesting a physical association between SS433 and the distorsion of the super nova remnant. Niell et al (1981) conducted extensive VLBI observations at 2.3 GHz at 12 epochs over 15 months. They found that the variation of the position angle of the elongated components has a sinusoidal signature corresponding to the optical period and that the position angle of this radio component lags that of the optical model by 17 days. These observations provide physical evidence for precessing jets, a model which has been suggested to explain the misalignment of the small and large scale structures of extragalactic sources. These authors also resolved the ambiguities in the two angles of the optical model of SS433 (see also the VLBI observations of Schilizzi et al 1981). From a subset of their VLBI observations, Niell et al (1981) found evidence for motion of individual radiating knots and combined the proper motion of these blobs with the absolute ejection velocity (0.26c) obtained from the optical data to determine the distance of SS433 (5.1 kpc). Vermeulen et al (these proceedings) have conducted six Mark III VLBI experiments at two-day intervals to study the properties of the individual knots and of the ambient medium in which they move.

4. ASTROMETRY WITH RADIO STARS

The primary celestial reference frame is now based on precise VLBI positions of extragalactic sources. This is the most stable system because quasars or galactic nuclei have no detectable proper motions. However, this system is not dense ($\sim$ 200 sources for the whole sky) and is not as accessible as the optical system which is based on the positions of several thousands stars. Neverthless, the optical system lacks stability and so it is beneficial to tie the two frames together. For example, this process directly removes the global rotation of the optical system due to the galactic rotation. The global angular offsets and rates can be determined by comparing the radio positions (unit-vector: $\boldsymbol{\sigma}_{vlbi}$) and radio proper motions ($\dot{\boldsymbol{\sigma}}_{vlbi}$) referred to the VLBI reference frame with the corresponding optical quantities ($\boldsymbol{\sigma}_{opt}$ and $\dot{\boldsymbol{\sigma}}_{opt}$) of radio stars. The RS CVn binaries and some X-ray stars are optically bright (magnitude < 11) and sufficiently active in radio to be used as link stars. The following relations hold for each link star:

$$\boldsymbol{\sigma}_{vlbi} = [R]\boldsymbol{\sigma}_{opt} \quad ; \quad \dot{\boldsymbol{\sigma}}_{vlbi} = [R]\dot{\boldsymbol{\sigma}}_{opt} + [\dot{R}]\boldsymbol{\sigma}_{opt}$$

The matrices $[R]$ and $[\dot{R}]$ are functions of the 3 angular offsets and rates. These parameters can be solved for with the above set of equations if enough positions of radio stars are measured.

Niell et al (these proceedings) have determined VLBI positions of 8 radio stars with astrometric uncertainties ranging from 2 to 40 milliarcseconds. These positions are derived from VLBI delays and fringe rates and their astrometric precision strongly depends on their flux densities. Lestrade et al (these proceedings) have used a phase reference VLBI technique to determine the relative position between a reference source and Algol in a very low state of activity (4 mJy) by coherently integrating 4.1 hours of VLBI data. The uncertainty on their relative position is estimated to be about twice the stochastic precision of 0.8 milliarcsecond (angular resolution / SNR). This technique makes at least a dozen radio stars always available for differential astrometry.

Linking the optical and radio celestial reference frames is crucial for the Hipparcos space mission which should provide a new optical celestial frame as precise as the VLBI system (i.e. milliarcsecond precision) in the 90's.

5. PROSPECTS

Is radio emission from RS CVn a local phenomena, occuring in a single coronal loop of the subgiant for instance, or is it a global phenomena, involving the two stellar components through connecting magnetic loops for instance ?. Klein and Chiuderi-Drago (1987) have simulated both models and concluded that a structure interconnecting regions of intense magnetic fields on both stars of the binary should be revealed by a double structure of the radio source (two hot spots) at high frequency ($>$ 10 GHz). These two possible models could be discriminated by high dynamics range VLBI maps. Another issue is how large is the active core where outbursts are triggered by some energy deposition mechanism ?.

The cores observed so far are unresolved on terrestrial baselines > 100 × 10^6 λ for UX Ari and Algol and even baselines > 200 × 10^6 λ for HR1099. High frequency or orbiting VLBI observations might provide new insights.

Ultra-precise differential astrometry (< 100 μas) repeated over hour, day, month and year intervals should separate between intrinsic motion of the active core, orbital motion, space motion of the whole system (proper motion) and trigonometric parallax. If the core is stable in a loop attached to the subgiant, its motion around the gravity center of the binary system could be used to measure the major axis by repeated VLBI astrometric measurements. Then, individual masses of the components could be determined. Mass is one of the most difficult parameter to measure in stellar physics.

VLBI investigations should be initiated on new classes of stars which are susceptible to exhibit milliarcsecond structures, e.g. dMe, FK Com and T Tauri stars.

Acknowledgements
The research described in this report was carried out, in part by the Jet Propulsion Laboratory of the California Institute of Technology, under contract with the National Aeronautics and Space Administration.

References

Abbott, D. C., 1985, in "Radio Stars", Workshop held in Boulder August 1984, edited by R. M. Hjellming and D.M. Gibson, 61.
Clark, B.G., Kellermann, K.I., Shaffer, D.B., 1975, *Ap. J.*, **198**, L123.
Clark, T.A. et al, 1976, *Ap. J.(Letters)*, **206**, L107.
Dulk, G.A., 1985, An. Rev. Astron. and Astroph, vol 23, 169.
Hall, D.S., 1975, in "Multiple Periodic Variable Stars", IAU Colloquium No 29, Budapest.
Hjellming, R.M., Johnston, K.J., 1985, in "Radio Stars", Workshop held in Boulder August 1984, edited by R. M. Hjellming and D.M. Gibson, 61.
Klein, K.L., Chiuderi-Drago, F., 1987, Astron. Astrophys. (in press)
Kwok, S., 1985, in "Radio Stars", Workshop held in Boulder August 1984, edited by R. M. Hjellming and D.M. Gibson, 79.
Lestrade, J.-F., Mutel, R.L., Preston, R.A., Scheid, J.A., Phillips, R.B., 1984a, *Ap. J.*, **279**, 184.
Lestrade, J.-F., Mutel, R.L., Phillips, R.B., Niell, A.E., Preston, R.A., 1984b,*Ap. J. (Letters)*, **282**, L23.
Lestrade, J.-F., Mutel, R.L., Preston R.A., Phillips, R.B., 1987, submitted Ap.J.
Little-Marenin, I.R., Simon, T., Ayres, T.R., Cohen, N.L., Feldman, P.A., Linsky, J.L., Little, S.J., Lyons, R., 1986, *Ap. J.*, **303**, 780.
Ma, C., et al, 1986, *A.J.*, **92**, 1020.
Melrose, D.B., Dulk, G.A., 1982, *Ap. J.*,**259**, 844.
Molnar, L.A., Reid, M.J., Grindlay, J.E., 1984, *Nature*, **310**, 662.
Molnar, L.A., 1986, in "Physics of Accretation onto Compacts Objects", held in Tenerife (Spain) April 1986, M. Watson K. Mason and N. White (eds)
Mutel, R.L., Doiron, D., Lestrade, J.-F., Phillips, R.B., 1984, *Ap. J.*, **278**, 220.
Mutel, R.L., Lestrade, J.-F., Preston, R.A., Phillips, R.B., 1985, *Ap. J.*, **289**, 262.
Mutel, R.L., Lestrade, J.-F. , 1985, , *A.J.*, **90**, 493.
Mutel, R.L., Morris, D.H., Doiron D.J., Lestrade, J.-F., 1987, *A. J.*, **93**, , 1220.
Niell A.E., Lockhart, T.G., Preston, R.A., 1981, *Ap. J.*, **250**, 248.
Owens, F.N., Spangler, S.R., 1977, *Ap. J.*, **217**, L41.
Pallavicini, R., 1986, Highlights of Astronomy IAU 1986, edited by J.P. Swings, 447.
Preston, R.A. et al, *Ap. J.*, **268**, L23.
Rodonò, M., 1986, Highlights of Astronomy IAU 1986, edited by J.P. Swings, 429.
Rogers, A.E.E. et al, 1983, *Science*, **219**, 51.
Uchida, Y., Sakurai, T., 1983, in "Activity in Red Dwarf Stars, IAU Colloq. 71, edited by M. Rodonò, Reidel, 629.
Schilizzi, R. T., Miley, G.K., Romney, J. D., Spencer, R.E., 1981, *Nature*, **290**, 318.
Vestrand, W. T., 1983, *Ap. J.*, **271**, 304.
Vogt, S.S., Penrod, G.D., 1983, *P.A.S.P.*, **95**, 565.
Walker, R.C., et al, 1981, Ap. J., **243**, 589.
Walter F.M., Gibson, D.M., Basri, G.S., 1983, *Ap. J.*, **267**, 665
White, N.E. et al, 1985, in "Cool Stars, Stellar Systems and the Sun", edited by D. Gibson and M. Zeilik.

EVIDENCE FOR A RADIO BRIGHTENING ZONE AROUND SS433

R.C. Vermeulen[1], R.T. Schilizzi[2], V. Icke[1], I. Fejes[2], R.E. Spencer[3]
[1] Leiden Observatory, Box 9513, NL-2300 RA, Leiden, NL
[2] NFRA, Box 2, NL-7990 AA, Dwingeloo, NL
[3] NRAL, Jodrell Bank, Macclesfield, Cheshire, SK119 DL, UK

SS433 is thought to be a binary system from which two antiparallel beams are ejected at 0.26c. An excellent review of this peculiar object has been given by Margon (1984). The so-called "kinematic model" describes the precessing motion of the beams. They can be observed over at least 8 orders of magnitude in distance: from 10^{12} cm to 10^{20} cm.

We have investigated the beams of SS433 on scales of 10^{15} cm. A series of 6 European VLBI Network observations at 5 GHz was made in May 1985 at two-day intervals, using Mark III recording in Mode E. In the 6 hybrid maps we have obtained, the dynamic range is typically 50 to 1. Maps 2 and 3 which are shown in Figure 1 display all the characteristic features of this series. We have drawn the projection on the sky of the beam trajectories as predicted by the 5-parameter kinematic model (Margon, 1984) by adopting a distance of 5 kpc. The predicted ejection angle was almost at right angles to the line of sight.

There is a strong (75 mJy), variable, unresolved core, which contains a large fraction of the total flux density. Sometimes the central structure is elongated along the precession locus; this may be related to subsequent unresolved ejection episodes. Additionally there are discrete emission patches (total flux density typically 20 mJy). They are unresolved perpendicular to the model trace, but much broader along it. In general, all maps are quite symmetric, which supports our choice of the brightest central knot as the core. There is no emission beyond 100 mas (500 AU) in any of our maps.

When the locus of the brightest points in each map is followed from map to map, the rate of divergence is found to be (at most) three quarters of that predicted. This is unlikely to be due to a low ejection velocity or a very deviant ejection angle: such large deviations have never been seen in earlier radio or optical data, and are not observed in simultaneous optical spectra (Vermeulen et al., in preparation). The symmetric flux density distribution is hard to reconcile with the differential Doppler boosting expected at deviant ejection angles. We believe it is much more plausible that we observe a series of blobs which pass through a zone where they brighten, after which they decay in flux density. We can decompose the emission patches along the model trace into substructures, which are unresolved, and which move in

M. J. Reid and J. M. Moran (eds.), The Impact of VLBI on Astrophysics and Geophysics, 275–276.

accordance with the kinematic model. The "brightening zone" is located at about 50 mas (4×10^{15} cm) from the centre. We do not observe the brightest spots of emission to be stationary in our series of maps since, when there is a succession of blobs rather than a continuous jet, each of these will dominate the appearance of the maps for some time.

The decay time by synchrotron radiation for the relativistic electrons in the observed blobs is some 10^6 years. Therefore, the observed flux density decay after the blobs have passed the brightening zone is probably due to adiabatic expansion. The stellar wind emanating from the normal star in the SS433 system, if moving at 2000 km/s, traverses 3×10^{15} cm in one precession period. This is therefore the largest distance at which any wind gas which may have been disturbed by a blob can be fully replenished before the next blob passes. Hence, expansion of the blobs beyond the brightening zone could be explained by the relative absence of confining material there.

The nature of the brightening zone is unclear. Because the beams are evidently composed of discrete clouds, a speculative explanation is available which needs no assumptions about the external medium. An ejected clump of gas develops a bow shock which points backward at the Mach angle. Because the ejection direction precesses, and since all blobs move at the same speed, each blob will overtake the bow shock of its precursor. When this happens, density enhancement, particle acceleration through turbulence, or compression of the magnetic field, could all lead to brightening.

Reference: Margon, B. Ann. Rev. Astr. Astrophys. 22, 507-536 (1984)

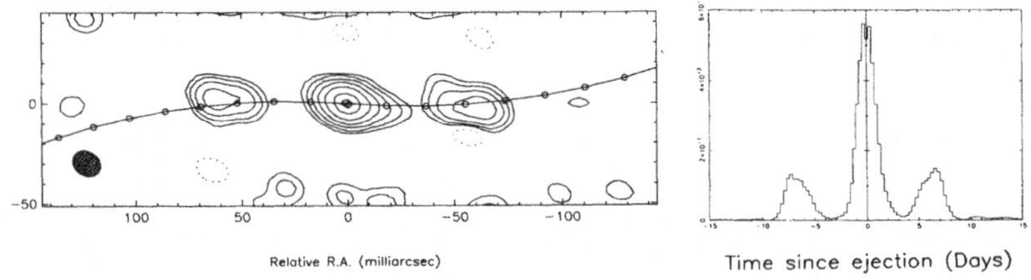

Figure 1: Maps 2 and 3 of the series of 6 EVN observations of SS433. The ellipse at lower left is the FWHM of the restoring beam. Contour levels are drawn at -2,2,4,8,16,32,64 and 90 percent of the peak flux level in each map. Marks along the kinematic model locus are present at two-day intervals. We display a crosscut of the maps along this locus, with the horizontal axis labelled in days since the assumed ejection.

RS OPHIUCHI - A RECURRENT NOVA WITH TWIN JETS

R.J. Davis & A.R. Taylor, NRAL, Jodrell Bank
M.F. Bode, Lancashire Polytechnic
R.W. Porcas, MPI Bonn

ABSTRACT An analysis of European VLBI data of nova RS Ophiuchi shows the presence of a linear jet-like structure which has a brightness temperature of 10^7K at 1.7 GHz. This ejection is shown to be two sided and the 1.7 GHz emission is non-thermal. The light curves show the presence of a second, thermal component emerging at higher frequencies at a later epoch.

1. THE VLBI OBSERVATIONS

The recurrent nova RS Ophiuchi went into outburst on January 26th 1985 (Morrison 1985). The radio light curve was monitored initially at Jodrell Bank at 4.9 GHz (Padin et al. 1985) and from March 1985 by the VLA at its four operating frequencies (Hjellming et al. 1986). The behaviour was clearly very different from a classical nova, and from the rapid development of the radio light curve it was clear that the brightness temperature was much in excess of 10^4K. Observations were subsequently carried out with four antennas of EVN at 1.7 GHz on 13th April 1985. We used the Mk III recording system with 28 MHz bandwidth and the noise level on Jodrell-Bonn, the most sensitive baseline, was 0.4 mJy in 13 min integration. By global fringe fitting around one triangle of telescopes, a set of closure phases were measured. The hybrid map so obtained is shown in Figure 1. Because of the weakness of the source and the sparceness of the u-v coverage simulations were made to check the three component structure of the source. No one-sided ejection model can be fitted satisfactorily to the data. The map enables brightness temperatures for the three components to be calculated and values of $1.0 \pm 0.1 \times 10^7$K were obtained.

2. INTERPRETATION AND DISCUSSION

We initially tried to fit these observations with thermal emission from 10^7K gas. However, the level of X-rays measured with the Exosat satellite rules out the presence of sufficient 10^7K gas to produce this radio emission. The 1.7 GHz flux density must therefore be non-thermal and

M. J. Reid and J. M. Moran (eds.), The Impact of VLBI on Astrophysics and Geophysics, 277–278.

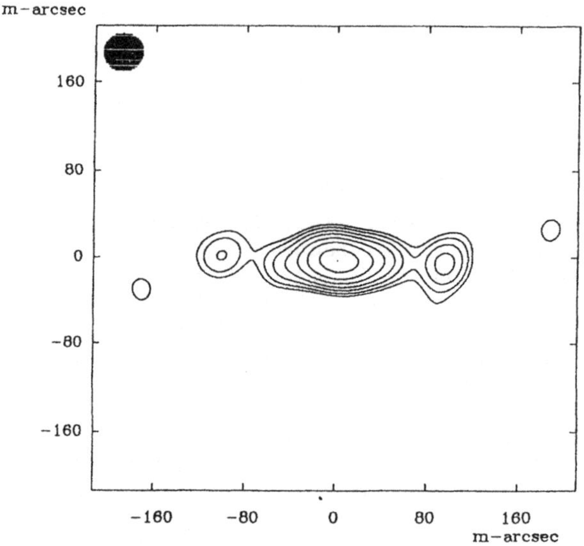

Figure 1. The EVN map made on 13 April 1985 at 1.7 GHz. The peak is 8.8 mJy and the resolution is 35 mas.

the synchrotron mechanism can easily fit the observations.

As the nova progresses a new spectral component emerges at high frequencies. The non-thermal flux extrapolated from 1.7 GHz has been subtracted from the other frequencies with a spectral index of $\alpha = -0.08$ and a new set of light curves were generated. These curves stay high ($\sim$23 mJy at 22 GHz) for $\simeq$100 days (cf. 30 days at 1.7 GHz) and initially have a spectral index of $\alpha \simeq +1$ thereafter decreasing. Thus there appears to be a two component spectrum, primarily non-thermal at 1.7 GHz with a thermal component emerging at higher frequencies.

Comparison of the total angular extent from the VLBI observations, and reinterpreted 15 GHz measurements made with the VLA but using the one dimensional model as seen in the VLBI map, indicate non-decelerating flow at 3800 km s^{-1}. This tangential velocity agrees well with the maximum radial velocity of 3500 km s^{-1} seen shortly after optical maximum.

REFERENCES

Hjellming, R.M., van Gorkom, J.J., Taylor, A.R., Seaquist, E.R., Padin, S., Davis, R.J. & Bode, M.F. (1986). Ap.J.Lett. 305,L71-L75.

Morrison, W. (1985). IAU Circ.No. 4030.

Padin, S., Davis, R.J. & Bode, M.F., (1985). Nature, 315, 306

VLBI Observations of an Expanding Jet in Cygnus X-3

Lawrence A. Molnar, M. J. Reid, and J. E. Grindlay
Center for Astrophysics, Cambridge, MA USA

The X-ray binary Cyg X-3 offers a rare opportunity to study the physical conditions of the formation of a relativistic jet. While the central engines of extragalactic jets are generally poorly understood, the infrared and X-ray data on Cyg X-3 allow detailed modelling of the conditions in the inner disk and corona, the source of the jet energy. Study of the periodic nature of particle injection into the jet may yield an important clue to the nature of the particle injection mechanism. And the time scale of jet evolution is much shorter than for extragalactic sources, allowing the possibility of observing a large number of flares to determine general characteristics.

The radio emission was discoverd in 1971 (Braes and Miley 1972) and attracted wide attention in September, 1972 when the radio flux density rose from ~10 mJy to ~20 Jy over a span of a few days (Gregory *et al.* 1972). The spectral evolution of this flaring event led to the interpretation of the radio waves as synchrotron emission from an expanding plasma of relativistic electrons. Measurements (Geldzahler *et al.* 1983) several weeks after another giant flare show the source to have become slightly resolved in the north-south direction, demonstrating expansion and suggesting a collimated (jet) morphology.

We have carried out a program of measurements of the spectral evolution of Cyg X-3 at times of low flux density that has shown the radio emission may at all times be described as a series of overlapping flares (Molnar, Reid, and Grindlay 1984, 1985). Flare amplitudes range from tens of milliJanskys to tens of Janskys (the giant flares), the larger flares being less common. While the amplitudes vary from one flare to the next, the time from the beginning of one flare to the beginning of the next is fixed at 4.95 h, some 3% longer than the orbital period. Each flare shows the spectral evolution expected of an adiabatically expanding synchrotron source: the flux density peaks sooner and at a greater value at higher frequencies (van der Laan 1966). We analyzed a typical flare under this hypothesis assuming the total energy in the flare was less than the energy emitted in the X-rays during the flare to predict a range of expansion velocities of $.05c$–$.25c$ (Molnar, Reid, and Grindlay 1984).

As a preliminary VLBI study we took 6 hours of data on each of 2 days, 5 and 8 February, 1985. We used a 5 element array at 1.3 cm wavelength recording with the MkIII recorders in Mode A for maximum bandwidth. Measurements of

M. J. Reid and J. M. Moran (eds.), The Impact of VLBI on Astrophysics and Geophysics, 279–280.

interstellar scattering at longer wavelengths implied that we would have to use the shortest wavelength possible to see a flare become larger than its scattering size before it was too faint to detect.

We fit all the data to a model in which the source is an elliptical Gaussian of fixed axial ratio, A, and orientation, with the FWHM major axis expanding at a constant rate, v, from zero size at flare onset, and convolved with a Gaussian scattering disk of FWHM θ_{scat}. The orientation (5°) was fixed by previous measurements and the flare onset by the light curve. The probability of obtaining a higher χ^2 than our minimum value was .20; the best fit parameters (v, A, θ_{scat}) were (.40 mas/hr, 2.2, .68 mas) with asymmetric statistical errors less than 15% (1-σ). Wilkinson *et al.* presented high accuracy scattering data at this meeting that predict a scattering size of $0.68 \pm .09$ mas, in agreement with the value we derive, supporting the validity of our model.

We chose the model above because our data are insufficient to make independent hybrid maps at each time step. The data and the model both suffice to determine the features of greatest interest: significant deviations from circularity and significant changes in size. Our main conclusions are that with greater than 5-σ certainty, the source is elongated and is expanding along its major axis. For our best values of the externally fixed parameters (orientation and time of flare onset), we can also conclude that with greater than 5-σ certainty the source is also expanding along its minor axis. Changing the externally fixed parameters within their allowed range, the first two conclusions remain unchanged but values can be found for which this third conclusion is not significant.

We have also fit a two-sided jet model that is mathematically similar to the model above and allows a direct comparison with our earlier prediction. It consists of two identical circular Gaussian sources that have a an apparent transverse motion from the center as well as an intrinsic expansion. For a distance of 10 kpc, the fit yields a transverse motion of .20c, similar to that of SS433, and an expansion velocity of .13c, within our predicted range of velocities.

We are currently working on a more extensive data base that will be able to confirm and extend the conclusions reached here. Only the short baselines show significant correlated flux because of the large scattering size. As short baselines in the current network do not tend to be connected, these data do not have any reliable closure phase measurements. In this regard the first few stations of the VLBA (Pietown, Los Alamos, and Kitt Peak) will allow significantly improved measurements because they, used together with the VLA, Goldstone, and Owens Valley, offer a large number of interconnected short spacings.

Braes, L. L. E., and Miley, G. K. 1972, *Nature*, **237**, 506.
Dickey, J. M. 1983, *Ap. J. (Lett.)*, **273**, L71.
Geldzahler, B. J., *et al.* 1981, *Ap. J.*, **251**, 230.
Gregory, P. C. 1972, *Nature*, **239**, 439.
Molnar, L. A. 1985, Ph.D. thesis, Harvard University.
Molnar, L. A., Reid, M. J., and Grindlay, J. E. 1984, *Nature*, **310**, 662.
Molnar, L. A., Reid, M. J., and Grindlay, J. E. 1985, *Radio Stars*, (R. M. Hjellming and D. M. Gibson, eds.), Astrophysics and Space Science Series, Reidel, Dordrecht, p. 329.
van der Laan, H. 1966, *Nature*, **211**, 1131.

VLBI AND VLA OBSERVATIONS OF SCO X-1

B. J. Geldzahler
Applied Research Corporation
Landover, MD

E. B. Fomalont
National Radio Astronomy Observatory
Charlottesville, VA 22903

N. L. Cohen
Boston University
Boston, MA 02215

We have measured the absolute and relative positions of the components of Sco X-1 (Fig. 1) with the VLA from 1980–1986 (Geldzahler, Fomalont, Hilldrup, and Corey, 1981, *A.J.*, **86**, 1036; Fomalont, Geldzahler, Hjellming, and Wade, 1983, *Ap.J.*, **275**, 802; Geldzahler and Fomalont, 1986, *Ap.J.*, **313**, 815). We have used these observations to test the dual-beam model of radio sources proposed by Blandford and Rees (1974). Our latest VLA results may be summarized as follows:

1. All radio components of Sco X-1 have an absolute proper motion that is nonzero and consistent with the proper motion of the optical star. Thus, there is no doubt of the physical association of the components with the x-ray emitting star.
2. The NE lobe is moving radially away from the core at a rate of 0.015 ± 0.002 arcsec yr^{-1}, corresponding to a transverse velocity of 35 ± 5 km s^{-1} at an assumed distance of 500 parsecs.
3. The relative proper motion of the core and the SW hot spot at the outer edge of the SW lobe is < 0.030 arcsec yr^{-1}.
4. The assumption of linear motion of all three components is consistent with the post-fit residuals.
5. The flux density of the core can vary by an order of magnitude in a period of an hour.
6. The NE lobe varies by about 10% over periods of a year or less.
7. The SW hot spot does not appear to vary within the errors. However, because it is faint, percentage variations as large as those of the NE lobe could be present and may be reflected in the actual flux density values of the hot spot.
8. The integrated flux density of the SW lobe varies by about 30% over time scales of less than one year, and variations are not localized to a small region in the lobe.
9. There may be a correlation between the flux density variation in the NE lobe, SW hot spot, and SW lobe.

M. J. Reid and J. M. Moran (eds.), The Impact of VLBI on Astrophysics and Geophysics, 281–282.

With 18 cm VLBI observations of the NE lobe in 1985, we measured a lobe size of ~ 7 mas, but the source is weak and the consequent noisiness of the data allows for $\sim 20\%$ halo. Our first VLBI study used NRL, HRAS, VLA (phased array), and OVRO.

Our second epoch 18 cm VLBI observations of Sco X-1 used Green Bank, HRAS, VLA (phased array, D configuration), and OVRO. We positioned the VLA phase center midway between the NE lobe and the core such that both sources were in the beam. The results of this study are: (1) the confirmation of the amazingly small size of the NE lobe, and (2) the measurement of the offset of the NE lobe and the core to milliarcsec accuracy. Thus, we have reduced the measurement errors by an order of magnitude over the VLA results. The measurement of the offset was achieved because we were successful in detecting the core, at a $S/N \sim 30$, through phase referencing of the core to the NE lobe. Subsequent VLBI observations of the relative offsets should be sensitive enough to look for nonlinear motions.

There are several problems that remain in understanding the radio results of Sco X-1. The measured size of the NE lobe, < 7 mas, is an upper limit so the true size of the component is unknown. Confinement of this lobe seems to require an exceptionally large magnetic field or dense ambient medium under conventional emission processes. The uncertainty of the distance to Sco X-1 causes uncertainties in the application of the dual-beam model. We plan to measure the distance with the Hubble Space Telescope.

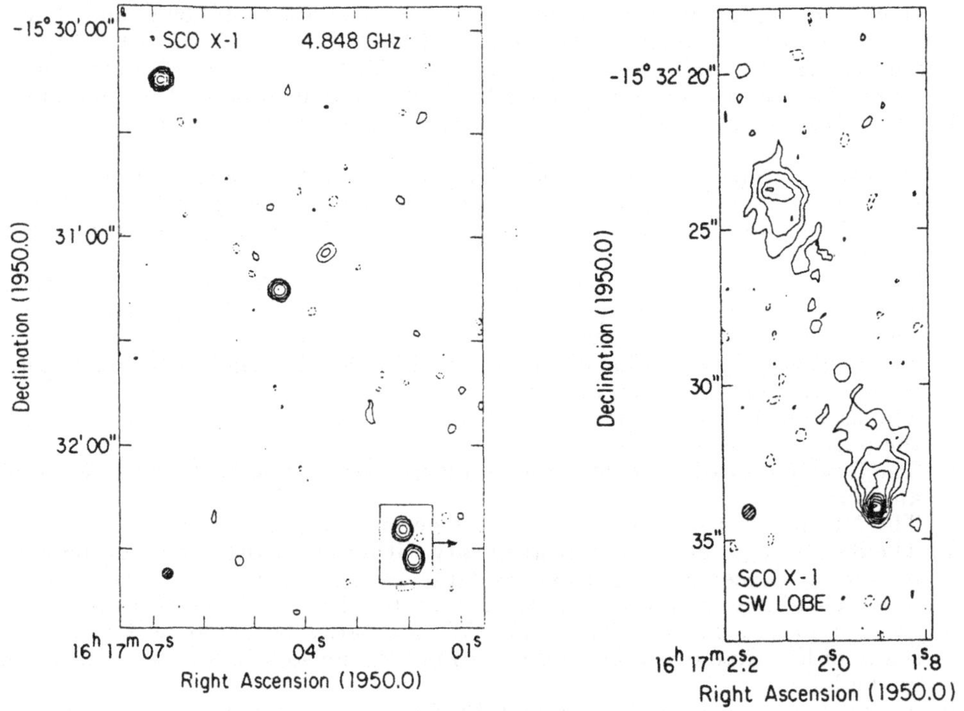

Figure 1

VARIATIONS OF VLBI STRUCTURE IN UX ARI

M. Catarzi[1], M. Felli[1], M. Massi[1], F. Palagi[2], R. Pallavicini[1] and G. Tofani[1]
[1] Osservatorio Astrofisico di Arcetri
Largo E. Fermi 5, 50125 Florence, Italy
[2] Gruppo Nazionale di Astronomia, CNR
Largo E. Fermi 5, 50125 Florence, Italy

Five stellar binary systems of RS CVn and Algol type have been observed on June 12, 1986, with the EVN (Effelsberg, Jodrell, Medicina, Onsala, Westerbork), and two US antennas (Green Bank, Haystack). Observations were carried out at 5 GHz, with 13 min of integration and with MARK III mode B.

Four of the sources (i.e. HR 1099, AR Lac, SZ Psc and Algol), were detected on short baselines only, while UX Ari was observed on all the 21 baselines, thus allowing a more detailed discussion. The observed visibility amplitudes are reported in Fig. 1 (triangles), versus the projected baselines. Since radio emission from UX Ari has been reported to range from less than 10 mJy to more than 250 mJy, the 93 mJy maximum flux level detected during our VLBI run clearly indicates that the source was observed during a flare event, as it was during the previous VLBI observation by Mutel et al. (1985). Due to the limited UV coverage, only gaussian models have been used to fit the data, leading to a single component with a flux density of 60 mJy and size of 1 mas. With this model the brightness temperature of the source is $T_b = 4.2\ 10^9$ K.

At the distance of UX Ari (50 pc), the separation between the binary components is 1.5 mas (16 Ro), and the sizes of the individual components are < 0.56 mas. Since the system was observed at phase 0.46 when the two stars are almost aligned, but not eclipsing, a source size comprized between the above values is fully consistent with the emission coming from an extended region comparable with the binary separation.

The visibility amplitudes observed by Mutel et al. are shown in Fig. 1 (dots). These have been interpreted as due to an unresolved core (size < 0.4 mas, flux density = 25 mJy), and a larger halo (3.2 mas, 120 mJy). Assuming that our single gaussian source coincides with the halo component, it appears to be of lower flux and smaller size than the halo component observed by Mutel et al.. The lower intensity and the absence of a core suggest that we are observing a flaring event when the rapidly decaying core had already disappeared and the halo was slowly fading away. The smaller size may be related to the different phase of the observation (0.46 versus 0.20).

Alternatively, we can interpret our observations as indicating a decreased intensity at the shorter spacings and a virtually unchanged

M. J. Reid and J. M. Moran (eds.), The Impact of VLBI on Astrophysics and Geophysics, 283–284.

component at the longest spacings. By assuming the components' size from Mutel et al., we derive a flux density of 50 mJy for the halo and 25 mJy for core. From our data alone we cannot distinguish between the two hypotheses. However from physical considerations, a single extended component is a more plausible interpretation.

The scatter of the visibility amplitudes with respect to the fitting curves is well outside the error bars. These "fringes" in the visibility function, which are clearly present in both our and Mutel's data, suggest a multi-component structure within the halo.

References

Mutel, R.L., Lestrade, J.F., Prenston, R. A. and Phillips, R.B.: 1985, Ap. J. 298, 262.

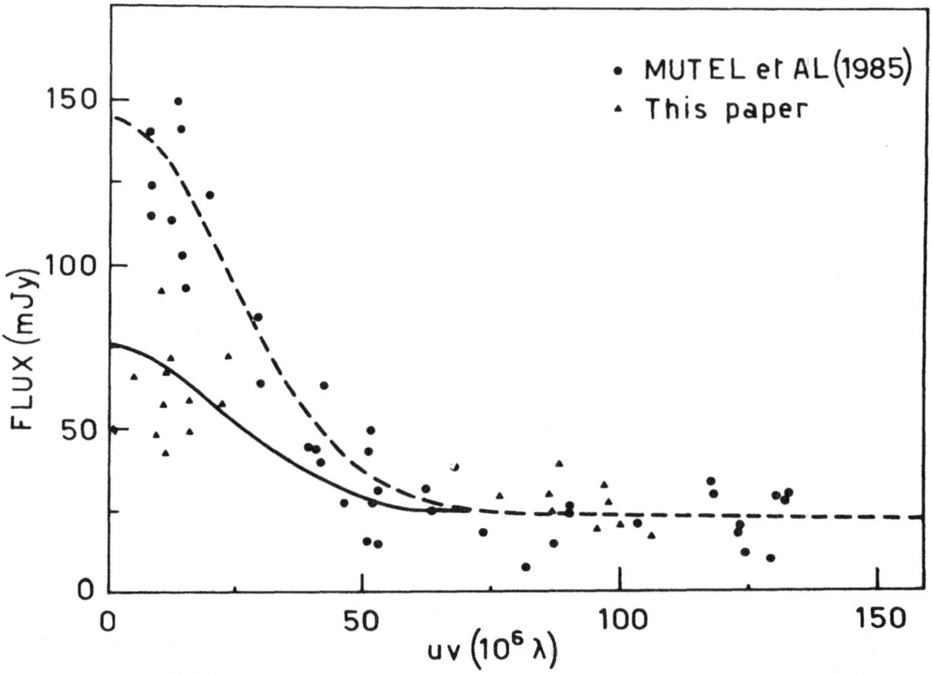

Fig. 1 - UX Ari Visibility Amplitudes

PULSAR INTERFEROMETRY WITH MICROARCSECOND RESOLUTION

A. Wolszczan
NAIC, Arecibo Observatory, Arecibo, PR 00613

J.M. Cordes
Astronomy Department and NAIC, Cornell University,
Ithaca, NY 14853

ABSTRACT. An 'interstellar interferometer' related to a double imaging of the pulsar PSR 1237+25 by refraction in the interstellar medium has been used to resolve the pulsar magnetosphere for the first time. With an interferometer baseline ~ 1 AU, we have measured a typical transverse separation between the emitting regions of PSR 1237+25 to be $\sim 10^8$ cm, or an angular separation ~ 0.02 μas.

Refractive effects can significantly modify the interstellar scintillation patterns arising from diffractive scatter of pulsar signals propagating through the interstellar medium (e.g. Hewish et al. 1985; Cordes and Wolszczan 1986). Strong refraction events sometimes split the apparent pulsar image into two or more distinct subimages. The resultant dynamic spectra (intensity as a function of time and frequency) are the superposition of separate patterns from each subimage combined with oscillatory cross terms caused by beating between subimages. The oscillatory terms are equivalent to fringes obtained with an interferometer baseline ~ 1 AU, sufficiently large to potentially resolve the pulsar magnetosphere (Cordes et al. 1986).

The dynamic spectra of PSR 1237+25 observed in December 1986 at 430 MHz with the Arecibo radiotelescope and with the 3-level, 40 MHz correlation spectrometer are shown in Fig. 1a. They display a strikingly periodic intensity modulation as a function of time and frequency. A model to explain this modulation (Wolszczan and Cordes 1987 and references therein) consists of a thin screen in which refractive phase variations are strong enough to give rise to a pair of diffracted subimages of the pulsar. Beating between subimages leads to oscillations (fringes) in the dynamic spectra, best characterized by the oscillation period in frequency, P_ν. The transverse displacement of a source by a distance δx_s shifts the observed fringes by approximately $\delta\phi \approx 3 \cdot 10^{-5} \nu \delta x_s [(D/D_s - 1)/DP_\nu]^{1/2}$ radians, where ν is the observing frequency, D is the pulsar distance and D_s is the distance between the pulsar and the screen. Fringe shifts may arise, if different pulse components are emitted from locations that are separated in the transverse direction (e.g. Cordes et al. 1983). We have resolved such separations by detecting fringe phase shifts in the dynamic spectra of PSR 1237+25 at different pulse longitudes (one pulsar period equals 360° longitude) as shown in Fig. 1b,c. The range of spatial separations of the emission regions calculated for $D = 2D_s$ is indicated on the right hand side of Fig. 1c. For a dipolar magnetic field and emission at a constant altitude,

M. J. Reid and J. M. Moran (eds.), The Impact of VLBI on Astrophysics and Geophysics, 285–286.

the fringe phase should be *symmetric* with respect to the pulse centroid and it should monotonically increase through the origin. This is in contradiction to the measurements, suggesting that the magnetic field of PSR 1237+25 may deviate strongly from a dipolar form in the emission region and emission altitudes may vary such that transverse separations are much larger for a given pulse longitude than implied by the dipole model. Alternatively, the refracting screen may be located much nearer the pulsar than the Earth ($D \gg D_s$), which would lead to transverse separations much smaller than for the screen that is midway ($D = 2D_s$).

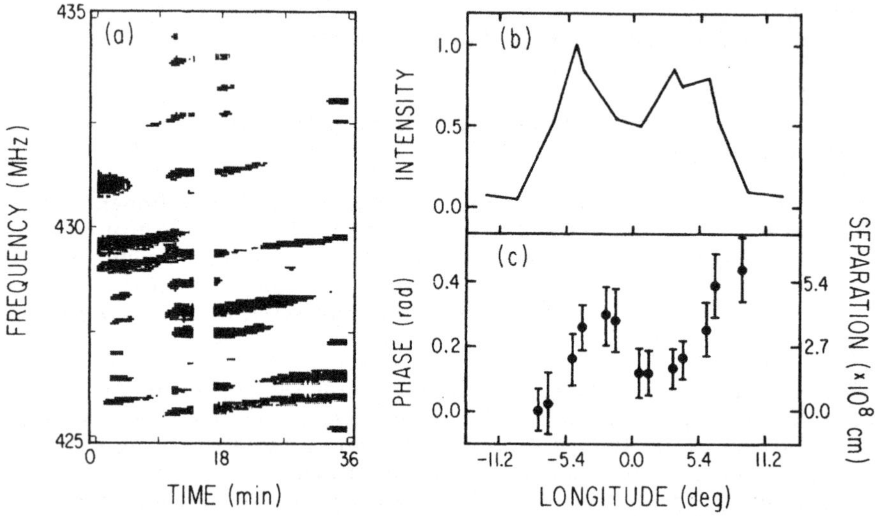

Fig. 1. PSR 1237+25: *(a)* Dynamic spectrum, *(b)* integrated pulse profile and *(c)* fringe phase as a function of pulse longitude.

In addition to resolving the pulsar magnetosphere, our measurements suggest that a discrete refracting structure of size > 1 AU crossed the line of sight, producing a predominantly double image of the pulsar with a splitting angle between subimages of about 3.3 mas. This episode may be similar in origin to recent caustic events seen in the light curves of compact extragalactic radio sources (Fiedler *et al.* 1987).

This work was supported by the Alfred P. Sloan Foundation, by the National Science Foundation through grant 85-20530 to Cornell University, and by the National Astronomy and Ionosphere Center at Cornell which operates the Arecibo Observatory under contract to the National Science Foundation.

REFERENCES

Cordes, J.M., Weisberg, J.M., and Boriakoff, V. 1983, *Ap.J.*, **268**, 370.
Cordes, J.M., and Wolszczan, A. 1986, *Ap. J. (Letters)* , **307**, L27.
Cordes, J.M., Pidwerbetsky, A., and Lovelace, R.V.E. 1986, *Ap. J.*, **310**, 737.
Fiedler,R.L., Dennison,B., Johnston,K.J., and Hewish,A. 1987, *Nature*, , in press.
Hewish, A., Wolszczan, A., and Graham, D. A. 1985, *M.N.R.A.S.*, **192**, 799.
Wolszczan, A., and Cordes, J. M. 1987, *Ap. J. (Letters)*, , submitted.

THE INFLUENCE OF PROPAGATION THROUGH THE IRREGULAR INTERSTELLAR PLASMA ON VLBI OBSERVATIONS.

B.J.Rickett and W.A.Coles,
Electrical and Computer Engineering Department,
University of California San Diego,
La Jolla, Ca 92093.

It has been known since pulsars were discovered that the interstellar medium (ISM) can cause angular broadening of radio sources. However, the amount of scattering along typical lines of sight to extra-galactic sources is not large enough to cause any loss of visibility, even on intercontinental baselines, unless the frequency is below about 100 MHz. Recent studies have shown that interstellar scattering (ISS) can have considerably stronger effects; this paper addresses their relevance to VLBI.

It is now known that there are two scales of fluctuation in the amplitude variations caused by ISS; a (smaller) diffractive scale and a (larger) refractive scale. For pulsars, which behave as effective point sources for ISS, both scales are observable, causing variability on times of the order of 10 minutes and 10 days, respectively. For extra-galactic sources, even the most compact have angular diameters larger by many orders of magnitude and only the longer term refractive fluctuations are detectable. Indeed such fluctuations are probably the major contribution to "low frequency variability", which had previously been ascribed to intrinsic variations (Rickett, Coles and Bourgois, 1984). Pulsar studies have been used to estimate the wavenumber spectrum of electron density in the ISM at scales of 10^8 to 10^{11} m as a function of galactic coordinates (Rickett, 1977; Cordes, Weisberg and Boriakoff, 1985 - CWB; Hewish, Graham and Wolszczan, 1985; Blandford and Narayan, 1985; Cordes and Wolszczan, 1986; Cordes, Lovelace and Pidwerbetsky, 1986; Coles et al. 1987). Three aspects are relevant to VLBI observations.

First, there are lines of sight through the Galaxy that suffer greatly enhanced scattering (by many orders of magnitude), particularly at low galactic latitudes and through the inner galactic plane. Second, the unexpectedly strong refractive amplitude variations imply that there are random refractive position variations, which are possibly larger than the scattered image itself. Third, the form of the scattered image may, on occasion, take the form of a few discrete beams.

The enhanced scattering on particular lines of sight have already been

M. J. Reid and J. M. Moran (eds.), The Impact of VLBI on Astrophysics and Geophysics, 287–294.

investigated in several VLBI onbservations, (e.g. Dennison et al. 1984). The physical site of the enhanced scattering is unknown. The evidence points to a low latitude distribution (± 100pc) of scattering "clouds", whose mean free path for intersection by a line of sight is about 1kpc (CWB). Specific structures may include the Cygnus loop (responsible for broadening CL4, Geldzahler and Shaffer, 1981). There is weak evidence for enhancements near other supernova remnants (Spangler et al. 1985). Also the source NGC 6334B is heavily scattered and lies in the same direction as a dense HII region and molecular cloud (Rodriguez et al. 1982). The enhanced scattering is to be compared with the background of a more uniform distribution modelled by a ±500pc disc (CWB). The influence of scattering is to cause an (extra-galactic) point source to be scattered into an angular spectrum that has an instantaneous width θ_{scat} (FWHM) and to be displaced from the true source position by an angle $\theta_{ref} \propto \lambda^2$, which varies slowly and randomly with time. The scattering angle θ_{scat} is the most important parameter in determining the ISS effects on VLBI observations. The uniform scattering distribution corresponds to a typical scattering path L= 0.5 (cosec b) kpc and a value:

$$\theta_{scat} \sim 0.7\ (f_{GHz})^{-2}\ (\text{cosec } b)^{0.5}\ \text{mas}. \qquad (1)$$

This equation should be compared with the data in Figure 1 from Cordes, Ananthakrishnan and Dennison (1984), from which we have also estimated equation (1) above. In using it to estimate scattering on a given latitude and longitude one must expect very large upward deviations as the line of sight passes closer toward the inner galactic plane. Equation (1) should be a reasonable estimate down to latitudes of 20 degrees; at lower latitudes an increasing longitude dependence is expected and an increasing chance of encountering enhanced scattering , raising θ_{scat} by more than 2 orders of magnitude; see also Rao and Ananthakrishnan (1985).

We now discuss the conditions under which ISS will be resolvable in VLBI (i.e. when ISS will cause loss of visibility on a given baseline). We follow the approach of Cohen and Cronyn (1974) in discussing whether there will indeed be a loss of fringe visibility in a given situation. The basic condition, of course, is that the baseline should be longer than the field "coherence scale" s_o. Expressed in angular coordinates; the angular resolution of the instrument θ_{res} should be less than $\theta_{scat} = (ks_o)^{-1}$, where $k=2\pi/\lambda$. With a long enough baseline, the amplitude and phase at the two antennas will be significantly different, differing by say 1 radian and 50% in amplitude over distance s_o . However for a point source, with no averaging due to bandwidth or integration time, the magnitude of the apparent visibility on a single baseline would be unity. There would also be a random phase. With a receiver bandwidth B > $2c/L(\theta_{scat})^2$, or integration time T > s_o/V or an intrinsic source diameter $\theta_{source} > s_o/L$, the visibility is reduced. Here L is an effective path length through the scattering medium, estimated by L = 500pc(cosec b) and V = 50 km/s is the velocity of the observer with respect to the diffraction pattern. The conditions are plotted in

Figure 2, which displays the variables (a) $\theta_{source}\, f_{GHz}\, L_{kpc}$; (b) $T_{min}\, f_{GHz}$; (c) $B_{MHz}\, L_{kpc}$ plotted against θ_{scat} . So for a given observation we need an estimate of the scattering angle, the path length and the instrumental quantities f_{GHz} ,B_{MHz} and T_{min}. There are separate vertical scales for each of the quantities (a), (b) and (c), which are each plotted as a point at the appropriate θ_{scat}. The interferometer visibility will be reduced when any of the points lie above and to the right of the corresponding lines. It is clear from the figure that the most likely cause of the loss of visibility is angular diameter smearing of the diffraction pattern by different (incoherent) parts of the source, based on angular diameters greater than (say) 0.1 mas.

With a single baseline one can predict whether ISS can reduce the visibility as discussed above. There is also a random phase slope or position shift. However, this is usually given as a position difference from a calibration source. With many baselines a brightness distribution can be estimated. In general the source visibility over baseline γ is multiplied by the visibility function due to scattering $\exp[-(\gamma/s_o)^{\beta-2}]$, where β is the power law exponent defined below. Thus the source brightness distribution will be convolved with the "point spread function" due to scattering. With approprate averaging this is the Fourier transform of the visibility given above; it depends on the wavenumber spectrum of the ISM density irregularities $P_N(K)$, where K is the wavenumber. A Kolmogorov spectrum has $P_N(K) \propto (K_o{}^2 + K^2)^{-\beta/2}$ with $\beta = 11/3$. In the absence of averaging, the detailed form of the image will show "speckles" down to the resolution limit. Any value of β between 2 and 4 should give an essentially circular image of the form given above. If the spectrum were steeper (β between 4 and 6) several differences arise. First is the prominence of random refractive shifts by angles larger than θ_{scat}; thus the form of the instantaneous scattering spread function is different from the long term average that includes the image wander. It is thought that the form given above corresponds to the "instantaneous" image, though the question has not been resolved formally. Goodman and Narayan (1985) argue that the image will show a slowly changing ellipticity and may be broken down into a hierarchy of sub-images; their analysis, however, does not appear to address questions of averaging over the source or bandwidth, nor the form of the instantaneous image in the absence of refractive displacement.

On the observational side, there is only little evidence on the scattered image itself. The compact source at the Galactic Center has been resolved and has a wavelength dependent diameter, corresponding to very highly enhanced scattering. In addition the image is elongated by about 1.8:1 (Lo et al. 1985). There are several other observations, which have shown a wavelength- squared dependence of diameter, but few have had a sufficient number of baselines to synthesize an image. See Spangler and Cordes, and Wilkinson, Spencer and Nelson (both in this volume) for recent measurements. Clues about the scattered image have also been assembled from observations of the dynamic ISS spectra of pulsars. The presence of multiple images has been inferred by Cordes and Wolszczan (1986). They suggest that there are times when the image consists of 2 or 3 discrete images, each broadened by scattering and separated from each other by several times their width. Such a situation is inconsistent with a simple Kolmogorov spectrum and requires either a steeper power spectrum (i.e. $\beta > 4$ or a cut-off due to an inner scale) or the chance focussing action of discrete structures in the ISM, superimposed on a background of diffractive scattering. The latter explanation has recently been advanced (Romani, Blandford and Cordes, 1987) following the discovery of unexpected "focussing events" by Fiedler et al. (1987). It would be very interesting to detect such structures with VLBI techniques. In this context, the most convincing proof of a plasma scattering origin for a multiple image would be a dependence of their separation on the square of the wavelength.

We now need an advance in scattering theory to predict the statistical form of scattered images when many antennas are used and self-calibration or other techniques are used to eliminate antenna-based phase errors. The critical questions concern the rate at which scattering distortions of an image will change with time (as the line of sight traverses different sections of the ISM) or with differing angles across the source. We will discuss below apparent position wander, under the simplifying assumption that sufficient baselines have been used with self-cal to construct an image. We proceed by also assuming that the point source image does not change its shape with time and look for departures from the isoplanatic condition on a scale larger than that instantaneous image (i.e. distortions of the separation between components in the source). We examine only the contribution to this distortion from the irregular phase change obtained on a straight line integration through the medium - call this the geometric phase $\phi(x,y)$, where (x,y) are coordinates transverse to the line of sight. The result will be reasonably accurate when the scattered image is well resolved; it is this case that we now examine.

Since a linear phase gradient is removed in self-cal (i.e. the image is centered) only position <u>differences</u> can be detected. Consider two point sources separated by $\Delta\theta$, we want to predict the magnitude of fluctuations in their apparent separation and how quickly such fluctuations change with time and with angle across the sky. This involves first and second order departures from a plane wavefront and depends on the form of the spectrum $P_N(K)$. Using the representation of a thin screen at distance L, we examine the second difference of geometric phase:

$$S(\boldsymbol{\gamma},\boldsymbol{\delta}) = \langle [\phi(\mathbf{r})-\phi(\mathbf{r}+\boldsymbol{\gamma}) - \phi(\mathbf{r}+\boldsymbol{\delta}) + \phi(\mathbf{r}+\boldsymbol{\delta}+\boldsymbol{\gamma})]^2 \rangle \text{, hence}$$
$$S(\boldsymbol{\gamma},\boldsymbol{\delta}) = 2D(\boldsymbol{\gamma}) + 2D(\boldsymbol{\delta}) - D(\boldsymbol{\delta}+\boldsymbol{\gamma}) - D(\boldsymbol{\delta}-\boldsymbol{\gamma}) \text{, where}$$
$$D(\boldsymbol{\gamma}) = \langle [\phi(\mathbf{r}) - \phi(\mathbf{r}+\boldsymbol{\gamma})]^2 \rangle$$

We consider the baseline magnitude = $\gamma > s_o$, so that the scattered image is resolved and $\delta = L\Delta\theta$. We find the error θ_{ref} in $\Delta\theta$ due to refractive distortion of the field using $k\theta_{ref} \sim$ (average difference over $L\Delta\theta$ in the phase gradient over baseline γ), which gives $\theta_{ref} \sim [S(\gamma, L\Delta\theta)]^{0.5}/k\,\gamma$. With the baseline limited to a maximum of 10^4 km and $L\Delta\theta > L\theta_{scat}$, it follows that $\gamma << \delta$.

When the spectrum has a power law exponent $2 < \beta < 4$, $S(\gamma,\delta) \sim 2D(\gamma)$ independent of δ. This gives $\theta_{ref} \sim \theta_{res}\,(\gamma/s_o)^{(\beta-2)/2} < \theta_{scat}$. In words, the refractive error in the separation of sources $\Delta\theta$ is larger than the angular resolution θ_{res} of the interferometer and smaller than the scattering angle. We can also analyze the rate at which this changes either across the sky or with time as the line of sight intersects differing portions of the ISM, and find that the decorrelation angle for θ_{ref} is approximately equal to θ_{ref} itself, and the decorrelation time would be $L\theta_{ref}/V$. With typical coherent integration times of minutes for VLBI, there would be detectable changes in component separation over times of the order months, but by angles smaller than the scatter broadening itself. As emphasized at the outset, in practice only the enhanced scattering is likely to be resolvable with earth-based interferometry and so too, refractive effects would only be important along such heavily scattered lines of sight.

For steeper power law exponents $4 < \beta < 6$, the quantity $S(\gamma,\delta)$ depends on both γ and δ, for $\gamma << \delta$; however, the phase gradient $[S]^{0.5}/\gamma$ is then independent of γ, giving $\theta_{ref} \sim \theta_{scat}(\Delta\theta/\theta_{scat})^{(\beta-4)/2}$. Thus the refractive errors in separation are larger than the instantaneous scattering angle and increase with separation $\Delta\theta$, (slowly for $\beta \sim 4.3$ as has been proposed by Goodman and Narayan, 1985). The angular decorrelation scale for the refractive errors would be $\Delta\theta$ and the decorrelation time would be $L\Delta\theta/V$. The refractive errors would saturate at separations such that $L\Delta\theta \sim$ (outer scale of the ISM irregularities). The third possible wavenumber spectrum for the ISM that has been proposed involves an inner scale at about 10^9 m (Coles et al. 1987). This scale is larger than s_o and is specified in such a way that at the distance of a typical pulsar, the inner scale will cause interference - that is it will "focus". Under these conditions, that pertain to a frequency somewhat below 1 GHz, the scattering angle and refraction angle are about the same.

We conclude that there is indeed a predicted difference between the spectral models in regard to refractive distortion in the field of view; it is possible that low frequency VLBI may be able to resolve such differences and so provide a constraint on the wavenumber spectrum for the ISM. The measurements of Gwinn et al. in this volume have provided an upper limit on the amount of angular wander of 20 μas over 9 months in the Sag B2 H_2O maser source, which has an apparently scatter-dominated diameter of about 0.3 mas. For the Kolmogorov type spectra we only predict that $\theta_{ref} < 0.5$ mas. The $\beta = 4.3$ model predicts $\theta_{ref} \sim \theta_{scat}\,(VT/L\theta_{scat})^{(\beta-4)/2}$. Taking an 8 kpc distance, the 9-month observing time gives a refractive error about equal to the scattering angle. However, the critical calculation is how quickly does the refractive error change, since the true positions are not known. This will be not be dealt with here. In addition the issue of the shape of the scattered image of a point source may be measurable, though the intrinsic source averaging may smear such effects.

REFERENCES

Blandford, R.D. and Narayan, R.:1985, Mon. Not. Roy. Astro. Soc., **213**, 591.

Cohen, M.H. and Cronyn, W.M.:1974, Astrophys. J., **192**, 193.

Coles, W.A., Frehlich, R.G., Rickett, B.J. and Codona, J.L.:1987, Astrophys. J., **315**, 666.

Cordes, J.M., Ananthakrishnan, S. and Dennison, B.:1984, Nature, **309**, 689.

Cordes, J.M., Pidwerbetsky, and Lovelace, R.V.E.:1986, Astrophys. J., **310**, 737.

Cordes, J.M. and Wolszczan, A.:1986, Astrophys. J. Lett., **307**, L27.

Cordes, J.M., Weisberg, J.M. and Boriakoff, V.:1984, Astrophys. J., **288**, 221.

Dennison, B., Thomas, M., Booth, R.S., Brown, R.L., Broderick, J.J and Condon, J.J.:1984, Astrophys. J., Astron. Astrophys., **135**, 199.

Fiedler, R.L., Dennison, B., Johnston, K.J. and Hewish, A.:1987, Nature, in press.

Geldzahler, B.J. and Shaffer, D.B.:1981, Astrophys. J., **248**, 132.

Goodman, J.J., and Narayan, R.:1985, Mon. Not. Roy. Astro. Soc., **214**, 519.

Hewish, A., Wolszczan, A. and Graham, D.:1985, Mon. Not. Roy. Astro. Soc., **213**, 167.

Lo, K.Y., Backer, D.C., Ekers, R.D., Kellerman, K.I., Reid, M.J. and Moran, J.M.:1985, Nature, **298**, 825.

Rao, A.P. and Ananthakrishnan, S.:1984, Nature, **312**, 707.

Rickett, B.J.: 1977, Ann. Rev. Astron. Astrophys., **15**, 479.

Rickett, B.J., Coles, W.A. and Bourgois, G.:1984, Astron. Astrophys., **134**, 390.

Rickett, B.J.: 1986, Astrophys. J., **307**, 564.

Rodriguez, L.F., Canto, J. and Moran, J.M., Astrophys. J.:1982, **225**, 103.

Romani, R.W., Blandford, R.D. and Cordes, J.M.:1987, submitted to Nature.

Spangler, S.R., Mutel, R.L., Benson, J.M. and Cordes, J.M.:1985, Astrophys. J., **301**, 312.

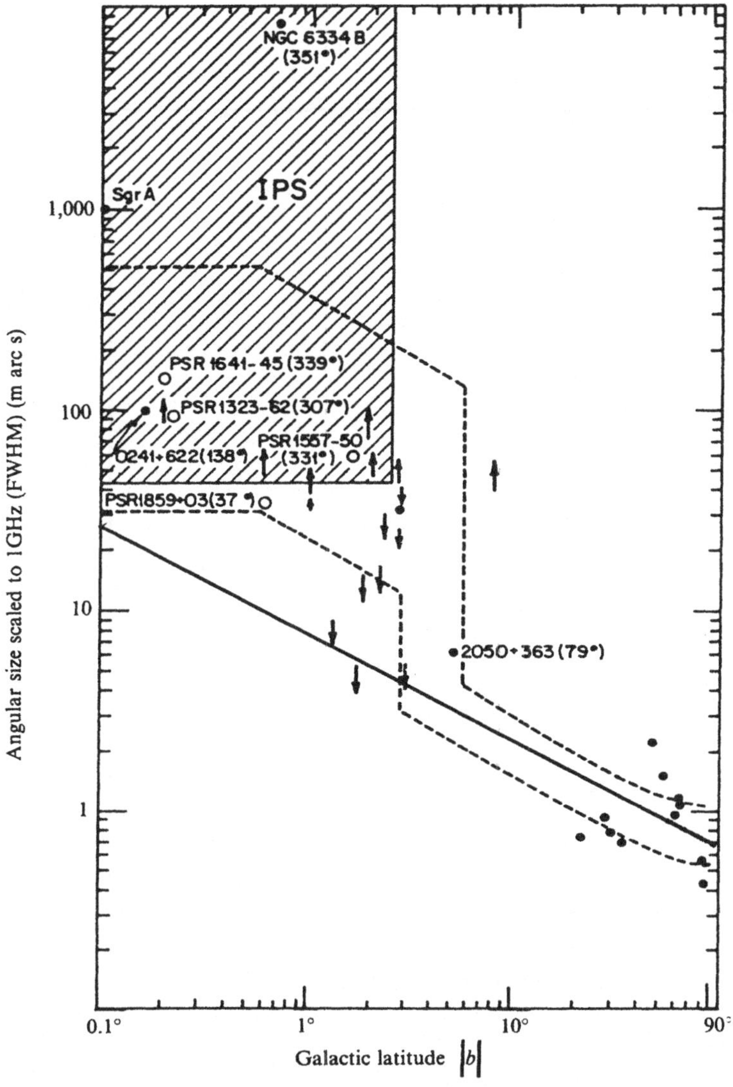

Figure 1. Scattering diameters, scaled to 1 GHz versus magnitude of galactic latitude for the inner galaxy, from Cordes et al. (1984). The filled circles and arrows represent values and limits from VLBI at 408 MHz by Dennison et al. (1984). The point for NGC 6334B is from Rodriguez et al. (1982). The open circles are implied angles for a plane wave source seen through 5-10kpc of scattering material, derived from particular pulsar ISS measurements and the dashed lines are from a survey of pulsar ISS (CWB). The shaded region is from interplanetary scintillation observations of sources toward the galactic center by Rao and Ananthakrishnan (1984). Note the large excursions upwards from the solid line that represents equation (1).

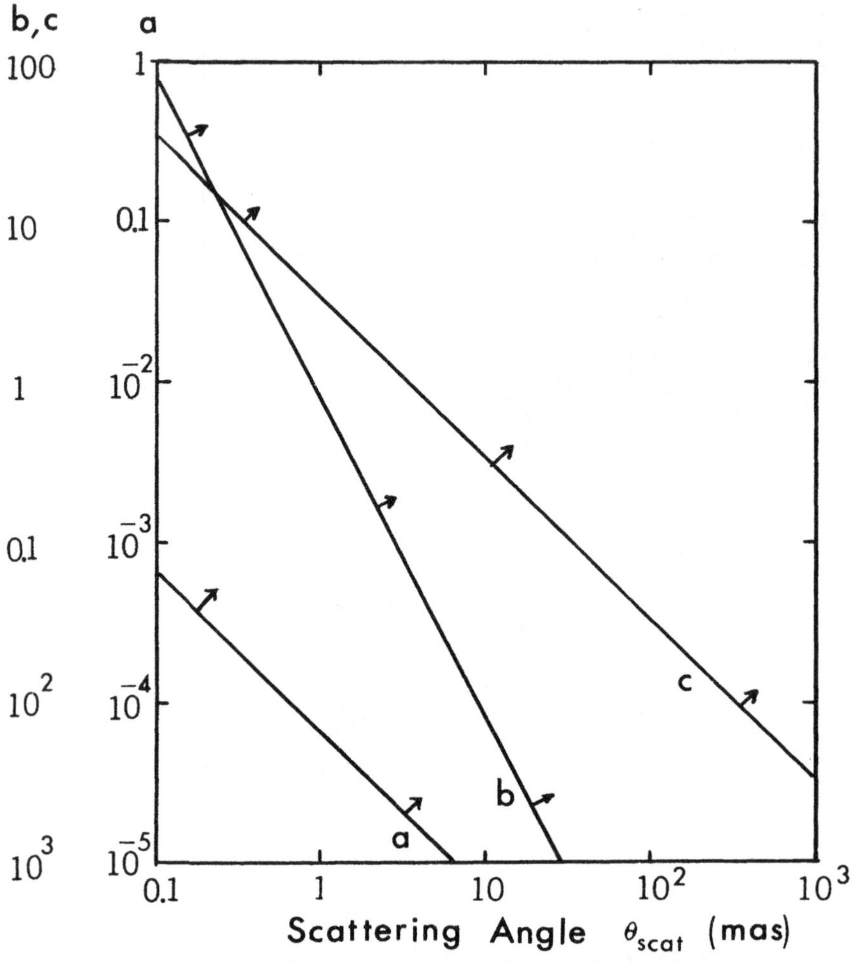

Figure 2. The various contributions to loss of visibility due to averaging of the ISS diffraction pattern, following the method of Cohen and Cronyn (1974). The baseline γ is assumed larger than the field coherence scale s_o. The observing conditions are plotted as 3 points versus the expected scattering angle. Using scale (a) the product $\theta_{source}\ f_{GHz}\ L_{kpc}$ (diameter in mas) is plotted; loss of visibilty occurs if the point lies above and to the right of line (a). Using scale (b) the product $B_{MHz}\ L_{kpc}$ is plotted; loss of visibility occurs if the point lies above and to the right of line (b). Using scale (c) the product $T_{min}\ f_{GHz}$ is plotted; loss of visibility occurs if the point lies above and to the right of line (c). Note that for most extra-galactic sources, source diameter averaging dominates and ensures that the visibility will be reduced by $\exp[-\gamma/s_o{}^{\beta-2}]$.

Limits on Refractive Interstellar Scattering of H_2O Masers

C.R. Gwinn, J.M. Moran, M.J. Reid and M.H. Schneps
Harvard-Smithsonian Center for Astrophysics

We limit the spectral index of density fluctuations in the interstellar plasma by comparing maser sizes, due to small-scale fluctuations, with wander of masers from constant velocity, possibly due to refraction by large-scale fluctuations. The wander of the masers about constant-velocity motion limits power-law spectra of the fluctuations to indices less than 4.0.

1. INTRODUCTION

Refractive effects in interstellar scattering have recently attracted much interest, motivated by the possibility that they may explain low-frequency variability of quasars and pulsars, and flicker of quasars at decimeter wavelengths (Rickett, Coles and Bourgois 1980). Large-scale ($\approx 10^{14}$ cm) density fluctuations could produce the observed effects by focusing or defocusing radiation from the sources. Such fluctuations should also act as prisms, producing an apparent time-variable deflection of the source. We place limits on such deflections.

2. OBSERVATIONS AND ANALYSIS

The data discussed here were obtained from three VLBI observations of the water masers in Sgr B2(N) over a period of six months. We used antennas at Haystack, Green Bank, the VLA and Owens Valley. The minimum fringe spacing was 700 microarcseconds (μas). We phase-referenced all frequency channels to a reference channel, and mapped the masers with fringe-rate mapping. We determined sizes and more precise positions by fitting in the (u, v) plane. We then removed differential precession and nutation, and fitted for proper motions. Since the beam was elongated in declination, our fits were much more accurate in right ascension, and we discuss that coordinate here. Estimated random and systematic errors were less than 10 μas.

The fitted sizes of the masers in Sgr B2(N) ranged upward from a sharp cutoff at 300 μas (FWHM), which we ascribe to interstellar scattering. It is consistent with sizes of OH masers in Sgr B2 at 18 cm and the scaling of size θ with wavelength λ as $\theta \propto \lambda^2$ expected for interstellar scattering (Moran 1968), and is larger than minimum sizes in closer maser clusters.

M. J. Reid and J. M. Moran (eds.), The Impact of VLBI on Astrophysics and Geophysics, 295–296.

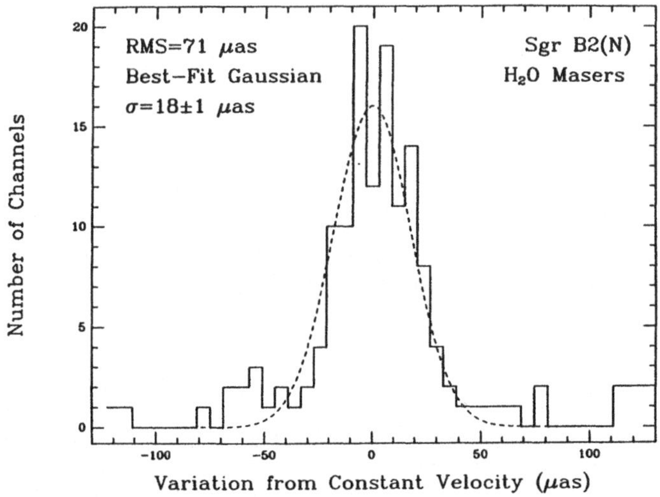

Figure The histogram shows residuals for the fits to constant-velocity motions for H_2O masers in Sgr B2(N); the dashed line shows the best-fitting Gaussian distribution.

We fit constant-velocity motions to positions of maser features in 49 frequency channels and positions on the sky, which appeared at all three epochs with flux densities of 3 Jy or more. The fitted motions are consistent with physical motions of the masing clouds (Reid *et al.* 1987). The figure shows a histogram of the residuals, corrected for the reduction in rms by the fit. The best-fitting Gaussian, which is sensitive to the width of the central peak, has a standard deviation of $18 \pm 1\mu$as. Maser wander due to refractive effects is generally taken to be a Gaussian process, as would result from the incremental effects of many scattering screens along the line of sight. We can place an upper limit of 20 μas on any such wander. The wander is expected to be of the order of the scattering disk size for power-law density fluctuation spectra with indices greater than 4.0 (Rickett 1986), so the index must be less than that. Comparison with quantitative theory favors the Kolmogorov index of 3.67 (Romani, Narayan and Blandford 1986).

3. CONCLUSIONS

The scattering size and observed wander for H_2O masers in Sgr B2 require any power law spectrum to have an index of less than 4.0, and are consistent with a Kolmogorov spectrum of density fluctuations.

4. REFERENCES

Moran, J. 1968, Ph.D. thesis, Massachusetts Institute of Technology.

Reid, M.J., Schneps, M.H., Moran, J.M., Gwinn, C.R., Genzel, R., Downes, D., and Rönnäng, B. 1987, in preparation.

Rickett, B.J. 1986, *Ap. J.*, **307**, 564.

Rickett, B.J., Coles, W.A., and Bourgois, G. 1980, *Astr. Ap.*, **134**, 390.

Romani, R.W., Narayan, R., and Blandford, R. 1986 *M.N.R.A.S.*, **220**, 19.

THE LOW FREQUENCY VARIABILITY OF EXTRAGALACTIC RADIO SOURCES: A RELATIVISTIC EFFECT OR GALACTIC SCINTILLATION ?

L. Padrielli[1], R. Fanti[1], A. Ficarra[1], L. Gregorini[1], F. Mantovani[1], S. Spangler[2]

1. Istituto di Radioastronomia, Bologna, Italy
2. Dep. of Physics and Astronomy, University of Iowa, USA

Results obtained on the Low Frequency Variability (LFV) phenomenon, by means of combined multifrequency observations of 50 sources, on a period of more than ten years on a frequency grid of 0.4, 2.3, 4.8, 8.0, and 14.4 GHz and two epoch VLBI observations at 18 cm can be summarized as follows:

1. 15-20% of variables appear to have variations consisting either of quasi-simultaneous outbursts at all frequencies or of bursts which drift to lower frequencies with time and decreasing amplitude. In our sample, we find five good cases: 3C 120, 0605-085, 1510-089, 3C 345, BL Lac. Three of these are famous superluminals; the other two show significant structural changes between our 18 cm VLBI measurements. The corresponding expansion rate for these five sources is in agreement with the γ's derived from LFV with the usual causality arguments. For the sources of this class, the observations are therefore in agreement with models that explain the phenomenon of the variability as synchrotron emission of relativistic electrons beamed in a direction close to the line of sight.

2. 35 % of variables show only low frequency (<1 GHz) variability and little or no intermediate high frequency variations. In DA 406, prototype of the category, no superluminal motions have been observed, even if the resolution of our VLBI observations should allow the detection of the structural change expected on the basis of intrinsic LFV. In this case we do not find direct evidence of relativistic motions associated with the LFV and the process is most easily explained if the variations are extrinsic (propagation effects through the interstellar medium as the slow refractive scintillation).

3. The remaining 40-45% of variables show uncorrelated high (>5 GHz) and low frequency variability with a minimum of activity at the intermediate frequencies. The explanation of the phenomenon is less clear. It could be attributed to intrinsic (superluminal) variations at high frequencies, coexisting with unrelated processes at low frequencies.

It is quite likely that both intrinsic and extrinsic processes are at work in the LFV.

M. J. Reid and J. M. Moran (eds.), The Impact of VLBI on Astrophysics and Geophysics, 297–298.

We have accepted as "bona fide" intrinsic variables the five sources of the first category; we ask whether the remaining sources are all explained by the scintillation model and what parameters are required by the interstellar medium.

The variability measurements are characterized by two parameters: the scintillation index (m=r.m.s./average flux), and the time scale (t). We statistically derived these parameters for the 50 LFV sources monitored at 0.4 GHz, by means of structure function and autocorrelation function. We have found that 30% and perhaps as many as 50% of the sources show two distinct time scales. The rms flux variability associated with each of the two time scales has been evaluated and two variability indices m short and m long have been computed.
In the light of the predictions of a slow refractive scintillation model (Rickett, 1986, Ap. J, 307, 564), a relation has been derived between the observable quantities (m, t, and galactic latitude (b) of the source) and the properties of the medium.

$$m\ t\ (\sin b)^{3/2} = f(\text{medium})$$

The quantity m*t is, to first order, independent of source characteristics.
The distribution of the m*t associated with the shorter time scales follows the expected $\sin b^{-3/2}$ law with a large scatter. Noticeable exceptions are present: DA 406 and 1117+14 vary too slowly or too much. Furthermore two sources with low galactic latitude in the direction of the galactic anticenter have very small m*t. This latter anomaly is not unexpected, owing to the finite size of the Galaxy.
The products m*t associated with the longer time scales do not show a clear correlation with the galactic latitude.

It could be argued that the large scatter results from merging all sources with the same absolute magnitude of b; the characteristics of the medium could be different for positive or negative b and for different galactic longitude. To circumvent this problem we have considered groups of sources near each other within a few degrees; the radiation of the sources should pass through a medium with similar characteristics and m*t should be similar. The data only weakly support this hypothesis, because the scatter does not decrease substantially. In particular DA 406 and 3C 345, 8 deg. close in the sky, show definitely different variability behaviour, even if their low frequency VLBI structures are very similar on a scale of 10 mas.

Finally we want to remark that the variability parameters related to propagation effects show m*t values less than those quoted by Rickett (1986). This requires some modification of the medium model.

CONCLUSION

15-20% of LFV sources give indications of intrinsic variability. A component of LFV can be due to propagation effects through the Galaxy. However several sources show a longer time scale variability, whose nature is not understood: they could define a new class of intrinsic variables.

VLBI AND EXTREME SCATTERING EVENTS

Brian Dennison
Physics Department
Virginia Polytechnic Institute and State University
Blacksburg, VA 24061 USA

R. L. Fiedler and K. J. Johnston
E. O. Hulburt Center for Space Research
Code 4130
Naval Research Laboratory
Washington, DC 20375 USA

ABSTRACT. Ray tracing numerical simulations of extreme scattering events result in light curves that display many of the qualitative features observed in such events. The apparent images of an affected source show strong and variable distortions, which could be readily discerned with VLBI observations.

An extreme scattering event (ESE) is evidently caused by an interruption of the line of sight to a compact source by a dense compact structure in the interstellar medium (Fiedler *et al.* 1987a,b). The resulting refraction produces strong fluctuations in the received flux density. An affected source should therefore exhibit major, and variable, distortions in its apparent structure, during the course of such an event.

To gain a qualitative understanding of the appearance of an affected source, ray tracing calculations were carried out for various types of intervening ionized clouds. The background source was assumed to be a circularly symmetric Gaussian, with a half-power diameter of 0.8 mas. The calculations were carried out in the limit of geometric optics, with the refraction angles calculated from the impulse approximation.

A model that qualitatively reproduces many of the observed features of the ESE seen in 0954 + 658 (Fiedler *et al.* 1987a,b) involves a highly irregular cloud, in which the two-dimensional column density profile was synthesized from 12 randomly chosen Gaussian distributions. The overall properties of the cloud are given in Table 1.

The apparent brightness distributions, and light curves (Figure 1), were computed for a range of impact parameters and at two frequencies (2.7 and 8.1 GHz). Not surprisingly, major distortions were present, including multiple imaging and elongations. In Figure 2, we show contour diagrams of the apparent brightness distribution at 8.1 GHz, for a range of offsets between the source and cloud center. At 2.7 GHz, the apparent brightness distribution consists of the source plus scattered radiation just before and after the event, and scattered radiation during the event.

We conclude that synoptic VLBI observations, conducted during the course of an event, can directly test the refractive focusing model for ESE's and provide important information on the structure of the refracting clouds.

M. J. Reid and J. M. Moran (eds.), The Impact of VLBI on Astrophysics and Geophysics, 299–300.

Table 1
Irregular Cloud Model

Mean column density $= 5 \times 10^{15}$ cm^{-2}	Distance = 1000 pc
RMS column density $= 1.4 \times 10^{16}$ cm^{-2}	Approximate Diameter = 6 A.U.

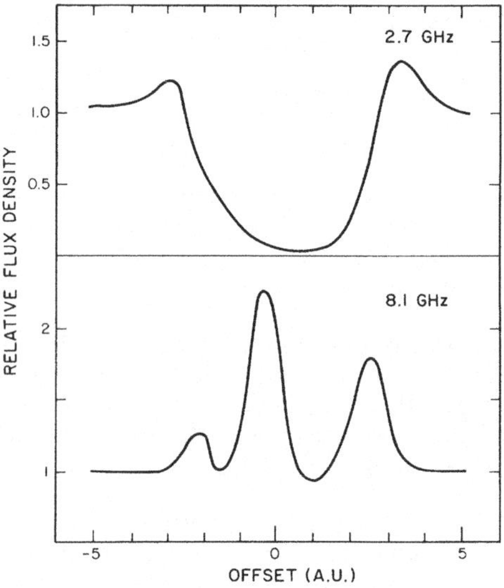

Figure 1. Light curves of a simulated ESE.

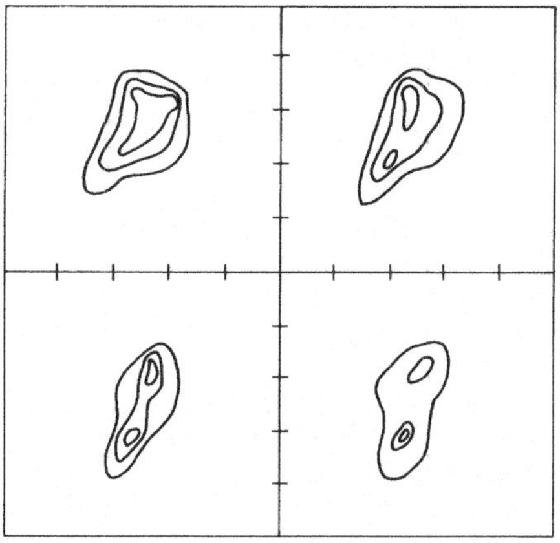

Figure 2. Contours of apparent brightness at 8.1 GHz, for source-cloud center offsets of (left to right from upper left) –0.5, 0, 0.5, 1.0 A.U. Spacing between tick marks equals 1 mas (equivalent to 1 A.U. at the cloud).

4. REFERENCES

Fiedler, R. L., Dennison, B., Johnston, K. J., and Hewish, A. 1987a, *Nature*, **326**, 675.

Fiedler, R. L., Dennison, B., and Johnston, K. J. 1987b, this symposium.

EXTREME SCATTERING EVENTS

Ralph Fiedler[1], Brian Dennison[2], and Kenneth Johnston[1]

[1] E. O. Hulburt Center for Space Research
Naval Research Laboratory, Washington, DC, USA

[2] Department of Physics, VPI&SU, Blacksburg, VA, USA

Abstract. Daily flux density measurements of 36 extragalactic radio sources over a seven year period, obtained by the Green Bank interferometer, reveal several unusual minima in the light curves that do not follow typical source variations (Fiedler et al. 1987). The most significant departure from typical source variability occurred at both frequencies in the quasar 0954+658 between 1980.95 and 1981.3. Refractive focussing by small scale inhomogeneities in an ionized structure in the interstellar medium appears to be the most likely explanation.

Unprecedented flux density variations have been observed in the radio light curves of 0954+658, 1502+106, and 1611+343. These sources are all between 40 and 50 degrees galactic latitude. The best case, 0954+658, is illustrated in Figure 1 where the data is plotted for the entire 7-year observing interval. It is clear from the figure that the typical variations of 0954+658 bear no relation to the event between 1980.95 and 1981.3.

The unusual variations appear to be indicative of an occultation event. The rapid decline of the 2.7 GHz flux density just prior to the period of low, constant flux, and the subsequent rapid rise together with a VLBI angular size (Pearson and Readhead 1981) provides information on the proper motion of the occulter relative to the line of sight. The proper motion is estimated to be 0.09 mas/day and implies a transverse velocity of 500c at the quasar. We consider as unlikely the possibility that the occulter is local to the source, or at a cosmological distance.

An occulter that is local to our galaxy with a transverse velocity $v < 200$ km/s seems more likely. When combined with the estimated proper motion we obtain a distance limit of $D < 1.3$ kpc. From the duration of the minimum we estimate an upper limit to the scale sizes of the 2.7 GHz footprint, at the earth. The velocity

M. J. Reid and J. M. Moran (eds.), The Impact of VLBI on Astrophysics and Geophysics, 301–302.

and the duration of the prolonged minimum, about 60 days, give s < 7 AU. This dimension should be comparable to the physical size of the occulter.

The irregular fluctuations observed at 8.1 GHz and the more regular variations at 2.7 GHz may be explained using refractive effects through irregularities in ionized gas density in the occulter. The estimated upper limit to the ionized particle density is 4000 cm^{-3}. If the irregularities are extended along the line of sight, then the density can be much lower.

Three events observed spanning 395 days in about 160 source-years implies a number density of 120 pc^{-3} f^{-1}, if the occulters are a halo population objects. This is about 1000 times the number density of stars in the galactic disk. The dimensionless factor, f, is included to allow for the possibility that the occulters may be filamentary, and is the ratio of the linear extent to the width.

Clearly, the nature of these objects still needs to be elucidated by observations of additional events.

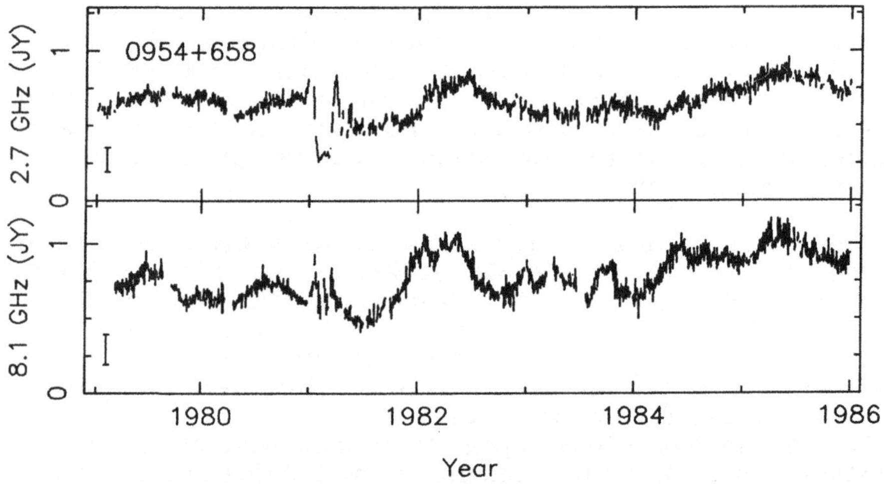

Figure 1. The 2.7 and 8.1 GHz flux densities of the quasar 0954+658 are plotted from 1979 through 1985. The error bars are twice the rms error in height. The unusual variations in the light curves between 1980.95 and 1981.3 may be an extreme scattering event caused by the passage of a cloud in our own galaxy across the line of sight to the quasar.

References:

Fiedler, R.L., Dennison, B., Johnston, K.J., and Hewish, A. 1987, Nature, **326**, 675.

Pearson, T.J. and Readhead, C.S. 1981, Ap.J., **248**, 61.

MEASUREMENTS OF INTERSTELLAR SCATTERING IN THE CYGNUS REGION

S. Spangler, A. Fey, and R. Mutel
Department of Physics and Astronomy
University of Iowa
Iowa City, IA 52242

ABSTRACT

We report progress on a study of interstellar scattering in the direction of the constellation Cygnus. A high quality, 1663 MHz VLBI image has been obtained of the radio source 2013+370, whose structure at this frequency is dominated by scattering. The visibility function indicates that the spectral index of the interstellar density irregularities is between 3.7 and 4.0, with the higher value being somewhat more likely. Another project consists of multifrequency observations of eight sources in this region. Pronounced changes in the scattering are observed for sources separated by only a few degrees.

1. GENERAL REMARKS

Many types of radioastronomical observations reveal the presence of plasma turbulence in the interstellar medium. Since angular broadening is one of the phenomena resulting from wave propagation through a random medium, VLBI observations can assist in the study of interstellar turbulence. A VLBI observation of angular broadening gives us a measurement of the path-integrated "strength of scattering". Comparison of different lines of sight therefore permits us to identify astronomical objects or regions characterized by particularly heavy scattering. This allows us to discriminate between turbulence distributed along the line of sight and that which is concentrated in one or a few regions. Furthermore, the interferometric visibility of a point source viewed through a turbulent medium is directly related to the structure function of the density irregularities. A careful measurement of a compact radio source may therefore permit us to deduce the spatial spectrum of the density irregularities. This, in turn, provides valuable information on the physical processes operative in forming the irregularity spectrum.

M. J. Reid and J. M. Moran (eds.), The Impact of VLBI on Astrophysics and Geophysics, 303–304.

2. VARIABILITY OF SCATTERING ON CLOSE LINES OF SIGHT

We are in the process of obtaining three-frequency scattering measurements for nine sources in the direction of Cygnus. For one of the sources (2013+370, see below) the observations are complete. For the remaining eight objects, observations at 5 and 1.6 GHz have been acquired and partially analysed. Observations at 0.61 GHz are awaiting scheduling. The nine sources are 1923+210, 1954+513, 2005+403, 2013+370, 2021+317, 2022+542, 2023+336, 2048+313, and 2113+293. We believe we have detected interstellar scattering for five of the sources: 2005+403, 2013+370, 2021+317, 2023+336, and 2048+313. Interstellar scattering is manifest by a λ^2 dependence of the angular size on wavelength, and a size which is much larger than observed for high-latitude sources with similar spectra and variability characteristics.

A striking result of our preliminary investigations is that the amount of angular broadening changes substantially for lines of sight separated by only a couple of degrees, suggesting that the scattering is dominated by a single region. However, the nature of these regions is uncertain. The sources 2013+370 and 2048+313 lie behind the Cygnus OB1 association and the Cygnus Loop supernova remnant, respectively, yet are less scattered than nearby sources whose lines of sight do not intersect such obvious "extreme Population I" objects.

3. SCATTERING OF THE RADIO SOURCE 2013+370

In previous observations it was found that the source 2013+370 displayed broadening due to interstellar scattering. We therefore undertook a six-station observation at 1663 MHz to obtain a good map of this object and to acquire information on the form of the density irregularity spectrum.

The image of the source revealed two striking attributes. The first is its asymmetry. A Gaussian fit to the Cleaned map yields a major axis of 17 milliarcseconds, a minor axis of 12 milliarcseconds, and a position angle of 14 degrees. Since this is the same position angle as observed for the intrinsic structure at a higher frequency, it seems likely that this asymmetry reflects underlying intrinsic structure. However, this statement is not completely certain. The second noteworthy feature of the image is its smoothness; there is no evidence of multiple images, as might be expected if the irregularity spectrum were quite steep, with attendant strong refractive effects. Comparison of the observed visibility with that given by power law turbulence indicates a best fit for a power law index of between 3.7 and 4.0.

INTERSTELLAR SCATTERING TOWARDS CYG X-3

P.N. Wilkinson, R.E. Spencer & R.F. Nelson
University of Manchester, NRAL, Jodrell Bank

MERLIN observations at 0.408 GHz of the December 1985 outburst of Cyg X-3 have enabled us to characterise the angular broadening, caused by interstellar scattering, very accurately. The scattering along the line of sight towards Cyg X-3 is exceeded by only two other lines of sight through the Galaxy: towards the Galactic centre (see e.g. Lo et al. 1985. Nature, 315, 124) and towards NGC 6334 (Rodriguez et al. 1982. Astrophys.J., 225, 103); as a result quite short baselines are needed to study the scattering disk at 73 cm wavelength (0.408 GHz). The projected MERLIN baselines for our observations cover the range from a few km (on which the source was unresolved and had a correlated flux density $\sim$3.6 Jy) to $\sim$130 km (on which the source was totally resolved i.e. a correlated flux density $\leq$0.030 Jy). The basic result from these observations is that over this range of baselines the scattering appears to be purely diffractive in character. The scattering disk is, to quite a good approximation, a circular gaussian and shows no evidence of fine scale substructure. Model fitting to the visibility amplitudes, assuming circular symmetry, yields $\theta_{0.408GHz} = 2.85\pm0.05$ arcsec (FWHM).

However a detailed examination of the data reveals deviations from this simple picture. There is a hint of dipolar distortion (axial ratio 0.93±0.03) in the N-S direction and a slight but significant deviation from a gaussian profile. We have characterised this latter effect quantitatively as follows: for a power law spectrum of turbulence the normalised fringe visibility $\gamma = \exp - (\text{baseline})^n$ thus a plot of $\log(-\ln(\gamma))$ against log (baseline) should yield a straight line with slope n. Here the baseline is measured in wavelengths and the usual power law index β corresponds to n+2 (see the review article by Rickett and Coles in this volume). The MERLIN 0.408 GHz data were plotted in this way out to 100,000 wavelengths (i.e. $\sim$73 km) at which point the fringe amplitude drops below the noise and the data begin to deviate sharply from a straight line. The data with good signal-to-noise ratio are well fitted by a line with slope $n = 1.88\pm0.05$ (i.e. $\beta = 3.88\pm0.05$). This is to be compared with $n = 1.67$ (i.e. $\beta = 11/3$) for the classical Kolmogorov spectrum.

On the basis of diffractive scattering theory the wavelength dependence of the scattering diameter is $\lambda^{\beta/(\beta-2)}$ (e.g. Romani et al.,

M. J. Reid and J. M. Moran (eds.), The Impact of VLBI on Astrophysics and Geophysics, 305–306.

1986, Mon.Not.R.astr.Soc., 220, 19) the predicted dependence is, therefore, $\lambda^{2.06\pm0.03}$. MERLIN observations of Cyg X-3 at 1.66 GHz enable us to measure this wavelength dependence directly even though we were only able to observe the source in its quiescent mode (flux density 0.057 Jy). Fitting a circular gaussian model to the visibility amplitudes yields $\theta_{1.66GHz}$ = 0.154±0.010 arcsec (FWHM) from which we infer that the exponent of the wavelength dependence is 2.08±0.05 i.e. consistent with the value derived from the shape of the scattering disk at 0.408 GHz alone. Note that the expected size at 1.66 GHz for the β = 4.00 case is 0.172±0.003 arcsec while that for β = 11/3 is 0.130±0.002 arcsec. Both are inconsistent with the present 1.66 GHz data, especially the β = 11/3 case.

Re-analysis of the Jodrell Bank data taken at the time of the 1972 outburst (Anderson et al. 1972, Nature Phys.Sci. 239, 117) yields $\hat{\theta}_{0.408GHz}$ = 2.7±0.3 arcsec. The size of the scattering disk has not changed by more than ~10% in the last 15 years.

MERLIN observations of Cyg X-3 during future outbursts should enable us to image the scattering disk directly at 0.408 and 1.66 GHz and to constrain β even more closely.

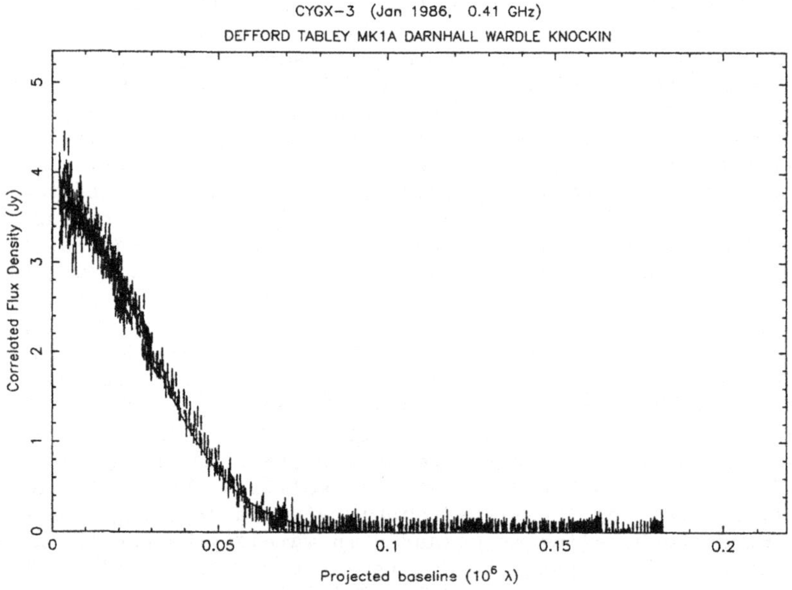

The MERLIN fringe amplitude at 0.408 GHz (in Jy) plotted against projected baseline (in Mλ). The larger scatter on the short baseline data is due to confusion from the Cygnus X region. The signal-to-noise ratio falls below unity just before 0.1 Mλ and so the right hand half of the plot represents only thermal noise. The solid line shows the Fourier transform of a circular gaussian of total flux 3.65 Jy and diameter (FWHM) 2.85 arcsec.

LIMITS TO VLBI RESOLUTION DUE TO SOLAR CORONAL SCATTERING

Stanislaw Gorgolewski
Torun Radio Astronomy Observatory
Nicolaus Copernicus University
Torun, Poland

ABSTRACT. Electron irregularities in the solar corona scatter the radiation from radio sources. During more than 30 years of coronal electron irregularity investigations, a very complex temporal and spatial behavior of the solar corona have been observed. In particular, the scattering dependence on distance from the Sun varies with both the phase of the cycle and the different solar maxima. The influence of scattering on the highest resolution observations in radio astronomy, i.e., those made with VLBI, is not negligible. It should be taken into account in planning of both the global VLBI as well as the future orbital VLBI stations such as the QUASAT and RADIOASTRON missions. The "normal" scattering dependence on the distance from the Sun is close to the inverse square law, but substantial deviations from this dependence have been found.

1. INTRODUCTION

The irregular structure of the solar coronal electron density was discovered in the early fifties, during the search for refraction during the occultation of Tau A by the solar corona. No refraction was found, but scattering, which causes an apparent increase of the observed radio source diameter, was detected. Interferometers of about 100 wavelength spacings were used in these measurements in the meter wavelength range. More recently, small diameter sources have been observed with proportionally longer baselines at decimeter waves. With the discovery of interplanetary scintillations, finer structure in radio sources could be observed; this led the way to the discovery of the pulsars. Since scattering is proportional to wavelength squared, long wavelength observations were preferred for higher sensitivity.

Vast observing materials have become available during the last three decades. Anisotropic scattering provided evidence for the existence of radial magnetic fields in the outer solar corona. With the achievement of high resolution by VLBI, solar coronal scattering gains new importance. It may soon become the chief limiting factor in highest resolution observations at long decimetric waves using the longest ground-based baselines, as well as space VLBI baselines.

M. J. Reid and J. M. Moran (eds.), The Impact of VLBI on Astrophysics and Geophysics, 307–308.

2. RESULTS

Scattering without refraction was found to be the main effect of the passage of radio waves through the solar corona at distances up to several tens of solar radii. These results refer to medium baseline interferometers at meter waves during solar maxima. Interplanetary scintillations can be observed on very small diameter sources at distances comparable to the astronomical unit, i.e., about 200 solar radii. The most surprising result was the discovery made in 1969 by both radio source occultations and interplanetary scintillations that solar coronal scattering did not follow the presumed correlation with the solar activity cycle. It was found in 1970 that the scattering was indeed stronger than ever before observed, i.e., in 1959. It existed only at rather small distances from the Sun. At the distance of 10 solar radii from the Sun, the scattering did not differ from that during the solar minimum. Another surprise came in 1979, when the scattering dependence on distance from the Sun followed an inverse 1.24 power law, the flattest dependence yet observed. One can thus see that the solar corona does not behave predictably. We must monitor the ever changing conditions of the Sun and its influence on the solar corona if we want to avoid surprises.

3. DISCUSSION

The simple idea of an 11-year solar cycle does not fully account for all observed scattering events as related to both the distance and phase of the solar cycle. One should take the 22-year period of solar activity as the full cycle. This total period of solar magnetic activity contains the full cycle of magnetic field evolution. During solar field reversal, the magnetic field lines are wrapped around the corona. Normally, the radial field lines permit the solar wind to flow from the Sun, but during the time of field reversal, the magnetic lines inhibit the flow from the solar regions. Should this pattern of events be the long-term solar behavior, the next most favorable observing period for highest resolution VLBI should come near the next solar maximum, i.e., near 1991. This period is near the time of the expected QUASAT and RADIOASTRON missions. The following solar maximum near the year 2002 may well be the time of the extended solar coronal irregularities—a time not very suitable for highest resolution VLBI.

I thank the I.A.U. for subsistence and a travel grant in the U.S.A. Other travel expenses as well as financial support for research were taken care of by the "Resortowy Program Badan Rozwojowych": RPBR NR. RR. I. 11/2.

4. BIBLIOGRAPHY

Erickson, W. C. 1964, *Ap. J.*, **139**, 1290.
Gorgolewski, S., and Hewish, A. 1960, *The Observatory*, **80**, 99.
Gorgolewski, S., and Krygier, B. J. 1977, *Interdiscipl. Cycle Res.* **8(3-4)**, 189.
Hewish, A. 1958, *M. N. R. A. S.*, **118**, 534.
Hewish, A., Dennison, P. A., and Pilkington, J. D. H. 1966, *Nature*, **209**, 1188.
Hewish, A., Wyndham, J. D. 1963, *M. N. R. A. S.*, **126**, 469.
Little, L. T., Hewish, A. 1966, *M. N. R. A. S.*, **134**, 221.
Okoye, S. E., and Hewish, A. 1967, *M. N. R. A. S.*, **137**, 287.
Slee, O. B. 1961, *M. N. R. A. S.*, **123**, 223.
Vitkevich, V. V. 1960, *A. J. U.S.R.R.*, **37**, 32.

CELESTIAL REFERENCE FRAMES : DEFINITIONS AND ACCURACIES

E M Standish
Caltech / Jet Propulsion Laboratory
Pasadena, CA 91109

ABSTRACT. The determination of a specific catalogue or ephemeris reference frame is a highly over-determined problem, depending on the particular selection of which coordinates, which objects and at what time(s) the determination is made. The consistency which various determinations exhibit is dependent upon the accuracy of the catalogue or ephemeris itself. This paper discusses the accuracies of the three most prominent celestial reference frames: stellar catalogues, the lunar and planetary ephemerides and the radio source catalogues.

The FK4 stellar catalogue contains known systematic errors amounting to a few tenths of an arcsecond; the FK5 will yield nearly an order of magnitude improvement; HIPPARCOS and Space Telescope expect 0".002 by the mid 1990's; optical interferometry should approach 0".01 within a couple of years, tens of micro(!)arcseconds after a couple of decades. Present-day lunar and planetary ephemerides have accuracies at the level of 0".01 for the moon and inner four planets; 0".1 for the outer planets. Further observational data will permit continued improvement. Radio source catalogues now show internal consistency of 0".001.

I. INTRODUCTION

A given celestial catalogue does not explicitly define its own reference frame. Neither does a given set of lunar and planetary ephemerides. The reference frame of a given catalogue is only implied by the actual coordinates of the catalogue. In general, only three independent coordinates at one specific instant of time are necessary and sufficient for defining a frame uniquely. However, when speaking of a reference frame for a catalogue, one envisions some global set of general coordinates, applied over an extended interval of time.

Any actual attempt to determine the orientation of a reference frame for a full catalogue automatically invokes a specific set of weighted coordinates over a specific set of time points. On the other

M. J. Reid and J. M. Moran (eds.), The Impact of VLBI on Astrophysics and Geophysics, 309–315.

hand, during a subsequent application of such a determined orientation, a different weighted selection of coordinates is used. The two sets may not be consistent. That depends upon the accuracy of the catalogue or ephemeris.

Section II presents the known systematic corrections for the FK4 stellar catalogue, estimates the uncertainties expected for the FK5 and mentions the accuracies expected from future improvements to the stellar reference catalogues. Section III gives realistic estimates of errors in modern lunar and planetary ephemerides along with a comparison of two relatively recent, but independently created, ephemerides. Section IV gives accuracies of radio source catalogues, derived from inter-catalogue comparisons.

II. THE FUNDAMENTAL STELLAR CATALOGUES, FK4 AND FK5

The fundamental stellar catalogue, the FK4 (Fricke and Kopff, 1963) has known systematic errors which have been estimated by Fricke and his colleagues at Heidelberg. These may be categorized as global errors, zonal errors and individual errors, both in the listed 1950 positions and in the listed proper motions. Application of these corrections, along with the incorporation of additional observational data, have been used to produce the FK5.

The global systematic corrections to the FK4 are Δp, a correction to Newcomb's value of the general precession in longitude; D, an offset to the declination system; E, the equinox offset at 1950 and $\dot{E}$, the equinox drift. Table I gives numerical values for these

Table I. Known corrections to and estimated uncertainties of the fundamental stellar catalogue, FK4.

Equinox Offset at 1950	E	$= 0''525 \pm 0''045$
Equator Orientation	D	$= 0''00 \pm 0''02$
Equinox Drift	$\dot{E}$	$= (1''275 \pm 0''150)$/cty
1976 Precession Correction	Δp	$= (1''10 \pm 0''15)$/cty
[VLBI Precession Correction]	Δp	$= (0''94 \pm 0''08)$/cty
Zonal uncertainties		$\pm 0''3$, $\pm 0''3$/cty
Individual uncertainties		= 2-3 X Zonal uncertainties

Application of Known Corrections to FK4/5-Based Optical Observations

astrometric : $\Delta\alpha = E + [\dot{E} - \Delta p_\alpha]\,T$; $\Delta\delta = D + [-\Delta p_\delta]\,T$

apparent : $\Delta\alpha = E + [\dot{E}]\,T$; $\Delta\delta = D$

$[\Delta p_\alpha = \Delta p\,(\cos\epsilon + \sin\epsilon \sin\alpha \tan\delta),\ \Delta p_\delta = \Delta p\,(\sin\epsilon \cos\alpha)]$

corrections along with their uncertainties. Included are also uncertainty estimates for the zonal and individual errors. It is seen that, in general, one may expect a few tenths of an arcsecond uncertainty in the FK4.

Subsequent determinations of the precession value from VLBI and from lunar laser-ranging have indicated that Fricke's correction is about 10-15% too high, implying that there is already a known systematic error in that catalogue.

Table I also shows how the known errors enter differently into both astrometric and apparent positions: different types of observations require different corrections when they are compared or reduced to the new J2000 system.

By the mid-1990's, observations from the HIPPARCOS astrometric satellite are expected to give another order of magnitude increase in accuracy, producing uncertainties as low as a few milli-arcseconds in relative positions and a few milli-arcseconds per year in relative proper motions. Space telescope is expected to reach fainter objects, allowing frame orientations using a number of sources available in the radio source catalogue. However, the establishment of inertial proper motions must unfortunately await a future second set of such astrometric missions. Within the next year or so, modest optical interferometers may produce positions of the brighter stars at the level of $0\overset{''}{.}01$ (see Johnston et al., 1987), an accuracy which may eventually reach a few micro-arcseconds by the turn of the century (see, e.g., Boyce and Reasenberg, 1984).

III. LUNAR AND PLANETARY EPHEMERIDES

Modern ephemerides of the moon and inner planets are determined primarily by the accurate ranging measurements to which they are adjusted. It has been shown elsewhere (Newhall et al., 1983; Standish, 1982a) not only how the relative positions and motions of these bodies are determined by ranging, but also how the ranging determines their mean motion with respect to inertial space. Furthermore, if desired, the orientation of the system can be adjusted to the dynamical equator and equinox of date, from only ranging measurements alone (see Standish, 1982b).

For purposes of comparing an ephemeris with a celestial catalogue, it is desirable to estimate the uncertainties in geocentric coordinates, namely, right ascension and declination, even though these are not the most accurately determined parameters of the ephemerides. The main strength lies in relative positions and motions. Table II, however, shows that even when expressed in right ascension and declination, modern lunar and planetary positions have achieved a remarkable degree of accuracy.

Table II. Estimated realistic uncertainties for present-day ephemerides: geocentric coordinates given with respect to the dynamical equinox. The table is valid over a two-decade time-span surrounding the mean epoch of the lunar-laser ranging data (~1978).

	Dist	R.A.	Dec.	Mean Motion
Moon	2 m	0."005	0."002	[0."5/cty**2]
Sun	200 m	0."005	0."002	0."01/cty
Mars	50	0."005	0."002	0."01/cty
Venus	100	0."005	0."01	0."06/cty
Mercury	100	0."02	0."03	0."14/cty
Jupiter	100 km	0."1	0."1	0."2/cty
Saturn	500	0."1	0."1	0."5/cty
Uranus	1000	0."2	0."2	1"/cty
Neptune	10000	0."2	0."2	1"/cty
Pluto	30000	0."5	0."5	2"/cty

The lunar laser-ranging data is sensitive to both the of-date equator (station locations) and to the ecliptic (solar perturbations). With these data now covering nearly the full 18.6-year period of the precession of the node, the alignment of the equator to the ecliptic approaches an accuracy of 0."001. However, a constant offset in longitude introduces uncertainties of 0."005 and 0."002 in right ascension and declination, respectively. Mars is determined relative to the earth-sun system by means of the accurate spacecraft ranging; Venus and Mercury less accurately by radar ranges. As described by Williams (1982), the planetary mean motions are affected most by uncertainties in the masses of many asteroids. The error of 0."01/cty applies only to the mean epoch of the accurate ranging data (~1978); it will grow to about 0."03/cty for epochs about a decade before or after this date. For epochs remote from present, the uncertainty of the lunar tidal deceleration in longitude produces an error for the moon which is quadratic in time. New data types for the outer planets are rapidly improving these bodies' ephemerides, now approaching the level of 0."1 or so throughout the present two decades.

The estimates of ephemeris errors in Table II can in some way be substantiated by comparing two different ephemerides, especially if these have been created independently. This is possible using the Center for Astrophysics' PEP740, created in 1983 and the Jet Propulsion Laboratory's DE118, created in 1981. With respect to the values quoted in Table II, both of these ephemerides may be considered somewhat obsolete because of the data to which they are adjusted: we now have the full span of the Viking Lander ranging data, additional radar-ranging, newer data types for the outer planets, nearly 18.6 years of lunar laser-ranging, improved modeling of the asteroid perturbations and improved values for the outer planet masses.

Table III. Geocentric comparison of the CfA ephemeris, PEP740, vs. the JPL ephemeris, DE118, over the time-span, 1969-1983. Comparison is made after the rotation, $R_x(-0\overset{''}{.}0017)\ R_y(-0\overset{''}{.}0058)\ R_z(+0\overset{''}{.}4648)$, was applied to the CfA ephemeris. The planetary coordinates were fit with a linear function, the lunar ones with a quadratic. Only significant terms are shown. The rate terms are centennial.

Body	Distance [km]			Right Ascension			Declination		
	const	rate	s.d.	const	rate	s.d.	const	rate	s.d.
Moon	-.0005	-	.0003	-0".0014 -0".4364/cty**2	0".016	0".0005	-	-	0".0016
Sun	-	-	.05	-	-0".005	0".0002	-	-	0".0001
Mar	-	-	.27	-0".0006	-0".008	0".0009	-	-0".010	0".002
Mer	-.27	-	.57	-	-	0".0021	-	-	0".006
Ven	-	-	.66	-	-	0".0025	-	-	0".005

Table III shows the result of comparing the two ephemerides in distance, right ascension and declination, after the application of a three-axis rotation to the PEP740 position vectors. This was determined by a least squares fit over the time interval of the comparison, 1969-1983, using the geocentric solar coordinates only. The rotation, given in the table, shows less than a 0".01 adjustment to the equator/ecliptic orientation (x- and y- rotations), but a substantial alignment necessary for the zero-point of right ascension: DE118 was aligned to the FK4 equinox through the incorporation of optical data; no specific alignment was attempted for PEP740.

The resulting ephemeris coordinate differences from the comparison were fit with linear functions in time (quadratic for the moon) and are given in the table. The slight mean motion differences for Mars and the sun are consistent with the uncertainties estimated in Table II. The quadratic term for the lunar motion demonstrates the uncertainty of the tidal deceleration. Comparisons for the outer planets show an order of magnitude greater error than what is now expected.

IV. RADIO SOURCE CATALOGUES

The relative positions of nearly 100 radio sources have now been determined with an accuracy of a milliarcsecond or less (Ma and

Shaffer, 1987). Even a comparison of relatively older catalogues has shown agreement to the level of 0".002, after an orientation adjustment had been performed (Ma, 1986; Sovers and Treuhaft, 1987). These catalogues are essentially absolute in declination at the epoch of observation; they inherit any precession error when transformed to a different epoch. Also, since no absolute right ascension information is present in the measurements, it has been the practice to adopt a somewhat artificial definition for the right ascension orientation. For this, the optically determined position of the source 3C273B (Hazard et al., 1971) has been transformed into the J2000 reference system and then adopted as a defining value.

For absolute positions aligned onto the equator and equinox of J2000, then, one expects an error of $\Delta p \sin\epsilon \cos\alpha\, T$ = 0".003 cos or so in declination, where the present uncertainty in precession is taken as 0".05/cty and the time, T is 0.15 centuries. The right ascension uncertainty is dominated by the expected error of about 0".2 in the position of 3C273B. It is argued by Ma and Shaffer (1987) that future catalogues could well abandon such an artificial orientation: 1) the main strength of the VLBI observations lies in relative angular separations, now well below the milliarcsecond level; 2) radio sources are assumed to have no proper motion and therefore could be expressed by time-invariant positions; and 3) a global alignment could be used which does not depend upon observations of a single source (3C273B shows time-dependent source structure).

V. CONCLUSIONS

The stellar reference system presently shows uncertainties up to nearly a tenth of an arcsecond. However, the future shows promise of dramatic increases in accuracy. Similarly, the lunar and planetary ephemerides continue to be improved, now below the level of 0".01. Radio source catalogues have internal consistencies of 0".001 or less; their orientations, however, are less certain.

ACKNOWLEDGMENT: This paper presents the results of one phase of research carried out at the Jet Propulsion Laboratory, California Institute of Technology, sponsored by the National Aeronautics and Space Administration.

REFERENCES

Boyce,P.B. and Reasenberg,R.D., eds.: 1984, "Proceedings of the Workshop on High Angular Resolution Optical Interferometry from Space", Baltimore.

Fricke,W.: 1971, "A Rediscussion of Newcomb's Determination of Precession", Astron. Astrophys. 13, 298-308.

Fricke,W.: 1982, "Determination of the Equinox and Equator of the FK5", Astron. Astrophys. 107, L13-L16.
Fricke,W. and Kopff,A.: 1963, "Fourth Fundamental Catalogue (FK4)", Veroff. #10, Astronomisches Rechen-Institut, Heidelberg.
Johnston,K.J.: 1987, this volume.
Ma,C.: 1986, "The Celestial Reference Frame of the NASA Crustal Dynamics Project", in "The Earth's Rotation and Reference Frame for Geodesy and Geodynamics" (D.McCarthy and A.Babcock, eds.), IAU Symposium #128.
Ma,C. and Shaffer,D.B.: 1987, "The Celestial Reference Frame Defined by VLBI", poster session, IAU Symposium #129.
Newhall,X X, Standish,E.M. and Williams,J.G.: 1983, "DE102: a numerically integrated ephemeris of the Moon and planets spanning forty-four centuries", Astron. Astrophys. 125, 150-167.
Sovers,O.J. and Treuhaft,R.N.: 1987, "Radio Reference Frame Stability from VLBI Data", poster session, IAU Symposium #129.
Standish,E.M.: 1982a, "The JPL Planetary Ephemerides", Cel. Mech. J. 26, 181-186.
Standish,E.M.: 1982b, "Orientation of the JPL Ephemerides, DE200/LE200, to the Dynamical Equinox of J2000", Astron. Astrophys. 114, 297-302.
Williams,J.G.: 1984, "Determining Asteroid Masses from Perturbations on Mars", Icarus 57, 1-13.

TOWARDS THE DEFINITION OF A UNIFIED CELESTIAL OPTICAL/RADIO REFERENCE FRAME

K. J. Johnston, J. Russell
E. O. Hulburt Center for Space Research, Naval Research Laboratory, Washington, DC 20375-5000
Ch. de Vegt, Hamburger Sternwarte
J. Hughes, United States Naval Observatory
D. Jauncey, G. White CSIRO
G. Nicholson, Hartebeestok, S. A.

ABSTRACT. An almost inertial celestial reference frame based upon extragalactic sources is in the process of being established. This reference frame is to be global with a minimum density of one source/100 square degrees. The source positions will be based upon radio observations and will allow optical reference frames to be related to this frame at the 0.03 arc second level.

1. INTRODUCTION

Radio astrometry has made great progress in the past ten years in making accurate positional measurements over large angles. The accuracy of these measurements now appears to be approaching a milliarcsecond. Objects such as extragalactic radio sources make excellent reference points for establishing a reference frame as they should display little or no detectable proper motions.

2. THE PRESENT STATUS OF CELESTIAL REFERENCE FRAMES

The optical reference frame, the FK5, contains about fifteen hundred bright stars located over the celestial sphere and some three thousand fainter stars. The positions of the brightest stars are believed to be known to 0.03 arc seconds at the mean epoch which is about 1950 for most of the stars. The uncertainty in proper motion is estimated at 0.001 arc seconds/year so that at 1987 the apparent positions should be accurate to 0.05 arc seconds.

The radio reference frame has yet to be established. There are various catalogs of from twenty-five to over one hundred extragalactic objects (Robertson et al. 1986; Ma et al. 1986; Fanselow et al. 1984) which report positional accuracies of less than a few milliarcseconds. The majority of the sources are located in the northern hemisphere with some objects having declinations as low as -40 degrees.

3. PROGRAM PLAN

The program is aimed at establishing a reference frame of four hundred or more objects giving a density of at least one source/100 square degrees. The objects will be extragalactic radio sources which display optical emission. The positions will be determined by the radio

M. J. Reid and J. M. Moran (eds.), The Impact of VLBI on Astrophysics and Geophysics, 317–318.

observations. These objects will be selected from the list of radio/optical candidates put foreward by IAU Commission 24 (Argue et al. 1984) and supplemented by other flat spectrum objects chosen from the Parks and Green Bank-MPI surveys.

Optical observations will be carried out using wide field astrograph cameras such as those located at Hamburger Sternwarte in Germany and Black Birch Astrometric Observatory in New Zealand and large 4 meter class prime focus photography. The astrographs will allow a large number of stars to be used to relate the position of the extragalactic radio source on the prime focus plate to the bright stars making up the optical reference frame. The accuracy of the relationship of the extragalactic radio source's position to the stars on the astrograph plate should be better than 0.03 arcseconds. For objects too faint for prime focus plates, CCD observations will be used as was the case for 1928+738 (de Vegt et al. 1987). In this way the positions of the extragalactic sources in the optical reference frame can be determined.

Observations will be obtained over a five year period beginning in July 1987. The radio observations from this campaign will be analyzed for self consistency with observations of all candidate objects obtained at least once a year in the northern hemisphere and observations obtained in the southern hemisphere when possible. The radio positions of sources down to declinations of -40 degrees should approach an accuracy of a milliarcsecond while those south of this declination may be degraded by factors of from 2 to 10. The optical observations will require a minimum of two prime focus plates per source and two wide field astrographic plates of the star field around the source. The optical observations should be on the optical reference frame to an accuracy exceeding 0.05 arcseconds globally.

The radio positions of the objects will be used to establish the reference frame and will be adopted as the positions of the objects. We are assuming that the optical and radio photocenters of the objects are coincident to the measurement accuracies of the reference which will be a few milliarcseconds. The relationship of this reference to the FK5 (or any other optical system, e.g. HIPPARCOS) frame can then be obtained to the accuracy of the optical plate transfer from the 4 meter prime focus plates to the astrograph reference stars (which will be of order less than 0.03 arcseconds). This should allow alignment of radio and optical images at this level which is about one half an Airy Disc of the Hubble Space Telescope.

4. REFERENCES

Argue, A. N. et al.: 1984, Astron. & Astrophys., **130**, 191.

de Vegt, C., Schramm, J., and Johnston, K. J.: 1987, Astron. J., **93**, 261.

Fanselow, J. L., Sovers, O. J., Thomas, J. B., Purcell, G. M., Cohen, E. J., Rogstad, D. M., and Skjerve, L. J.: 1984, Astron. J., **89**, 987.

Ma, C. et al.: 1986, Astron. J., **92**, 1020.

Robertson, D. S., Fallon, F. W., and Carter, W. E.: 1986, Astron. J., **91** 1456.

SOLAR SYSTEM VLBI ASTROMETRY: THE VENUS BALLOON EXPERIMENT

R. A. Preston, J. Ellis, S. G. Finley, C. E. Hildebrand, G. H. Purcell, Jr., C. T. Stelzried, Jet Propulsion Laboratory; R. Z. Sagdeev, V. M. Linkin, V. V. Kerzhanovich, L. R. Kogan, V. I. Kostenko, L. I. Matveenko, R. R. Nazirov, S. V. Pogrebenko, I. A. Strukov, Space Research Institute, Moscow, USSR; J. E. Blamont, Centre National d'Etudes Spatiales, Paris, France; N. Armand, Yu. N. Alexandrov, Institute of Radio Electronics, Moscow, USSR; L. Boloh, G. Laurans, Centre National d'Etudes Spatiales, Toulouse, France; G. Petit, Institut Geographique National, Saint-Mande, France; A. F. Bogomolov, Energy Institute, Ministry of High Education, Moscow, USSR; I. G. Moiseev, Crimean Astrophysical Institute, Simeiz, USSR; F. Biraud, A. Boischot, A. Ortega-Molina, C. Rosolen, Observatoire de Paris-Meudon, Meudon, France; P. Kaufmann, Itapetinga Radio Observatory, Instituto de Pesquisas Espaciais, Sao Paulo, Brazil.

ABSTRACT. A global array of 20 radio observatories was used to measure the three-dimensional position and velocity of the two meteorological balloons that were injected into the equatorial region of the Venus atmosphere by the VEGA spacecraft.

Previous Venera and Pioneer Venus probes into the Venus atmosphere have determined essentially instantaneous vertical profiles of the wind, showing a wind speed ranging from about 100 m/sec at the visible top of the clouds (about 65 km altitude) to near zero at the surface. The VEGA balloons provided a means of measuring variations in the wind and in situ parameters at a nearly constant altitude below the tops of the clouds.

At least three antennas had to observe simultaneously each balloon and its associated flyby spacecraft at 1.7 GHz to determine a complete set of three-dimensional position and velocity components relative to Venus. Plane-of-sky balloon velocity information was obtained by VLBI observations of the balloon carrier signal, where the measured quantity was the difference of the phase rates of the signals received at different sites. The line-of-sight velocity component was derived from measurements of the carrier Doppler shift at a single antenna. Measurements of altitude by a pressure sensor provided an independent means of determining the radial velocity component, and allowed a calibration of the frequency of the onboard oscillator which controlled the balloon transmissions. Plane-of-sky balloon

M. J. Reid and J. M. Moran (eds.), The Impact of VLBI on Astrophysics and Geophysics, 319–320.

position was obtained by VLBI observations of two coherent tones transmitted by the balloons, where the measured quantity was the received phase difference between the two tones, differenced again between sites. The radial component of position was derived from the pressure measurements.

To obtain both components of transverse position and velocity it was necessary to perform the VLBI observations on at least two baselines with large orthogonal components. For these VLBI measurements, simultaneous observations were also made of identical signals from the flyby spacecraft (in the same antenna beam as the balloon). When the balloon and flyby measurements were differenced, the effects of several error sources cancelled, providing precise measurements of balloon position and velocity relative to the flyby. An independent determination of the trajectory of the flyby relative to Venus was used to infer balloon position and velocity relative to a Venus-centered reference frame.

The total array of 20 ground antennas allowed simultaneous observations with good baseline geometry for 22 hours each day. Velocity measurements were made during each transmission from the balloons (every half hour or hour), while position measurements were possible only every 2 hours. For a single transmission, the typical one-sigma accuracies are about 1 m/sec and 15 km, respectively.

Initial analysis of the VLBI, Doppler and in situ data has provided preliminary scientific results. Each balloon was blown westward about 11,500 km (8,000 km on the night side) by zonal winds with a mean speed of about 70 m/sec at the float altitude of 54 km. The wind speed varied as much as several meters per second on time scales of tens of seconds to hours. Correlation of vertical velocity and temperature variations revealed an upward convective mean heat flow of 70 watts per square meter and the existence of gravity waves. There was also possible evidence of a solar-fixed atmospheric feature and topography-induced wind turbulence. Future analysis is expected to provide more complete descriptions of the dynamical processes that drive the atmosphere and the transports of heat and momentum. An understanding of these processes is essential to an adequate explanation of the high temperatures of the lower atmosphere and the high wind velocities of the upper atmosphere.

A SPACEBORNE MULTI-ARM INTERFEROMETER FOR VLF GRAVITATIONAL WAVE DETECTION. (THE SMILE PROJECT)

Allen Joel Anderson
Planetary Geodesy and Geophysics
The University of Uppsala
S-755 90 Uppsala
Sweden

ABSTRACT. This project would be the next step in our ability to detect very low frequency (VLF) gravitational waves and the first committed spaceborne designed experiment. Present Deep Space spacecraft tracking experiments are severely limited in their detection capability. It is proposed to construct a spaceborne multi-arm microwave interferometer using current elements of design applicable for the detection of VLF gravitational waves. The elements are outlined with particular emphasis placed on the utilization of small inexpensive get away special (GAS) modules currently under development at JPL for launch in the 1990's.

1. INTRODUCTION

The search for gravitational waves of astrophysical origin is an outstanding challenge to present day physics and is being persued by eminent experimental teams in several parts of the world. Most efforts aim at the detection of high frequency, broadband pulses emitted by catastrophic event during the collapse of supernovae in our own or nearby galaxies.

Another important class of sources of gravitational waves are primordial events, collapses of nuclei of quasars and galaxies, and a full set of sources during the early epochs of the cosmos. This very low frequency (VLF) component, the cosmic background of gravitational radiation, bathes the solar system isotropically with an unknown amount of energy. This background produces a stationary spectrum in which the sensitivity of the experiment increases with the square root of the observing time. More important, one can exploit the fact that the correlation function of the observable is negative at the round trip light time in the detector.

At present we have been searching for VLF gravitational radiation using standard Deep Space tracking procedures (1). Basic limitations of the current systems are well understood and severely limit detection capabilities.

M. J. Reid and J. M. Moran (eds.), The Impact of VLBI on Astrophysics and Geophysics, 321–322.

2. THE SMILE PROJECT

One way around the current limitation is to construct a totally space based system. A spaceborne multi-arm interferometer contains elements which have recently been addressed through laboratory implementation, laboratory experimentation and in recent space experiments. The study underway proposes to implement a multi-arm microwave interferometer which will have a resolution of a few parts in 10^{17} in spatial resolution operating in the period range of 100 to 1000 seconds. This would allow measurements of G-wave background at 10^{-4} that of cosmic closure density for these wavelengths.

2.1. UTILIZATION OF CURRENT MICROWAVE TRACKING TECHNOLOGY

The design is based upon current proven techniques and would give improvement in sensitivity of over 4 orders of magnitude in power spectral density compared with current space experiments. The design uses 1 meter dish beacon transponders which are carried into position using special small ion drive propulsion modules (2).

2.2. MICROWAVE LINK MARGIN AND SYSTEM CHARACTERISTICS

The study has shown that Ka and Ku frequency bands (16 and 32 Ghz) perform with adequate link margins. Work is now preceeding to design a base module which will have two identical transmit/receive systems and make the carrier phase comparison in the interferometer. Mechanical stability of this module must be monitored to 100 μm for the time periods of 1000 seconds. Rotational stability is similarly monitored using a differential 3 component accelerometer similar in design to one component of the current ESA gravity gradiometer project.

3. CONCLUSIONS

We describe that by using current techniques it is possible to construct a spaceborne microwave multi-arm interferometer suitable for the detection of VLF gravitational waves. Spaceborne laser interferometer systems have been proposed with theoretically greater sensitivities, which however are far more difficult to build and operate than the system described here which is based upon current technology. Furthermore we suggest that the next constructive step to detect gravitational waves in space would be to deploy these small inexpensive modules. This would open up a new range of experimental sensitivities to VLF gravitational waves and allow us to overcome practical difficulties created by even more challenging experiments in the future.

4. REFERENCES

(1) R.W. Hellings, J.D. Anderson, P.S. Callahan and A. Moffet. Phys. Rev. D 23, 834-843, 1981.

(2) Lunar Get Away Special Ion Drive feasibility study. JPL D-4181, March 4, 1987.

ASTROMETRY OF MILLIJANSKY SOURCES USING A PHASE REFERENCE VLBI TECHNIQUE

Jean-François Lestrade
Bureau des Longitudes, Paris, France
and Jet Propulsion Laboratory

Alan E. E. Rogers and Arthur E. Niell
Haystack Observatory, Westford, MA 01886

Robert A. Preston
Jet Propulsion Laboratory, Pasadena, CA 91109

Atmospheric fluctuations and hydrogen maser clock instabilities limit the coherence time (T_{coh}) and hence the signal-to-noise ratio of VLBI data ($SNR \propto \sqrt{T}_{coh}$), to a few minutes at GHz frequencies. However, T_{coh} can be increased up to many hours if a phase reference is established throughout the experiment by using a strong source observed alternatively with the program source. Weak celestial radio sources with flux densities in the milliJansky range at centimeter wavelengths, such as radio stars, fast pulsars and faint components of extragalactic sources, can then be reliably detected by this VLBI technique, simultaneously providing accurate astrometry relative to the reference source.

An arbitrarily small signal from the program source can be tracked by the VLBI correlator over successive scans of the experiment if fringe search is replaced by a fringe filter. The fringe filter is implemented by restricting the "search" to a fixed delay and fringe rate (i.e. a zero width window search) determined from the reference source fringe search for each scan. For the program source signal to be within the fringe filter passband, the uncertainty in the program source delay and rate must be smaller than $1/b$ sec in delay and $1/T_{int}$ Hz in fringe rate or, for example, $<$ 35 nanosec and $<$ 16 mHz for a bandwidth b = 28 MHz (Mark III mode B) and a scan integration time T_{int} = 60 sec. Such small uncertainty is not usually achieved in VLBI because the clock offset and rate are larger and are a priori unknown. However, if the clock parameters are first derived from the reference source observed throughout the experiment, then the fringe processing parameters for the program source scans can be properly set.

As a second step of the procedure, all the "noisy" complex visibilities of the program source must be coherently added together to built up the final SNR. This coherent addition can be carried out with the differenced phases between the program and reference sources computed by a precise astrometric model. If the uncertain parameter of this model is only the position of the program source, a grid of possible positions on the sky can be searched to obtain the position which optimizes the coherent addition.

The stellar system Algol and the nearby reference source MW 0309+41, separated by 70 arcminutes, were observed for 6 hours at 5 GHz at both left and right circular polarizations on 1984 October 11. The total flux density of Algol was approximately 4 milliJansky or about 1/6 the detection threshold provided by each 2-minute scan on the

M. J. Reid and J. M. Moran (eds.), The Impact of VLBI on Astrophysics and Geophysics, 323–324.

two baselines between Greenbank and the VLA (GY) and between Owens Valley and the VLA (OY). The coherent addition was carried out over all the scans of Algol acquired on the two baselines in both senses of circular polarization. The resulting effective coherent integration time is 4.2 hours, providing a 10-fold increase in sensitivity compared to the original 2-minute scan. The coherent addition of all the complex visibilities over a grid of possible positions of Algol yields the "dirty" map in Fig. 1. High side lobes are apparent in this map because the u-v coverage (Fig. 2) is poor. However, it is sufficient to unambiguously locate the peak of the map to provide the astrometric position of Algol.

Several data sets were formed from the two baselines GY and OY and the two senses of circular polarization L and R. There are 4 independent estimates of the position of Algol relative to MW 0309+41 (Table 1). Uncertainties in station locations, earth orientation parameters, ionospheric and tropospheric effects and errors in the extrapolation of the reference source phase, can cause systematic errors as large as 1.5 milliarcsecond according to our analysis. The stochastic errors (angular resolution/SNR) are 0.8 and 1.2 milliarcsecond in α and δ, respectively, for an SNR of 6. Hence, the spread of 1.5 milliarcsecond in α and 1.7 milliarcsecond in δ found in Table 1 is consistent with the expected precision.

Baseline/ Polarization	T_{coh} hours	$\Delta\alpha$ 4^m	$\Delta\delta$ $22'$	SNR
OY/L	1.1	$51^s.83081$	40".8348	6
OY/R	1.0	$51^s.83079$	40".8350	7
OY/I	2.1	$51^s.83079$	40".8348	9
GY/L	1.0	$51^s.83071$	40".8360	6
GY/R	1.1	$51^s.83073$	40".8365	6
GY/I	2.1	$51^s.83078$	40".8360	9
OY+GY/L	2.1	$51^s.83076$	40".8360	8
OY+GY/R	2.1	$51^s.83076$	40".8362	9
OY+GY/I	4.2	$51^s.83076$	40".8360	12

Table 1: VLBI relative position of MW 0309+41 minus Algol. The symbol I is for the sum of the LCP and RCP data. If the reference source MW 0309+41 is at $\alpha = 3^h\ 13^m\ 1^s.9615$ and $\delta = 41°\ 20'\ 1".202$ then Algol is at $\alpha = 3^h\ 8^m\ 10^s.1307 \pm 0^s.0001$ and $\delta = 40°\ 57'\ 20".366 \pm 0".0017$. The positions are in the J2000.0 system and neither the parallax nor the proper motion effects have been removed.

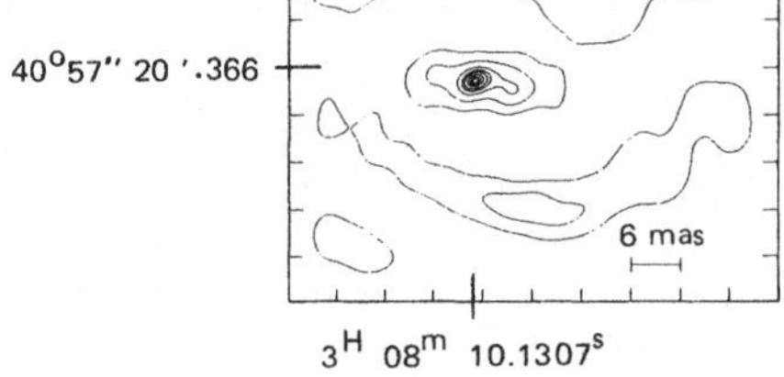

Figure 1: Coherent addition of visibilities over a grid of possible positions for Algol. Contour levels: 10,50,75,85,90,95,99 % peak

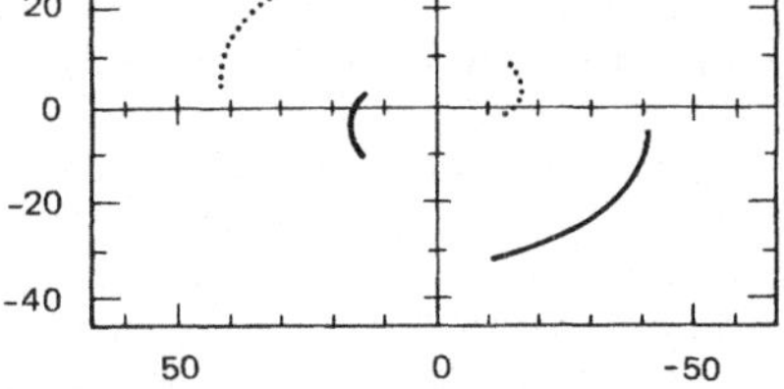

Figure 2: u-v coverage of the VLBI observation of Algol and MW 0309+41

THE CELESTIAL REFERENCE FRAME DEFINED BY VLBI

C. Ma
Goddard Space Flight Center
Geodynamics Branch 621.9
Greenbelt, Maryland 20771 USA

and

D. B. Shaffer
Interferometrics Inc.
8150 Leesburg Pike
Vienna, Virginia 22180 USA

ABSTRACT. VLBI currently produces the most accurate positions of celestial objects. From 1979 to 1987, 114 extragalactic radio sources have been observed with dual-frequency Mark III VLBI as part of the NASA Crustal Dynamics Project and the NGS POLARIS/IRIS program. The formal statistical errors of conventional celestial coordinates are as small as 0.3 milliarcseconds. The fundamental quantity measured by VLBI is the arc length between radio sources. Thus, we suggest that VLBI be used to establish a coordinate reference frame based solely on radio positions, and that this system not necessarily be coupled to right ascension and declination.

The Mark III VLBI system was developed by the Crustal Dynamics Project (CDP) of the National Aeronautics and Space Administration (NASA) to support research in geodynamics. This system is described in Clark *et al.* (1985). The Mark III is also used in the POLARIS and IRIS networks coordinated by the National Geodetic Survey (NGS). Since the quasi-inertial reference frame in which terrestrial motions are to be measured with centimeter precision is that defined by extragalactic radio sources, it is necessary to have correspondingly precise positions of these fiducial points.

More than 150,000 dual-frequency observations have been made of 114 radio sources to determine their positions. These observations come from more than 400 experiments, each covering approximately one day and using up to seven stations simultaneously. A large majority of the sessions are from the POLARIS/IRIS networks, but the observations are roughly equally divided between CDP and POLARIS/IRIS. About half of the sources have been used as geodetic reference sources. The others have been observed in survey sessions designed to enlarge our reference catalog. There is a wide variation in the number of observations for various sources. Some of our standard sources have been observed more than 15,000 times while some of the survey sources have been observed less than 10 times. Formal errors for the positions of the sources range from about 0.3 milliarcseconds (mas) for the most frequently observed sources, up to a few mas for sources with fewer than 50 observations.

The stations and sources are almost entirely in the northern hemisphere, and the observing baselines have predominantly east-west orientation. Recent observations to South Africa have enabled us to observe more southerly sources (down to -30 degrees declination) and to reduce the declination errors on low declination sources.

M. J. Reid and J. M. Moran (eds.), The Impact of VLBI on Astrophysics and Geophysics, 325–326.

The data were analyzed in a single, sequential least-squares solution as described in Ma *et al.* (1986). The astronomical and geophysical models generally followed the MERIT standards. However, these standards do not specify the theoretical model for the VLBI observables. In our analysis, as in all Mark I and Mark III geodetic work described and published by the VLBI groups at Goddard, NGS, and the Center for Astrophysics, the theoretical model was that of Robertson (1975). Further discussions of our VLBI radio source position measurements can be found in Ma *et al.* (1986) and Ma (1987).

The fundamental celestial quantity measured by VLBI is the arc length between radio sources, now being measured with sub-milliarcsecond accuracies. A comparison of Mark III positions with Mark II VLBI positions derived by JPL from the Deep Space Network (O. Sovers, private communication) shows a weighted mean arc length difference of 2 mas for the approximately 60 sources in common. The HIPPARCOS satellite will also measure arc length differences, of similar accuracy, for many bright stars. Neither the VLBI nor HIPPARCOS measurements require any reference to the complicated dynamics of the Earth.

Values of right ascension and declination depend on knowledge of the orientation of the earth and the definition of the origin of the right ascension system. (VLBI declinations are nominally absolute, but only with respect to the instantaneous rotation axis of the earth.) Hence, we suggest that consideration should be given to the establishment of a coordinate reference frame based solely on (relative) positions of radio sources. The relative positions of the radio sources would define a very nearly inertial frame. Positions of celestial objects would be invariant with time, except for true proper motions.

The system would be defined by the assigned positions of several sources. The relative orientations of these sources would have to be well measured and repeatable at any frequency. They should be located such that no degeneracies occur when determining the relative position of any other objects. This system could be defined independently of any consideration of earth orientation, but it should be straight forward to determine RA/Dec-like quantities for observation purposes. Thus, precession would not "move" the sources, but rather indicate where the earth is with respect to the celestial frame. The celestial frame could be aligned with some large scale structure of the universe (the ecliptic, the Galactic equator, or some super-cluster equator) for conceptual purposes, but this is not necessary.

REFERENCES

Clark, T. A. *et al.* (1985). *IEEE Trans. Geoscience & Remote Sensing* **GE-23**, 438.

Ma, C. *et al.* (1986). *Astronomical Journal* **92**, 1020.

Ma, C. (1987), *IAU Symposium #128* (in press).

Robertson, D. S. (1975). Ph.D. thesis, Massachusetts Institute of Technology.

VLBI POSITIONS OF EIGHT STELLAR SYSTEMS

A.E. Niell (1), J.-F. Lestrade (2,3), R. A. Preston (2),
R.L. Mutel (4), and R. B. Phillips (1)
(1) Haystack Observatory, Westford, MA 01886
(2) Jet Propulsion Laboratory, Pasadena, CA 91109
(3) Bureau des Longitudes, Paris, France F-75014
(4) University of Iowa, Iowa City, IA 52242

ABSTRACT. Radio positions of eight stellar systems have been derived with accuracies of 3 to 300 milliarcseconds from MkIII VLBI observations conducted with multi-station arrays. The best accuracy was obtained during a relatively strong outburst of the RS CVn system HR 5110. The epoch J2000.0 positions are obtained in the Jet Propulsion Laboratory VLBI reference frame of extragalactic radio sources.

1. Radio star astrometry and the Hipparcos Satellite

Observations of optically bright stars using VLBI are being made to determine accurate positions (uncertainties of a few milliarcseconds) and proper motions for about 15 objects well distributed over the sky. The extragalactic radio sources which form the reference frame for these positions will then provide an inertial frame for the optical positions of 100,000 stars determined by the Hipparcos satellite. The accuracy expected for the satellite observations is 2 mas in position and 2 mas/yr in proper motion.

2. Observations and Results

Survey work has shown that only the class of stars called RS Canum Venaticorum and other close binary systems (e.g. Algol, CYG X-1, LSI61 303) are at times sufficiently strong and compact in their radio emission to be useful for the astrometric requirements. During three sessions in 1983, using arrays of 5 or 6 antennas, we determined positions for eight stars with formal uncertainties of 0\"003 to 0\"3 (see tables 1 and 2). No correction has been made for proper motion or parallax. The precision was limited primarily by the strengths of the sources, which ranged from 9 to 400 mJy at the times of the observations.

We have compared these results with radio positions obtained with the VLA and with optical positions after transforming the VLA and

M. J. Reid and J. M. Moran (eds.), The Impact of VLBI on Astrophysics and Geophysics, 327–328.

optical positions from epoch B1950.0 to J2000.0. For 5 sources the mean (and rms) of the differences (VLA-VLBI) are 0″.00 (0″.06) and -0″.06 (0″.06) in right ascension and declination, respectively. For seven stars, excluding Algol, the mean (and rms) of the differences with the optical (OPTICAL-VLBI) are 0″.03 (0″.08) and -0″.01 (0″.09). The good agreement with the optical in right ascension is fortuitous since the origin of right ascension for the radio frame has an uncertainty of 0″.2 with respect to the optical (Hazard et al., 1971). However, it does indicate that the JPL VLBI radio reference frame is aligned with the preliminary FK5 frame at a level of better than 0″.1. This uncertainty is limited by the systematic errors over the sky of 0″.1 to 0″.2 in the optical positions.

Table 1. Dates of observations

	Date	Freq	Stations
I	83/03/20	8.4 GHz	DSS14,DSS13,OVRO,Ft.Davis,Westford
II	83/07/26	5.0 GHz	Bonn,Haystack,NRAO140",Ft.Davis,VLA,OVRO
III	83/10/16	1.7 GHz	Bonn,Haystack,NRAO140",Ft.Davis,VLA,OVRO

Table 2. Positions of eight radio stars. No correction has been made for parallax or proper motion.

Star	Session	Right Ascension (J2000.0) hr mn sec	+/-	Declination (J2000.0) dg mn sec	+/-	Flux Density (mJy)
LSI61 303	III	02 40 31.686	+/-0.008	61 13 45.56	+/-0.07	40
Algol	II	03 08 10.1308	0.0003	40 57 20.359	0.005	45
UX Ari	II	03 26 35.3375	0.0002	28 42 56.026	0.005	145
HR 1099	I	03 36 47.3240	0.0002	00 35 18.66	0.02	400
HR 5110	II	13 34 47.6893	0.0002	37 10 56.859	0.003	165
SigCrB	II	16 14 41.2011	0.0007	33 51 32.47	0.01	16
CygX-1	II	19 58 21.6804	0.0004	35 12 05.887	0.009	15
SZ Psc	II	23 13 23.7645	0.0005	02 40 31.31	0.04	38

REFERENCES:

Hazard,C, Sutton, J., Argue, A.N., Kenworthy, C.M., Morrison, L.V., and Murray, C.A., 1971, Nature, 233,89.

MERIDIAN & ASTROLABE OBSERVATIONS OF RADIO STARS: A CONTRIBUTION TO THE CONNECTION OF RADIO AND OPTICAL REFERENCE SYSTEMS

M. Sarasso,[1] G. Chiumiento,[1] and A. Poma[2]
[1] Osservatorio Astronomico di Torino
I-10025 Pino Torinese (TO), Italy
[2] Stazione Astronomica Internaz. di Latitudine
Via Ospedale 72
I-09100 Cagliari, Italy

ABSTRACT. Optical positions of eight radio stars observed with the Photoelectric Transit Instrument (PTI) at Torino and with the Danjon Astrolabe (AST) at Cagliari are presented here. The values of observed right ascensions have been computed at the Epoch J2000.0.

Precise positions of optical counterparts of radio sources are needed for the connection of optical and radio reference frames.

The optical counterparts of eight galactic radio sources—from the list proposed by the IAU Commission 24 Working Group on the identification of Radio-Optical Astrometric Sources—have been regularly observed with the Photoelectric Transit Instrument at Torino Observatory. The radio source HR 5110 has also been observed, at west transit only, with the Danjon Astrolabe at Cagliari Observatory.

The observations, obtained in different periods, have been reduced according to the MERIT Standards, that is, on the basis of the IAU (1976) System of Astronomical Constants, the IAU (1980) Theory of Nutation, and the FK5 System.

The optical positions of seven of the above galactic radio stars—at the epoch J2000.0—are presented in Table 1 as well as a comparison with right ascensions obtained with Carlsberg Meridian Circle at La Palma (1986).

In particular, with reference to the radio source HR 5110, the values of right ascensions, computed at Torino and Cagliari at the epoch J2000.0, are presented in Table 2 as well as a comparison with optical positions obtained with the observations for the Astrolabe General Catalogue (Billaud *et al.*, 1977), with Transit Instrument (TI) at Bratislava (Hefty *et al.*, 1987) and the Automatic Meridian Circle (AMC) at Bordeaux Observatory (Lestrade *et al.*, 1985).

Finally, a preliminary comparison with VLBI radio position of this source, measured with the NASA-JPL Deep Space Network and the US VLB Network antennas, is also presented.

M. J. Reid and J. M. Moran (eds.), The Impact of VLBI on Astrophysics and Geophysics, 329–330.

Table 1: OPTICAL POSITIONS OF GALACTIC RADIO SOURCES OBSERVED AT TORINO AND LA PALMA OBSERVATORIES

Source	DM	m	sp	α(O-C)	α obs	α calc	δ calc	N	Ep.	Cat	
b Per	49°1150	4.56	A2	$-0^s.0379$	$4^h18^m14^s.604$	$14^s.642$	50°17'44".05	20	84.37	SAO	(1)
				±0.0049	14.522			7	85.11		(2)
54 Cam	57 1118	6.49	G0	-0.0193	8 2 35.809	35.828	57 16 24.67	22	83.79	SAO	(1)
				±0.0044	35.842			6	85.73		(2)
ZZ Boo	26 2508	6.80	F0	0.1076	13 56 9.505	9.397	25 55 7.41	6	82.46	SAO	(1)
				±0.0055	9.622			7	85.04		(2)
β Lyr	33 3223	3.4-4.3	Bp	0.0001	18 50 4.802	4.802	33 21 45.71	4	81.74	FK4	(1)
				±0.0059							
P Cyg	37 3871	4.77	Bp	0.0479	20 17 47.219	47.171	38 1 58.85	21	83.73	SAO	(1)
				±0.0038		47.211		6	85.69		(2)
HD 216489	16 4831	5.84	K0	-0.0056	22 53 2.291	2.297	16 50 27.86	37	83.76	FK4S	(1)
				±0.0035		2.292		6	85.52		(2)
HR 8752	56 2923	4.96	G0	0.0691	23 00 5.102	5.033	56 56 43.12	4	82.07	FK4S	(1)
				±0.0101		5.112		6	85.55		(2)

(1) Photoelectric Transit Instrument - Torino Observatory - (2) Carlsberg Meridian Circle - La Palma
N = Number of observ. Ep = Mean epoch of observ. (1980+) Cat = Optical Ref. adopted Catalogue

Table 2: OPTICAL AND RADIO POSITIONS OF RADIO SOURCE HR 5110 = DM 37 2426 = FK4 502 m=4.96 Sp. F0

Station	α(O-C)	ε	α obs	α calc	δ obs	δ calc	N	Ep.
Torino PTI	$0^s.0009$+/-$0^s.0020$		$13^h34^m47^s.803$	$47^s.802$		37°10'56".70	19	82.67
Cagliari AST	0.0105	0.0036	47.812	47.802		56.70	21	84.09
G.C.A. AST	0.0021	0.0007	47.804	47.802				64.50
Bratislava TI	0.0004	0.0035	47.802	47.802				76.20
Bordeaux AMC	0.0090		47.695	47.686	37°10'56".92	56.92		83.32 *
VLBI Network			47.689		56.86			83.57 *

* To compute this positions, proper motions and trigonometric parallaxes have not been taken into account, between the epoch of observation and J2000.0

References

1) BILLAUD et Al.: 1977, Astron. Astrophys. 63, 87
2) LESTRADE et Al.: 1985, Proc. Coll. 'Scientific Aspects of the Input Catalogue preparation'. Aussois, 3-7 June 1985, 251 (ESA SP-234)
3) CARLSBERG MERIDIAN CATALOGUE - La Palma, No. 2, 1986
4) HEFTY et Al.: 1987, Bull. Astr. Inst. Czechosl. 38, 24

RADIO REFERENCE FRAME STABILITY FROM VLBI DATA

O. J. Sovers and R. N. Treuhaft
Jet Propulsion Laboratory, California Institute of Technology
4800 Oak Grove Drive
Pasadena, California 91109

ABSTRACT. Intercontinental radio interferometric measurements between NASA's Deep Space Network stations yield a catalog of positions of 106 extragalactic radio sources, uniformly distributed between $-45°$ and $+85°$ declination. Almost all of the source positions have formal uncertainties between 0.5 and 3 milliarcseconds. Estimates of three independent sets of 32 source coordinates at average epochs 1980.0, 1983.2, and 1984.5 assess the level of stability of the reference frame over a five-year period. Comparisons of the three sets of source coordinates show a number of cases of $> 2\sigma$ differences, particularly betwen the 1984-85 and 1981-83 periods. No sources, however, show 2σ shifts among both pairs of epochs. Upper limits of the order of 1 mas/yr can be placed on the time rates of change of the 32 source coordinates.

1. OBSERVATIONS AND ANALYSIS

A total of 48 VLBI observing sessions were carried out during 1978-85 on the two DSN intercontinental baselines [California (C) to Spain (S) and to Australia (A)]. The majority were of 24-hr duration, equally divided between CS (470 hr) and CA (415 hr). We employed JPL Mark II data acquisition systems and a bandwidth synthesis technique with channels of 2 MHz bandwidth, spanning about 40 MHz in each radio frequency band. Only dual-frequency measurements were used, to permit elimination of ionospheric delays. Each source was observed a number of times in a given session to cover the full range of mutual visibility. Many different schedules were used on each baseline, but each included observations of 3C 273, the right ascension reference source.

Modeling was performed in Solar System barycentric coordinates defined in terms of the mean equator of J2000.0. We attempted to adhere to the Project MERIT standards (Melbourne et al., 1983) for astronomical constants and Earth models. Tropospheric zenith delays were mapped with the function of Lanyi (1984). The usual value of $12^h29^m6^s.6997$ at J2000.0 was adopted for the right ascension of the reference source 3C 273. A multiparameter diagonally weighted least-squares fit (the "standard fit") determined approximately 1300 parameters from the 6800 delay and delay rate pairs. The fit included session-specific clock and station location parameters, and a troposphere zenith delay at each station for every 3 hour period. The only "global" parameters common to all sessions were the right ascension and declination of each source. A number of variants of the standard fit were also performed, including a search for mismodeling by solving for independent positions of 32 sources in three subsets of the 1978-85 time span.

M. J. Reid and J. M. Moran (eds.), The Impact of VLBI on Astrophysics and Geophysics, 331–332.

2. SOURCE POSITIONS, COMPARISONS, AND STABILITY TEST

The standard fit to 1978-85 DSN data produced positional coordinates of 106 sources, which form the JPL 1986-3 catalog. Root-mean-square uncertainties are approximately equal for arc lengths in RA and declination: 2.1 and 2.0 mas, respectively, with the range 0.5 to 3 mas including nearly all sources.

Two recent extragalactic source catalogs [GSFC (Ma, 1986), and IRIS (Robertson et al., 1986)] have 55 (of 101) and 19 (of 26) sources in common with JPL 1986-3, respectively. Since these VLBI programs employ completely different hardware, baselines, observing schedules, and correlation and parameter estimation software, comparison of results was thought to be valuable in identifying residual systematic errors in the VLBI technique. For the common sources the rms formal coordinate uncertainties are substantially smaller than 2 mas for both catalog pairs, indicating that the quality of the three catalogs is essentially comparable. Right ascension rms differences are of the order of 2 mas, declination rms differences are considerably larger (6 mas for GSFC-JPL), and the χ^2_ν values (1.2 to 2.3) indicate that one or both sets of formal uncertainties are underestimates of the true errors.

The sizable discrepancies between the catalogs led us to perform a number of tests on the DSN data that were designed to uncover systematic problems. One of these was a fit estimating independent positions in each of three time spans for sources with at least 10 intercontinental observations in each period. The 1978-85 DSN observations fall into three time periods : 1978-80 (average epoch 1980.0), 1981-83 (1983.2), and 1984-85 (1984.5). Along with three sets of coordinates of 32 sources, the estimated parameters included a single set of positions for all other sources, and the nutation angles for each intercontinental observing session. Differences of the three sets of source coordinates derived in this manner were calculated, with error estimates taking into account correlations between coordinates at different epochs. While there are four cases of $> 2\sigma$ differences, no source shows 2σ shifts between the two pairs of epochs. Of these four sources, only for DA 55 and P 0202+14 are the directions of the shifts consistent between the two epoch differences. Root-mean-square differences are approximately 2 mas for both pairs of epochs, and χ^2_νs are close to 1 with the exception of that for 1984.5 minus 1983.2 right ascension differences, which becomes 1.03 when one outlier is omitted. From these results, upper limits of the order of 1 mas/yr can be placed on the time variation of the 32 source coordinates. We conclude that the marginally significant positional differences between JPL and other catalogs may arise from unidentified systematic errors.

3. ACKNOWLDEGMENT

The research described in this report was carried out by the Jet Propulsion Laboratory, California Institute of Technology, under a contract with NASA.

4. REFERENCES

Lanyi, G. E. (1984). *Proceedings of the International Symposium on Space Techniques for Geodynamics* , Sopron, Hungary, July 9-13, 1984, vol. 2, p. 184.

Ma, C. (1987). *Proceedings of IAU Symposium 128, "The Earth's Rotation and Reference Frames for Geodesy and Geodynamics", 1986*, ed. A. K. Babcock, in preparation.

Melbourne, W., Anderle, R., Feissel, M., King, R., McCarthy, D., Smith, D., Tapley, B., and Vicente, R. (1983). USNO Circular No. 167, Washington, D.C.

Robertson, D. S., Fallon, F. W., and Carter, W. E. (1986). Astron. J. **91**, 1456.

ASTROMETRY IN LOCAL REFERENCE FRAMES FOR DEEP SPACE NAVIGATION

R. N. Treuhaft
Jet Propulsion Laboratory, California Institute of Technology
4800 Oak Grove Drive
Pasadena, California 91109

ABSTRACT. Future deep space missions will benefit from submilliarcsecond angular navigation accuracy (< 25 km projected distance at Neptune). Conventional Very Long Baseline Interferometry (VLBI) measurement strategies entail observations of a radio source (RS) and spacecraft pair. To increase the 3–5 milliarcsecond accuracies currently attainable, instrumental improvements could be coupled with either of two methods: 1) Use improved calibration measurements with the single RS as the spacecraft reference, or 2) use a reference frame established by a local network of RS's near the spacecraft to reduce the contribution of sky-dependent error sources. The demanding baseline vector (earth orientation) and tropospheric calibrations needed to realize submilliarcsecond accuracy with the single-source strategy would not be necessary with the multi-source strategy. The two strategies are compared via a sample covariance analysis. In this analysis, the single-source method yields 1.5 milliarcseconds for the navigation accuracy, while the multi-source method yields 0.5 milliarcseconds.

Deep space angular navigation is currently achieved with VLBI measurements of a spacecraft and a radio source (RS), typically within 10 degrees of each other. Demonstrated accuracies for this technique have been in the range of 3–5 milliarcseconds for the projection of the unmodelled spacecraft position on the baseline vector (J. S. Border et. al., 1982). Improvements in instrumentation in the Deep Space Network (DSN), particularly the increased bandwidth and bit rates attainable with the Mark III VLBI system (Rogers et. al., 1983), will allow rapid observations of a few RS's near the spacecraft. These additional RS observations could be used to calibrate errors which are a function of the direction in which the antennas point, called sky-dependent errors. The local reference frame technique described here is an extension of the technique used to measure relativistic bending (Fomalont and Sramek, 1975). Below, an example of a navigation covariance analysis is given, describing the parameterization of the sky-dependent errors, the observational configuration, and the results for the conventional and local network techniques.

The conventional and local network techniques can be viewed as estimations of an unmodelled spacecraft delay parameter in the presence of other unmodelled effects. With the conventional technique, assuming that the observation sequence is RS-spacecraft-RS, only station clock epoch and clock rate are estimated along with the desired spacecraft delay. All other errors which cause delay differences between the observations of the two objects propagate into the spacecraft delay; they will be called "consider" errors, as

M. J. Reid and J. M. Moran (eds.), The Impact of VLBI on Astrophysics and Geophysics, 333–334.

opposed to "explicit" errors, which are explicitly solved for in the estimation procedure. In the proposed local network technique, the two largest sky-dependent errors, earth orientation and tropospheric uncertainties, become explicit parameters. Therefore, to the extent that they are correctly modelled, the effects of earth orientation, troposphere, and clocks do not need to be externally calibrated.

Table 1 below describes the covariance results for the two navigation methods. The levels of assumed errors are shown along with the level at which they affect the spacecraft angular determination. The DSN California–Australia baseline (~ 10000 km) was used with five radio sources from the DSN catalog forming the local net. With RS1 being radio source #1, the local network technique entailed seven VLBI observations in the sequence RS1–RS2–RS3–spacecraft–RS4–RS5–RS1. For the conventional technique, the closest source, RS3, was assumed to be observed in the sequence RS3–spacecraft–RS3.

TABLE 1. Comparison of Conventional and Local Net Errors

Error	Assumed Magnitude	Conventional Error (μas)	Local Net Error (μas)
Earth Orient.	10 mas per component	1352	explicit
Static Trop.	4 cm	476	explicit
Fluctuating Trop.	1 cm daily zen/station	299	180
Baseline Length	3 cm	79	10
Celestial Pole	5 mas long. 5 mas obl.	0.3	13
System Noise	30 psec spacecraft 30 psec RS	213	427
RSS		1482	464

The results suggest that the local network technique is a promising alternative to conventional navigation, with minimal external calibration required. The earth orientation error, assumed at the approximate level of that required for the Galileo mission to Jupiter, dominates the conventional technique error for the RS–spacecraft configuration studied. It should be noted that the exact parameterization of earth orientation and troposphere effects in the local network technique is preliminary, and the contribution from system noise may be reduced by a more judicious parameter selection. In this analysis, sources were considered as perfectly known points; effects of source structure and position errors must be considered in future analyses.

The research described in this report was carried out by the Jet Propulsion Laboratory, California Institute of Technology, under a contract with NASA.

REFERENCES

Border, J. S., et. al., *AIAA Paper #82-1471*, August 1982.
Fomalont, E. B., and Sramek, R. A., *Astrophysical Journal*, **199**, 749, August 1975.
Rogers, A. E. E. et. al., *Science* **219**, 51, January 1983.

USING THE HUBBLE SPACE TELESCOPE TO RELATE THE HIPPARCOS AND EXTRAGALACTIC REFERENCE FRAMES

P. D. Hemenway and R. L. Duncombe
Department of Astronomy and Center for Space Research
University of Texas at Austin
Austin, Texas 78712

ABSTRACT. The HIPPARCOS satellite will produce positions, motions and parallaxes of celestial objects with previously unattained accuracy. This HIPPARCOS Instrumental System, however, will have an unknown solid body rotation with respect to an inertial reference frame. One aspect of our program of astrometric observations with the Hubble Space Telescope is to determine the rotation of the HIPPARCOS reference frame with respect to an extragalactic reference system.

The European Space Agency plans to launch an astrometric satellite (HIPPARCOS) to establish the most accurate relative stellar coordinate system of positions and motions ever determined. However, the HIPPARCOS system must be tied to an extragalactic (VLBI) reference frame to be used for studies of physical motions of objects and for the studies of relativistic effects.

The Hubble Space Telescope (HST) will provide a means of accurately measuring the relative positions and motions of bright objects with respect to faint ones in a small field of view. The expected accuracy (0.002 arcsec, rms) is an order of magnitude better than the accuracies obtained by current optical techniques for the same measurements.

Ninety one extragalactic optical objects (EGOs) in the vicinity (angular separation less than 18 arcminutes) of 160 HIPPARCOS stars (mostly SAO stars brighter than 10.5), evenly distributed over the sky, have been selected to be used to tie the HIPPARCOS system to the Extragalactic Reference Frame. The pairs are to be observed with the HST Fine Guidance Sensors (FGS), sometimes in conjunction with the HST Planetary Camera, in order to determine the relative positions and motions of the individual HIPPARCOS stars with respect to the EGOs.

In order to insure that the observations will be made and will be of use to the HIPPARCOS project, several conditions must be met:

A) The stars have to be within one FGS field of view of an extragalactic object (about 18 arcminutes along the longest dimension). Therefore, a region around each prospective EGO was searched for Smithsonian Astrophysical Observatory (SAO) stars, and for other, non-catalogued suitable anonymous stars.

M. J. Reid and J. M. Moran (eds.), The Impact of VLBI on Astrophysics and Geophysics, 335–336.

B) The extragalactic objects must be observable by HST. The limiting magnitude of the FGSs is nominally 17th. However, by using a mode of observation which incorporates Planetary Camera exposure of the "targets" and with the astrometric information from the Guide Stars, fainter EGOs may be reached. From more than 500 candidates, for which accurate positions, flux densities, optical identifications, and radio visibilities from the VLA and VLBI were available, 91 EGOs were finally chosen.

C) The EGO-HIPPARCOS star pairs must be evenly distributed on the sky. To this end, many "old favorite" radio sources were not included because an EGO-HIPPARCOS star pair in the same region of the sky was deemed more suitable.

D) The EGOs should be optically identified radio sources with compact cores which allow them to be included in the radio reference frame derived from VLBI observations. When possible, the EGOs were selected from the IAU list of recommended Radio Source Position Standards. In some cases, the only object bright enough to observe in a particular part of the sky was a radio quiet EGO. Such objects were included because of their usefulness in determining the rotation of the HIPPARCOS system independently of the radio reference frame.

E) The stars must be bright enough to be observed with HIPPARCOS (i.e. brighter than about 12.0, and brighter than 10.5 for full accuracy). The initial selection was made from the SAO catalogue, using a computer search around candidate EGOs. Other EGOs were measured on the UTRAO Laser Measuring Engine and positions of candidate stars determined from glass copies of the National Geographic Society, Palomar Sky Survey, at the University of Texas.

F) The stars must be included in the HIPPARCOS INPUT CATALOGUE (INCA) in order to be observed by HIPPARCOS. An initial selection of 414 stars was submitted to the HIPPARCOS project in 1982. This program was given the highest priority by the HIPPARCOS project; the southern extension was a major undertaking in support of the HST program. We have worked closely with various elements of the HIPPARCOS project to determine which of the stars will be included in the final catalogue.

G) Since the FGSs are interferometers, the objects must be small enough in angular extent to be observed with the FGSs. A massive program of Speckle Interferometry in the northern and southern hemispheres on as many of the 414 stars as possible has been conducted. These observations have shown that most of the stars in the final list of 160 are not multiple at the 0.04-0.08 arcsec level.

The final output from this work is the inclusion of the program objects in both the HST and the HIPPARCOS programs of observation.

Using this technique, we hope to make the most accurate determination possible of the motion of the HIPPARCOS system with respect to the extragalactic reference system.

We are indebted to J. Kovalevsky, C. Turon, N. Argue, O. Franz, H. McAlister and B. Morgan for observations and consultation. This work was supported by NASA grant NAGW-233.

THE CRUSTAL DYNAMICS PROJECT

Robert J. Coates
Crustal Dynamics Project
Code 601, Goddard Space Flight Center
Greenbelt, Maryland, 20771

ABSTRACT. The Crustal Dynamics Project has been developing, deploying, and operating very-long-baseline interferometry (VLBI) systems and satellite laser ranging (SLR) systems for highly accurate geodetic measurements of global plate motion, plate stability, regional crustal deformation, and earth rotation/polar motion. Over the past 10 years, the measurement accuracies of these systems have been improved by a factor of 10 to the cm level. Plans are to continue these developments to reach mm level accuracies. The present deployment of the VLBI systems is primarily in the Northern Hemisphere. This network has produced measurements of the relative plate motion between the North American, Eurasian, and Pacific plates; the stability of the same plates; and the regional deformation at the North American/Pacific plate boundary in California and Alaska.

1. GEODETIC VLBI DEVELOPMENT

For the past 10 years, the National Aeronautics and Space Administration (NASA) has been the primary sponsor of the development of very-long-baseline interferometry (VLBI) for very accurate geodetic measurements over long distances. VLBI experiments, in the 1970s, demonstrated feasibility of the technique at the decimeter precision level. Next, the Mark III VLBI system was designed to provide improved accuracy for geodesy. By 1980, the Mark III system had been implemented in a few stations, and baseline measurement precisions of 5 cm had been achieved. At the same time, similar results had been obtained with satellite laser ranging (SLR) developments. Thus, NASA established the Crustal Dynamics Project (CDP) to further develop and implement geodetic VLBI and SLR for measurements of the deformations in the plate boundary regions in western North America; the contemporary relative plate motions of the North American, South American, Nazca, Eurasian, and Australian plates; the internal deformation of the plates, with emphasis on North America and Pacific; the rotational dynamics of the earth; and the regional deformations in other areas of high earthquake activity.

M. J. Reid and J. M. Moran (eds.), The Impact of VLBI on Astrophysics and Geophysics, 337–338.

The CDP developed several improvements in the VLBI system, and with the cooperation of many organizations in several countries, established a global VLBI network for crustal dynamics measurements. The current network measurements of baselines have a precision of about 1 cm. Further developments are under way with the goal of millimeter precisions in the future.

2. GLOBAL NETWORK FOR PLATE MOTION

The global network consists of fixed antennas equipped with the Mark III VLBI system. Some stations are dedicated geodetic stations, but most are multi-use stations scheduled occasionally for VLBI. The stations used for the CDP plate motion measurements are located on the North American Plate at Westford, MA, Ft. Davis, TX, Mojave, CA, Owens Valley, CA, Hat Creek, CA, and Fairbanks, AK; on the Pacific plate at Vandenberg, CA, Kokee Park, HI, and Kwajalein Is.; and on the Eurasian plate at Kashima, Japan, Onsala, Sweden, and Wettzell, FRG.

3. NORTH AMERICAN NETWORK FOR REGIONAL DEFORMATION AND PLATE STABILITY

For regional deformation and plate stability, the CDP uses the fixed stations, listed above, as base stations for measurements to mobile VLBI stations that occupy a large number of sites in the region to determine baselines crisscrossing the faults. The CDP developed two mobile VLBI stations which are deployed in western North America at sites located at Black Butte, CA, Deadman Lake, CA, Ft. Ord, CA, Mammoth Lakes, CA, Monument Peak, CA, Palos Verdes, CA, Pasadena, CA, Pearblossom, CA, Pinyon Flat, CA, Point Reyes, CA. Presidio, CA, Quincy, CA, Santa Paula, CA, Flagstaff, AZ, Yuma, AZ, Platteville, CO, Ely, NV, Vernal, UT, Cape Yakataga, AK, Kodiak, AK, Sand Point, AK, Sourdough, AK, Algonquin, Canada, Penticton, Canada, Whitehorse, Canada, and Yellowknife, Canada.

4. MEASUREMENT SCHEDULE

Since VLBI measures the vector baselines between two or more stations, repeated measurements over a period of time are needed to determine motions. The VLBI determination of velocities of centimeters/yr with a precision of millimeters/yr requires measurements to be made over several years.

The first measurements were started in 1980 with Westford, Ft. Davis, and Owens Valley in North America and Onsala in Europe. By 1983, the second European station was added at Wettzell, North American stations were added at Mojave, Vandenberg, and Hat Creek, and the two Mobile VLBI systems were deployed to many sites in California. The Pacific, Alaskan, and Canadian stations completed the network with measurements beginning in 1984. Thus, there are 3 to 6 years of measurements of these baselines. The CDP plans to continue these VLBI measurements for several more years in order to achieve the mm/yr precision in velocities of the sites.

NASA/CRUSTAL DYNAMICS PROJECT RESULTS: Tectonic Plate Motion Measurements with Mark-III VLBI

J. W. Ryan and T. A. Clark (for the CDP VLBI Group)
NASA/GSFC
Code 621.9
Greenbelt, MD 20771

The NASA Crustal Dynamics Project (CDP) has been using VLBI on intercontinental baselines to measure tectonic plate motions since 1979. We report on measurements between sites on the North American plate (Haystack/Westford, MA; Owens Valley and Mojave, CA; Ft. Davis, TX and Gilmore Creek, AK), the Eurasian plate (Onsala, Sweden; Wettzell, West Germany, and Shanghai, China), the Pacific plate (Kauai, HI; Kwajalein in the Marshall Islands, and Vandenberg AFB, CA), the African plate (Hartebesthoek, RSA), and Japan (Kashima).

The Mass.–Sweden baseline has been measured over 100 times as part of the CDP, MERIT, POLARIS, and IRIS projects. Since 1984, measurements to Europe have also included a new 18-m antenna at Wettzell, FRG. In 1984, the CDP began GAPE (the Great Alaska and Pacific Experiment) to measure motions of sites on the Pacific plate using our existing 9-m VLBI antenna at Vandenberg AFB plus new VLBI installations at the 9-m NASA antenna on Kauai and on a 26-m antenna at Kwajalein. At the same time, we also implemented a dedicated Alaskan VLBI station on a 26-m antenna at Gilmore Creek, near Fairbanks; the GAPE measurements have also included Mojave, Kashima, Westford, Onsala, and Wettzell. Selected CDP VLBI results are summarized in the following table along with a comparison to predictions of the Minster–Jordan RM-2 relative plate motion model.

M. J. Reid and J. M. Moran (eds.), The Impact of VLBI on Astrophysics and Geophysics, 339–340.

Baseline	Observed Rate (mm/yr)	Observed Sigma (mm/yr)	RM-2 Prediction (mm/yr)
North American–European Baselines			
Westford–Onsala	16	3	17
Westford–Wettzell	13	5	19
North American–Pacific Baselines			
Gilmore Creek–Vandenberg	–56	6	–60
Gilmore Creek–Kauai	–53	9	–52
Gilmore Creek–Kwajalein	–36	7	–23
Mojave–Vandenberg	20	7	22
Mojave–Kauai	8	7	16
Mojave–Kwajalein	–2	13	23
Baselines to Japan			
Gilmore Creek–Kashima	–15	8	–7*
Mojave–Kashima	–20	12	–9*
Kauai–Kashima	–78	11	–87*
Vandenberg–Kashima	–61	26	–51*
Kwajalein–Kashima	–77	9	–94*
Selected Intraplate Measurements			
Gilmore Creek–Mojave	–12	4	0
Kauai–Vandenberg	–15	9	0
Kauai–Kwajalein	–7	8	0
Vandenberg–Kwajalein	–20	18	0
Westford–Owens Valley	2	2	0
Westford–Mojave	0	4	0
Mojave–Owens Valley	–3	2	0

* Assuming Kashima is on the Eurasian plate.

Beyond Plate Tectonics: Looking at Plate Deformation with Space Geodesy

THOMAS H. JORDAN

Department of Earth, Atmospheric and Planetary Sciences
Massachusetts Institute of Technology, Cambridge, MA 02139

J. BERNARD MINSTER

Institute of Geophysics and Planetary Physics
Scripps Institution of Oceanography
La Jolla, CA 92093

ABSTRACT. We address the requirements that must be met by space-geodetic systems to place useful, new constraints on horizontal secular motions associated with the geological deformation of the earth's surface. Plate motions with characteristic speeds of about 50 mm/yr give rise to displacements that are easily observed by space geodesy. However, in order to improve the existing plate-motion models, the tangential components of relative velocities on interplate baselines must be resolved to an accuracy of < 3 mm/yr. Because motions considered small from a geodetic point of view have rather dramatic geological effects, especially when taken up as compression or extension of continental crust, detecting plate deformation by space-geodetic methods at a level that is geologically unresolvable places rather stringent requirements on the precision of the measurement systems: the tangential components on intraplate baselines must be observed with an accuracy of < 1 mm/yr. Among the measurements of horizontal secular motions that can be made by space geodesy, those pertaining to the rates within the broad zones of deformation characterizing the active continental plate boundaries are the most difficult to obtain by conventional ground-based geodetic and geological techniques. Measuring the velocities between crustal blocks to ± 5 mm/yr on 100-km to 1000-km length scales can yield geologically significant constraints on the integrated deformation rates across continental plate-boundary zones such as the western United States. However, baseline measurements in geologically complicated zones of deformation are useful only to the extent that the endpoints can be fixed in a local kinematical frame that includes major crustal blocks. For this purpose, the establishment of local geodetic networks around major VLBI and SLR sites in active areas should receive high priority.

1. INTRODUCTION

Deformation of the earth's lithosphere covers a broad spectrum of temporal and spatial scales, from seconds to aeons and from mineral grains to planetary dimensions. Table 1 categorizes a subset of lithospheric motions that cause geologically and geophysically significant deformations. It is convenient to discriminate secular motions persisting on

M. J. Reid and J. M. Moran (eds.), The Impact of VLBI on Astrophysics and Geophysics, 341–350.

geological time scales of thousands to millions of years from transients associated with, for example, seismic and volcanic events. Practical research is more concerned with the transients, because they tend to disturb human activities. Secular motions also warrant vigorous study, however, since they provide the kinematical framework for describing transients and understanding their driving mechanisms.

TABLE 1. Types of motions observed at the earth's surface.

	SECULAR	TRANSIENT
HORIZONTAL	Plate motions Boundary-zone tectonics Intraplate deformation	(Pre,co,post)-seismic Fault creep Stress redistribution
VERTICAL	Tectonic motions Thermal subsidence Diapirism Crustal loading Post-glacial rebound Cratonic epeirogeny	(Pre,co,post)-seismic Magma inflation Tidal loading

The most significant long-term deformations are those related to plate tectonics. Although local tectonic movements near plate boundaries display large vertical components and time-dependent behavior, the net motions between the stable interiors of large blocks are forced by viscous damping and gravity to be nearly steady and horizontal. The characteristic tangential velocity of the plate system is about 50 mm/yr, which gives rise to displacements easily measured by geodetic methods. Horizontal secular motions have been observed both by ground-based networks [e.g., *Savage,* 1983] and by space-geodetic systems [e.g., *Christodoulidis et al.,* 1985; *Herring et al.,* 1986]. Though their application to geodesy is relatively new, space-based techniques -- Very Long Baseline Interferometry (VLBI), Satellite Laser Ranging (SLR), Global Positioning System (GPS) -- have already revolutionized the science of terrestrial distance measurement. They are contributing new information about active tectonics, particularly on the planetary scales previously inaccessible to ground-based surveys (Figure 1).

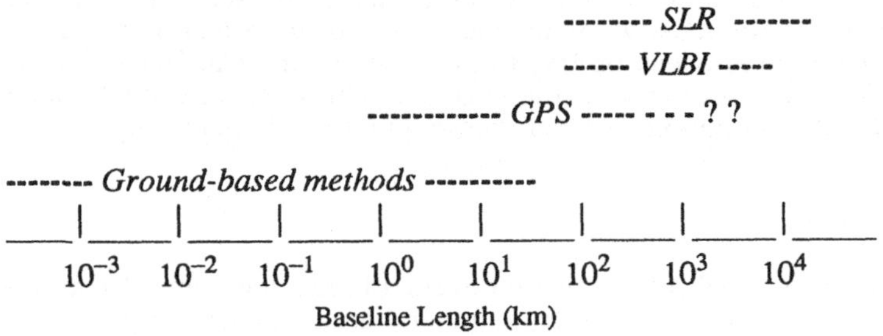

Figure 1. Spatial scales sampled by various geodetic methods.

In this paper, we examine the space-geodetic observations needed to solve geologically interesting problems. Specifically, we are concerned with the measurement accuracy required to place useful, new constraints on models of global plate velocities and related horizontal secular motions.

2. PLATE MOTIONS

In the ocean basins, most of the deformation related to horizontal secular motion occurs in well-defined, narrow zones that are the boundaries of a dozen or so large lithospheric plates. The current plate velocities are constrained by three basic types of data collected along these submerged boundaries: (1) spreading rates on mid-ocean ridges from magnetic anomalies, and directions of relative motion from (2) transform-fault azimuths and (3) earthquake slip vectors. The first self-consistent global models were synthesized soon after the formulation of plate tectonics [*Le Pichon*, 1968], and significant refinements were made throughout the next decade [*Chase*, 1972; *Minster et al.*, 1974]. Third-generation models were published by 1978 [*Chase*, 1978; *Minster and Jordan*, 1978] and are still in use. The velocity field for the RM2 model of *Minster and Jordan* [1978] is depicted in Figure 2.

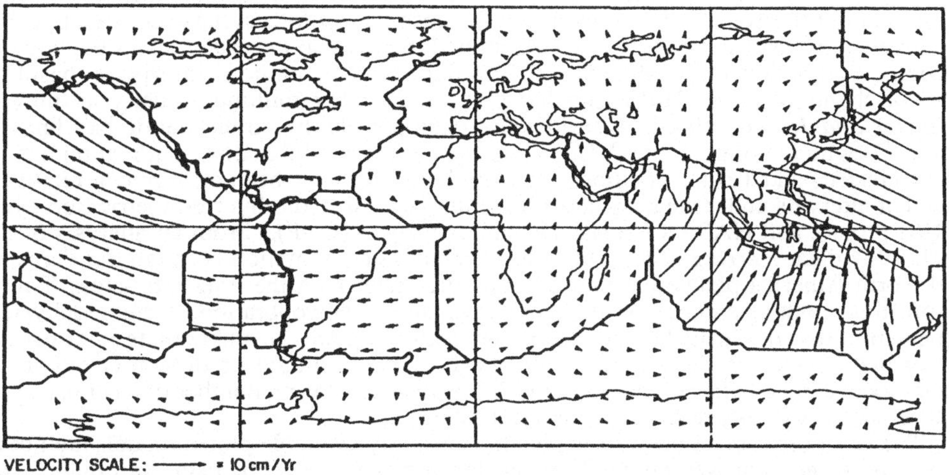

Figure 2. The plate geometry and velocity field for the eleven-plate RM2 model of *Minster and Jordan* [1978] in the AM1-2 (hotspot) reference frame (map courtesy of A. Forte and W. Peltier).

Since the reference frame is arbitrary, the angular velocity vectors describing the instantaneous relative motions among M rigid plates are specified by $3(M - 1)$ independent components. The 30 numbers representing the eleven-plate RM2 model were derived by least-squares inversion of a carefully selected, globally distributed data set comprising 110 spreading rates, 78 fault azimuths, and 142 slip vectors. The magnetic anomalies employed in this analysis average the rates over the last 2-3 million years, about the shortest time span for which good spreading rates can be obtained on a global basis.

Although the interval sampled by RM2 is hardly "instantaneous" from a geodetic point of view, it is geologically brief, and the small plate displacements that take place during it are well described by infinitesimal (as opposed to finite) rotations.

TABLE 2. Comparison of extension rates, in mm/yr, for two transatlantic baselines.

Baseline	VLBI*	RM2†
Haystack-Onsala	17 ± 2	17 ± 3
Ft.Davis-Onsala	20 ± 7	16 ± 3

* CDP/IRIS data from *Herring et al.* [1986].
† Calculated for NOAM-EURA plate pair.

It will probably be some time before the global plate-tectonic models can be significantly improved by space geodesy. Because the geological data sets are large and the inverse problem is strongly overdetermined, the formal uncertainties in the angular velocity components are already fairly small. RM2 predicts interplate baseline rates-of-change with standard errors that are typically a few millimeters per year (mm/yr) [*Minster and Jordan*, 1983]. Consider, for example, the two transatlantic baselines in Table 2, which connect the VLBI antenna at Onsala, Sweden, with antennas in Massachusetts and Texas. Positions between these sites have been measured at relatively frequent intervals under the auspices of the International Radio Interferometric Surveying (IRIS) project and NASA's Crustal Dynamics Project (CDP). From an analysis of the 1980-1984 data, *Herring et al.* [1986] have found rates of motion between the Eurasian and North America plates that are in good agreement with the RM2 estimates. This is comforting, both as a check on the techniques and as corroboration of the geophysical expectation that the instantaneous velocities between points in stable plate interiors are dominated by secular plate motions. However, if the 3-mm/yr prediction errors of the plate model can be believed, high-quality VLBI data sets over a number of such baselines will be needed to really refine the geological estimates.

There are other plate pairs for which the currently available global models are not as good, and some improvement could be made from space-geodetic data. In Figure 2, for example, Southeast Asia is assumed to be part of the large Eurasian plate, but the active tectonics of China imply it should be moving as a separate entity. Because it is completely surrounded by complex zones of deformation, its motion relative to Eurasia will be difficult to quantify without space-geodetic observations. Closer to home is the case of the Pacific-North America plate pair, whose relative rate of motion can be directly measured only on a tiny ridge segment in the mouth of the Gulf of California. From a reanalysis of the magnetic anomaly profiles across this ridge by *DeMets et al.* [1987], it appears that previous estimates of the spreading rate made by us and others were biased high by about 15-20%, and standard models like RM2 overestimate the relative plate speed. Since this rate is critical to models of deformation in the western United States (see below), a geodetic check on the Pacific-North America angular velocity will be very valuable.

With these exceptions and a few others, however, the global networks of VLBI and

SLR stations are just too sparse to control plate motions as tightly as the geological observations. Work is nearing completion at Northwestern University on an improved fourth-generation plate-motion model (R. Gordon, personal communication, 1987), which should remedy the few problems thus far identified with RM2. The impact of geodetic observations on the development of these plate-motion standards will be relatively minor, at least for the next few years.

Of course, this does not say that interplate observations made by space-geodetic methods will not reveal new and interesting phenomena associated with other categories of motion listed in Table 1; particular attention should be focused on time-dependent signals, including the possibility that plate speeds and directions have changed significantly during the 2-My averaging period of the geological data. Our point is to emphasize that the exciting issues for space geodesy lie beyond the now-classical descriptions of major plate motions.

3. Deformation of Plate-Boundary Zones

In particular, space geodesy can be used to quantify models of the complex deformation that characterizes all major continental plate boundaries. In the western United States, the interaction between the North America plate and northwestward moving Pacific plate is spread out over a 1000-km wide zone of deformation, stretching from the Rocky Mountains to the continental margin. Although California's San Andreas Fault is the major locus of movement on the plate boundary, the likelihood that significant crustal deformation is occuring both east and west of the fault has long been recognized by geologists. Reliable constraints on the rates of deformation integrated across regions with characteristic dimensions of 100 km or more are critical for understanding fundamental tectonic processes and assessing seismogenic potential. Rates of motion in continental regions are notoriously difficult to pin down by geological methods alone, and at this scale, ground-based geodetic surveys are expensive and inaccurate. Space geodesy appears to be the most feasible methodology.

VLBI measurements have already established significant constraints on geological motions in the Basin and Range Province, a peculiar region of diffuse seismicity and ubiquitous vulcanism which encompasses the Great Basin of Nevada and western Utah. A variety of geological and geophysical data indicate that the Basin and Range is stretching along an axis oriented approximately N60°W [*Zoback and Zoback,* 1980], but the rate of opening has been the subject of a long-standing controversy. In some cases the slip rates on individual faults have been inferred from detailed paleoseismic mapping [*Wallace,* 1984], but very few have been studied well enough to establish more than crude bounds. Geological estimates integrated across the entire province range from as low as 1 mm/yr to as high as 22 mm/yr, with a probable value in the range 4-12 mm/yr [*Minster and Jordan,* 1984, 1987]. Unlike the case of global plate motions discussed above, the assumptions that go into deriving these estimates are questionable, and the data redundancy is low.

In a recent analysis of western U.S. deformation [*Minster and Jordan,* 1987], we have used 4-yr VLBI rates for seven baselines spanning the Basin and Range to derive the vector motion $\mathbf{v}_{CN}$ of the aseismic Sierra Nevada-Great Valley block relative to stable North America. We express this motion by a vector $\mathbf{v}_{CN}$ obtained by extrapolating the relative velocity field of the block to a fiducial point C, fixed at a particular (though arbitrary) location on the San Andreas Fault in central California; our conventional choice, adopted here, is at 36°N, 120.6°W. The VLBI data alone gives an estimate of 9 ± 4 mm/yr, N48±17°W at point C (Figure 3, left panel). When the geological constraints on the azimuth of the opening are combined with the VLBI rates, this estimate becomes 10 ± 2 mm/yr at N56±10°W (right panel). Although the formal errors of the estimate should be regarded with appropriate caution, the precision achieved from such a short observing period is impressive.

The VLBI value of $\mathbf{v}_{CN}$ can be combined with the slip vector $\mathbf{v}_{SA}$ across the San Andreas, known from ground-based geodetic data and geological mapping, and the total Pacific-North America plate motion $\mathbf{v}_{PN}$ to derive a residual vector $\mathbf{v}_{CP} = \mathbf{v}_{CN} + \mathbf{v}_{SA} - \mathbf{v}_{PN}$, which measures the deformation west of the San Andreas. (All of these velocities are computed at the fiducial point *C*.) The geometry of this vector closure equation is shown in Figure 4. If RM2 is used to estimate $\mathbf{v}_{PN}$, the magnitude of $\mathbf{v}_{CP}$ is 15 mm/yr (Figure 3), implying considerable deformation across the California margin. The revised Pacific-North America rate of *DeMets et al.* [1987] reduces this discrepancy to about 8 mm/yr, a smaller but nevertheless substantial value. The vector $\mathbf{v}_{CP}$ can be resolved into two parts: a component of right-lateral shear parallel to the California coastline, which is presumably taken up on the largely offshore San Gregorio-Hosgri fault system, and a component of compression perpendicular to the coastline, which may be distributed across a region extending from the Great Valley east of the San Andreas to the edge of the continental margin [*Gawthrop,* 1978; *Crouch et al.,* 1984; *Minster and Jordan,* 1984, 1987]. If this model of the deformation is correct, the seismogenic potential of the California margin is likely to be higher than previously thought.

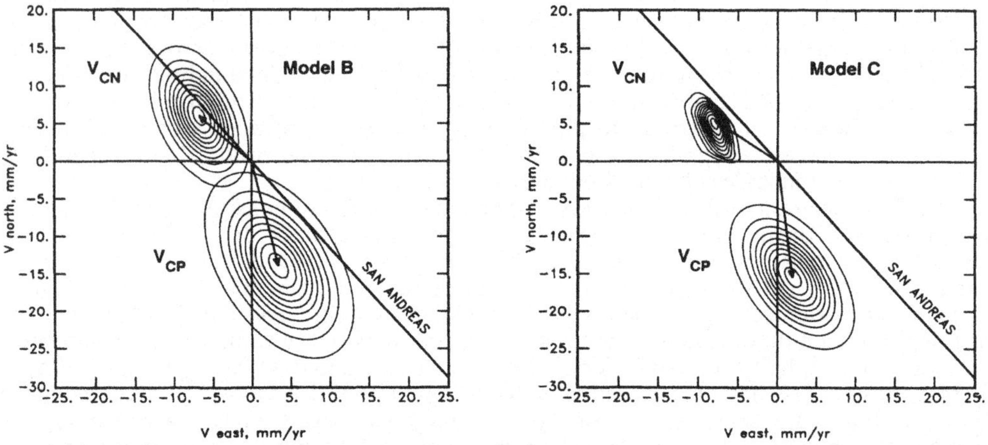

Figure 3. Maximum-likelihood estimates of the Basin and Range opening-rate vector $\mathbf{v}_{CN}$ derived from VLBI rates-of-change on seven baselines (left panel) and from a combination of VLBI rates and geological azimuths (right panel). Contours delimit the 5% through 95% confidence regions in 10% increments, calculated by the Bayesian procedure of *Minster and Jordan* [1987]. The residual vector $\mathbf{v}_{CP}$ describes deformation west of the San Andreas and is computed as the sum $\mathbf{v}_{CN} + \mathbf{v}_{SA} - \mathbf{v}_{PN}$, where $\mathbf{v}_{SA}$ is the San Andreas slip vector (34 ± 3 mm/yr, N41 ± 2°W) and $\mathbf{v}_{PN}$ is the RM2 Pacific-North America velocity (56 ± 3 mm/yr, N35 ± 2°W). All vectors correspond to velocity fields extrapolated to a fiducial point *C* at 36°N, 120.6°W.

In view of the large population and critical facilities that could be affected by earthquakes near the coast, tests of such speculations by direct geodetic measurements are urgently needed. VLBI and SLR stations along the California coast are beginning to yield baseline rates with the precision required to evaluate competing hypotheses [*Lyzenga et al.*, 1986; *Clark et al.*, 1986; *Christodoulidis et al.*, 1987], but the small number of sites leaves serious gaps in the coverage and does not provide adequate redundancy. The best way to remedy this deficiency is to densify the VLBI and SLR networks with portable GPS instruments. To this end, a consortium of universities is cooperating with several government agencies to set up a GPS network in southern and central California [*Agnew et al.*, 1987]. Included in this network are a number of islands in the Santa Barbara Channel and Southern California Borderlands, which will permit the observation of deformation associated with offshore faulting.

In the western U.S., geologically useful information has been derived from space-geodetic data with rate uncertainties for individual baselines on the order of ± 5 mm/yr. The kinematical complexities and seismic hazards of the Pacific-North America plate-boundary zone are by no means unique. The Mediterranean, Middle East, and of course China all display broad belts of tectonic activity, reflecting the relatively low yield strength of the upper continental lithosphere. These and other regions provide ample opportunities to employ the current capabilities of space geodesy as a tool for geological and geophysical investigation.

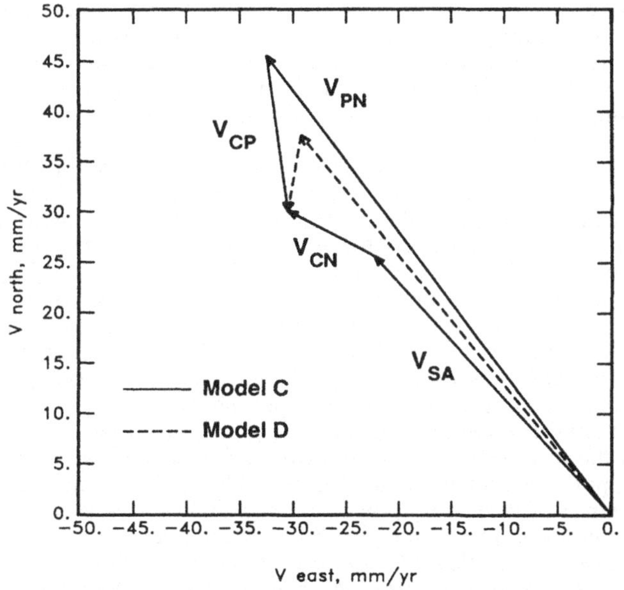

Figure 4. Vector closure diagram for western U.S. deformation showing the contributions of San Andreas slip ($\mathbf{v}_{SA}$) and deformation east ($\mathbf{v}_{CN}$) and west ($\mathbf{v}_{CP}$) of the San Andreas to Pacific-North America relative motion ($\mathbf{v}_{PN}$). The solid line uses the RM2 estimate of $\mathbf{v}_{PN}$, whereas the dashed line employs a relative velocity consistent with a revised spreading rate [*DeMets et al.*, 1987] and fault azimuths [*Goff et al.*, 1987] in the Gulf of California.

4. Intraplate Deformations

Plate tectonics is only an idealization. If wide zones of deformation extend away from continental plate boundaries, what about deformations within the supposedly stable plate interiors? An obvious application of space geodesy, and a stated objective of the Crustal Dynamics Project, is to evaluate the accuracy of the rigid-plate approximation. In designing plate-stability experiments, however, it should be realized that motion considered small from a geodetic point of view has rather dramatic geological effects, especially when it is taken up as compression or extension of continental crust. Detecting plate deformation by geodetic methods at a level that is geologically unresolvable will places rather stringent requirements on the precision of space-based measurement systems.

An example of this fact is the spectacular Rio Grande Rift of New Mexico, formed by Neogene extension related to Basin-and-Range opening on the other side of the Colorado Plateau. The rates of deformation across this obviously active, 50-km-wide feature have been estimated by both geological and ground-based geodetic techniques, and they are low. *Woodward* [1977] suggests a rate of only 0.3 mm/yr for the expansion of Albuquerque-Belen basin over the last 27 My, and *Golombeck* [1981] estimates an even lower rate for the opening of the Velarde graben since 1.1 My ago. *Savage et al.* [1980] did not resolve any strain across the rift near Socorro, New Mexico, during the 1972-1979 trilateration surveys, placing an upper bound of 1 mm/yr (95% confidence level) on the integrated strain rate. Hence, the motion across the Rio Grande Rift is small enough that it is usually ignored in discussions of how the Pacific-North America plate motion is partitioned across the western U.S. zone of deformation [*Minster and Jordan,* 1984].

Compression at this rate will lead to even more impressive features. Because continental crust is buoyant and cannot easily be remixed into the mantle, a secular horizontal shortening will cause it to pile up; 1 mm/yr across a width of 50 km thickens the crust at a rate of about 5 km every 10 million years, enough to build a sizable topographic expression in (geologically) short order. In fact, the uplift rate of the high Himalayas is only about an order of magnitude greater than this value; i.e., the compression rate which is building the main range of the highest mountains on earth is only about 10 mm/yr [*Lyon-Caen and Molnar,* 1985]. Similar rates are estimated across the Tien Shan, another major Asian mountain range [*Molnar and Deng,* 1984]. (Plate models tell us that the Indian subcontinent is moving northward into Asia at a speed of ~50 mm/yr, but most of this motion is taken up by the sideways escape of China [*Lyon-Caen and Molnar,* 1985].) Therefore, outside regions of substantial seismicity and/or young topography, intraplate deformation due to horizontal secular motions probably lies at or below the resolution of current space-geodetic experiments.

Zones of substantial intraplate seismicity are more intriguing. One is in the central United States near the town of New Madrid, Missouri, just south of the confluence of the Ohio and Mississippi Rivers, where faulting is active beneath the thick sedimentary cover. In 1811-12, a sequence of three very powerful earthquakes took place; these events had surface-wave magnitudes greater than eight and were felt over the entire eastern half of the continent [*Nuttli,* 1983]. The New Madrid seismic zone lies directly between the VLBI stations at Ft. Davis, Texas, and Westford, Massachusetts, and VLBI data on this baseline show an average closure rate of 5-10 mm/yr [*Herring and Shapiro,* 1985]. As *Herring and Shapiro* [1985] note, systematic measurement errors or local antenna motions may be responsible, but the observation has nevertheless prompted informal speculation that the seismic and geodetic observations are both manifestations of the same intraplate tectonics. The maximum compressive stress is indeed oriented in a northeast-southwest direction [*Zoback and Zoback,* 1980], and some estimates of the strain rates within the New Madrid seismic zone [*Johnson and Nava,* 1985] are high enough to make a causal relation plausible, but these rates depend critically on a poorly constrained earthquake recurrence time. The tectonics of the New Madrid region are controlled by very old (Precambrian) structures reactivated by intraplate stress [*Braile et al.*, 1982]. Because the rate of topographic change associated with these features appears to be low, we estimate that the closure rate on the Westford-Ft.Davis baseline due to horizontal secular motion distributed across the eastern U.S. is probably less than 1 mm/yr.

5. CONCLUSIONS

Many geological and geophysical problems related to the various types of surface motion in Table 1 are currently being addressed by space-geodetic experiments. Space geodesy can be an effective tool for solving these problems only if critical measurements are made at accuracies not feasible by more conventional (and less expensive) techniques. In this paper, we have addressed the requirements for observing the subset of motions that are tangential to the earth's surface and persist for geologically significant intervals. Our conclusions can be summarized as follows:

- To contribute to the improvement of existing models of present-day motion among the major plates, the tangential components of relative velocities on interplate baselines must be resolved to an accuracy of < 3 mm/yr.

- Measuring the velocities between crustal blocks to ± 5 mm/yr can place geologically useful constraints on the integrated deformation rates across continental plate-boundary zones, such as the western U.S.

- To detect the horizontal motions associated with the secular deformation of stable plate interiors, the tangential components on intraplate baselines must be resolved to an accuracy of < 1 mm/yr.

With regard to the second bullet, we wish to emphasize that baseline measurements in geologically complicated zones of deformation are useful only to the extent that the relationship of the endpoints to geologically significant crustal blocks is understood. Some antennas have a long history of participation in VLBI experiments, so that their motions in the VLBI reference frame are becoming well known; but they lie within complex zones of faulting, and their motions in kinematical frames fixed to local geology are not at all known. For example, the baseline rates for the Owens Valley Radio Observatory (OVRO) in California relative to the Westford, Massachusetts, and Ft. Davis, Texas, antennas have been measured to a precision of about 2 mm/yr (T. Clark, personal communication, 1986). In our analysis of Basin and Range extension, we have assumed OVRO moves with the Sierra Nevada-Great Valley block, but unfortunately, it is separated from the Sierra Nevada by a major system of faults, one of which broke in the great 1872 Owens Valley earthquake. Until the position of OVRO is regularly resurveyed in a local geodetic network which includes stations planted firmly on the Sierra block, the geological implications of the VLBI data will remain in doubt. The establishment of local geodetic networks around major VLBI and SLR sites in active areas should receive high priority.

Acknowledgements. We thank C. DeMets and J. Goff for preprints and T. Clark, R. Gordon, T. Herring, and J. Sauber for helpful discussions. This research was sponsored by the Crustal Dynamics Project of the National Aeronautics and Space Administration under contracts NAG5-459 and NAS5-28572.

6. REFERENCES

Agnew, D. C., et al., GPS measurements in central and southern California (abstract), *Eos Trans. AGU*, **68**, 282, 1987.

Braile, L. W., G. R. Keller, J. W. Hinze, and E. G. Lidiak, An ancient rift complex and its relation to contemporary seismicity in the New Madrid seismic zone, *Tectonics*, **1**, 225-237, 1982.

Chase, C. G., The *n*-plate problem of plate tectonics, *Geophys. J. R. Astr. Soc.*, **29**, 117-122, 1972.

Chase, C. G., Plate kinematics: the Americas, East Africa, and the rest of the world, *Earth Planet. Sci. Lett.*, **37**, 353-368, 1978.

Christodoulidis, D. C., D. E. Smith, R. Kolenkiewicz, S. M. Klosko, M. H. Torrence, and P. J. Dunn, Observing tectonic plate motions and deformations from satellite laser ranging, *J. Geophys. Res.*, **90**, 9249-9264, 1985.

Christodoulidis, D. C., D. E. Smith, S. M. Klosko, J. W. Robbins, P. J. Dunn, and M. H. Torrence, Tectonic motion in western USA from satellite laser ranging, preprint, 1987.

Clark, T., J. Ryan, C. Ma, E. Himwich, D. Gordon, and A. Mallama, Geodesy by radio interferometry: measurements of vector site motions in the western U.S. and Alaska (abstract), *Eos Trans. AGU*, **67**, 906, 1986.

Crouch, J. K., S. B. Bachman, and J. T. Shay, Post-Miocene compressional tectonics along the central California margin, in *Tectonics and Sedimentation along the California Margin*, edited by J. K. Crouch and S. B. Bachman, *Pac. Sect. Soc. Econ. Paleontol. Mineral.*, **38**, 37-54, 1984.

DeMets, C., R. G. Gordon, S. Stein, and D. F. Argus, A revised estimate of Pacific-North America motion and implications for western North America plate boundary zone tectonics, *Geophys. Res. Lett.*, submitted, 1987.

Gawthrope, W. J., Seismicity and tectonics of the central California coastal region, in *San Gregorio-Hosgri Fault Zone, California*, edited by E. A. Silver and W. R. Normark, *Calif. Div. Mines Geol. Spec. Rept. 137*, 45-56, 1978.

Goff, J. A., E. A. Bergman and S. C. Solomon, Earthquake source mechanisms and transform fault tectonics in the Gulf of California, *J. Geophys. Res.*, in press, 1987.

Golombeck, M. P., Geometry and rate of extension across the Pajarito fault zone, Espanola Basin, Rio Grande Rift, northern New Mexico, *Geology*, **9**, 21-24, 1981.

Herring, T. A., and I. I. Shapiro, Geodesy by radio interferometry: stability of the North American plate, *Eos Trans. AGU*, **66**, 848, 1985.

Herring, T. A., et al., Geodesy by radio interferometry: evidence for contemporary plate motion, *J. Geophys. Res.*, **91**, 8341-8347, 1986.

Johnson, A. C., and S. J. Nava, Recurrence rates and probability estimates for the New Madrid seismic zone, *J. Geophys. Res.*, **90**, 6737-6753, 1985.

Le Pichon, X., Sea-floor spreading and continental drift, *J. Geophys. Res.*, **73**, 3661-3697, 1968.

Lyon-Caen, H., and P. Molnar, Gravity anomalies, flexure of the Indian plate, and the structure, support and evolution of the Himalaya and Ganga Basin, *Tectonics*, **4**, 513-538, 1985.

Lyzenga, G. A., K. S. Wallace, J. L. Fanselow, A. Raefsky, and P. M. Groth, Tectonic motions in California inferred from very long baseline interferometry observations, *J. Geophys. Res.*, **91**, 9473-9487, 1986.

Minster, J. B., T. H. Jordan, P. Molnar, and E. Haines, Numerical modelling of instantaneous plate tectonics, *Geophys. J. R. Astr. Soc.*, **36**, 541-576, 1974.

Minster, J. B., and T. H. Jordan, Present-day plate motions, *J. Geophys. Res.*, **83**, 5331-5354, 1978.

Minster, J. B., and T. H. Jordan, PMOTION: interactive software for computing RM2-predicted relative motions and uncertainties between CDP sites, NASA Crustal Dynamics Project, unpublished technical report, April, 1983.

Minster, J. B., and T. H. Jordan, Vector constraints on Quaternary deformation of the western United States east and west of the San Andreas Fault, in *Tectonics and Sedimentation along the California Margin*, edited by J. K. Crouch and S. B. Bachman, *Pac. Sect. Soc. Econ. Paleontol. Mineral.*, **38**, 1-16, 1984.

Minster, J. B., and T. H. Jordan, Vector constraints on western U.S. deformation from space geodesy, neotectonics and plate motions, *J. Geophys. Res.*, **92**, 4798-4804, 1987.

Molnar, P., and Q. D. Deng, Large earthquakes and average rate of deformation in Asia, *J. Geophys. Res.*, **89**, 6203-6228, 1984.

Nuttli, O. W., Average seismic source-parameter relations for mid-plate earthquakes, *Bull. Seismol. Soc. Am.*, **73**, 519-535, 1983.

Savage, J. C., Strain accumulation in western United States, *Ann. Rev. Earth Planet. Sci.*, **11**, 11-43, 1983.

Savage, J. C., M. Lisowski, W. H. Prescott, and A. R. Sanford, Geodetic measurements of horizontal deformation across the Rio Grande Rift near Socorro, New Mexico, *J. Geophys. Res.*, **85**, 7215-7220, 1980.

Wallace, R. E., Patterns and timing of Late Quaternary faulting in the Great Basin province, and relation to some regional tectonic features, *J. Geophys. Res.*, **89**, 5763-5770, 1984.

Woodward, L. A., Rate of crustal extension across the Rio Grande Rift near Albuquerque, New Mexico, *Geology*, **5**, 269-272, 1977.

Zoback, M. L., and M. Zoback, State of stress in the conterminous United States, *J. Geophys. Res.*, **85**, 6113-6156, 1980.

CURRENT PLATE MOTIONS

Richard Gordon, Charles DeMets, Seth Stein, Don Argus, and Dale Woods: Department of Geological Sciences, Northwestern University, Evanston, Illinois 60201

The standard against which VLBI measurements of continental drift and plate motions are compared are self-consistent global models of "present-day" plate motions determined from geophysical data: marine magnetic anomalies at oceanic spreading centers, azimuths of transform faults, and orientations of earthquake slip vectors on transform faults and at subduction zones. Past global plate motion models have defined regions where the assumption that plates behave rigidly has apparently lead to systematic misfits, sometimes exceeding 10 mm/yr, of plate motion data. Here, we present some of the results from NUVEL-1, a new, self-consistent global model of present-day relative plate motions determined from a compilation and analysis of existing and new geophysical data. These data and new techniques have allowed us to eliminate nearly all statistically significant systematic misfits identified by earlier models, suggesting that the rigid-plate assumption is an excellent approximation when plate motions are averaged over several million years. Beside improving estimates of the motion on previously identified plate boundaries, we have also identified and determined motions on other boundaries whose subtle morphologies, lack of seismicity, and very slow ($<$ 10 mm/yr) relative motions have made them difficult to detect. Here we focus on the application of VLBI measurements to help resolve plate tectonic problems and then briefly outline our results for Pacific-North America motion and plate motions in the Indian Ocean.

What important plate tectonic problems can VLBI measurements address? Since the NUVEL-1 model, like earlier models, averages plate motions over 3 Ma, it has nothing to say about short-term plate motion episodicity, something that geodetic techniques such as VLBI can measure directly. Second, conventional geophysical data can directly determine the motion only between two plates which share a boundary. Estimates of motion between plates lacking a common boundary or separated by a subduction zone can only be made by invoking plate circuit vector closure constraints based on the assumption that the plates are rigid (see Figure 1). In contrast, VLBI measurements can yield the rate and direction of motion between any two plates. Comparison between rates measured by VLBI to those predicted by closure of the plate motion model at subduction zones and between plate pairs without a common boundary thus tests the extent to which plates act rigidly.

Plate motion models have long suggested that the rate of motion between the Pacific and North American plates is ~60 mm/yr, in contrast to the well-documented ~35 mm/yr of motion along the San Andreas fault. This discrepancy, termed the San Andreas discrepancy, presumably must be reconciled by extension in the Basin and Range and by strike-slip motion on other California faults parallel to the San Andreas. However, estimates of the rate and direction of Basin and Range extension fall at least

M. J. Reid and J. M. Moran (eds.), The Impact of VLBI on Astrophysics and Geophysics, 351–352.

10 mm/yr short of the expected Pacific-North America rate [Minster and Jordan, 1984]. A compilation and analysis of the available marine magnetic profiles from the southern Gulf of California, where the long-term measure of the Pacific-North America rate can be found, suggests spreading in the Gulf has averaged 48 mm/yr since 3 Ma, 20% lower than estimates of the Gulf spreading rate [Larson et al., 1968] used in prior models [e.g. Chase, 1978, Minster and Jordan, 1978].

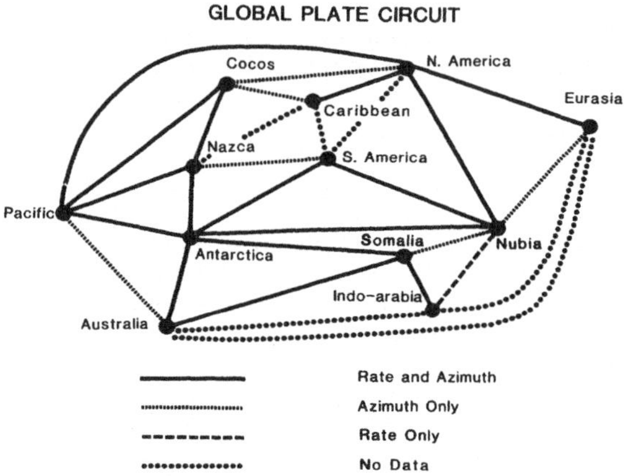

Figure 1

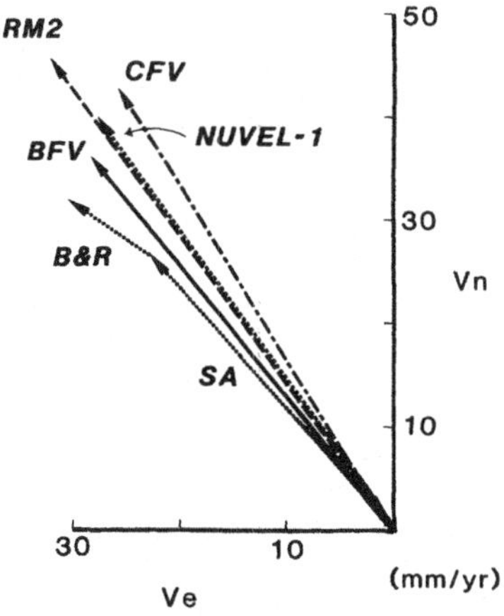

Figure 2

Independently, when direct Pacific-North America data are deleted from the NUVEL-1 global dataset, we predict rates of 49 mm/yr in the southern Gulf and in central California, implying that the "missing motion" along the San Andreas is 30% less than estimated from prior models. Figure 2 shows that the observed relative motions across the San Andreas Fault and the Basin and Range province fail to sum to the Pacific-North America velocity vector predicted by NUVEL-1. Both NUVEL-1 and Minster and Jordan's [1978] RM2 model predict comparable (~9 mm/yr) shortening across the plate boundary zone; however, NUVEL-1 predicts only 4 mm/yr of additional slip parallel to the San Andreas, ~60% less than predicted by RM2.

The Indian Ocean plate motion circuit between the Antarctic, African, and Australian plates has posed problems for prior global relative plate motion models and therefore provides an important test of the rigid-plate hypothesis. The Indian Ocean is often cited as the type-location of oceanic intraplate deformation because of the non-closure found in prior models, high levels of seismicity and heat flow, mappable reverse faults, and undulations in topography and gravity. The three spreading centers, the Central, Southwest, and Southeast Indian ridges are all long and have well-distributed observations of azimuths and rates, thereby guaranteeing strong constraints on the individual Euler vectors and a strong test of circuit closure. Our evaluation of Indian Ocean plate circuit closure [DeMets et al., 1987], using a statistical test of closure [Gordon et al., 1987], yields no evidence for non-rigid plate behavior at even the 2-3 mm/yr level, a result mainly attributable to new spreading rate data from the Southeast Indian Ridge and a revised geometry for the Indian and Arabian plates [Wiens et al., 1985]. In contrast to earlier models, we place India and Australia on separate plates and predict the distance between the continents is decreasing at 12 mm/yr. This could be tested using VLBI between stations in India and Australia.

CONSTRAINTS ON NORTH AMERICAN–PACIFIC PLATE BOUNDARY DEFORMATION IN CENTRAL CALIFORNIA FROM VLBI AND GROUND-BASED GEODETIC DATA

Jeanne Sauber, Thomas H. Jordan, Gregory C. Beroza
Department of Earth, Atmospheric, and Planetary Sciences
Massachusetts Institute of Technology
Cambridge, MA 02139

Thomas A. Clark
NASA–Goddard
Greenbelt, MD 20771

Michael Lisowski
U.S.G.S.
Menlo Park, CA 94025

To accommodate the relative motion across the North American–Pacific plate boundary predicted by global plate solutions, significant deformation on faults other than the San Andreas is necessary. In central California, this deformation is thought to include distributed compression perpendicular to the San Andreas as well as right-lateral strike-slip motion parallel to the San Andreas on faults such as the San Gregorio/Hosgri system. A self-consistent set of VLBI observations from experiments beginning in October 1982 is used to determine the vector rate of change of station position at central California VLBI sites Ovro, Mojave, Vandenberg, Fort Ord, Presidio, and Point Reyes. To estimate VLBI station positions, a procedure is used that minimizes the uncertainties in defining a reference frame by including a priori geologic and geodetic information. The vector rate of change of station positions provides constraints on the integrated deformation rates between stations. Geologic and geophysical data suggest that the rate and mode of deformation varies on both local and regional scales. Thus, the VLBI derived results are interpreted in the context of an overall tectonic framework by examining geologic and ground-based geodetic data.

M. J. Reid and J. M. Moran (eds.), The Impact of VLBI on Astrophysics and Geophysics, 353.

GEODESY BY RADIO INTERFEROMETRY: DETERMINATION OF VECTOR MOTIONS FOR SITES IN THE WESTERN UNITED STATES

David Gordon
Science Applications Research
Code 621.9; NASA/Goddard Space Flight Center
Greenbelt, Md. 20771

Since the late 1970's, NASA's Crustal Dynamics Project has been using Mark III VLBI to study plate tectonic motion, plate boundary deformation and earth dynamics. One major thrust has been the study of crustal motions in the western continental United States. For this effort, several relatively large, fixed antennas (Mojave, Owens Valley, Hatcreek, Ft. Davis and Vandenberg) have been used along with two small, highly mobile VLBI systems which have periodically visited sites of tectonic interest.

For the work reported here, all dual frequency (S and X band) mobile VLBI experiments conducted between October, 1982 and February, 1987, and some fixed station only experiments dating back to 1979 were used. Each experiment was analyzed using weighted least squares. For each baseline, three orthogonal coordinates (length, transverse and vertical) were determined. The transverse component (in the horizontal direction at either site and perpendicular to the baseline vector), being very sensitive to earth orientation, was used only for data after January 1984, when the four station IRIS program (Carter et al., 1985) began making highly accurate measurements of the earth's orientation. The vertical component was not used because of its greater error due to uncertainties in the atmospheric delay. The length and transverse components for each baseline were solved via least squares for their rates-of-change. Though primarily a study of the western U.S., we have included several baselines to Westford and Fairbanks to provide some information on the rigidity of the North American plate. The length and transverse rates were used in a least-squares analysis of site motions. There were 83 length and 74 transverse determinations. Mojave was selected as the fixed reference point, and Westford was additionally held fixed to anchor the coordinate system. The results are shown in Figure 1, where velocity vectors and their one sigma error ellipses are plotted. For greater details of the experiments and analysis procedures, see Clark et al. (1987) and references therein.

Eastward motion is seen at Ft. Davis and Plattville, indicative of the expected spreading across the Basin and Range. Also as expected, the six sites west of the San Andreas show N.W. motions essentially parallel to the local trace of this fault. Pearblossom and Presidio,

M. J. Reid and J. M. Moran (eds.), The Impact of VLBI on Astrophysics and Geophysics, 355–356.

just east of the San Andreas, also show similar N.W. motions. None of the Pacific plate sites, however, shows the full predicted RM2 (Minster and Jordan, 1978) velocity of 56 mm/yr (now thought to be too high by about 20%). Interestingly, Fairbanks is required to move 14 mm/yr (1 sigma = 4 mm/yr) to the S.E in this solution. Thus, with respect to Fairbanks, Vandenberg and Ft. Ord do show the full RM2 motion and Mojave shows N.W. motion of about 14 mm/yr. Thus, inasmuch as Fairbanks can be considered representative of stable North America, there is evidence that significant N.W. motion is occurring well east of the San Andreas. This possibility is also supported by the southerly motion seen at Ft. Davis, Plattville, Ely, Flagstaff and Yuma.

REFERENCES

Carter, W. E., Robertson, D. S., and Mackay, J. R., 1985, J. Geophys. Res., **90**, p. 4577.
Clark, T. A., Gordon, D., Himwich, W. E., Ma, C., Mallama, A., and Ryan, J. W., 1987, submitted to J. Geophys. Res.
Minster, J. B. and Jordan, T. H., 1978, J. Geophys. Res., **83**, p. 5331.

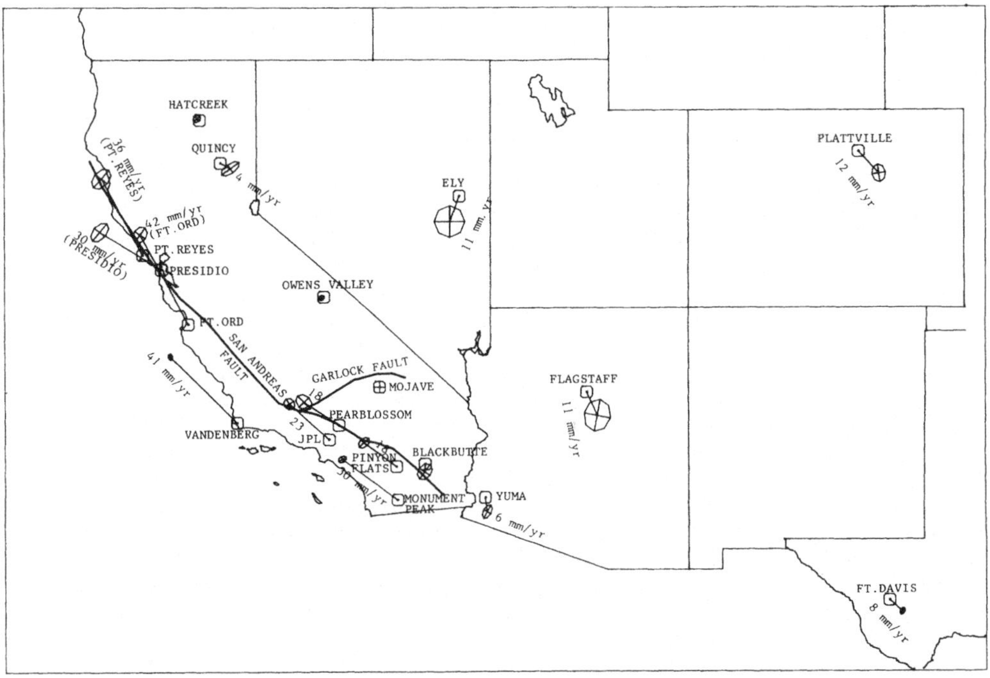

Figure 1: Western U.S. geodetic VLBI sites and their velocities in mm/yr with respect to a fixed Mojave. Also shown are the San Andreas and Garlock faults. The velocities of Owens Valley, Hatcreek and Blackbutte are essentially zero with respect to Mojave.

NASA/CRUSTAL DYNAMICS PROJECT GEODETIC DATA ANALYSIS

W. E. Himwich
Interferometrics Inc.
8150 Leesburg Pike, Suite 1400
Vienna, VA 22180

ABSTRACT. The VLBI group in NASA's Crustal Dynamics Project (CDP) maintains an integrated system for analyzing geodetic VLBI data. This system includes: CALC, calibration programs, SOLVE, GLOBL, and the Data Base System. CALC is the program which contains the models used to calculate the theoretical delay. SOLVE is used to analyze single experiments. GLOBL is used to analyze large groups of experiments. The Data Base System is a self-documenting data storage system used to pass data between programs and archive the data. Kalman filtering is being investigated for operational use in geodetic data analysis.

1. Description

The general flow of geodetic data is from the observatories where the data are recorded, to the correlator for correlation, and finally to the analysis centers, such as Goddard Space Flight Center (GSFC) for analysis. At GSFC, the data pass through CALC, calibration programs, and several housekeeping programs. SOLVE is then used by an analyst to examine the data in detail and develop models for the clock and atmosphere behavior. Once the models have been developed, the data are passed on to GLOBL where they are used in larger solutions.

CALC calculates the geometric delay for the observations including the effects of the best understood geophysical and astronomical models. Currently, the models include: J2000 precession, 1980 IAU Nutation Series, UT1/PM BIH *a priori*, solid earth tides, pole tide, Yoder UT1 tidal variations, ocean loading, gravitational deflection of light, and the PEP numerical ephemeris. CALC also calculates the partial derivatives used for estimation of station and source positions and model parameters.

SOLVE is an interactive weighted least-square analysis program used to analyze single day experiments. The most difficult part of this analysis is developing models for the atmosphere and clock behavior during experiments. Accordingly, SOLVE includes a powerful interactive plotting utility to aid the analyst in developing good

M. J. Reid and J. M. Moran (eds.), The Impact of VLBI on Astrophysics and Geophysics, 357–358.

polynomial models for these effects. Usually 5 to 15 clock or atmosphere coefficients per station must be estimated. The use of these polynomial models does not completely account for the systematic variation of the atmosphere and clocks during the experiment. In order to account for this 'noise,' SOLVE calculates 'Error Constants' that are used to modify the observation weights. SOLVE also allows adjustment of station and source coordinates, UT1/PM, daily nutation offsets or nutation model coefficients, solid earth tides, precession constant, and gamma.

GLOBL is an extension to SOLVE used to estimate parameters from large data sets. Typical large solutions use 470 experiments, 180,000 observations, and 12,000 parameters. To make this tractable, a technique called 'arc-parameter elimination' is used. This technique takes advantage of the large scale structure of the normal equations to reduce the matrix manipulation by a factor of more than 10,000 compared to a direct solution. Parameters can be estimated independently for each experiment or globally from all the experiments. The two most common types of solutions are UT1/PM and Baseline. In a UT1/PM solution, the station and source coordinates are estimated globally and UT1/PM are estimated independently for each experiment. The result is a UT1/PM series. In a Baseline solution, sources coordinates are adjusted globally, station coordinates are estimated independently for each experiment. The produces a history of baseline evolutions. Model parameters can be adjusted as part of either solution.

The Data Base System is the glue that holds the system together. It provides a standard, secure, and permanent means of passing data between programs and archiving the data. One of the goals of the Data Base System is to maintain the accountability of the data analysis. A record is kept of each step in the data analysis process, along with comments by the analyst. The Catalog System keeps track of all the data bases and where they reside on disk and tape. The Data Base System is designed to be as portable as possible and has been used successfully on HP, UNIVAC, IBM 360, and DEC VAX computers.

Kalman filtering is being investigated for use as an operational technique for data analysis. Filtering Mark III geodetic data was pioneered by Tom Herring and Jim Davis at CfA. It uses statistical descriptions of clock and atmosphere behavior to model their variation from one observation to another. The statistical model of the clock and atmosphere used by the filter is physically more realistic than the polynomials used by SOLVE. This improvement effectively removes the need to adjust the observation weights to account for unmodeled 'noise.' Because much of the analyst time in SOLVE is spent parametrizing the clocks and atmosphere, the filter has the potential to free analyst time for other work. The filter requires about two to three times as much computation as least-squares.

2. Acknowledgements

This analysis system is the result of the combined efforts of the GSFC/CDP VLBI Group over a period of more than 10 years.

FIRST RESULTS OF THE VLBI-INVESTIGATION OF SOURCES FROM GEODETIC IRIS-EXPERIMENTS

C.J. Schalinski, W. Alef, A. Witzel
Max-Planck-Institut für Radioastronomie, Bonn, F.R.G.

J. Campbell, H. Schuh
Geodätisches Institut der Universität Bonn, F.R.G.

The geodetic *IRIS*-network is a transatlantic 5-station VLBI-array dedicated to monitor variations of the earth's rotation (e.g. Carter et al.,1985). In a 5 day cycle 14 extragalactic radio sources are observed in regular intervals, covering a GST-range of up to 24 hours, using the MkIII-system in Mode C at 3.6cm (X-band) and 13cm (S-band) wavelength. As this database provides a great potential to monitor source structures and their changes with mas-resolution (the sample contains at least 7 "superluminals"), we have initiated a project to obtain radio maps at both frequencies and to determine, how accurate physical parameters like component flux densities and separations can be measured. Early results are published in Schalinski (1985) and Schalinski et al. (1986).

In addition to astrophysical applications, we investigate the influence of structure phases on phase- and group-delay corrections (s. Campbell et al.,this vol.).

Since the amplitude calibration of geodetic VLBI-data at present is a major problem due to insufficient radiometry, and self-calibration techniques are of only limited use for snap-shot data, the instalment of a consistent calibration is of great importance. We selected the BL-Lac objects 0212+73 and 1803+78, for the following reasons:

1. Due to the fact that these sources are circumpolar and are thus observable during the whole session, these two objects are the best observed sources among the sample with the most regular uv-coverage.
2. The mas structures of these sources are well-known from VLBI-experiments at different frequencies and epochs (e.g. Eckart et al.,1986), so that the calibration of the *IRIS*-data can be checked independently (partly the data were taken at the same epoch).
3. The source morphology (core-jet type) is such that these sources can be mapped adequately with four station networks; components or structural variations can be easier identified at different epochs than in maps of sources with complex structure. In the following we summarize the main results: {1} We obtained a consistent calibration for both sources at S- and X-band by incorporating models at 1.6 and 5 GHz (details s. Schalinski, 1985). 1803+78 serves as a "calibrator", because it's structure basically consists of "stationary" components, separated by about 1.4 mas (X-band) and 5 mas (S-band) under a position angle of almost 260 (s.Fig.1). {2} We made *hybrid*-maps with the *HYBER*-software

M. J. Reid and J. M. Moran (eds.), The Impact of VLBI on Astrophysics and Geophysics, 359–360.

package of the MPIfR for these sources at the following epochs: 1983.4, 1983.9,1984.3,1984.5,1984.6,1985.8,1986.2,1986.5 and 1986.9. Up to 7 iterations were sufficient for convergence. The dynamic range in the VLBI-maps reaches 50:1. Furthermore, we demonstrated the applicability of the calibration scheme by mapping the quasar 4C39.25 at 2 epochs (s.Fig.2: Schalinski,Alberdi et al.,this vol.). The procedures used to calibrate and map the sources allow to reduce more data in a reasonable amount of time and thus enable us to extend the mapping to the other *IRIS*-sources. This work is in progress. (3) From modelfitting of the 1803+78- and 0212+73-data we determined the errors on component flux densities and separations to be on the order of 20%. Preliminary analysis of spectral indices for 0212+73 showed that the *core* has an inverted spectrum with a median of α [2.2/8.4GHz]=0.3 +/- 0.1, in good agreement with the value obtained for a statistically complete sample of flat spectrum radio sources (0.4 +/- 0.2:Eckart et al.,1986).
Further improvements of the calibration and dynamic range of the maps require flux density monitoring during each *IRIS*-run - at least one measurement of the "calibrator" 1803+78.

REFERENCES

Carter W.E.,Robertson D.S.,Mackay J.,1985: Journ.of Geophys.Res.,90,4577
Eckart A.,Witzel A.,Biermann P.,Johnston K.J.,Simon R.,Schalinski C.J., Kühr H., 1986 : Astron. & Astrophys., 168, 17
Schalinski C.J.,1985: Diploma thesis, Universität Bonn
Schalinski C.J.,Alef W.,Campbell J.,Schuh H.,Witzel A., 1986: *Die Arbeiten des SFB 78 Satellitengeodäsie der TU München 1984 und 1985*, ed.M. Schneider(München : Bayerische Akademie der Wissenschaften), 292

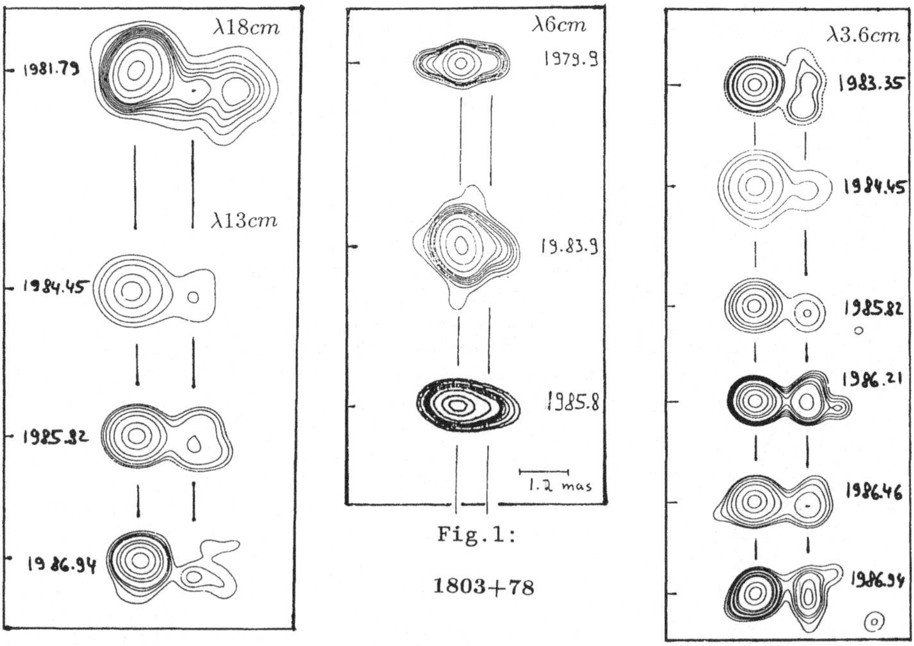

Fig.1:

1803+78

COMBINING VLBI AND GPS TO PRODUCE A NATIONAL CRUSTAL MOTION NETWORK IN THE UNITED STATES

W. E. Strange and G. L. Mader
National Geodetic Survey
Charting and Geodetic Services
National Ocean Service, NOAA
Rockville, Md. 20852

ABSTRACT. A National Crustal Motion Network for the United States using VLBI and GPS technologies is being developed. Implementation of both the VLBI and GPS components of the network is underway.

The many geophysical requirements for crustal motion monitoring in the United States result in the need for a National Crustal Motion Network (NCMN) consisting of hundreds of stations which are reobserved at varying intervals over extended time periods. Differential station positions must be determined with an accuracy of ± 1 to 3 cm. Fixed and mobile VLBI systems have demonstrated their ability to monitor crustal motion with the required accuracy. However, cost and logistical considerations prevent the use of mobile VLBI to establish and monitor the large number of stations constituting an NCMN. Comparisons of differential positions obtained during Global Positioning System (GPS) receivers with those obtained using mobile VLBI demonstrate that GPS results can provide equivalent accuracies of 1 to 2 cm over baselines up to about 500 km in length. Because a GPS measurement is about two orders of magnitude less expensive than a mobile VLBI measurement and logistically much simpler (hand portable equipment as compared to two large tractor-trailers for mobile VLBI), GPS will be used for establishing and monitoring most of the NCMN.

The replacement of mobile VLBI by GPS for measurements over shorter baselines heralds an era where VLBI will provide global and regional frameworks for the more detailed GPS measurements. The NCMN will use a regional VLBI reference frame consisting of about 30 VLBI sites. The remaining stations of the NCMN will be established using GPS geodetic receivers. These receivers will be referenced to the VLBI network through permanent co-location of GPS receivers at about 5 fixed VLBI sites in North America to provide GPS satellite orbit information in the VLBI reference frame and temporary occupation of fixed or mobile VLBI sites with GPS in the immediate vicinity of GPS observation campaigns while they are being carried out.

Implementation of the NCMN has begun. To the maximum extent possible

M. J. Reid and J. M. Moran (eds.), The Impact of VLBI on Astrophysics and Geophysics, 361–362.

the establishment of the NCMN is being carried out as a cooperative effort between NGS and other Federal agencies, state, and local agencies, and university groups. The fixed VLBI stations to be used in the regional VLBI network will consist of the three existing U.S. stations of the IRIS network, other existing stations at Mojave, Owens Valley, and Hat Creek, California, Maryland Point, Maryland and Green Bank, West Virginia, and the Very Long Baseline Array (VLBA) stations as they are implemented. Mobile VLBI sites wil consist of nine stations previously established as part of the NASA Crustal Dynamics Project (CDP) in California, Arizona, Nevada, Utah, and Colorado and stations established by the National Geodetic Survey (NGS) in the States of Washington, Texas, Oklahoma, Georgia, and Indiana during the 1986-87 time period specifically to support the NCMN. Permanent GPS stations for orbit determination are currently being operated at VLBI stations in Massachusetts, Texas, and California as a joint activity by NGS, the University of Texas and the Texas Department of Highways. NGS will begin operation of a permanent GPS station at its Richmond, Florida, VLBI site in August 1987.

In support of the NASA CDP, NGS currently operates its mobile VLBI systems to monitor a 15 station network in California and some five additional stations in the western United States. Mobile VLBI measurements over the California network have shown that movement across the San Andreas and associated fault systems is 3.5 to 4.5 cm/yr, less than the estimated long term North American-Pacific plate motion of 4.8 to 5.6 cm/yr. GPS measurements over the southern California mobile VLBI network were carried out by NGS in June 1986, and repeated in January 1987. These measurements demonstrated the ability of GPS to replace mobile VLBI over this network provided two to three stations continue to be monitored with mobile VLBI and represents the initiation of a transition from mobile VLBI to GPS monitoring over the next 2 to 3 years. In conjunction with other organizations additional GPS stations were established to densify the NCMN in southern California and extend it to the offshore islands. By the end of 1987 additional GPS activities to extend and densify the NCMN will include: (1) a 43 station GPS strain network reaching from the eastern slope of the Rocky Mountains to the Atlantic Ocean, (2) interconnection of the Hawaiian Islands, (3) repeated GPS occupations of VLBI sites in Alaska, (4) detailed GPS networks in the Imperial Valley of California, the Hegben Lake-Yellowstone area, the Anchorage area, Coastal Maine, the State of Tennessee, southern Arizona and the Houston-Galveston area of Texas, (5) a 20 to 30 station regional network in northern California and (6) a regional network in northwest Washington State.

PLATE MOTION AND EARTH ORIENTATION

A. Mallama and M. Kao
Science Applications Research
4400 Forbes Boulevard
Lanham, Maryland 20706

ABSTRACT. Earth orientation series are linked to the terrestrial reference frame in which the observing site locations are measured. The effect of tectonic motion is a simple rotation for any given plate, but the overall effect depends on the distribution of sites. The magnitude of this motion is large enough to be evident in the data. For example, the coefficient of rotation for the North American plate around the Earth's Y-axis is -0.8 millarcseconds per year in the AM0-2 plate motion model of Minster and Jordan. The VLBI analysis system at NASA/GSFC for computing earth orientation series has recently been enhanced by including the Minster and Jordan model for a priori tectonic effects. Tests indicate that the weighted-root-mean-square residual of observations to the solution is decreased by using this model.

1. INTRODUCTION AND DATA DESCRIPTION

NASA's Crustal Dynamics Project (CDP) developed and began using Mark III VLBI during the 1970's to study earth orienation and plate tectonic motion. Earth orientation is the instantaneous position of the crust of the earth relative to its spin axis (i.e., the two orthogonal components of polar motion), and the phase of the earth's rotation (i.e., Universal Time). Since the earth's crust is composed of tectonic plates that are in motion relative to one another, it is clear that plate motion is connected to the earth orientation series.

The VLBI data reported on in this paper were collected in three series of experiments, CDP, IRIS, and MERIT. The CDP experiments are managed by NASA/GSFC. They are a heterogeneous set taken on a network of about 20 sites that span the North American, Eurasian, and Pacific plates. The IRIS experiments are managed by NGS for the purpose of monitoring earth orientation. This system provides UT1 and both components of

M. J. Reid and J. M. Moran (eds.), The Impact of VLBI on Astrophysics and Geophysics, 363–364.

polar motion at 5 day intervals. The MERIT data are from 14 experiments conducted during 1980 that serve to link the CDP and IRIS data.

2. PLATE MOTION MODEL AND RESULTS

The model of Minster and Jordan, 1978, has been chosen to account for plate motion in the VLBI data. It is derived from a geophysical data set consisting of hundreds of spreading rates, transform fault azimuths, and earthquake slip vectors. The results from our VLBI data when solved for site motions agree in an overall way with the Minster and Jordan model. The absolute motion model AMO-2, which corresponds to no net rotation, has been implemented in our VLBI analysis system.

The result that best illustrates an improvement in the data fit comes from the comparison between sets of experiments involving sites on the Pacific and North American plates, which have a large relative motion. The wrms fit in the solution is 93.2 psec without the plate motion model and 89.6 psec with it. The improvement with the model included indicates that it has successfully accounted for site motion. In the case where all the CDP, IRIS, and MERIT data are combined in a global solution the fit is 88.0 psec without the plate motion model and 87.0 psec with it. In an rss sense the model removes 13 psec of noise.

3. DISCUSSION

In our long-term effort to refine the VLBI technique we have included many calibrations such as precession and nutation, ionospheric delay, and ocean loading. The addition of the plate motion model to our system is in the category of a geophysical calibration. The results of our tests show that the application of the model improves the fit between observed and calculated data. We expect that a plate motion model will also improve the consistency between different sets of earth orientation data taken on entirely different networks, VLBI and otherwise.

4. REFERENCE

Minster, J.B. and Jordan, T.H., 'Present Day Plate Motions,' J. Geophysical Research, 83, 5331-53, 1978.

DISTRIBUTION OF RELATIVE PLATE MOTION ALONG THE PACIFIC-NORTH AMERICAN PLATE BOUNDARY DETERMINED FROM MOBILE VLBI MEASUREMENTS

P. M. Kroger, G. A. Lyzenga, K. S. Wallace, and J. M. Davidson
Jet Propulsion Laboratory, California Institute of Technology
4800 Oak Grove Drive, MS 238-640
Pasadena, California 91109

1. INTRODUCTION

The problem of understanding the deformation occurring along the Pacific-North American plate boundary in the western United States depends upon understanding the forces which drive the plates in this region. One of the primary sources of our knowledge concerning these forces lies in their manifestation as relative displacements which occur throughout the broad zone of deformation surrounding the San Andreas fault system. It is information concerning the spatial and temporal distribution of these motions which will be of greatest benefit in helping to determine which of several possible mechanisms is responsible for driving contemporary plate motions in this region.

The resolution of the motions in this region is itself proving to be a problem of considerable magnitude. One of the principal questions involving the present day motions is the apparent discrepancy between the relative plate rate of 5.6 cm/yr predicted by rigid plate models and the slip rates across the San Andreas fault of 2.5–3.5 cm/yr which have been determined from geologic and earth-based geodetic measurements. There is a strong possibility that this "missing motion" is occurring west of the San Andreas fault, probably in association with offshore faults. Measurements using the Global Positioning System (GPS) satellites have recently extended the coverage of geodetic observations to island sites off the southern California coast, but it will require at least several years of such measurements before accurate velocities to these offshore sites can be inferred.

2. RESULTS

We have used over 6 years of mobile VLBI baseline measurements in this region to determine the velocities of 11 sites relative to North America (see Figure 1.). We find these velocities to be generally consistent with those determined from geologic data and contemporaneous satellite laser ranging measurements made in the same region.

The most striking feature of the results presented in Figure 1 is the contrast between the local distribution of velocities in northern and southern California. In southern California the velocities of sites west of the San Andreas fault increase gradually with increasing distance west of the fault reaching a maximum value of 4.1 cm/yr at Vandenberg AFB on the Pacific coast approximately 100 km west of the fault. In northern

M. J. Reid and J. M. Moran (eds.), The Impact of VLBI on Astrophysics and Geophysics, 365–366.

California, on the other hand, velocities of 4.6 and 4.5 cm/yr are obtained for the sites at Ft. Ord and Pt. Reyes located within 15 km of the fault. It is noteworthy that even the westernmost of the VLBI sites in southern California falls significantly short of the full 5.6 cm/yr predicted by rigid plate models.

3. DISCUSSION

We find that once transient cyclic deformation associated with the earthquake cycle is taken into account, the observed geodetic site motions are consistent with the activity of known subsidiary faults to the San Andreas and with the 5.6 cm/yr plate motion rate.

These conclusions are reached through consideration of a finite element model of elastic deformation in the lithosphere. The aim of the modeling is to attempt an explanation of the apparent variations in strain distribution along the strike of the San Andreas system. In particular, a suite of two-dimensional models of time dependent response in the coupled lithosphere-asthenosphere system is considered. These models provide synthetic displacement profiles across three typical cross sections of the plate boundary.

The three "slices" perpendicular to the faults are differentiated by the number and location of active faults with respect to the plate boundary. While the results of these calculations are successful in reproducing both classical geodetic and VLBI results, limitations in the coverage and precision of the available data preclude the placing of strong constraints upon the parameters of these theoretical models. This fact points out the need for a much denser spatial network of observing sites if more sophisticated models of interplate deformation in this region are to be successfully tested.

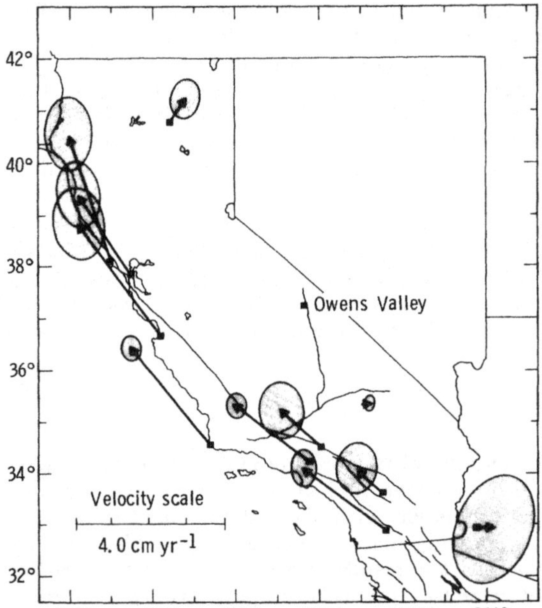

Figure 1. Velocities of VLBI sites relative to the site at the Owens Valley Radio Observatory which is assumed fixed. Error ellipses represent one standard deviation.

VLBI geodesy: 2 parts-per-billion precision in length determinations for transcontinental baselines

J.L. Davis, T.A. Herring, and I.I. Shapiro
Harvard-Smithsonian Center for Astrophysics
Cambridge, Massachusetts 02138

ABSTRACT. We have used very-long-baseline interferometry (VLBI) to make twenty-two independent measurements, between September 1984 and December 1986, of the length of the 3900-km baseline between the Mojave site in California and the Haystack/Westford site in Massachusetts. These experiments differ from the typical geodetic VLBI experiments in that a large fraction of observations are obtained at elevation angles between 4° and 10°. Data from these low elevation angles allows the vertical coordinate of site position, and hence the baseline length, to be estimated with greater precision. For the sixteen experiments processed thus far, the weighted root-mean-square scatter of the estimates of the baseline length is 8 mm. We discuss these experiments, the processing of the data, and the resulting baseline length estimates.

1. INTRODUCTION

In recent years, the precision of estimates of baseline length obtained from the analysis of Mk-III very-long-baseline interferometry (VLBI) data has passed below the level of 10 parts-per-billion. The limiting source of error continues to be the atmosphere. The effects of atmospheric errors on estimates of baseline length are discussed in Lanyi [1984], Davis *et al.* [1985], Davis [1986], Herring [1986], and Treuhaft and Lanyi [1987]. The typical treatment of atmospheric propagation delay in the analysis of geodetic VLBI data involves the use of surface meteorological measurements (and possibly radiometric measurements; see Elgered *et al.* [1987]) to determine an *a priori* value for the zenith propagation delay, and the use of a mapping function. (The mapping function describes the elevation-angle dependence of the propagation delay, *i.e.*, the number of air masses traversed.) Corrections to the *a priori* value of the zenith delay are then estimated, along with the relevant geodetic parameters, from the VLBI data.

The unique elevation-angle dependence of the atmospheric propagation delay allows a precise estimate to be made of the zenith-delay corrections. However, the sensitivity of the VLBI group-delay measurement to a change in the zenith delay is highly correlated with the sensitivity of the group delay to a change in the vertical coordinate of site position for small ranges of air mass. By observing sources at low elevation angles, we can decrease the correlation of the estimates of the zenith-delay corrections with those of the vertical coordinate of site positions, and thereby obtain a more precise estimate of those (and other) parameters. Nevertheless, many schedules for geodetic VLBI experiments include no observations below about 10° elevation at any site. This omission is primarily intended to *avoid* the effects of errors in the mapping function used, since such errors are generally worse at lower elevations. (It is harder to predict atmospheric properties at greater horizontal distances from the site.) We therefore pose the question: What is the "optimum" minimum elevation angle to minimize errors in baseline-length estimates?

M. J. Reid and J. M. Moran (eds.), The Impact of VLBI on Astrophysics and Geophysics, 367–368.

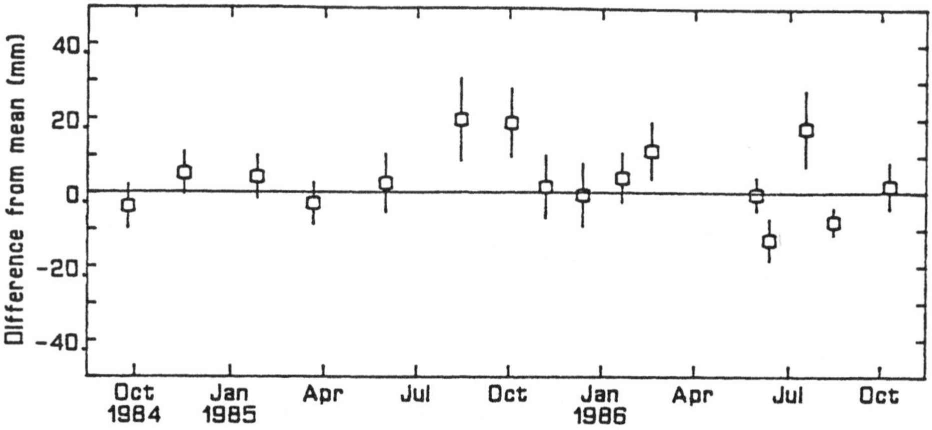

Figure 1. Estimates of the length of the baseline from Haystack/Westford to Mojave (see text). A mean value of 3,904,144,248 mm has been subtracted from the values shown.

2. DESCRIPTION OF EXPERIMENTS and RESULTS

In order to study this question, we designed VLBI experiments known as "low-elevation" experiments in reference to their scheduling strategy: the attempt is made to observe low in the sky as frequently as possible, within the limits of the antennas used. For the standard set of low-elevation experiments, we used the Mojave site ($\epsilon_{min} = 8°$) in California and the Haystack/Westford site ($\epsilon_{min} = 4°$) in Massachusetts.

The processing of the data from these experiments differed in several respects from the processing which we employed several years ago (see, *e.g.*, Clark *et al.* [1985]) for geodetic VLBI data. A new mapping function for the dry atmosphere, developed by Davis *et al.* [1985], and accurate to about 10 mm at 5° elevation for a wide range of atmospheric conditions, was used. Another improvement was the use of a Kalman filter to estimate clock offsets and zenith delay corrections. Each of these corrections is modeled as a time-varying stochastic process, the statistics of which are deduced from other sources; the values of the corrections for the epochs of the VLBI observations are estimated by the Kalman filter. Finally, we used radiometric data to estimate the "wet delay" when these data are available.

In Figure 1, we present the estimates of the Haystack-Mojave baseline lengths as a function of experiment date. The error bars represent the standard deviation of the estimate of the baseline length, deduced from the propagation of the stochastic errors involved. These errors include the the measurement errors of the group-delay observations and the stochastic behavior of the clocks and atmospheres. The weighted root-mean-square scatter of the estimates about their weighted mean is 8 mm, representing a fractional repeatability of 2 parts-per-billion.

This work was supported by Air Force Geophysics Laboratory contract F-19628-86-K-0025; NASA grant NAG5-538; and NSF grants EAR-83-06380 and EAR-86-18989.

Clark, T.A. *et al.*, *IEEE Trans. GARS* **GE-23**, 438–449, 1985.
Davis, J.L., T.A. Herring, I.I. Shapiro, A.E.E. Rogers, G. Elgered, *Radio Science* **20**, 1593, 1985.
Davis, J.L., U.S. Air Force report AFGL-TR-86-0243, 1986.
Elgered, G., J.L. Davis, T.A. Herring, I.I. Shapiro, this volume, 1987.
Herring, T.A., *J. Geophys. Res* **91**, 9177–9182, 1986.
Lanyi, G.E., *Proc. Int. Symp. Space Tech. Geodyn.*, IAG/COSPAR,**2**, 184–195, 1984.
Treuhaft, R.N., G.E. Lanyi, *Radio Science* **22**, 251–265, 1987.

THE LONG TERM STABILITY OF VLBI EARTH ORIENTATION MEASUREMENTS

T.M. Eubanks and J.A. Steppe
Jet Propulsion Laboratory
California Institute of Technology
Pasadena, California 91109

ABSTRACT. Tectonic motions will, in general, change the orientation as well as the length of baselines used in Very Long Baseline Interferometry (VLBI), and will thus cause slow divergences between Earth orientation results obtained with different VLBI networks, as well as between VLBI results and those obtained by Satellite Laser Ranging (SLR) and Lunar Laser Ranging (LLR). Such drifts (on the order of a milliarcsecond / year) are inherently interesting as well as being significant in combinations of orientation results from different sources. The geodetic study of tectonic motions is also closely connected to research into the nature and causes of systematic errors in data from the modern techniques of space geodesy. We describe both a special coordinate system found to be of use in the analysis of VLBI data and tectonic motion estimates for a VLBI baseline stretching from California to Australia.

1. Geometrical Nature of Errors in Geodetic VLBI

Understanding the distribution of errors in single baseline geodetic VLBI is an important step in the validation of global geodesy. We have found a special coordinate system, the Transverse, Vertical, Length (TVL) coordinate system, to be very useful in analysis of VLBI data. This system is defined by the baseline vector (the length component) and the cross product between the vectors from the center of the Earth to the telescopes at the ends of the baseline (the transverse component), with the vertical component completing a right handed orthogonal system. The transverse is defined so that a positive transverse motion moves one end of the baseline in a counterclockwise fashion as seen from above the other.

Only a small region of the sky is visible from both ends of a intercontinental baseline, forming a narrow sector oriented along the transverse direction. Since the geometrical distribution of observations largely determines the inherent accuracy of baseline vector estimates, the transverse component is relatively well determined even for very long baselines, while the accuracy of the length and vertical components decreases with increasing length. In addition, errors in modeling the atmospheric propagation delay at an observatory, probably the largest source of systematic errors in geodetic VLBI (Herring et al. 1986), are highly correlated with the local vertical component of the position of the observatory. Such "local vertical" errors will cause errors in the baseline length and vertical components, but not in the baseline transverse, since this component is in the horizontal plane at both ends of the baseline. Similar errors are present in multi-baseline VLBI data and the effects of unmodeled local vertical errors should be accounted for in the design and analysis of observations from global VLBI networks.

M. J. Reid and J. M. Moran (eds.), The Impact of VLBI on Astrophysics and Geophysics, 369–370.

2. The relative tectonic motion of Australia and California

The Australia - California baseline of the Deep Space Network (DSN) has a large transverse tectonic velocity and is thus quite sensitive to tectonic motions. Figure 1 shows the difference between VLBI baseline transverse measurements for this baseline and a Kalman smoothing of LLR, SLR and non-DSN single baseline VLBI; only measurements with formal errors < 5 milliarcseconds are shown for clarity. (The data used are described in the BIH Annual Report for 1986, and the Kalman filter in Morabito et al. 1987). The baseline vector is defined as pointing from California (longitude 243°E and latitude 35°N) to Australia (149°E and -35°N); the baseline length is ~10,589 km. The observed motion is in excellent agreement with the AM1-2 model of Minster and Jordan (1977), with a drift rate estimate of 9.6 times the formal error. The scatter of the residuals is clearly time correlated, but the observed drift would be significant at the 99.5% level even assuming the ~50 day correlation time of Herring et al. (1986).

ACKNOWLEDGEMENTS. The work described in this paper was supported by the Jet Propulsion Laboratory, California Institute of Technology, under contract with the National Aeronautics and Space Administration.

REFERENCES

Bureau International de l'Heure, Annual Report for 1986, in press.
Herring, T.A. et al., 1986, J. Geophys. Res., 91, 8341-8347.
Minster, J.B. and Jordan, T.H., 1978, J. Geophys. Res., 83, 5331-5354.
Morabito, D.D., Eubanks, T.M., and Steppe, J.A., 1986, Proceedings IAU Symposium 128, in press.

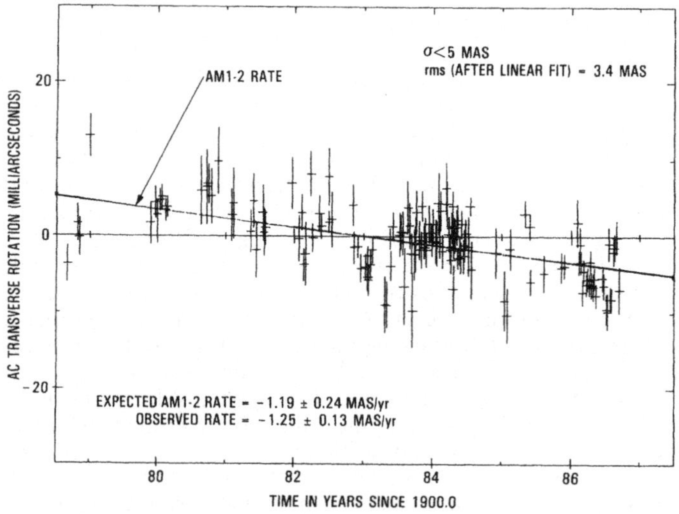

Figure 1. Estimates of the transverse orientation change of the DSN Australia - California baseline minus a smoothing of other data, together with the Minster and Jordan AM1-2 tectonic rate estimate.

VLBI studies of the nutations of the earth

T. A. Herring, Harvard–Smithsonian Center for Astrophysics
60 Garden Street
Cambridge, MA. 02138

ABSTRACT. The application of very–long–baseline interferometry (VLBI) to the study of the nutations of the earth has yielded unprecedented accuracy for the experimental determination of the coefficients of the nutation series. The analysis of six years of VLBI data has yielded corrections to the coefficients of the seven largest terms in the IAU 1980 nutation series with periods of one year or less, with accuracies approaching the truncation error of this nutation series (0.1 mas). The nutation series coefficients computed from the VLBI data, and those obtained from theoretical considerations (the IAU 1980 nutation series), are in excellent agreement. The largest corrections are to the coefficients of the retrograde annual nutation [2.0 $\pm$ 0.1 mas], the prograde semiannual nutation [(0.5 - $\imath$ 0.4) $\pm$ 0.1 mas], and the prograde 13.7 day nutation [-0.4 $\pm$ 0.1 mas]. (The imaginary term for the semiannual nutation represents a term 90° out–of–phase with the arguments of the nutation series.) The geophysical implications of these results are currently under active investigation. We discuss the methods used to extract the nutation information from the VLBI data, the calculations of the uncertainties of the resultant corrections to the coefficients of the nutation series, and the current research into the nutations of the earth.

1. INTRODUCTION

The determination of the coefficients of the nutation series using very-long-baseline interferometry data has been discussed recently in a number of papers (*Herring et al.*, 1983; *Gwinn et al.*, 1984; *Herring et al.*, 1985; *Eubanks et al.*, 1985; *Gwinn et al.*, 1986; *Herring et al.*, 1986a; *Herring et al.*, 1986b; and *Himwich et al.*, 1986). A detailed description of the analysis techniques used for determining the corrections to the coefficients of the nutation series is given in *Herring et al.*, 1986a. The interpretation of the results given in *Herring et al.*, 1986a is discussed in detail by *Gwinn et al.*, 1986. The interpretation of these results has also been discussed by *Wahr and Bergen et al.*, 1986; *Yoder and Ivinus*, 1986; and *Yoder and Ivinus*, 1987. Here we review briefly the procedures used to estimate the corrections to the coefficients of the nutations, and we present the results obtained from the analysis of 412 VLBI experiments obtained during the 6.5 year period between July 1980 and February 1987.

M. J. Reid and J. M. Moran (eds.), The Impact of VLBI on Astrophysics and Geophysics, 371–375.

2. DATA ANALYSIS TECHNIQUE

The techniques we have used to estimate corrections to the IAU 1980 nutation series have been discussed in *Herring et al.*(1985) and *Herring et al.*(1986a). For each VLBI observing session (typically of 24 hours duration), we estimate corrections, $\delta\Delta\psi$ and $\delta\Delta\epsilon$, to the nutation angles computed from the IAU 1980 nutation series. In a post–processing operation, we use these nutation–angle corrections to estimate corrections to the coefficients of certain terms in the nutation series. An alternative technique is to directly estimate the corrections to coefficients of the nutation series from the VLBI data themselves. Results obtained using this technique have been presented by *Himwich et al.*, 1986.

The nutation–angle corrections are not only used to estimate the corrections to the coefficients of the nutation series, but they are also be used to study the noise characteristics of the nutation estimates (see *Herring et al.*,1986a), and to study phenomena not explicitly included in the standard nutation series such as the "free core–nutation" (see *Herring et al.*, 1986b). In particular, we will discuss the power–spectral density (PSD) function of the nutation–angle corrections. The calculation of the PSD function from the nutation–angle corrections have been discussed in *Herring et al.*, 1985. The technique we adopt is to "lightly" smooth the nutation–angle corrections using a Gaussian filter with a full–width–at–half–maximum (FWHM) of 5 days, and to then compute the PSD function from these uniformly spaced smoothed values (5 day spacing). The (assumed uncorrelated) noise in the nutation–angle corrections is propagated through the smoothing and the fast fourier transform (FFT) algorithms to compute the 99.5% confidence limit for the PSD function.

3. RESULTS

The VLBI data we have analyzed are from 412 observing sessions carried out by the NASA Crustal Dynamics Project and the NGS IRIS program between July 1980 and February 1987. This data set is an extension of those discussed in *Herring et al.*(1985) and *Herring et al.*(1986a and b). In Table 1, we give the estimates of the amplitudes of the circular nutations that can be resolved with the limited temporal range of the VLBI data. The values for $\delta\Delta\epsilon$ and $\delta\Delta\psi \sin\epsilon_o$ we used to obtain these amplitude estimates are shown in Figure 1. The PSD function computed from the nutation–angle corrections shown in Figure 1 is shown in Figure 2. The three largest peaks in the PSD function correspond to the three largest corrections to the coefficients of the nutation series given in Table 1. There is also a large peak corresponding to the retrograde frequency closest to 1 cycle per sidereal day. Any corrections to the long period nutations (18.6 and 9.3 years) would all appear in this term. With only 6.5 years of data we can not yet reliably estimate corrections to the nutation series terms with these long periods. The remainder of the PSD function appears to be consistent with the white–noise statistics used to compute the 99.5% confidence interval.

The results given in Table 1 are very consistent with previously published results within the uncertainties of these earlier analyses. The correction to the retrograde annual nutation has been interpreted as being due to the effects of the earth's fluid core being more strongly coupled to the mantle than was inferred for the calculation of the IAU 1980 nutation series (*Gwinn et al.*, 1986). The corrections to the prograde semiannual term are partly due the coupling of the fluid core to the mantle and partly due to the effects of ocean tides on the nutations (*Sasao and Wahr*, 1981). The correction to the 13.7 day prograde term has not

TABLE 1. Estimates of Corrections to Amplitudes of Circular Nutations of IAU 1980 Nutation Series (*Wahr*, 1981).

Period in inertial space, days	Observed Amplitude, mas	Corrections* In-Phase δa_r, mas	Out-of-Phase δa_i, mas
182.6 prograde	-548.62	0.45	-0.41
182.6 retrograde	-24.63	-0.11	-0.07
13.7 prograde	-94.45	-0.37	-0.00
13.7 retrograde	-3.63	-0.03	0.05
365.3 prograde	25.71	0.05	0.18
365.3 retrograde	-33.13	-2.07	0.36
27.6 prograde	14.50	-0.02	0.06
27.6 retrograde	-13.79	0.03	-0.01
121.7 prograde	-21.56	-0.08	-0.03
121.7 retrograde	-0.95	-0.04	-0.03
9.1 prograde	-12.53	-0.10	0.00
9.1 retrograde	-0.43	-0.01	-0.02
31.8 prograde	3.19	-0.01	-0.03
31.8 retrograde	-3.02	0.07	0.00
433.2 prograde	0.06†	-0.06	-0.00
433.2 retrograde FCN	0.33†	0.32	-0.04

*The standard deviation of each correction is estimated to be 0.10 mas. The standard deviations are obtained using the techniques described in *Herring et al.* (1986), with the root–sum–squares addition of a contribution of 0.07 mas to account for (i) the effects of the truncation of the IAU 1980 nutation series to the nearest 0.1 mas, (ii) the possible effects of atmospheric excitation of the nutation, and (iii) the effects of ocean and earth tide modeling errors in the analysis of the VLBI data. See also *Herring et al.* (1986) for definitions of the quantities listed.

†The amplitudes of these terms are the root–sum–squares of the amplitudes of the real and imaginary components.

been satisfactorily explained. Since this nutation term is the one most affected by the elasticity of the earth, the correction may arise from the treatment of elasticity or the neglect of the effects of anelasticity in the IAU nutation series. The root–mean–square scatter of the remaining 25 corrections given in Table 1 is 0.06 mas, which is consistent with our estimated uncertainties of 0.10 mas.

4. CONCLUSION

The quality of the VLBI estimates of the corrections to the nutation series will improve as further VLBI experiments are carried out. The VLBI data from the international earth rotation service will continue to provide high quality measurements of the nutation angles on a regular basis (once every 5 days). These data and those obtained by the NASA Crustal Dynamics Project will ensure that the coefficients of selected terms in the nutation series will be determined with precisions of less than 0.1 mas. However, the current nutation series is truncated at 0.1 mas, and thus, will soon be the limiting error source in the interpretation of the corrections to the nutation series. Since all modern nutation series are obtained by convolving the response function of a "realistic" model earth with the nutations of a rigid earth, we will soon start developing a rigid–earth nutation

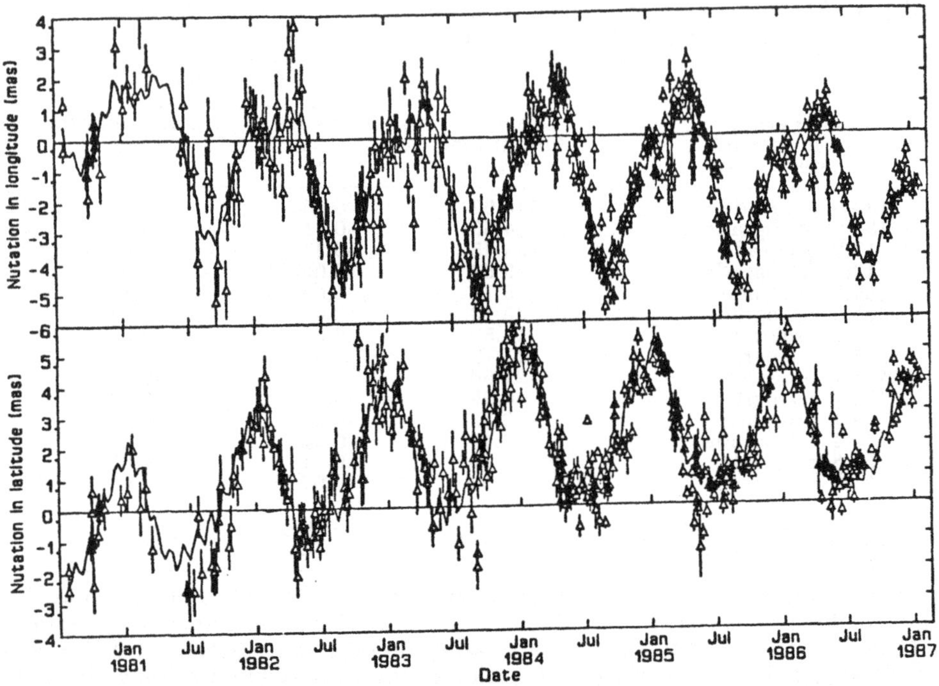

Figure 1. Estimates of $\delta\Delta\epsilon$ and $\delta\Delta\psi \sin\epsilon_o$ from the VLBI data analyzed in this paper. The solid line in each figure is computed from the corrections given in Table 1 and from long-period terms (periods $\geq$ 18.6 years) which are not given in the table because their values are highly correlated and the estimates are not considered reliable.

series with an accuracy of 0.01 mas. This new nutation series will provide an accurate framework in which the observed nutations of the earth can be compared with that computed from geophysical theory.

5. ACKNOWLEDGMENTS

This work was supported by the Air Force Geophysics Laboratory, contract F-19628-86-K-0025; NASA grant NAG5-538; and NSF grants EAR-83-06380 and EAR-86-18989.

6. REFERENCES

Eubanks, T.M. *et al.*, in "Proc. of the international conference on earth rotation and the terrestrial reference frame", Ohio State University, 326–340, 1985.

Gwinn, C.R. *et al.*, *EOS*, **65**, 859, 1984.

Gwinn, C.R. *et al.*, *J. Geophys. Res.*, **91**, 4755–4765, 1986.

Herring, T.A. *et al.*, *EOS*, **64**, 674, 1983.

Herring, T.A. *et al.*, in "Proc. of the international conference on earth rotation and the terrestrial reference frame", Ohio State University, 307–325, 1985.

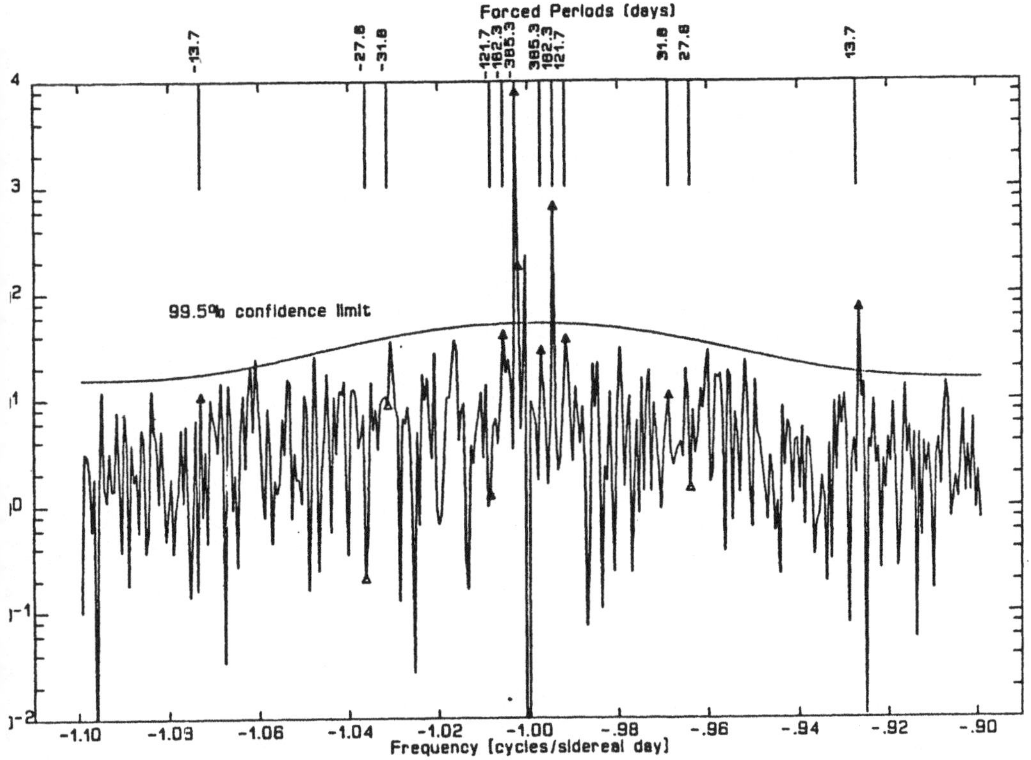

Figure 2. The power–spectral–density (PSD) function of the nutation–angle corrections shown in Figure 1. The decrease in power in the PSD function and in the 99.5% confidence limit with increasing frequency away from -1 cycle per sidereal day (cpsd) is due to use of a (5 day FWHM) Gaussian filter. The frequency spacing in the PSD function is 1/1830 cpsd. At the top of the figure we have shown the frequency positions of some of the major terms in the nutation series. We have also marked the frequency components in the PSD function closest to these frequencies with triangles.

Herring, T.A. *et al.*, *J. Geophys. Res.*, **91**, 4745–4754, 1986a.
Herring, T.A. *et al.*, in "Proc. of IAU Symp. No. 128", 1986b.
Himwich, W.E. *et al.*, in "Proc. of IAU Symp. No. 128", 1986.
Sasao, T. and J.M.Wahr, *Geophys. J. Roy. Astr. Soc.*, **64**, 729–746, 1981.
Wahr, J.M. and T.Sasao, *Geophys. J. Roy. Astr. Soc.*, **64**, 747–765, 1981.
Wahr, J.M. and Z. Bergen, *Geophys. J. R. Astr. Soc.*, **87**, 633-668, 1986.
Yoder, C.F. and E.R. Ivinis, in "Proc. of IAU Symp. No. 128", 1986.
Yoder, C.F. and E.R. Ivinis, in "Proc. of IAU Symp. No. 129", 1987.

PROPERTIES OF THE CORE-MANTLE BOUNDARY

Bradford H. Hager
Seismological Laboratory
California Institute of Technology
Pasadena, CA 91125 USA

The core-mantle boundary (CMB), separating the molten metallic core from the overlying solid silicate mantle, marks the largest discontinuity in mechanical properties within the Earth. The $\sim$ 200 km thick region just above the CMB, named D'' by Bullen (1950), is characterized by an anomalous gradient in seismic velocity versus depth. D'' was originally interpreted as a region with a strong compositional gradient due to the accumulation of dense material at the base of the mantle. Subsequently, the anomalous gradient was interpreted as the result of a strong temperature gradient in a hot thermal boundary layer at the base of the mantle, an interpretation motivated by the requiremnet that heat involved in generating the geodynamo must be transported out of the core and through the mantle by convection.

Determining the nature of D'' is important to understanding the geochemical evolution of Earth, in addition to understanding mantle and core dyanmics. If D'' is compositionally distinct from the mantle, it represents a reservoir of material not yet considered in modeling the composition of the mantle. Since direct sampling is impossible, we must resort to indirect methods to determine its properties.

Mapping the topography fo the CMB is one way of constraining the nature of D''. Due to the high temperature near the CMB, mantle materials respond to stresses by creeping flow; any topographic relief on the CMB must be dynamically maintained by convection in the overlying mantle. If D'' were a separately convecting layer, this topography would be much smaller than if D'' were the thermal boundary layer of the (entire) convecting lower mantle. CMB topography can be probed in a number of ways. Nutational coupling of the core and mantle depends upon CMB ellipticity. Torques caused by the interaction of fluid flow in the core and CMB topography would cause changes in the length of day. The predicted dynamic topography of the CMB caused by lower mantle density contrasts inferred from seismic tomography can be calculated from fluid mechanical models. The variations in travel times of seismic waves that reflect off or are transmitted through the CMB can be mapped into models of CMB topography.

The results from a number of recent models of CMB topography coming from various branches of geophysics are not in agreement. Nutation and length of day studies indicate CMB topography of order hundreds of meters, while seismic models have topography of over 10 km. Dynamic models that include D'' as a chemically distinct layer are in accord with the former estimates.

M. J. Reid and J. M. Moran (eds.), The Impact of VLBI on Astrophysics and Geophysics, 377.

IMPROVED ANALYTIC NUTATION MODEL

C. F. Yoder and E. R. Ivins
Jet Propulsion Laboratory
Calfironia Institute of Technology
4800 Oak Grove Drive
Pasadena, California 91109

Models describing earth's nutations can be separated into two major components, plus several small perturbations, of which ocean tides and solid friction are the most significant. The primary element of a successful model is an accurate rigid earth nutation theory. The effect of mantle elasticity and differential nutation of the fluid core is then obtained as a correction factor which multiplies the rigid body result. Wahr's (1981) theory is certainly the most sophisticated elastodynamic response model. However, we have found that the simple model of Sasao et al., (1980) differs from Wahr's theory term by term by less than 0.3mas if a modern earth structure model (1066B) is used to evaluate the nutation structure constants.

The complex version of Euler's equations is

$$A\quad (D-ie\Omega)m + (D+i\Omega(A_f m_f+Ac_{21}) = -iAe\Omega\phi\ .$$

$$Dm + \left[D+i(1+e_f)\right]m_f + Dc_{21}^{f} = 0$$

where $D=\frac{d}{dt}$, $e=(C-A)/A$, $e_f=(c_f-A_f)/A$, $\Omega=2\pi/\text{day}$, $m=(\Omega_x+i\Omega_y)/\Omega$ and $i=\sqrt{-1}$. The rigid polar motion m_r due to a periodic external potential variation $\theta=\phi_o\ \exp i\ (n-\Omega)$ is obtained in the limit n=0; $m_r=(e/[1-e])\ \phi_o$, and the scale factor is simply $m(n)/m_r$. The off-diagonal variations in the moment of inertia of whole earth c_{21} and fluid core c_{21}^f depend on four structure constants κ, ζ, γ and β; $c_{21}=\kappa(m-\phi)+\zeta m_f$; & $c_{21}^f=\gamma(m-\phi)+\beta m_f$. The values obtained from a standard elastic theory analysis are $e=3.259\text{x}10^{-3}$, $e_f=2.555\text{x}10^{-3}$, $A_f=0.1139A$, $\gamma=1.963\text{x}10^{-3}$, $\zeta=2.253\text{x}10^{-4}$, $\kappa=1.048\text{x}10^{-3}$ and $\beta=6.214\text{x}10^{-3}$. The FCN frequency obtained (-458.5d) is in excellent agreement with Wahr's value of -460d.

The effect of oceans has also been estimated using a similar model in which the effective c_{21} and c_{21}^f corrections due to the combined ocean mass and load terms are: $\kappa^o=0.066f\kappa(i-1)$ $\zeta^o=0.0061f\kappa(1-i)$, $\gamma^o=0.029f\gamma(1-i)$ and $\beta^o=0.0020f\gamma(i-1)$. From an intercomparison of Laplace ocean models and Lageos data, we find f=1 for the O1 tide (equivalent to 13.66d prograde nutation) and ≅0.8 for the P1 (semi-

M. J. Reid and J. M. Moran (eds.), The Impact of VLBI on Astrophysics and Geophysics, 379–380.

annual nutation) and K_1 tides. Although f is probably accurate to ±25% or better, ocean currents contribute an additional source of angular momentum variation which could be as large as the ocean mass effect and which may add 0.3mas to the 13.66d nutation, based on a related study of the M2 tide's effect on UTI (Baader et al., 1983).

Solid friction can be modelled by multiplying $\{\gamma,\beta\}$ and $\{\kappa,\zeta\}$ by $(1+i/Q_c)$ and $(1+i/Q_t)$, respectively. We find that $Q_c \approx Q_t$ to within ~20%. For a plausible Q_t=100, solid friction adds ~0.2mas to a few terms. Ocean and solid friction equally contribute to a FCN damping time of ~ 50 yr.

Herring et al.'s latest VLBI solution indicates that the annual residual is -2.2mas relative to Wahr (1981). If the FCN frequency is reduced to -430.5d by changing e_f to 2.681×10^{-3}, the resulting VLBI residuals (minus ocean and solid friction effects) relative to the improved SOS theory are shown below.

	OCEAN CORRECTIONS			SOLID FRICTION		VLBI RES.
	WAHR	OUR RESULTS		CORE:	TIDE:	Herring
Period	SASAO	MASS	LOAD	Q_C=100	Q_t=100	et al. 1987)
13.66d	0.02 (j)	-0.22 (j)	0.24 (j)		0.09i	-0.36 - 0.06i
182.62	0.61 (j)	-0.07 (j)	0.69 (j)	0.01i	0.30i	-0.31 - 0.10i
-365.26	0.17 (j)	<0.01	0.21 (j)	0.12i	0.12i	0.30i
						±0.10mas

Table 1: Oceanic and tidal corrections to improved SOS theory (with FCN=-430d), compared to Herring et al., (1987) VLBI solution. Here j=1-i, $i=\sqrt{-1}$

Acknowledgement

The research in this publication was carried out at the Jet Propulsion Laboratory, California Institute of Technology, under a contract NAS7-100 with the National Aeronautics and Space Administration.

References

Baader, H. -R., P. Brosche, W. Hovel, J. Geophys., 52, 140-142, 1983.

Herring, T.A., C.R. Gwinn, B.A. Buffett and I.I. Shapiro, to appear in IAU Symp. 128: Earth Rotation & Reference Frames for Geodesy, 1987.

Sasao, T.S., S. Okubo and M. Saito, in Proc. IAU Symp. No. 78, Nutations and Earth Rotation, Kiev, 1977, (eds. E.P. Federov, M.L. Smith, P.L. Bender), Reidel, Dordrecht, 165-183, 1980.

Wahr, J.M., Geophys. J. Roy. astr. Soc., 64, 705-727, 1981.

Wahr, J.M. and T.S. Sasao, Geophys. J. Roy. astr. Soc., 64, 747-765, 1981.

THE THEORY OF THE EARTH'S ORIENTATION, WITH SOME NEW RESULTS FOR NUTATION

John M. Wahr
Department of Physics and
Cooperative Institute for Research in Environmental Sciences
Campus Box 390
University of Colorado
Boulder, Colorado 80309

ABSTRACT. The rotation of the earth is variable at a number of time scales, from a few days to thousands of years and longer. Variations occur in the rotation rate, in the geographical position of the rotation axis (referred to as polar motion), and in the position of the axis relative to inertial space (referred to as nutation and precession). The interpretations of the various observations have implications for the dynamical behavior and structure of the earth's deep interior, and for various aspects of meteorology and oceanography. These are reviewed below. Also included is an Appendix describing a model of the diurnal resonance in nutation for a non-hydrostatically pre-stressed earth.

1. INTRODUCTION

The earth's rotation is not constant. Instead, there are variations both in the rate of rotation and in the position of the rotation axis. Variations in the rate of rotation affect the length of a day (lod), where "a day" is defined in terms of the overhead transits of celestial objects. The time integral of the lod variability is proportional to perturbations in Universal Time.

To describe motion of the rotation axis, it is useful to differentiate between (1) geographical motion: motion with respect to fixed points on the earth's surface; and (2) inertial space motion: motion with respect to celestial objects. Geographical motion of the axis is referred to as polar motion or, sometimes, wobble. Inertial space motion is called nutation or, in a special case, precession. In a sense, this differentiation is artificial because there cannot be inertial space motion without geographical motion, and vice versa. Still, the distinction is useful because most disturbances of the pole can be characterized as either predominately nutation or predominately polar motion, depending on the frequency of the disturbance.

For example, any perturbation of the axis with a period much longer than one day as seen from space, will produce motion which is mostly nutation with very little polar motion: the rotation axis and fixed terrestrial points move almost together at long periods. This means that the external gravitational torques on the earth from the sun and moon will cause motion which is predominately nutation, since those torques vary with the various orbital periods characterizing the lunar and solar orbits, and those

M. J. Reid and J. M. Moran (eds.), The Impact of VLBI on Astrophysics and Geophysics, 381–390.

periods are all much longer than one day. Note that to an observer fixed to the rotating earth, the periods of those torques, and so of the corresponding axis motion, will be close to one day. On the other hand, disturbances which perturb the pole at periods much longer than one day as seen from the rotating earth, cause mostly polar motion with very little nutation: the rotation axis doesn't move much in inertial space, but fixed terrestrial points do. This means that processes which originate within the earth and which have time scales much longer than a day, cause mostly polar motion.

I will discuss, below, the observed variability in the lod, in polar motion, and in nutation, together with the potential geophysical implications. I will concentrate on the shorter period variability (periods of a few years and less), with the rationale that because of the relatively short data span of high quality VLBI observations, those observations through the near future will mostly impact our understanding of the short period variability. More detailed discussions can be found in Munk and MacDonald (1975), Lambeck (1980), and Rochester (1984).

2. LENGTH OF DAY

Let the unperturbed rotation vector of the earth be in the $\hat{z}$ direction, with amplitude Ω equal to one cycle per sidereal day. Suppose the rotation rate of the mantle is perturbed, so that its time-dependent amplitude is $\Omega(1+m_3(t))$. Suppose we apply an external torque, $L_3(t)$, to the mantle along the $\hat{z}$ axis. Let C be the unperturbed value of the earth's principal moment of inertia along the $\hat{z}$ axis. Suppose we perturb the moment of inertia, so that its value at time t is $C+\Delta c(t)$. Then the angular momentum equation for the mantle reduces to

$$\frac{d}{dt} m_3 = \frac{L_3}{C\Omega} - \frac{1}{C}\frac{d}{dt}\Delta c \tag{1}$$

As equation (1) demonstrates, the earth's rotation rate can be perturbed either by applying an external torque to the earth, or by changing the earth's inertia tensor.

2.1. Long Period Variability

Historic eclipse records indicate that the lod has been increasing by about 2 msec per century over at least the last 2500 years. This linear increase is mostly due to lunar and solar torques (L_3 in (1)) on the earth which tend to slow the earth's rotation and which arise because of dissipation of tidal energy in the oceans. However, these tidal torques are about 25% larger than those implied by the eclipse data. The discrepancy is due to the effects of post-glacial rebound. Mass continues to flow laterally into the region of the mantle beneath Hudson's Bay as the earth adjusts to the melting of the Canadian ice load at the end of the last ice age. As a result, the earth's polar moment of inertia, Δc, is decreasing and this tends to decrease the lod, partially offsetting the effects of tidal dissipation.

Telescope data over the last few hundred years demonstrate that there are also irregular fluctuations in the lod with periods of a few decades and amplitudes of up to 4-5 msec. These are presumably caused by the exchange of angular momentum between the earth's mantle and fluid core. The torques responsible for the exchange of angular momentum could be due either to electromagnetic or to topographic coupling. As the VLBI data span becomes longer so that VLBI is able to resolve the long period variability, the results may help increase our understanding of the dynamical processes and the material and structural properties of the core and lower mantle.

2.2. Short-Period Variability

There are two important excitation sources which cause variations in the lod at periods of a few years and less: tides, particularly in the solid earth; and the exchange of angular momentum between the earth's mantle and the atmosphere.

Tidal deformation in the earth causes the principal polar moment of inertia, and so the lod, to vary at the periods of the long period tides. Particularly important perturbations occur at periods of 1 year, 6 months, 1 month, and 2 weeks. The amplitudes of these lod variations depend on how easily the earth deforms at these periods. The deformation can be calculated by solving a set of partial differential equations which represent the dynamical response of the earth to tidal forcing. Any such model requires, as input, a set of numerical parameters describing the earth's material properties. The parameters which describe the elastic behavior of the earth are well known from seismology. But, the parameters which describe the departure of the earth from perfect elasticity are not well constrained. Seismic observations are sensitive to anelastic effects, but those effects are apt to be substantially different at the long tidal periods than at the much shorter periods (seconds to minutes) characteristic of seismic phenomena. In fact, one of the important problems in geophysics is to understand more fully the physical processes responsible for the earth's anelasticity. Tidal lod observations can help in this regard, since they offer information on anelasticity in a unique frequency regime.

The 1 month and 2 week tidal variations in the lod have amplitudes of about .2 msec and .4 msec, respectively. The effects of anelasticity on the amplitudes are probably 1%-5% of these numbers (Wahr and Bergen, 1986), so they perturb the lod at the .01 msec level. Merriam (1985) used BIH optical results for the tidal variations to constrain the anelasticity. Results based on VLBI observations should give improved estimates within the near future, although the interpretation of any data is limited by the accuracy with which the effects on the lod of ocean tides and atmospheric fluctuations at these periods can be removed from the data. In fact, this is why the observed 6 month and 1 year amplitudes can not be used to constrain mantle anelasticity: the seasonal effects of the atmosphere are large, comparable to the tidal effect at 6 months and about 10 times larger at 1 year.

Atmospheric effects occur both at seasonal and at more irregular periods. One of the impressive results obtained from the improved lod data of the last decade is excellent agreement between the short period rotation results after subtraction of the tidal effects, and the atmospheric excitation estimated from meteorological data (see, for example, D.A. Salstein, this volume). The atmospheric effects are estimated from global atmospheric wind and pressure data. These data are integrated over the globe to find the total atmospheric angular momentum as a function of time. The negative time derivatives of these integrated results should be equal to the torque on the mantle, L_3 in (1), and so can be used to find m_3.

Most of the variability in the atmospheric angular momentum is due to changes in zonal winds. The effects of the redistribution of atmospheric mass on the moment of inertia, as inferred from surface pressure observations, are only about 10% of the effects of winds. There are also likely to be effects due to seasonal variations in the circum-polar ocean current around Antarctica. Those effects are probably smaller than the effects of atmospheric pressure, although the estimates are reasonably uncertain.

The most important implication of the good agreement between the lod and atmospheric data is that it suggests existing atmospheric data adequately reflect real global disturbances in the atmosphere. For example, Langley, et al. (1981) found that the inferred atmospheric angular momentum data and the observed lod have a well-correlated 50 day oscillation in common. This has led to atmospheric modelling efforts to understand the source of the oscillation.

3. POLAR MOTION

The most pronounced features of the observed polar motion are oscillations at 12 months and 14 months, called the annual wobble and Chandler wobble, respectively. The annual wobble is due to seasonal forcing of the earth by the atmosphere, particularly by the large seasonal pressure variations over central Asia. Since the earth is elliptical rather than spherical, those pressure variations cause a seasonal torque on the earth about an equatorial axis, and that tips the earth causing polar motion. There is also sizable seasonal excitation due to similar torques caused by global changes in ground water storage (Van Hylckama, 1970). The agreement between the observed annual wobble amplitude and the equatorial torques inferred from global pressure and ground water data is good (see Merriam, 1982; Wahr, 1983).

The Chandler wobble is a rotational normal mode of the earth, equivalent to the free nutation of a top. For a rigid earth, the period would be roughly 10 months. The increase of the period to 14 months is due to the ability of the earth to deform : solid earth deformation adds 4 months, oceanic displacements add 1 month, the earth's fluid core subtracts 1 month. Observations of the period have been used to constrain mantle anelasticity at a period of 14 months (see, for example, Smith and Dahlen, 1981). The observed damping time (several decades) has also been used to constrain anelasticity.

The primary excitation source for the Chandler wobble is, however, not well understood. Although there is some disagreement, it appears as though neither earthquakes (through induced changes in the earth's inertia tensor; see eg. Dahlen, 1973) nor the atmosphere and oceans (Wilson and Haubrich, 1976; Wahr, 1983) were the primary excitation source during this century. It may be that the excitation is due to fluid pressure in the core against the core-mantle boundary, as proposed by Le Mouel, et al. (1985). Or, perhaps it is due to irregular fluctuations in global ground water storage. Future VLBI data may help resolve this problem through accurate observations of individual excitation events.

4. NUTATIONS

The earth's nutational motion is caused by the gravitational attraction of the sun and moon. The motion can be separated into a discrete sum of periodic terms with frequencies, as seen from the diurnally rotating earth, of one cycle per day modulated by the lunar and solar orbital frequencies.

The earth is believed to have a rotational normal mode, called the free core nutation (FCN), with an eigenfrequency within the diurnal band of nutation frequencies. To understand the dynamics of the FCN, suppose the fluid core and solid mantle were tipped about an equatorial axis in opposite directions, and then released. If both the core's internal density distribution and the core-mantle boundary were spherical, there would be no restoring torque, and the core and mantle would remain tipped relative to each other. But, because the real core/mantle boundary is elliptical and the earth is elliptically stratified, there are, instead, restoring pressure and gravitational torques between the core and mantle. As a result, the core and mantle execute periodic twisting motion with respect to each other. This motion is the FCN, and its frequency is one cycle per day plus a factor dependent on the strength of the restoring torque.

In the Appendix, it is shown that under very general conditions (including the possibility that the mantle is not hydrostatically pre-stressed, but not including the effects of a solid inner core) the FCN eigenfrequency is

$$\omega_o = \Omega \left[1 + \frac{A}{A_m} (e_f - \beta) \right] \tag{2}$$

where A and A_m are the equatorial principal moments of inertia for the entire earth and the mantle alone; $e_f \cong (C_f - A_f)/A_f$ is the dynamical ellipticity of the core, where C_f and A_f are the principal moments of the fluid core about the polar and equatorial axes, respectively; and β is a numerical factor which represents the effects of deformation and is effectively independent of any aspherical stratification within the earth. The quantity e_f depends on the Y_2^0 spherical harmonic term of the core internal density field and on the Y_2^0 component of the shape of the core-mantle boundary. It does not depend on any other spherical harmonic component of the structure. For a hydrostatically prestressed earth, β is about 25% of e_f (Sasao, et al, 1980), and $\omega_o \cong (1+1/460)$ cycles/day (Wahr, 1981).

The free nutational motion at the period ω_o has never been unambiguously observed. Evidently, it is not excited sufficiently by any interior process. But, because ω_o is so close to the frequencies of the luni-solar nutations, the amplitudes of those nutations are affected by the presence of the mode by up to 20 or 30 milli-arcseconds, a perturbation which can be readily detected even by conventional optical observations of nutation. These effects of the FCN are included in the standard forced nutation model adopted by the IAU, which is based on rigid earth values of Kinoshita (1977) and corrections for non-rigidity from Wahr (1981), the latter assuming a hydrostatically pre-stressed earth.

Because VLBI is tied directly to an inertial reference system, it is superior to any of the other new techniques in resolving the forced nutations. As discussed by T.A. Herring in this volume, the VLBI results disagree with the IAU adopted theory by almost 2 mas, particularly at the annual frequency: $\omega = \Omega(1+1/365.25)$ cycles/day. This is the forcing frequency closest to ω_o, and it suggests that the FCN frequency may be somewhat larger than expected, close to $\omega_o = \Omega(1+1/435)$ cycles/day (Gwinn, et al, 1986).

Equation (2) suggests that ω_o can be modified by changing either β or e_f. Wahr and Bergen (1987) (see, also, Dehant, 1987) considered the effects of mantle anelasticity on β, and concluded that the corresponding perturbation of ω_o is too small and, more importantly, has the wrong sign. Although the effects of anelasticity are, in principle, observable using the current VLBI observations, the larger discrepancy between observation and theory must first be resolved before the VLBI results can be used to learn about diurnal anelasticity.

Probably the most likely explanation for the discrepancy is, instead, uncertainty in e_f, as postulated by Gwinn, et al. (1986). Recent results based on seismic tomography (eg. Richards and Hager, 1984; Creager and Jordan, 1986; Morelli and Dziewonski, 1987) suggest that the shape of the core-mantle boundary may diverge appreciably from hydrostatic equilibrium. It is not straightforward to compute the effects on e_f given only the shape of the core-mantle boundary. A perturbed boundary and an aspherical density distribution in the mantle will cause perturbations in the internal density surfaces in the core, and these will affect e_f. But, current models of the boundary based on seismic data suggest values for the non-hydrostatic portion of e_f which are as large or larger than the result inferred from the nutation observations. In fact, the VLBI nutation observations are proving to be a valuable independent constraint on the seismic models.

Acknowledgements

The work described in the Appendix was supported by Air Force Geophysical Laboratory contract F19628-86-k-0011, and NASA grant NAG5-485.

REFERENCES

Alterman, Z., H. Jarosch, and C.L. Pekeris, 1959; *Proc. Roy. Soc. London, Ser. A.,* **252**, 80.

Creager, K.C. and T.H. Jordan, 1986; *Geophys. Res. Lett.,* **13**, 1497.

Dahlen, F.A., 1973; *Geophys. J. Roy. Astr. Soc.,* **32**, 203.

Dehant, V., 1987; *Phys. E. Planet. Int.*, in press.

Gwinn, C.R., T.A. Herring, and I.I. Shapiro, 1986; *J. Geophys. Res.,* **91**, 4755.

Kinoshita, H., 1977; *Cel. Mech.,* **15**, 277.

Lambeck, K., 1980; *The Earth's Variable Rotation: Geophysical Causes and Consequences,* Cambridge University Press.

Langley, R.B., R.W. King, I.I. Shapiro, R.D.Rosen, and D.A. Salstein, 1981; *Nature,* **294**, 730.

Le Mouel, J.L., C. Gire, and J. Hinderer, 1985; *C.R. Acad. Sc. Paris,* **301**, 27.

Merriam, J.B., 1982; *Geophys. J. Roy. Astr. Soc.,* **70**, 41.

Merriam, J.B., 1985; *J. Geophys. Res.,* **90**, 9423.

Munk, W.H., and G.J.F. MacDonald, 1975; *The Rotation of the Earth,* Cambridge University Press.

Morelli, A. and A.M. Dziewonski, 1987; *Nature,* **325**, 678.

Richards, M.A., and B.H. Hager, 1984; *J. Geophys. Res.,* **89**, 5987.

Rochester, M.G., 1984; *Phil. Trans. R. Soc. Lond. A,* **313**, 95.

Sasao, T., S. Okubo, and M. Saito, 1980; in *Proceedings of IAU Symposium no. 78 "Nutation and the Earth's Rotation"*, eds. Fedorov, E., M. Smith, and P. Bender; pub. D. Reidel.

Smith, M.L., and F.A. Dahlen, 1981; *Geophys. J. Roy. Astr. Soc.*, **64**, 223.

Van Hylckama, T.E.A., 1970; in *Int. Assoc. Sci. Hydrology, Proc. Reading Symp. World Water Balance, Publ. 92*, AIHS-UNESCO.

Wahr, J.M., 1981; *Geophys. J. Roy. Astr. Soc.*, **64**, 705.

Wahr, J.M., 1983; *Geophys. J. Roy Astr. Soc.*, **74**,451.

Wahr, J., and Z. Bergen, 1986; *Geophys. J. Roy. Astr. Soc.*, **87**, 633.

Wilson, C.R., and R.A. Haubrich, 1976; *Geophys. J. Roy. Astr. Soc.*, **46**, 707.

Woodhouse, J.H., and F.A. Dahlen, 1978; *Geophys. J. Roy. Astr. Soc.*, **53**, 335.

APPENDIX: FCN FREQUENCY FOR A NON-HYDROSTATICALLY PRESTRESSED EARTH

All existing nutation models assume the earth is hydrostatically pre-stressed. Among the implications of this assumption which could be important for nutation calculations are that, in the unperturbed state, the core-mantle boundary is a surface of constant potential, and that the core-mantle boundary and all internal density surfaces in the core and mantle are elliptical: ie. their departure from spherical symmetry can be described with the single spherical harmonic $Y_2^0(\theta,\lambda)$, where θ and λ are co-latitude and eastward longitude, respectively.

Recent results based on seismic tomography suggest that surfaces of constant density in the mantle and surfaces of discontinuity, such as the core-mantle boundary, have shapes which are not consistent with the hydrostatic pre-stress assumption. The Y_2^0 components of these surfaces differ from the hydrostatic values, and there are large components with other Y_l^m angular dependence. Since the FCN is critically dependent on pressure coupling acting across the core-mantle boundary (and, to a lesser extent, gravitational coupling between the core and mantle), the FCN eigenfrequency is apt to be sensitive to this sort of detail about the surface shapes.

In this Appendix, the FCN eigenfrequency is modelled under only minimal assumptions about the state of pre-stress in the earth. The core is assumed to be hydrostatically pre-stressed (it is, after all, a fluid even at seismic periods), but the mantle is allowed to possess a non-zero deviatoric pre-stress. As a result, the core-mantle boundary and the internal density surfaces in the mantle can have arbitrary shape. Constant density surfaces in the core are uniquely determined by the shape of the core-mantle boundary and by the gravitational potential from the mantle acting on the core. But, the constant density surfaces in the core need not coincide with the surfaces expected for an everywhere hydrostatically pre-stressed earth. The derivation, below, extends the semi-analytical model of Sasao, et al. (1980) to account for an initially non-hydrostatic mantle and core-mantle boundary, and to more fully justify a number of the approximations. In this derivation, the inner core is assumed to be fluid.

Consider a reference frame fixed to the mantle and rotating with it. For nutational motion at frequency ω_o, the rotation vector of the frame is $\Omega\hat{\mathbf{z}} + \Omega m_o(\hat{\mathbf{x}}+i\hat{\mathbf{y}})e^{i\omega_o t}$, where the nutational motion of the mantle is described by m_o, and $m_o \ll 1$. For the FCN, $(\omega_o-\Omega)/\Omega$ is first order in ϵ, where ϵ is some measure of the departure of the earth from spherical symmetry. The equations of motion in the hydro-statically pre-stressed core for the case of no external forcing are

$$\rho[-\omega_o^2\mathbf{s}+2i\omega_o\mathbf{\Omega}\times\mathbf{s}] = -\rho\nabla\phi_1-\nabla P_E-\rho_1\nabla\Phi-\Omega^2\rho m_o[(\hat{\mathbf{x}}+i\hat{\mathbf{y}})z(\Omega-\omega_o)+ +\hat{\mathbf{z}}(x+iy)(\Omega+\omega_o)] \tag{A1}$$

$$\nabla^2\phi_1 = 4\pi G\rho_1$$

where $\mathbf{s}$ is the displacement field in the core; ϕ_1, P_E, and ρ_1 are the Eulerian perturbations in gravitational potential energy, pressure, and density, respectively; $\mathbf{\Omega}=\Omega\hat{\mathbf{z}}$ represents the mean rotation of the earth; and ρ and Φ are the initial density and total potential energy (gravitational plus centrifugal) and can have arbitrary angular dependence, except that surfaces of constant ρ must be surfaces of constant Φ in the core. There are, also, equations relating ρ_1 and P_E to derivatives of $\mathbf{s}$, which are not shown here. And, there are similar, although more complicated, differential equations describing displacements in the mantle, which are also omitted here. The mantle equations explicitly include the non-hydrostatic pre-stress (see Woodhouse and Dahlen, 1978, for both the mantle and core equations).

For nutational motion, the displacement field in the core has the form

$$\mathbf{s}(\mathbf{r}) = \Omega\theta_o(\hat{\mathbf{x}}+i\hat{\mathbf{y}})\times\mathbf{r} + \mathbf{s}_d(\mathbf{r}) \tag{A2}$$

where θ_o represents the mean nutational motion relative to the mantle, and $\mathbf{s}_d$, which represents the deformation of the core, is first order in ϵ compared with the θ_o term. Taking $(\hat{\mathbf{x}}-i\hat{\mathbf{y}})\cdot[\mathbf{r}\times(\text{eqn.(A1)})]$ and integrating through the core, and using (A2) for $\mathbf{s}$, gives an angular momentum balance equation for the core:

$$2i\Omega^2 A_f m_o + 2\omega\Omega^2 A_f\theta_o\frac{\Omega-\omega_o}{\Omega} + i\omega_o\Omega c_-^f = (\hat{\mathbf{x}}-i\hat{\mathbf{y}})\cdot\int_{\text{core}}\mathbf{r}\times\left[-\rho\nabla\phi_1-\nabla P_E-\rho_1\nabla\Phi\right] \tag{A3}$$

where A_f is the core's principal moment of inertia in the equatorial plane; and $c_-^f=c_{13}^f-ic_{23}^f$, where the c_{ij}^f represent the perturbation in the core's inertia tensor due to the motion. The terms on the right hand side of (A3) include the pressure and gravitational torques on the core from the mantle.

All terms in (A3) are second order in ϵ or smaller, given that θ_o is zero order. For example, ϕ_1 is first order since it is a deformation term, and the integral of $\rho\mathbf{r}\times\nabla\phi_1$ is 0 no matter what ϕ_1 is, unless either the core shape or the internal density in the core is aspherically symmetric. As another example, $\Omega\omega_o\cong\Omega^2$ is a first order quantity (the unperturbed centrifugal force is order ϵ times the unperturbed gravitational force) as is $(\Omega-\omega_o)/\Omega$. Thus, the θ_o term in (A3) is second order. As a final example, m_o is first order compared with $\Omega\theta_o$ (m_o represents motion of the mantle rotation axis as seen from the mantle and is of order $(\Omega-\omega_o)\theta_o$) so that $\Omega^2 m_o$ is second order.

A comparable angular momentum equation for the entire earth, after dropping all terms third order or smaller, is

$$2iAm_o = 2(\omega_o-\Omega)A_f\theta_o \tag{A4}$$

where A is the principal moment of inertia for the entire earth. (This result is consistent with the assumption that m_o is first order.)

Next, c^f_- and the right hand side of (A3) can be related to θ_o. This is done by using (A2) in (A1), dropping all terms second order or smaller in ϵ which occur in (A1) and in the corresponding mantle differential equations and in all boundary conditions, and then solving the entire system on a computer. In the core, for example, the first order (spheroidal) equations for ϕ_1, P_E, and ρ_1 reduce to

$$\rho_0\nabla\phi_1+\nabla P_E+\rho_1\nabla\Phi_0 = \left(i\theta_o\Omega^3\sqrt{8\pi/15}\right)\rho_o\nabla(r_2Y_2^1) \tag{A5}$$

where ρ_0 and Φ_0 are the spherically symmetric parts of ρ and Φ. No m_o terms are included in (A5), because those terms are second order.

The corresponding first order differential equations in the mantle are the usual set of equations describing deformation of a spherical, non-rotating, static, and hydrostatically pre-stressed mantle (see, for example, Alterman, et al., 1959), with no first order apparent force corresponding to the θ_o term in (A5). Although there is an apparent force in the mantle which depends on m_o, that force is second order. All first order boundary conditions within the core and the mantle and at the outer surface are also equivalent to the boundary conditions for a spherical, non-rotating, hydrostatically pre-stressed earth.

In effect, then, all deformation-related quantities can be computed by solving the static equations of motion for a spherical, non-rotating, hydrostatically pre-stressed earth, subject to an apparent force proportional to $\rho_0\nabla(r^2Y_2^1)$ in the core and equal to 0 in the mantle. Note that, as a consequence, neither the non-hydrostatic pre-stress in the mantle nor any of the earth's aspherical structure enters explicitly into any of the first order deformation equations. Their effects are included only through the integrals on the right hand side of (A3).

Because the apparent force in the core is proportional to $\rho_0\nabla(r^2Y_2^1)$, and because the first order deformation equations are spherically symmetric, ρ_1, P_E, and ϕ_1 have $Y_2^1(\theta,\lambda)$ angular dependence. Using this angular dependence, (A5) yields directly

$$P_E(r,\theta,\lambda) = \rho_0(r)\left[\left(i\Omega^3\theta_o\sqrt{8\pi/15}\right)r^2Y_2^1(\theta,\lambda)-\phi_1(r,\theta,\lambda)\right] \tag{A6}$$

$$\rho_1(r,\theta,\lambda) = \frac{\partial_r\rho_0(r)}{\partial_r\Phi_0(r)}\left[\phi_1(r,\theta,\lambda)-\left(i\Omega^3\theta_o\sqrt{8\pi/15}\right)r^2Y_2^1(\theta,\lambda)\right]$$

Using (A6) in (A3) and doing the integrals gives a result accurate to second order of

$$2i\Omega^2m_o+2\Omega^2\theta_o(\Omega-\omega_o)+\frac{ic^f_-\Omega^2}{A_f} = -2\Omega^3\theta_oe_f \tag{A7}$$

where e_f is the dynamical ellipticity of the core and is equal to $(C_f-A_f)/A_f$, where C_f is the polar moment of inertia of the core.

Define the dimensionless, real parameter β so that

$$c^f_- = \beta 2A_fi\Omega\theta_o \tag{A8}$$

Here, β can be determined by solving the deformation equations on a computer. Using

(A4) and (A8) in (A7) and solving for the FCN eigenfrequency, ω_o, gives

$$\omega_o = \Omega\left[1+\frac{A}{A_m}(e_f-\beta)\right] \tag{A9}$$

where A_m is the principal moment of inertia of the mantle.

Although the result (A9) was derived here without assuming a hydrostatically pre-stressed mantle, it is identical in form to the hydrostatic result. The dynamical ellipticity, e_f, depends on the Y_2^0 component of the core-mantle boundary shape, and on the Y_2^0 terms in the density stratification inside the core. There is no dependence, to this order of approximation, on any other Y_l^m terms in the aspherical structure. The factor β represents the effects of deformation and is insensitive, to this order, to aspherical structure. For a hydrostatically pre-stressed earth, β is about 25% of e_f, and $\omega_o \cong (1+\frac{1}{460})$ cycles per day.

EARTH ROTATION FROM THE IRIS PROJECT

D. S. Robertson, W. E. Carter, and F. W. Fallon
National Geodetic Survey, Charting and Geodetic Services
National Ocean Service, NOAA
Rockville, Maryland 20852

ABSTRACT. Project IRIS (International Radio Interferometric Surveying) was set up under the IAG and COSPAR to provide an operational system that would employ Very-Long-Baseline Interferometry (VLBI) techniques to monitor variations in the rotation of the Earth. Currently the IRIS-A network, with stations at Westford, MA, Ft. Davis, TX, Richmond, FL, and Wettzell, FRG, conducts 24-hour observing sessions every five days to produce determinations of pole position, UT1, and nutation in addition to other parameters of geophysical and astrometric interest. The resulting Earth orientation parameters (EOP) have been shown to have an accuracy of 1 to 2 milliseconds of arc in pole position, and 0.05 to 0.1 milliseconds of time in UT1. In order to observe the relatively large higher frequency variations in UT1, daily 45-minute observing sessions are conducted using the single baseline between Westford and Wettzell. Intercomparison of the UT1 values from the daily and the 5-day series indicates that the accuracy of the daily values is better than 0.1 millisecond of time.

The longer term objectives of the IRIS project include improving the monitoring of Earth orientation by increasing the sampling rate and accuracy of the observations. In April, 1987, the IRIS-P network, with stations in Kashima, Japan, Fairbanks, AK, Ft, Davis, TX, and Richmond, FL began monthly 24-hour observing sessions, and a second series of daily UT1 observing sessions was begun using the stations in Richmond and Bologna, Italy. The additional networks will provide redundancy that will improve the reliability of the system and allow the accuracy of the EOP values to be estimated.

The IRIS UT1 time series provides, for the first time, sufficient accuracy and temporal resolution to look for the few percent increase in k/C caused by the anelastic response of the mantle. Initial results presented here suggest that improved methods of accounting for the dynamics of the oceans and atmosphere may be required before the intertwined variations in UT1 can be fully separated.

M. J. Reid and J. M. Moran (eds.), The Impact of VLBI on Astrophysics and Geophysics, 391–400.

I. INTRODUCTION

By the late 1970's it was clear that VLBI techniques had advantages over all of the new high precision geodetic techniques being developed for observing the rotation of the Earth. A National Geodetic Survey (NGS) review of candidate technologies and instrumentation in 1977 concluded that very long baseline interferometry (VLBI) using the NASA/Haystack MARK III system was clearly the system of choice for developing a modern polar motion and UT1 monitoring service [Carter, 1978; Carter and Strange, 1979]. Project POLARIS (POLar-motion Analysis by Radio Interferometric Surveying) was subsequently implemented jointly by the NGS, the National Aeronautics and Space Administration (NASA), and the United States Naval Observatory (USNO) [Bossler, 1982].

Under project POLARIS, three permanent geodetic VLBI observatories have been developed: Westford, in Massachusetts; the George R. Agassiz Station (GRAS), in Texas; and Richmond, in Florida. Even before the completion of the POLARIS network, the Federal Republic of Germany (FRG) began constructing a dedicated geodetic VLBI observatory at their Satellite Observation Station in Wettzell, Bavaria. To exploit the improved capabilities of the combined POLARIS/Wettzell network the responsible geodetic agencies in the FRG and NGS jointly initiated project IRIS (International Radio Interferometric Surveying) in 1982 [Carter and Robertson, 1984]. The International Association of Geodesy (IAG) and the Committee on Space Research (COSPAR) subsequently approved the formation of a permanent joint IRIS subcommission; organizations in several additional nations have agreed to participate, or expressed interest in participating, in IRIS activities. The Onsala Space Observatory, Sweden, is regularly participating in one IRIS observing session per month, and the VLBI observatories in Bologna, Italy, Kashima, Japan, and Fairbanks, Alaska have recently begun observing sessions under the IRIS project.

The IRIS observations are designed to monitor all of the important components of the rotation of the Earth, including motion of the rotation axis in space (nutation), motion of the Earth's crust and mantle relative to the rotation axis (polar motion), and the angle of rotation about the rotation axis (UT1), in addition to baseline components, radio source coordinates, and other parameters of geophysical and astronomical interest.

II. POLAR MOTION AND NUTATION

Although this paper will focus on the IRIS determinations of UT1, the UT1 estimation process is so critically dependent on accurate polar motion and nutation values that we felt it would be useful to display the IRIS determinations of those quantities before describing the UT1 observing sessions.

Figure 1 shows the pole position determinations from the IRIS observations and Satellite Laser Ranging (SLR) measurements between January, 1984, and March, 1987. The close agreement (2 milli-arcseconds RMS difference) between the two completely independent methods of observation places a tight bound on the total errors in both sets of determinations [Robertson *et al.*, 1985a].

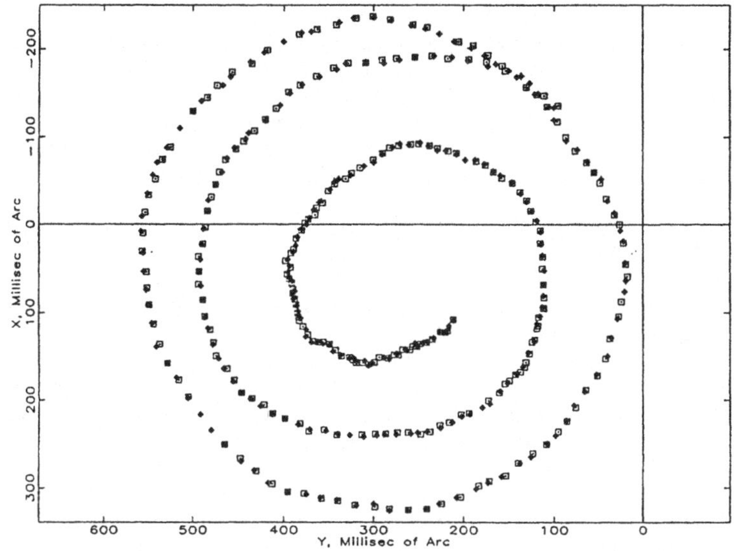

Figure 1. Pole position determinations from IRIS VLBI observations (rectangles) and SLR observations (diamonds).

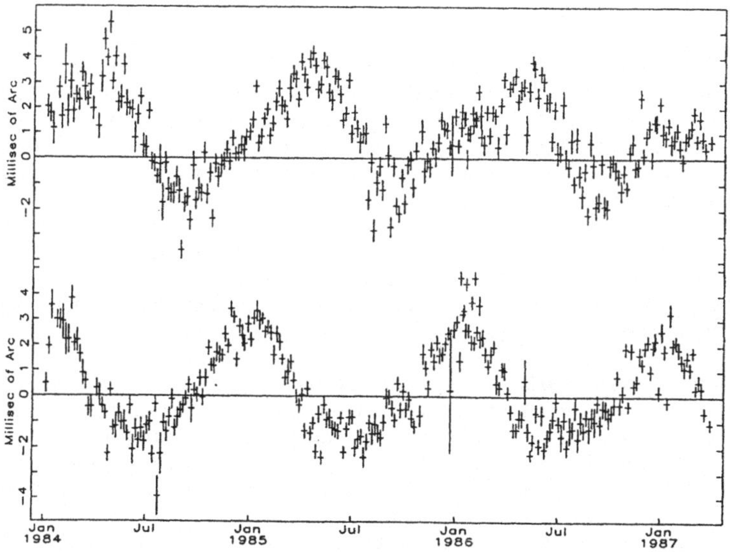

Figure 2. Corrections to the IAU 1980 nutation series as determined from IRIS VLBI observations. The top plot shows nutations in longitude, scaled by the sine of the obliquity, and the bottom plot shows nutation in obliquity.

The IRIS determinations of the nutations of the Earth's spin axis in space are shown in figure 2, displayed as residuals to the IAU 1980 nutation model [Wahr, 1981]. The clear annual signature in these residuals has been interpreted by Herring and his colleagues at the Center for Astrophysics as resulting from an error in the value for the core-mantle coupling assumed by Wahr in calculating the standard nutation model [Herring et al., 1986; Gwinn et al., 1986]. This analysis will be discussed in another paper at this conference [Herring, 1987].

III. UT1 DATA ANALYSIS

Determinations of UT1 have long been hampered by the fact that UT1 has more high-frequency time variations than the other components of Earth orientation. To try to understand these time variations the IAU/IAG project MERIT [Wilkins, 1980] scheduled intensive observing campaigns in 1984 and 1985 with a goal of obtaining daily determinations of UT1. Although there was no technical reason why the IRIS project could not have conducted its 24-hour multi-station observing sessions every day instead of every 5 days, the costs of such an increase in duty cycle were prohibitive. However for monitoring only UT1 a much less costly observing program was found to be sufficient. In principle UT1 could be estimated from a single delay measurement of an equatorial source using one equatorially oriented baseline if the baseline coordinates, radio source coordinates, pole position, nutation, and clock parameters were well enough known. In practice, all of these parameters except the clock error parameters are sufficiently well known from the regular IRIS observations. Therefore all that is needed to monitor UT1 is a brief observing session designed to separate UT1 from clock errors parameters. This can be accomplished by observing a set of sources that cover a wide range in declination. In practice, the observing sessions employed for daily UT1 determinations use the Westford-Wettzell baseline and make eight observations on four sources that span about 70 degrees in declination. These observing sessions take about 45 minutes each day.

It would be nice to be able to verify the accuracy of these observations by intercomparison with determinations from observations based on a different technique, as the VLBI and SLR pole position values were intercompared. Unfortunately there is no independent observing technique that can match the accuracy, stability and time resolution of the VLBI determinations. We can, however, intercompare the UT1 determinations from the daily measurements with the corresponding results from the 5-day, multi-station IRIS observing sessions. Such an intercomparison consistently shows agreement between the observing sessions at the level of about 0.1 millisecond of time [Robertson et al., 1985b]. This value represents a bound on the combined error in both series, plus any real UT1 variations with periods less than about 10 days.

The important geophysical parameters that can be estimated from these high precision UT1 determinations relate to the UT1 variations caused by changes in the Earth's moment of inertia under the effects of the solid Earth tides. Yoder et al. [1981] describe a detailed model for such variations in an elastic Earth with liquid oceans and core. The resulting model is proportional to the elasticity parameter k/C. Wahr has discussed the possibility of using the tidal variations in UT1 to determine the anelastic behavior of the mantle [Wahr, 1987; Wahr and Bergen, 1986]. Qualitatively, the anelastic effects are not difficult to understand: an anelastic Earth is a bit "softer" than a perfectly elastic one, and its response to tidal perturbations is slightly greater than the comparable effects for an elastic Earth. Quantitatively, the magnitude of the amplification is about 2% in k/C. Yoder's estimate for k/C is about 0.944, and Wahr's corrections for anelastic effects indicate that the value for fortnightly terms should be in the range of 0.97 to 1.06, and for monthly terms in the range of 0.96 to 1.00. This corresponds to an increase in the tidal UT1 variations by about 0.02 milliseconds of time. The need to measure such a small effect places a severe requirement on the measurement accuracies.

Figure 3 shows the IRIS daily UT1 time series from May, 1985 through March, 1987, after removing a linear drift, three long-period sinusoids (6,12 and 24 month periods, respectively), and the effects of the tabulated variations in atmospheric angular momentum as compiled by the National Meteorological Center [Rosen and Salstein, 1983]. The high frequency tidal variations in UT1 are easily seen in these data. How accurately do the tidal variations seen in Figure 3 match the Yoder's theoretical model? Figure 4 shows a periodogram of these data together with a periodogram constructed from the Yoder model evaluated at one-day intervals over the span of the VLBI observations. The tidal peaks at 27 and 14 days (13.5 and 26.1 cyc/year) are seen to match nicely, as do the lesser peaks at 9 and 31 days (11.8 and 40.6 cyc/year). Notice that the 9-day peak is outside the Nyquist limit for 5-day sampling, and can be observed only with the daily VLBI observations.

In order to use these data to evaluate the k/C tide coefficients with standard least-squares techniques it was necessary to remove the low-frequency variations. To do this the data were low-pass filtered using a filter with an absolute cut-off at 10 cycles/year (period of 36.524 days). The resulting function, shown in figure 3, was subtracted from the data. About 5 points at each end of the data set were not fit closely by this function, and were deleted in the ensuing analysis. A six-month sample of the differences, together with values calculated from the Yoder model, are shown in figure 5. To these data we fit three amplitude coefficients for the Yoder model, the first applied to the first 12 terms (with periods about 9 days), the second applied to the next 13 terms (with periods about 14 days), and the third applied to the next 16 terms (with periods about 27 days). The results of this fit are tabulated in table 1. The formal errors for the 14 and 27 day coefficients are seen to be substantially smaller than the precision required to determine the anelastic effects. Of course the formal errors are not the same as the total errors in the determinations. Rather, they represent the limiting accuracy which might be obtained in the absence of unmodeled systematic error. Their principal significance in this context is to demonstrate that the measurement errors are not a significant fraction of the true error budget.

To estimate the magnitude of the total error in the determinations of k/C we divided the data into subsets in two different ways and repeated the solutions, as shown in table 1. The resulting differences in the k/C

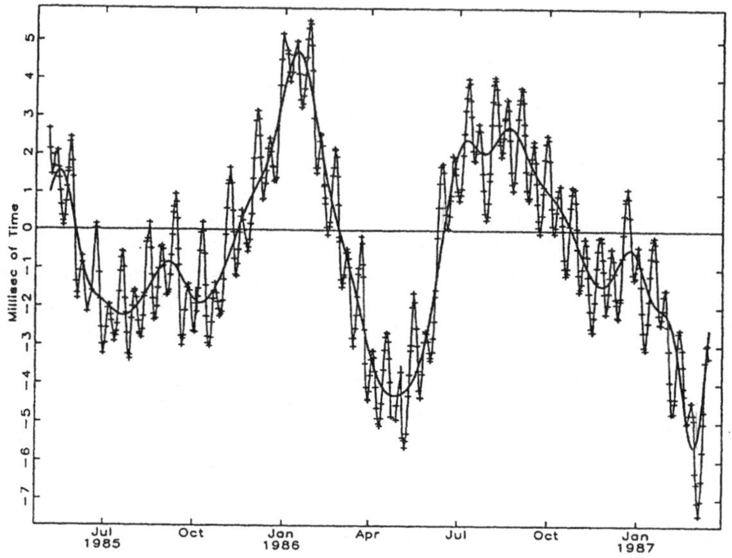

Figure 3. Daily determinations of UT1, after removal of a linear trend and long-period sinusoids. The heavy line was generated by smoothing the observations with a low-pass boxcar filter with a cut-off at 10 cyc/year.

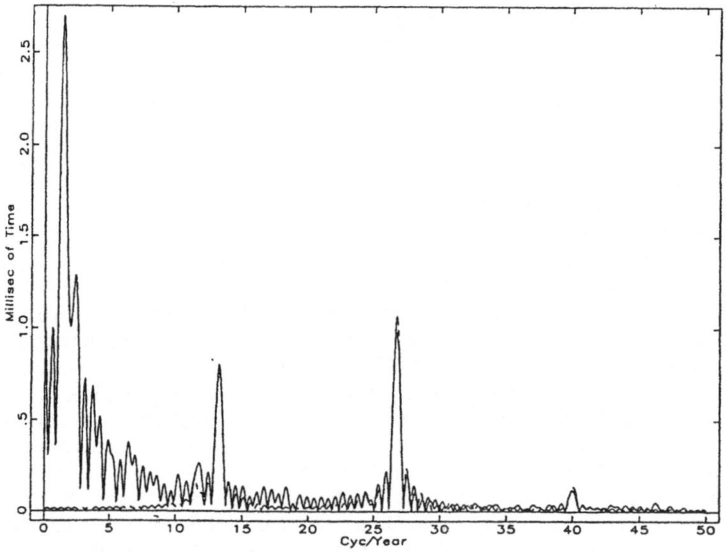

Figure 4. The solid line shows a periodogram of the data shown in Figure 3, and the dashed line is a similar periodogram of a daily tabulation of Yoder's tide model, evaluated over the same time interval as the VLBI data.

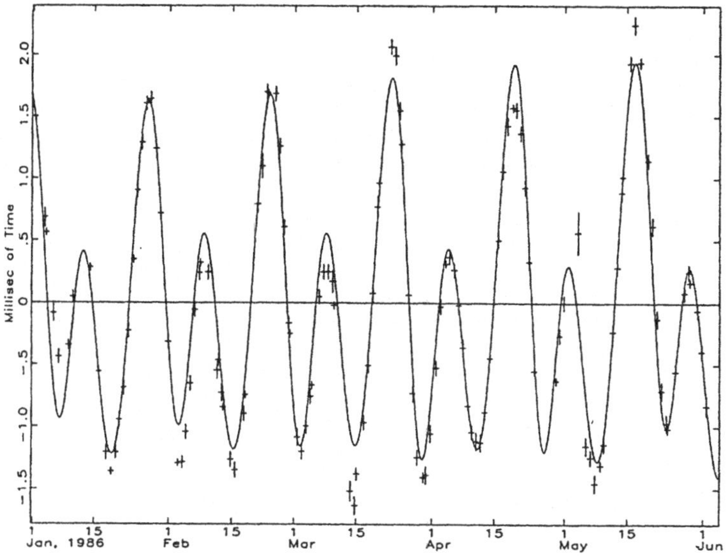

Figure 5. Sample of the data shown in figure 3 after removal of the smooth curve from that figure. The solid line shows the Yoder tidal variation model.

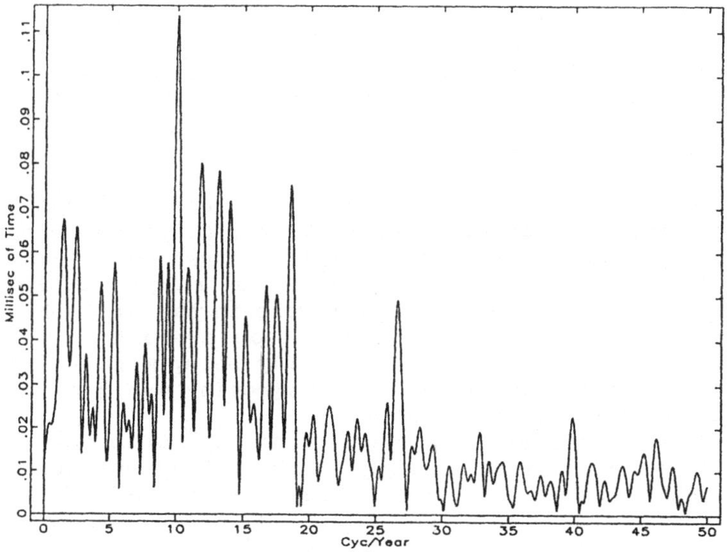

Figure 6. Periodogram of the residuals of the UT1 determinations after fitting three k/C amplitudes as described in the text.

Table 1. k/C determinations. The column labeled "sigma" tabulates the formal standard errors. Under "differences," the first column tabulates differences from interleaved subsets, and the second tabulates differences from the first and second halves of the data set.

Period	K/C	sigma	Differences 1	Differences 2
9 day	0.898	0.02	0.09	0.24
14 day	0.948	0.003	0.02	0.03
27 day	0.986	0.004	0.04	0.09

determinations are generally much larger than the formal errors of the individual determinations. These differences provide much better estimates of the true uncertainties in the k/C determinations than do the formal errors. For the 27 day term and especially the 14-day term these uncertainty estimates approach the level needed for determining the mantle anelasticity. Clearly, however, there is some unmodeled source of systematic error or process noise interfering with the estimations of k/C. If it could be eliminated or substantially reduced then we might be very close to being able to observe the effects of the mantle anelasticity.

The set of possible sources of the observed process noise can be conveniently divided into 5 categories according to their distance from the center of the Earth: extra-terrestrial, atmospheric, oceanic, crust-mantle, and liquid-core effects. The two extreme categories can be largely dismissed out of hand: extra-terrestrial effects (luni-solar torques) are unlikely to hold surprises at any significant level, and the liquid core is unlikely to be able to generate any response at the <30 day periods needed. Of the other three effects, the ones that relate to the crust and mantle are the very ones we are trying to observe. That leaves the oceans and atmosphere as the principal problem areas.

The known effects of atmospheric winds and pressure variations have already been removed from the data, but the error levels in the wind values are poorly known. If there are large errors in the atmosphere data at fortnightly and monthly periods then they could prove to be a serious problem. The large effort that is currently being made to study these atmospheric effects holds some hope for improving our knowledge of the atmospheric variations in the future.

The remaining likely source for errors in estimating k/C is oceanic effects. The oceans respond strongly at the tidal frequencies, and because of local resonances they are commonly significantly out-of-phase with the driving torques. The periodogram of the residuals to the UT1 data after fitting the k/C values, shown in figure 6, shows peaks at the tidal frequencies, exactly as might be expected from such out-of-phase oceanic effects. (There are, of course, other peaks that are not at the tidal frequencies, which may or may not be related to oceanic effects.) Furthermore, Yoder calculates that the ocean tides will cause UT1 variations at semi-diurnal and diurnal frequencies with amplitudes ranging from 0.02 to 0.07 milliseconds [Yoder et al., 1981]. Since this power is above the Nyquist frequency for the daily sampling employed here, it must alias into the frequency band that we are able to observe. There is no ready analytic cure for the process noise introduced by the oceans, but there may be ways to deal with it numerically. For example, given an analytic model for the ocean tide displacements such as the Schwiderski model [Schwiderski, 1980], it should be possible to numerically integrate the variations in moment of inertia of the oceans as a function of time. The implied variations in UT1 could then be compared with the observed residuals. Whether this approach would reduce the scatter in the estimates of k/C to the level that would make possible a useful determination of anelastic effects in the mantle remains to be seen. If the existing level of accuracy of ocean tide models is insufficient to reduce the scatter in the k/C estimates and the corresponding residuals in the UT1 observations, then these UT1 observations may have a role to play as a constraint on future ocean models.

If the semi-diurnal and diurnal tides cannot be modeled with sufficient accuracy to remove their effects from the UT1 values, it may be necessary to increase the sampling rate of the UT1 observations to at least the Nyquist frequency for the semi-diurnal tides, i.e., 4 observing sessions per day. VLBI is the only observing technique that can currently achieve such a sampling rate, and, indeed, sampling rates up to several times higher than this pose no particular technical problems.

IV. CONCLUSIONS:

The extraordinary power of VLBI for monitoring Earth rotation has been demonstrated by the quality of the IRIS results. The observed nutation corrections have placed a constraint on the interactions between the core and the mantle which is unavailable from any other observational techniques. The UT1 measurements have reached a level where the observational uncertainty is not a major part of the error budget, and the geophysical information that can be recovered is limited by our understanding of the complex interactions between the mantle, the oceans and the atmosphere. As these interactions are unraveled we can expect that the VLBI observations will play an important role in measuring anelastic effects in the mantle, and possibly in constraining models of ocean tide motions.

V. REFERENCES:

Bossler, J.D., Federal Implementation Plan for the Application of Space Technology to Crustal Dynamics and Earthquake Research, prepared by the Interagency Coordinating Committee for Geodynamics, 113 pp., 1982.

Carter, W.E., Modern Methods for the Determination of Polar Motion and UT1, Proceedings of Tenth Annual Precise Time and Time Interval Applications and Planning Meeting, NASA Memorandum #80250, 1978.

Carter, W.E. and W.E. Strange, The National Geodetic Survey Project "POLARIS," Tectonophysics, 52, 39-46, 1979.
Carter, W.E. and D.S. Robertson, IRIS Earth Rotation and Polar Motion Results, in Proc. International Symposium on Space Techniques for Geodynamics, Sopron, Hungary, 1984.
Gwinn, C.R., T.A. Herring and I.I. Shapiro, Geodesy by Radio Interferometry: Studies of the Forced Nutations of the Earth, Part II: Interpretation, J. Geophys. Res., 91, 4755-4765, 1986.
Herring, T.A., C.R. Gwinn, and I.I. Shapiro, Geodesy by Radio Interferometry: Studies of the Forced Nutations of the Earth, Part I: Data Analysis, J. Geophys. Res., 91, 4745-4754, 1986.
Herring, T.A., VLBI Studies of the Nutations of the Earth, this volume, 1987.
Robertson, D.S., W.E. Carter, B.D. Tapley, B.E. Schutz, R.J. Eanes, Polar Motion Measurements: Sub-Decimeter Accuracy Verified by Intercomparison, Science, 229, 1259-1261, 1985a.
Robertson, D.S., W.E. Carter, J.A. Campbell, and H. Schuh, Daily UT1 Determinations from IRIS Very Long Baseline Interferometry, Nature, 316, 424-427, 1985b.
Rosen, R. D., and D. A. Salstein, Variations in Atmospheric Angular Momentum on Global and Regional Scales and the Length of Day, J. Geophys. Res., 88, 5451-5470, 1983.
Schwiderski, E.W., On Charting the Global Tides, Rev. Geophys. and Sp. Phys., 18, 243-268, 1980.
Wahr, J.M., The Forced Nutations of an Elliptical, Rotating, Elastic and Oceanless Earth, Geophys. J. R. Astron. Soc., 64, 705-728, 1981.
Wahr, J.M., Earth Orientation and Nutation, this volume, 1987.
Wahr, J.M., and Z. Bergen, The Effects of Mantle Anelasticity on Nutations, Earth Tides, and Tidal Variations in the Rotation Rate, Geophys. J. R. Astron. Soc., 87, 633-668, 1986.
Wilkins, G.A., (ed.), Project MERIT: A Review of the Techniques to be Used During Project MERIT to Monitor the Rotation of the Earth, Royal Greenwich Observatory, Herstmonceux, 77 pp., 1980.
Yoder, C.F., J.G. Williams, and M.E. Parke, Tidal Variations of Earth Rotation, J. Geophys. Res., 86, 881-891, 1981.

ATMOSPHERIC EFFECTS ON EARTH ROTATION AND POLAR MOTION

David A. Salstein
Atmospheric and Environmental Research, Inc.
840 Memorial Drive
Cambridge, MA 02139 USA

ABSTRACT. The variability in the earth's rotation rate not due to known solid body tides is dominated on time scales of about four years and less by variations in global atmospheric angular momentum (M), as derived from the zonal wind distribution. Among features seen in the length of day (Δl.o.d.) record produced by atmospheric forcing are the strong seasonal cycle, quasi-periodic fluctuations around 40-50 days, and an interannual signal forced by a strong Pacific warming event, known as the El Niño. Momentum variations associated with these time scales arise in different latitudinal regions. Furthermore, winds in the stratosphere make a particularly important contribution to seasonal variability.

Other related topics discussed here are (i) comparisons of the M series from wind fields produced at different weather centers, (ii) the torques that dynamically link the atmosphere and earth, and (iii) longer-term non-atmospheric effects that can be seen upon removal of the atmospheric signal. An interesting application for climatological purposes is the use of historical earth rotation series as a proxy for atmospheric wind variability prior to the era of upper-air data. Lastly, results pertaining to the role of atmospheric pressure systems in exciting rapid polar motion are presented.

1. INTRODUCTION

The earth's variable rotation rate and polar motion have been measured accurately in recent years by a number of techniques, including VLBI. Such observations have been accompanied by renewed interest in the geophysical causes of these variations, including those originating within atmospheric winds and pressure systems. Earlier reviews of the subject by Munk and Macdonald (1960) and Lambeck (1980) pointed to the atmosphere for explanations of the observed seasonal variability in earth rotation. With the advent of modern atmospheric data in combination with new geodetic data sets, studies of the earth/atmosphere momentum budget have become possible on a much finer time scale. At

M. J. Reid and J. M. Moran (eds.), The Impact of VLBI on Astrophysics and Geophysics, 401–410.

operational agencies such as the U.S. National Meteorological Center (NMC) and the European Centre for Medium Range Weather Forecasts (ECMWF) regularly gridded atmospheric analysis fields are produced on a twice-daily basis as a by-product of their weather forecasting mission; furthermore, these fields have been determined to be sufficiently accurate for geodetic applications.

Recent studies have demonstrated that the expected proportionality between changes in the length of day (Δl.o.d.) not due to known tides and changes in global atmospheric angular momentum (M) exists on time scales of less than a few years (Hide et al., 1980; Rosen and Salstein, 1983). Thus many of the variations found in the earth's rotation are due to phenomena of meteorological origin. In addition, the atmosphere appears to play a role in the excitation of polar motion, due largely to shifts of mass associated with pressure systems.

2. TEMPORAL VARIABILITY IN M AND ΔL.O.D.

If the solid earth and the atmosphere form a two-body closed dynamical system, then changes in the momentum about the polar axis of one component must be matched by equal, but opposite changes in the momentum of the other. Based on this simple model, Rosen and Salstein (1983) derived the following linear relationship between ΔM and Δl.o.d.:

$$\Delta \text{l.o.d.} = 1.68 \times 10^{-29} \, \Delta M$$

where Δl.o.d. is in units of seconds and ΔM is in kg m^2 s^{-1}. The quantity M is given by the following expression:

$$M = \frac{2\pi a^2}{g} \iint_{\text{whole atmosphere}} [u] \cos^2\phi \, d\phi \, dp$$

where a is the mean radius of the earth (6.37×10^6 m), g is the acceleration due to gravity (9.81 m s^{-2}), ϕ is latitude, p is pressure and [u] the zonal mean zonal (eastward) wind. In calculations of M, the vertical extent of the pressure integration has varied in different data sets. The most complete data set has incorporated levels between the the 1000 and the 1 mb level, or 99.9% of the atmosphere. Since [u] values are not routinely available higher than 100 or 50 mb, winds at the high stratospheric levels required special processing.

Figure 1 shows the striking agreement between M and Δl.o.d. series for a four-year period. Evident in both series is the strong seasonal cycle; indeed the angular momentum of the atmosphere varies by nearly a factor of two between its minimum in July/August and its maximum in January/February. The annual signal is largely driven by a summer to winter contrast in tropospheric zonal winds which are greater in the Northern Hemisphere than in the Southern Hemisphere. A semiannual signal is also present, related in part to the presence of that frequency in the tropical stratosphere (Rosen and Salstein, 1985).

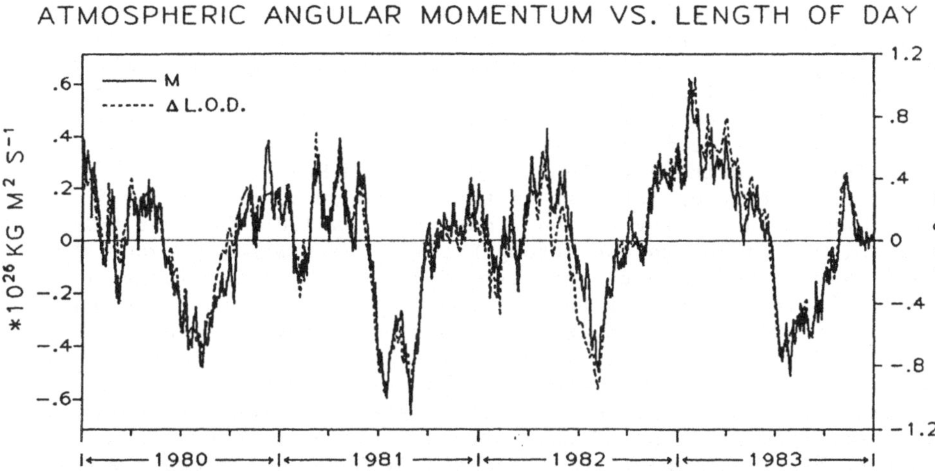

Fig. 1 Time series of daily values of atmospheric angular momentum between 1000 and 1 mb (solid line; scale on left) and three-day means of changes in Δl.o.d., minus tidal terms. The atmospheric series is based on NMC analyses for 1000-100 mb plus those obtained for 100-1 mb as described by Rosen and Salstein (1985). The Δl.o.d. series was produced by Morgan et al. (1985). The mean value of each series has been removed.

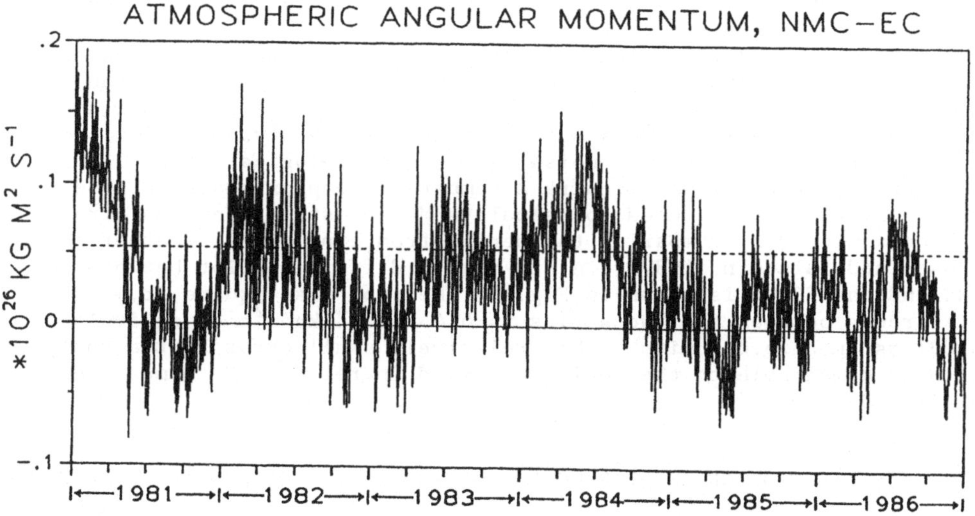

Fig. 2 The difference between time series of daily values of M from the NMC and ECMWF analyses. Values are integrated between 1000 and 50 mb. The dashed line indicates the r.m.s. difference between the two series.

Another pervasive feature in the curves of Fig. 1 arises from the quasi-periodic fluctuations in the atmosphere with periods of around 40-50 days (Langley et al., 1981). These fluctuations are the topic of considerable meteorological study and appear to arise in the tropical band, possibly related to convective activity (Madden and Julian, 1972). Furthermore, evidence exists for the propagation of these features into midlatitude regions (Anderson and Rosen, 1983).

A third feature of interest is the interannual variability in M and Δl.o.d., and in particular the peak during early 1983 that was related by Rosen et al. (1984) to the strong Pacific Ocean warming event known as the El Niño. Such events are part of one phase of the large-scale Southern Oscillation, a shifting of atmospheric mass across the breadth of the Pacific Ocean with an irregular period of several years. During the 1983 El Niño the westerly winds in mid-latitudes were observed to be exceptionally strong and the tropical easterlies weaker than normal; the strong peak in M was the result. A correlation between El Niño/Southern Oscillation (ENSO) events and the earth's rotation rate has been observed both over the past three decades by Chao (1984) and over a century and a quarter using historical data by Salstein and Rosen (1986).

3. CALCULATION AND COMPARISON OF M SERIES FROM THE WEATHER CENTERS

The complex system in place at the NMC (Dey and Morone, 1985) is similar to that found at other major weather centers. Meteorological fields are produced on a 2-1/2° x 2-1/2° latitude-longitude grid at 12 levels in the vertical by assimilating a large collection of unequally spaced weather observations every 12 hours. This analysis procedure can also incorporate values forecasted by the numerical prediction model and it weights them heavily in data-sparse areas. The analyzed fields are further processed to reduce noise of non-meteorological origin.

Because the systems require a number of complex steps to produce the zonal wind field, used to calculate M, an estimate of the error in M can be achieved by comparing the series from a number of weather centers (Rosen and Salstein, 1987). A six-year M difference plot between the NMC and the ECMWF is shown in Fig. 2; the root mean square (r.m.s.) difference between the two series is 5.4×10^{24} kg $m^2 s^{-1}$, equivalent to 0.091 ms in Δl.o.d. units. This relatively small error in the analysis of M is comparable to that being achieved in relating M to measurements of Δl.o.d.

4. DISTRIBUTION OF ANGULAR MOMENTUM VARIABILITY WITHIN THE ATMOSPHERE

To study regional contributions to the total variability in angular momentum, we have subdivided the atmosphere into a set of 46 equal-area latitudinal belts. In each belt the 10 year series of angular momentum was calculated from zonal winds and separated into three broad frequency bands: interannual, seasonal, and high frequency. The covariance of these series with the global M series, also similarly separated in frequency, was calculated, so that at each frequency the sum of the 46 covariance values yields the total variance in M. The results are displayed in Fig. 3.

This figure depicts the strong out-of-phase seasonal signal of the subtropical jet streams of the two hemispheres, with a stronger cycle in the Northern Hemisphere. Both the high frequency and interannual signals are relatively evenly contributed by the whole region between 30°S and 30°N.

5. TORQUES THAT LINK THE EARTH AND ATMOSPHERE

Angular momentum has been demonstrated to be conserved in the earth-atmosphere system by comparing the M and Δl.o.d. series kinematically. However, the physical mechanism, i.e. the torque that dynamically couples the earth and atmosphere is beginning to be studied also, though this problem is less straightforward to formulate. This torque has been computed as the contribution from two separate processes: that from surface friction and the more macroscopic interaction called "mountain torque." The friction torque has been modeled by the use of drag coefficients at the earth's surface while the mountain torque involves knowledge of the atmospheric pressure distribution around topographic features, as well as their elevation.

Recent estimates of the torques remain somewhat inconclusive. On monthly time scales, Wahr and Oort (1984) have found the friction torque to be the more dominant, though they discuss the large uncertainties involved in their computation. On less than monthly time scales, results by Swinbank (1985) and Wolf and Smith (1987) have isolated mountain torques as being the dominant mechanism for rapid transfer of momentum between earth and atmosphere.

6. HISTORICAL EARTH ROTATION DATA AS A PROXY FOR ATMOSPHERIC BEHAVIOR

The result that changes in M and Δl.o.d. are well correlated on time scales of up to several years enabled Salstein and Rosen (1986) to examine the feasibility of using a historical record of the earth's rotation as a proxy for year-to-year changes in the global zonal wind field. Such an earth rotation series has been prepared by Morrison (1979) and further analyzed by McCarthy and Babcock (1986), and is based mainly on telescopic observations of the occultation of stars by the moon.

The time series of fluctuations in l.o.d. contains small enough errors to relate to the atmosphere since 1860, and was filtered to remove the large variations on decadal time scales not of meteorological origin. The resulting filtered time series is shown in Fig. 4; also marked are the years in which a well-defined ENSO warm event occurred, according to a compilation by van Loon and Shea (1985). A student's t-test confirms (at the 95% level) that winters in years following ENSO events have "longer days" than otherwise. Since this connection was

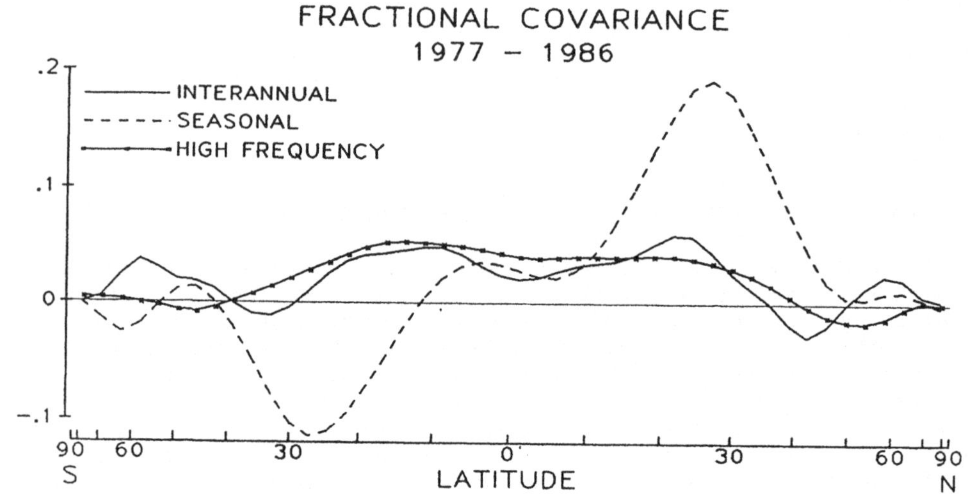

Fig. 3 Fractional part of the variance in the NMC M series for 1977-86 explained by each of 46 equal-area belts for three frequency bands, given by covariance (m_b, M) ÷ variance (M), where m_b is the belt angular momentum. Each belt's central latitude is at the dots on the high frequency curve.

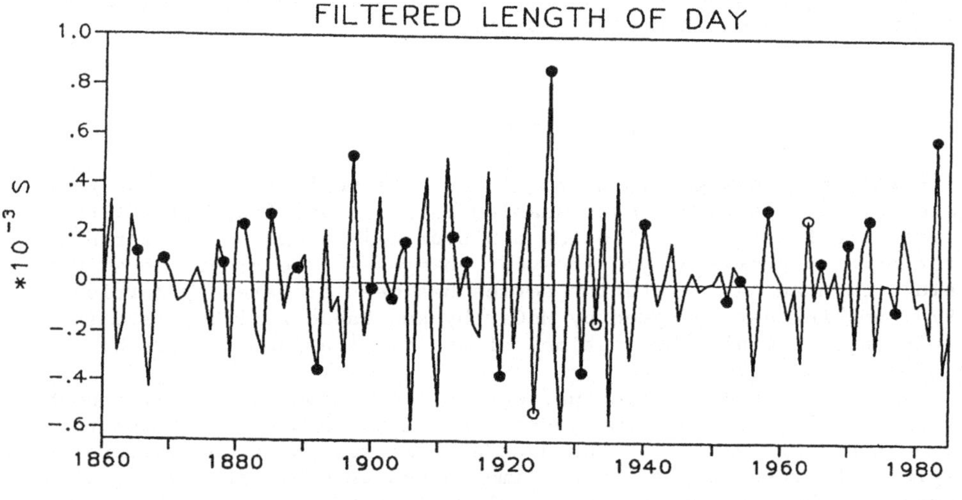

Fig. 4 High-pass filtered values of Δl.o.d. for northern winters during 1860-1985. Solid dots mark those winters that follow a year containing a well-defined ENSO warm event; open dots indicated weak ENSO events.

expected, we have at least partially justified the use of this historical l.o.d. series in meteorological studies. We then used the results in Fig. 4 to examine the intra-decadal variability of the implied northern winter global M values. A large contrast was found, for example, between high variability during the 1920s and low variability of the 1940s, possibly associated with different temperature regimes during these decades. Based on the interannual curve in Fig. 3, these circulation changes apparently result from zonal wind anomalies in the region between 30°S and 30°N.

7. ATMOSPHERIC EXCITATION OF POLAR MOTIONS

As in the model whereby M and Δl.o.d. are closely connected, changes in the equatorial component of atmospheric angular momentum can be related to equatorial changes in the earth rotation vector, that is, the wobble of the polar axis. In this case, however, calculations have indicated that more than the atmosphere needs to be accounted for to produce the observed polar motion.

The largest signals in the polar motion time series are associated with the (approximately) 14-month resonant Chandler wobble and an annual wobble. For quite some time, the annual wobble has been assumed to be largely excited by seasonal fluctuations in the atmosphere's mass (Munk and Macdonald, 1960); however, the forcing mechanism of the Chandler wobble has not been discovered yet, but may also derive some of its power from the atmosphere (Eubanks, personal communication).

Rapid polar motions on the time scale of 15 to 45 days, with smaller amplitudes than the Chandler and seasonal wobbles, have been observed from recent high precision data (Robertson et al., 1985). Using daily NMC analyses and a formulation for computing the atmospheric excitation functions by Barnes et al. (1983), Eubanks et al. (1987) demonstrated that these rapid polar motions are largely driven by surface air pressure changes, as partly modified by the response of sea level to atmospheric loading. Hide (1984) also concluded from the use of ECMWF atmospheric analyses that an overall atmospheric forcing of polar motions was occurring. It should be added here that the Barnes et al. development comprises the effects on polar motion of both the winds and the atmospheric surface pressure distribution modified by the elastic rotational and surface loading deformations of the earth.

We have separated the atmospheric surface pressure excitation values from NMC into a number of equal-area sectors in order to determine which regions are most responsible for behavior on high frequencies of the global excitation. To do so we have calculated the covariance between the regional series and the global series filtered in the 15-45 day high frequency band, and the results are displayed in the map in Fig. 5 (Salstein and Rosen, 1987). As shown in the figure, the main regions contributing to the excitation of rapid polar motions are located over the oceanic storm track regions of the North Pacific and North Atlantic and the midlatitudes of the Southern Hemisphere.

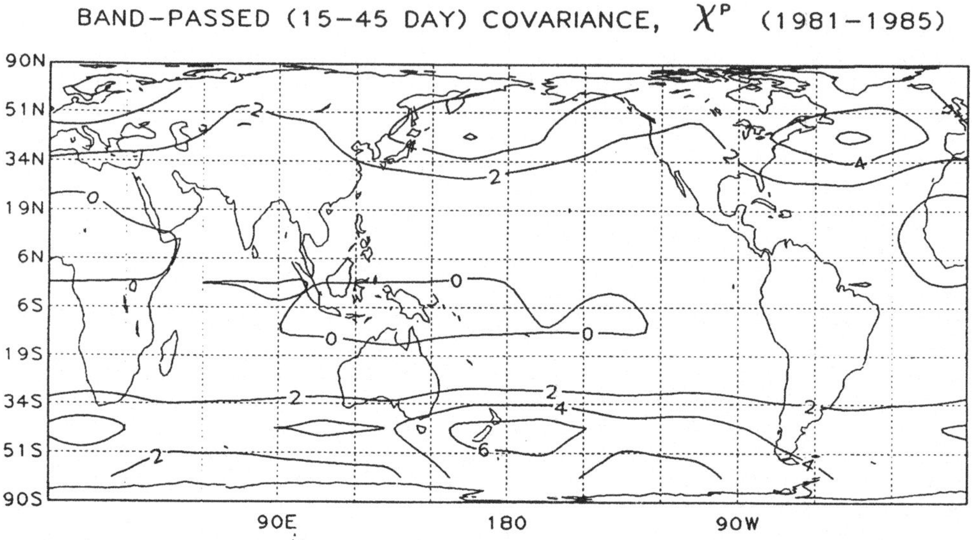

Fig. 5 Covariance between the global atmospheric excitation function for polar motion associated with surface pressure fluctuations (χ^P) between ~15-45 days and the contribution made to χ^P in each of 108 equal-area sectors, bounded by the dashed lines. Results are based on NMC analyses for 1981-85, in non-dimensional units of 10^{-16}.

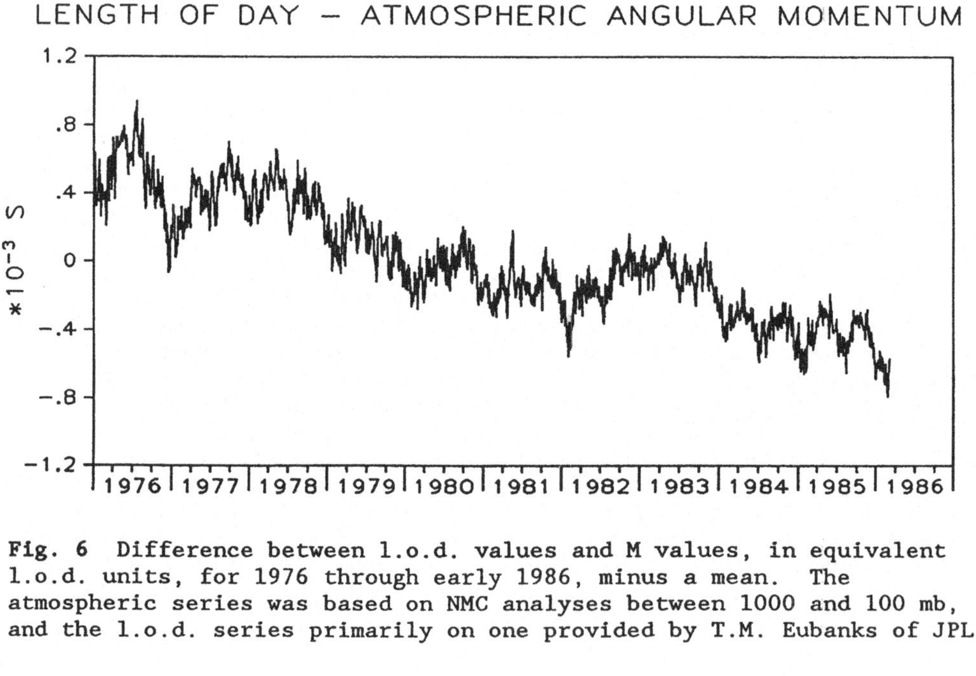

Fig. 6 Difference between l.o.d. values and M values, in equivalent l.o.d. units, for 1976 through early 1986, minus a mean. The atmospheric series was based on NMC analyses between 1000 and 100 mb, and the l.o.d. series primarily on one provided by T.M. Eubanks of JPL.

8. AFTERWORD

I have focused here on a number of recent issues and results regarding the role of the atmosphere in both earth rotation and polar motion. The modern atmospheric and geodetic data sets have enabled these studies to progress rapidly.

Knowledge of the atmospheric signal in earth rotation allows a more reliable calculation of the role of non-atmospheric mechanisms, and in particular, the influence of the earth's core. For example, subtraction of the equivalent M series from the Δl.o.d. series for the 1976-86 period shows a long period signal which may be modeled by the alternation of relatively rapid decreases with plateau periods (Fig. 6). From such results, a better understanding of the core-mantle interaction may emerge.

A number of other challenges in this field lie ahead including better determination of the Chandler wobble excitation, of the role of the oceans and the hydrological cycle, and improved calculations of earth-atmosphere torques. Advances in both data quality and theoretical formulations will undoubtedly help provide more answers.

9. ACKNOWLEDGMENTS

Most of my research reported here has been performed in close collaboration with Dr. Richard D. Rosen of AER Inc., to whom I am much indebted. Support came from the Geodynamics Program of NASA, most recently under contract NAS5-28195.

10. REFERENCES

Anderson, J.R. and R.D. Rosen, 1983: 'The latitude-height structure of 40-50 day variations in atmospheric angular momentum.' J. Atmos. Sci., **40**, 1584-1591.

Barnes, R.T.H., R. Hide, A.A. White and C.A. Wilson, 1983: 'Atmospheric angular momentum fluctuations, length-of-day changes and polar motion.' Proc. R. Soc. London, **A387**, 31-73.

Chao, B.F., 1984: 'Interannual length-of-day variation with relation to the Southern Oscillation/El Niño.' Geophys. Res. Lett., **11**, 541-544

Dey, C.H. and L.L. Morone, 1985: 'Evolution of the National Meteorological Center global data assimilation system: January 1982 - December 1983.' Mon. Wea. Rev., **113**, 304-318.

Eubanks, T.M., J.A. Steppe, J.O. Dickey, R.D. Rosen and D.A. Salstein, 1987: 'Rapid polar motions associated with atmospheric pressure changes and oceanic response.' Manuscript in preparation.

Hide, R., N.T. Birch, L.V. Morrison, D.J. Shea, and A.A. White, 1980: 'Atmospheric angular momentum fluctuations and change in the length of the day.' Nature, **286**, 114-117.

Hide, R., 1984: 'Rotation of the atmospheres of the Earth and planets.' Phil. Trans. R. Soc. Lond., **A313**, 107-121.

Lambeck, K., 1980: The Earth's Variable Rotation. Cambridge University Press, Cambridge, 449 pp.
Langley, R.B., R.W. King, I.I. Shapiro, R.D. Rosen and D.A. Salstein, 1981: 'Atmospheric angular momentum and the length of day: a common fluctuation with a period near 50 days.' Nature, **294**, 730-732.
Madden, R.A. and P.R. Julian, 1972: 'Description of global-scale circulation cells in the tropics with a 40-50 day period.' J. Atmos. Sci., **29**, 1109-1123.
McCarthy, D.D. and A.K. Babcock, 1986: 'The length of day since 1656.' Phys. Earth and Planet. Inter., **44**, 281-292.
Morgan, P.J., R.W. King and I.I. Shapiro, 1985: 'Length of day and atmospheric angular momentum: a comparison for 1981-1983.' J. Geophys. Res., **90**, 12645-12652.
Morrison, L.V., 1979: 'Re-determination of the decade fluctuations in the rotation of the earth in the period 1861-1978.' Geophys. J. R. astr. Soc., **58**, 349-360.
Munk, W.H. and G.J.F. Macdonald, 1960: The Rotation of the Earth. Cambridge University Press, Cambridge, 323 pp.
Robertson, D.S., W.E. Carter, B.D. Tapley, B.E. Schutz, R.J. Eanes, 1985: 'Polar motion measurements: subdecimeter accuracy verified by intercomparison.' Science, **229**, 1259-1261.
Rosen, R.D. and D.A. Salstein, 1983: 'Variations in atmospheric angular momentum on global and regional scales and the length of day.' J. Geophys. Res., **88**, 5451-5470.
Rosen, R.D. and D.A. Salstein, 1985: 'Contribution of stratospheric winds to annual and semiannual fluctuations in atmospheric angular momentum and the length of day.' J. Geophys. Res., **90**, 8033-8041.
Rosen, R.D., D.A. Salstein, A.J. Miller and K. Arpe, 1987: 'Accuracy of atmospheric angular momentum estimates from operational analyses.' Mon. Wea. Rev., **115**, in press.
Rosen, R.D., D.A. Salstein, T.M. Eubanks, J.O. Dickey and J.A. Steppe, 1984: 'An El Niño signal in atmospheric angular momentum and earth rotation.' Science, **225**, 411-414.
Salstein, D.A. and R.D. Rosen, 1986: 'Earth rotation as a proxy for interannual variability in atmospheric circulation, 1860-present.' J. Climate Appl. Meteor., **25**, 1870-1877.
Salstein, D.A. and R.D. Rosen, 1987: 'Regional contributions to the atmospheric excitation of rapid polar motions,' manuscript in preparation.
Swinbank, R., 1985: 'The global atmospheric angular momentum balance inferred from analyses made during the FGGE.' Quart. J. R. Met. Soc., **111**, 977-992.
van Loon, H. and D.J. Shea, 1985: 'The Southern Oscillation. Part IV: The precursors south of 15°S to the extremes of the oscillation.' Mon. Wea Rev., **113**, 2063-2074.
Wahr, J.M. and A.H. Oort, 1984: 'Friction- and mountain-torque estimates from global atmospheric data.' J. Atmos. Sci., **41**, 190-204.
Wolf, W.L. and R.B. Smith, 1987: 'Length of day changes and mountain torque during El Niño.' J. Atmos. Sci., in press.

AN ELASTO-VISCOUS MODEL OF THE EARTH

S. Losito*, B. Pernice, D. Picca, G. Verrone
*IESI-CNR
Dipartimento di Fisica
Via Amendola 173
70126 Bari, Italy

ABSTRACT. A two-symmetric-rigid-rotators model of the Earth has been studied, under the hypothesis of elasto-viscous coupling. The free Eulerian equation of motion has been solved in the linear approximation related to small wobbling amplitudes. Under these hypotheses, polar motion is stable, and the angular velocity of the Earth is the sum of three vectors rotating with different frequencies and damped amplitudes. One of these terms turns out to be retrograde with a quasi-diurnal frequency and could be identified with a similar term appearing in liquid core models of the Earth. The other two terms are identified with the Chandler wobble and the annual term according to observational data. The elastic coupling produces, in the time variation of L.O.D., a periodic term whose frequency is about one year. It could be hypothesized that the "decade fluctuation" could be partially attributed to the free oscillation of the Earth.

1. THE MODEL

The problem of polar motion has not been well modeled until now. We have studied an elasto-viscous model of the Earth under some simplifying hypotheses. Detailed calculations will appear elsewhere. Here we outline the main results. Two rigid symmetric rotators are considered, the outer for the crust-mantle and the inner for the core. We adopt two reference frames, with the same origin, $(\bar{x}_1, \bar{x}_2, \bar{x}_3)$ for the mantle and $(\bar{a}, \bar{b}, \bar{c})$ for the core. It is assumed also that the elastic coupling between the two rotators can be expressed as $T = \sigma \cdot \alpha$ where σ is a positive constant and represents the angular deviation of the core from its equilibrium position with respect to the mantle.

The viscous coupling is assumed to be proportional to the difference between the angular velocity of the core $\bar{\omega}$ and the mantle $\bar{\Omega}$, via a second order tensor of inertia $\overleftrightarrow{u}$. The Eulerian equations of motion for the mantle and the core give rise to four differential equations for $\dot{\bar{\Omega}}$, $\dot{\bar{\omega}}$, $\dot{\bar{c}}$, $\dot{\bar{a}}$. The total torque considered is the sum of the elastic torque, the viscous torque, and a residual torque, which include all the external interactions of the systems.

M. J. Reid and J. M. Moran (eds.), The Impact of VLBI on Astrophysics and Geophysics, 411–412.

2. FREE MOTION SOLUTIONS

To solve this set of differential equations, we adopt for $\bar{\Omega}$, $\bar{\omega}$, $\bar{c}$, $\bar{a}$ the hypothesis of small wobbling amplitudes, well supported by observations.

The Free Motion Solution, that is, the equation with no external interactions, can be split into two new equations; one differential equation is referred to as the free equation of L.O.D. owing to the fact that its most representative term is the parameter m_3 related to $\Delta\Omega/\Omega_\circ$ where $\Omega_\circ = 1$ rev/day is the mean angular velocity of the Earth.

The second equation is referred to as the Free Polar Motion Equation (FPME), owing to the fact that it contains the parameters m_1 and m_2 of the angular velocity of the Earth in the equatorial plane $(\bar{x}_1, \bar{x}_2)$ that is the wobbling components.

From the FPME one finds that the $m_1(t)$ and $m_2(t)$, parameters of the wobbling components of $\bar{\Omega}$, are the sum of three rotating vectors, each in the equatorial plane $(\bar{x}_1, \bar{x}_2)$, each with its frequency and each with damped amplitudes.

The first two complex eigenvalues of the solution are identified respectively with the annual term and the Chandler term of the Earth rotation; the third eigenvalue is related to a term of period $T_d = -.999991$ days and damped time $\tau_d = .4693713$ years. That is to say the third term turns out to be a retrograde diurnal term.

It is also possible to obtain an estimate of the nondimensional parameters related to the elastic and viscous coupling. The main term $m_3(t)$ of the F.E. of L.O.D. contains oscillating terms with frequency depending on the viscous parameter. Setting this parameter to zero, one obtains a frequency = 19.1910 rad/day with a period of 327 days.

3. SUMMARY

The annual term is usually related to the interaction of the Earth with the atmospheric angular momentum. In this work it is assumed as a characteristic of the Free Motion. The quasi-diurnal retrograde term has been found for liquid core models. In this work it is derived from a solid core model.

The annual term variation in the L.O.D. is an indication that the "decade fluctuations" in the L.O.D. can be partly attributed to the viscous coupling between the core and the mantle.

4. REFERENCES

Flodmark, S., and Davstad, K., "Core-Mantle Interaction and Chandler Wobbles," Earth's Rotation: Solved and Unsolved Problems, NATO Advances Research Workshop, Chateau de Bonas, Gers, France, 1985.

Kubo, Y., "An Explanation of the Polar Motion by a Rigid Core-Mantle Model," in *Rotation of the Earth*, Melchior, P., and Yumi, S., eds., copyright IAU, 1982, p. 182.

Lambeck, K., *The Earth Variable Rotation: Geophysical Causes and Consequences*, Cambridge University Press, Capt. I, 1980.

Rochester, M. G., "The Earth's Rotation," *EOS (Trans. Am. Geophysical Union)*, **54**, 769-780, 1973.

ON THE ROTATION MODEL IN VLBI GEOPHYSICAL ANALYSIS

Y. F. Xia
Astronomy Department
Nanjing University
People's Republic of China

The basic observables of VLBI for geophysics are the delay and the delay rate. These two basic observables depend on relative geometric relations between the baseline vector and the source vector. The transformation of baseline orientation from the terrestrial coordinate system, in which the stations are located, to the celestial coordinate system, in which VLBI observables are calculated, involves a number of models. One of them is the rotation model.

In this paper, we first describe the relations among various reference poles for the deformable earth and the liquid core earth; then the rotation model in VLBI geophysical analysis is discussed. It points out:

1. If the nutation given is the nutation of the angular momentum pole, and if the diurnal spin matrix rotates about the pole, then the rotation model can be represented as

$$\mathbf{B}_{sf} = PNSDW\ \mathbf{B}_{bf} \quad , \tag{1}$$

   where $\mathbf{B}_{sf}$ and $\mathbf{B}_{bf}$ represent the baseline vector given in the space-fixed and body-fixed coordinate system, respectively, and D is the diurnal polar motion matrix. It is defined as the transformation in the body-fixed coordinate system from the angual momentum pole to the angular momentum pole with force-free polar motion. Its amplitude is that of the forced polar motion of the angular momentum pole.

2. If the diurnal spin matrix rotates about the angular momentum pole with force-free polar motion, then the rotation model can be represented as

$$\mathbf{B}_{sf} = PNO_pS'W\ \mathbf{B}_{bf} \quad , \tag{2}$$

   where O_p is called the Oppolzer term transformation matrix. Its amplitude is that of the nutation Oppolzer term of Celestial Ephemeris Pole (CEP). O_p can be predicted.

3. If the force nutation of the mean figure pole is given, then the rotation model can be represented as

$$\mathbf{B}_{sf} = PN'S'W\ \mathbf{B}_{bf} \quad , \tag{3}$$

M. J. Reid and J. M. Moran (eds.), The Impact of VLBI on Astrophysics and Geophysics, 413–414.

which include neither the diurnal polar motion matrix nor the Oppolzer term transformation matrix.

4. For the liquid core earth, the matrix that transforms the angular momentum pole with force-free polar motion into a Chandler angular momentum pole is neglected. Its amplitude is that of the nearly diurnal free polar motion of the angular momentum pole in the body-fixed coordinate system and is that of the free core nutation of the mean figure pole in the space-fixed coordinate system.

The research of liquid core earth will provide more perfect rotation model for VLBI geophysical analysis.

ACKNOWLEDGMENT

The author wishes to acknowledge the help of Professor W. Cannon in discussions of reference system.

ON THE ERROR EQUATION FOR DETERMINING EARTH ROTATION PARAMETERS WITH VLBI NETWORK

Y. F. Xia
Astronomy Department
Nanjing University
People's Republic of China

1. INTRODUCTION

In this paper, the partial derivative matrix in the error equation for determining earth rotation parameters (ERP) with the VLBI network is described, and the effect of the functional parameter constraint on solving ERP is discussed.

2. THE COMPUTATION OF THE PARTIAL DERIVATIVE MATRIX FOR DELAY AND DELAY RATE WITH RESPECT TO PARAMETERS

The delay for computing partial derivatives may be written as

$$\tau = \tau_o \left[1 - (\dot{\mathbf{r}}_2)_s \cdot \mathbf{S} + ((\dot{\mathbf{r}}_2)_s \cdot \mathbf{S})^2\right] \quad , \tag{1}$$

where

$$\tau_o = -\frac{1}{c}\, \mathbf{B}_s \cdot \mathbf{S} = -\frac{1}{c} PN'S'W\mathbf{B}_b \cdot \mathbf{S} \quad . \tag{2}$$

The following expression of the delay in the celestial coordinate system is given:

$$\tau = -\frac{\mathbf{B}_s}{c}\left[\sin\delta_B \sin\delta + \cos\delta_B \cos\delta \cos(\alpha_B - \alpha)\right] \quad . \tag{3}$$

Obviously, α_B and δ_B are the right ascension and declination of the baseline. If the Celestial Ephemeris Pole is adopted as the uniform pole for the space-fixed and body-fixed coordinate system, then $\alpha_B = S_G + \lambda_B$.

If we directly utilize equation (3) to compute the partial derivatives of the delay and delay rate with respect to the parameters and mix up the corresponding reference system, then the correct ERP cannot be calculated. The expressions of the delay and delay rate for computing partial derivatives should be in explicit function form of source position parameters in the space-fixed coordinate system and baseline parameters in the body-fixed coordinate system.

3. SOLVING FOR THE ERROR EQUATION

In determining the ERP, the radio source is observed simultaneously at all stations of the VLBI network. At that time, the constraint condition can sometimes be imposed in least squares adjustment.

M. J. Reid and J. M. Moran (eds.), The Impact of VLBI on Astrophysics and Geophysics, 415–416.

The error equation of determining ERP with VLBI network is

$$\mathbf{L} = F\,\Delta\mathbf{X} + \mathbf{e} \ , \tag{4}$$

where $\mathbf{L}$ is the residual vector of the delay and delay rate, F is the partial derivative matrix of the delay and delay rate with respect to the parameters estimated, $\Delta\mathbf{X}$ is the corrected vector of the a *priori* parameters, and $\mathbf{e}$ is the vector of observational error.

By means of functional parameter constraints, we can derive the solution of $\Delta\mathbf{X}$ (Mikhail 1970, 1976)

$$\Delta\mathbf{X}_{lsr} = \Delta\mathbf{X}_{ls} + \left(Z^T - \Delta\mathbf{X}_{ls}^T Q^T\right) \left(QDQ^T\right)^{-1} QD \ , \tag{5}$$

where Q is the coefficient matrix of $\Delta\mathbf{X}$ under the constraint condition, and Z is the conditional vector constrained.

The variance covariance matrix for parameters estimated is

$$D_R = D - DQ^T(QDQ^T)\,QD \ . \tag{6}$$

If the VLBI network is composed of three stations, then the general constraint condition is (Davidson 1980)

$$\sum_{m=1}^{3} B_m = 0 \ , \qquad \sum_{m=1}^{3} a_m^k = 0 \ , \tag{7}$$

where m is the baseline number, and k is the order of clock polynomial coefficients.

In order to examine the effect of the functional parameter constraint on the precision of earth rotation parameters estimated, we take three baselines composed of three stations: Shanghai (China), Effelsberg (FRG), and Algonquin (Canada). The positions of 10 radio sources whose correlation flux densities are greater than 0.5 Jy are taken from the catalogue given by Preston *et al.* (1978).

The altitude h and the azimuth A of these radio sources observed from the station are computed by means of the positions of the stations and the radio sources, once every 15 minutes of Greenwich sidereal time. The observed altitude is greater than 20 . Therefore the observational schedule can be made.

The clock model takes a linear form. The standard deviations, σ_τ and $\sigma_{\dot\tau}$, of the delay and delay rate observed, are assumed to be equal to 0.5 ns and 0.1 ps s^{-1}, respectively.

The computation results show that the functional parameter constraint imposed in the least squares adjustment may reduce the correlation of ERP with the baseline parameters and clock polynomial coefficients. All of these parameters can be solved together, and the precision of solving ERP can be increased if the functional parameter constraint is used.

4. REFERENCES

Davidson, D. A. 1980, *A Least Square Adjustment for LBI*, Univ. of New Brunswick, Technical Report No. 71.

Mikhail, E. M. 1970, *Photogrammetric Engineering*, **36**, 1277.

Mikhail, E. M. 1976, *Observations and Least Squares* (New York: IEP).

Preston, R. A., *et al.* 1978, DSN Progress Report 42–46.

STABILITY of DETERMINING the EARTH ROTATION PARAMETERS (ERP) with VLBI

Shifang Luo and Dawei Zheng
Shanghai Observatory, Academia Sinica
Shanghai, People's Republic of China

Abstract

By using the observations of IRIS network, the stability of determinang ERP with VLBI is studied. It is concluded that the uncertainties from initial values of ERP, the errors of other parameters are at the same level as the formal errors in determination of ERP. The geometric effect on determination of ERP is important and appears as systematic errors. Geometric uncertainty on polar motion is greater than that on UT1. and specially much worse for the continenal network. The stability of determining ERP with VLBI can be improved either by increasing new stations at reasonable location in a VLBI network or by increasing new networks.

I. Introduction

In 1977 the project of POLARIS was operated in NGS (NOAA, USA)(Carter et al.,1979). In recent years, POLARIS has been developed to IRIS (the International Radio Interferometric Surveying).IRIS consists of 5 stations which are distributed in northern America and Europe. The IRIS network is sensitive to determination of ERP (Carter et al, 1984).

By using the analysis and comparison of several series obtained by resolving the observations from IRIS network, the stability of determining ERP is studied, which includes the uncertaintied from the ERP initial values errors, the errors of unadjusted parameters and geometric errors from network.

II. Data

1.ERP series from IRIS Bullitin with Code of ''Q''(QR) and the Definite Results (DR) from nov. 1984 to May 1985.

2.(a).resolved ERP series from Nov.1984 to May,1985 by the single day's session and ERP from Circular D of BIH as initial values, same clock pattern and atmospheric zenith height parameters as in QR, and fixing positions of sources and stations.

(b).ERP series from re-structured 3 VLBI networks by IRIS observations. The IRIS observations is composed of those in two Euro-America networks and an American network as shown in Fig.1.

M. J. Reid and J. M. Moran (eds.), The Impact of VLBI on Astrophysics and Geophysics, 417–420.

The parameter for the 3 networks are shown as follows:

European-American networks:

(I)E-F-W

baseline	L	F	length
	o	o	(km)
E-F	197	-20	3134.9
E-W	64	5	5998.3
F-W	49	11	8417.6

(II) E-R-W

baseline	L	F	length
	o	o	(km)
E-R	66	-50	2044.5
E-W	64	5	5998.3
R-W	65	16	7588.4

American network

(III) E-F-R

baseline	L	F	length
	o	o	(km)
E-R	66	-50	2044.5
E-F	197	-20	3134.9
F-R	-9	-12	2362.6

E-Westford,Mass.; Westford Observatory
F-Ft.Davis,Texas; Harvard Observation Station
R-Miami, Fla.; Richmond POLARIS Observatory
W-Wettzell, Federal Republic of Germany

The background for solving ERP from the observations of these networks is the same as in (a). In the new solution, the same computer and the software are used as in processing the data of IRIS Bulletin.

III. Comparison and Conclusion

1. uncertainty from the initial values of ERP

Comparing the series between QR and 2.(a) in section II., we obtained the uncertainty from the errors of initial values of ERP, which are shown on the first line of Table 1. The maximum difference of ERP between Rapid Service and Circular D of BIH during the studied period is 5 ms for UT1 and 10 mas for x,y components, but both the systematic errors (Mean) and random errors(RMS) in ERP series caused by these are almost equal to the internal errors in magnitude (formal errors, 0.05ms for UT1 and 1 mas for x,y).

Table 1. uncertainties from initial values of ERP (A) and other unadjusted parameters (B)

		DUT1		Dx		Dy	
Code	Number	Mean	RMS	Mean	RMS	Mean	RMS
		(0.01ms)		(0.1mas)		(0.1mas)	
(A)	33	-2	±5	+2	±6	-1	±6
(B)	35	-7	5	+6	6	-2	6

2. uncertainty from other unadjusted parameters

This type of uncertainty is obtained by comparing the series between DR and 2.(a) in section II. and shown on the second line of Table 1. It is obvious that the uncertainty from the unadjusted parameters is greater than that from initial values of ERP, but still is almost equal to the formal errors in magnitude.

3. Geometric uncertainty of IRIS network

The geometric uncertainty of IRIS network on determining ERP are studied by comparing the series in 2.(b) with the one in 2.(a) in section II. and shown in Table 2 and Fig 2.

From Table 2 and Fig.2, the following arguments can be drawn:

There is a systematic error about 3 mas in x component for (E-F-W) network.

There is a systematic error about 2 mas in x and y components for (E-R-W) network.

But for the American network (E-F-R), there are big systematic error (Mean) and statistical error (RMS) in y component, and a strong correlation exists between x components and UT1 because of the location of stations.

Table 2. Geometric uncertainty of VLBI network on determining ERP

		Formal Errors			External Accuracy					
network	number	SUT1	Sx	Sy	DUT1		Dx		Dy	
					Mean	RMS	Mean	RMS	Mean	RMS
E-F-W	35	± 5	±11	± 8	+8	± 4	-28	±13	0	± 7
E-R-W	31	6	13	11	+3	9	+23	14	+17	16
E-F-R	31	16	18	31	-22	26	+20	28	-72	37

The systematic errors in x component for 2 European-American networks are nearly equal in magnitude, but opposite in sign (Fig.2).

So it is suggested that a new station should be set up in IRIS network in order to reduce the uncertainty from the geometry of network when the observations fail in some stations. On the other hand, some realistic and new intercontinental networks should be set up to determine ERP regularly in order to improve the stability of ERP by combining the results from different networks.

IV. Acknowledgements

The work reported here was supported by a grant from the National Geodetic Survey Division (NGSD),Carting and Geodetic Survice, National Oceanic and Atmospheric Administration. We thank the many people at the observatories, the correlator center, the analysis center, and the support facilities for their hard work.

References

1. Carter,W.E.,and W. E. Strange,
The National Geodetic Survey Project ''POLARIS''
Tectonophysics 52.(1979) p30-46

2. Carter,W.E. and D.S. Robertson,
IRIS Earth Rotation and Polar Motion Results
Proceedings of the International Symposium on Space Techniques for Geodynamics Sopron, Hungary (1984)

Fig. 1. Restructured 3 VLBI networks

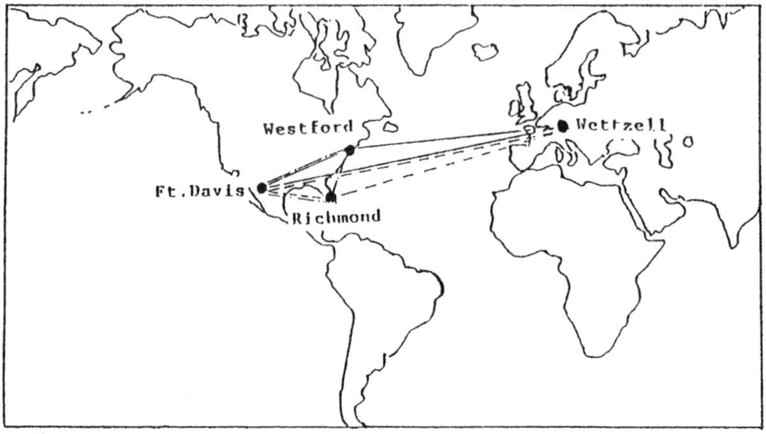

Fig. 2. Geometric effects on UT1, x and y

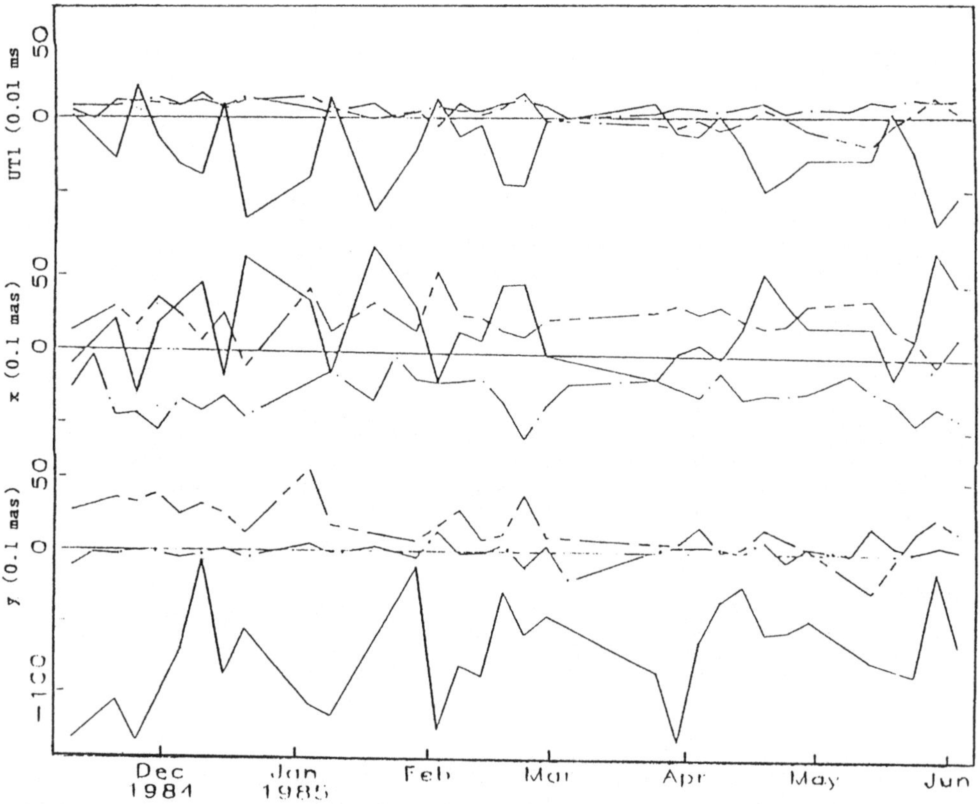

JOINT USES OF VLBI WITH ASTRONOMICAL OPTICAL INSTRUMENTS FOR STUDYING VERTICAL CHANGES

Li Zhi-sen, Zhang Guo-dong, and Han Yan-ben
Beijing Astronomical Observatory
Academia Sinica
Beijing, People's Republic of China

The description of the gravitational field at the surface of the Earth requires two quantities: the absolute value of the gravitational acceleration and the gravitational direction (deviation from vertical direction). At present, the various gravimeters measure the former quantity, and there is no effective instrument for monitoring the latter. This shortcoming seriously affects the comprehension and further knowledge of the gravitational field.

It is well known that VLBI can be applied to measure earth rotation parameters. VLBI observations have great precision, and the measurements are referred to the inertial frame of the radio sources. They will soon replace optical instruments and other techniques that monitor earth rotation parameters.

Astronomical optical observations in support of geodesy have a history of about 80 years. They give numerous data of angle values between the rotation's axis and the vertical (e.g., plumb line) at some places. By analysis of their residuals, we have found that the optical instruments can detect changes in the local vertical, at least the violent vertical changes that occur before major earthquakes. The Soviet astronomer N. T. Mironov has also studied the variations of the directions of plumb lines from data of astronomical observations. The plumb line at a given place not only has tidal variation (about $0''.03$) but also changes due to the motions of ground mass, such as ground water before major earthquakes. These changes can exceed $0''.1$.

VLBI is a geometrical measurement technique, and it can be used to measure the geodetic coordinates. Therefore, its joint use and coordinated observation with optical measurements at a certain place could provide valuable data on vertical changes. For this purpose, portable VLBI equipment should be developed.

M. J. Reid and J. M. Moran (eds.), The Impact of VLBI on Astrophysics and Geophysics, 421.

CONVENTIONAL TERRESTRIAL SYSTEM BY VLBI: A KINEMATIC APPROACH

H. B. Papo and J. Saleh
Technion, Israel Institute of Technology
Technion City, Haifa 32 000
Israel

The paper examines the potentials of VLBI time delay observables for the establishment and maintenance of a Conventional Terrestrial System (CTS). The CTS is defined in 2-D by the standard epoch positions and velocities of a network of control points located on a spherical reference surface.

VLBI time delay observables are ideally suited for the establishment of a CTS due to their sensitivity to the rotational motion of the observing stations with respect to inertial space (CIS). The rotational motion of a VLBI station with respect to the CIS is partitioned into individual or regional motion with respect to a discrete terrestrial coordinate frame and into global motion of the terrestrial coordinate frame with respect to the CIS. The global motion is mathematically described by the models and parameters of general precession, nutation and diurnal spin. The unmodelled part is defined as polar motion and spin-rate variations. The above three models together with the unmodelled polar motion and spin-rate variation determine at a given epoch the orientation of the CTS coordinate frame with respect to the CIS.

A kinematic model of analysis of the CTS resembles the system of a fundamental star catalogue. Each CTS control point defines the CTS by its two spherical coordinates at the standard epoch T_0 and by the coordinate time rates. The only type of measurement considered in this paper is VLBI time delay, observed between points of the CTS. However, VLBI time delay measurements alone are inadequate for defining completely the CTS in 2-D at a given epoch or for monitoring its rotational motion. There is a need for additional data sets and models (derived from non-VLBI sources) that would convey a conventional datum to the CTS. The method of analysis that incorporates these data sets should be designed so as to preclude any distortion of the measurements.

The datum of a kinematic network becomes a problem when the observational material used for estimating coordinates and velocities of the individual points is deficient in content. The datum problem encountered in most large geodetic projects is solved by introducing additional and subjective information – the datum basis. It has the form of a set of point-coordinates, velocities or other relevant parameters with a common property of being somewhat less consistent as compared to the "inadequate" but objective observations. The subjective datum content of the basis is enforced without disturbing the consistency of the observations, through the application of free net constraints. The resulting solution has two desirable properties:

- the coordinate frame defined by the datum basis is maintained, i.e., the esti-

M. J. Reid and J. M. Moran (eds.), The Impact of VLBI on Astrophysics and Geophysics, 423–424.

mated coordinates define an identical coordinate frame; and

- the corrections to the measurements are independent of the specific datum basis. They reflect only the measurement noise.

The CTS treated as a kinematic entity needs two datum bases:

- the first datum basis (a set of point coordinates) defines the CTS frame at the standard epoch;
- the second basis consists of a set of velocities that refers to the same datum points.

An obvious choice for the first datum basis is the BIH coordinates of a selected number of coordinated observatories. Such a choice would insure a smooth transition from the present BIH/CIO terrestrial system to the CTS. For the second datum basis we recommend a set of point velocities derived from the parameters of a plate tectonics model. The advantages of adopting the second basis are:

- reduced sensitivity to CTS point distribution;
- minimization of the motion of the asthenosphere as represented by a network of hot spots;
- velocities are minimized using weights that reflect the quality of the plate tectonics model parameters.

We consider an extended period of time (years) during which all CTS control stations participate in VLBI measurements. A small subset of three to four stations observes continuously while the remaining majority observes from time to time. The parameters that can be estimated are grouped as follows: (1) earth orientation parameters; (2) CTS station standard epoch coordinates; (3) CTS station velocities; (4) radio source CIS coordinates; and (5) station clock error model parameters.

A simulation study in which four CTS stations participated in VLBI measurements over five years for a total of only five days resulted in the following standard deviations:

Date of observation	Accumulated days	Earth orie. (cm)	CTS pos. (cm)	CTS velo. (cm/year)
1–1–81	1	8 – 10	5 – 23	–
1–6–81	2	6 – 9	4 – 17	100. – 300.
1–1–82	3	6 – 8	3 – 15	1. – 3.5
1–1–83	4	5 – 8	1 – 14	.7 – 1.7
1–1–86	5	5 – 7	1 – 12	.3 – .7

We believe that the idea of treating the Conventional Terrestrial System as a kinematic entity should be adopted by the scientific community as a standard method of analysis. Although crude and unsophisticated in technical detail, our studies demonstrate the high effectiveness of VLBI for the establishment and maintenance of a kinematic CTS. The main characteristics of the proposed kinematical approach to CTS analysis are:

- point velocities are estimated directly without having to resort to multi-stage adjustment techniques with the inevitable but dubious assumptions on error propagation;
- CTS datum determination and maintenance are explicitly associated with discrete data sets;
- the adjustment procedure is sequential in the true sense. CTS parameters are updated continuously at the rate of the incoming data.

PHASE CONNECTION FOR GEODESY: RESULTS FROM A 245-KM BASELINE

J. R. Ray
Interferometrics Inc.
GSFC/NASA
Code 621.9
Greenbelt, MD 20771

ABSTRACT. In an effort to greatly improve the measurement precision of VLBI observables used to determine geodetic baselines, the potential of utilizing fringe phase has been investigated. The 245-km baseline between Mojave and Owens Valley (in CA) has been most thoroughly studied for this purpose. In trying several different experimental designs it appears that source scheduling is not the critical factor in determining when phase connection can be established. Normally, instrumental instabilities are also not limiting (although nominal functioning of the instrumentation is essential). The most serious factor in limiting opportunities for phase connection seems to be atmospheric variability. However, when the observation SNR is made large (> 50) to give very precise group delays (and hence enhance the chance of resolving phase ambiguities), the improvement in baseline uncertainty obtained using phase delays is only marginal compared with the corresponding group delay solution. For this case, the baseline error budget is dominated by stochastic variations in clock and atmospheric delay contributions rather than by observation noise.

1. INTRODUCTION

Normally, geodetic VLBI measurements are made using bandwidth synthesis group delay observations of quasars, the fundamental precision of which is limited by the radio bandwidth that can be spanned. A large gain in measurement precision is possible if, instead, phase delay can be used as the observable. For the frequency sequences used by NASA's Crustal Dynamics Project (CDP), the improvement in measurement precision is roughly a factor of 50. The overwhelming obstacle to using phase delays is the resolution of the cycle (two-pi) ambiguities inherent in phase measurements. Effective utilization of fringe phase requires that those ambiguities be resolved to a common lobe for all observations of a given experiment session. Recent CDP research efforts to use the precise phase observable have concentrated on developing observing and analysis techniques that will enhance the probability for maintaining phase connection for one-day intervals over

M. J. Reid and J. M. Moran (eds.), The Impact of VLBI on Astrophysics and Geophysics, 425–426.

baseline lengths of 100 to 1000 km. The goal is to measure the lengths of such baselines to within a few millimeters (i.e., a precision of about 0.01 ppm or better). Potential applications include refined studies of tectonic activity in regions such as southern California. Recognizing that phase connection over baselines of such lengths may not be a routine occurrence, a secondary interest is the identification of those processes which set the limit for phase stability. In particular, the characterization of instrumental variations has played a significant role in the effort.

2. METHODS & RESULTS

The baseline most intensively investigated for this study is the 245-km baseline between the 12-m Mojave Base Station antenna (in the Goldstone complex near Barstow, CA) and the 40-m dish at the Owens Valley Radio Observatory (OVRO). Several different observing strategies have been attempted, including scheduling observations of sources within a few tight clusters and more conventional geodetic schedules with longer slews between scans. Each approach has yielded successful phase connections at least once, indicating that schedule design need not be the critical factor in determining phase connectability. In addition, several analysis techniques have been applied to the problem of phase ambiguity resolution: shifting of the phases to the lobe nearest to the corresponding group delay (after dual-band correction for ionospheric dispersion); comparison to refined theoretical delays formed by parameter adjustments using the group delays. Both weighted least-squares and Kalman filtering have been used for the parameter estimation. Experience indicates that the most powerful approach for ambiguity resolution is the use of Kalman filtering of group delays made with SNR values of 50 or greater.

The CDP experiment most successful in its phase resolution was run on 1986 October 18 and, in addition to Mojave and OVRO, involved the 26-m antenna at Hat Creek, a 5-m mobile station located at Quincy, and a 4-m mobile at Monument Peak (all in CA). Kalman filter analysis of the group delays from all the stations except Hat Creek was used to generate refined theoretical delay values which in turn were used to produce phase residuals. The phases for the Mojave-OVRO baseline were found to be readily and reliably resolvable, whereas those from the remaining baselines were not. The ambiguity resolution was performed independently for both the X- and S-band data, ionospheric correction values were recalculated from the phases, and a Kalman solution was run for the Mojave-OVRO baseline. Comparison of the phase delay baseline solution with that using the group delays shows, at best, only marginal improvement in the determination of the baseline length and the three-dimensional station coordinates. The reason for this is that estimation of the stochastic clock and atmospheric variations, rather than measurement noise, dominates the error budget for the baseline results. Nonetheless, the effort has proven useful in revealing subtle instrumental variations in the Mark III system that had not previously been detected.

ON THE COMPUTATION OF GROUP DELAY CORRECTIONS CAUSED BY RADIO SOURCE STRUCTURE

J. Campbell, H. Schuh, G. Zeppenfeld
Geodetic Institute
University of Bonn
Nussallee 17
D-5300 Bonn

ABSTRACT: We developed a structure algorithm for the computation of phase- and group delay corrections to geodetic VLBI observations. For some of the sources, i.e. 3C454.3, 3C273B and 3C345 the corrections are in the range of several hundred picosecs of group delay.

1. INTRODUCTION AND CONCEPT

In geodetic VLBI one of the tacit assumptions is that the observed sources are pointlike objects serving as ideal benchmarks in a virtually perfect realisation of an inertial reference system. In reality however, the strongest of the compact radio sources which are used in geodetic VLBI are also among those displaying both complex structure and high variability. In a companion paper by Schalinski et al. it has been shown that it is possible to use the geodetic observations of the IRIS and CDP programs to map the structure as well as to monitor its temporal changes. Here we use typical hybrid mapping results to derive phase and group delay corrections for geodetic VLBI.

Our work is based on a detailed study by Thomas (1980) on source structure effects. Using the concept of "effective position" which is the position obtained for one particular observation when the structure effects vanish, THOMAS arrives at the position of the "centroid" as a weighted mean associated with all observations of a given source in a multiparameter baseline fit.This centroid is obtained if no source corrections are applied. If we consider to use a structure map, which describes the features of a source in terms of relative coordinates, we are free to choose a reference point for the structure corrections. There appear to be two alternatives:

a. if we choose the centroid as origin of our structure map, we can compute the delay corrections needed to modify our observations as if they were produced by a point source at the location of the centroid. The potfit rms will decrease, but the problem is not solved because the centroid changes its absolute position if the brightness distribution changes.

M. J. Reid and J. M. Moran (eds.), The Impact of VLBI on Astrophysics and Geophysics, 427–428.

b. if we choose a prominent feature (e.g. the brightest component) as origin and use the same procedure as above we obtain corrected observations which correspond to this component.

In view of what is known today about the physical nature of compact radio sources we may have hopes that features with a stable position in the absolute sense really exist (e.g. the core of a core-jet source).

2. PROCEDURE AND FIRST RESULTS

We have developed a software module STRUC which can be linked to our geodetic VLBI program BVSS and which contains a package of subprograms dealing with the different steps in the generation of structure corrections. At the first observation of a particular source STRUC loads the delta map (in a standardized format), transforms it to the desired new origin and performs an FFT to obtain the complex visibility function which is then saved on a temporary memory. In the observation loop the visibility function is read from this memory and after computing the baseline projections is interpolated to derive the phase and its derivatives with respect to the u-, v-coordinates, i.e. to frequency. Then the group delay corrections are formed and returned to the main program. The delay rate corrections are calculated by numerical differentiation. For first numerical tests we used the structure maps of 1803+784, 0212+735, 3C454.3 (see fig.1), 3C273B, 3C345 and applied the corrections to the CDP-experiment May, 5/6, 1983 (baseline Effelsberg-Haystack). The results show a significant influence on the residuals of the solution as well as on the geodetic parameters but more extensive testing is needed to establish the tendency of the effects.

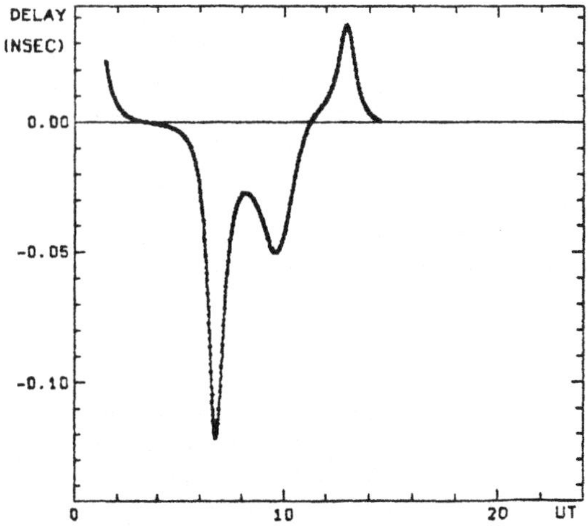

Fig.1. Group delay corrections for 3C454.3, Haystack-Effelsberg

In the future a combined package of source imaging and structure correction software will be included in regular geodetic VLBI-processing. Then, a reanalysis of the existing VLBI-databases solving for the position of the stationary component of each observed radio source will become a feasible task.

Ref.: Thomas, J.B. An Analysis of Source Structure Effects in Radio Interferometry Measurements. JPL Publ. 80-84,1980

EFFECTS OF SOURCE STRUCTURE ON ASTROMETRY AND GEODESY

James S. Ulvestad
Jet Propulsion Laboratory
California Institute of Technology
Pasadena, California 91109

ABSTRACT. VLBI hybrid maps of a number of radio sources have been used to make calculations of the effects of source structure on group delay and phase delay for baselines of different lengths. The effects on the two types of delay are of comparable magnitude and have a slightly stronger than linear dependence on baseline length. For sources selected to have only modest structure, the r.m.s. delay error caused by structure at 8.4 GHz is on the order of 30 to 50 picoseconds for intercontinental baselines.

A limit to the accuracy of astrometric and geodetic observations of compact radio sources is provided by the non-pointlike structures of some such sources. The nature of the source structure "error" is somewhat different in group and phase delay. At (u,v) points where the fringe visibility is low, the group delay caused by the source structure may be very large, while it is relatively small at high visibility points. The phase delay, on the other hand, will typically undergo large *changes* at low visibility points. In a region of high fringe visibility, the phase delay caused by structure may be fairly constant, but may be quite different from that in another high-visibility region in the (u,v) plane.

Source structure effects at 8.4 GHz (X band) have been calculated using 5-GHz maps of 26 sources from the first epoch of the Caltech VLBI survey (Pearson and Readhead 1984; Pearson, Readhead, and Barthel, this volume). These calculations have been made by considering each source to be a collection of point sources, as given by the model components in a hybrid map.

Calculations of structure effects on delay have been made every three minutes along the (u,v) tracks for each source on three different baselines. The (u,v) tracks for the real (northerly) declinations of the sources have been used. The three baselines used in the computations are Goldstone to Madrid (8390 km), Fort Davis to Haystack (3140 km), and Green Bank to Haystack (840 km). For the purposes of the calculations, the 8.4-GHz structures have been assumed to be identical to the 5.0-GHz structures. Therefore the calculations for individual sources are not correct in detail, but the computations are used only for statistical analysis and not for detailed corrections for individual source structures. Since the structure that would actually be seen at 8.4 GHz would be smaller and more core-dominated than at 5 GHz, the structure effects are probably overestimated by a factor of 1.5 to 2.

M. J. Reid and J. M. Moran (eds.), The Impact of VLBI on Astrophysics and Geophysics, 429–430.

Although the two shorter baselines were actually used in making the hybrid maps, the Goldstone-Madrid baseline was not. However, almost all the maps used had data from the baseline between Owens Valley and Bonn, which is not very different from the Goldstone-Madrid baseline. The major source of uncertainty is the interpolation to unsampled parts of the (u, v) plane because the frequency used for calculations was not the same as that used in making the maps.

For the full set of 26 sources used, the r.m.s. errors due to unmodeled source structure are misleadingly large because of the inclusion of sources that have so much structure that they generally would not be present in astrometric or geodetic VLBI catalogs. Therefore, a subset of 16 "good" astrometric sources has been constructed, from which 3C 84 and all sources classified as compact doubles or steep-spectrum cores by Pearson and Readhead (1984) have been eliminated.

TABLE 1. Calculated RMS Structure Delays for 16 "Good" Sources

Baseline	Length	Group Delay	Phase Delay
Goldstone-Madrid	8390 km	83 psec	56 psec
Haystack-Fort Davis	3140 km	28 psec	19 psec
Haystack-Green Bank	840 km	5 psec	3 psec

Table 1 shows the r.m.s. delays caused by structure of the 16 sources on the three baselines studied. The group delay is calculated relative to the flux density centroid of each source, while the phase delay is calculated relative to the mean value for each source. The r.m.s. group delays range from 5 picoseconds on the shortest baseline to 83 picoseconds on the longest baseline, while the r.m.s. phase delays are about 30–40% lower. As mentioned above, these calculated values are probably overestimates by factors of 1.5 to 2. The higher r.m.s. for the group delays is probably caused by the extended tail of the structure delay distribution, which in turn is related to the behavior of the group delay near visibility minima. Power law fits of delay *vs.* baseline length give exponents of 1.19 for the group delay and 1.28 for the phase delay over the factor of 10 in baseline length studied here. The approximate scaling with baseline length indicates that uncorrected source structure will cause astrometric errors at the level of a several tenths of a milliarcsecond for the range of baselines considered. Preliminary indications are that power laws with similar indices continue down to baselines of several hundred kilometers.

I thank Tim Pearson for supplying the hybrid maps used in this work.

REFERENCES

1. Pearson, T. J., and Readhead, A. C. S.: 1984, IAU Symp. No. 110, p. 15.
2. Pearson, T. J., Readhead, A. C. S., and Barthel, P. D.: 1987, IAU Symp. No. 129, this volume.

MONITORING OF EXTRAGALACTIC RADIO SOURCES VIA GEODETIC VLBI OBSERVATIONS

Guoqiang Tang and Bernt Rönnäng
Onsala Space Observatory
Chalmers University of Technology
S-439 00 Onsala
Sweden

1.INTRODUCTION

It was pointed out already in the early stage of the development of the Mark-III VLBI that the structure of the quasars could affect VLBI results for geodesy or astrometry (Cotton, 1980; Robertson, 1981). As errors from other sources are gradually reduced, this effect can not be ignored any longer. We should monitor the structure variations of the sources and study their internal kinematics closely. 8GHz maps made from geodetic VLBI observations may give us important information.

2.MULTI-EPOCH MAPS AT 8 GHz

Mark-III geodetic VLBI observations are used to map some compact extragalactic radio sources. Our maps are made from global geodetic VLBI experiments, e.g. MERIT campaign, CDP and IRIS projects etc., in which Onsala is involved. Six maps of 4C39.25 from 1980-1985 have been made(Tang, Rönnäng and Bååth, 1987). In our maps, a third component became visible since 1984.30. The component B moves towards A (or vise versa) at an apparent projected velocity of $v/c=3.1\pm0.4$ ($z=0.699$, $q_0=0.05$ and $H_0=100$km/s/Mpc). Five maps of 3C345 at 8 GHz are obtained from the same geodetic VLBI experiments. The maps show the characteristics of this most well-studied quasar. The component C3.5(Biretta et al, 1986) is confirmed in the map of 1982.96. The expansions of jets C3, C4 and C5 from the core are clearly indicated.

3.CALCULATIONS OF SOURCE STRUCTURE EFFECTS ON GEODETIC VLBI

The group delay correction due to the radio source structure is given by,

$$T_{str} = \frac{\mathrm{Re}(V)\frac{\partial}{\partial w}[\mathrm{Im}(V)] - \mathrm{Im}(V)\frac{\partial}{\partial w}[\mathrm{Re}(V)]}{|V|^2} \qquad (1)$$

where the visibility $V(u,v)$ is the Fourier transform of the source brightness distribution $B(x,y)$, which can be properly represented by a number of Gaussians. The Gaussian models are obtained from X-band hybrid maps made from geodetic VLBI data.

The delay corrections on the baseline Onsala-Westford due to the structure of 3C345 and 4C39.25 are calculated during the CDP experiments 15-18 December 1982 (Fig.1). It is shown that the delay correction

M. J. Reid and J. M. Moran (eds.), The Impact of VLBI on Astrophysics and Geophysics, 431–432.

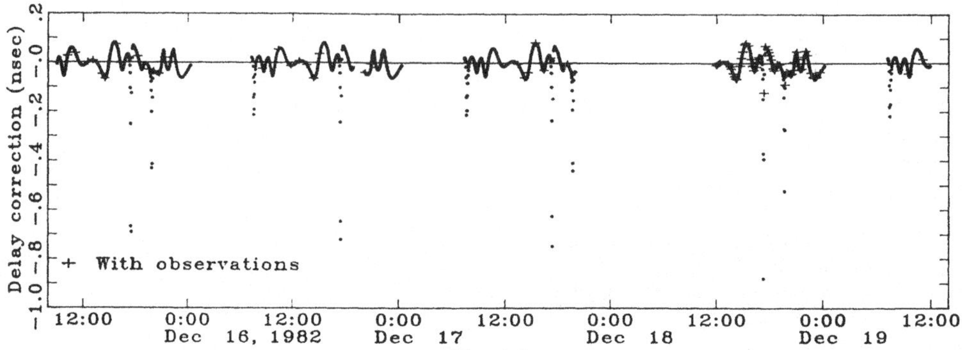

Fig. 1-1

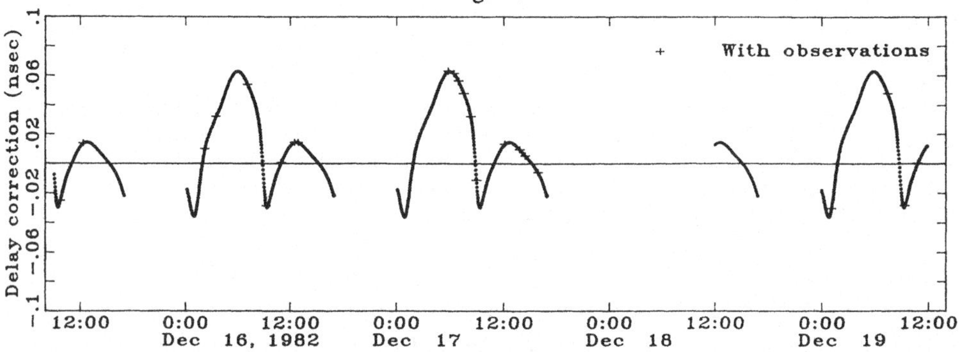

Fig. 1-2

Figure 1. The group delay corrections due to the radio source strucure on the Westford-Onsala baseline during the global geodetic VLBI experiment, Dec. 15-18, 1982. The calculations are based on the Gaussian models obtained from the 8GHz maps which are made from the same observations. Fig. 1-1 shows the structure effect of 3C345, and Fig. 1-2 shows that of 4C39.25.

due to the source structure varys periodically from a few hours to daily. In general, this error effect falls in the range of current measuring accuracy (±3cm) of geodetic VLBI. However, for some large scale one-sided core-jet sources, it can change dramatically.

Hybrid maps made from geodetic VLBI data can provide valuable information on the internal kinematics of compact extragalactic radio sources and they are certainly important for estimation of the source structure effect on geodetic VLBI. Source monitoring will become more significant if it is based on a long term basis. Therefore the mapping will be continued and expanded. Further investigations are needed to study the source structure effect on geodetic VLBI, especially the long term influence, using source maps obtained from geodetic VLBI observations.

REFERENCES

Biretta, J.A., Moore, R.L. and Cohen, M.H.:1986, Astrophys. J., **308**, 93.

Cotton, W.E.:1980, in *Radio Interferometry Techniques for Geodesy*, NASA Conference Publication, No.2115 (NASA, Washington, DC), pp.193-198.

Robertson, D.S.:1981, in *Reference Coordinate Systems for Earth Dynamics*, eds. E.M. Gaposchkin and B. Kolaczek (Reidel, Dordrecht), pp.205-216.

Tang, G., Rönnäng, B. and Bååth, L.:1987, Astron. Astrophys. (in press).

THE RADIOASTRON PROJECT

N. S. Kardashev and V. I. Slysh
Space Research Institute, U.S.S.R. Academy of Sciences
Moscow, U.S.S.R.

ABSTRACT. The RADIOASTRON mission (Andreyanov *et al.* 1986) is designed to achieve angular resolutions as fine as 6 microarcsec and will be used to study radio sources with very high brightness temperature. It will form an orbiting radio interferometer between a satellite radio telescope 10 m in diameter and several large ground-based radio telescopes. The orbit of the satellite makes it possible to have baselines from several thousands to 80,000 km. The satellite will be equipped with four dual polarization receivers at 327, 1665, 4830, and 22235 MHz, with local oscillators phase-locked to a ground-based hydrogen maser frequency standard via an S-band microwave link. The IF signals from the receivers will be transmitted to the ground by an X-band high data rate link and recorded on magnetic tape in the VLBA format. Participation of large radio telescopes from many countries is envisaged and encouraged.

1. INTRODUCTION

The angular resolution of ground-based VLBI systems such as the European VLBI Network (EVN) or the Very Long Baseline Array (VLBA), which is under construction, is limited by the dimensions of the Earth; the baseline of any ground-based interferometer cannot exceed 12,000 km, which limits the angular resolution to about 0.1 milliarcsecond at 6 cm. There are two ways of increasing the angular resolution further: one is to go to millimeter and submillimeter wavelengths; another is to go to space to increase the baseline length. The former way encounters certain problems caused by atmospheric phase fluctuations; the latter way is more straightforward and presents no limitations in principle to the baseline length. Microarcsecond angular resolution will allow us to study the following phenomena:

- flare activity of nearby stars;
- planets around nearby radio emitting stars;
- measurement of distance to pulsars and other compact galactic sources through their parallax and proper motion;
- light cylinder zone of pulsars and the ejection of relativistic particles;
- structure of binary systems, neutron stars, or black hole-stars and the ejection of relativistic particles from them;
- structure and dynamics of maser radio sources in star-forming regions; tur-

M. J. Reid and J. M. Moran (eds.), The Impact of VLBI on Astrophysics and Geophysics, 433–440.

bulent structure of interstellar matter and scintillation of radio sources;
- candidate large black hole at the galactic center;
- acceleration and collimated ejection of relativistic particles by supermassive AGN's and QSO's;
- super-Compton emission and superluminal motion;
- cosmological evolution of compact extragalactic radio sources;
- determination of fundamental cosmological parameters and of the nature of hidden mass.

Since the flux density sensitivity of a space interferometer and a ground-based interferometer is approximately the same, the space interferometer will be sensitive to much higher brightness temperatures of radio sources, because this sensitivity scales as baseline length squared. Although brightness temperatures of synchrotron emission sources must be limited to 10^{12} K (Kellermann and Pauliny-Toth 1969, 1981), there is evidence of Doppler boosting, which increases the brightness temperature of AGN's and QSO's by a factor of $\gamma^{3+\alpha}$, where γ is the Doppler factor, and α is the spectral index. For $\gamma \sim 10^3$, $\alpha \sim 0$, the brightness temperature may be as high as 10^{21} K. Nonsynchrotron radio sources such as OH and H_2O masers and pulsars may also have very high brightness temperatures. Some OH masers have been measured to have brightness temperatures as high as 4×10^{12} K, while H_2O masers reach 10^{16} K. The brightness temperature of pulsars is estimated to reach 10^{30} K.

Another advantage of the space radio interferometer over a ground-based array is its ability to measure much smaller angular motions related to superluminal motion in quasars. As shown in Figure 1, angular resolution from space at 1.35 cm might be high enough to discriminate between different cosmological models through measurement of proper motion of large samples of superluminal radio sources with large redshifts (Cohen *et al.* 1987).

2. ORBITAL PARAMETERS AND U-V PLANE COVERAGE

The RADIOASTRON project is aimed at the creation of a space-Earth radio interferometer that will be able to realize angular resolution as fine as 6 microarcseconds. It will consist of one radio telescope on a satellite circling around the Earth on a highly eccentric orbit and several large ground-based radio telescopes operating in the standard VLBI mode. The orbital parameters are given in Table 1.

Table 1. Orbital parameters of the RADIOASTON satellite

Period	$P = 24^h$
Semimajor axis	$a = 42,000$ km
Semiminor axis	$b = 26,000$ km
Eccentricity	$\varepsilon = 0.8$
Inclination	$i = 67°$
Perigee radius	$r_p = 9,000$ km
Apogee radius	$r_a = 75,000$ km

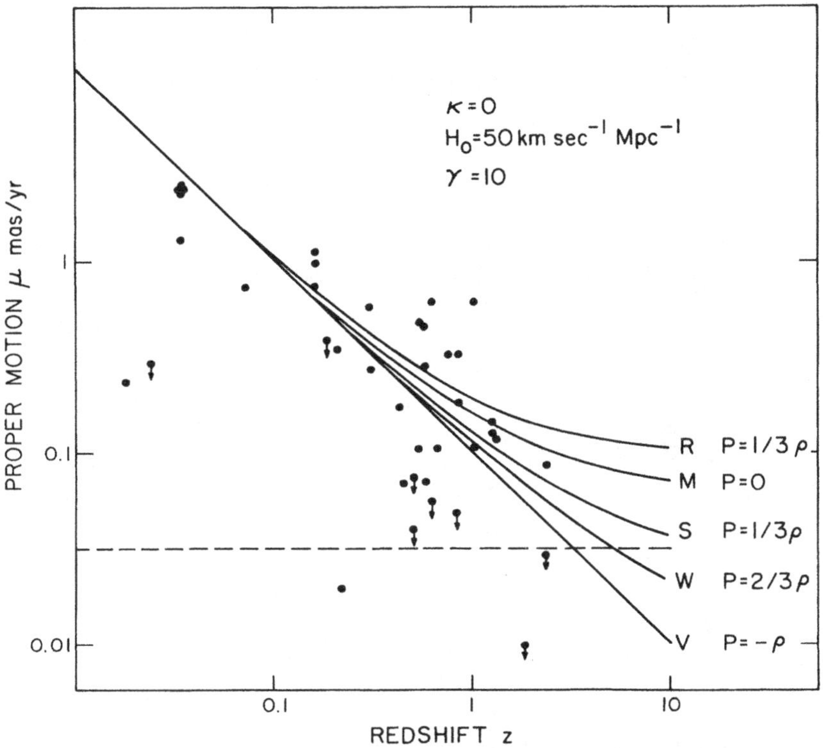

Figure 1. Proper motion of superluminal components versus redshift. The data points are from Cohen *et al.* (1987). Solid lines show the proper motion-redshift relation for different cosmological models: *R*—cosmological model with dominating relativistic matter (equation of state indicated on the right side); *M*—the same for nonrelativistic matter (pressure equals zero); *S*—the same for infinite strings; *W*—the same for infinite walls; *V*—the same for vacuum. The curvature, the Hubble constant, and Lorentz-factor used for plotting the solid lines are 0, 50 km s^{-1} Mpc^{-1}, and 10, respectively. The dashed line gives the sensitivity of RADIOASTRON at 1.35 cm wavelength.

The orbit is schematically shown in Figure 2. uv-plane coverage of such an interferometer may be rather poor for certain projections, as shown in Figure 3, and is not as well suited for construction of images as the VLBA system or the QUASAT project. New data processing algorithms must be designed to deal with this poor uv-coverage.

3. BASIC COMPONENTS OF THE RADIOASTRON SYSTEM

An overall schematic diagram of the RADIOASTRON system is shown in Figure 4. We describe briefly some of the major components.

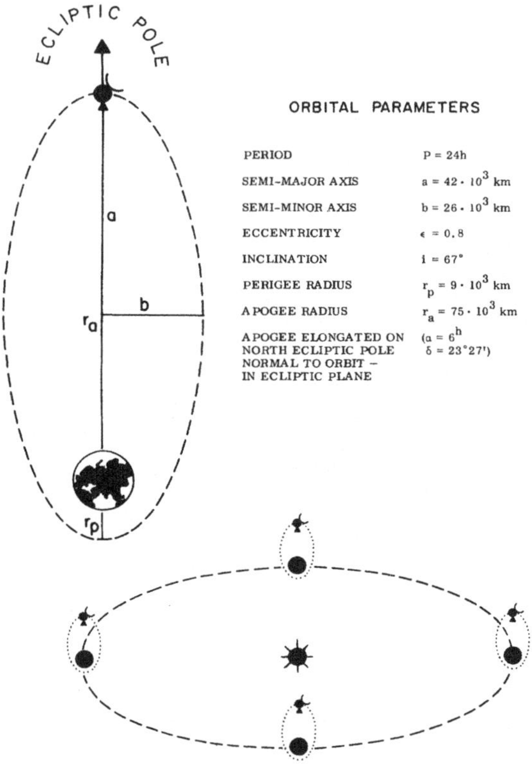

Figure 2. Proposed orbit of the RADIOASTRON satellite.

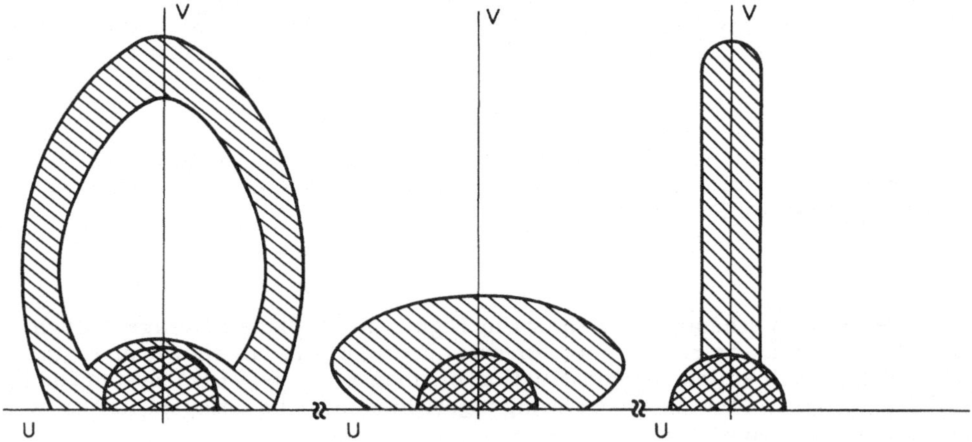

Figure 3. uv-plane coverage for a VLBI network including RADIOASTRON. *Left*: for a source lying in the direction normal to the orbital plane. *Center*: the same for a direction to the source making an angle of 75° with the major axis. *Right*: the same for a direction to the source coinciding with the minor axis. The double cross-hatched area near the origin corresponds to uv-plane coverage of the ground-based VLBI network without RADIOASTRON.

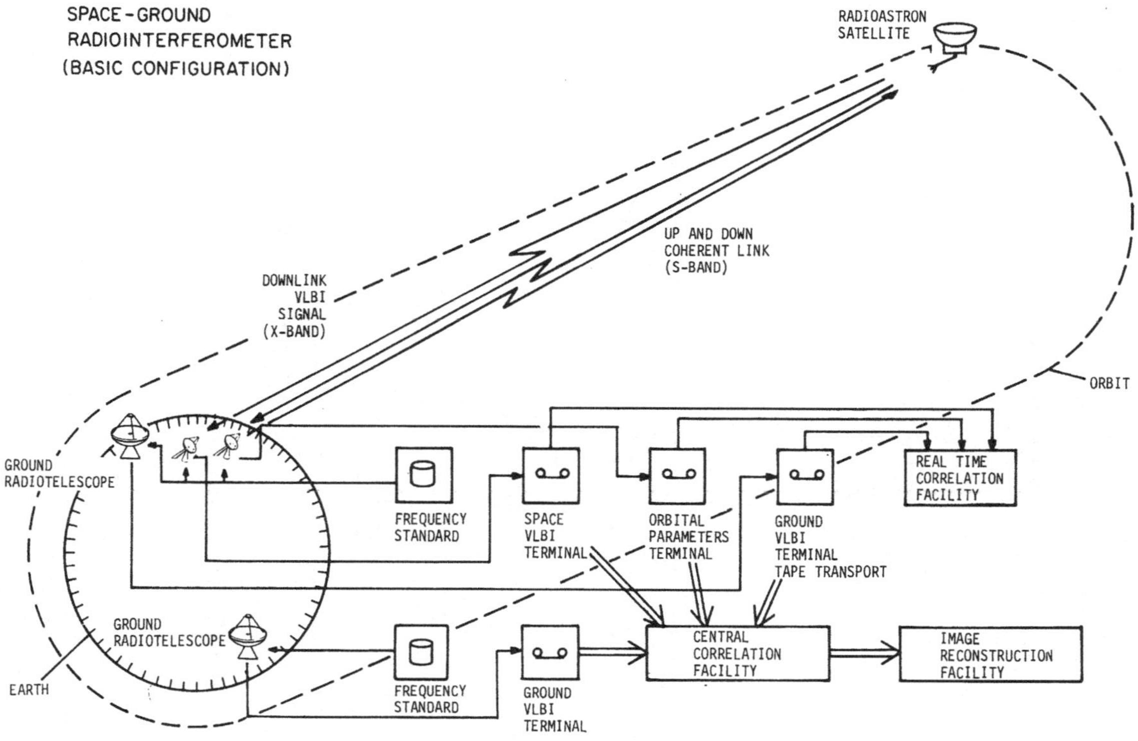

Figure 4. The space-ground radio interferometer.

3.1. Antenna

The satellite radio telescope, shown in Figure 5, will be a prime focus system with a deployable parabolic reflector 10 m in diameter. The reflector has a fixed inner part of 3-m diameter and 24 unfoldable panels. The focal package is attached to the inner 3-m part of the reflector with four focal support legs. Feeds and receiver front-ends are located in the focal package. The reflector is made of reinforced carbon fiber and has an rms surface accuracy of 0.5 mm.

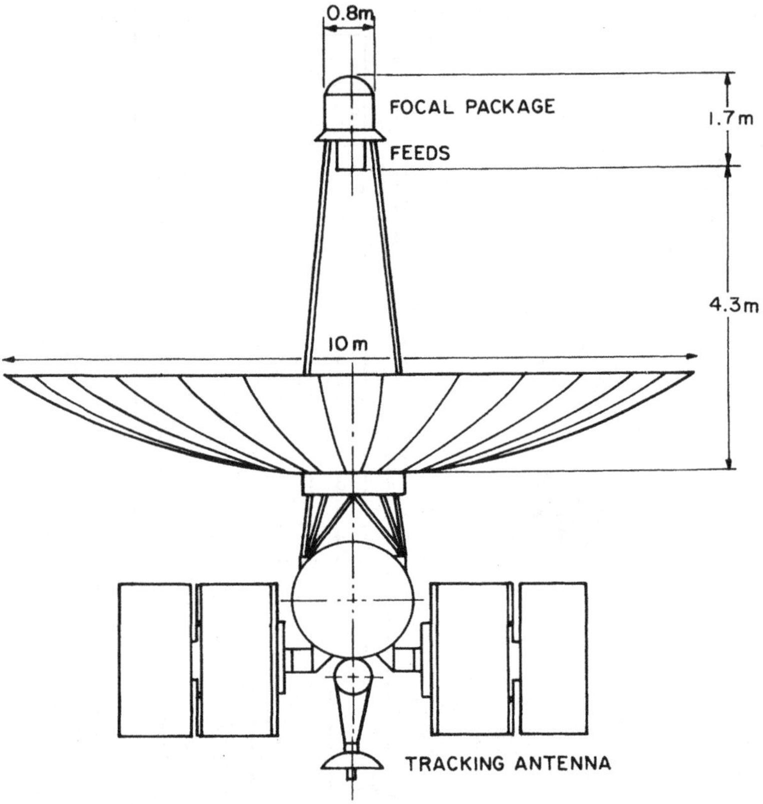

Figure 5. The RADIOASTRON satellite.

3.2. Receivers

Four dual circular polarization receivers will be provided on the satellite. Table 2 shows the basic parameters of the receivers and the expected sensitivity and angular resolution of the RADIOASTRON satellite in conjunction with a 70-m diameter ground-based radio telescope.

The receivers will have cooled low noise transistor amplifiers as front-ends (except at 327 MHz, where the receiver will be uncooled). The refrigeration system

Table 2. Parameters of the RADIOASTRON system

Frequency (MHz)	Bandwidth (MHz)	System Noise Temperature (K)	Flux Density Sensitivity (mJy)[1]	Angular Resolution (μas)[2]
327	4	80	10	400
1665	32	50	3.5	80
4830	32	50	3.5	24
22235	32	70	4.0	6

[1] $5 \times$ rms, averaged over two circular polarizations, 18 hours integration time.
[2] $0.2\lambda/B_{max}$, $B_{max} = 80,000$ km. This criterion is used throughout the paper. It assumes a moderate signal-to-noise ratio such that details smaller than the nominal diffraction-limited beam size of $\sim 0.5\lambda/D$ can be discerned.

relies on radiation cooling to 150 K and a closed-cycle cooling machine, which will cool the front-ends to 80 K. Each receiver will have its own feed in the focal plane of the antenna with polarization splitters and calibration system. The coma losses due to the offset positions of the three feeds (22 GHz feed will be placed at the focus) will not exceed 0.2 db. The cooled front-ends and polarization splitters will be located in a separate focal plane assembly connected by RF cables and waveguides to the focal package where the receivers are located.

3.3. Local oscillator

All local oscillator frequencies will be derived from a coherent frequency synthesizer, which will be phase-locked to a ground-based hydrogen maser frequency standard by a two-way (up and down) S-band link that utilizes an omnidirectional antenna on the satellite. As a backup, an on-board rubidium frequency standard with a stability of 10^{-12} will be available.

3.4. IF and video

The receivers have standard 500 MHz IF amplifiers and IF-to-video converters. Two independent channels for both circular polarizations are converted to the baseband and then sampled. The sampling rate is 64 mbits/sec in each channel, which allows for two-level sampling of two 32 MHz bandwidth channels or four-level sampling of two 16 MHz bandwidth channels.

3.5. Down-link

The down-link to the ground station is at X-band (8.2 GHz) with a two-channel transmitter and high-gain parabolic antenna. The antenna for the down-link has a diameter of 1 m and can be pointed to the ground within 0–120° from the antisolar direction in one plane and ±30° for the perpendicular plane, with a pointing error less than 0.5°.

3.6. Ground-based station

The ground-based station of the RADIOASTRON mission consists of one transmitting (command and S-band local oscillator up-link) antenna and two receiving antennas (S-band local oscillator down-link and X-band high-data rate down-link). The 70-m antenna will be used as a radio telescope to receive signals from a radio source. Both X-band data and 70-m radio telescope outputs will be recorded on two tape recorders in the VLBA format. Simultaneous observations by telescopes outside the Soviet Union will be recorded in a VLBA-compatible format; participation of large radio telescopes from Australia, Canada, Europe, India, Japan, and the U.S. is expected. Tapes of the data from the satellite and all participating radio telescopes will be processed on the NRAO VLBA correlation facilities, on the European VLBI Network correlator, and on the facilities of the Space Research Institute in Moscow.

4. SUMMARY

The RADIOASTRON mission is scheduled to fly in 1991 and will have a lifetime of two years. Realization of this project will be possible with broad international cooperation, both through a combined effort to construct the satellite radio telescope and through participation of as many large radio telescopes as possible. It will be possible to achieve sufficient sensitivity even with the modest 10-m diameter satellite radio telescope to undertake many important investigations. An increase of angular resolution by an order of magnitude compared to ground-based VLBI is only a first step for future improvements of many orders of magnitude using space interferometers. However, this first step will permit the solution of many of the astronomical problems mentioned earlier.

REFERENCES

Andreyanov, W., *et al.* 1986, *Astron. Zh.*, **63**, 850.

Cohen, M.H., Barthel, P.D., Pearson, T.J., and Zensus, J.A. 1987, *The Impact of VLBI on Astrophysics and Geophysics*, Proc. of IAU Symposium 129 (this volume).

Kellermann, K.I., and Pauliny-Toth, I.I.K. 1969, *Ap. J. (Letters)*, **155**, L71.

Kellermann, K.I., and Pauliny-Toth, I.I.K. 1981, *Ann. Rev. Ast. Astrophy.*, **19**, 373.

QUASAT

R.T. Schilizzi
Netherlands Foundation for Radio Astronomy
Dwingeloo
The Netherlands

ABSTRACT: QUASAT was proposed to ESA as a joint ESA-NASA medium scale mission for space VLBI in December 1982. At the time of writing, four and a half years later, QUASAT has passed through an Assessment Study in ESA, a pre-Phase A study in NASA and has just entered an ESA Phase A study in European industry, supported by JPL and Australian and Canadian space interests. If QUASAT is selected to fly at the end of the Phase A study, it will be launched in 1995-1996. The scientific goals, mission concept and current status are reviewed this paper.

1. INTRODUCTION

The quest for higher angular resolution has taken radio astronomy a long way since the first two element interferometric observations were made in 1946 with baselines of a few hundred meters. Current VLBI networks and the planned VLBA in the USA utilize the whole Earth as a radio telescope to achieve not only sub-milli-arcsecond angular resolution, but also good quality images. The next obvious step for the interferometry technique is to go beyond the confines of the Earth and position an element or elements in Earth orbit.

Apart from the considerable technical challenge of going into space, there are sound scientific reasons for doing so. Foremost amongst these is high quality imaging at even greater angular resolution than on the Earth to investigate those sources and source components still unresolved on Earth baselines. In addition, prospects for imaging in the equatorial region of the sky, and in the southern hemisphere, are significantly enhanced by the presence of a space element to provide adequate U-V coverage for the first time. "Rapid" imaging on timescales of a few hours for variable galactic sources is also made possible, again because of enhanced U-V coverage compared to ground based arrays.

QUASAT was proposed to ESA as a joint ESA-NASA space VLBI mission in November 1982 by an international group of astronomers. An ESA Assessment Study was carried out from October 1983 to June 1985 in

M. J. Reid and J. M. Moran (eds.), The Impact of VLBI on Astrophysics and Geophysics, 441–446.

parallel with a pre-Phase A study in NASA. QUASAT was approved for Phase A in Europe in February 1986.

The aftermath of the Challenger crash in January 1986 forced NASA to delay consideration of new projects in the Explorer program, including QUASAT, and caused ESA to re-evaluate the cooperative framework under which QUASAT had been studied up to that time. This re-evaluation resulted in a mission concept with ESA as the leading agency but with substantial support from JPL, and Australian and Canadian space interests (ESA SCI(86)8). This new concept was approved for an ESA Phase A in February 1987.

In the meantime in 1986, a succesful demonstration of space VLBI was made at the first attempt using the NASA TDRSS satellite (Linfield et al., these Proceedings). This result emphasised that space VLBI is a viable, and potentially very valuable, space astronomy discipline.

2. SCIENTIFIC GOALS

With QUASAT we expect to obtain high quality images at resolutions of ~60 micro arcsec at 22 GHz in the "final" orbit and lower quality images at angular resolutions of 40 micro arcsec in the "initial" orbit (see discussion on orbits in the next section). The primary goal will be to probe the nuclei of radio galaxies and quasars more deeply and in greater detail than is possible with ground-based networks alone (see Table 1). The scope of applications of VLBI demonstrated by the contributions at this Symposium is evidence that the scientific goals of QUASAT will be wide-ranging. One other area is worth mentioning as potentially very important. This is the use of astrometric techniques on H_2O masers in nearby galaxies to directly measure distances to those galaxies (Reid et al, these Proceedings).

The main scientific goals are listed in Table 2 and described in detail in the Assessment Study Report ESA SCI(85)5 and in the QUASAT Workshop Proceedings ESA SP-213.

Table 1: Spatial resolution at 1.35 cm wavelength. (H_o=100 km^{-1} Mpc^{-1}, q_o=0.5)

Object	Distance	Linear size for 60 micro-arcsec beam	
Galactic Center	7.1 kpc	6.7×10^{12} cm	0.4 A.U.
Centaurus A	5 Mpc	4.6×10^{15}	1.8 light days
NGC 1275 (3C84)	53 Mpc	4.6×10^{16}	0.6 light months
3C273 (z=0.158)	491 Mpc	3.4×10^{17}	4 light months
3C345 (z=0.595)	1992 Mpc	7.4×10^{17}	0.8 light year

Table 2: QUASAT Scientific Goals

Measurement of high brightness temperature phenomena in:

- **Active Galactic Nuclei**
  Quasars, Radio Galaxies, BLLacs/Blazars, Seyfert I+II, Liners
  Images on scales of accretion disk, broad line region, narrow line region
- **Newly-formed stars**
  masers, flow patterns, magnetic fields, densities
- **Evolved stars**
  masers, mass-loss rates, densities
- **Binary flare stars**
  images of giant flares, energy generation
- **X-ray binary stars**
  images of mini-jets, collimated flow

Extragalactic masers
properties of neutral gas in active nuclei

Scattering properties of the interstellar medium

Distances within our galaxy

Distances to nearby galaxies - Hubble Constant

---

3. MISSION CONCEPT

3.1. General

QUASAT is a free-flying satellite carrying a 15 m class radio telescope in elliptical orbit around the Earth. The radio telescope will be deployed in space, and used to make interferometric observations of radio sources in conjunction with the major ground-based VLBI arrays in Europe, USA, Australia and the USSR. The mission design lifetime is 2 years, but an operational lifetime of 5 years is expected.

The space-borne antenna will be capable of observing in both hands of circular polarization simultaneously at any two of the frequency complement of 22, 5, 1.6 and 0.3 GHz, and will relay the received signals via a digital or analogue link directly to telemetry stations (NASA DSN and/or ESA) on the ground (see Figure 1). The bandwidth of the dual polarization link will be at least 32 MHz in each hand of polarization. A phase/frequency reference for the antenna in space, stable to about 1×10^{-14}, will be based on hydrogen maser oscillators on the ground and relayed directly to the satellite via a two-way link from the telemetry stations in turn. All communication with the space element will be through one or more telemetry stations in the network.

processing facility of the European or US VLBI array for correlation with similar tapes from the ground VLBI arrays. After correlation and calibration the data will be sent to the principal investigators for further analysis.

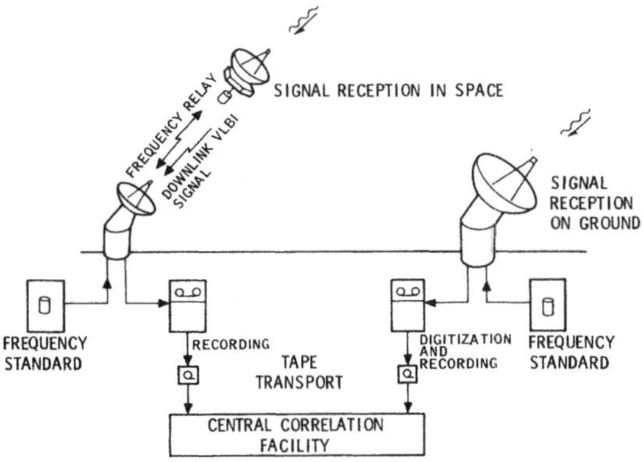

Figure 1. Schematic diagram of a space/ground VLBI system.

3.2. Orbits

Two orbital situations have been considered:

1) an "imaging" orbit, optimized for very high quality imaging with the European and US ground arrays. The perigee altitude is 5700 km, the apogee altitude 12500 km, and the inclination 63°. This orbit would require a dedicated launch since no passengers to this orbit have been identified. Reduced levels of funding projected for the next medium scale mission in ESA, make the cost of getting to this orbit unattractive.
2) a "two-tier" orbit to provide high angular resolution with lower quality imaging, the "initial" orbit, before bringing the satellite down to lower altitudes for higher quality imaging at somewhat lower angular resolution, the "final" orbit, after perhaps 1 or 2 years. The initial orbit has perigee and apogee altitudes of 5000 and 36000 km respectively, an inclination of 30° and a period of $12^h.2$. The final orbit has its apogee altitude reduced to 22000 km and the period to $7^h.75$. The advantage of this concept is that QUASAT would share a launch to geostationary transfer orbit, and thus share the costs. Figure 2 depicts the two-tier orbit.

Computer simulations of the imaging potential of QUASAT in the "imaging" orbit have demonstrated that superb results are obtained (ESA SCI(85)5). Simulation of the imaging potential of the "initial" and "final" orbits is in progress at Jodrell Bank as part of the Phase A study. Table 3 lists the angular resolution and sensitivity expected for QUASAT in the "final" orbit at its observing wavelengths of 1.35, 6, 18 and at 92 cm.

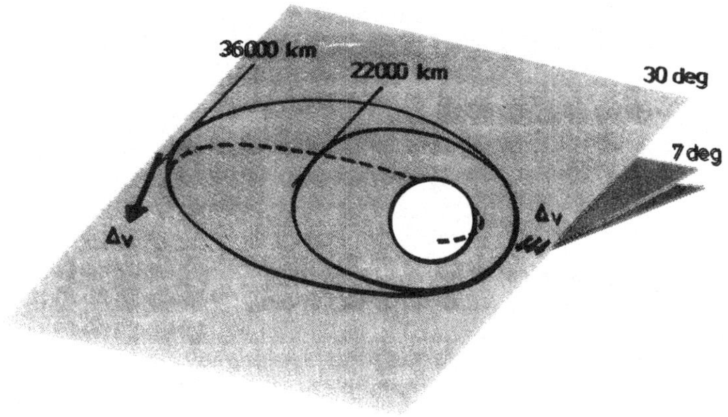

Figure 2. The orbit injection sequence, and two-tier orbit.

Table 3: Angular resolution and sensitivity of QUASAT

λ (cm)	Angular resolution* (micro-arcsec)	RMS noise per resolution element (continuum)**	RMS noise per spectral channel+	Frequency (GHz)
1.35	60	1 mJy	35 mJy	22.0-22.50
6	270	0.5	-	4.72-5.02
18	800	0.3	20	1.60-1.72
92	4000	3	-	0.324-0.327

* this is the dimension of the synthesized beam. Information on size and position angle is in fact available on scales 2 to 3 times smaller than this.

** for a 48 hour observation assuming there is a sufficiently strong feature (~150 mJy at 1.35 cm, ~50 mJy at 6 and 18 cm and 300 mJy at 92 cm) present in the map to allow self-calibration.

\+ for a 48 hour observation assuming there is a spectral feature present of sufficient strength (~20 Jy at 1.35 cm and ~5 Jy at 18 cm) to allow self-calibration.

3.3. Science Payload

The Assessment Studies have shown that the overall mission, which is based on current VLBI and spacecraft engineering practice, is technically feasible. As far as the spacecraft is concerned the major new aspect is the deployable antenna. Two antenna configurations have been studied; in ESA a 15 meter diameter inflatable space rigidized structure developed by Contraves, Switzerland (see Figure 3), and in

NASA a 15 meter wrap-rib deployable mesh structure developed by the Lockheed Corporation, California. Both appear to be satisfactory for the mission. Other areas of development are the radio astronomy feeds and an active cooling system for the radio astronomy receivers. An industrial study of the feeds has shown good performance for a compact three frequency co-axial feed at prime focus which can also accommodate a dipole array for 92 cm. The radio astronomy receivers will be based on HEMT amplifiers, cooled actively by Stirling cycle coolers.

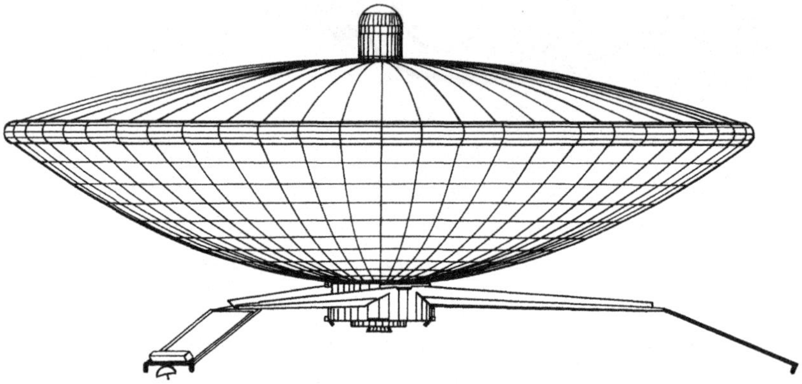

Figure 3. ESA concept for the orbiting antenna.

4. CURRENT STATUS AND FUTURE DEVELOPMENT

The industrial Phase A study finally began on 12 May 1987 and is being conducted by a European consortium led by Aeritalia in Turin, Italy. NASA has at the same time allocated funds to JPL for support of the ESA Phase A, particularly in the area of antenna technology, links, tracking, and operation. Details of the Australian and Canadian participation are currently under discussion.

The "science payload" - feeds, receivers, coolers, synthesizers, etc, but excluding the antenna - is to be studied by the radio astronomy community in parallel with the Phase A, with the aim of establishing the costs and effort required in constructing space qualified payload elements. Institutes in Europe, USA, Australia and Canada have been identified for the various elements.

QUASAT will enter a competitive selection round in ESA at the end of 1988. Assuming it is selected, the project will then move into a detailed design phase in 1989-90, followed by the construction and testing phase from 1990-1994. Final integration of the spacecraft will occur in 1995 with launch in that same year or in 1996.

ANALYSIS OF THE OFFSET QUASAT ANTENNA

M. Catarzi and G. Tofani
Osservatorio Astrofisico di Arcetri
L. E. Fermi 5
50125 Florence Italy

ABSTRACT. In the ESA phase A of the Quasat project the use of a 10 m offset antenna, has been suggested (1). In the present work the use of the multifrequency coaxial feed, developed by ESA for the symmetric configuration, is analyzed with a multireflector configuration. The coaxial feed has the advantage of a more compact configuration. However reduced efficiency and higher cross polarization are expected at the lower frequency, due to the scattering on the subreflector surfaces.

THE OFFSET REFLECTOR AND THE COAXIAL FEED

The major disadvantage of the offset configuration is the depolarization effect which introduces beam squint (2). In a prime focus configuration multifrequency operation would imply the use of a complex feed (i.e. cluster of corrugated horns or dichroics). Multireflector may be used in order to compensate depolarizing effects and to change the feed configuration. In the assessment phase, ESA has commissioned to CSELT the development of a multifrequency feed prototype for the prime focus symmetric reflector. The system is composed of an innner circular waveguide feed at 22.2 GHz surrounded by two waveguide feeds at 4.9 and 1.6 GHz. Aperture and spillover efficiencies of about 0.7 are obtained for symmetrical parabolic reflectors, having f/D=0.43.

THE MULTIREFLECTOR OFFSET ANTENNA

Ray tracing has been used in order to match the sequence of quadric surface reflectors to the radiation pattern of the coaxial feed. A possible configuration, based on clear aperture, is shown in fig.1. An ellipsoid is used in order to concentrate the feed broad beam on the hyperboloid of the Cassegrain antenna. Physical optics analysis (3) has been applied to the quadric surface multireflector antenna. The scattered field of the feed-ellipsoid system has been analyzed at 1.6 GHz. The result is shown in fig 2a. Examples of hyperboloid and primary reflector scattered fields are reported in figs 2b and 2c. A

M. J. Reid and J. M. Moran (eds.), The Impact of VLBI on Astrophysics and Geophysics, 447–448.

clear aperture multireflector design offer a superior performance compared to the standard symmetric system in terms of blockage efficiency. The shaping of the surface by means of geometrical optics synthesis analysis (4) can improve the aperture and spillover efficiency. In Table 1 a budget, limited to some of the total efficiency parameters, is shown (5).

TABLE 1

Efficiency	Symmetric quadric	Offset quadric	Offset shaped
Illumination	0,83	0,83	0,99
Spillover	0,94	0,94	0,99
Blockage	0,85	1,00	1,00

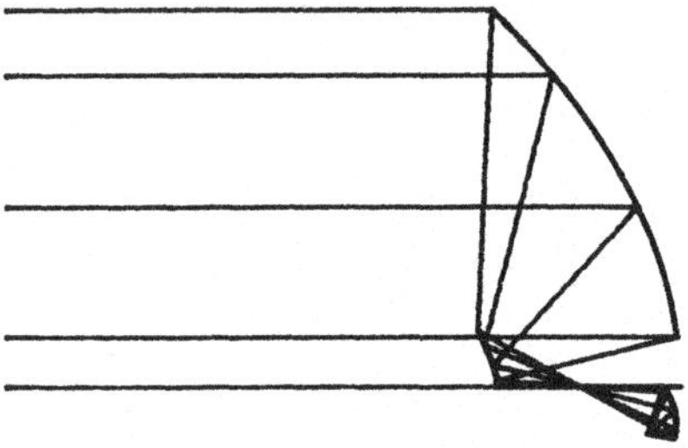

Figure 1. Clear aperture configuration.

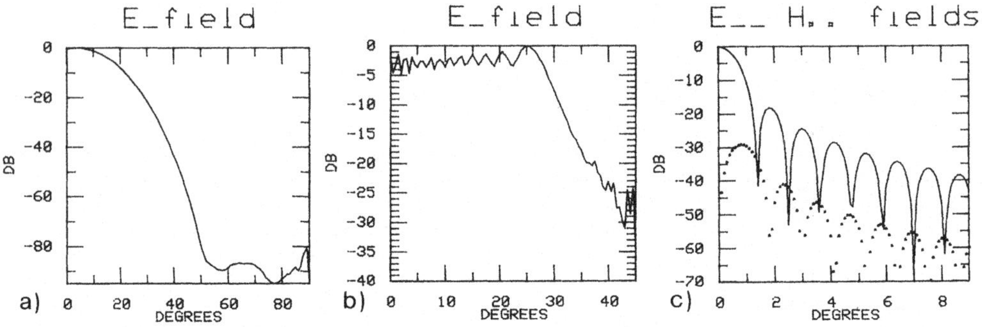

Figure 2. Scattered fields from:
2a) Ellipsoid; 2b) Hyperboloid; 2c) Primary Reflector.

REFERENCES

(1) QUASAT: A space VLBI satellite. Update report, 1987, ESA Sci (86), 8.

(2) Rudge A.W.: Offset parabolic reflector antennas in Satellite communication antenna technology, Ed. Mittra R., Imbriale W.A., Maanders E.J., 1987, North Holland.

(3) Cha A.G.: Physical optics analysis of NASA/JPL DSN antennas, 1984 JPL Techn. Rep. D-1853.

(4) Galindo Israel V., Mittra R., Cha A.G.: Aperture amplitude and phase control of offset and reflectors, 1979, IEEE Trans. AP 27, 154.

(5) Cha A.G., Bathker D.A., Williams W.F.: Advanced design concepts in ground station antennas, 1981, in Proc. 9th European Microwave Conference, Brighton, England, 147.

VLBI ACTIVITIES IN JAPAN AND A PROJECTED SPACE-VLBI PROGRAM

H. Hirabayashi
Nobeyama Radio Observatory
Tokyo Astronomical Observatory
Nobeyama, Minamimaki,
Nagano, 384-13, Japan

ABSTRACT. There are three institutes in Japan that are equipped with VLBI facilities. Of these, Nobeyama Radio Observatory started VLBI experiments with astrophysical and astrometric interests. The Institute of Space and Astronautical Science, the Nobeyama Radio Observatory, and the Radio Research Laboratory collaborated with JPL and other institutions for this first space-VLBI test experiment, generating fringes on ground-TDRSS satellite baselines at S band. A working group formed by scientists of the above three Japanese institutes proposed to ISAS to launch a space-VLBI satellite. The project, called VSOP, for VLBI Space Observatory Program, is described here.

1. INTRODUCTION

VLBI was first developed in the field of radio astronomy. It is now a basic tool for astrometry and for geodynamics. In Japan, three institutes have VLBI facilities for different purposes. Also, space-VLBI is becoming a reality with the recent successful demonstration of the first space-VLBI test experiment using the TDRSS (Transfer and Data Relay Satellite System) satellite and ground stations. The Russian RADIOASTRON project (Sagdeev 1984; Kardashev and Slysh 1987) is funded, and the Euro-US QUASAT (Schilizzi *et al.* 1984; Schilizzi 1987) project has been extensively studied. In Japan, researchers at the Nobeyama Radio Observatory (NRO), the Institute of Space and Astronautical Science (ISAS), and the Radio Research Laboratory (RRL) have been discussing the feasibility of a Japanese space-VLBI project (Hirabayashi 1984; Nishimura *et al.* 1986). For space-VLBI to become operational, coordination of observing, tracking, and data transfer and processing is needed. This paper reports the Japanese activities in VLBI with special attention to space-VLBI.

2. VLBI STATIONS IN JAPAN

In Japan, there are several institutes that are now using VLBI technology or in which related projects are considered.

M. J. Reid and J. M. Moran (eds.), The Impact of VLBI on Astrophysics and Geophysics, 449–456.

Figure 1. The 64-m antenna of the Usuda Deep Space Center of the Institute of Space and Astronautical Sciences.

RRL operates a 26-m telescope in Kashima. It is equipped with two hydrogen masers, two K3 terminals, and a two-baseline correlator. The K3 terminal and the correlator are Mk III compatible. This equipment was primarily developed by RRL. The RRL group has been actively involved in the Crustal Dynamics Project (CDP) with NASA and is engaged in other projects with many institutes. They operate at S and X bands.

The Geographic Survey Institute (GSI) has a 5-m quasi-mobile antenna, primarily stationed at Tsukuba, for obtaining basic data for geographic mapping purposes. S and X band frontends, a hydrogen maser, and a K3 terminal are installed here. The baseline between two points in Japan has already been measured using this station with the collaboration of the RRL group.

NRO operates a 45-m radiotelescope that is equipped with a hydrogen maser, a Mk III terminal, and two Mk II terminals. The VLBI research activities conducted here will be treated in more detail in the next section.

ISAS has a 64-m deep space station at the Usuda Deep Space Center (UDSC) located in Usuda (Nomura *et al.* 1985). This antenna, shown in Figure 1, was used for the first space-VLBI demonstration experiment with the TDRSS satellite. For this historical experiment with the Jet Propulsion Laboratory (JPL), NRO and RRL collaborated with ISAS. This experiment will be described in more detail in Section 4. X band VLBI experiments with Kashima and with telescopes abroad have been done by the NRO group here. Mk II terminals, a Mk III terminal, and a rubidium frequency standard were shipped from Nobeyama. ISAS is preparing a K3 terminal, which should be ready by the end of 1987 (T. Nishimura, private communication).

The International Latitude Observatory of Mizusawa (ILOM) proposed project VERA, an acronym for VLBI for Earth Rotation Study and Astrometry, which is a two-station VLBI observing system for earth rotation study and astrometry (Fujishita 1987). Two antennas of about 35-m and 15-m diameter and related backends are proposed here. ILOM may be combined with the Tokyo Astronomical Observatory within a year.

Figure 2. An aerial view of the Nobeyama Radio Observatory.

3. VLBI AT THE NOBEYAMA RADIO OBSERVATORY

NRO is a radio astronomical observatory open to outside users (Akabane *et al.* 1983). NRO has two major mm-wave radiotelescopes: a 45-m radiotelescope, and a five-element 10-m super-synthesis interferometer. These instruments are shown in Figure 2. For the interferometer, the so-called "FX" architecture was adopted. For this instrument, we developed a 1024-channel spectro-correlator, which has advanced performance compared to conventional cross-correlation type spectrometers (Chikada *et al.* 1986). This scheme is applicable to a new generation of multi-baseline VLBI spectro-correlators.

For VLBI, the 45-m radiotelescope is used. This telescope is a large-aperture, high-precision millimeter-wave radiotelescope with an rms surface accuracy of about 0.15 mm and a pointing accuracy of 0.001 degree. A VLG-11 hydrogen maser made by the Smithsonian Astrophysical Observatory, a Mk III recording terminal, and two Mk II terminals are installed. VLBI experiments started in 1983 (Hirabayashi *et al.* 1985). The VLBI activities include domestic X band experiments, international centimeter and millimeter wave experiments, and space-VLBI related activities. Domestic VLBI experiments have been done with the Kashima 26-m antenna of RRL for precise coordinate determinations and for VLBI polarimetry. The data tapes were analyzed with the two-baseline correlator at Kashima (Kunimori *et al.* 1986). The first three baseline experiments were done among the telescopes at Kashima, Nobeyama, and Tsukuba (5-m antenna of GSI) in 1985.

The 22 and 10 GHz bands can be used for VLBI with the 45-m radiotelescope in the microwave region. The 1.4/1.6 and 5 GHz bands are not used any more because the primary focus cabin was removed in order to improve the pointing accuracy of the telescope. The 22 GHz water vapor maser line emission was observed with U.S. and European telescopes both in galactic and extragalactic sources.

Millimeter-wave VLBI is well suited to the 45-m radiotelescope because of its high performance in the mm-wave region. Due to the high gain of the telescope and the good uv-coverage afforded by Nobeyama's participation, high quality images can be obtained of the very compact cores of radio galaxies and quasars and of compact line emission regions (Hirabayashi 1981). A 43 GHz experiment with U.S. and European stations generated the first fringes to Nobeyama at millimeter wavelengths in an experiment conducted in May 1986. The Nobeyama-Haystack baseline in this experiment provided a fringe spacing of 0.16 mas, one of the smallest ever obtained. Nobeyama participated in an 89 GHz experiment with four U.S. stations, but no fringes were detected on Nobeyama baselines. The local oscillator chains have been extensively checked and upgraded for mm-wave VLBI.

The first space-VLBI test experiment using the 64-m telescope with the TDRSS satellite proved successful with NRO's full participation with the Usuda Deep Space Center of ISAS (Levy *et al.* 1986). The NRO group has been interested in space-VLBI, and years of discussion with the scientists in ISAS are continuing. Space-VLBI matters will be discussed in more detail later.

X band VLBI mapping experiments were done with international telescopes with the X band front and the VLBI backend of NRO, and the 64-m antenna at Usuda made available by ISAS.

4. THE TDRSS-OVLBI DEMONSTRATION EXPERIMENT

Scientists at NRO, ISAS, and RRL are interested in space-VLBI and joint discussions are underway (Hirabayashi 1984; Nishimura *et al.* 1986). When the idea of the TDRSS-OVLBI (Transfer and Data Relay Satellite System – Orbiting VLBI) demonstration experiment was proposed by JPL, the above scientists were all interested in the experiment. The TDRSS-OVLBI experiment generated close collaboration at the inter-institute level in Japan.

In the early calibration experiments among the ground stations, Mk II terminals were used. Just before that time, the NRO Mk II terminal was completed, and a test experiment for the terminal was done using the 45-m telescope and U.S. telescopes. At the Usuda station, there were no VLBI facilities, and two Mk II terminals, a rubidium clock, a Mk III recording terminal, and other miscellaneous instruments of NRO were shipped to Usuda. Scientists and technicians of NRO installed these instruments on the 64-m antenna and collaborated with the staff of ISAS throughout the experiment.

The Kashima group of RRL participated in the experiment using both the 26-m antenna for reception of the cosmic signals and the K3 correlator for processing the data. They also played an important role in the calibration of the ground stations to phase the telescopes for VLBI use. Also, throughout the TDRSS-OVLBI experiments, the Kashima group observed the sources and recorded tapes.

Fringes were successfully obtained among three stations (TDRSS 4.9-m, Usuda 64-m, and Tidbinbilla 64-m antennas) for three quasars in the July–August 1986 session, but not for the TDRSS-Kashima baseline. The maximum projected baseline obtained was 17,800 km (1.4 times the earth diameter) by the

TDRSS-Usuda baseline (Levy *et al.* 1986). Here, the feasibility of space-VLBI was demonstrated and the scheme of transmitting the reference signal from the ground to the space antenna proved useful. Figure 3 shows the concept of the TDRSS-OVLBI demonstration experiment.

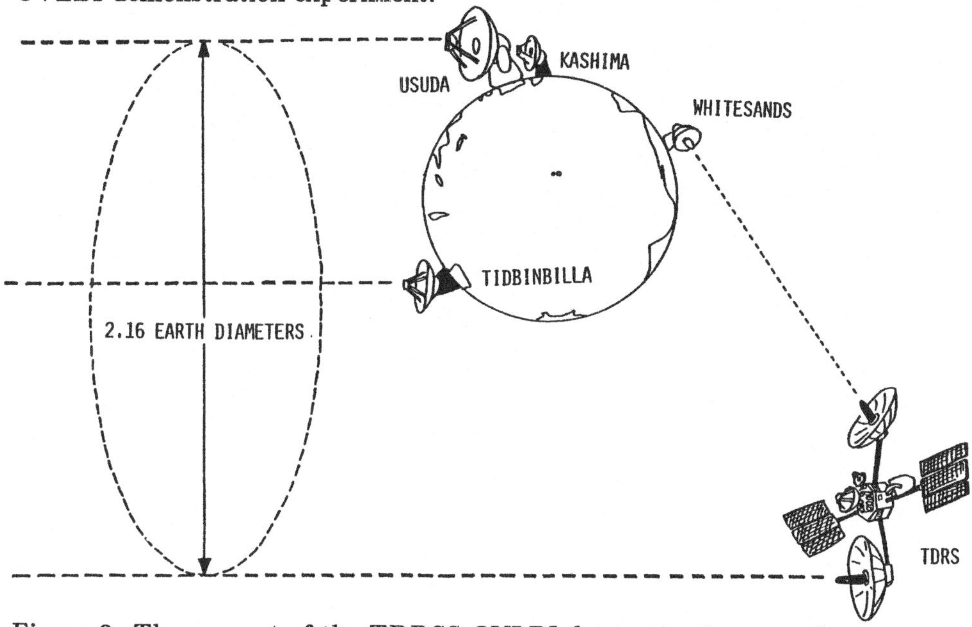

Figure 3. The concept of the TDRSS-OVLBI demonstration experiment.

In the January 1987 session of the experiment, we detected fringes on 23 radio sources out of 25 observed, and in spite of the small gain of the telescopes, the TDRSS (4.9 m)-Kashima (26 m) pair generated fringes for at least 3C279. The maximum projected baseline obtained was extended to 2.2 times the earth diameter (Linfield *et al.* 1987). The above experiments were done in S band. The possibility of observations at 15 GHz is under discussion. Participation of the 45-m radiotelescope at that frequency is possible.

5. THE VSOP JAPANESE SPACE-VLBI MISSION

A working group formed in ISAS by scientists from ISAS, NRO, and RRL proposed a Japanese space-VLBI mission, called VSOP, VLBI Space Observatory Program, to ISAS in March 1987 after a two-year feasibility study. The specifications are not yet fixed. Figure 4 shows a rough sketch of the VSOP satellite. For space-VLBI to become operational, cooperation is needed because Japan cannot develop all the items independently of the RADIOASTRON and QUASAT projects.

The best candidate for the observing frequency bands are 5 and 22 GHz. The 5-GHz band was selected because it is one of the most active bands for ground-based VLBI. The 22-GHz band was selected because of the importance of water vapor maser line emission, to map the stars of special phase, to obtain the astronomical distance scale, and to measure the rotation of galaxies.

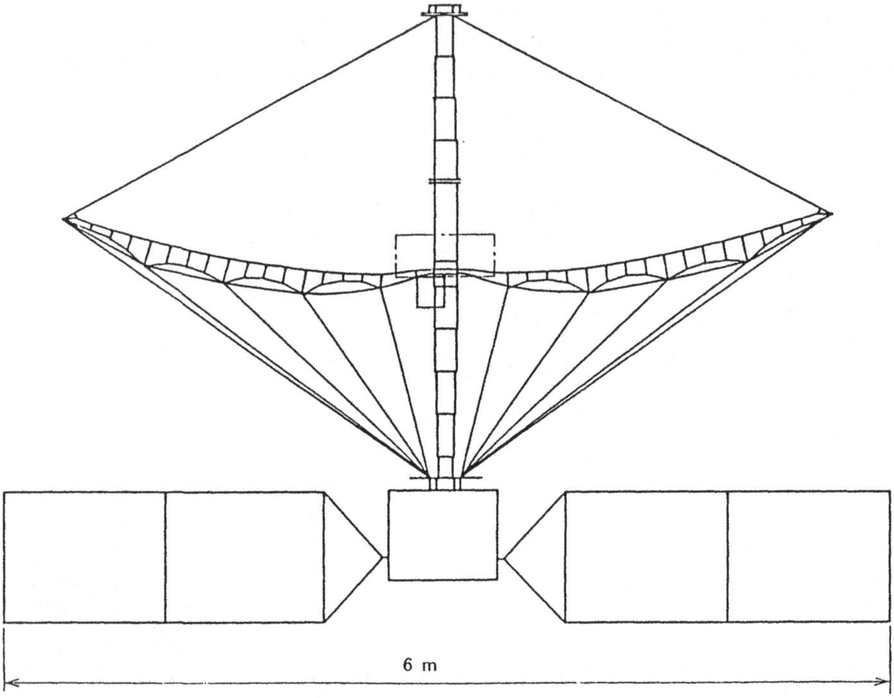

Figure 4. A possible configuration of the VSOP satellite.

We assume that an ISAS rocket is available to launch the VSOP satellite. The M3S-II rocket is the largest existing launching rocket in ISAS and is the same type used for launching the two Japanese Halley probes (Hirao and Itoh 1986). ISAS is considering a more powerful rocket, but it has not been approved yet. Due to the severe limitations of the launching capability, parameters of the payload mass and the satellite orbit were traded off roughly as below. Note that this is a case study where a M3S-II rocket is assumed. The parameters are:

inclination	=	31°
perigee altitude	=	1000 km
apogee altitude	=	10000 km
eccentricity	=	0.379
satellite mass	=	300 kg

The M3S-II rocket would be modified to have the extra apogee-kick capability. The payload weight includes all subsystems: the observing and data link, power supply, thermal control, receiving antenna, pointing, satellite structure, and integration system. An effort will be made to make the diameter of the antenna as large as possible, at least larger than 5 m, with the limitation that the mass of the antenna must be less than 100 kg.

The inclination of 31° is the same as the latitude of the Uchinoura Launching Station of ISAS and was chosen to maximize the payload mass (Tsuboi and

Hirabayashi 1985). If we fix the orbit as above, the variation of the longitude of the ascending node (Ω) is $-1.32°$/day, or $-480°$/year. The variation of the argument of the periapsis (ω) is $2.07°$/day, or $750°$/year. In the QUASAT proposal, the former parameter is $-0.22°$/day, and the latter is set to zero by choosing an inclination of 63°. When a more powerful rocket is available in ISAS, then a larger space-VLBI satellite can be considered. The VSOP satellite introduced here is too limited in mass, and a more powerful launching rocket is highly desirable.

A deployable antenna structure is considered. Figure 4 shows one example of the antenna mounted on the satellite, and Figure 5 shows a related example of the satellite in the launching rocket, with the antenna folded up.

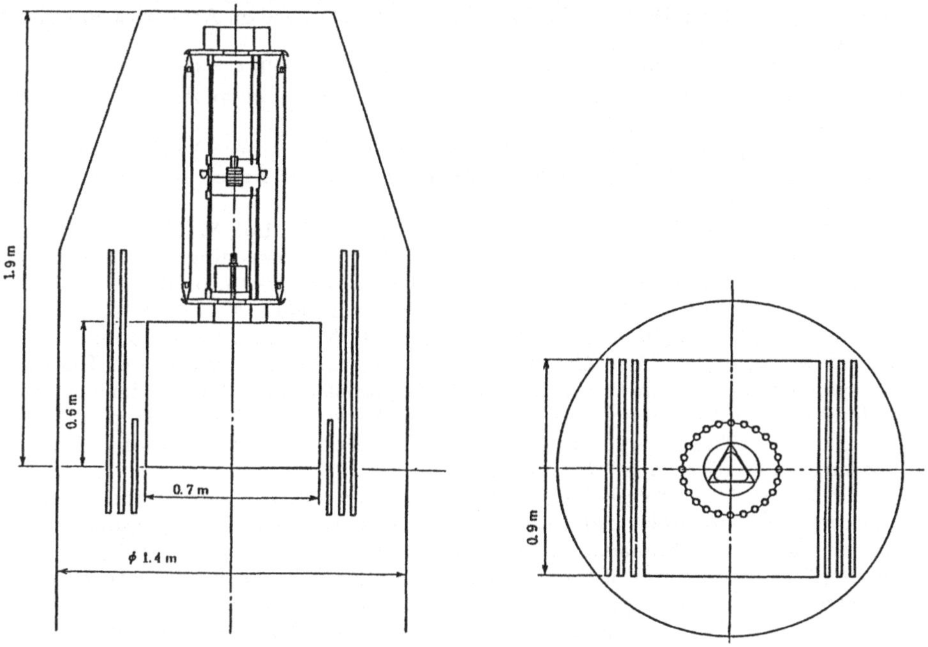

Figure 5. An example of the VSOP satellite folded in the launching rocket.

6. INTERNATIONAL COLLABORATION AND VSOP

To make the space-VLBI satellite an effective mission, an array of ground stations for the reception of terrestrial signals from the radio sources as well as other antennas for tracking, for the local signal link, and for the information downlink are needed. Without these ground stations, sensitivity and uv-coverage are limited. These ground stations must be coordinated for the effective utilization of the VLBI satellite. At present, Japan has neither a world-wide ground network for satellite tracking nor a VLBI ground network. Collaboration at the international level is an important item to be discussed. Effective sharing of the orbit and compatibility with other possible space-VLBI missions should also be considered to maximize the efficiency of the satellites.

7. CONCLUSION

VLBI activities and the status of space-VLBI in Japan have been described. Discussion both inside Japan and at the international level is needed to promote an efficient space-VLBI program.

8. REFERENCES

Akabane, K., Morimoto, M., Kaifu, N., and Ishiguro, M. 1983, "The Nobeyama Radio Observatory," *Sky and Telescope*, Dec. 1983, p. 495.

Chikada, Y., *et al.* 1986, "A Very Fast FFT Spectrum Analyser for Radio Astronomy," in Proceedings of IEEE-IECEJ-ASJ International Conference on Acousties and Signal Processing, April 7-11, 1986, p. 2907.

Fujishita, M., and Hara, T. 1987, "Japanese VLBI Project – VERA," this volume.

Hirabayashi, H. 1981, "International VLBI Array," in *Ann. Tokyo Astron. Obs.*, 2nd Series, **18(2)**, p. 82.

Hirabayashi, H. 1984, "An Introductory Review for Japanese Space-VLBI Mission," in Proceedings of the Meeting on Space-VLBI, Institute of Space and Astronautical Science, May 26, 1984.

Hirabayashi, H., Inoue, M., Iwashita, H., Morimoto, M., and Tsuboi, M. 1985, "Very Long Baseline Interferometry by 45 m Radiotelescope," in Proceedings of International Symposium on Antennas and Propagation, Kyoto, **ISAP-85**, p. 975.

Hirao, K., and Itoh, T. 1986, "The Planet-A Halley Encounters," *Nature*, **321**, 294.

Kardashev, N. S., and Slysh, V. I. 1987, "RADIOASTRON Project," this volume.

Kunimori, H., *et al.* 1986, "The Geodetic Results of VLBI Experiment on Kashima-Nobeyama Baseline," in Proceedings of the Symposium on Applications of Space Technologies to Astronomy and Astrophysics, Tokyo, Jan.–Feb. 1986, p. 138.

Levy, G. S., *et al.* 1986, "Very Long Baseline Interferometry Observations Using an Orbiting Radio Telescope," *Science*, **234**, 187.

Linfield, R. P., *et al.* 1987, "Very Long Baseline Interferometry Observations Using the Tracking and Data Relay Satellite as an Orbiting Telescope," this volume.

Nishimura, T., Takano, T., Yamada, T., and Hirabayashi, H. 1986, "Japanese Space VLBI Mission," in Proceedings of the Symposium on Applications of Space Technologies to Astronomy and Astrophysics, Tokyo, Jan.–Feb. 1986, p. 152.

Nomura, T., Hayashi, T., Nishimura, T., Hirosawa, H., and Ichikawa, M. 1985, "Usuda Deep Space Station with 64 m Diameter Antenna," in Proceedings of 36th IAF Symposium, Stockholm, **IAF-85**, 381.

Sagdeev, R. Z. 1984, "Some Prospects of Space VLBI," in *QUASAT – A VLBI Observatory in Space*, Proceedings of a Workshop held at Gross Enzersdorf, Austria (European Space Agency, Noordwijk, The Netherlands), p. 19.

Schilizzi, R. T. 1987, "QUASAT," this volume.

Schilizzi, R. T., Burke, B. F., Jordan, T. F., and Hawkyard, A. "The QUASAT Mission: An Overview," in *QUASAT – A VLBI Observatory in Space*, Proceedings of a Workshop held at Gross Enzersdorf, Austria (European Space Agency, Noordwijk, The Netherlands), p. 13 (see also p. 27).

Tsuboi, M., and Hirabayashi, H. 1985, "Toward the Optimum Selection of the Orbit of a Japanese Space-VLBI Satellite," *Ann. Tokyo Astron. Obs.*, 2nd Series, **20(3)**, p. 191.

VERY LONG BASELINE INTERFEROMETRY OBSERVATIONS USING THE TRACKING AND DATA RELAY SATELLITE AS AN ORBITING RADIO TELESCOPE

R. P. Linfield[1], G. S. Levy[1], J. S. Ulvestad[1], C. D. Edwards[1], J. F. Jordan, Jr.[1], S. J. DiNardo[1], C. S. Christensen[1], R. A. Preston[1], L. J. Skjerve[1], L. R. Stavert[1], B. F. Burke[2], A. R. Whitney[3], R. J. Cappallo[3], A. E. E. Rogers[3], K. B. Blaney[4], M. J. Maher[5], C. H. Ottenhoff[6], D. L. Jauncey[7], J. E. Reynolds[8], T. Nishimura[9], T. Hayashi[9], T. Takano[9], T. Yamada[9], H. Hirabayashi[10], M. Morimoto[10], M. Inoue[10], M. Tokumaru[11], N. Kawaguchi[11], J. Amagai[11]

[1]Jet Propulsion Laboratory, Pasadena, CA 91109
[2]Massachusetts Institute of Technology, Cambridge, MA 02139
[3]Haystack Observatory, Westford, MA 01886
[4]Goddard Space Flight Center, Greenbelt, MD 20771
[5]Bendix Field Engineering Corporation, Seabrook, MD 20706
[6]TRW Space and Technology Group, Las Cruces, NM 88004
[7]Division of Radiophysics, Commonwealth Scientific and Industrial Research Organization, Canberra, ACT 2601, Australia
[8]Mount Stromlo Observatory, Canberra, ACT 2606, Australia
[9]Institute for Space and Astronautical Science, Tokyo, 153, Japan
[10]Nobeyama Radio Observatory, Nobeyama, 384-13, Japan
[11]Radio Research Laboratory, Kashima, 314, Japan

ABSTRACT. An antenna in geostationary orbit was used for VLBI observations at 2.3 GHz, in combination with ground antennas in Australia and Japan. 23 of the 25 observed sources were detected on orbiter-ground baselines, with baseline lengths as large as 2.15 earth diameters. Brightness temperatures between 10^{12} K and 4×10^{12} K were measured for 10 sources.

1. OBSERVATIONS

VLBI observations using a 4.9 m diameter antenna of the Tracking Data and Relay Satellite System were first performed in July and August 1986 (Levy *et al.*, 1986). In January 1987, additional observations were made, again using 64 m diameter ground antennas in Tidbinbilla, Australia and Usuda, Japan. The observing frequency was 2.3 GHz. 25 sources were observed, with a total observing time of 20 hours. The sources ranged in declination from 0° to −30°, and were observed at spacecraft hour angles of 180° ± 22° (i.e. nearly towards the earth).

M. J. Reid and J. M. Moran (eds.), The Impact of VLBI on Astrophysics and Geophysics, 457–458.

2. DATA REDUCTION

The data were correlated at Haystack, using the Mk IIIA correlator. Due to software limitations, it has not yet been possible to obtain closure phase values. The coherence fell significantly at short integration times, and then dropped more slowly. Measured values were 0.94 at 30 s., 0.86 at 100 s., 0.78 at 200 s., and 0.68 at 600 s. These are mean values; the standard deviations were 3% at 30 s., 9% at 100 s., 12% at 200 s., and 19% at 600 s. The reason for such a time-variable coherence is not known, but instabilities in the spacecraft electronics are suspected.

3. RESULTS

Of the 25 sources which were observed, 23 were detected, with data from one other lost due to equipment problems. The maximum baseline for each source depended primarily on declination, and ranged from 0.94 to 2.15 earth diameters. Visibilities on the maximum baseline for each source ranged from 0.11 to 0.68. The most compact source was 1519−273, with a visibility of 0.68 (1.1 Jy correlated flux) on a baseline of 2.02 earth diameters. With a fitted size of 0.33 milliarcseconds, the brightness temperature is 3.4×10^{12} K $(1+z)$ (the redshift for this source is not known). Other sources with very high brightness temperatures were: 0336−019 (2.4×10^{12} K), 0420−014 (3.3×10^{12} K), 1921−293 (3.8×10^{12} K), and 2216−038 (3.4×10^{12} K). The errors on these values range from 10% to 35%.

4. DISCUSSION

Baseline lengths greater than 1 earth diameter are necessary to measure the structure of numerous compact radio sources. Two (1519−273 and 1921−293) of the three sources observed on baselines greater than 2 earth diameters displayed correlated flux greater than 1 Jy on those baselines. However, none of the sources were completely unresolved. Baselines of 1–3 earth diameters, available only with orbiting VLBI, promise to reveal a great deal about the structure of compact radio sources.

The measured brightness temperatures for 10 sources exceed 10^{12} K. If incoherent synchrotron radiation is the emission mechanism in these sources, the Inverse Compton process will limit the brightness temperature to $< 10^{12}$ K in the source rest frame. Bulk relativistic motion in the direction towards us, with a Doppler factor of at least 1.5–2, is required to explain the observations. This provides additional evidence that many (if not most) compact radio sources are significantly Doppler boosted in flux.

5. REFERENCES

Levy, G. S. *et al.* 1986, Science, **234**, 187.

THE LOW FREQUENCY SPACE ARRAY (LFSA)

K.W. Weiler[1], B.K. Dennison[2], K.J. Johnston[1], R.S. Simon[1], J.H. Spencer[1], W.C. Erickson[3], M.L. Kaiser[4], H.V. Cane[4], M.D. Desch[4], and L.M. Hammarstrom[1]
[1]E.O. Hulburt Center for Space Research, Naval Research Laboratory, [2]Virginia Polytechnic Institute, [3]University of Maryland, [4]NASA Goddard Space Flight Center

At the lowest radio frequencies (≤30 MHz), the Earth's ionosphere transmits poorly or not at all. This relatively unexplored region of the electromagnetic spectrum is thus an area where high resolution, high sensitivity observations can open a new window for astronomical investigations. Also, extending observations down to very low frequencies brings astronomy to a fundamental physical limit where the Milky Way becomes optically thick over relatively short path lengths due to diffuse free-free absorption.

To obtain data at these frequencies requires the difficulty and expense of placing radio telescopes into Earth orbit, but the scientific rewards of a space mission are likely to be great. Even without considering the serendipitous discoveries which have always accompanied the opening of a new realm of frequency, resolution, or sensitivity in astronomy, a low frequency telescope in space can:

1. map the entire sky, with emphasis on the galactic background non-thermal emission, with high resolution and sensitivity;
2. determine the distribution of galactic diffuse ionized hydrogen by surveying its absorption to discrete background sources;
3. study of the interstellar plasma by investigating the origin, distribution, and magnitude of its scattering and refraction;
4. study individual source spectra for energy production mechanisms and such processes as synchrotron self-absorption, Razin-Tsytovich effect, HII absorption, inverse Compton scattering, and synchrotron losses;
5. study the origin of the correlation between low frequency steep spectrum clusters of galaxies and their enhanced x-ray emission;
6. search for "fossil" radio components in "radio quiet" objects and extend the counts of sources to the low frequencies where synchrotron lifetimes approach the age of the universe;

M. J. Reid and J. M. Moran (eds.), The Impact of VLBI on Astrophysics and Geophysics, 459–460.

7. image individual sources with high resolution (~10′ at ~1 MHz to ~5" at ~30 MHz) to investigate spectral index changes across source components and to search for extended halos;
8. study the impulsive emission from Jupiter and the Sun and search for similar radiation from other Solar System bodies; and
9. search for the coherent radiation found from Solar System bodies but undetected so far in larger systems.

A scientific justification and technical description has been developed for construction of a Low Frequency Space Array (LFSA) to carry out full sky mapping and individual source imaging between 1.5 and 26.3 MHz.

Artist's conception of the Low Frequency Space Array (LFSA) with four array elements.

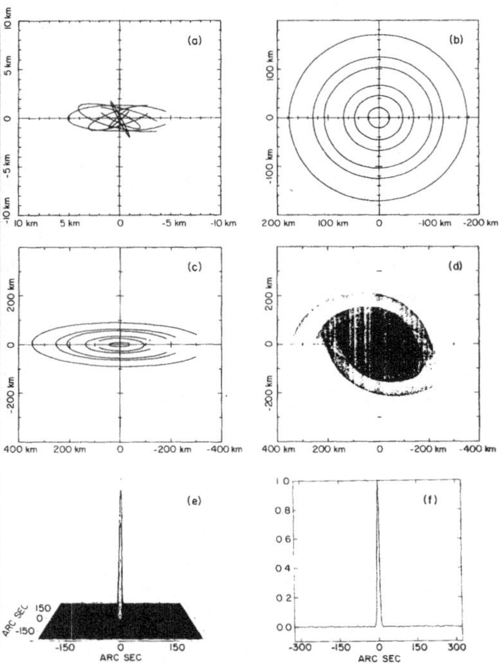

Aperture plane coverage by a 4 element array for a source at RA = 6^h, Dec = 45° for one orbit on (a) Day 1, (b) Day 180, and (c) Day 360. The total coverage during 1 year is given in (d) and the resulting beam pattern in (e). For a cross cut through the beam pattern better illustrating the sidelobe levels see (f).

THE VERY LONG BASELINE ARRAY

Jonathan D. Romney
National Radio Astronomy Observatory
Edgemont Road
Charlottesville VA 22903-2475

ABSTRACT. The Very Long Baseline Array (VLBA) will be the world's first large-scale dedicated VLBI facility. It is planned as a multi-purpose instrument, supporting continuum, spectroscopic, astrometric/geodetic, multi-frequency, bandwidth synthesis, polarization, and pulsar measurements -- with a variety of bandwidths, sampling rates, quantization schemes, and multi-band modes. The Array will consist of ten new 25-m precision antennas, in an optimized geographical configuration. Extremely low-noise receivers will support observations with dual polarization in nine frequency bands from 330 MHz to 43 GHz, including almost all conventional VLBI bands. A wideband, high-density recording system will allow unattended operation at a sustained data rate of 128 Mbit/s for 24 hours, and peak rates up to 512 Mbit/s. The VLBA correlator, planned as a 20-station system to support processing of all observations involving the Array, will be by far the largest ever built for VLBI. The Array will be operated remotely, from a new operations and data-reduction center to be built in Socorro, NM.

INTRODUCTION

The present Symposium is devoted to the impact of VLBI on various scientific disciplines. The Very Long Baseline Array, currently under construction by the National Radio Astronomy Observatory (NRAO), will have a major impact on the conduct of almost every area of research discussed here this week. This paper reviews the capabilities planned for the Array when it is completed early in the 1990's.

The shortcomings of the present ad-hoc VLBI arrays will be apparent to most participants in this meeting. Existing antennas are poorly located for adequate sampling of the aperture plane; many are poorly instrumented over the range of frequencies desired; most cannot be calibrated more than marginally despite heroic efforts; and few perform adequately at high frequencies. Notwithstanding the impressive volume of work in evidence at this symposium, available observing time is insufficient, and proper monitoring of the many structural

M. J. Reid and J. M. Moran (eds.), The Impact of VLBI on Astrophysics and Geophysics, 461–468.

changes seen with VLBI resolution is all but impossible. The advent of the narrow-track Mark 3A system has shifted the limitation on wideband VLBI observations from tape supply to correlator capacity; current systems cannot cope with the arrays required for imaging with high dynamic range. And despite recent improvements, scheduling remains laborious and must be done well in advance.

In 1982, after seven years of careful study and preliminary design, NRAO proposed the Very Long Baseline Array to alleviate these deficiencies and provide a dedicated national VLBI instrument. A good general description of the VLBA for the non-specialist can be found in the article by Kellermann and Thompson (1985); more detailed technical information is contained in a "VLBA Book" and several series of general and specialized memoranda.

CONFIGURATION

The VLBA configuration has been optimized to provide both high resolution and a large field of view, with the most uniform possible two-dimensional aperture coverage even at low declinations. A degree of condensation, centered on the VLA, extends satisfactory coverage over a broad range of scale sizes. Site selection emphasized high, dry locations, and freedom from RF interference. All sites are on US territory.

The first antenna was recently completed at the Pie Town site, on the Continental Divide west of the VLA; this highest continental site will become operational late this year. Selection of the final site is still underway, with the search concentrated in western Massachusetts, New Hampshire, and Vermont. Table I lists all VLBA sites in the planned order of construction, and shows the current status and planned first operation of each.

Table I. VLBA Configuration and Site Status

Code	Location	N. Latitude	W. Longitude	Elevation	Status	Operational
PT	Pie Town, NM	34° 18′ 03″.61	108° 07′ 07″.24	2371m	Outfitting	87/11
KP	Kitt Peak, AZ	31 57 22.39	111 36 42.26	1916	Erection	88/6
LA	Los Alamos, NM	35 46 30.33	106 14 42.01	1967	Developed	88/9
NL	N. Liberty, IA	41 46 17.03	91 34 26.35	241	Developed	88/12
FD	Fort Davis, TX	30 38 05.63	103 56 39.13	1615	Development	89/5
BR	Brewster, WA	48 07 52.80	119 40 55.34	255	Development	89/9
SC	St. Croix, VI	17°.76	64°.58	15	Acquisition	90/3
OV	Owens Valley, CA	37.23	118.28	1204	Acquisition	90/6
ML	Mauna Loa, HI	19.58	155.49	2735	Acquisition	90/12
NE	"New England"				Selection	91/5

ANTENNA

The VLBA antenna is a new, advanced design developed by the NRAO engineering division. It employs a wheel-and-track azimuth structure for superior pointing in the presence of wind and temperature differentials. A cone-shaped transition section from the elevation structure supports a conventional radial-rib reflector back-up on a ring of 20 equal attachment points; the subreflector support legs are mechanically isolated from the main reflector. These features provide enhanced performance under gravitational deformation. Shaped, high-precision main and subreflector surfaces afford high efficiency at frequencies up to 43 GHz. The fast slew capability will allow rapid sequential observation even of widely separated sources.

The main reflector panels are specified at 0.125 mm RMS surface accuracy, with a total error budget of 0.282 mm (RSS) under the primary "precision operating conditions" of moderate, relatively constant and uniform temperature, low wind, and no precipitation. Operation can continue, with reduced surface accuracy, over a wide range of secondary environmental conditions. The Cassegrain subreflector is specified at 0.15 mm surface accuracy, with a design goal of 0.10 mm. These characteristics will allow satisfactory operation at frequencies as high as 43 GHz, and use at 89 GHz under exceptional conditions.

NRAO has contracted with Radiation Systems, Inc., of Sterling, Virginia, to manufacture the ten VLBA antennas. Three antennas have been fabricated, and a fourth is nearing completion; the fifth and sixth antennas were recently authorized.

RECEIVERS & FEEDS

The Array will support observations in nine frequency bands from 330 MHz to 43 GHz. Details of the frequency ranges, expected total aperture efficiency, and receiver performance, are summarized in Table II. Three "optional" bands, shown in square brackets in the table, are foreseen as possible eventual additions to the VLBA but are not included in the construction project; one such addition would extend the frequency coverage to 89 GHz.

The last two columns in Table II show the planned installation schedule. To facilitate a thorough system evaluation, the frequency bands shown with check marks in column PT will be implemented at the Pie Town station by the end of 1987. These include eight of the nine standard VLBA bands, plus the optional 10.7-GHz band. The 43-GHz band will be deferred to allow further receiver development. All subsequent stations will be instrumented initially only in the 1.5-, 4.8- and 23-GHz bands shown by the asterisk symbol in the "Array" column; other frequencies will be introduced starting in the years shown.

Table II. VLBA Frequency Bands

Band Designation		Frequency Range [GHz]	Aperture Efficiency	Receiver Temp. [K]	System Temp. [K]	Installation PT	Installation Array
330	P	.312 → .342	.50	25	99	✓	'89
610		.580 → .640	.57	31	65	✓	'89
1.5	L	1.35 → 1.75	.57	12	30	✓	*
2.3	S	2.15 → 2.35	.71	15	35	✓	'89
4.8	C	4.6 → 5.1	.72	20	35	✓	*
6.1		⟦ 5.9 → 6.4 ⟧	.71	20	35		—
8.4	X	8.0 → 8.8	.69	28	49	✓	'88
10.7		⟦ 10.2 → 11.2 ⟧	.69	36	52	✓	—
15	U	14.4 → 15.4	.68	57	75	✓	'90
23	K	21.7 → 24.1	.66	60	87	✓	*
43	Q	42.3 → 43.5	.51	40	75		'88
89	W	⟦ 86? → 92? ⟧	.18	?	?		—

Receivers for the 1.5-GHz and higher-frequency bands generally employ cooled (20K) FET front ends, fed by horns at off-axis Cassegrain foci arranged in a ring. HEMTs (high-electron-mobility transistors) will be substituted as available; the tabulated noise temperatures are for FETs. The 43-GHz receiver will probably use an SIS mixer operating at 3.5K. Weinreb, Norrod, and Pospieszalski (1987) describe the preceding receivers in greater detail. The two lowest-frequency receivers operate at ambient temperature, with a crossed-dipole feed at the prime focus.

Simultaneous observations in orthogonal senses of circular polarization are supported in each individual band. In addition, dichroic optics provide a capability for simultaneous dual-frequency observations in the 2.3- and 8.4-GHz bands. The feed circle is designed to facilitate future dual-frequency band pairing of 4.8 with 15 and 23 GHz, and 15 with 43 and -- eventually -- 89 GHz.

CHANNELIZATION & DIGITIZATION

Each VLBA receiver will produce two independent IFs, in the 500-1000 MHz range. Four such signals -- two from each of any two receivers -- are transmitted to the control building. These inputs can be converted to as many as 16 baseband channels, generated as upper and lower sidebands with respect to 8 independent local oscillators, each tunable in 10 kHz steps. Each such channel can be filtered to bandwidths ranging from 16 MHz down to 62.5 kHz. The filtered signals are sampled at rates ranging from 2 to 32 Msmp/s, allowing oversampling except at the widest bandwidths. Sample quantization is supported at both two and four levels.

The aggregate throughput of these subsystems is thus 256 MHz, 512 Msmp/s, and 1024 Mbit/s. This equipment, developed for NRAO by Haystack Observatory, is designed to allow expansion by a factor of 2 in numbers of baseband converters, channels, and throughput.

RECORDING & PLAYBACK

Central to any VLBI instrument is the data recording and playback technology. The VLBA recording system, currently under development by Haystack Observatory, will represent an extension of the existing Mark 3A system to higher data densities (Webber and Hinteregger 1987), and will incorporate as well numerous technical improvements.

Since the VLBA stations will be controlled remotely, one of the fundamental specifications is the requirement for recording wideband data, without operator intervention, for a 24-hour period. The initial scheme will be capable of sustained recording at 128 Mbit/s for at least 12 hours on a single tape, with a peak rate of 256 Mbit/s. Each station will be equipped with two such recorders, to reach the 24-hour requirement as well as to allow peak data rates of 512 Mbit/s.

Preliminary specifications foresee using the 38-micrometer Mark 3A headstacks, recording up to 16 hour-long passes of 32 data tracks, at 50,000 bits per inch, on 27,000-foot lengths of 0.55-mil tape. Each tape would have a capacity of 7.4 Terabits; further enhancements of data capacity are anticipated by the time the Array is complete.

The recording system will support a flexible, programmable data format. During early VLBA operations, when the first several Array elements are used in conjunction with stations of the existing VLBI networks, and data reduction occurs on existing correlators, one of the standard Mark 3 formats will be written. This cannot support the VLBA's wider bandwidths and four-level sampling capabilities, however. An advanced format, incorporating more robust error-detection, flexible track assignment, and transparent framing, will be substituted when the Array nears completion and the correlator begins routine operation. Since the programmable format requires a commensurately capable decoder/deformatter, it will remain possible to read tapes recorded in Mark 3 format (although this will restrict the observations to relatively narrow aggregate bandwidths).

NRAO recognizes, and indeed urges, the importance of standardization of data recording to future large-scale VLBI ventures. A number of communications have already been received from observatories interested in obtaining VLBA-compatible recording equipment. We expect to be in a position to provide reliable cost and availability information to potential users upon completion of the pre-production phase of our contract with Haystack, currently planned for mid- to late 1988.

CORRELATOR

The VLBA correlator, intended to support processing of all observations involving the Array, will be the first large, wideband VLBI correlator ever built. Its design recognizes that the VLBA will be joined frequently by existing antennas to form a global array. With 20 station inputs, it will be able to accommodate large observations of this type, as well as to support extreme wideband measurements using two recorders simultaneously at each of 10 stations, or simultaneous reduction of two normal 10-station programs. A large, high-resolution system on this scale imposes a new cost/performance logic upon the design. After an extensive study, the spectral-domain or FX correlator algorithm was adopted because of the significant cost efficiencies achievable (especially for a large spectroscopic system), as well as a number of important scientific and technical advantages.

The FX scheme (Chikada, et al. 1984) differs from the conventional lag correlator essentially in the sequence of operations performed to calculate baseline cross-power spectra from the input station samples. Individual station signals are first Fourier transformed; correlation is then replaced by a much more economical pairwise cross-multiplication of the station spectra. Exploiting the efficient FFT algorithm, and organizing as much processing as possible on a station basis, this approach minimizes the number of arithmetic operations required. The reduction is sufficient to realize a significant advantage in basic signal-processing hardware, compared to a lag correlator, despite the greater complexity of the operations involved. The comparison is particularly favorable for many-station and/or high-resolution systems -- for the VLBA the factor is about three.

Further, the weak logarithmic dependence of the FFT's complexity on the number of samples transformed allows the FX correlator to provide superior spectroscopic performance. The preceding cost comparison applies in fact to a lower-performance lag system, which achieves full spectroscopic resolution only for a half-size array. Other advantages inherent in the FX architecture include the feasibility of true station-based fringe rotation, without loss of sensitivity or spurious correlation, and straightforward incorporation of an exact correction for fractional-sample delay error, again with no loss of sensitivity. Finally, the FX algorithm allows a spectrally-dependent de-dispersing gate for pulsar observations, and is easily and inexpensively adaptable to tracking fringes from orbiting VLBI "stations" such as Radioastron and Quasat.

The VLBA FX correlator will provide 160 separate FFT engines, enough to accommodate 8 baseband channels at each of 20 stations. When fewer channels are required (as for typical spectroscopic observations), or if only 10 station inputs are active, these engines are available to support one or more of several multi-processing modes, which serve mainly to obtain higher sensitivity from oversampled data or (for line spectra) to counter the effects of the segmentation inherent in the FX algorithm. Independently of these modes, each FFT engine can operate at resolutions ranging in binary steps from 32 to 1024 spectral points.

The cross-multiplier section imposes virtually no restrictions, but the long-term accumulator and output data rate ceiling limit the maximum available resolution. The correlator will support 1024-point resolution in one channel for a 20-station array, without polarization cross-products. In binary steps, this limit decreases for 2, 4, or 8 channels, or for polarization -- and increases, subject to a limit of 1024 points (512 for polarization), for 14- or 10-station arrays.

Final design of the VLBA correlator is currently in progress. A major element is the gate array which will serve both as a radix-4 FFT stage and as a combined cross-multiplier/accumulator. Correlator construction is expected to begin when funding becomes available at the beginning of 1988, with a 7-station, 2-channel subset system scheduled for completion two years later. Expansion to the final 20-station 8-channel configuration will then take a further two years.

POST-PROCESSING

The output from the correlator will be available to Array users in the standard FITS format. No specialized VLBA software is planned for post-processing (calibration, editing, fringe-fitting, imaging, etc.) of these data, because almost all the functions required are available within the AIPS system used for the VLA, for much of current VLBI work, and for many other applications. Instead, a number of extensions to AIPS are under development to support the few functions specific to VLBA (and other VLBI) observations.

A careful evaluation of the computing capacity required to support Array imaging was included in NRAO's original VLBA proposal, and the construction project will include adequate facilities to ensure that most observations can be processed. Because of the present extremely rapid evolution of computer technology, however, these facilities will neither be specified nor procured until full-time Array operation begins. The use of transportable AIPS software is expected to make available a broad range of potential post-processing hardware.

OPERATIONS

Efficient operation of a distributed instrument such as the VLBA will require that the individual stations be operated remotely from a single center, and all station equipment has been designed to allow unattended operation much of the time. One shift a day is planned, to cover such tasks as changing and shipping tapes, routine maintenance, and module replacement.

Central control will be exercised through polled multidrop communications over dedicated telephone lines. Dial-up lines will be available as a backup, and may be used exclusively during early operation. Much of the traffic over this network will be monitor data reported from the stations to the control center, where it will be logged for later use (including correlator control, calibration, and diagnosis), and analyzed to detect malfunctions. Samples of the received signal will also be transmitted for near-real-time fringe verification. Sufficient buffering will be provided in both directions to allow operation to continue for several hours in the event of a communication outage.

NRAO plans to operate both the VLBA and the existing VLA facility from a new operations center to be built on the campus of the New Mexico Institute of Mining and Technology in Socorro, NM. In addition to array operations, the center will house the VLBA correlator, imaging facilities for both arrays, and VLBA maintenance laboratories, plus other associated scientific, electronics, and computing activities. The building will be funded partially from the VLBA construction project and partially by the State of New Mexico. Bids for the first phase of construction have been solicited; occupancy is scheduled for July 1988.

REFERENCES

Chikada, Y., et al., 1984, in Indirect Imaging, ed. J. A. Roberts, Cambridge University Press, p. 387.
Kellermann, K. I., and Thompson, A. R., 1985, Science, **229**, 123.
Webber, J. C., and Hinteregger, H. F., 1987, this Symposium.
Weinreb, S., Norrod, R., and Pospieszalski, M., 1987, this Symposium.

NEW DEVELOPMENTS IN EUROPEAN VLBI

R.T. Schilizzi
Netherlands Foundation for Radio Astronomy
P.O. box 2
7990 AA DWINGELOO
The Netherlands.

ABSTRACT: Five major areas of development in European VLBI are reviewed: new telescope millimeter-wavelength VLBI; Mk2 and Mk3 correlators, the European Data Processing Facility for the 1990's and a 5 GHz receiver for the RADIOASTRON space VLBI program.

1. INTRODUCTION

The European Consortium for VLBI was established in 1980 with the member institutes listed in Table 1. The Consortium is governed by a Board of Directors consisting of the Institute Directors, which delegates the operation of the European VLBI Network (EVN) to the Program Committee which meets three times a year to select observing proposals, the Technical Working Group which coordinates instrumental developments in the EVN, and a network of "friends of VLBI" at each telescope to carry out the observing programs. Scheduling of the EVN observing sessions and coordination of the activities of the friends of VLBI during the sessions is the responsibility of the Chairman of the Program Committee. The Board of Directors recently appointed a Network Scientist to look into means for improving the scientific performance of the EVN, and a Project Manager to oversee efforts to develop new generation processing facilities.

In the seven years of its existence the EVN has seen observing time for VLBI more than double to about 65 days per year, the number of proposals more than double to about 80 per year, and the number of scientists involved in proposals also more than double to about 150 per year.

A number of initiatives have been undertaken in recent years to ensure a leading role for European VLBI in the next decade. These include a proposal for an advanced data processing facility, the construction of new large diameter telescopes, a proposal for equipping the millimeter-wave telescopes at Pico Veleta and Plateau de Bure with VLBI equipment, the agreement with the Space Research Institute in Moscow to construct two 5 GHz receivers for the RADIOASTRON space VLBI program (Kardashev and Slysh, these Proceedings), and the proposal for

M. J. Reid and J. M. Moran (eds.), The Impact of VLBI on Astrophysics and Geophysics, 469–473.

QUASAT, the European-led western proposal for space VLBI (Schilizzi, these Proceedings).

Table 1: Member institutes of the European Consortium for VLBI

Institute	Membership Status	Telescope Location	Telescope Diameter	Comment
Nuffield Radio Astronomy Lab, Jodrell Bank, UK	F	Jodrell Bank	76 m 26 m	
Paris Observatory, Meudon, France	A	Nancay	94 m	(equivalent diameter)
Netherlands Foundation for Radio Astronomy, Dwingeloo, NL	F	Westerbork	93 m	(phased array)
Max-Planck-Institut for Radio Astronomy, Bonn, FRG	F	Effelsberg	100 m	
Institute of Radio Astronomy, Bologna, Italy	F	Medicina Noto Sardinia	32 m 32 m $\geq$32 m	 new new
Chalmers Technical Univ Goteborg, Sweden	F	Onsala	26 m 20 m	
Technical University of Munich/University of Bonn, FRG	A	Wettzell	20 m	
Nicolaus Copernicus University, Torun, Poland	A	Torun	15 m 32 m	 new
Space Research Institute, Moscow, USSR	A	Simeis Suffa	22 m 70 m	 new

F : full member
A : associate member

2. NEW TELESCOPES

Four of the telescopes in Table 1 are noted as "new". The Noto, and Suffa telescopes are currently under construction, and should be operational in 1989 and 1990 respectively. The engineering design study for the Torun telescope has been completed. The Sardinia telescope is approved as part of a long term development plan for southern Italy, the details of which are still under discussion. A further 32m telescope has been proposed at Cambridge as an extension of the MERLIN system, which at the same time provides crucial intermediate baselines between the current MERLIN and the EVN. These telescopes will add significantly to the overall sensitivity and U-V coverage of the EVN.

Two other developments in the area of telescopes are worth mentioning: the resurfacing of the 25m Mk2 telescope at Jodrell Bank and the outfitting of the 11m telescope at Nsukka (Nigeria) for VLBI. The Mk2 telescope at Jodrell now has overall surface accuracy of 0.6mm rms which will allow high aperture efficiency operation at 22 GHz. The Nsukka telescope will operate at 18 cm initially and provide important intermediate baselines from the EVN to Hartebeesthoek.

3. MILLIMETER-VLBI

Onsala and Effelsberg have conducted experimental observations at 43 GHz as a single baseline and also as part of global observations with US stations. A proposal to expand European facilities for mm-VLBI has been prepared by MPIfR and IRAM, and will be submitted in Summer 1987. It envisages hydrogen masers and broad-band recording systems at Pico Veleta in Spain and Plateau de Bure in France. Table 2 lists the European telescopes capable of participating in a mm-VLBI network.

Table 2: European millimeter-wave telescopes

Location	Diameter	λ min
Pico Veleta (Spain)	30 m	0.8 mm
Onsala (Sweden)	20 m	2.6 mm
Yebes (Spain)	14 m	2.6 mm
Metsaehovi (Finland)	14 m	2.6 mm
Plateau de Bure (France)	3 x 15 m	0.8 mm

4. Mk2 AND Mk3 CORRELATION FACILITIES

An agreement between Bologna and Caltech will see the JPL/CIT 5-station Mk2 correlator move to Bologna late in 1987 following check out of the new Block II correlator at Caltech.

The MPIfR Mk3 correlator will be upgraded from 4 to 5 stations in 1987 and to the higher track density system by early in 1988.

Support has come for the tape density upgrade throughout the EVN from the Commission of the European Communities in Brussels via a grant from their Technology Stimulation Program.

5. EUROPEAN DATA PROCESSING FACILITY FOR THE 1990'S

To meet the obvious scientific challenges of the 1990's, the EVN will need to provide better image quality, higher spectral resolution, higher frequency capability, and be able to handle data from a space element. The European Consortium is seeking funding for an augmentation of the EVN facilities in the 1990's through the European Community's "Framework Programme of Technological Research and Development (1987-1991)".

The proposal has two major components: the Data Processing Facility (DPF) including the Data Playback System (DPS), and the Data Acquisition System (DAS). The current concept for the processing facility centers around a correlator capable of processing up to 256 Mbits/sec data in one polarisation from 16-stations simultaneously. The correlator is of the "X-F" type and will employ state of the art 2μ CMOS technology. A number of the possible correlation options are listed in Table 3.

The DAS and DPS will follow the US VLBA design for obvious compatibility reasons. The 24 acquisition terminals and 18 playback terminals envisaged are to be built in Europe.

The total cost is expected to be 16.4 MECU (18 M$), comprising 10.2 MECU for the DPF including the DPS, 4.6 MECU for the DAS and 1.6 MECU for contingency.

The NFRA in Dwingeloo will carry the overall project responsibility and will construct the DPF. The MPIfR in Bonn will manage the construction of the DAS and DPS.

6. 5 GHz RECEIVERS FOR RADIOASTRON

The European Consortium is jointly financing the construction of two space-qualified, coolable, 5 GHz receivers for RADIOASTRON. The NFRA in Dwingeloo is managing the overall receiver project and building the mixer/IF unit; the MPIfR is building the low noise amplifier. It is hoped that ESA/ESTEC will provide advice during the course of the project, as well as the use of environmental test facilities for testing the prototypes. The receiver project kickoff meeting took place in February 1987; delivery of the two flight units is expected in March 1990.

Table 3: Basic correlator configurations for the proposed European Data Processing Facility.

# Stations	# Pol	Continuum bandwidth (MHz) 2-Bit	1-Bit	# Spectral points per interferometer baseline
16	2	64	128	128
	1	128	256	256
12	4	64	128	128
	2	64	128	256
	1	128	256	512
8	4	64	128	256
	2	64	128	512
	1	128	256	1024
6	4	64	128	512
	2	64	128	1024
	1	128	256	2048

CHINESE VLBI NETWORK PROJECT

Wan Tongshan   Qian Zhihan
Shanghai Observatory, Academia Sinica
Shanghai, China

1. Introduction

The Chinese VLBI Network (CVN) project was proposed early in 1979, and since then VLBI research has been actively developed in Shanghai Observatory. However, only recently the whole project was accepted by the Astronomy Committee, Division of Mathematics and Physics, Academia Sinica.

2. The Sites

The selected sites of the network are Shanghai, Kunming and Urumqi. In addition, there are other two stations for radio astronomy. In Miyun near Beijing there is an aperture synthesis radio telescope with an equivalent diameter of 42-m in the meter waveband. And in Delingha in the west of China, a 13.7-m antenna in the millimeter waveband has been erected already. These two stations will also be involved in VLBI observation in the future.

3. Scientific Objectives

The Project is aimed at both geodynamics and astronomy applications. It is considered that the Chinese VLBI network will run in two modes of operation, i.e., domestic and international collaboration.

(i) Domestic mode

Three or four antennas can be used for VLBI survey to map large samples of the radio sources. Measurements of source positions and maser source study can also be made in this mode of operation. Also of interest are the determinations of Earth Rotation parameters (ERP) and the deformation of the crust on the territory, which is related to earthquake study and Earth science.
The Shanghai-Miyun-Urumqi-Kunming forms a meter waveband VLBI array.

M. J. Reid and J. M. Moran (eds.), The Impact of VLBI on Astrophysics and Geophysics, 475–476.

In this mode of operation, we can operate the network on our own initiative and lay down the groundwork for international collaboration.

(ii) International collaboration

The combination of the CVN with the EVN and Japanese radio telescopes would produce excellent image quality particularly on the northern sources. In the Asia-Pacific region, the Chinese VLBI network would play an important role. The Chinese meter waveband VLBI array can be largely extended with Australian antennas and the powerful Ooty radio telescope, India.
The CVN will make positive contributions to the investigation of global crustal motions and ERP, and the establishment of the terrestrial reference frame and its maintenance using VLBI. The Shanghai radio telescope will join regular observations in the IERS (International Earth Rotation Service) for the determination of ERP in early 1988. And Shanghai Observatory will become one of the VLBI data analysis centers in IERS project.

4. Major Facilities

Table 1 below lists the radio telescopes in China. Receivers, narrow-band terminals and hydrogen masers will be constructed in China. The CVN center is being set up in Shanghai. Investigation is being made for the construction of a wide-band processor for the CVN. The first station of CVN, Shanghai station, will conduct its first VLBI experiment in June, 1987. The completion of the whole project is scheduled for 1992.

Table 1 Chinese VLBI Network Project

Observatory	Antenna Diameter (m)	Wavelength (cm)	Terminal	Frequency Standard	Remarks
Shanghai	25	92,50,21,18, 13,6,3.6,2.8,1.3	Mk-2, Mk-3	H maser	completed
Urumqi	25	92,50,21,18, 13,6,3.6,2.8,1.3	Mk-2, Mk-3 or VLBA	H maser	planned
Yunnan	10	92,50,21,18, 13,6,3.6,2.8	Mk-2, Mk-3 or VLBA	H maser	antenna available
Miyun	42(equiv.diam.)	92	Mk-2	Rb	id.
Delingha	13.7	1.3	(future plan)		id.

DESIGN OF CHINESE SECOND 25-m VLBI ANTENNA

Xu Pei-yuan
Shanghai University of Science and Technology
Shanghai, People's Republic of China

The first Chinese 25-m VLBI antenna has just been completed at the She-Shan Station of the Shanghai Observatory. In accordance with the plan of the Chinese VLBI Network, the second 25-m VLBI antenna will be established at the Urumqi VLBI Station. It is a shaped symmetric dual reflector Cassegrain antenna and will be equipped with nine feeds from 0.3 GHz to 23 GHz. The surface accuracy of the main reflector is 0.65 mm, and the 14-m central part has a better surface accuracy (0.42 mm) to provide the possibility of making VLBI observations at 44 GHz. This antenna has a semihomology design, and its subreflector has a mechanism to focus it in three dimensions. Crossed dipole feeds for the 0.32 and 0.61 GHz bands are located in the middle of the subreflector, and the focussing mechanism will retract it to bring the dipoles to the prime focus. For other frequencies, a corrugated horn will be mounted at the Cassegrain focus. To obtain dual frequency operation at X and S bands, a flat dichroic reflector and an elliptical reflector will be mounted over the on-axis S band horn and off-axis X band horn, respectively. The antenna has a wheel and track mounting, and the pointing will be controlled by an IBM PC/XT microcomputer. The pointing and tracking accuracy is 16 arcsecond. The estimated aperture efficiencies are in the range of 45–65%. The estimated antenna temperatures are in the range of 35–100 K.

M. J. Reid and J. M. Moran (eds.), The Impact of VLBI on Astrophysics and Geophysics, 477.

18 cm VLBI NETWORK

L. I. Matveyenko,[1] V. I. Kostenko,[1] V. V. Timofeev,[1] L. R. Kogan,[1]
B. Z. Kanevskii,[1] I. G. Moiseev,[2] R. L. Sorotchenko,[3]
R. M. Martirosyan,[4] M. V. Golovnya[5]
[1] Institute for Space Research, Moscow, U.S.S.R.
[2] Crimea Astrophysical Observatory, Crimea, U.S.S.R.
[3] Lebedev Physical Institute, Lebedev, U.S.S.R.
[4] Radiophysical Institute, Riga, Latvia, S.S.R.
[5] Main Astronomical Observatory, Kiev, U.S.S.R.

The Soviet VLBI network includes parabolic 70-m antennas located near Usuriisk and Eupatoria, a 64-m antenna in Bear Lake, 22-m antennas in Pushino and Simeiz, and a 25-m antenna near Ulan-Ude. The maximum baseline length in the E–W direction is equal to about 7000 km, and in the N–S direction is equal to 1300 km. The minimum baseline length is equal to about 100 km.

Each 70-m antenna has a Gregorian system of illumination, the 22-m antenna has a Cassegrain system at 1.35 cm and a prime focus system at 18 cm, and the 25-m and 64-m antennas have Cassegrain systems. Each antenna has been equipped with a hydrogen frequency standard (H) or a rubidium standard (Rb), a low noise cooled 18-cm parametric or transistor preamplifier, and a 2-MHz digital recording system with a video recorder. The system noise temperatures (T_{sys}) of the radio telescopes at 18 cm are 50–70 K, and the aperture efficiencies (η) are equal to 0.45–0.57 (see Table 1).

Table 1

Station	Antenna Diameter (M)	$\lambda = 18$ cm		$\lambda = 1.35$ cm		Frequency Standard
		η	T_{sys} (K)	η	T_{sys} (K)	
Usuriisk	70	0.55	60	(40)	(60)	H
Ulan-Ude	25	0.45	65	–	–	Rb
Bear Lake	64	0.55	50	–	–	H
Pushino	22	0.49	65	0.30	100	H
Eupatoria	70	0.57	50	0.40	60	H
Simeiz	22	0.56	70	0.48	65	H

The antennas in Simeiz, Eupatoria, and Pushino have been equipped with 1.35-cm radiometers having maser preamplifiers. The fractional-frequency stability of the local oscillators is $< 5 \times 10^{-14}$ for 1000 sec. The system noise temperature and the efficiency of the radio telescopes are shown in Table 1 (Matveyenko *et al.* 1983). A three-station processor is used for the processing of continuum sources,

M. J. Reid and J. M. Moran (eds.), The Impact of VLBI on Astrophysics and Geophysics, 479–480.

and a two-station processor is used for spectral line objects. The processor has 96 scalar channels, and the integration time is in the interval 0.2 to 2.0 sec.

In 1985, balloon probes in the Venus atmosphere were observed with the global VLBI network, which included our antennas. The preliminary results of the balloon motion for June 11 and June 15 – Vega 1 and Vega 2, respectively – are shown in Figure 1. The power of the balloon transmitter was ~ 5 watts and of the spacecraft was ~ 45 watts. The trajectories of balloons are measured relative to the spacecraft on the antennas at Usuriisk, Ulan-Ude, and Eupatoria.

We are now observing radio sources. The fringe spectrum of the Seyfert galaxy NGC1275 on the Eupatoria–Usuriisk baseline is shown in Figure 2.

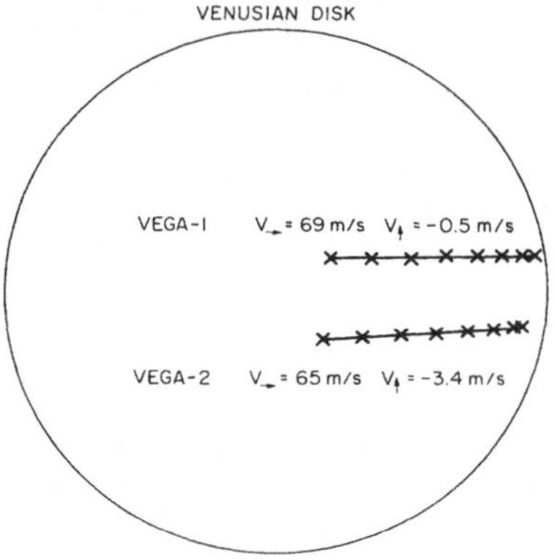

Figure 1. Tracks of the Vega 1 and Vega 2 probes projected on the Venusian disk. The time between data points is 4 hours.

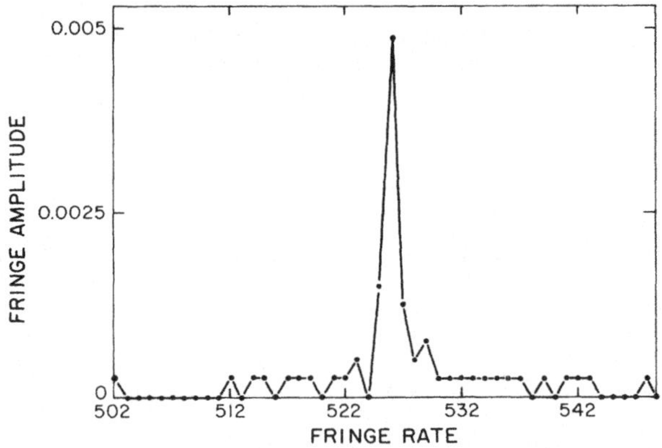

Figure 2. Fringe rate spectrum of NGC1275 on the Eupatoria–Usuriisk baseline. The fringe rate resolution is 4.8 mHz.

REFERENCES

Matveyenko, L., *et al.* 1983, *Sov. Letter Astron. J.*, **9**, 415.
Matveyenko, L., *et al.* 1986, *Sov. Letter Astron. J.*, **12**, 59.

URAN, A DECAMETER BAND VLBI SYSTEM

S. J. Braude, L. N. Litvinenko, A. V. Megn
Institute of Radio Astronomy
Kharkov, U.S.S.R.

Until recently, regular interferometric measurements were not performed at decameter wavelengths. The reasons are well known, i.e., effects of the ionosphere and the interplanetary plasma, the inherently high level of interference and of the galactic background, and the necessity to have very large antennas. Yet, because of the considerable interest in such observations, the Ukrainian SSR Academy of Sciences has started a project aimed at the development and construction of a system of decameter band interferometers (project URAN, from the "Ukrainian Radio Interferometers of Academy of Sciences (Nauk)"). By 1989, the system will involve four interferometers with baseline lengths between 40 and 900 km. The basic instrument of all these is the N–S arm of the large UTR-2 radio telescope operated by the Institute of Radio Astronomy in Kharkov. The antenna is an electrically controlled phased array of broad band dipoles (total length about 1800 m). At the other sites, the antennas are smaller phased arrays (maximum length about 230 m) consisting of crossed dipoles.

The shortest baseline is 42.2 km, achieved with the URAN-1 interferometer near Kharkov. It is oriented along the E–W direction and has been used over a decade for connected element interferometry at 25, 20 and 16.7 MHz. New techniques for measuring the magnitude γ of the visibility function have been developed, with an account of the special observation conditions at these frequencies. γ has been measured for about 30 sources of radio emission such as supernova remnants, radio galaxies and quasars. Some of the galaxies and quasistellar objects (e.g., 3C196 or 3C84A) have shown extended components of tens seconds of arc in size, characterized by "steep" power law spectra, that do not seem to have been observed before at higher frequencies. Compact components with steep spectra have been detected and investigated in supernova remnants (3C144 and 3C461). Measurements of the angular structure of the radiation from extended sources like radio galaxies or unidentified objects (e.g., 3C123 or 3C134) are in agreement with the meter and decimeter wavelength data.

Preliminary observations of 3C144, 3C196 and 3C461 were performed in 1986/87 with the new VLB interferometer URAN-3 (baseline length 616 km). We have found that high quality VLB records can be obtained at decameter wavelengths in autumn and winter during the night time, and γ can be estimated to within 5 or 10 percent.

M. J. Reid and J. M. Moran (eds.), The Impact of VLBI on Astrophysics and Geophysics, 481.

JAPANESE VLBI PROJECT - VERA

Mitsumi Fujishita and Tadayoshi Hara
International Latitude Observatory of Mizusawa
Hoshigaoka-cho, Mizusawa-shi,
Iwate-ken, 023
JAPAN

ABSTRACT. The Japanese VLBI project VERA (VLBI for the Earth Rotation study and Astrometry) is being designed. It will be dedicated to regular monitoring of the Earth's rotation and astrometry. VERA will be composed of two antennas spanning about 2300km and a correlator of multi-station capability. The diameters of the antennas are 35m and 15m.

1. TARGET OF THE VERA

The International Latitude Observatory of Mizusawa (ILOM) and the Tokyo Astronomical Observatory (TAO) designed the main points of the Japanese VLBI network VERA (VLBI for the Earth Rotation study and Astrometry). It will be dedicated to regular monitoring of the Earth's rotation in an international VLBI network like the current IRIS (IAG and COSPAR Sub-Commission; International Radio Interferometric Surveying). The accuracies of the Earth rotation parameters are expected to be remarkably improved by the participation of the VERA (Yokoyama and Manabe, in press). VERA also aims at intensive observations of UT1 and nutation using its own single baseline. It will also be used radio astrometry purposes, for example, establishment of the celestial reference frame and measurements of radio-source distances. In addition, it is expected to be used both for astrophysics and geophysics in cooperation with other fixed stations, mobile systems, SLR stations and GPS receivers. VERA will be open for any users, in principle.

2. OUTLINE OF THE VERA

VERA is planned to be composed of two antennas equipped with the K-3 or the Mark III system. One antenna of 35 m diameter is expected to be located in the vicinity of the ILOM which is in the northern part of the Japanese main island. The other antenna with the diameter of 15 m will be located in one of the South-West Islands. The baseline spans about 2300 km. Antennas will be built on granite bedrocks to

M. J. Reid and J. M. Moran (eds.), The Impact of VLBI on Astrophysics and Geophysics, 483–484.

avoid irregular local displacements. The slew rate of the antennas will be over 1 degree per second for raising the efficiency of the observation. A receiver of 22 GHz will be available, in addition to those of S and X bands. 43 GHz band is expected to be received, because RMS error of the surface accuracy of the antennas is under 0.5mm. The correlation center with a multi-baseline correlator will be located in the present complex of the ILOM. VERA will be brought into operation in three years.

3. PROMOTION OF THE VERA

The committee for promoting the VERA was organized early 1986 after the agreement of the Japanese group of astrometry and geodesy for constructing a domestic single baseline VLBI network. The committee members come from the ILOM, the TAO, the Radio Research Laboratory, the Geographical Survey Institute, the Earthquake Research Institute and the Hydrographic Department. Two VERA symposia were already held to discuss hardware and software systems of the VERA. Scientific proposals for the use of the VERA were also made at the symposium from researchers of various fields for discussing scientific targets and main points of the VERA. The ILOM is a responsible institute for the realization of this project. System design is being made at the ILOM in collaboration with other relevant institutes.

REFERENCE

K. Yokoyama and S. Manabe : in press, "The Role of Japan in the New International Earth Rotation Service (Expected Contribution of the Planned Japanese VLBI System VERA)", in Earth's Rotation and Reference Frames for Geodesy and Geodynamics.

THE AUSTRALIA TELESCOPE LONG BASELINE ARRAY

R.P.Norris
CSIRO Division of Radiophysics
PO Box 76, Epping, NSW 2121,
Australia

ABSTRACT

Progress on the Australia Telescope, due for commissioning in 1988, is on schedule. When complete, the Australia Telescope will consist of a 6-km compact array and a Long Baseline Array which will, in collaboration with the University of Tasmania and NASA, use baselines up to 1400 km. Here I focus on the novel features of the Long Baseline Array, which is planned to use satellite linked local oscillators and VLBA data acquisition terminals to provide a VLBI network capable of expansion both within Australia and overseas.

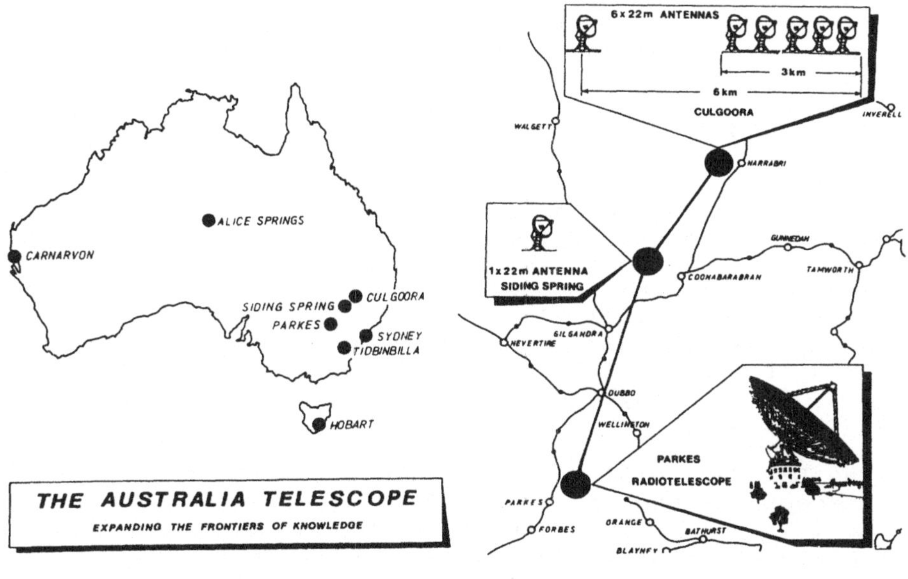

Figure 1: AT and Australian VLBI antennas

1. INTRODUCTION

The Australia Telescope (AT) will comprise a 6-km compact array (CA) at Culgoora, NSW, and a long baseline array (LBA) which will, in collaboration with the University of Tasmania and NASA, use baselines up to 1400 km, and be capable of further expansion in the future. The CA will use six movable 22-m antennas on a rail track to allow high dynamic range

M. J. Reid and J. M. Moran (eds.), The Impact of VLBI on Astrophysics and Geophysics, 485–486.

maps to be made at wavelengths extending eventually down to 3mm. The compact array is notable for its large bandwidth (up to 256 MHz with full polarisation information) and verstaile spectral-line capability (up to 2048 channels per Stokes parameter). The LBA will use the CA in a tied configuration together with the 64-m Parkes antenna, and a new 22-m antenna at Siding Spring. It will also be used in conjunction with the NASA DSN Tidbinbilla antennas and the 25-m antenna owned by the University of Tasmania.A recent review of the AT is given by Whiteoak (1986). Here I concentrate on the novel features of the LBA,the location of which is shown in Fig.1.

2. LOCAL OSCILLATOR DISTRIBUTION

Present plans are to use the AUSSAT geosynchronous satellites to distribute a master local oscillator from Culgoora. At each station of the LBA, a two-way AUSSAT link will be used to compare the phase of the local oscillator with that of the master at Culgoora, and the difference used to correct the phase of the data. The detailed design of this system is not yet complete, and a series of experiments using AUSSAT II will shortly be started to compare the performance of various configurations.

3. DATA TRANSMISSION AND PROCESSING

For various reasons, primarily the requirement for compatibility with the CA, fringe rotation for the LBA will be done at the antennas. This will be controlled by a local computer at each antenna which will communicate continuously with the central site computer, probably using a commercial packet-switched network. It is planned to record data at each antenna of the LBA using a standard Data Acquisition System purchased from the VLBA in the U.S. We also plan to purchase a number of Data Playback Systems which will then interface to our correlator at Culgoora.

The LBA data will be compatible with VLBA and other VLBI data, except for the provision of fringe tracking at the antennas. This will be overcome as follows: (i) For VLBI experiments to be processed overseas, fringe rotation at each antenna will be disabled. (ii) For VLBI experiments to be processed at the AT central site, our provisional plans are to implement a Hilbert Transform phase rotator at the correlator. As well as ensuring our ability to participate in international VLBI, this compatibility is intended to facilitate future expansion and use of QUASAT and RADIOASTRON.

REFERENCE

Whiteoak, J.B., 1986, Proc. Astr. Soc. Aust., **6**, 290.

THE PARKES - TIDBINBILLA INTERFEROMETER

R.P.Norris, M.J.Batty, M.J.Kesteven,
K.J.Wellington
CSIRO Division of Radiophysics
PO Box 76, Epping, NSW 2121,
Australia

ABSTRACT. The Parkes - Tidbinbilla Interferometer is a radio-linked interferometer using the 64-m telescope at Parkes together with one of the NASA antennas (34-m or 64-m) at Tidbinbilla. With a baseline of 275km, it is currently the world's longest real-time interferometer, and is usable at frequencies of 1.6, 2.3, and 8.4 GHz to give angular resolutions of 0.13, 0.09, and 0.03 arcsec respectively, with a sensitivity of 1-2 mJy rms in 5 minutes at a bandwidth of up to 10MHz.

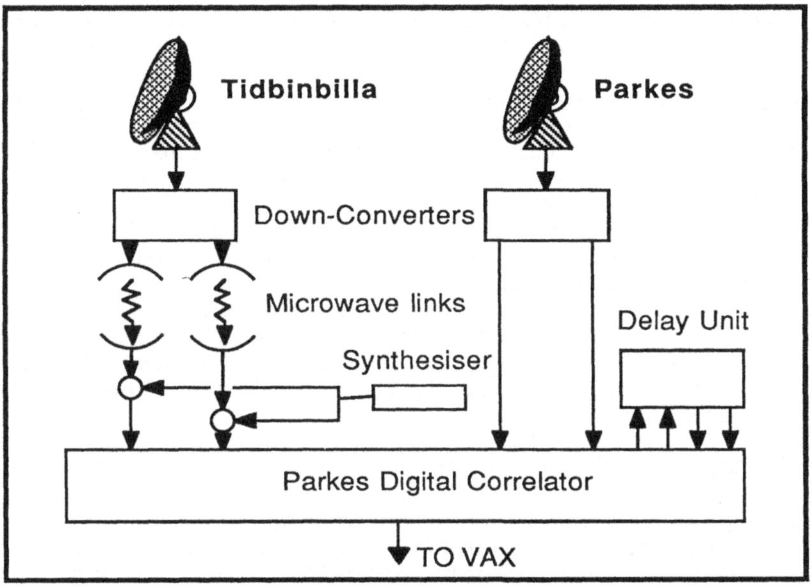

Figure 1. The Parkes-Tidbinbilla Interferometer

1. INTRODUCTION

The Parkes - Tidbinbilla Interferometer (PTI), was conceived largely as a prelude to the Australia Telescope (AT), and as such is useful both as a test-bed on which to tackle some of the problems that we will encounter on the AT, and also as a first step towards the construction of a frame

M. J. Reid and J. M. Moran (eds.), The Impact of VLBI on Astrophysics and Geophysics, 487–488.

of reference of calibration sources for the AT. In addition it constitutes a powerful instrument with which to tackle some current astronomical problems. First fringes were obtained on the PTI in a limited mode in June 1985 (Norris et al., 1985). Since then the system has been upgraded so that it uses two channels, each of which can be used at bandwidths up to 5MHz. Local oscillators at Parkes are locked to a Rubidium standard.

2. THE INSTRUMENT

A novel feature of the PTI is that many of the interferometer functions which have traditionally been implemented in hardware (before correlation) are instead software-based (after correlation). Only coarse phase and delay tracking necessary to prevent decorrelation in the 2-s integration time need be applied before correlation.

The IF signal from the Tidbinbilla antenna is sent to Parkes on microwave links installed by NASA for the Voyager-Uranus encounter. At Parkes, the two signals from Tidbinbilla are mixed with a 0.15 MHz tone derived from a Rockland Model 5100 synthesiser, which is adjusted every second by the on-line software to accomplish a first-order phase tracking. To prevent random long-term drifts,the phase of the synthesiser is set to zero at the start of each 'scan' (typically 10 minutes). The resulting phase steps, together with fine phase variations, are subsequently removed in software.

Coarse delay tracking is achieved by a circulating- RAM delay unit with a resolution of several μs. The two cross-correlation functions (XCFs) are then allowed to drift throught the 1024-channel correlator, and the online software selects two 256-channel XCFs, thereby tracking the delay to a resolution of 1 delay channel. The resulting XCF is then Fourier transformed into the frequency domain, where fine delay tracking is performed by applying a phase gradient to the complex spectrum.

3. THE SCIENCE

Current projects being tackled on the PTI include:

1. Observations at 2.3 and 8.4GHz of SN1987A
2. Astrometric measurements of calibrator sources
3. Determination of pulsar proper motions
4. Studies of OH maser sources
5. Searches for compact nuclei in active galaxies and flat-spectrum quasars.

REFERENCE

Norris, R.P., Batty, M.J., Kesteven, M.J., & Wellington, K.J., 1985, Proc. Astr. Soc. Austr., **6**, 137.

THE CANADIAN GEOPHYSICAL LONG BASELINE INTERFEROMETER

J.L.Yen, P.Leone, G.A.Watson, R.Wiedfeldt, J.Zao
University of Toronto, Toronto, Ontario, Canada
W.H.Cannon, P.Mathieu, H.Tan
York University, Toronto, Ontario, Canada
J. Popelar
Department of Energy, Mines & Resources, Ottawa, Ontario, Canada
J.A. Galt
Dominion Radio Astrophysical Observatory
Penticton B.C., Canada

A new VLBI system has been developed and put into operation. The system has three distinctive characteristics as compared with current VLBI practice, namely, the use of a wavefront clock at each station, the capability of a burst mode operation with high speed sampling, and the use of multiple 12 Mb/s data recording channels on commercial VCR's. The main use of the system is to be in geophysical investigations where these new features may prove of value.

A wavefront clock at each station is used to generate the local oscillator for down conversion of RF signal to baseband and to sample the baseband signal for recording. The clock depends on the reference source position as well as the station location. At each station a received wavefront from the reference source position is tagged with the time of its arrival at the earth's centre. Thus signals received at different stations tagged with the same time came from the same wavefront. This provides station oriented fringe rotation and delay tracking as opposed to the commonly used baseline oriented corrections. The more complicated microprocessor controlled clock at each recording station is balanced by the simplicity of the processor required to correlate the data.

The use of the wavefront clock to tag the received signal removed the constraint of sampling the received signal uniformly at each station. Since signals tagged with the same time received at different stations belong to the same wavefront and hence are correlated, as long as signals with the same times are recorded at each station, they can be correlated without regards to how these samples are distributed in time. This makes possible the burst mode operation of the system. Sampling at each station can be performed at high speed in bursts, stored in a buffer memory, and then recorded at a lower rate. This means bursts of signals of larger bandwidth than that dictated by the

M. J. Reid and J. M. Moran (eds.), The Impact of VLBI on Astrophysics and Geophysics, 489–490.

recording bit rate can be recorded. The width of the correlation function is thereby reduced. A ten to one expansion in signal bandwidth is readily achieved. In addition, to observe impulsive phenomena such as pulars, the entire record would contain useful signal. It should be noted that since the total number of samples recorded in a time interval is given by the recording bit rate, the signal to noise ratio obtained in each recording channel remains unchanged.

Each recording channel makes use of a VHS type VCR operating with a bit rate of 12 Mb/s. This is made possible by the use of partial response signaling, an adaptive equalizer and a viterbi decoder on playback. Currently the VCR industry is witnessing advances in higher quality video. This would be directly translated into higher channel bit rate. The electronics of the recording system is designed to operate at higher rates to accommodate anticipated enhancements without modifications. The small size and simplicity of VCR's as well as the potential for enhancement makes the system attractive in many situations.

MILLIARCSECOND IMAGING WITH A LARGE OPTICAL ARRAY

M. Shao, M. Colavita, SAO
K. Johnston, R. Simon, D. Mozurkewich, D. Hutter NRL
D. Staelin, B. Hines, MIT
J. Hughes, G. Kaplan, J. Hershey, USNO
C. Townes, UC Berkeley

Abstract. Considerable progress has been made over the last few years in ground based optical interferometery. The Mark III optical interferometer at Mount Wilson which at this time is operating with a 12 meter north south baseline is capable of measuring amplitude and phase. Experience gained from this facility will allow for the development of a large optical array. Preliminary plans are described for a facility that will be capable of milliarcsecond imaging of celestial objects at wavelengths from 0.4 to 10 microns. If funded this array will be in operation by 1993.

1. Introduction

There is the need for high resolution measurements at optical wavelengths to measure precisely the position and diameters of stars in our galaxy as well as compact collapsed objects in external galaxies. Interferometeric techniques offer an increase in resolution of an order of magnitude over other techniques using a single large aperture.

Optical interferometry is advancing rapidly as is evidenced by the development of independent aperture interferometers for amplitude interferometry by Labeyrie in France and Davis in Australia as well as interferometers that measure both amplitude and phase (Shao and Staelin) in the visible and Townes in the IR.

2. Mark II/Mark III Optical Interferometers

A series of stellar interferometers have been built on Mt Wilson over a period of years to develop the technology for aperture synthesis at optical wavelength. After an initial demonstration of phase tracking on Polaris, the Mark II was built on the principle that as much of the complexity of a stellar interferometer be placed in software. The Mark II

M. J. Reid and J. M. Moran (eds.), The Impact of VLBI on Astrophysics and Geophysics, 491–492.

demonstrated the technology to reliably and quickly find stellar fringes from stars all over the sky.

The Mark III is presently operating with a 12 m N-S baseline and 9 m E-S baseline. From measurements obtained in 1986 with just the N-S baseline, it has measured the position of stars to a formal accuracy of 0.02 arc sec (Mozurkewich et al), determined the diameter of several stars (Shao, et. al. 1987b) and made measurements characterizing the turbulence of the atmosphere (Colavita et al). Plans are underway in 1987 to operate with the two baselines and a laser metrology system consisting of 15 laser interferometers. Additional information on IR operation will be obtained from a two element 10 micron interferometer, to become operational in late 1987.

3. The Large Optical Array

The Mark III will be upgraded over the next two years in several ways to test aperture synthesis techniques. One planned modification is to increase the size of the collecting optics to approximately 1 meter. Another is to extend the baseline to 40-50 meters so that a much larger number of stellar disks are resolved. Another is to add the capability for simultaneous operation of three interferometers for closure phase/ triple correlation measurements. The development of these additions to the Mark III will give us the experience needed to build the large optical array. The preliminary design of this array which will image objects at wavelengths between 0.4 and 10 microns consists of approximately thirteen elements with apertures between 1.0 and 1.6 meters with baselines up to 200 meters at visible wavelengths to almost a kilometer at IR wavelengths. Table 1 gives the preliminary parameters for the array.

Table 1: Large Optical Array Instrument Parameters

# elements	7 (medium sized telescopes) 0.75 meter 6 1.65 meter telescopes one delay line per telescope.
baselines	5 meter to >200 meters (site dependent)
wavelength	0.4 to 10 microns
resolution	0.0002 arc sec at 0.4 micron 0.005 arc sec at 10. micron

4. Status

We are in the process of seeking funding for this program. If we are successful we hope to initiate this program in fiscal year 1989 and have initial operation of the array in 1993.

THE DEVELOPMENT OF THE TORUN VLBI STATION

Stanislaw Gorgolewski
Torun Radio Astronomy Observatory
Nicolaus Copernicus University
Torun, Poland

ABSTRACT. The Torun VLBI Station is situated in Piwnice near the site of the Torun Radio Astronomical Observatory on the Nicolaus Copernicus University Land Estate, about 10 km north of Torun. The VLBI station development started in 1976, when the rubidium frequency standard was ordered and the ground was broken for the foundation of the 15-m equatorially mounted dish. The construction of the Mark II-C VLBI system began in 1977, and the completed VLBI station was first tested in 1980. The first fringes were obtained on May 31, 1981, with the Effelsberg 100-m dish. This experiment started our work with the European VLBI Network (EVN), and later in 1984 with the Global VLBI Network. We now have five radiometers and have received funding for the construction of a new VLBI station, with a 32-m dish, which is now in the final design stage.

1. INTRODUCTION

The Torun VLBI team is an outgrowth from our first radio astronomical research dating from 1958 in the field of monitoring of the Sun during the IGY. Two British Council scholarships for research students at Cavendish Laboratory in Cambridge, England, brought us into the field of solar coronal investigations and radio source structure studies. This research involved the use of interferometric techniques and was the stepping stone to VLBI. We began using VLBI after several years of intense instrumental work that was supported by the European and the U.S. VLBI communities.

2. MATERIALS AND METHODS

The first VLBI radiometers came from the Westerbork Synthesis Radio Telescope (50 and 6 cm) and the Max Planck Institut für Radioastronomie (2.8 cm). New local developments were the 18, 92, and 21 cm radiometers as well as the data quality analyzer/decoder and coherence test systems for use at all VLBI bands covered in Torun, i.e., 6, 18, 21, 50, and 92 cm wavelengths. The 15-m dish, designed with generous Westerbork advice and experience, was built in Poland. The computer control system of the 15-m telescope underwent extensive modernization. It is now automatically running the VLBI observing. It is also used to plot the source

M. J. Reid and J. M. Moran (eds.), The Impact of VLBI on Astrophysics and Geophysics, 493–494.

maps from the data obtained from another computer that analyzes the correlated signals after primary data reduction in Bonn or Caltech.

The new VLBI project, to be completed by 1990, involves the construction of a new 32-m precision radio telescope and the acquisition of a hydrogen maser frequency standard and a VLBA compatible recording system. The funding of this enterprise started in 1986 and enabled us to nearly complete the final design of the 32-m dish and to obtain a PDP-11/45 look-alike as well as two IBM PC-compatible computers. The 32-m telescope is needed to ensure a far better sensitivity than that of the 15-m dish. It shall also be useful for the QUASAT and RADIOASTRON missions, thus enabling us to make another important step into space VLBI. The 32-m dish will also be used with the existing 15-m dish for a local interferometer on a 200-m E–W baseline.

3. ACKNOWLEDGMENTS

The development of the Torun VLBI station was the result of nearly 10 years of concentrated effort of our group, which had nearly 20 years of previous experience. The first impulse for radio astronomy in Torun came from Prof. W. Iwanowska, who in turn was inspired by Martin Ryle, whose genius she had already recognized in the mid-fifties. Capital investment in the Torun Radio Astronomy Observatory (TRAO) has been provided by the Ministry of Science and Higher Education through the Nicolaus Copernicus University. Operational costs of the VLBI projects were shared by the university, the Satellite Geodesy Division of the Space Research Centre, and the Copernicus Astronomical Centre, both in Warsaw. Crucial instrumental, educational, and technical support has been generously provided by the European and U.S. communities. Without this support, we would never have been able to begin VLBI work in the early eighties if ever at all. This work was made possible by the generous support from the I.A.U., which awarded me a travel grant within the U.S. and subsistence for participation during I.A.U. Symposium 129. Travel expenses to the U.S. and back to Poland as well as the support for our research in this field since 1986 have been borne by the "Resortowy Program Badan Rozwojowych" (The Ministry of Science and Higher Education): RPBR NR. RR. I. 11/2.

THE ITAPETINGA CONTRIBUTION ON NORTH–SOUTH HIGH RESOLUTION VLBI EXPERIMENTS

P. Kaufmann, Z. Abraham, Y. Bakor, C. B. Georges, E. Scalise, Jr., and R. E. Schaal
Itapetinga Radio Observatory, INPE
Instituto de Pesquisas Espaciais
12200 San Jose Dos Campos, Brazil

Since the very beginning of the Itapetinga 45-ft antenna operations in 1971, there were plans to use it as a southern hemisphere VLBI terminal for high north-south resolution. The first successful VLBI tests were performed on water vapor lines in cooperation with Haystack Observatory in 1978 in one of the last runs using the MkI system. This cooperation led to the construction of a MkII terminal for Itapetinga years later (1981–1983). Astrophysical VLBI runs have been carried out since 1984, with the cooperation of Caltech, producing maps of unique resolution on stronger quasars, especially on 3C273 whose evolution has been followed since then. Recently, Bonn joined the collaboration in astrophysical VLBI programs. Itapetinga also participated successfully in geodetic surveys, and in the Vega-Venus mission, in collaboration with JPL and French organizations CNES, Observatoire de Paris and IGN. Plans for continuation and expansion of the Itapetinga VLBI participation in international efforts depend strongly on support to maintain and upgrade instrumentation at the Brazilian Radio Observatory.

M. J. Reid and J. M. Moran (eds.), The Impact of VLBI on Astrophysics and Geophysics, 495.

PERFORMANCE OF THE MEDICINA RADIOTELESCOPE AND THE NEW S/X RECEIVER FOR GEODYNAMICS VLBI

G. Grueff[1,3] G. Comoretto[2], F. Mantovani[3] and P. Tomasi[3]

1. Dipartimento di Astronomia, Università di Bologna, Italy
2. Osservatorio Astrofisico di Arcetri, Firenze, Italy
3. Istituto di Radioastronomia del C.N.R., Bologna, Italy

INTRODUCTION

At the beginning of 1980 the Istituto di Radioastronomia of the Consiglio Nazionale delle Ricerche (C.N.R.) started the Italian VLBI Project.

The first antenna was completed in October 1983 (see Table 1 for telescope characteristics). The first VLBI Mark2 experiment was performed in March 1984. In December 1984 we had the first Mark3 experiment. From that time our Institute is officially member of the European VLBI Network (EVN) and since January 1986 it is also Associate Member of the US Network.

Table 1

Diameter	32 m
Surface Accuracy	1 mm rms
Mount	Az-El
Pointing Accuracy	10 arcsec rms
Slew Rate Az/El	48/30 deg/min
Limits Az	± 270 deg
El	0.5/109 deg
HPBW (5 GHz)	8 arcmin

1. RECEIVER CHARACTERISTICS

Freq. GHz	Tsys K	Bandwidth MHz	Peak Ant. Gain K/Jy	Polariz. Channels
22.3	180	400	0.08	1
10.7	100	300	0.12	2
5.0	55	500	0.18	2
1.6	90	100	0.12	2 +
1.4	90	100	0.12	2 +
2.2	180	140	0.15	1 +
8.2	210	400	0.15	1 +

+ uncooled (room temperature)

M. J. Reid and J. M. Moran (eds.), The Impact of VLBI on Astrophysics and Geophysics, 497–498.

All the receivers listed above, are MesFet-GaAs receivers, fully developed at the Institute. Half of them are at present cooled at 20K, using CTI, closed circuit helium cryogenic pumps. Feeds and receivers are located in the cassegrain focus. Only the S/X receiver is set-up for the primary focus.

2. GEODYNAMIC PROGRAMS

We joined Wettzell and Hartrao for our first geodynamic experiment (called SARG), from January 31st to February 7th 1987. Twentyone tapes were recorded during seven days of that experiment dealing with Polar Motion.

We have also observed two IRIS 24 hours sessions (No. 389, April 3rd and No. 395, May 3th) and, from April 10th to May 8th we took part to an ad hoc Intensive Campaigne for UT1 monotoring with the baseline Medicina-Richmond. The aim of that experiment was to repeat with a second independent baseline previous measurements done with the baseline Westford-Wettzell, in which UT1 glitches on short time scale were observed. Data analysis for both SARG and Intensive Campaigne is in progress.

Calc and Solve have been installed on the HP1000 E-Series of the Medicina Station. Computer facilities also include a 404 MByte disk that allows to analyze VLBI geodynamic data.

It may be of interest to state the plans we have for the next future related to geodynamic activities. They can be shortly summarized as follows:

- cooling of the S/X dual band receiver (expected System Temperature 60K and 80K in the S and X band respectively);
- construction of a platform for Mobile Satellite Laser Ranging;
- partecipation, on a regular basis, to the IRIS campaigne;
- partecipation to ad-hoc VLBI Geodynamic experiments.

It is also worth mentioning that construction of the second antenna started last April, near the town of Noto, Sicily. The antenna is scheduled to be in operation for both Radio Astronomy and Geodynamic in late 1988.

COMPACT CRYOGENIC RECEIVERS FOR THE 1.3 TO 43 GHZ RANGE

S. Weinreb, R. Norrod, and M. W. Pospieszalski
National Radio Astronomy Observatory*
2015 Ivy Road
Charlottesville, Virginia 22903

A series of front-ends utilizing small closed-cycle refrigerators and very low-noise, high-electron-mobility transistor (HEMT) amplifiers [1] have been developed for use in the Very Long Baseline Array (VLBA). The frequency bands, amplifier noise temperatures, expected system temperatures, and current status are shown in Table I. The receivers are designed to be light weight (~ 55 pounds except for 105 pounds at 1.5 GHz) for ease of installation and maintenance, are easily remotely controlled and monitored, and provide dual-channel circular polarization capability. Detailed descriptions of some of the front-ends are given in VLBA technical reports [2,3,4].

The HEMT amplifiers utilized in the receivers have been under development for several years. At 1.5 GHz a feedback, lumped-element amplifier has been improved by utilization of the Fujitsu FHR01FH HEMT. At 2.2 GHz through 15 GHz, coaxial designs [5] utilizing the same transistor are planned (other transistors giving still better performance may become available). At 23 GHz a hybrid integrated circuit amplifier utilizing chip transistors has been developed.

REFERENCES

[1] M. W. Pospieszalski, S. Weinreb, "FET's and HEMT's at Cryogenic Temperatures -Their Properties and Use in Low-Noise Amplifiers," to be presented at IEEE 1987 Int. Mic. Symp., Las Vegas, June 1987.
[2] S. Weinreb, H. Dill, and R. Harris, "Low-Noise, 8.4 GHz, Cryogenic GASFET Front-End," VLBA Technical Report No. 1, National Radio Astronomy Observatory, Charlottesville, VA, Aug. 29, 1984.
[3] R. Norrod, "Model F103, 1.5 GHz Cryogenic Front-End," VLBA Technical Report No. 2, National Radio Astronomy Observatory, Charlottesville, VA, Sept. 10, 1986.
[4] R. Norrod, "Model F105, 4.8 GHz Cryogenic Front-End," VLBA Technical Report No. 3, National Radio Astronomy Observatory, Charlottesville, VA, Dec. 18, 1986.
[5] G. Tomassetti, S. Weinreb, and K. Wellington, "Low-Noise, 10.7 GHz, Cooled GaAs FET Amplifier," Electronics Letters, vol. 17, no. 25/26, pp. 949-951, Dec. 10, 1981.

*The National Radio Astronomy Observatory is operated by Associated Universities, Inc. under contract with the National Science Foundation.

M. J. Reid and J. M. Moran (eds.), The Impact of VLBI on Astrophysics and Geophysics, 499–500.

TABLE I. VLBA Cryogenic Receivers

Freq. Range GHz	Amp. Noise K	Sys. Noise K	Status March 1987	Inp. WG Diam. cm
1.35-1.75	4*	22	3 complete	16.33
2.15-2.35	5	25	Under design	9.75
4.6-5.1	9	24	3 complete	4.490
5.9-6.4	10	24	No work	3.578
8.0-8.8	12*	29*	10 on VLA	2.602
10.2-11.2	14*	32	FET prototype	2.045
14.4-15.4	22	40	FET prototype	1.487
22.2-24.6	50*	77	8 on VLA	0.931
42.3-43.5	70	105	HEMT and SIS under devel.	0.526

*Measured result, all other figures are expectations.

Fig. 1. Photo of typical VLBA front-end. Width is 30 cm.

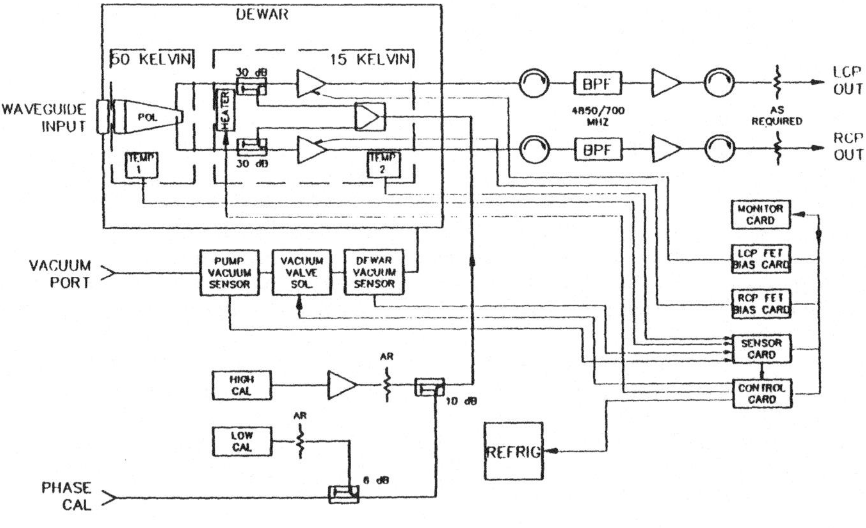

Fig. 2. Block diagram of typical VLBA front-end.

Mark IIIA and VLBA High-Density Recording

John C. Webber and Hans F. Hinteregger
M.I.T. Haystack Observatory, Westford, MA 01886, USA

ABSTRACT. The wideband Mark III recorder has been improved by the development at Haystack Observatory of a narrow-track recording system. Using 56 MHz bandwidth, more than 3 hours of VLBI data may be recorded on one tape. The same headstacks with new, thinner "digital video" tape allow recording at 128 Mbit/sec for 16 hours on one reel for the VLBA. New electronics for control, signal switching, and record/playback have been designed. Up to 2 Gbit/sec could be recorded by modular expansion.

1. Introduction

Haystack Observatory under NASA sponsorship has developed a narrow-track headstack which, using standard video tape, allows up to a 16-fold improvement in the area density of bits written with the Mark III VLBI wideband recorder. This permits operation with correspondingly fewer tape changes and greatly reduced costs of handling and shipping. The groundwork has also been layed for further improvements, including the recording system for the VLBA.

2. The Mark IIIA System

The Mark III wideband recorder writes 28 tracks of data at rates up to 8 Mbit/sec/track longitudinally on 1-inch videotape. With modern tape, the signal-to-noise ratio is much higher than required for good performance in a VLBI system. Haystack therefore undertook to increase the bit density by retrofitting with narrow-track headstacks and positioner. The headstacks were originally to be developed commercially, but failure after two years forced Haystack to undertake the development and participate in the manufacture of these critical components.

The crucial head tip trimming operation is performed at Haystack using a dicing saw custom-fitted with a laser interferometer position readout, a 2000× microscope for edge measurement, and intimate computer control. Ferrite "gapped bars" are precision-trimmed to an average edge placement accuracy of 0.5 μ; the tips are only 38 μ wide. The head tip array is fastened to a base containing the ferrite coils which complete the magnetic circuits.

M. J. Reid and J. M. Moran (eds.), The Impact of VLBI on Astrophysics and Geophysics, 501–502.

In addition, a precision positioning mechanism has been built to move the headstack between passes and control the position to better than 1 μ accuracy. This incorporates a piezoelectric motor, a linear variable differential transformer position sensor, and the necessary electronics. A headblock containing the headstack, actuator, sensor, and intimate interface electronics substitutes for the fixed wide-track Honeywell head assembly. The positioner and new, electrically cleaner power supplies and shields mount in the tape transport rack. This upgrade constitutes a "Mark IIIA" recorder.

This system typically provides >25 dB SNR at band-edge frequency for a Mark IIIA recording. Typical error performance is $< 3 \times 10^{-4}$ fraction of bits discarded due to synchronization loss, and a residual bit error rate of 5×10^{-5}, which is quite satisfactory for VLBI. The integrity of the VLBI scientific results is not easily compromised even when performance is seriously degraded by tape defects, clogged heads, electronic interference, *etc.*

3. The VLBA system

The same head and tape technology is being applied by Haystack under NRAO sponsorship to meet VLBA requirements. These include 100 Mbit/sec average record rate and 24-hour unattended recording. The VLBA recorder has all 36 heads connected instead of the 28 used by Mark IIIA, providing 32 data tracks and 4 system tracks which may substitute for dead tracks or record transverse parity. The VLBA recorder uses a new VME-based controller using a 68000 microprocessor for controlling transport, headstacks, and signal switching. The analog data electronics are new designs. The VLBA recorder must be reliable; MTBF $> 10^4$ hours appears achievable. The flexible formatting and decoding of the VLBA system will allow compatibility of the recordings made on either system in Mark IIIA mode.

The VLBA will probably use the new "digital video" tape which permits 50% higher longitudinal bit density and is half the present standard thickness. Packaged on a 16-inch reel, one volume will hold 8 *km* of tape and run 16 hours at 128 Mbit/sec. Each site will have 2 recorders, providing a maximum recording rate of 512 Mbit/sec. The system is expandable in a modular fashion and with additional headstacks and modules could record up to 2 Gbit/sec.

Even higher densities may be achieved by using "metal-in-gap" heads and metal particle tape. 12-hour continuous recordings at 1 Gbit/sec on one reel is projected if this new consumer head and tape technology is adapted for the VLBA.

4. Acknowledgements

We thank the more than 20 Haystack staff who have worked on this project, particularly J. W. Crowley, P. G. Bolis, J. I. Levine, P. M. Chizinski, H. L. Sellers, W. T. Petrachenko, and R. J. Cappallo.

THE MARK IIIA CORRELATOR SYSTEM

Alan R. Whitney
MIT Haystack Observatory
Westford, MA 01886
USA

ABSTRACT. The Mark IIIA correlator system, developed at Haystack Observatory, is a second-generation version of the original Mark III correlator which adds significant new capabilities, such as double-speed operation, longer integration periods, and improved internal modelling. One Mark IIIA correlator is currently operational at the U.S. Naval Observatory in Washington, D.C., and another is being readied to replace the Mark III correlator at Haystack.

The Mark IIIA correlator can simultaneously process up to 10 baselines of 24-MHz BW data from 5 stations. Experiments including more than 5 stations may be processed with multiple passes through the correlator. The architecture of the correlator allows a future expansion to a maximum of 16 stations. Due to a simple modular design and the low cost of required computer-support equipment, expansion is straightforward and relatively economical.

Both astronomy and geodetic data processing are supported by the Mark IIIA system, including spectral line and pulsar gating. In addition, recent work has led to support of space-based VLBI observations; this capability was critical in supporting the first successful space-based VLBI astronomy observations between ground-based antennas and an orbiting satellite in August 1986.

1. INTRODUCTION

In 1984, four U.S. government agencies (NASA, NGS, USNO, NRL) contracted with Haystack Observatory to develop and build a next-generation Mark III correlator to replace the original Mark III correlator in service since 1979. This correlator, dubbed the Mark IIIA, was to have significantly improved capabilities in several areas, primarily in modelling and data-throughput improvements. This new correlator system was initially installed at the USNO in January 1986, and has since been expanded to 5-station operation.

M. J. Reid and J. M. Moran (eds.), The Impact of VLBI on Astrophysics and Geophysics, 503–504.

2. SYSTEM DESCRIPTION

The Mark IIIA processor, like the Mark III processor, is based on a straight-forward modular concept utilizing one hardware correlator module per baseline track-pair, where a tape track contains the data from a single RF "window". Thus 14 modules are required to process one baseline of 14 tracks. Each module operates as a completely autonomous unit, but all are controlled by a common overseeing computer (HP1000 minicomputer). The module performs the functions of de-coding the two data streams (one from each station forming the baseline), buffering and correctly aligning the data streams to remove the effects of recorder jitter. Each module can process continuous data streams up to 8 Msamples/second.

Each Mark IIIA correlator module incorporates an internal microprocessor to maintain a high-precision a priori model for several seconds, even under conditions of very high delay rates or acceleration. Sample-by-sample fringe-rotation and cross-correlation is performed over a specified number of lags (up to 32/module) for each of a number of several-second integration periods over the duration of an observation. The correlation results are then communicated to the control computer, which performs the necessary Fourier transforms to create cross-spectral and delay functions for each integration period. Data from all integration periods are then combined to make estimates of various observables (correlation amplitude, phase, group delay, phase-delay rate, etc.). Special algorithms are used for bandwidth-synthesis measurements, which typically span RF bandwidths of several hundred Megahertz.

3. CAPABILITIES

Up to 5-stations (10 baselines) of 14-track data may be correlated through the ensemble of 140 identical correlator modules. Software-selectable signal-path selection allows considerable freedom with regard to configuring the correlator for processing other types of data. For example, up to 5 baselines of 28-track data may be processed sumultaneously; or all polarization combinations from a 3-station 14-track (7 track/polarization) may be processed.

For spectral-line processing, 4480 available lags (140 modules times 32 lags/module) may be divided among one or more baselines as desired.

References:

1. T.A. Clark et al, 'Precision Geodesy Using the Mark-III Very Long Baseline Interferometer System", IEEE Transactions on Geoscience and Remote Sensing, GE-23,438-449 (1985).
2. A.R. Whitney, 'The Mark III Correlator: Design, Operation, and Productivity', Proceedings of IAG Symposium No. 5: Geodetic Applications of Radio Interferometry, 152-162 (1982).
3. G.S. Levy et al, "Very Long Baseline Interferometric Observations Made with an Orbiting Radio Telescope', Science, 234,187-189 (1986).
4. A.R. Whitney, "Capabilities and Performance of the Upgraded Mark III Correlator System", Proceeding of IAU Symposium 128: The Earth's Rotation and Reference Frame for Geodesy and Geodynamics, in press.

HYDROGEN MASER SUPPORT OF VLBI FOR THE NASA CRUSTAL DYNAMICS PROJECT

R. E. Price, M. J. Chandler, B. R. Schupler and P. R. Dachel
Bendix Field Engineering Corporation
One Bendix Road
Columbia, Maryland 21045
USA

For the NASA Crustal Dynamics Project VLBI network, signals from quasars are recorded simultaneously at widely separated antennas. It is well known that hydrogen maser frequency standards provide the stable frequency reference used to precisely measure the difference in arrival time of the radio signals at the different antennas, enabling the determination of precise distances between the antennas. This paper reviews the practical requirements for maser support of VLBI for the Crustal Dynamics Project and describes the means used to meet these requirements for a network of eight fixed and three mobile stations which participate in approximately 200 VLBI experiments per year at locations in North America and the Pacific.

The masers must provide a stable reference signal with frequency stability better than 1E-14 for tau's of 100 to 1000 seconds. The masers used in the Project have a stability of approximately 2E-15 for these tau's.

The extensive schedule of observations demands exceptional reliability of the network of hydrogen masers. Each maser must have a long mean time between failures and short down times for preventive maintenance and any component failures. The NR masers were developed to fulfill those needs, and BFEC has established an extensive system of communications, data acquisition and analysis, design, and logistics support to ensure this performance.

The NR maser is equipped with a microprocessor which actively monitors 64 voltage, current, temperature and signal levels. Failures are often anticipated and prevented by thorough monitoring and analysis of important performance parameters over time. For data acquisition and analysis, BFEC has established a comprehensive network of communications with the masers. At the stations, data is retrieved by a small computer through a maser RS-232 port and then relayed to BFEC through electronic mail or direct phone lines. In the mobile VLBI data systems, a lap-top computer is used for data retrieval and then taken to a nearby telephone to access electronic mail.

With data for each maser taken an average of twice a week, a large data base has been compiled over the past four years. With such an extensive history on eleven NR masers, many failures can be predicted

M. J. Reid and J. M. Moran (eds.), The Impact of VLBI on Astrophysics and Geophysics, 505–506.

and nearly all have been prevented from occurring during VLBI operations.

The frequency performance of a maser is vital in determining the quality of the standard and also in the early detection of problems. To compare the remote masers with a common reference, time position measurements are made relative to the Global Positioning System. On the mobile sites and at fixed stations, the portable computer utilized for maser data transmission receives a digital output from a GPS timing receiver. Averaged time positions are sent to BFEC headquarters via electronic mail and a data base is collected for each maser time position. From this data, frequency offsets for each maser are determined and corrected.

Frequency offsets can develop as a result of drifts in a maser's cavity thermal control point due to aging of components in the control circuitry. These aging rates cause a second order time component, a change in the maser operating frequency over time. In addition, the frequency stability of the maser degrades as changes in the thermal control point cause small changes in the physical size of the cavity. The NR maser can be programmed to offset the cavity drift by controlling the cavity heater current over time. The cavity drift of a maser must be characterized through frequent flux-tuning with a second maser as reference. Cavity tuning is performed in the BFEC laboratory and can be done elsewhere when two of more masers are located at a common site.

Cavity tuning and characterization, as well as frequency stability measurements, are performed when the mobile VLBI systems are stationed at the Mojave Base Station. The frequency difference measurements utilize a measurement system contained within each NR maser, and the hydrogen flux variations and measurements are controlled and recorded by the masers' microprocessors. This configuration is also used to measure the masers' hydrogen line quality factor (line Q). The variation in line Q for each maser is studied to detect deterioration, if any, over the years. Masers with a line Q of less than 1.0 E9 have been returned to the BFEC laboratory for improvement.

All efforts are made to assure optimum maser performance through remote data monitoring coupled with preventive maintenance. But, in addition to thorough analysis and prevention, there is a need for rapid response hydrogen maser support. The Crustal Dynamics Project requires that BFEC "fix 95% of the maser problems within 72 hours or less" for a network of eleven observatories in North America and the Pacific. Rapid response support is achieved through communications, logistics, and maintenance techniques developed over the past five years. Maser problems are answered by phone and electronic mail seven days a week, 24 hours a day. More serious problems prompt rapid shipment of replacement components or field maintenance. When necessary, a replacement maser is shipped out with all systems operating. An extensive network of air cargo and truck transport arrangements has been formed to handle maser shipments on short notice.

HYDROGEN MASER CLOCKS AND DEEP SPACE SYSTEMS INVOLVING CLOCKS:
Ultra-Long Baseline Interferometry Definition of an Inertial Frame, Searches for Pulsed Gravitational Radiation

R.F.C. Vessot, Center for Astrophysics, 60 Garden Street,
Cambridge, MA 02138

The present status of research and development in atomic hydrogen maser clocks gives stability of 3×10^{-16} in $\Delta f/f$ for intervals of several hours. Long-term drift rates are in the order of 1×10^{-15}/day and uninterrupted operating lifetimes of more than 4 years are fairly common with SAO's VLG-series masers. Experiments with improved techniques for separating atoms in the desired state confirm that stability can be achieved with existing technology at levels below 1×10^{-16}. A four-fold improvement beyond this level appears feasible with more smoothly applied storage volume coatings of fluorinated materials. A recent breakthrough in technology has been made by SAO/Harvard, M.I.T. and, University of British Columbia, by operating hydrogen masers at 0.4K using storage coatings of superfluid liquid helium. This promises stability better than 1×10^{-17} in $\Delta f/f$ for intervals longer than 30 seconds. Features of a hydrogen maser operating at room temperature, developed at SAO for long-term ($\sim$7 years) unattended operation in space, are described. Possible applications of such masers with stability $\sim1\times10^{-16}$ for intervals 10^4-10^5 seconds in a deep space network of 4 (or more) radio astronomy antennas interconnected by time correlated one-way and two-way Doppler systems are discussed. By comparing the phase of signals propagated in both senses around the four (or more) closed triangular paths, changes in the magnitude and direction of the baselines can be determined with reference to inertial space defined by the constancy and isotropy of the velocity of light. The system is also sensitive to pulsed gravitational radiation at wavelengths less than the inter-terminal distances. Sensitivity is limited by the stability of the clocks and the fluctuations of ionization in the paths between stations. Multiple station time correlation of Doppler fluctuation in the signals between the six (or more) inter-terminal paths can be used to determine the direction and polarization of the pulsed gravitational radiation.

M. J. Reid and J. M. Moran (eds.), The Impact of VLBI on Astrophysics and Geophysics, 507.

DEVELOPMENTS IN CONTINUUM IMAGING

P.N. Wilkinson and J. Conway
University of Manchester, NRAL, Jodrell Bank

J. Biretta
Harvard-Smithsonian Center for Astrophysics

ABSTRACT. We review recent progress in the practical aspects of making VLBI images from continuum data.

1. INTRODUCTION

Aperture synthesis with intercontinental baselines came of age with the publication of the Wilkinson *et al.* (1977) map of 3C147. The subsequent development of 'hybrid mapping' or 'self-calibration' was rapid and by 1981 it had largely reached its present, widely used, form. Several accounts of this development are now available (Ekers 1983; Cornwell and Wilkinson 1984; Pearson and Readhead 1984). However innovations are still being made, most notably in the 'difference-mapping' technique, a brief outline of which is given by Muxlow *et al.* (this volume).

There were very few presentations of fits to data or even of (u, v) curves during this meeting, clearly VLBI arrays can now be regarded as tools with which to do astrophysics. The number of resolution elements in the images and their dynamic range has increased by more than 100 since the early days, and as a result of all this the astronomical output has mushroomed. It is easy to forget how much we now take for granted about central engines, beams, bulk relativistic motion etc. which would still be mere conjecture if VLBI images had not been available.

Software developments have played only a part in this success story. The development of the U.S. and European Networks and their joint scheduling, the accessibility of the correlators, and the widespread availability of the latest data reduction programs have all played a role. And last, but not least, MERLIN has blazed the trail ahead, showing how much can be achieved with a well-understood sparse array.

It goes without saying that there are many astronomical reasons why we should try to achieve images of ever greater resolution, sensitivity and overall quality. The astrophysical payoff has only just begun.

2. DATA ACQUISITION

The quality of the images depends fundamentally on the quality of the original data. As one of us has outlined elsewhere (Wilkinson 1987) there is no reason why

M. J. Reid and J. M. Moran (eds.), The Impact of VLBI on Astrophysics and Geophysics, 509–518.

images of the sensitivity and quality now produced by the VLA should not be achieved by VLBI arrays in the 1990's. The major problems involved are to acquire data good enough (residual phase errors $< 0\overset{\circ}{.}01$, residual amplitude errors $< 0.02\%$) to yield a dynamic range of $10^5 : 1$; and, for weak sources, to calibrate the phase accurately (see section 3.6). To achieve the quality goals many current sources of error will have to be overcome.

2.1 Current Problems

Primary amplitude calibration is poor. Even at 1.6 GHz the initial calibration of an array is never better than a few percent rms (see also Jones *et al.* 1986). At other frequencies errors of tens of percent are quite common. Not enough systematic investigations have been made (or at least reported) of calibration errors and how time-variable they are. Clearly they must be affecting the quality of our images, but it is hard to say by how much.

There are baseline-related offsets of several percent in amplitude and several degrees of phase in many data sets. Such effects are the enemy of self-calibration algorithms (Wilkinson 1983; Perley 1986) and there are many ways in which they can arise (see *e.g.* Cornwell 1986a); specific to VLBI is poor recording and playback. With Mk2 systems significant amplitude errors can certainly be introduced in this way although the situation is improving as new correlators come into use (S. Unwin, private communication). We do not know the facts about Mk3 recording/playback repeatability. Excellent quality control throughout the data transmission path is a *sine qua non* if our $10^5 : 1$ goal is to be achieved.

The final passband-defining filters must be well-matched or baseline offsets will occur (see *e.g.* Thompson 1980). In the recent 'world array' observations the amplitude and phase responses of the final 1.8 MHz video filters showed a wide variation in performance relative to the ideal. Calculations of the effect on the correlation coefficients (J. Benson, private communication) showed baseline offsets ranging from 0% to 8% in amplitude and 0° to 13° of phase, with the errors predominantly at the low end of these ranges. We compared these calculated offsets with those determined empirically (see below) from the "world array" data on M87 (W. Junor and T. Muxlow, private communication). There is a reasonable correlation between the phase offsets but no correlation between the amplitude offsets - it appears that non-matched passbands are only part of the problem in this data set.

Polarisation impurity in the front-ends is another potential source of baseline-related errors. If LCP channels from the two antennas contributing to a baseline contain an RCP impurity *i.e.* $L_1 = L_1 + aR_1$ and $L_2 = L_2 + bR_2$ then $L_1L_2 = I(1 + ab) + mI(a + b)$. If the impurities are very bad *i.e.* $a = b = 0.1$ and the compact source is overall 10% polarised (as is commonly observed, see the review by Wardle in this volume) then $\sim 1\%$ complex gain errors are introduced per baseline. These will vary with parallactic angle and hence with time. This represents very much a worst case but this effect may become important as other sources of error are eliminated.

Finally not much has been reported about hardware problems (*e.g.* Wilkinson 1983; Cornwell 1986a) which may give rise to offsets in VLBI correlators. Overall we stress that much more attention should be given to identifying and eliminating sources of non-closing errors in VLBI data.

3. DATA CORRECTION

After correlation many stages of data correction are needed before the final map is produced. Typically this now involves delay/rate estimation (global fringe fitting), data editing, self-calibration and a final deconvolution step. Some observers even iterate around this loop. A case study has recently been presented by Jones *et al.* (1986).

3.1 Global Fringe Fitting

The idea of estimating residual fringe rate and delay offsets on a station-by-station basis was introduced by Schwab and Cotton (1983, SC). More recently Alef and Porcas (1986, AP) have described a simpler approach which may yield comparable results in practice. These algorithms are intended to overcome the problem of non-closing delay estimates, and hence non-closing phase errors, which arise when fringe fitting is performed on a baseline-by-baseline basis as was the case for the first 17 years of VLBI.

Utilising current algorithms to their best advantage is, by common consent, quite tricky. There is no simple recipe which works under all circumstances given the several different timescales and signal-to-noise ratios which are inherent to the problem. A good introduction to the practical aspects of using the SC algorithm has been given by Walker (1986), while Jones *et al.* (1986) also pass on some useful experience. M. Cohen (private communication) has compiled (and is continually updating) a "users guide" to the AIPS implementation of the SC algorithm.

3.2 Data Editing

For high dynamic range work one should ruthlessly discard bad data including whole baselines where there is no signal to be seen above the $\sim 1\sigma$ level. Such data seem only to add unwanted noise to the image (see for example Jones *et al.* 1986, although other observers have come to the same conclusion). Ruthless editing means point-by-point editing on each baseline and not merely "global" editing via, for example, "UVSUB" and "CLIP" in AIPS package. The latter approach is clearly inadequate. R.C. Walker (private communication) reports that the inclusion of data from UTs where there is no closure information (typically at the start and finish of intercontinental runs) can also degrade the final image.

3.3 Self-calibration

Self-calibration is basically the reason for the success of VLBI imaging. Useful practical discussions regarding its use have recently been given by Cornwell (1986a) (in a VLA context) and by Walker (1986). Despite its success there are circumstances in which imprudent application of the method can lead to an imperfect image—even to the extent of inventing spurious components and/or eliminating real components. However we stress that such cases are a) rare and b) invariably associated with data sets possessing limited closure information or low signal-to-noise ratio or large amplitude errors or a combination of all three.

Spurious symmetrisation can occur in a one-sided source if the starting model is symmetric (*e.g.* a point source). Usually the algorithm will converge to the correct solution (see for example the test presented by Readhead and Wilkinson 1978) but one should always be suspicious of "ghost" components on the opposite side of a bright core. Linfield (1986) points out that symmetrisation is most likely to be a problem when the array is not well-mixed, for example if it contains an outrigger antenna. Linfield describes a lengthy procedure to help convergence in tricky cases, however it may by quicker to use a different starting model. The "difference mapping" technique is ideal for rejecting suspicious components and hence overcoming this particular problem.

Poorly-mixed arrays can also cause problems with amplitude self-calibration. Since the amplitude of the visibility function falls off as a function of baseline length, part of the true structure can be "telescope factorisable" and it is possible to "lose" real flux from extended components by injudicious (too early, too vigorous) amplitude correction. Always examine the fit to the *observed* data as well as to the corrected data after the "final" map has been obtained. If the fit on the short baselines is systematically low the intermediate maps have been insufficiently cleaned and hence the trial maps used in the data correction stage do not contain enough flux. If the array is sparse and poorly mixed the algorithm can systematically "correct" (downwards) the amplitude scale on telescopes involved in shorter baselines. A simple trick to preserve the *overall* flux scale is to rescale the mean of the amplitude correction factors to unity before each data correction stage.

Amplitude correction is notoriously more tricky than phase correction. Excluding negative image components always helps to correct phase errors (antisymmetric) but is less effective in reducing amplitude errors (symmetric) which can be largely positive. Careful image plane windowing can drive the algorithm towards the correct solution but great care is needed if the windows are the strongest constraints. As an example, the data from MERLIN at 151 MHz can be badly affected by radio frequency interference which in turn causes major (tens of percent) amplitude calibration problems. Because MERLIN is a sparse, relatively ill-mixed array, obtaining good maps at 151 MHz demands great care on the part of the astronomer, even though the source is apparently simple enough for the method to converge easily when the amplitude errors are small. However, many reliable MERLIN maps have been made at 151 MHz using the "difference mapping" technique.

3.4 Removal of baseline-related errors

Empirical correction of simple baseline-related offsets is now becoming common. The transform of the "almost-final" image is compared with the self-calibrated data; the offsets in amplitude and phase are summed for each baseline and the mean subtracted from the data. This is an inherently dangerous procedure which can be justified pragmatically by the reduction of spurious sidelobes after application. It should be used very sparingly and it is obviously crucial that the "almost final" image be a good representation of the true source.

3.5 Example

In Figs. 1, 2, and 3, we show images made by one of us (JB) at different stages of the data correction cycle. The data are those from the 'world-array' at 1.66 GHz on M87 (see also Biretta *et al.*, this volume). The data reduction was performed mostly with the NRAO AIPS package, save for the detailed editing, for which the CIT VLBI package was used.

Fig. 1 shows the result of a standard ASCAL+MX iteration where the data were weighted by the square root of the standard weights. (There is conflicting opinion about the usefulness of such weights). The dynamic range in Fig. 1 is 800:1. To produce the image in Fig. 2, negative CLEAN components were deleted prior to self-calibration, the normal AIPS weights were used and smaller windows were employed in CLEAN. The dynamic range in Fig. 2 is 1500:1. Finally, to produce Fig. 3 the baseline-related errors were removed by the empirical method described above. The effect on the north-south sidelobe level is clear and the dynamic range at this stage of the reconstruction is 2300:1. The noise level in Fig. 3 is $\sim$ 0.4 mJy/beam, and is roughly equal to that expected from thermal noise (*cf.* Wilkinson 1983).

3.6 Phase referencing

For self-calibration the signal-to-noise ratio on individual baselines must typically be of order unity or greater in a coherence time. This places a lower limit on the flux density of the target source even if it is compact. The weakest sources so far imaged with the EVN using fringe-fitting (AP) and self-calibration have peak flux densities of $\sim$ 6 mJy per synthesised beam (with the thermal noise level ranging from $<$ 1 mJy to $\sim$ 10 mJy in a 60 second integration depending on the baseline). Using a nearby source to determine the telescope-related errors is becoming possible as sensitivities improve to the extent that suitable reference sources (*e.g.* flat spectrum, $S >$ 100 mJy at 5 GHz) can be found within a few degrees of any target source. Alef and Lestrade *et al.* (this volume) indicate what is currently being achieved. Note that neither do the source positions need to be known perfectly, nor does the reference source need to be unresolved for the technique to be useful for structure work (*e.g.* Wilkinson 1983). One can expect to see many useful images of mJy sources in the next five years.

4. DECONVOLUTION

Non-linear deconvolution algorithms (see *e.g.* Cornwell 1986b) attempt to interpolate across the unmeasured parts of the (u, v) plane or, equivalently to suppress the sidelobes in the 'principal' solution. As has been pointed out many times (*e.g.* Wilkinson 1983, 1987; Dulk *et al.* 1984, Perley 1986) the on-source fidelity of the resulting image can be very much less than suggested by the dynamic range. The tests described by Dulk *et al.* and Readhead (1984) show that the on-source errors induced by CLEAN can be $>$ 10 times worse than the thermal noise level especially on and around a bright core. Note that no work has been done in quantifying image fidelity when using self-calibration. There will inevitably be an interaction between deconvolution-induced artefacts and the determination of the complex gains.

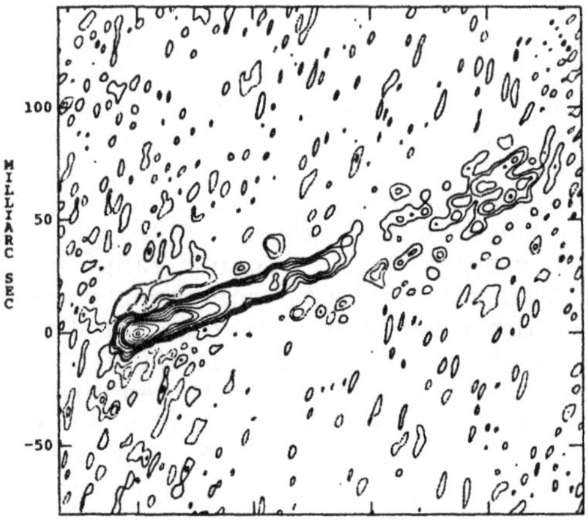

Fig.1 M87 at 1.6 GHz

(for explanation see section 3.5)

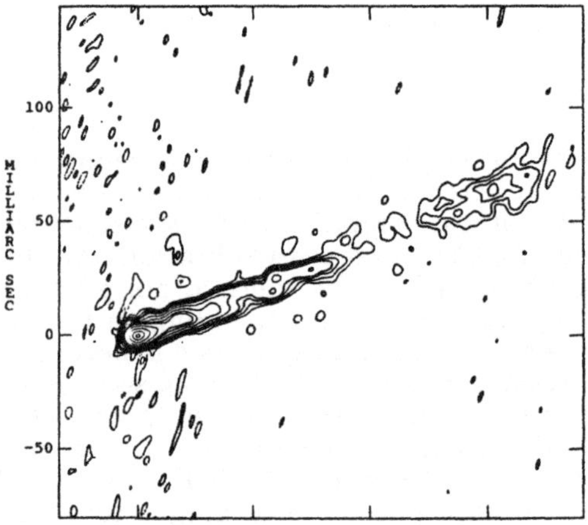

Fig.2

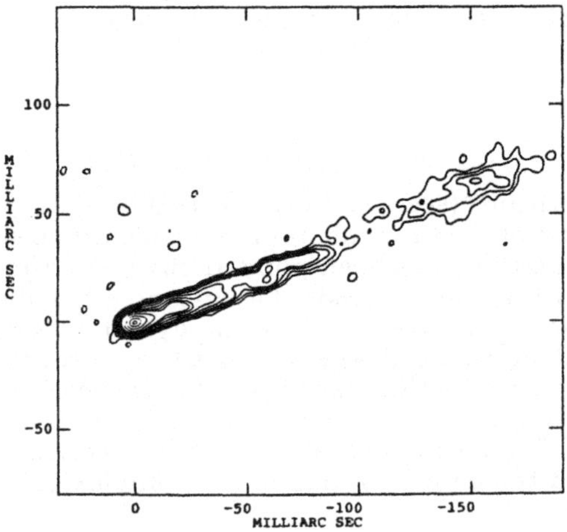

Fig.3

Recently the MEM algorithm, of Cornwell and Evans (1985) has been used to reconstruct VLBI images from the 'world array' (see ***e.g.*** Muxlow *et al.* this volume) ***after*** self-calibration (it is too slow to use in the loop) and subjectively superior images to those produced by CLEAN have resulted. We expect that, in line with VLA practice, steadily more use will be made of MEM for this final deconvolution step. For current VLBI data sets it can be tricky to get the algorithm to converge reliably (see Cornwell 1986b). "Pre-convolution" of the dirty map, removal of strong point sources with CLEAN, and a good knowledge of the expected noise level in the outer parts of the map are often required (T. Muxlow, private communication).

5. THE STATE OF THE ART

The weakest source for which a true image (by which we mean one utilising phase information) has been made is probably the nucleus of NGC 4151 (total flux density $\sim$ 15 mJy; Preuss *et al.* 1987). Like many of the images of weak objects now being produced this was made with the large EVN telescopes and Mk3 recording terminals; it still involved self-calibration and not phase-referencing. As far as dynamic range is concerned the current record is probably the $\approx$ 8000 : 1 (rms noise to peak flux) achieved in the map of 3C273 made by Unwin and Davis (this volume). The most complex images made so far are certainly those resulting from the 18 station 'world array' (M87: Muxlow *et al.* and Biretta *et al.*; 3C236: Schilizzi *et al.*; 3C120: Walker *et al.*, all in this volume). However the maps of 3C236, for example, are rather similar in quality to the 6 station MERLIN map of 3C249.1 at 408 MHz produced in 1981 (Lonsdale and Morison 1981). There is still much untapped potential in current VLBI arrays. Higher quality data is the key to releasing it.

6. MULTI FREQUENCY SYNTHESIS

The only way to improve the coverage of the aperture plane for a given VLBI array is to observe at different frequencies. This idea has already been used at the VLA (*e.g.* Braun *et al.* 1987) to improve the image of Cas A. Even for arrays consisting of 8 stations observations in different frequency bands, covering a range $\Delta\nu = \pm 10\%$ often fill in the synthesised aperture very satisfactorily.

The fact that the source structure is different in each frequency channel has so far been ignored. However, following Cornwell's (1984) pioneering analysis of the problem, a new technique has been developed which takes account of spectral variations if the intensity follows a power law at each pixel (Conway, Cornwell and Wilkinson, in preparation). To a good approximation the "dirty map" (DM) derived from multi-frequency data is a superposition of two distributions. One of them is the true distribution at a specific "reference frequency" ν_0, convolved with a "composite dirty beam" obtained in the usual way from all the (u, v) points assuming that there are no differential spectral index variations across the face of the source. The second distribution contains the effects of these variations. It is the true distribution (weighted by a differential spectral index at each pixel) convolved with a "spectral dirty beam" which is the Fourier transform of the aperture coverage but now with each (u, v) point weighted by $(\Delta\nu/\nu_0)$. These weights are positive or negative depending on the sign of $\Delta\nu$.

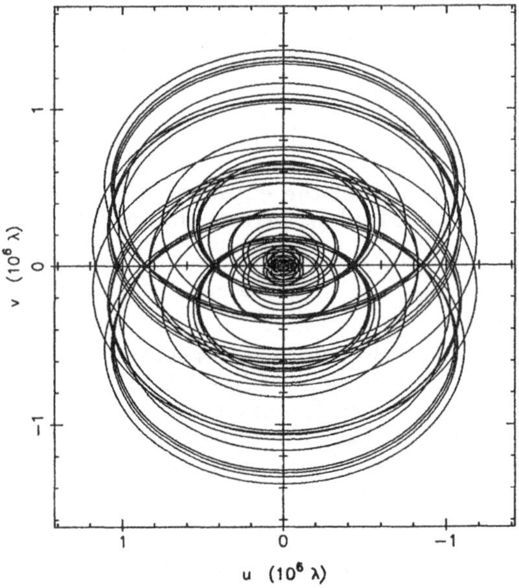

Fig.4

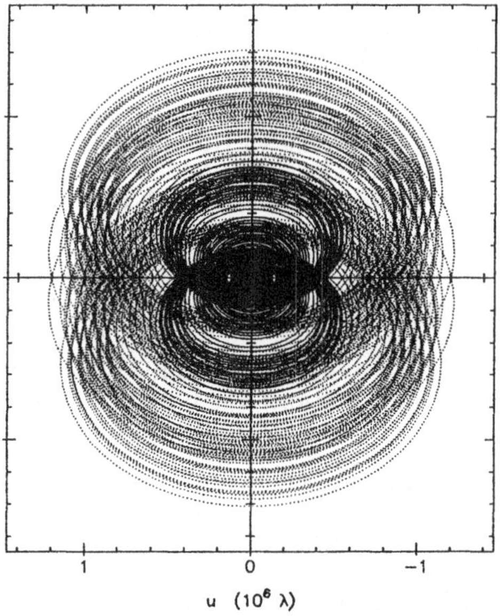

Fig.5

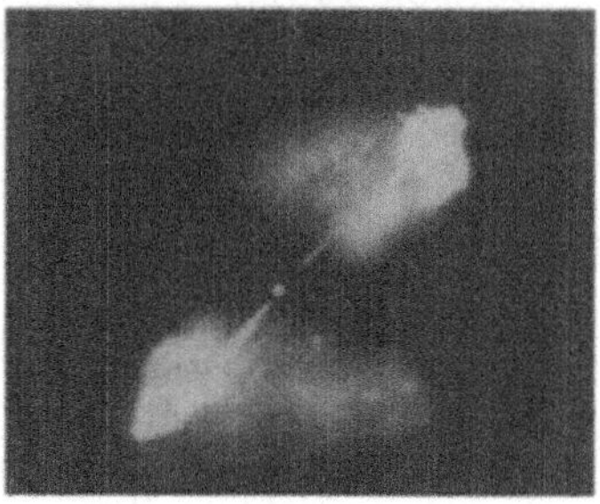

Fig.6

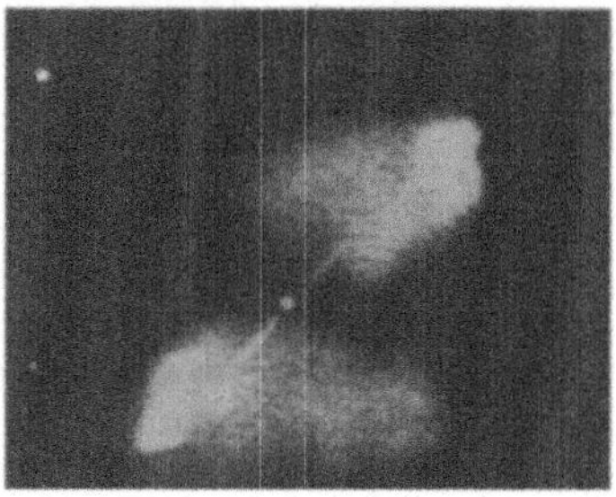

Fig.7

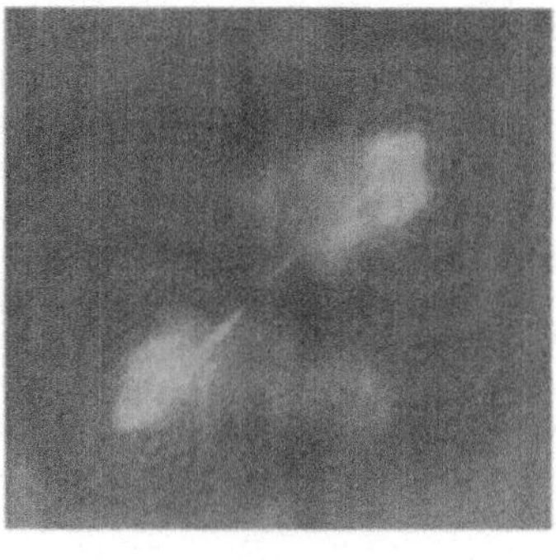

Fig.8

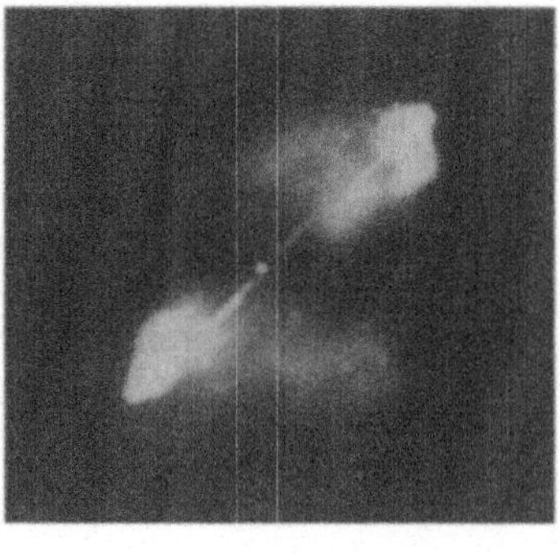

Fig.9

One of us (JEC) has developed a scheme, dubbed "double deconvolution," to separate out these two distributions. Here we show the results of one of the tests which have been run to prove the method. Fig. 4 show the aperture plane coverage of MERLIN at $\delta = 50°$ as it would be if the current 6 telescope array were enhanced by the addition of telescopes in Cambridge and Chilbolton. Fig. 5 shows the greatly improved coverage which would result from observing in 5 separate bands over a $\pm 12\%$ range in frequency (*viz.* 1350 to 1710 MHz). The effect is similar to increasing the number of MERLIN telescopes to 17. Fig. 6 shows a model intensity distribution; not shown is the assumed spectral index distribution which varies from -0.5 in the "hot spots" to -2 in the "lobes." Fig. 7 shows the reconstruction of the source from simulated data at only a single frequency—note in passing the obvious CLEAN artefacts in the faint "lobes" and the subsequently poor on-source fidelity in these regions (*cf.* section 4). Fig. 8 shows the effect of using five-frequency data without taking account of the spectral index variations— note that the clear "spectral sidelobes" are primarily associated with the "hot spots." Finally Fig. 9 shows the image after the application of double deconvolution—virtually all of the subtle details in the model are recovered. The technique appears to be a very promising way of improving VLBI maps in the next few years. The major problems will be amplitude calibration and radio frequency interference.

7. POSTSCRIPT

There is now an awareness that thermal-noise-limited imaging should be the norm in VLBI. To achieve it regularly more understanding is urgently needed regarding: i) baseline-related errors; ii) the utilisation of arrays in which the baselines have markedly different sensitivities; iii) the effect of poor initial amplitude calibration on an image—this is important for mm-wavelengths and we may need larger arrays than we think merely to ensure reliable self-calibration.

ACKNOWLEDGEMENTS

We thank the many people who supplied information for this review; in particular Walter Alef, Richard Porcas, Steve Unwin, Marshall Cohen, Craig Walker, John Benson, Tom Muxlow, Bill Junor, Paddy Leahy and Robin Conway.

REFERENCES

Alef, W. and Porcas, R. W., 1986 *Astr. Ap.*, **168**, 688.
Braun, R. *et al.* 1987. *Nature*, **327**, 395.
Cornwell, T. J., 1984. VLBA Array Memo. No. 324.
Cornwell, T. J. and Wilkinson, P. N., 1984. Proc. URSI/IAU Symp. 'Indirect Imaging' p.207. ed. J. A. Roberts Cambridge Univ. Press.
Cornwell, T. J. and Evans, K. F., 1985. *Astr. Ap.*, **143**, 77.
Cornwell, T. J., 1986a. Course notes from NRAO Summer School 1985 'Synthesis Imaging' p.137. ed. R. A. Perley, F. R. Schwab, A. H. Bridle.
Cornwell, T. J., 1986b. ibid, p.109.
Dulk *et al.* 1984. (see Cornwell and Wilkinson 1984) p.355.
Ekers, R. D., 1983. Proc. NRAO Workshop 'Serendipitous Discoveries in Radio-astronomy' p.154. ed. K. Kellermann and B. Sheets.
Jones, D., *et al.* 1986. *Ap. J.*, **305**, 684.
Linfield, R. P., 1986. *A. J.*, **92**, 213.
Lonsdale, C. J. and Morison, I., 1983. *M. N. R. A. S.*, **203**, 833.
Pearson, T. J. and Readhead, A. C. S., 1984. *Ann. Rev. Astr. Ap.*, **22**, 97.
Perley, R. A., 1986. (see Cornwell 1986a) p.161.
Preuss, E. *et al.* 1987, Proc. IAU Symp. No. 121, Byurakan, eds. E. Y. Khachikian, K. Fricke, J. Melnick, Reidel, Dordrecht.
Readhead, A. C. S. and Wilkinson, P. N., 1978. *Ap. J.*, **223**, 25.
Readhead, A. C. S., 1984. Proc. ESA Workshop on QUASAT, Gross Enzersdorf 1984. p.101, ESA SP-213.
Schwab, F. R. and Cotton, W., 1983 *A. J.*, **88**, 688.
Thompson, A. R., 1980. VLA Electronics Memo. No. 192.
Walker, R. C., 1986. (see Cornwell 1986a) p.189.
Wilkinson, P. N., *et al.* 1977. *Nature*, **269**, 764.
Wilkinson, P. N., 1983. Proc. Int. Conf. on VLBI Techniques, CNES Toulouse 1982, p.375, ed. F. Biraud, Editions Cepadues.
Wilkinson, P. N., 1987. Proc. Workshop on Superluminal Sources, eds. T. J. Pearson and A. Zensus, Cambridge Univ. Press.

MAPPING WITH A SINGLE-SPACING INTERFEROMETER

B. A. Doubinsky
Institute of Radioengineering and Electronics, U.S.S.R. Academy of Sciences
Moscow, U.S.S.R.

N. S. Kardashev
Space Research Institute, U.S.S.R. Academy of Sciences
Moscow, U.S.S.R.

Recent VLBI development provides opportunities for aperture synthesis, or super synthesis, with terrestrial interferometer antennas, and also in the near future with orbiting antennas. In this connection it is useful to consider the potential mapping capability of a single interferometer. Usually one would assume the use of image restoration procedures to be indispensable for observations with a single interferometer, because the *uv*-plane is covered only on an elliptical track, and is very much unfilled. High and distant sidelobes of the synthesized beam, caused by gaps in the coverage of the *uv*-plane, are usually considered to be influenced only to a small degree by any weighting function by which the visibility function is multiplied in the Fourier image formation algorithm. It has been shown that the mean-squared value of the point-source response function is invariant with respect to the weighting function (Doubinsky 1983). However, this does not prohibit the modulation of the point-source response by the visibility weighting function. This study shows that one can suppress the sidelobes in two symmetrical sectors of the point-source response with the help of such modulation. Thus for a line-shaped source model (that is, linear within the angular resolution of the synthesized beam) one could map without excessive distortion by Fourier transformation alone (Doubinsky 1979). The exclusion of restoration procedures, which are largely empirical, increases the reliability of the image when the source structure is extended and the signal-to-noise ratio is low. It also saves much computer time.

The linear form of the model is a significant restriction. Nevertheless, in radio astronomy there are many objects that appear to be linear: for example, a radio galaxy with fairly compact radio emitting regions symmetrically located on each side. Many unresolved sources may be shown to be of similar type by future VLBI observations. Such a one-dimensional image can be analyzed as a strip brightness distribution when the width of the *uv*-track in a direction orthogonal to the long axis of the source is small enough so that the source is not resolved in that direction.

As a simple illustration of the principles involved, consider a source that is linear in the east-west direction. Then the visibility is a function only of u, and in the v direction remains constant out to values beyond the projected spacings of the interferometer. If the source is observed continuously over a sufficient portion of the *uv*-track, the visibility is determined continuously as a function of u from zero to the maximum spacing in the u direction, with no gaps in the u-coverage. Then if the source is also well resolved by the maximum spacing in the u direction, the east-west profile can be accurately determined. The point-source response function,

M. J. Reid and J. M. Moran (eds.), The Impact of VLBI on Astrophysics and Geophysics, 519–520.

with which the source profile is convolved, can be controlled by introducing a visibility weighting function equal to the Fourier transform of the desired point-source response. The same considerations can be applied to a linear source at any position angle by appropriate rotation of axes. The linear-model analysis has been applied to the source 3C263, in which four components are widely spaced in a linear manner. The visibility data used were obtained at the Mullard Radio Astronomy Observatory, and the brightness profile derived is in strict agreement with the two-dimensional map of the source.

The more general case of a curvilinear source has also been considered. If the minimum radius of curvature of such a source is large compared with the width of the synthesized beam, then the source can be treated as a number of linear components.

In the case where the form of the source is initially unknown, a two-dimensional map can first be made by two-dimensional Fourier transformation of the visibility data. The resulting beam will contain large sidelobes, but the map should reveal whether any major component of the source is of linear form, and if such a component exists, the position angle can be estimated. The linear-model analysis can than be applied as a second step.

In conclusion, the authors have demonstrated a simplified approach to the analysis of interferometer observations of sources of linear form. This work has been stimulated by the prospect of increased synthesis mapping with the use of both terrestrial and space-based VLBI antennas. The authors wish to express their thanks to Dr. G. G. Pooley who kindly provided the data for 3C263 for two different interferometer baselines.

REFERENCES

Doubinsky, B. A. *Astronom. Letters*, 1979, **5(10)**, 557-560 (in Russian).
Doubinsky, B. A. *Radiophysics*, 1983, **26(11)**, 1493-1496 (in Russian).

GLOBAL FRINGE FITTING FOR SPACE INTERFEROMETRY

Dayton L. Jones
Mail Code 238-700
Jet Propulsion Laboratory
4800 Oak Grove Drive
Pasadena, CA. 91109

The technique of global fringe fitting has proven to be very valuable for very-long-baseline interferometry (VLBI). It allows weaker fringes to be detected than is possible with conventional single-baseline fringe fitting algorithms, and thus improves the sensitivity of present VLBI arrays. Global fitting was developed for VLBI by Schwab and Cotton (1983, *Astron. J.* **88**, 688), and has been incorporated into the NRAO Astronomical Image Processing System as program VBFIT. As Schwab and Cotton point out, this is a generalized from of self-calibration in which closure relations for residual fringe delay and rate (frequency) are satisfied in addition to those for fringe phase and amplitude. These additional closure relations allow station-dependent delays and rates to be solved for, using data from all baselines with sufficiently strong fringes. Then it is possible to calculate where fringes will be on any less sensitive baselines between these telescopes, so that very narrow search windows in delay and rate can be used.

A generalized version of VBFIT has been developed for use with VLBI arrays that include orbiting radio telescopes. This program, VBGFF, can accomodate the rapidly-changing geometric delays and fringe rates on baselines to orbiting telescopes. It requires at least one ground-space baseline with enough sensitivity to produce strong fringes in integration times which are shorter than the time it takes the actual baseline motion to differ from a smoothly-varying model of its motion through the (u, v) plane.

Let $\phi(i,j)$ be the source phase on the baseline between antennas i and j, which is initially unknown. To get around this problem, we require a model of the source brightness distribution from which the $\phi(i,j)$ can be calculated. As in normal hybrid mapping, a more accurate source model will improve convergence of the fringe search, although in practice a point source model is often sufficient.

The observed visibility phase is $P(i) - P(j) + \phi(i,j)$, where $P(i)$ is the phase error associated with antenna i. The partial derivative of the visibility phase with respect to frequency is the baseline delay $D(i,j)$ and the partial derivative with respect to time is the baseline fringe rate $R(i,j)$. The location of the maximum amplitude of the Fourier transform of the observed visibility function in delay-rate space gives the values of $D(i,j)$ and $R(i,j)$. Standard fringe fitting uses this method to find $D(i,j)$ and $R(i,j)$ for all i and j during each integration interval. The values of D and R for a given baseline

M. J. Reid and J. M. Moran (eds.), The Impact of VLBI on Astrophysics and Geophysics, 521–522.

are then used to correct the observed baseline phases for the effects of $P(i) - P(j)$ and hence allow coherent integrations of the visibility function for periods during which $\phi(i,j)$ remains nearly constant. This can be as long as several minutes.

The problem with this approach is that it only works on baselines where the Fourier transform of the observed visibility function has a large enough maximum amplitude. There is no way to use the $D(i,j)$ and $R(i,j)$ for other baselines to calculate D and R for a baseline on which fringes are very weak, since the D and R have not been separated into antenna-dependent $D(i), D(j), R(i)$, and $R(j)$. Global fringe searching solves for antenna based $P(i), D(i)$, and $R(i)$ using a set of $\phi(i,j)$ calculated from the source model. Once the antenna-based quantities have been determined (by least-squares, for example), they can be used to find fringes on all baselines.

The VBGFF program allows searching for fringes which vary over much wider ranges in delay and rate for some antennas than for others. This is done by setting different widths for the delay and rate search windows for the various antennas. This is useful when an array includes both ground and space based antennas, or when equipment problems or bad weather cause some ground based antennas to have less stable delay and rate solutions than others. The delay and rate search windows are defined by adding a "penalty function" to the quantity being minimized to estimate the antenna-based parameters. The penalty increases with increasing absolute value of the residual $R(i)$ and $D(i)$. VBGFF uses a more steeply-rising function than VBFIT does.

An additional modification to VBFIT allows different solution intervals to be used for the initial Fourier transform fringe search and the final least-squares search. Now one can use a relatively long time interval to determine initial values for the least squares part of the program, but use a shorter interval for the least squares fit so it can more accurately follow changes in the delays and rates. This avoids the need to run VBFIT twice – once to detect fringes using a very long solution interval and a second time to get more accurate telescope solutions with a shorter solution interval. Using a short solution interval can be particularly helpful for high frequency experiments, where bad weather can cause rapidly changing telescope solutions.

A final modification to VBFIT concerns the way in which the delay and rate solutions are edited or smoothed prior to being applied to the input data. The initial solutions found by the FFT search can be boxcar smoothed over an arbitrary time interval before being used to "center" the visibility data for a second fringe fit. The final solutions can still be edited and smoothed in the same way as in VBFIT before being applied.

Several limitations remain, including a requirement that search windows be centered at zero. This requirement will eventually be removed. In addition, it is assumed that antenna amplitude errors will remain constant during the period of the fringe search (up to several minutes), and that the amplitude of the source visibility will also remain constant during this period. These assumptions are not always valid.

This research was carried out by the Jet Propulsion Laboratory, California Institute of Technology, under contract with the National Aeronautics and Space Administration.

Test of Phase-reference Mapping for Switched Observations

Walter Alef
Max-Planck-Institut für Radioastronomie
Auf dem Hügel 69
D-5300 Bonn 1
West Germany

ABSTRACT. An observing scheme and software for single frequency switched phase-referencing observations have been developed and tested on a source pair separated by 0.5° (0248+430, 0249+436). The phase connection was done utilizing closure properties of phase and phase-rate. A comparison of the phase-reference map with a hybrid map shows the success of the method.

1. Motivation for Phase-reference Mapping

Self calibration algorithms for the interferometer visibility have proven to work well in the high signal-to-noise case. When the signal-to-noise is low, however, the interferometer phase cannot be determined unambiguously. And below a flux limit imposed by relatively short coherence times no signal can be detected at all.

To be able to map very weak radio sources we wanted to explore the feasibility and the limitations of phase-reference mapping for switched observations. Our aim was to develop a technique for single frequency observations, which would be easy to use, and could hopefully work without the need for additional atmospheric/ionospheric calibration measurements.

2. Switched Observation of 0248+430 and 0249+436

We conducted a first test of the phase-reference mapping technique for single frequency switched observations in the spring of 1984. The source pair 0248+430 (~1.2 Jy at 6 cm) and 0249+436 (~0.2 Jy at 6 cm) with a separation of 33′ was observed with the radio telescopes in Onsala, Effelsberg, Westerbork, Green Bank, and Owens Valley. We chose the observing frequency of 5 GHz in order to minimize the combined atmospheric and ionospheric effects. The data were recorded using the Mk3 system in mode E, so that one tape could record four 13-minute intervals of data.

In order to remove structural phase effects from the reference source 0248+430 we mapped it with high dynamic range (see fig. 1). The phases of the reference source were connected using closure properties of the phase and the phase rate to eliminate ambiguities. The remaining instrumental phase variations are due only to atmospheric and oscillator instabilities. Subtracting these from the visibilities of the weak source provided an interferometer signal, approximately free from phase corruptions, which contains information about the source structure and position only. The data were gridded, inverted, and CLEANed using standard mapping programs. The resulting phase-reference map is displayed in fig. 2.

M. J. Reid and J. M. Moran (eds.), The Impact of VLBI on Astrophysics and Geophysics, 523–524.

A comparison with a hybrid map of the same source (fig. 3) shows that both maps agree well down to the 10% level. The higher noise level in the phase-reference map and its seemingly more complex structure are due to additional phase noise which is introduced by interpolating the reference phase (errors of the order of ±30°), and systematic errors which are caused by inaccuracies of the correlator models for the interferometer geometry, atmosphere, earth rotation, etc. The largest systematic errors are introduced by not modelling the behaviour of the ionosphere and the troposphere, which show up as spurious structures in a phase-reference map. From the low level of such false components we can conclude that our phase-reference mapping test was not dominated by systematic phase errors. However, it is difficult to give an estimate of their magnitude. We believe that typical phase offsets are of the order of 30° to 50°.

0248+430 HYBRID MAP OF REFERENCE SOURCE

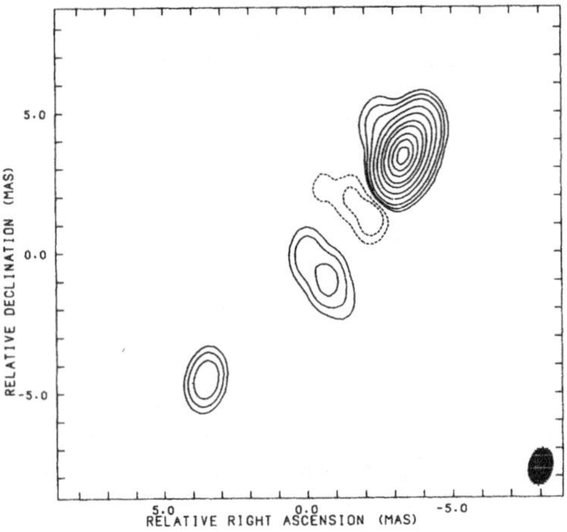

0249+436 HYBRID MAP OF SIGNAL SOURCE

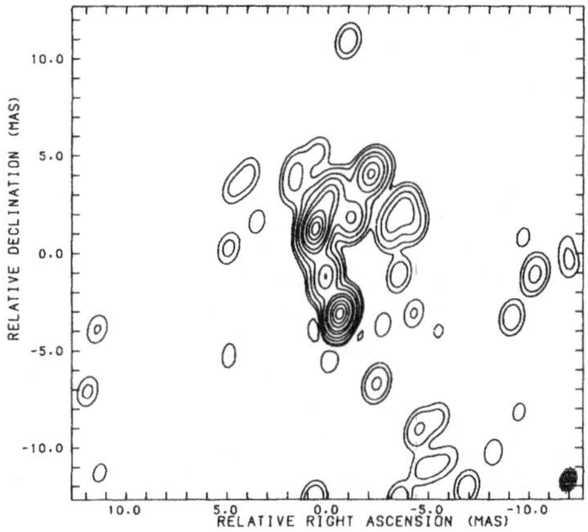

0249+436 PHASE-REFERENCE MAP

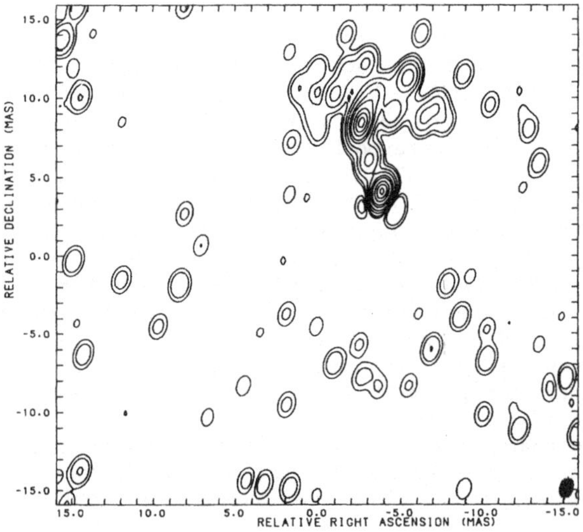

Contour levels:

In Reference source:
0.5, 1, 2, 5, 10, 20,
30, 50, 70, 90% of peak
0.64 Jy/beam

Signal source both maps:
1, 2, 5, 10, 20, 30,
50, 70, 90% of peak
0.031 Jy/beam

An Optimal Identification of Clock Behaviour Model for VLBI

Dawei Zheng and Shifang Luo
Shanghai Observatory, Academia Sinica

J.R.Mackay
National Geodetic Survey Division
Charting and Geodetic Survices
National Ocean Service

1. Introduction

The accuracy of VLBI measument has reached centimeters(Carter et al.,1985). But in VLBI observation the observable are broken sometimes because of the variations of the clock phases or frequencies. This is the first problem met in VLBI data processing.

Recently, the interactive mode is adopted to determine the intant of the clock discontinuities and the clock behaviour model of the seperate sections. By using this mode the results will not only be effected by personal cause (Robertson,1975), but also it takes a lot of computer time.

In this paper, we will discuss the estimates of clock models and the discontinuities in VLBI data processing by information criteria in order to improve the results and the procedure of VLBI data processing.

2. The Optimal Estimate Method of Clock Behaviour Model

According the physical character of atomic clock, the clock behaviour model in one session of VLBI can be described as

$$OCt=\sum_{j=1}^{L}(a_j + b_j t + c_j t^2) + \varepsilon_t^j \qquad (1)$$

where t is the argument of time, ε_t^j is the noise in clock. When L=1, it means that the clock has no discontinuities in this session and only one clock behaviour model is needed. If $L \geqslant 2$, the clock has some variations and two or more models are needed to describe the clock behaviour in this session.

Then the followed Information Criteria presented by Akaike (1973) is adopted:

$$AIC(\hat{B},\hat{T},\hat{K})= \min\sum_{j=1}^{L}(Nj'Ln(RSS(Kj)/Nj')+2(Kj+1)) \quad L=1,2,...,Lm \qquad (2)$$

$\hat{B}$, $\hat{T}$ and $\hat{K}$ are the optimal estimates for the number, the instant of the clock discontinuities and for the orders of clock behaviour model respectively. Where Lm is the maximum number of clock behaviour model existed possiblly in the observations. RSS is the sum of the residural square after fitting jth model and Nj' is the freedom for jth model.

M. J. Reid and J. M. Moran (eds.), The Impact of VLBI on Astrophysics and Geophysics, 525–528.

From the formula (2), the optimal estimates of B, T and K must simultaneous satisfy the condition of minimum of AIC.

3. The Results

In addition to the simulated data, the observations data of single baseline in POLARIS project in 1983 are also selected to estimate the B,T and K, then to compute the site coordinates. The estimates of clock bahaviour models are listed in Table 1 and Table 2 with that ones obtained by manual mode in NGS. The results for the code of OC137 and OC144 are shown in Fig.1.

Table 1. the Estimates of Clock Behaviour Models by AIC and Manual

Code	Date			AIC						Manual					
				k1	k2	instant of disc.			RMS	k1	k2	instant of disc.			RMS
	y	m	d			d	h	m	ns			d	h	m	ns
OC122	1983	8	1	1	2	2	1	24	±.511	2	2	2	6	20	±.544
OC137		10	22	2	2	23	10	16	.351	1	2	23	10	20	.378
OC144		11	26	2	1	27	10	1	.211	2	2	27	6	0	.216
OC146		12	6	2	1	7	5	42	.205	2	2	7	6	15	.205
OC147		12	11	2	1	12	8	56	.296	2	2	12	4	15	.303

Table 2. Differences of the Site Coordinates caused by Estimates of the Clock Behaviour Models with AIC and Manual

Code	X (m)	Y (m)	Z (m)
OC122	-0.0144	0.1043	-0.0831
OC137	-0.0013	-0.0265	0.0291
OC144	-0.0031	-0.0074	0.0082
OC146	-0.0030	-0.0372	0.0312
OC147	-0.0011	-0.0050	-0.0024

The numbers of discontinuities obtained by AIC and Manual are compared as shown in Table 3.

Table 3. the Effects on Site Coordinates from the Numbers of Clock Discontinuity Estimated by AIC and Manual

Code	Date			number of clock disc.		DX	DY	DZ
	y	m	d	AIC	Manual	(m)	(m)	(m)
OC104	1983	3	28	1	2	0.0019	0.0320	-0.0390
OC113		5	31	1	2	0.0209	-0.2380	0.2064
OC119		7	11	0	2	-0.0058	-0.0433	0.0433
OC130		9	17	1	0	0.0042	-0.0238	0.0192
OC151		12	31	1	0	0.0010	0.0252	-0.0213

The results obtained by AIC are different a few centimeters from that ones by interactive mode generally, and are approximately similar to the results of Clark et al.(1985). But the differences will reach about 10 to 20 centimeters sometimes when the observations are more dispersive. It should not be negletcted in VLBI survey with so high accuracy.

By using AIC mode, the identification of clock models will base on a rigorous mathematical criterion and the effects on the results of VLBI from personal cause will essentially be elimited. Furthermore it will improve the efficiency of VLBI data processing because the clock model is automatically selected by computer.

4. Acknowledgements:

The work reported here was supported by a grant from the National Science Foundation as part of the Senior Visiting Scientist Program of the National Geodetic Survey Division, Charting and Geodetic Survices, National Oceanic and Atmospheric Administration.

References:

1.Akaike,H., (1973), Information theory and an extension of the maximum likelihood principle, 2nd Int.Symp. on Inf. Theory (B.N.Petry and F. Csaki),p267-281. Budapest: Akademiakiado.

2.Robertson,D,S., "Geodetic and astrometric measurments with very-long baseline interferometry", 1975, M.I.T. (Ph.D.Thesis)

3.Carter,W.E., D.S.Robertson and J.R.Mackay, "Geodetic radio interometric surveying: Applications and Results", J.Geophys.Res., 90, No.B6, p4577-4587,1985

4.Clark,T.A. C.Ma, J.W.Ryan and A.Mallama, A comparison of Earth Orientation Data obtaied by Different Space Techniques, Proceedings of International Conference on Earth Rotation and Terrestraal Reference Frame, Ohio, USA,(1985).

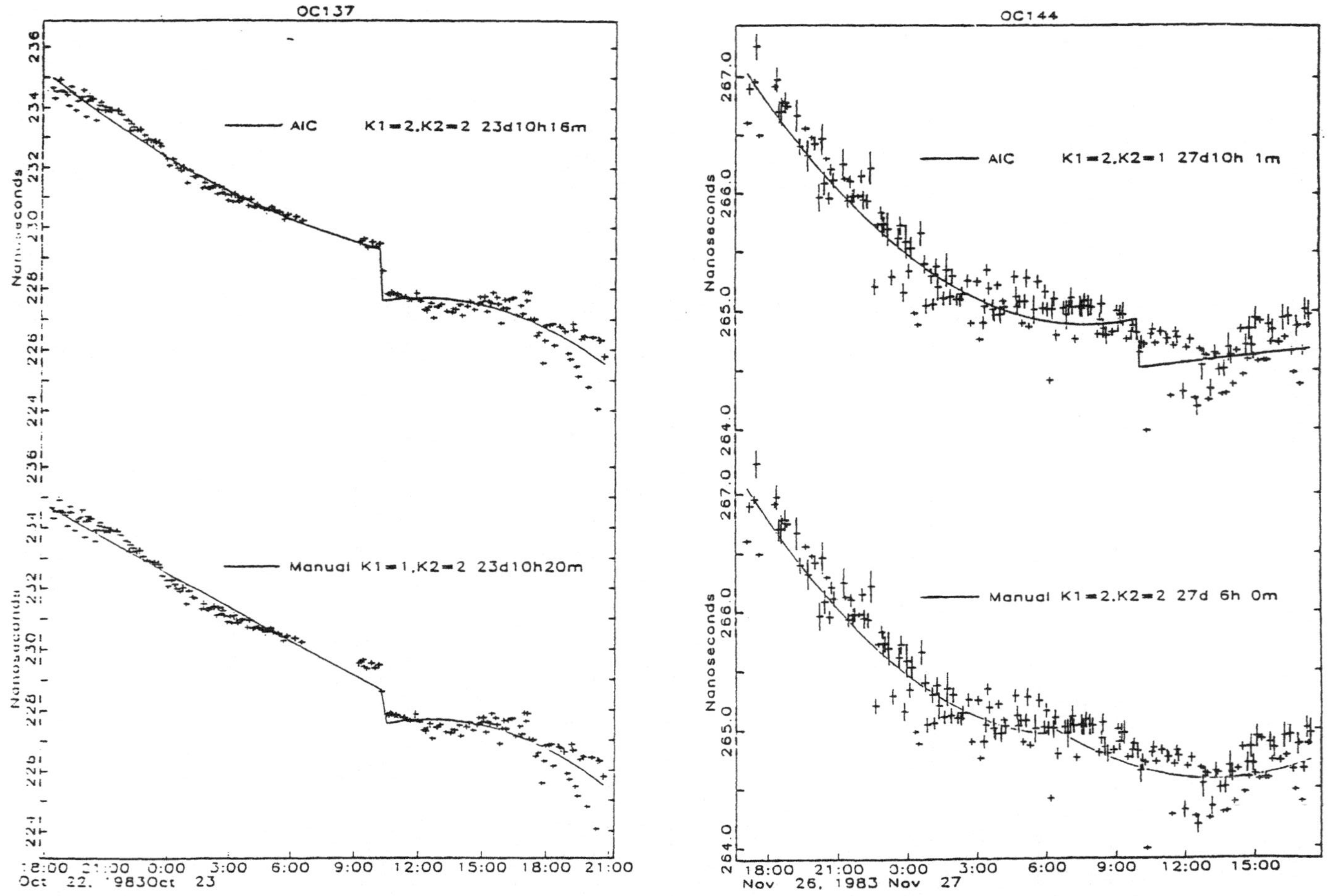

Fig. 1. The estimates of clock behaviour models for VLBI observation with code of OC137 and OC144 by AIC and Manual metheds. The time and K1,K2 represent respectively the broken time and the orders of models.

A Novel Method for Suppression of Aliasing Ghosts

Yoshihiro Chikada
Nobeyama Radio Observatory
Minamisaku, Nagano-Ken 384-13,
Japan

ABSTRACT. This paper proposes a "half-sample delay switching" to suppress aliasing ghosts caused by finite frequency sampling without the clumsiness of a sharp anti-aliasing filter.

Nyquist theorem says that signals beyond $R_s/2$ will be overlayed over in-band signals, where R_s is the sampling rate. Any anti-aliasing filter leaks undesired out-band supurious signals. And "good" sharp-cut filters are usually big and not simple. Are there no ways to suppress aliasing-ghosts without the clumsiness of big anti-aliasing filters? Yes. It is "delay switching".

Consider a cross-power spectrum analyzer for a two-element interferrometer, and assume stationary signals and complex-samplers (Figure). When we put a delay δ before one of the samplers, there is phase shift $\Delta\phi$ in the output phase spectrum which is proportional to the input frequency. When the delay δ is a half of the sampling period $1/f_s$, the phase shifts have a difference of 180° between two signals which are:
(1). In-band signal of frequency f.
(2). Aliasing signal of frequency f_a, where $f_a = f + f_s$.

$$\Delta\phi(f) = 2\pi f \delta = \pi f/f_s.$$

$$\Delta\phi(f_a) = 2\pi f_a \delta = \pi(f+f_s)/f_s = \Delta\phi(f) + \pi.$$

The 180° phase shift difference $\Delta\phi(f_a)-\Delta\phi(f)$ is not affected by delay compensation of phase slope $\exp(-\pi i\ f/f_s)$ at the output of the analyzer, because both in- and out-band signals appear in the same frequency channel. Therefore, by adding the compensated output of the delay-on state to the output of delay-off state, we get aliasing-free spectrum. To get precise suppression, put half-sample clock skew circuit (0° /180° switching of sampling clock, Figure) instead of analog half-sample delay in the signal path. Then we have:
(1). Precise half-sample delay by simple circuit in any sampling frequency.

M. J. Reid and J. M. Moran (eds.), The Impact of VLBI on Astrophysics and Geophysics, 529–530.

(2). No gain/phase change between on- and off-state of the delay switch.

Half-sample delay switching does not discriminate the aliasing ghost whose frequency is separated from the in-band signal by $k \times f_s$ where k is an even number. These ghosts can be discriminated by m/n sample delay switching, for an example, 0/3-1/3-2/3 sample delay switching.

Ghost caused by unbalance in a pair of real- and imaginary-part ADC is an aliasing ghost from the negative frequency band. Therefore it can also be suppressed by the delay switching. In the extreme case of the unbalance, aliasing in a case of real-sampling can also be suppressed by the delay switching.

New ADC system of the Nobeyama FX will have this half-sample clock skew switching facilities in 1987.

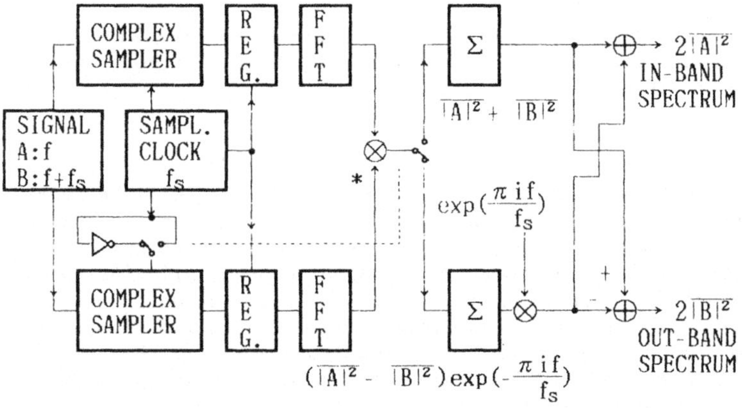

Figure. Block diagram of the half-sample delay (1/2 f_s) switching to eliminate aliased out-band spectrum.

FFT PROCESSOR AS A DIGITAL LENS IN GRID ARRAY VLBI

T. Daishido, K. Asuma, K. Nishibori, S. Inoue
Department of Science, School of Education, Waseda University,
1-6-1 Nishiwaseda, Shinjuku-ku, Tokyo

ABSTRACT. Correlator type processors have been currently used in VLBI, in which the number of correlators increases as N(N-1)/2 where N is the number of antennas. In a large elements VLBI array in future, this will introduce difficulty in constructing the processor. If we took the grid array configuration of antenna, FFT processor could be used in VLBI processing instead. FFT processor is used as a digital lens for a directional finding facility of the arrival waves. In FFT, the number of multiplications increases only as N log N. Although it gives maximum redundancy of Fourier components, the next step of the research on crustal motion will require measuring the velocities as a vector fields. Higher S/N ratio is also obtained in the VLBI maps.

The present processor, which was originally constructed for a pilot system of the radio patrol camera at Waseda University, is formed by a FFT Processor and eight Complex Amplitude Equalizers(1,2). When it is used as an VLBI processor, phase and amplitude fluctuations due to local oscillators and/or propagation effects are absorbed by the quick control of the Complex Amplitude Equalizers.

Here, we show examples of the computer controlled phase adjustment using the Equalizers(3). Our telescope is an 8 elements Very Short Baseline Interferometer(only 2m) at 10 GHz. Fig.1 is the block diagram of the present system. In VLBI system the recording facilities should exist between A/Ds and the Equalizers. The FFT processor is an error free ideal lens, and from the sampling theorem it makes eight independent beams in our system. Fig.2 is the dirty beams before removing phase errors, and Fig.3 the beams after the error removed. Both are 180 deg beam switched, while non switched in Fig.4.

Parameters of the processor are follows.

FFT Processor and Complex Amplitude Equalizers

Input/Output	8 complex
Dynamic range	8 bit
Band width	20 MHz/antenna (-10MHz to +10MHz)
Clock	20 MHz
Multipliers	CMOS LSI (SONY CX7997)

M. J. Reid and J. M. Moran (eds.), The Impact of VLBI on Astrophysics and Geophysics, 531–532.

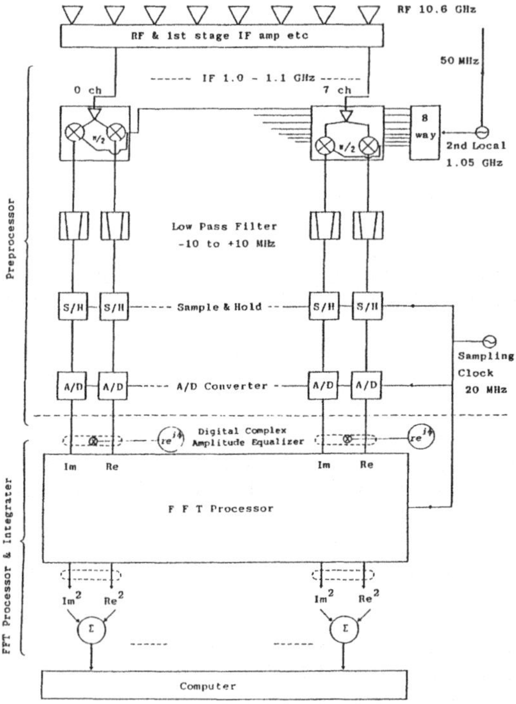

Fig.1 Block Diagram

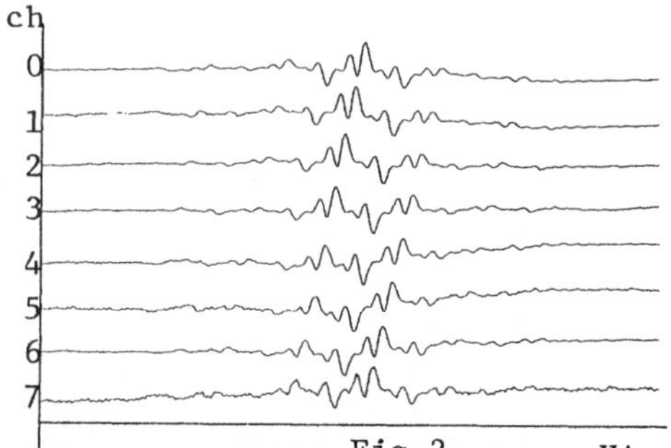

Fig.2

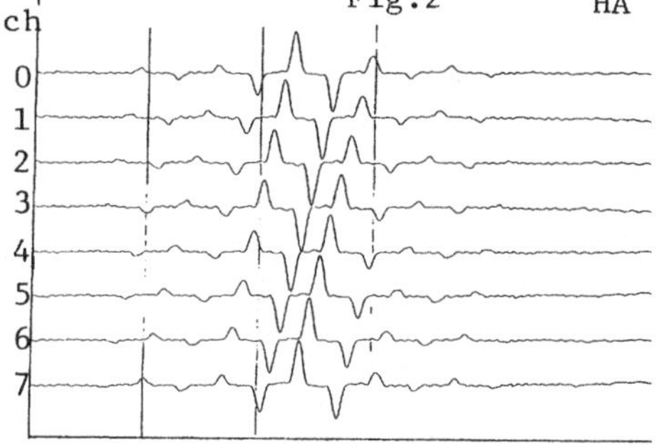

Fig. 3

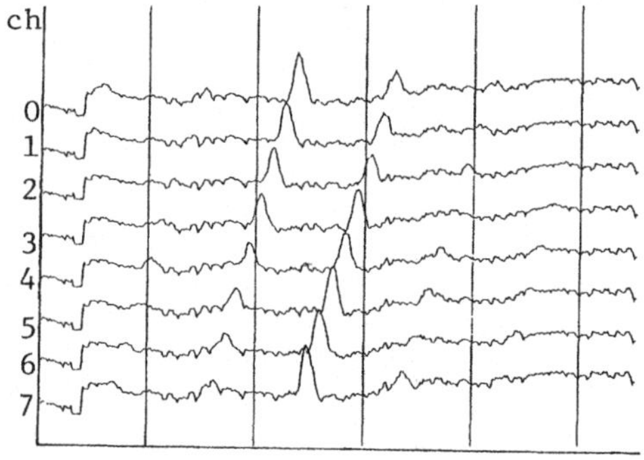

Fig. 4

Fig. 2,3,4 are obtained by scanning the sun. Channels 0-7 indicate the outputs of eight independent beams, which are used for the observations of CMB and transient radio sources.

References:

1. Daishido, T., Ohkawa, T., Yokoyama, T., Asuma, K., Kikuchi, H., Nagane, K., Hirabayashi, H., and Komatsu, S., 1984, in "Indirect Imaging", J. A. Roberts, Ed., Cambridge U P, pp81
2. Daishido, T., Asuma, K., Ohara, H., Komatsu, S., and Nagane, K., 1986, IEEE/ICASP86, Tokyo, 53.4.1, 2855
3. Asuma, K., et al., 1987(in preparation)

ATMOSPHERIC LIMITS: A REVIEW OF THE EFFECT OF PATH LENGTH VARIATIONS ON THE COHERENCE AND ACCURACY OF VLBI

Alan E.E. Rogers
Haystack Observatory
Westford, MA 01886

ABSTRACT. Interferometer phase fluctuations produced by path length variations in the troposphere limit the coherence to between 100 and 1000 seconds at a 3 millimeter wavelength. An expression for the coherence is given using an Allan variance to characterize the atmospheric phase fluctuations. Methods of optimizing the fringe detection threshold and calibrating the fringe visibilities under poor conditions of poor coherence are outlined. The astrometric and geodetic accuracy of VLBI is limited by our ability to calibrate the atmospheric path. Atmospheric "self-calibration" techniques which use the elevation dependence of interferometer delay to solve for the highly variable "wet" component of troposphere are discussed. Various models for the elevation dependence of atmospheric path or "mapping function" are reviewed. The accuracy limits imposed by the atmosphere are discussed.

1. INTRODUCTION

In early Very Long Baseline Interferometry (VLBI) experiments the phase fluctuations were largely due to instabilities in the frequency standards used to control the local oscillators. In more recent experiments at short wavelengths which used hydrogen maser frequency standards whose stability is 10^{-14} or better, the phase fluctuations are largely the result of atmospheric path variations. These path variations are non-dispersive and thus result in phase fluctuations which are proportional to the observing frequency. Experiments at wavelengths of 10 centimeters and longer start to be influenced by the ionospheric phase fluctuations which increase at a rate which is proportional to the observing wavelength. The atmospheric path length variations at wavelengths in the radio range are thought to be dominated by variations in the "wet" component or that due to water vapor which typically produces about 10 centimeters out of the approximately 2 meters of the total "wet" plus "dry" delay at the zenith. The "wet" component fluctuations, which are absent at optical wavelengths, constitute one of the few disadvantages of radio interferometry. However the development of Water Vapor

M. J. Reid and J. M. Moran (eds.), The Impact of VLBI on Astrophysics and Geophysics, 533–541.

Radiometers (WVR) may eventually eliminate the problems of the "wet" component by providing accurate corrections for both the constant and time variable components of "wet" path delay.

2. EXAMPLES OF ATMOSPHERIC PHASE FLUCTUATIONS

Typical good weather interferometer phases are shown in Figure 1 below for an experiment at 3 millimeter wavelengths [Rogers et al, 1984].

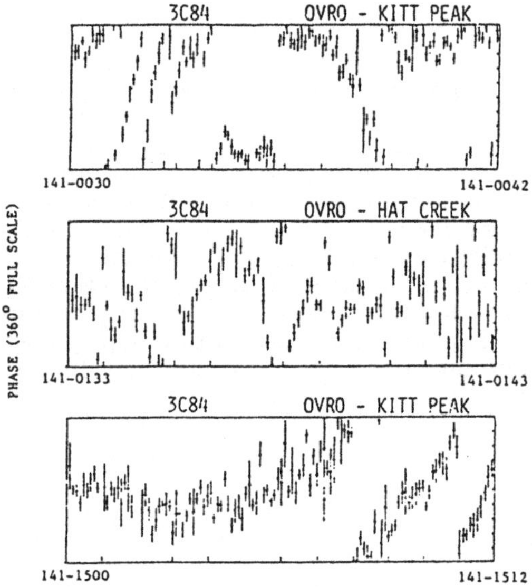

Figure 1. Typical 89-GHz interferometer phases as a function of time from observations made in May 1982 using a 10.4-m antenna of the Owens Valley Radio Observatory (OVRO), a 6-m antenna of the Hat Creek Observatory and the 12-m antenna of the National Radio Astronomy Observatory on Kitt Peak. Scans are 12 min long with phases shown for 5-s segments.

The typical r.m.s. path length variations are about 1 millimeter for coherent averaging time of 5 seconds. These variations result in a significant loss of coherence for longer coherent integration as shown in Figure 2. The range of coherence loss for experiments of good weather observing is typically about 4 to 1, the very best scans allowing a coherent integration of up to 700 seconds without large coherence losses.

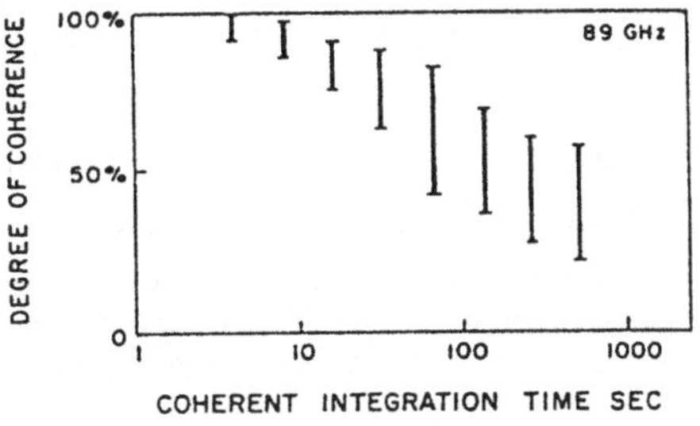

Figure 2. The range of normalized correlation coefficient as a function of coherent integration time for observations at 3 mm wavelength [Rogers et al, 1984].

3. COHERENCE AND ALLAN STANDARD DEVIATION

The coherence time T at observing frequency ω (radians/sec) and Allan standard deviation $\sigma(T)$ is approximately given by

$$\omega T \sigma(T) = 1 \tag{1}$$

A more precise relation has been developed by Rogers and Moran [1981] who starting with the following definition of a coherence function C(T)

$$C(T) = \left|\frac{1}{T}\int_0^T \exp\left[i\varphi(t)\right] dt\right| \tag{2}$$

and using the method outlined by Rutman [1978] derived the following integral equation

$$\langle C^2(T)\rangle^{1/2} = \Big[\frac{2}{T}\int_0^T \exp\left[-\frac{\omega^2\tau^2}{4}(\sigma^2(\tau) + \sigma^2(2\tau) + \cdots)\right](1-\tau/T)d\tau\Big]^{1/2} \tag{3}$$

Evaluation of the integral is simple in the case of white phase noise (Allan standard deviation power law exponent = -1) and white frequency noise (exponent = -0.5) resulting in 15 and 8 percent coherence loss when

$$\omega\sigma(T)T = 1. \tag{4}$$

The integral was also numerically evaluated to derive the coherence at several frequencies and integration times using the Allan standard deviation estimates for atmospheric phase fluctuations. The results are not significantly different, however, from the simple relationship derived from equation 1 which gives the maximum observing frequency for a coherence loss of approximately 10 to 20% as follows:

$$\omega_{max} \simeq 1/(\sigma(T)T) \quad rad/s \tag{5}$$

This simple relation predicts a maximum observing frequency of 100 GHz for a coherent integration time of 16 seconds and Allan standard deviation of 10^{-13}. Allan standard deviation values derived from VLBI data [Rogers et al, 1984] along with the Allan standard deviation for the hydrogen maser is shown in Figure 3 below.

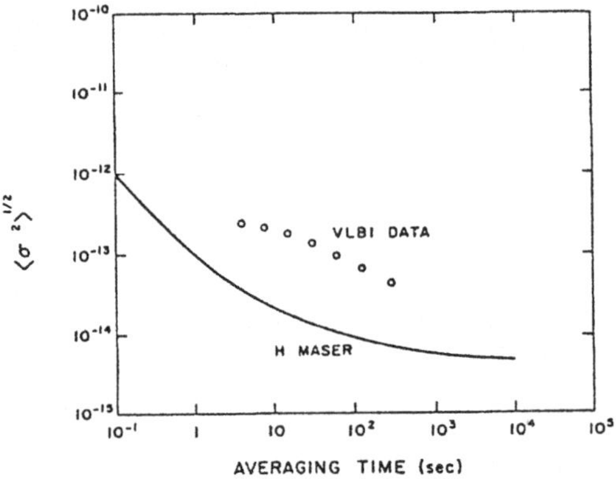

Figure 3. Allan standard deviation of atmospheric path length variation for an average of high SNR scans in April 1983. The solid curve is the Allan standard deviation observed between the Smithsonian Astrophysical Observatory (model VLG10) and Goddard Space Flight Center (model NP3) maser during tests in 1975 at Haystack Observatory. The masers used in the millimeter experiments were either of more recent design or upgraded versions for which the curve shown is a very conservative estimate of performance. (Allan standard deviation is given for difference between two sites and should be reduced by $2^{1/2}$ for path variations to a single site.)

4. ANALYZING DATA WITH SEVERE PHASE FLUCTUATIONS

In order to reduce the effects of phase noise on the estimation of correlation amplitude and closure phase, special data averaging methods are used. For a single baseline, the correlation amplitude is estimated by segmenting an observation. In this case the amplitude can be estimated by incoherently averaging N segments. An estimate of the correlation amplitude is given by Moran [1976], as

$$A = \left[\frac{1}{N} \sum_{i=1}^{N} a_i^2 - 2 \right]^{1/2} \qquad (6)$$

where a_i is coherent amplitude or signal to noise ratio of each segment in units of the sigma noise level. The subtraction of 2 is needed to remove the bias that results from the noise. As long as the phase variations of the signal within each segment are small, the amplitude A will not be significantly reduced. Thus it is desirable to make the segments as short as possible to contain all the fringe energy while maintaining a reasonable signal to noise ratio for A. When the SNR for each segment is much greater than one, the SNR of A increases with the square root of N. In this case the incoherent averaging produces no significant loss in SNR when compared wth coherent averaging. When the SNR of each segment is much less than one, a threshold proportional to the fourth root of N sets the detection limit. In this case the incoherent averaging process becomes relatively inefficient because the SNR will decrease with a reduction in segment length.

If fringes are detected on two baselines of a triangle, the coherent integration time can be extended on the third baseline by using the phases on the two stronger baselines to take out the phase variations. This technique requires a correction for the amplitude on the third baseline be made for phase noise on the two stronger "phase reference" baselines. The loss factors are multiplicative, or add when measured in decibels. The loss factor (Rogers el al. [1984]) depends on SNR and is greater than 0.9 (loss$<$ 10%) for SNR $>$ 2.5 for each data segment.

5. STATISTICAL MODELS OF PHASE FLUCTUATIONS

Important statistical properties of the phase fluctuations are the r.m.s. variation with baseline length, source elevation (or path length) and time scale (Allan standard deviation). Armstrong and Sramek [1982] and Treuhaft and Lanyi [1987] have shown how the power law exponents are related for these statistical quantities assuming that "frozen" turbulence is moving over the observing sites with the wind.

Statistical function	model	value	meas.	Reference
Spatial Power Spectrum(short)	-α	-3.67		
Spatial Power Spectrum (long)	-α	-2.67		
r.m.s. vs baseline (short)	$\alpha - 2$	1.67	1.4	Armstrong and Sramek [1982]
" (long)	"	0.67	0	see text
Allan standard dev. ($\tau < 30$)	$(\alpha - 4)/2$	-0.17	-0.25	Rogers et al. [1981]
" ($\tau > 30$)	"	-0.67	-0.50	"
" (long)	"	-0.67	-0.63	Hogg et al.[1981]
" (short)	"	-0.17	-0.07	Thompson et al.[1975]
" (long)	"	-0.67	-0.54	"
r.m.s. vs path length		0.50	0.50	Treuhaft and Lanyi [1987]

TABLE OF POWER LAW EXPONENTS

The table above shows the predicted and measured power law exponents for a model of Kolmogorov turbulence with two-dimensional spatial power exponent range of -11/3 for scales shorter than the atmospheric scale height and -8/3 for longer scales. The value of -11/3 corresponds to fully developed isotropic turbulence while -8/3 is the value expected for larger scales for which the structure effectively becomes two dimensional. The r.m.s. phase and phase rate variation with source elevation should depend on the square root of the path length at least for scales shorter than the scale height. Measured values of the power laws agree fairly well with the theory except that the theory is not adequate to correctly characterize the very long scales and correctly predict the observed independence of the r.m.s. fluctuations with baseline length for baselines longer than about 100 km. A "saturation" scale length like that introduced introduced by Treuhaft and Lanyi [1987] is needed to refine the model.

The strength of the phase fluctuations or magnitude of the turbulence varies (at 10 seconds) from an Allan standard deviation of 10^{-14} under the best meteorological conditions to 10^{-12} under the worst conditions. Most of the phase variations are thought to be due to the "wet" component but while Resch et al [1984] show encouraging results in their experiment to use a WVR to correct interferometric phases at the VLA it seems probable that perhaps as much as 30% of the variations are from the dry component. Some experiments are planned in which both the wet and dry components can be measured along a line from the site to an aircraft using multi-wavelength optical and radio transponders.

6. TROPOSPHERIC PATH DELAY MODELS

While the atmospheric path variations discussed in the previous sections limit the coherent integration and add noise to the measurements an incomplete knowledge of the average atmospheric path over the coherent integration period will limit the accuracy of the astrometric and geodetic parameters estimated from measurements of phase and group delay. In this section I review the path delay models.

The "dry" component of path delay at the zenith in a static atmosphere depends only on the ground pressure. The path delay variation with elevation ϵ or "mapping" fuction $m(\epsilon)$ can be predicted by a "ray trace" through an atmospheric model. A simple model for the atmospheric being one made of spherical shells of gas of known constituents and refractivity obeying the gas equations and in hydrostatic equilibrium. Additional model parameters are the ground temperature, lapse rate and height of the trospospause. For many years two simple analytic approximations to the ray tracing results were used in the analysis of VLBI data. The simplest model was that of Chao [1972].

$$m(\epsilon) = 1/[\sin\epsilon + [0.00143/(\tan\epsilon + 0.0445)]] \qquad (7)$$

and contains only fixed constants for "average" conditions including an average contribution from the "wet" component. A somewhat more sophisticated model was that of Marini [1974] (and modified for the radio range of frequencies in an unpublished note)

$$m(\epsilon) = 1/[\sin\epsilon + [(B/(A+B))/(\sin\epsilon + 0.015)]] \qquad (8)$$

whose constants A and B are dependent on ground measurements of pressure, temperature and humidity. After it was suspected that neither of these models were quite good enough Davis et al [1985] developed a model known as CfA-2.2 as follows:

$$m(\epsilon) = 1/[\sin\epsilon + [a/(\tan\epsilon + [b/(\sin\epsilon - 0.009)])]] \qquad (9)$$

where a = 0.001185 + correction terms
and b = 0.001144 + correction terms.

The terms added to a and b are linearized corrections for the deviations of the partial pressure of water vapor, ground temperature, pressure, atmospheric lapse rate and height of the tropopause from the nominal average values for which the constants a and b are valid without the added terms.

CfA-2.2 fits the ray tracing results to within 2 mm down to 5 degrees elevation whereas the simpler functional forms of Chao and Marini deviate from the numerical ray tracing by several centimeters or more at elevations below 15 degrees.

7. TESTS OF THE ACCURACY OF ATMOSPHERIC MODELS

Errors in the atmospheric model will result, primarily, in an error in the estimate of the local station height derived from inteferometric delays to a reference point whose path delay has been accurately corrected. For long baselines the error in height or radial component of either site Δr leads to an error in baseline length of

$$\Delta b \simeq \frac{b\Delta r}{2r_e} \tag{10}$$

where r_e is the radius of the earth. The first suggestion of the significant errors in the Chao and Marini models came from a test made by Davis et al [1985] of the dependence of baseline length on the elevation cutoff used in the least squares analysis. The elevation cutoff test showed a baseline length change of about 0.5 cm per 1000 km (fractional error of 5×10^{-9}) for a change in elevation cutoff between 5 and 15 degrees. The use of CfA-2.2 has made changes in baseline length with elevation cutoff undetectable at the present level of measurement uncertainty. With the improved atmospheric models the repeatability of baseline length measurements are at the few centimeter level for baselines longer than 3000 km, about 1 centimeter for intermediate baselines and at the millimeter level for baselines under 300 km. However even with the improved "mapping" function it is still necessary to either solve for the zenith path delay at each site every few hours, estimate the zenith path continuously using Kalman filter analysis or make highly accurate pressure and WVR measurements at each site to calibrate the atmospheric path.

8. SUMMARY

Atmospheric phase fluctuations are present at the level of approximately 10^{-13} (Allan standard deviation) and limit coherent integration of 3 mm VLBI to a few hundred seconds. These fluctuations are independent of baseline length for baselines longer than 100 km and increase with the square root of the path length thereby being approximately doubled at 15 degrees elevation. Accurate models for the elevation dependence of the atmospheric path length are needed to remove systematic errors in astrometric and geodetic results in order to reach a fractional accuracy of better than 5×10^{-9}.

References

Armstrong, J.W., and R.A. Sramek, 'Observations of tropospheric phase scintillations at 5 GHz on vertical paths', Radio Sci., **17**, 1579-1586, 1982.

Chao, C. C., 'A model for tropospheric calibration from daily surface and radiosonde balloon measurements.' *Tech. memo Cal. Inst. technol. Jet Propulsion Lab.,* 391-350, 1972.

Davis, J. L., T. A. Herring, I. I. Shapiro, A. E.E. Rogers, and G. Elgered, 'Geodesy by radio interferometry: Effects of atmospheric modeling errors on estimates of baseline length', Radio Sci., **20**, 1593-1607, 1985.

Hogg, D. C., F. O. Guiraud, and W. B. Sweeney, 'The short-term temporal spectrum of precipitable water vapor', Science,**213**, 1112-1113, 1981.

Moran, J. M., 'Very long baseline interferometric observations and data reduction', in *Methods of Experimental Physics*, edited by M. L. Meeks, vol. 12, part C, pp. 228-260, Academic, New york, 1976.

Marini, J. W., 'Correction of satellite tracking data for an arbitrary tropospheric profile', Radio Sci.,**7**, 223-231, 1972. and unpublished note 1974.

Resch, G. M., D. E. Hogg, and P. J. Napier, 'Radiometeric correction of atmospheric path length fluctuations in interferometric experiments', Radio Sci.,**19**, 411-422, 1984.

Rogers, A. E.E., A. T. Moffet, D. C. Backer, and J. M. Moran, 'Coherence limits in VLBI observations at 3-millimeter wavelength', Radio Sci.,**19**, 1552-1560, 1984.

Rogers, A. E.E., and J. M. Moran, 'Coherence limits for very-long-baseline interferometry', IEEE Trans. Instrum. Meas., IM-30(4), 283-286, 1981.

Rutman, J., 'Characterization of phase and frequency instabilities in precise frequency sources: Fifteen years in progress', Proc. IEEE,**66**, 1048-1075, 1978.

Thompson, M. C., L. E. Wood, H. B. James, D. Smith, 'Phase and amplitude scintillations in the 10 to 40 GHz band', IEEE Trans. Antennas Propag.,**AP-23**, 792-797, 1975.

Treuhaft, R.N. and G.E. Lanyi, 'The effect of the dynamic wet troposphere on radio interferometric measurements', Radio Sci., **22**, 251-265, 1987.

METHODS OF CORRECTION FOR THE "WET" ATMOSPHERE IN ESTIMATING BASELINE LENGTHS FROM VLBI

G. Elgered, J. L. Davis, T. A. Herring, and I. I. Shapiro
Center for Astrophysics
60 Garden Street
Cambridge, MA 02138

ABSTRACT. The error in VLBI estimates of baseline length caused by unmodelled variations in the propagation path through the atmosphere is greater for longer baselines. We present and discuss series of estimates of baseline lengths obtained using different methods to correct for the propagation delay caused by atmospheric water vapor. The main methods are use of data from a water-vapor radiometer (WVR) and Kalman-filtering of the VLBI data themselves to estimate the propagation delay. Since the longest timespan of WVR data associated with geodetic VLBI experiments was obtained at the Onsala Space Observatory in Sweden, we present results for the following three baselines: (1) Onsala–Wettzell, FRG (920 km), (2) Onsala–Haystack/Westford, MA (5600 km), and (3) Onsala–Owens Valley (7914 km).

1. INTRODUCTION

This paper addresses the utility of water-vapor radiometers (WVR) in geodetic radio-interferometry experiments. The WVR data are providing *a priori* information about the wet delay. It is also possible to estimate a correction to any *a priori* delay during post-processing. The estimate can be of a mean zenith bias for the entire experiment or of values of samples of, say an assumed random (Markov) process. We have analyzed 77 experiments (made during 1980–1985) several times, each time with a different method to correct for the wet delay at Onsala, but with the atmospheric delays for all other sites (as well as the clocks) modeled as Markov processes.

2. RESULTS AND DISCUSSION

The results are presented in Table I as weighted root-mean-square (WRMS) scatters of baseline lengths about estimated slopes. The WRMS value is a measure of repeatability. We also present one solution where observations made at elevation angles lower then 25° ($\epsilon < 25°$) at Onsala were deleted since low elevation angle observations are not important when no delay corrections are estimated (*Herring* 1986). The WRMS for the Wettzell–Onsala baseline is given with respect to its mean value since no baseline change is expected. For this baseline, when the WVR data are used in place of the Markov process, the WRMS decreases from 5.2 to 4.0 mm. If we assume that atmospheric delay errors are not correlated with other errors, and that the wet atmospheres over Wettzell and Onsala are statistically similar, the WRMS would be 2.2 mm were a WVR installed at Wettzell. The results for the Haystack/Westford–Onsala baseline are presented for each of the former antennas separately since the WRMS is significantly different in the two cases. This

M. J. Reid and J. M. Moran (eds.), The Impact of VLBI on Astrophysics and Geophysics, 543–544.

TABLE I. BASELINE LENGTH REPEATABILITY

Baseline Onsala to	Method used to correct for the wet delay at Onsala: *A priori*	Adjustment	Mean baseline[5] +919660 m (mm)	WRMS about mean (mm)
Wettzell[1]	None	Offset	999 ± 1	5.8
Wettzell[1]	None	Markov	1000 ± 1	5.2
Wettzell[1]	WVR	Offset	999 ± 1	5.1
Wettzell[1]	WVR	None	996 ± 1	4.0
Wettzell[1]	WVR ($\epsilon > 25°$)	None	996 ± 1	4.4
			Slope[5] (mm/year)	WRMS about slope (mm)
Haystack[2]	None	Offset	17.8 ± 1.4	14.1
Haystack[2]	None	Markov	19.0 ± 1.2	12.8
Haystack[2]	WVR	Offset	16.1 ± 1.2	12.5
Haystack[2]	WVR	None	14.2 ± 1.5	15.0
Haystack[2]	WVR ($\epsilon > 25°$)	None	16.6 ± 1.1	11.5
Westford[3]	None	Offset	15.9 ± 2.5	21.6
Westford[3]	None	Markov	17.8 ± 2.1	18.3
Westford[3]	WVR	Offset	14.5 ± 2.5	21.8
Westford[3]	WVR	None	13.7 ± 2.8	23.1
Westford[3]	WVR ($\epsilon > 25°$)	None	15.8 ± 2.3	18.8
Owens Valley[4]	None	Offset	12.8 ± 4.0	39.1
Owens Valley[4]	None	Markov	12.7 ± 3.2	31.4
Owens Valley[4]	WVR	Offset	10.3 ± 3.2	31.3
Owens Valley[4]	WVR	None	9.0 ± 6.1	57.6
Owens Valley[4]	WVR ($\epsilon > 25°$)	None	11.4 ± 3.6	35.4

[1] 25 experiments [2] 36 experiments [3] 45 experiments [4] 29 experiments
[5] The sigmas are scaled so that reduced $\chi^2 = 1$ (*Herring et al.* 1986).

difference indicates that the accuracy of the Westford–Onsala baseline estimates is not limited by atmospheric errors. It is also clear that low elevation observations do not improve the accuracy of baseline length estimates when WVR data are used and no adjustment to the *a priori* delay is estimated. Finally, for the longest baseline (Owens Valley–Onsala) it appears better to estimate a correction to the zenith delay inferred from the WVR data rather than to discard low elevation observations, because a bias affects the accuracy of the vertical position which is more important for estimates of length of longer baselines.

3. CONCLUSIONS AND SUGGESTIONS FOR FUTURE WORK

The differences in the rates obtained with the different methods are small compared to 1 cm/year. It is important to minimize possible biases in the atmospheric delays inferred from WVR data. A main source of such bias is the uncertainty of the attenuation coefficients due to water vapor. Another important task is to find the "best" elevation cut-off angle when WVR data are to be used.

ACKNOWLEDGMENTS
This work was supported by the Air Force Geophysics Laboratory, contract F–19628–86–K–0025; NASA grant NAG5–538; and NSF grants EAR–83–06380 and EAR–86–18989.

REFERENCES
Herring, T.A., *J. Geophys. Res.*, **91**, No. B9, pp. 9177–9182, 1986.
Herring, T.A. et al., *J. Geophys. Res.*, **91**, No. B8, pp. 8341–8347, 1986.

ATMOSPHERIC EFFECTS ON INTERFEROMETRIC OBSERVATION

Han Wen-jun
Beijing Observatory
Academia Sinica
Beijing, People's Republic of China

In recent years, the influence of atmospheric irregularities on electronic systems has been recognized. How these irregularities affect electronic systems is related to the propagation environment, the scales of the irregularity, and parameters of the equipment. The phase deviation caused by the atmosphere is important to the interferometer. However, there are some filter effects, namely antenna aperture filter, Fresnel filter, and baseline filter, which must be considered. The first two filter out the effects of small scale irregularities, and the last one filters out larger ones.

The atmospheric influence on centimeter wavelength and meter wavelength radio waves is mainly due to the troposphere and ionosphere, respectively. Concerning the excess propagation path, these two effects on observation have opposite signs. Taking the operating frequency to be about 2200 MHz under normal conditions, these two effects cancel each other to first order. In the meter waveband, because of the larger beam of the antenna, the ground reflected wave must be considered at large zenith angle, especially in aperture synthesis systems. The phase deviation caused by this effect is more serious, since the antenna heights for each set referred to its reflecting plane are in practice not equal.

The effect of atmospheric inhomogeneities on the interferometer is caused by horizontal gradient of inhomogeneity. If the baseline is not longer than about 1 km, the phase deviation caused by average atmosphere (quiet condition) is not larger than about 20° at 2 GHz, but that caused by irregularities (disturbed condition) is serious. The irregularities of the troposphere are due to the uneven distribution of water vapor in the form of clouds, fronts, etc., and those of the ionosphere are due to travelling ionospheric disturbance (TID), spread F and sporadic E. A model of sporadic E is suggested, and the effect of this on meter wave aperture synthesized beam is computed. The following conclusions are drawn: the gain of the main lobe of the synthesized beam decreases about 10 percent, and a right ascension error of about 0.6 second is introduced.

M. J. Reid and J. M. Moran (eds.), The Impact of VLBI on Astrophysics and Geophysics, 545.

WATER VAPOR RADIOMETRY AT THE ONSALA SPACE OBSERVATORY FROM 1980 TO 1987

J. M. Johansson, G. Elgered, and B. O. Rönnäng
Onsala Space Observatory
S–43900 Onsala
Sweden

ABSTRACT. Variations in atmospheric water vapor are difficult to correct for in geodetic, astrometric, and mm-wave astronomy VLBI. The use of a water vapor radiometer (WVR) has given promising results. The algorithm which relates the sky brightness temperatures observed by the WVR to the delay caused by atmopheric water vapor is discussed. Examples of WVR measurements made at the Onsala Space Observatory from 1980 to 1987 are presented.

1. INTRODUCTION

The main reason for the development of the water vapor radiometer (WVR) at the Onsala Space Observatory was to have a method to correct for the atmospheric delay due to water vapor during geodetic radio interferometry experiments. This delay is a major source of error when geodetic parameters such as baseline lengths are estimated.

The instrument commonly known as a WVR is a microwave radiometer which measures the emission from the sky at two (or more) well separated frequencies around the water vapor emission line at 22.235 GHz. The WVR observable at the frequency channel close to the center of the line depends on the amount of water vapor along the line of sight. The second frequency channel of the WVR is needed in order to correct for emission caused by liquid water, which occasionally can be even stronger than the vapor contribution. Since the wet delay also depends on the amount of atmospheric water vapor, we can formulate a (non-unique) algorithm that will allow us to infer the wet delay from the WVR observables.

The WVR at the Onsala Space Observatory has two channels; one centered at 21.0 GHz and the other at 31.4 GHz. System temperatures are typically around 600 K at both frequencies. The bandwidth of each channel is 1 GHz at the RF-port and typical integration times are around one second. The instrument is fully steerable and has a maximum slewing speed of 1.7° per second in both azimuth and elevation. Each channel is internally calibrated by a liquid nitrogen load and a load at room temperature. External calibration is performed by elevation-scans (tip-curves). The root-mean-square (RMS) repeatability over time periods of days is believed to be better than 1 K in each channel.

The algorithm relating the WVR observables to the wet path delay is in no way unique. Therefore, it is possible to make many rather different approaches to the problem. Empirical parameters in the algorithm show depencence on meteorological conditions such as vertical profiles of atmospheric temperature. Algorithms can therefore be optimized to a particular site. A detailed study has been made by *Johansson et al.* (1987) where algorithms including as well as excluding ground meteorology parameters are derived for about 20 geodetic VLBI sites. The RMS error

M. J. Reid and J. M. Moran (eds.), The Impact of VLBI on Astrophysics and Geophysics, 547–548.

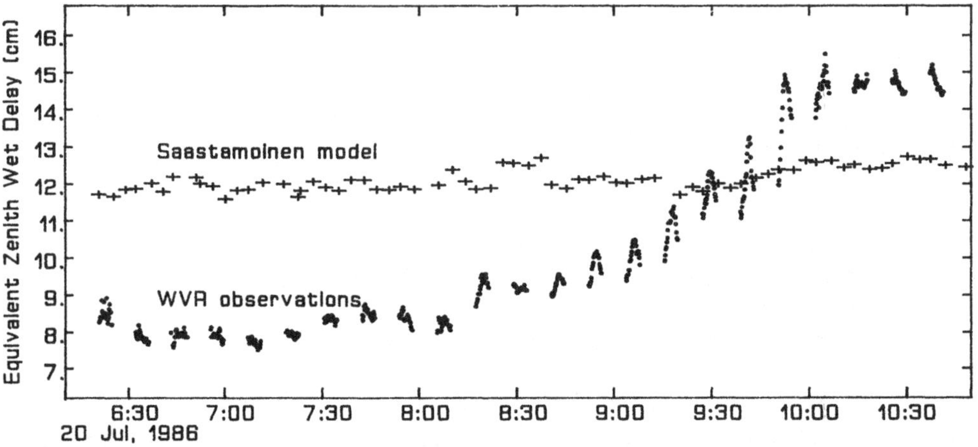

Figure 1. Equivalent zenith wet path delays measured with the WVR scanning at an elevation of 30°. Each data group consists of a scan from the east over the south and west and ending in the north direction. We conclude that a more humid air-mass was approaching from the west.

introduced by approximations in the algorithm is typically 2 mm in the zenith direction. When the algorithm is derived we assume that the attenuation due to water vapor can be perfectly calculated from *in situ* meteorological parameters. However, the uncertainties of these formulas are around 5 %, implying that the inferred wet delay has an uncertainty of the same order.

2. RESULTS.

Results obtained when WVR data have been used in the processing of geodetic VLBI data are reported by *Elgered et al.* (1987). These results suggest that it is important to improve the accuracy of formulas for the attenuation due to water vapor mentioned above.

Other results have been obtained from comparison of simultaneous WVR and radiosonde measurements at Göteborg-Landvetter Airport, 1980, and at Onsala Space Observatory, 1983. At both these occasions the RMS zenith difference of the wet delay was just below 8 mm.

Finally, in Figure 1, we present a typically unfavourable day for the prediction of wet delays using a ground based model. The wet delay increases with almost a factor of two over a two-hour period. However, the delay predicted by the model (*Saastamoinen* 1972) (approximately proportional to the absolute humidity at the ground) shows no variations above the 10% level.

REFERENCES.

Elgered, G., J.L. Davis, T.A Herring and I.I. Shapiro, 'The effect of the wet atmosphere on estimates of baseline lengths from VLBI,' *IAU Symposium No. 129*, Cambridge, MA, 1987.

Johansson, J.M., G. Elgered, J.L. Davis, 'Geodesy by radio interferometry: Optimization of wet path delay algorithms using microwave radiometer data,' *Res. Rept. No. 152*, Onsala Space Observatory, Chalmers Univ. of Tech., 1987.

Saastamoinen, J., 'Atmospheric correction for the troposphere and stratosphere in radio ranging of satellites,' in *The Use of Artificial Satellites for Geodesy, Geophys. Monogr. Ser.*, **15**, ed. S.W. Henriksen et al., pp. 247–251, American Geophysical Union, Washington, D.C., 1972.

CHINESE DUAL FREQUENCY WATER VAPOR RADIOMETER FOR VLBI

Xu Pei-yuan
Shanghai University of Science and Technology
Shanghai, People's Republic of China

A Chinese VLBI Network Project is under construction in China. The first 25-m antenna and associated VLBI system have just been completed at the She-Shan Station of the Shanghai Observatory. Both geodetic and astrophysical VLBI programs will be undertaken with the Network. A dual frequency water vapor radiometer is being developed to determine the tropospheric path delay correction. This WVR has the following characteristics:

Operating frequencies	20.60 and 31.65 GHz
IF bandwidth	50–500 MHz
Noise figure	≤ 6 dB
Integration time	Software selectable in steps of 1 sec
Antenna configuration	Two mechanically coupled elevation steerable 90 offset paraboloid mounted on an azimuth pedestal
Antenna beamwidth	2.4 deg
Beam efficiency	90% for ± 5 deg around beam center
Pointing accuracy	0.5 deg azimuth; 0.1 deg elevation
Control & data acquisition	Microcomputer IBM PC/XT
Calibration	Waveguide load at 313 K, noise diode, tipping curves
Reference stability	Waveguide load 0.05 K/hour, diode 0.2%/week
Antenna temperature accuracy	0.5–1.0 K

M. J. Reid and J. M. Moran (eds.), The Impact of VLBI on Astrophysics and Geophysics, 549.

INFLUENCE OF THE IONOSPHERE MODEL ON THE SOLUTION OF A VLBI GEODETIC EXPERIMENT

G. Petit
Institut Géographique National
2, Avenue Pasteur 94160 Saint-Mandé, France
J.F. Lestrade, BDL France; C. Boucher, IGN France;
F. Biraud, Meudon Obs. France; A. Rius, Madrid Univ. Spain;
A. Nothnagel, Hart. RAO S. Africa; P. Kaufmann, INPE Brazil

ABSTRACT. The GRIG-2 geodetic VLBI experiment was conducted in 1985, linking for the first time South America, Europe and Africa. At the single frequency band of 1.66 GHz which was used, the monitoring of the ionosphere is a critical aspect and several predictions of Total Electron Content (TEC) were used. One of them is derived from dual band Doppler observations of TRANSIT satellites, which were simultaneously conducted. The influence of these models on the solution is presented, with comparisons with other VLBI solutions. Decimetric accuracy has been achieved.

1. THE EXPERIMENT

The participating stations were Atibaia (Brazil), Hartebeesthoeck (S.Africa), Madrid (Spain, Deep Space Station 63), Nançay (France, a quasi meridian antenna of 94m equivalent diameter), and Onsala (Sweden). Some constraints in the available equipment led to the following set-up : L-band observations (1.66 GHz), with Mark II recording of two channels 18 MHz apart and switched at 1 pps to construct BWS delay. To minimize the ionospheric effect in taking advantage of the common night time at all stations, two 6.5 hour sessions were conducted starting June 29. 1985 at 20h45 UT and July 4, 1985 at 21h00 UT.

2. THE IONOSPHERE MODELS

The post-correlation processing was done with the JPL software MASTERFIT [1]. It allows to introduce ionosphere models in the form of lists of zenithal TEC values versus time. This value is then mapped to the relevant longitude with an hour angle dependence, and to the direction of the source. Five different models were used to derive the TEC values. They are thereafter represented by the symbols I0, to I4.

- I0 is the absence of any model
- I1 is the default MASTERFIT model
- I2 is the Bent model [2]
- I3 is derived from dual band Doppler observations (see below)
- I4 incorporates ionosonde data from South Africa

M. J. Reid and J. M. Moran (eds.), The Impact of VLBI on Astrophysics and Geophysics, 551–552.

3. TEC FROM DUAL FREQUENCY DOPPLER

The NNSS TRANSIT satellites transmit two frequencies around 150 and 400 MHz. Dual frequency receivers allow to derive the ionospheric delay in the direction of the satellite, in fact its variation with time. The absolute value of the delay is then computed either from a theoretical assumption on the time variation of the delay or, when two stations observe the same satellite, without this uncertainty. Several authors have presented the method, for exemple [3]. A program developed at the Geodetic Institut in Bonn [4] has been adapted to derive the TEC model I3.

4. COMPARISON OF TEC MODELS

The I3 model has been derived from Doppler data collected by TRANET stations or by receivers specially operated for the experiment. However, only the South African zone could be covered with two stations. The I4 model has been derived from ionosonde measurements in South Africa, with Dudeney model up to F2 layer plus Chapman Alpha model for top-side ionosphere. These two models agree reasonably well over the South Africa zone : they differ by about 20 to 30%. They have also the same kind of agreement with model I2. Thus, the confidence in the ionosphere model for the Hartebeesthoeck station is important.

In comparison, there are serious doubts on the TRANET data, which model on the European zone seems to degrade the results. It has not been used in the final I3 solution.

5. COMPARISON OF RESULTS

Three out of the five stations (DSS63, Onsala and Hartebeesthoeck) have already participated in high precision VLBI experiments. The GSFC solution [5] has been used as a "fiducial" one to compare our different solutions : we have adjusted a single translation between each GRIG-2 solution and the GSFC one. The RMS of the residuals of the coordinate differences is then a good estimate of the quality of the GRIG-2 solution. All I2, I3 and I4 solutions achieve RMS of residuals in the range I0-15cm.

REFERENCES

[1] J.L. Fanselow, O.J.Sovers, JPL Publication, **83-39**, 1985, Rev.1
[2] S.K. Llewellyn, R.B. Bent, Rep. **AD-772-773**, Nat. Tech. Inf. Serv., Springfield, Va. 1973
[3] R. Leitinger, G. Schmidt, A. Taurianen, J. Geophys. **41**,201-213,1975
[4] F.J. Lohmar, Dr.-Ing. Thesis, Bonn, 1984
[5] C. Ma, W. Himwich, A. Mallama, M. Kao, in Annual Report of the BIH for 1986, 1987 (in press)

CONCLUDING REMARKS

I. I. Shapiro
Harvard-Smithsonian Center for Astrophysics
Cambridge, MA, USA

This symposium was the first interdisciplinary conference devoted solely to applications of VLBI. Even its most jaded practitioners could hardly help being impressed by the broad sweep of these applications to astrophysics and geophysics. In 20 years, we have advanced from bare detections of the brightest compact radio sources to making inferences on the shape of the core-mantle boundary within the earth, monitoring the motions of the earth's continents, mapping the radio emission around stars, measuring the relative positions of compact sources across the sky with uncertainties under one milliarcsecond, and producing detailed polarization information on the radio emission from the hearts of quasars. Superluminal motion, merely a gleam in Martin Rees' eyes 20 years ago, is now a household word, with VLBI being used to follow virtually every twist and turn in the motion of superluminal jets that appear to be ejected from the cores of quasars. The sheer enormity of the current VLBI research discussed at this symposium defies simple summary. As a corollary, the number of people using VLBI as a research tool has also undergone nearly exponential growth; no longer do any of us know all of the others. By its very nature VLBI, of course, requires cooperation and the trend in all sciences towards increased numbers of authors per paper has always been much in evidence with VLBI. A newer trend, evidenced in this symposium, perhaps for the first time, is toward an ever larger number of papers per author; surely ten papers at one symposium co-authored by a single person must be a record.

Although VLBI, along with the rest of astronomy (and biology and physics and ...) is in a "golden age," the end is by no means in sight; one could predict with reasonable confidence a host of new extensions. When we or our descendants gather here, in 2007, for the fortieth anniversary of VLBI, the sophistication of the applications may compare to ours about as the Concorde compares to the horse and buggy. I hesitate to comment on the quantity.

M. J. Reid and J. M. Moran (eds.), The Impact of VLBI on Astrophysics and Geophysics, 553.

Discussion of the Papers Presented at the Symposium

Discussion of the paper by ZENSUS (p. 7)

Morimoto: What roughly is the velocity in the majority of sources?

Zensus: Most sources have apparent velocities between $\sim 2c$ and $\sim 10c$.

Schilizzi: What limits can you place on interblob emission in superluminal sources? Are these sources really jets or a sequence of bullets or plasmons being fired out from the nucleus?

Biretta: In 3C345, the region between the superluminal blobs is roughly ten times fainter than the knots, or hundred times fainter than the core.

Discussion of the paper by BARTHEL (p. 17)

Porcas: How well aligned are the mas and arcsec jet in 4C34.37? The alignment/misalignment statistics give another constraint on how to adjust the "standard" model.

Barthel: They are misaligned by 5°. I agree. Naively, one would expect beaming effects to be dominant in the plane defined by mas jet and observer, which would result in the generally observed alignment.

Preuss: With regard to core prominence in the large double-lobed quasars you mentioned, I would like to point out that the largest radio galaxies known also tend to have relatively strong cores.

Barthel: I am fully aware of that but did not want to include radio galaxies in the present discussion since *on average* radio galaxy cores are one order of magnitude less luminous than those in quasars. Particularly interesting in this respect is the giant radio galaxy 3C236, the largest known object in the universe. We are currently investigating the presence of possible superluminal motion in its core.

Discussion of the paper by BIRETTA (p. 19)

Ulvestad: When comparing the internal pressure of superluminal components with external pressure in the narrow line region, you should be careful about the pressure assumed for the narrow line region. In nearby active galaxies, there is some evidence for stratification in density in the narrow line region. Therefore,

M. J. Reid and J. M. Moran (eds.), The Impact of VLBI on Astrophysics and Geophysics, 555–580.

the density and pressure of that region within a few parsecs of the core may be considerably higher than in the material hundreds of parsecs from the core.

Biretta: We simply used an upper limit to the NLR density of 10^6 cm^{-3} in computing the external pressures and compared this to knots 20 to 200 pc from the nucleus. A more detailed model for the NLR densities that includes stratification would be useful.

Chu: Have you any intention to make the VLBI polarization observations so as to confirm the magnitude of magnetic fields?

Biretta: Others are doing this.

Marcaide: Do you have any reason to believe that the core position shift with frequency should not also be found in other superluminals? Note: This shift was first found for 1038 + 528A with 3.6/13 cm observations and a proper registration, leading to the result that 1038 + 528A is a mild superluminal.

Biretta: No. It seems likely that similar shifts will be found in other superluminals, as a similar shift was found (by you) in the quasar pair 1038 + 528 using phase reference VLBI.

R. J. Davis: Are you worried by the chance that the initial orientation is 1° to the line of sight?

Biretta: This is a question of statistics. It would be worrisome if many sources required an alignment within 1° of the line of sight, but it is not a serious problem in any one source. Further, if C4 has an intrinsic acceleration, the initial angle to the line of sight could be much larger than 1°.

Niell: If you calculate the speed of the components along the track (arclength), wouldn't the acceleration be significantly larger? (Upon further thought: If components approach an asymptotic direction, then accounting for arclength will reduce the increase of v/c.)

Biretta: The motion of C4 is initially nonradial and later becomes nearly radial. Computing the proper motion along the path would give a slightly larger initial speed and slightly less acceleration than when only the radial component of motion is considered.

vanBreugel: You mentioned two possible explanations for the curvature observed in 3C345: bending by external gas or internal jet stabilities. Could the curvature also not be due to precession of the nucleus?

Biretta: Precession cannot work in its simplest form since then the superluminal expansion would not be constant.

vanBreugel: But don't you think that there could be an external medium in addition to precession?

Biretta: Yes, but then any precession model could work.

vanBreugel: That was exactly my point.

Shaffer: How confident are you that the apparent shift in position of the core (with frequency) is not due to optical depth effects in the jet components?

Biretta: It is possible that the position shifts occur in the knots. However, the shifts are about 1/3 of the knot diameters, so that very strong spectral index gradients would be required.

Barthel: Was the acceleration of C4 accompanied by component brightening?

Biretta: Yes. The acceleration occurred near epoch 1982, and the flux of C4 doubled between 1981 and 1983. The acceleration and brightening might be related kinematically, though the flux of C4 will be affected by intrinsic evolution as well as by kinetic changes.

Discussion of the paper by *SIMON* (p. 21)

Porcas: How colinear are the three VLBI components?

Simon: The three VLBI components are colinear to within the measurement errors. The stationary component is at $PA = 118° \pm 1$, while the superluminal component is at $117° \pm 4$. The arcsecond component is at $PA \sim -50°$, within 8 degrees of the axis defined by the VLBI components.

Barthel: Since the superluminal motion relies heavily on the 1979.8 map, is it possible that you missed the moving component at that epoch?

Simon: No. The superluminal component was clearly seen in the 1979.8 map and in each of the subsequent maps. It has faded somewhat with time, but that has been compensated for by the recent maps having slightly higher dynamic range.

Cohen: Superluminal motion always is jittery, and a slowdown is expected here if a new component is coming out of the core. You measure from the centroid of (core + new component).

Simon: This might explain these observations, but the discrepancy is larger than a beam diameter. There is no evidence of any extension to the core away from the superluminal component, as might be expected if the change in speed is a resolution effect.

Cotton: If the apparent superluminal motion decreased by about 10, then the apparent brightness of that component should have decreased by a factor of about 1000. Doesn't this cause difficulty to the interpretation of this component?

Simon: Yes, if the moving component were assumed to be a simple plasmoid. Obviously, the moving component in 3C395 is more complicated than that. A random theoretician could probably cobble together at least two models before breakfast; more observations are clearly needed.

Discussion of the paper by *SALVATI* (p. 47)

deBruyn: Although the x-ray intensity in 3C273 fits the SSC model predictions, its slope certainly doesn't. Doesn't this worry you?

Salvati: The *local* slope in any given spectral band depends on the electron energy distribution. This is treated very crudely in the present version of the model; in fact, it predicts *monoenergetic* electrons and very bumpy synthetic spectra. Until the electron energy distribution is treated in a more realistic way, the model will only give a *global* slope over several spectral bands.

Biermann: For our Bonn sample of flat spectrum radio sources, we have observed all sources with the 30-m mm-telescope at a wavelength of 1 mm. All sources have a flat or nearly flat spectrum to 1 mm. One source, the BL Lac 1803 + 78, still has nearly 1 Jy at 60 μ (IRAS). Thus, we confirm that the turnover in the overall spectrum is near 10^{12} Hz, consistent with your picture.

In your initial expression for the total luminosity, the Klein-Nishina cutoff limits the Compton orders to about 2 at most. From the mm-x-ray correlation, the Lorentz factor can be estimated to about 10^3, limiting you indeed to 2 orders.

Salvati: The Lorentz factor can be lower and the number of Compton orders larger if one allows for an intermediate order between submm and x-ray wavelengths. Obviously, very low Lorentz factors are ruled out by synchrotron opacity limits.

Discussion of the paper by *LIND* (p. 55)

vanBreugel: Given the many different parameters that can determine the observed flux density variabilities and velocities in "real" jets (shocks, obstacles, opacity effects, viewing angles, etc.), isn't it surprising that the superluminal expansion speeds that are observed seem to be rather constant in time? Could this not be used to constrain your "more realistic" models?

Lind: The constancy for an individual component would provide some limit on the evolution of patterns within a jet, e.g., shocks do *not* slow down appreciably. Similarities between a number of components within a given source suggest that the ejection of the *brightest* features is fairly uniform in time. So, constancy of one component indicates limited *component* evolution; constancy of characteristics of different components indicates limited *central engine* evolution and/or precession, over time scales of years.

Discussion of the paper by *READHEAD* (p. 65)

Wilkinson: Now that MERLIN has been enhanced with the link into Cambridge, one can observe at a low frequency, 408 MHz, with a resolution of ~ 0.5 arcsec. It may well be that one will be able to map the weak, steep-spectrum, extended emission in the very core-dominated objects in your sample that are impossible to map with the VLA at ~ 5 GHz. This would help your statistics on the core–extended structure position angle differences.

Readhead: This would be well worth doing. It could almost double the number of objects that we could include in our misalignment analysis.

Porcas: Istvan Fejes and I have EVN data at λ18 cm and λ50 cm that indicate that the VLBI jet turns anticlockwise toward the direction of the southwest arcsecond scale component.

Readhead: Thank you. We shall use this result in calculating the misalignment in this object.

Discussion of the paper by SCHALINSKI (p. 71)

vanBreugel: Of the four objects without redshifts, are these highly polarized? Are they BL Lac objects?

Schalinski: Polarization observations, both at optical and radio frequencies, are consistent with the identification of the four sources as BL Lac type objects.

Discussion of the paper by MUTEL (p. 73)

Lind: Since the momentum transfer surface between the jet and the surrounding medium can be considerably wider than the jet or the terminal hot-spot (if the terminal hot-spot corresponds to the terminal mach disk in numerical simulations), then one can have considerably slower velocities of expansion than estimated in the ram-pressure calculations.

Porcas: Have you managed to find any central components yet?

Mutel: No!

Neff: What are the optical properties of the galaxies in which the CD's are found? You mentioned that all CD's were in galaxies or EF's, not QSO's, but there are different sorts of galaxies. What are the infrared properties of the systems? It would be useful to know if the optical spectra show evidence for an AGN (BLR, outflow, etc.).

Mutel: In general, not much is known about the optical properties of CD sources, except that the few galactic identifications appear to be elliptical. I have checked the IRAS point source catalog for all known CD sources. There are no CD sources listed.

Preuss: With regard to the evolutionary sequence you propose: Do the statistics of the source types you consider as evolutionary stages agree with what you would expect from the characteristic time scales?

Mutel: The relative number of CD, SSC, and FRII doubles at a given luminosity depends greatly on the time dependence of the expansion speed and the luminosity evolution, neither of which are known.

Discussion of the paper by <u>*MARR*</u> (p. 91)

Shapiro: Can you comment on the apparent change in orientation of the "symmetry axis" of the core from $PA \sim 30°$ in 1981 (April and December) to $PA \sim 0°$ in 1984–85?

Marr: The core is the dominant component in our maps at 1.3 cm and at 2.8 cm. The structure in this region is consistent in 1981 at both wavelengths. Unfortunately, the time interval between our images is insufficient to follow the evolution of the core accurately. We have no physical explanation for these structural changes. Precession, or curved paths of particles ejected from the core, are possibilities. We plan to study the core structure in more detail in the future.

Discussion of the paper by <u>*WRIGHT*</u> (p. 93)

Biermann: The mm-x-ray connection suggests one more strong argument for mm-VLBI. With mm-VLBI and x-ray observations we will be able to test all the nice SSC models we discussed this morning. The German x-ray satellite ROSAT is scheduled now to go up about 1990. Germany has just finished a report on the future of astronomy (15 years). In this report, mm-VLBI gets very strong support as a national goal. So we hope to equip the European mm-dishes with VLBI equipment, increasing the sensitivity dramatically.

Wright: Unpublished Einstein data provides evidence for a 30% to 50% increase in the x-ray intensity between February 1979 and February 1980, suggesting inverse Compton emission at the time of the millimeter flare. This is consistent with the observed size of the core of 3C84 at $\lambda 3$ mm. We can look forward to a world array at $\lambda 3$ mm.

Porcas: How confident are you in the reality of your detection of the 10 mas knot component?

Wright: All the components in the model fit are present in the uncalibrated visibility data. There is evidence for compact structure ~ 10 mas to the south of the core of 3C84 at all epochs. Model fitting determines the flux and location of this component.

Langston: Can you use your proposed multifrequency observations to trace atmospheric phase variations and then lengthen to coherent integration time?

Wright: Observations at three frequencies across the $\lambda 3$ mm window would not be simultaneous and cannot be used to increase the coherent integration time. On the other hand, we do have simultaneous observations at 5 GHz that might be used to track atmospheric fluctuations and provide a phase reference to extend the coherent integration at $\lambda 3$ mm.

Preuss: You expressed your confidence in the "bright future" of mm-VLBI. How bright is it actually going to be within, say 5-10 years from now, in terms of sensitivity and image quality?

Wright: We have demonstrated that there are no major technical problems with VLBI at $\lambda 3$ mm. With rather modest improvements in sensitivity and a few additional antennas, we should be able to make images with dynamic ranges of ~ 10. Use of the larger telescopes at Nobeyama, Onsala, and IRAM will improve the sensitivity and resolution. The MkIII density upgrade and the use of VLBA antennas and recorders will allow more frequent sampling at higher bandwidth so that in 5-10 years we can expect to see high quality maps with a flux limit $<\sim 100$ mJy.

Discussion of the paper by *FANTI* (p. 111)

Menon: I had shown in the Bologna Symposium Proceedings that for quasars the low frequency turnover suggests that the linear size of compact sources depends on redshift. Is it possible that the size difference between galaxies and quasars that you find is also partly at least a redshift effect since the quasars are in general at higher z than galaxies?

Fanti: The difference we find between galaxies and QSS's is mainly in their morphology. I think the redshift has little relevance since CSS radio galaxies and CSS QSS's in our sample have rather similar redshifts.

Discussion of the paper by *VAN BREUGEL* (p. 115)

Birkinshaw: Roger Davies and I have been using the KPNO 4-m telescope to image a large sample of radio galaxies and a companion sample of nonradio galaxies. There are no obvious, dramatic differences between the samples. The strangest galaxies so far are nonradio emitting.

vanBreugel: When discussing galaxy peculiarities, it is important to specify at which level, and of what kind (tidal tails, dust lanes, etc.), they are. Furthermore, (radio) galaxy peculiarities seem to depend on radio power and redshift. For example, the peculiarities found by Heckman *et al.* *(Ap.J.)* and Hutchings (1987, preprint) are at a much higher surface brightness level than in the lower luminosity radio galaxies. Also, at very high redshifts, galaxies have peculiarities at even higher surface brightness levels (e.g., McCarthy *et al.* 1987).

Barthel: In high z quasars where we can study the associated ($\Delta v \lesssim 3000$ km s^{-1}) CIV absorption, it has been found that the occurrence of such systems in steep spectrum QSR's is much higher than in flat spectrum QSR's (Foltz *et al.*, *Ap.J.*, 1986; Barthel *et al.*, in preparation). Such differing environmental properties argue against the unified scheme.

vanBreugel: History shows that ignoring the importance of one's environment can be hazardous.

Wrobel: You seem to want to link optical peculiarities with the presence of powerful radio sources. Have you looked for such peculiarities in radio-quiet ellipticals?

vanBreugel: A large optical imaging survey of radio-quiet ellipticals has been made recently by Djorgovski (1986, thesis, UC, Berkeley). Surface photometry was obtained of several hundred bright ellipticals. Very few showed evidence for dramatic peculiarities at the same surface brightness levels as found by Heckman *et al.* (1986). (See also my answer to Dr. Birkinshaw.)

Discussion of the paper by *__BÅÅTH__* (p. 117)

Jauncey: Are you monitoring the 930 MHz flux density history of your sample sources, as it would seem that such close monitoring should be an essential part of your program?

Båååth: The flux densities were measured at Arecibo during the VLBI observations only. The flux density history of these sources is monitored by other groups at nearby frequencies, though.

vanBreugel: What are the optical properties of the sources in your sample?

Bååth: The sources were selected for low frequency variability and in a declination range suitable for Arecibo ($\sim 20°$). Many of the sources have not been optically identified and thus have no known redshift.

Cohen: Could you comment on the possibility that some variety of interstellar scattering is responsible for the expansions?

Bååth: The components are moving linearly with the same position angle at all three epochs. Scattering should cause the components to wander around, shifting the position angle. Also, the jet in CTA102 seen with the VLA (R. Rusk, this meeting) has a position angle in excellent agreement with the direction of the motion of the component we see in our VLBI maps.

Discussion of the paper by *__MUXLOW__* (p. 131)

Romney: Did you find it necessary, or even possible, to apply baseline-based editing in producing images from your large VLBI arrays? VLBI observations are approaching the multiplicity of the VLA, where this sort of processing is generally regarded as impractical.

Muxlow: Both global and baseline-related editing are necessary and were indeed performed on these data. Software development to improve the response time for baseline-related editing is underway.

Discussion of the paper by *__WARDLE__* (p. 143)

Pooley: With what accuracy can you place the polarization maps relative to the total intensity?

Wardle: The antenna phases needed to make the polarization map are the same antenna phases found by the hybrid-mapping procedure for the total intensity map. To the extent that these phases are determined correctly, the maps are perfectly registered. In practice, we find that the maps are registered to better than 0.2 mas.

deBruyn: The region that is most likely to contribute most to the Faraday depolarization is the transition zone between the BLR and the NLR. We know little about this transition region. Can your data set any limits on the product of $n_e B$ in this region?

Wardle: The limit depends on the size and distribution of the clouds, which we do not know. Values of $n_e B$ in the range 0.1 to 10 cm^{-3} seem reasonable.

Inoue: I would like to comment on the penetration of polarized emission. Last year we found a synchronous variation in polarization angle between optical and radio frequencies in OJ287. They varied more than 90° within a week. So, at least at that time, both polarized regions are associated with each other.

Wardle: Other sources also exhibit episodes where the optical and radio polarizations change together. VLBI polarimetry will be able to determine the location of the varying region (i.e., in the jet or in the core).

Ekers: A smooth and uniformly polarized jet that has a changing position angle of polarized emission can appear double in a barely resolved polarized intensity image (beam depolarization). Consequently, it is unlikely that the solution for the separation of "apparent" components will be unique and capable of giving reliable estimates of component motion.

Wardle: A uniformly polarized jet with a changing position angle can be distinguished from a point double in the u-v plane. We do our model fitting in the u-v plane.

Discussion of the paper by REID (p. 169)

Rickett: Carl Gwinn showed earlier that the position errors due to refractive effects in the ISM are (surprisingly) small, $< 20\ \mu$as. Evidently, you have considered this as a source of error in your distance estimate to Sgr B2. I would like to point out that if the ISM has a "steep spectrum vs. wavenumber" (as opposed to the "Kolmogorov" spectrum), there is an additional source of error in the proper motion vectors. Have you considered this?

Reid: The limits quoted by Gwinn are from the Sgr B2 data themselves. As such, they place a limit on refractively induced apparent motion about $< 20\ \mu$as yr^{-1} or < 1 km s^{-1}. Perhaps with a very steep fluctuation spectrum and a certain pattern speed, the 15 km s^{-1} "turbulent" velocity residuals could be explained by refractive effects. However, even if this were the case, the distance estimate comes from the systematic motions (expansion) and would not be biased in any way.

Discussion of the paper by *__BARTEL__* (p. 175)

Kylafis: 1) If the expanding shell is spherically symmetric, how do you explain the observed polarization? 2) Electron scattering in an expanding shell shifts the spectral lines and gives you a false reading of the expansion velocity. Does that affect your results?

Bartel: 1) Since we have not yet mapped any supernova, we allow, in our estimates of uncertainties, for a (moderate) deviation from spherical symmetry. However, I do not know of any detection of polarized emission of the supernovae I discussed here. 2) Computations of models with electron scattering in the atmosphere show that mostly the shape of the P-Cyg line profiles are changed but hardly so the values for the expansion velocity of the photosphere and for the maximum expansion velocity. Therefore, electron scattering is unlikely to significantly affect our results.

Chu: The value of H_0 that Cohen inferred from superluminal movement measurements is likely to be 100 km s^{-1} Mpc^{-1}, while from your direct measurements it is 65 km s^{-1} Mpc^{-1}. I would like to know which value of H_0 is more accurate or more believable.

Bartel: Since the inference of H_0 from superluminal sources does not (yet) include an estimate of the uncertainty, it is hard to say which of the two values is more accurate. As to the second part of your question, I tend to regard, in general, direct measurements as more believable.

Pacini: I feel that it is dangerous to infer from the radio portion of SN1987A that the radiosphere always expands faster than the photosphere. The emission of SN1947A was a minor prompt event taking place outside the photosphere. The emission of radio supernovae like SN1979C occurs one year or so later, and it may be due to different causes and involve different physical processes.

Bartel: Alternatively, the delay may be due to a circumstellar medium much denser than that of SN1987A. The radio light curve and the evolution of the radio spectrum of SN1987A are very similar to the light curves and spectral evolutions of other supernova; only the time between the supernova explosion and the onset of radio emission is much shorter and the radio luminosity much weaker. I think that it is fair to say that our VLBI results on SN1987A are consistent with our assumptions for other supernova about the location of the radiosphere with respect to the photosphere.

Discussion of the paper by *__WILKINSON__* (p. 187) *(presented by DeBruyn)*

Molnar: Have the historic plates from the 1940's been searched for an optical supernova?

deBruyn: As far as I know, they have not. We have not taken any action yet but will, in a paper soon to be submitted, draw attention to this possibility.

Pacini: What is the minimum energy content of the SNR?

deBruyn: It is very comparable to that of most young galactic SNR's, but because of the ten times smaller linear size, it radiates much more efficiently.

Discussion of the paper by TURNER (p. 191)

Rogers: Wouldn't the microlensing or "Einstein" lensing of objects within our galaxy be a very transient phenomenon owing to differential galactic rotation?

Turner: Yes, there would be a transient signature. I haven't thought about this yet. (Answer from Art Niell: About 8 milliarcseconds/year.)

vanBreugel: For an assumed range of $H_0 = 50 - 100$ km s^{-1} Mpc^{-1}, what would the expected time delay be between the gravitationally lensed images in $0957 + 561$?

Turner: The answer is somewhat model-dependent, but the best current estimate is one to two years, with longer delays corresponding to smaller H_0 values.

Gwinn: 1) Why is high dynamic range necessary to observe gravitational lensing of masers by stars? 2) Does the brighter image remain nearly undeflected?

Turner: 1) It is not strictly necessary, but it allows one to detect large impact parameter (i.e., poor alignment) lensing events. These are the most probable, of course. 2) Yes.

Kylafis: What is the typical time delay for two images if the gravitational lens is a star?

Turner: They are of order of the light travel time across the Schwarzschild radius for a black hole with the mass of the lensing object, i.e., $\sim 10^{-5}$ sec for a solar mass object.

Burke: We have been following the flux variations in $0957 + 561$ for the past six years using the VLA. We have a result, but it is not yet sufficiently accurate to give a cosmologically interesting answer. The time delay is probably in the range 400–700 days, with A varying first. If B were to vary first, the time delay would have to be greater than six years, although this case is not generally predicted in current models, which favor A varying first.

Discussion of the paper by GORENSTEIN (p. 201)

Backer: On what time scale would you expect changes in the VLBI structure from transverse motion of the line of sight with respect to the lensing system (velocities up to 500 km s^{-1})?

Gorenstein: Transverse motions yielding apparent displacements of about $0.1''$ or more will yield detectable changes in image magnification. This angular scale is set by the $\sim 1''$ size of the lensing galaxy. Transverse velocities of ~ 500 km s^{-1} yield $0.1''$ displacements in about 10^6 years at distances of 10^{10} lyr.

Discussion of the paper by BRIGGS (p. 211)

Flatters: Would it be possible to use these techniques to investigate gas close to the quasar itself?

Briggs: Perhaps. High column densities of cool neutral gas ($\gtrsim 10^{20}$ H/cm^2) are necessary to create a detectable 21-cm absorption line. If the HI spin temperature is raised, either by collision in a hot gas or by bathing the gas in the 21-cm continuum of a nearby quasar, the optical depth will be reduced. One certainly does not expect 21-cm absorption by the quasar emission line region, for instance.

Neff: 1) Do the 21-cm absorption lines change with time? 2) Do you see absorption lines in the optical spectrum? If so, are they at the same redshift as the HI?

Briggs: 1) Two spectra have been obtained at Arecibo about a year apart. Within the errors, they are consistent with no change. 2) There is a rich optical absorption spectrum at the same redshift. In fact, we selected this object for study at radio wavelengths on that basis (see Wolfe *et al.* 1985, *Ap. J. Lett.*, **294**, L67).

Discussion of the paper by KYLAFIS (p. 223)

Norris: Have you yet had a chance to look at the pumping of the 12-GHz methanol masers?

Kylafis: No, not yet. So far, I have concentrated on water masers.

Reid: You asked why no beautiful SiO maser maps have been made. Basically, we do not have the critical short VLBI baselines at the high frequencies of the SiO transition. The VLBA will help alleviate this deficiency.

Discussion of the paper by HASCHICK (p. 233)

Shaffer: How well do you know that the H_2O maser region is coincident with the nucleus of the galaxy? How well are relative optical, radio continuum, and H_2O positions known?

Haschick: The optical positions are good to a few arcseconds. The radio continuum and H_2O positions are coincident to within $0\rlap{.}''05$.

Claussen: NGC1068 has a nuclear radio continuum triple source. The water masers fall on top of part of that triple. The radio continuum towards NGC4258 shows only one weak continuum source. The maser source falls on top of it.

Discussion of the paper by LESTRADE (p. 265)

Spencer: Could the decrease in the size of Algol with frequency as measured with VLBI be due to interstellar scattering rather than an intrinsic effect?

Lestrade: The distance to Algol is 27 parsecs and so interstellar scattering is very small and would not broaden the radio image of Algol.

Kaufmann: I have comments that are somewhat independent of your very interesting talk. It might be dangerous to make a direct physical association between solar and stellar activities. Sunspots hardly affect a fraction of a percent of solar irradiance, while "starspots" must be huge and of entirely different nature (if they really exist). Gyrosynchrotron masering produces very fast pulses in the Sun (milliseconds), while flares in stars are considerably longer. I finally wish to suggest VLBI experiments on the closest star, the Sun. The physics of active regions and flaring processes would greatly benefit from this high spatial resolution diagnostics.

Lestrade: I agree that the solar analogy must be used with care and that the stars that are 10^4 to 10^5 more radio luminous than the Sun might involve different rather than scaled-up emission mechanisms. This cyclotron masering is a short time scale phenomenon (millisecond), but multiple masers might occur and last much longer, as suggested by our observations.

Preuss: The registration of the radio image of UX Ari with the optical star looks plausible. Do you consider that objects of this type are particularly promising candidates for tying radio and optical reference frames together?

Lestrade: Actually, our VLBI project is motivated by the potential of VLBI observations of radio stars to tie astrometrically the optical and radio reference frames. The coincidence of the optical and radio components of the RS CVn binaries is an important question that should be answered by tackling the astrophysical problem of understanding the radio emission mechanism in these binaries.

Discussion of the paper by VERMEULEN (p. 275)

Biretta: In your beautiful sequence of maps, the radio emitting knots are ejected simultaneously in opposite directions. Is the ejection always simultaneous in the two directions?

Vermeulen: These maps indeed show blobs being ejected simultaneously to the east and to the west. No other such sequence of hybrid maps exists, but model fitting to an earlier sequence of observations by Schilizzi *et al.* (IAU Symposium 110, 1984, p. 280) indicated rather one-sided emission.

Ball: As a rule, one cannot do earth–rotation interferometry on an object that changes significantly with respect to the synthesized beamwidth in a day's time. This would seem to apply to SS443. Are your maps perhaps confused by this effect?

Vermeulen: We are fully aware of the 4.5-mas proper motion ($\sim 1/3$ of the synthesized beam FWHM), which takes place during each 12-hour synthesis. Subdivision of each observation was, however, not permitted by the data quality. The effects of this *predictable* motion have to be investigated more closely, but the lateral smearing that will result does not affect our conclusions about the existence of a brightening zone, derived from this sequence of independent maps.

Discussion of the paper by R. DAVIS (p. 277)

Simon: Since you have both an *angular* expansion velocity and (presumably) an optical *radial* expansion velocity, can you say anything about the distance to this nova? Is the distance consistent with other distance estimates?

R. Davis: The distance is measured as 1.6 kpc, which gives an expansion velocity of 3800 km s^{-1}. This velocity is consistent with the initial optical line widths. However, the line widths become narrower with time, but the expansion appears to continue to 3800 km s^{-1}.

Molnar: Which of the three components in your map do you identify with the separate thermal and nonthermal components?

R. Davis: The thermal contribution to the map is calculated to be 0.1 mJy, and thus all the components are nonthermal. The thermal component presumably emerges in the central region.

Shapiro: Is it possible that your two ("double-sided") jets are a ring of material viewed edge-on?

R. Davis: Yes, indeed, A. R. Taylor is investigating just such a possibility. I am investigating an SS433 type of jet.

Discussion of the paper by MOLNAR (p. 279)

Schilizzi: Is there any evidence for changes in the orientation of the expanding components that might suggest precession such as observed in SS433?

Molnar: Not yet. For precession similar to SS433, one might expect the position angle to vary over a range of $\sim 40°$. VLA and MERLIN measurements following giant flares in September 1982, September 1983, and October 1985 are all consistent with the same position angle (5° errors on each measurement).

R. Spencer: We have made continuous observations of Cyg X-3 over a 60-hour period with the Jodrell Bank Mk1A–MkII short baseline interferometer at λ6 cm, in August 1986. The most dominant period is 9 hours, with well separated flares. The individual flares last $\sim 4 - 6$ hours.

Molnar: As flare amplitudes vary from one to the next, the best way to search for the 4.95^h period is to look at the times at which each flare begins rather than at a Fourier transform of the entire data set.

Johnston: The flaring events in Cyg X-3 may be due to beaming. The orbital period of the binary system is 4.8^h. How does one rationalize your 4.95^h period for flaring events with the orbital period if the variations that are called flares are due to beaming?

Molnar: The observed radio period, 4.95^h, is not consistent with the orbital period, 4.79^h. As the observed spectrum and VLBI size of flares both indicate that the physical size increases linearly with time since the beginning of a flare, it seems that the observed variation is due to a physical change of the source and not due to a change of the angle from which we view the source due to rotation. We therefore do not expect a priori to see the orbital period as we would from a geometrical model like beaming. The similarity of the radio period to the orbital period does suggest that the timing mechanism is connected to the orbital motion but mixed with some other frequency.

Discussion of the paper by RICKETT (p. 287)

Diamond: With regard to your diagram of scattered size vs. galactic latitude, Kemball, Diamond, and Mantovani have new VLBI observations of OH masers at 1.66 GHz that show that some masers at $|b| < 2°$ have sizes between 0.8 and 2.5 mas. Is this an embarrassment for the size vs. $|b|$ diagram?

Rickett: This result is interesting, and it appears to emphasize the *wide* disparity in level of scattering in lines of sight toward the galactic plane.

Discussion of the paper by GWINN (p. 295)

Ekers: Given the very large variations in scattering from source to source that were just discussed by Rickett, how do you justify the assumption of a large number of scattering centers?

Gwinn: The level of scattering varies widely in the Galactic Plane on scales of a degree or less; our results are consistent with uniform scattering over the extent of the cluster of maser features, a few arcseconds. Even if a single cloud is responsible for all the scattering, its extent along the line of sight should be sufficient to produce many independent deflections, yielding Gaussian statistics for the wander, if it is more than 10 A.U. deep, or about 10^{-4} its breadth.

Discussion of the paper by PADRIELLI (p. 297)

Ananthakrishnan: Did you say that you find the variability percentage falling as a function of latitude in the case of short time variable sources?

Padrielli: There is a tendency for the *mt* products to be larger for galactic latitudes less than 20°. However, there are noticeable exceptions to the trend with the galactic latitude: DA406 and 1117 + 14 (single time scale variables) vary too slowly or too much, and some sources with low galactic latitude in the direction of the galactic anticenter show *mt* products much smaller than expected.

Fiedler: I would like to point out that DA406 displayed an extreme scattering event from 1985.3 to 1985.6. Do you have any data in this time frame?

Padrielli: We have monitored the source monthly until September 1985. The source shows a secondary peak of ~ 0.3 Jy in ~ 1985.1, and then it shows a monotonic decrease until the end of our observations.

Discussion of the paper by *<u>STANDISH</u>* (p. 309)

Shapiro: It is also necessary to define a space–time geometry in order to make use of any direction prescription of a reference frame.

Burke: One hopes that radio astronomers will start to appreciate the advantages of giving positions referred to J2000 rather than B1950. No problems with source names (generally defined by position nowadays) arise if IAU conventions are adhered to.

Bartel: Did you find any time dependence of the rotation parameters you estimated for the relation between DE118 and PEP740R?

Standish: I did not explicitly solve for any. However, had I done so, the angular rotation about the z-axis would have reflected the $0\overset{''}{.}005$/cty shown in the difference of the (geocentric) solar mean motion in Table III.

Discussion of the paper by *<u>JOHNSTON</u>* (p. 317)

Preuss: Will it be possible in the framework of your program to measure the coincidence of a milliarcsecond radio feature with the nucleus of a bright Seyfert galaxy?

Johnston: This program is aiming at accuracies at optical wavelengths of $0\overset{''}{.}03$. Thus, if the radio and optical reference frame can be aligned to this accuracy, then this is the accuracy of the coincidence. HIPPARCOS, if successful, will be accurate to $0\overset{''}{.}003$. This should allow the coincidence of radio/optical features to be aligned on mas scales, again if the relationship of the optical and radio reference frames are defined at the mas level.

Cannon: In view of the fact that the quasars have very small proper motions (if any at all), do you have any estimate of the rate of degradation of the accuracy of the radio reference frame?

Johnston: This effect should be very small and has yet to be observed. Changes in source structure, if not monitored, will have a larger effect. I estimate this effect at the 0.1 mas level.

Morimoto: I have several comments: precession does not randomly change RA and Dec; the optical-radio reference frame tie must be heavily affected by differential rotation of the galaxy; at an accuracy of $0\overset{''}{.}01$, galactic rotation (not differential rotation) can be measured in $\sim 2-3$ years.

Hemenway: We are coming to the point where the ability to determine the differences between reference frames (e.g., radio and optical) will be limited by our knowledge of the structures and their differences (in frequencies and time) of the objects used to define the frames. In response to a previous question, I would like to add that in finding the relationship between a radio map and an optical map (images), if the radio map has absolute coordinates at the $0\overset{''}{.}01$ level (from HIPPARCOS), then the Hubble Space Telescope may be used to point to 0.01

accuracy if the guide stars are determined with that accuracy. A camera image would then be relatable to the radio image at the 0.″01 or 0.″02 level.

Discussion of the paper by COATES (p. 337)

Preuss: In your diagram "accuracy vs. time (years)," there is a dotted line advancing into the mm accuracy regime for the next decade. Is this meant to be a (serious) goal of your operations?

Coates: Yes. NASA's goal of millimeter accuracy is based on the requirements of the scientists to get crustal motion velocities to an accuracy of 1 mm/year. This is discussed in T. Jordan's paper. It will take a lot of work to reach millimeter accuracies.

Chu: In your first photograph, What is the ring disk beneath your antenna? Is it for reducing the ground radiation?

Coates: It is for aerodynamic purposes.

Discussion of the paper by GORDON (p. 351)

Shapiro: What are the causes of the difference between the rate of motion of the Pacific with respect to the North American plate in your model and the RM2 models? Why is this difference apparently not covered by the error bar on this rate in the RM2 model?

Gordon: The difference in the direct estimate of the rate is probably because, until a few years ago, the magnetic data from the Gulf of California were too sparse to determine spreading rates reliably. The difference in the indirect rate reflects systematic differences between the NUVEL-1 and RM2 data sets along the East Pacific Rise. Part of the difference is caused by a revision of the geomagnetic reversal time scale a few years ago. The cause of the rest of the difference is unclear but might be due in part to slightly different time-averaging intervals between the two models.

Jordan: I have a comment on Shapiro's question. The differences between RM2 and NUVEL-1 Pacific–North America motions are evidently due to bias in the RM2 data set. It is interesting to note that this discrepancy, perhaps the largest between the two models, amounts to only about 8–10 mm/year, compared to a formal *one-sigma* uncertainty of ±3 mm/year for RM2.

Sauber: In your abstract, you referred to variability in plate motions on time scales of 0.5–1 million years. You did not discuss this in your talk; do you have any comments?

Gordon: Some changes of about 10% are seen. For the Pacific–North American plate motion, the rate based on recent data (700,000) suggests a spreading rate of 52 mm/year vs. 48–50 mm/year (averaged over 3 million years).

Discussion of the paper by *KROGER* (p. 365)

Whitney: Can you explain the commonality between the data sets and between your analysis and that of J. Sauber? At what point do the analyses become independent?

Kroger: Our data set spans the period January 1980 to May 1986. I believe that Sauber is using a data set spanning the period October 1982 through the early part of 1987. Since there are very few measurements before October 1982, we differ in the measurements made after 1986, which are not included in our analysis.

Wright: Owens Valley seems to be used as the reference station for measurements in California. What is the evidence that OVRO is stationary with respect to the North America plate? In particular, note that if OVRO is moving south, then the discrepancy vector between the San Andreas fault motion and the model goes away, and that this discrepancy is exactly the motion of Hat Creek, presumably with respect to OVRO.

Kroger: The Owens Valley–Westford vector does not change.

Wright: The discrepancy vector is orthogonal to this vector; what are the errors on the orthogonal component?

Kroger: ± 1 cm/year.

Wright: This is the same magnitude as the discrepancy; I propose that Owens Valley may be moving south by 1 cm/year.

Discussion of the paper by *J. DAVIS* (p. 367)

R. Gordon: Is the very small motion (-3 ± 2 mm/year) motion between Haystack and Mojave stations consistent or inconsistent with the large motion (~ 15 mm/year) we have heard about in other talks inferred between Mojave and other sites from presumed-stable North America?

J. Davis: Yes. The Goddard group gets -3 ± 3 mm/year using an independent set of experiments.

Clark: Measurements continue with Alaska and Ft. Davis.

Discussion of the paper by *EUBANKS* (p. 369)

Tang: 1) Since you usually use very long baselines, I suppose that the source structure effects play some role as an error source. 2) Did you see some rotation of the reference frame in your data analysis?

Eubanks: 1) I think you are probably right, although we hope that source structure effects will appear random and thus will be accounted for by the additive noise. 2) The reference frame clearly rotates with respect to the quasars. Since the effects of Earth rotation must be removed empirically, it is thus only possible to estimate relative rotations with this method.

Discussion of the paper by HERRING (p. 371)

Ulvestad: The 0.1 milliarcsecond errors that you quote for the nutation terms are very small. Could you comment on the possibility that systematic errors may be occurring at a larger level?

Herring: We have carried out a number of studies of this data set (for example, processing the data set as disjoint sets, and interleaved but independent sets) and have detected no changes that are inconsistent with the (66% confidence interval) one-sigma standard deviation estimates. Also, the power-spectral density function shows no anomalous peaks.

Cappallo: Might not decade-timescale source structure changes make it difficult to estimate reliably long-period nutation terms? Also, as the source pool slowly changes, it seems as though there will be a source-frame random walk, with a similar effect.

Herring: These are problems that will need to be addressed as more data are obtained and as the accuracy of estimates of the nutation series coefficients improve. However, the effects of changing source structure can be corrected, although we do not currently do this.

Caporali: The power spectral densities you showed give information on the amplitudes of the various terms. Spectral analysis also gives the phases of each term. In addition to amplitude corrections, are there also terms that need to be corrected only in phase, or both in phase and amplitude?

Herring: When we estimate the corrections to the coefficients of the nutation series, we use the phase information. The deviations of the phase from the values predicted by the IAU nutation series are given by the out-of-phase components in Table 1 of this paper.

Discussion of the paper by WAHR (p. 381)

Gwinn: Morelli and Dziewonski use seismic techniques to obtain Y_2^0 core–mantle boundary topology consistent with our interpretation of the nutations, and sub-km errors.

Cappallo: Since the excitation spectrum of the 14-month Chandler wobble seems to have well-defined lines, it seems likely that the amplitudes of the driving terms may not be smooth across the Chandler response bandpass. Wouldn't this call into question the use of a smoothed half-width of the response in order to infer a characteristic decay time and thus a long-term Q for the solid earth?

Wahr: Yes, results would be improved if the excitation source could be identified and its effects removed.

Discussion of the paper by ROBERTSON (p. 391)

Popelar: Are there any plans to increase the frequency of observations to obtain better data for determination of diurnal nutation?

Robertson: We are attempting to arrange special UT1 observing sessions using Richmond and Bologna at a rate of four observing sessions per day.

Herring: The so-called "diurnal polar motion" is incorporated into our nutation determination.

Babcock: The IRIS data set is certainly very impressive, particularly the high-frequency UT1 data. But you mentioned that polar motion information is required in order to determine UT1 from single-baseline experiments. Would you comment on the source of polar motion information used for these daily UT1 measurements?

Robertson: The pole position and nutation data used to reduce the daily observing sessions was obtained by interpolating the IRIS values for those parameters.

R. Gordon: Could you comment on the accuracy of the determination from VLBI data of the secular trend in polar motion, which you discussed in your *Scientific American* article last year?

Robertson: The problem of determining a secular drift is not limited by the accuracy of the pole position determination but by the relatively brief ($\sim$ 5 years) duration of the observations. It is difficult to distinguish short-term trends from truly secular trends.

Roberts: You should try observing at *random* times each day, rather than at a given time each day. This will remove your aliasing problem and permit you to extract information at frequencies far beyond 1/2 (minimum data spacing). This is described in Roberts, Lehar, and Dreher (*A. J.*, **97**, 968, 1987).

Robertson: We have some constraints on our observing times, but this is an idea that should definitely be investigated.

Discussion of the paper by LOSITO (p. 411) *(presented by Verrone)*

Herring: What was the γ quantity in your polar motion differential equations?

Verrone: The γ factor in the Free Polar Equation is related to γ_1 and γ_2, which are basically direction cosines of the $\bar{c}$ axis, axis of rotation of the core, relative to the reference frame of the mantle. $\gamma = \gamma_1 + i\gamma_2$. $\bar{c} = \gamma_1\bar{x}_1 + \gamma_2\bar{x}_2 + (\gamma_3 + 1)\bar{x}_3$.

Discussion of the paper by CAMPBELL (p. 427)

Readhead: The spectral index is a safer criterion to use in trying to identify the stationary component than the flux density of individual components. There are a number of objects in which the core is not the brightest component. In addition, the relative brightness often changes between epochs.

Campbell: Yes, this is an important piece of information. In our best computations, we used the brightest component as a first viable approach, knowing that each source has its own particular problem. This matter can only be dealt with by close cooperation between astrophysicists and geodesists.

Ratner: Along the line of the previous comment (about using spectral indices rather than flux density as criterion to identify reference points in a brightness distribution), in the case of 3C345 one cannot identify the stationary point with the brightest point in a map. This has not worked in the past.

Campbell: Yes, we know that before the final reprocessing of all geodetic data, we have to look at the structure history of each source. But here also lies the advantage of the proposed method: by using the regular geodetic VLBI campaigns, such as IRIS and CDP, we will have this structure history.

Marcaide: There may be cases where you will not be able to use either the spectral index as a guide to identify the stationary feature, as just pointed out by Readhead, nor the brightest component for it. An example is 4C39.25.

Shapiro: The criteria for reference points in the quasar will perforce be complicated because the sources are. In particular, the "brightest" component can change identity because the "same" component does not necessarily remain brighter than the other component(s) in the source.

Potash: In correcting geodetic measurements for ionospheric effects, the low frequency (S band) measurements factor in with a relatively low weight, about 1/13. The effect of S band source structure is, however, somewhat more important than this low weight suggests. This occurs because S band source structures tend to be larger than those at X band, and the magnitude of the source structure corrections scale, on average, with source size.

Eubanks: Could I conclude from your delay error plots that delay rate errors introduced from source structure effects will also be significant?

Campbell: We also compute the rate corrections, but the rate observations normally have a much smaller weight in geodetic solutions.

Tang: If you apply δ-functions as the source model, it is not necessary to map the source. You can fit directly the δ-functions to visibilities. However, we do see significant differences in estimations of source structure effect between using δ-functions and a Gaussian model. Therefore, we apply the more precise model – Gaussians – for our calculations.

Discussion of the paper by *KARDASHEV* (p. 433)

Romney: I think one of your slides mentioned, in Russian, a two-year lifetime for the spacecraft. Is that right?

Kardashev: Yes, two years.

Shapiro: How is the phase integrity of the signals to be assured in RADIOASTRON?

Kardashev: The phase stability will be assured by accurate measurements of the orbit by a ground-based network. Also, one of the space–earth interferometer's antennas is outside the atmosphere/ionosphere, which disturbs the phase.

Discussion of the paper by SCHILIZZI (p. 441)

Marcaide: The dual polarization option in QUASAT must be certainly tagged to the decision to use, or not to use, an offset antenna (with direct illumination, that is, without subreflector) since instrumental cross-polarization is expected to be a problem.

Schilizzi: Catarzi and Tofani (these proceedings) suggest that polarization measurements with an offset antenna will in fact be possible.

Chu: It seems to me that the offset illumination will cause considerable trouble in polarization measurements. Also, the spurious polarization caused by offset illumination is difficult to compensate for. Since the ground VLBA antenna is 25 m in diameter, could one use an umbrella-type antenna to increase the diameter of the space antenna and the sensitivity of the interferometer pairs?

Schilizzi: For QUASAT, we are studying offset and center-fed inflatable antennas as well as deployable mesh antennas. Umbrella-type antennas are not presently under consideration. I should note that not all of the antennas on the ground are 25 m in diameter; we will certainly use the larger telescopes (for example, some elements of the EVN) as well to increase the sensitivity.

Biermann: In Germany, we have proposed a 50-m high frequency antenna – included in the national report on the future of astronomy – to support space VLBI.

Schilizzi: Very good news!

Discussion of the paper by LINFIELD (p. 457)

Barthel: Did your Gaussian models agree with earth-based data obtained at higher frequency?

Linfield: We have not yet checked.

Bartel: I am worried that your values for the brightness temperature are model-dependent. If you had, for example, estimated the full-width of a uniform sphere model for the sources' brightness distribution rather than the FWHM of a Gaussian, your brightness temperature values would come down by a factor of ~ 3 and would drop at least in some cases below the magic temperature value of 10^{12} K.

Linfield: Although the fitted FWHM of a uniform sphere might be significantly different, the resulting brightness temperature would be only slightly different than for our Gaussian models. The formula for converting FWHM to T_B depends on the type of source model.

Cannon: Is the inverse Compton brightness temperature limit of 10^{12} K so firmly established that you can say with confidence that a brightness temperature of 3×10^{12} is a "strong" violation of this limit?

Linfield: Yes, inverse Compton losses vary as a large power of the temperature, and 3×10^{12} K is excluded for the source rest frame.

Discussion of the paper by *ROMNEY* (p. 461)

Wielebinski: Will the correlator be built in one stage or do you expect to do the work in sections?

Romney: A subset correlator of six or seven stations and two channels will be built first. This is intended to be adequate for thorough verification of correlator performance, as well as to support the first stand-alone operation of the VLBA. Expansion to the full 20-station 8-channel processor will then follow immediately.

Kaufmann: How will data from different antennas reach the central processing facility in Socorro? Are there any possible real time operations?

Romney: Recorded data tapes will be shipped from each station to the central processing facility by air or surface freight. A near-real-time correlation capability is planned but is intended only for technical verification, rather than scientific purposes.

Marscher: How will the multifrequency capability be implemented, i.e., how many frequencies will be able to be observed at once, and how rapidly will one be able to switch frequencies?

Romney: *Simultaneous* multifrequency observations will only be possible at S and X bands, unless one or more of the dual-frequency options, mentioned above, are implemented subsequently. Switching between frequencies (or dual-frequency pairs) is accomplished by rotating the subreflector and is expected to require on the order of 10 seconds.

Chu: For the purpose of site selection of the Chinese VLBI Network, we have calculated the u-v coverage with the VLBA. It seems that the u-v coverage with the previous D2 configuration is more uniform. (The coverage of short spacings cannot be seen in the u-v diagram.)

Walker: The u-v coverage with the D2 array did not include many short spacings. The high value of short spacings was given more weight in the later configuration studies. The current VLBA configuration has better coverage of short spacings than array D2. Since the number of VLBA telescopes was fixed early in the project, improving short spacing coverage required some degradation of long spacing coverage. I suspect that you are seeing this effect.

Discussion of the paper by WAN (p. 475)

Shaffer: How and when can we apply for observing time on the Chinese array?

Wan: Please write to Dr. Wan Tongshan, Division of Radio Astronomy, Shanghai Observatory, Shanghai, People's Republic of China.

Discussion of the paper by NORRIS (p. 485)

Shapiro: What has been your experience with the phase stability when using fiber feeds for transmitting LO signals?

Norris: Experiments have failed to attain the required stability with fibers, and so we are distributing the local oscillators via coax cable and using the optical fibers for IF transmission only.

deBruyn: Do you plan to bring the "compact array" to a very small configuration when doing VLBI (to have an acceptably large fanbeam)?

Norris: We have not yet established a policy on this, but we are aware of the problem. Probably we will decide on the basis of individual experiments and whether a trade-off between sensitivity and field-of-view can be made. Certainly moving the elements to a small configuration will be possible when required.

Discussion of the paper by YEN (p. 489)

Simon: What is the physical principle of your wavelength clock?

Yen: The local oscillator frequency and station delay are adjusted continuously to remove the station-dependent frequency offset and delay based on an assumed reference position on the sky and station location. The arrival time and frequency is thereby referred to a fictitious station at the center of the earth. Hence, no first-order fringe rate and delay tracking are required in the processor.

Discussion of the paper by ALEF (p. 523)

Shapiro: Just a historical comment: Our group first succeeded with elimination of fringe-phase ambiguities in switched observations between two sources about $0\overset{\circ}{.}5$ apart about five years before we were lucky enough to find two sources that would be in all telescope beams simultaneously.

Alef: I agree. However, I am not aware of any phase-reference map before Marcaide 1982.

Bartel: What is the uncertainty of your determination of the separation between components in the two sources?

Alef: The error in the measurement of the separation is dominated by the errors in the source structure. Our estimate is better than 0.5 mas.

R. Davis: Can you say whether your phase stability is limited by spatial or temporal fluctuations in the atmosphere?

Alef: At a source separation of 0°5, we are certainly limited by temporal fluctuations.

Discussion of the paper by <u>*ROGERS*</u> (p. 533)

Treuhaft: You mentioned that the rms delay fluctuation is proportional to the square root of the air mass. This is only true at time scales less than a few hundred seconds. Beyond that, the rms fluctuation is approximately proportional to the air mass.

Rogers: I took the figure from your paper and the air mass dependence fit the square root fairly well.

Treuhaft: But that was given for 200 seconds only. For longer times, the fluctuation dependence on air mass increases.

Rogers: That's correct.

Readhead: At mm wavelengths, the atmospheric coherence time is 10 sec in many cases. If the noise in closure phase measurements over each 10-second period is $\gtrsim 2\pi$ radians, then we cannot use the closure phase determined from 10-second coherent averages to determine the mean closure phase since, for low SNR, 2π ambiguities in the closure phase bias the mean toward zero. This difficulty can be avoided by using the bispectrum $\Gamma^3(x_1, x_2) = \Gamma(x_1)\Gamma(x_2)\Gamma(-x_1, -x_2)$ of each 10-second period and averaging this. The argument of the bispectrum is the closure phase.

Rogers: We "vector" average the closure phase for many 10-second segments, which overcomes the problem of 2π ambiguities and results in a SNR that continues to improve with the square root of the total averaging time. We do in fact use the "bispectrum" although we have called it the "vector" closure phase.

Discussion of the paper by <u>*ELGERED*</u> (p. 543)

Eubanks: What factors could give rise to a zenith troposphere bias in the WVR data?

Elgered: There are several factors that may be significant. One is instrumental errors, i.e., a bias in any or both of the WVR channels. Another factor, depending on how the algorithm relating WVR observables and the wet delay is derived, is uncertainty in the attenuation coefficients of water vapor or biases of independent calibration methods. In addition to this, since the "wet" and the "dry" delays cannot be separated in the estimation process, a bias in the zenith delay can also arise from possible errors in the pressure readings at the site.

Author Index

Source Index

(References are to first page of papers)

Subject Index

(References are to first page of papers)

The manufacturer's authorised representative in the EU is Springer Nature Customer Service Centre GmbH, Europaplatz 3, 69115 Heidelberg, Germany. If you have any concerns regarding our products, please contact ProductSafety@springernature.com

Printed and bound by CPI Group (UK) Ltd, Croydon, CR0 4YY
30/09/2026
02225505-0019